# Cord Blood Stem Cells Medicine

# Cord Blood Stem Cells Medicine

*Edited by*

**Catherine Stavropoulos-Giokas**
**Dominique Charron**
**Cristina Navarrete**

AMSTERDAM • BOSTON • HEIDELBERG • LONDON
NEW YORK • OXFORD • PARIS • SAN DIEGO
SAN FRANCISCO • SINGAPORE • SYDNEY • TOKYO

Academic Press is an imprint of Elsevier

Academic Press is an imprint of Elsevier
32 Jamestown Road, London NW1 7BY, UK
525 B Street, Suite 1800, San Diego, CA 92101-4495, USA
225 Wyman Street, Waltham, MA 02451, USA
The Boulevard, Langford Lane, Kidlington, Oxford OX5 1GB, UK



ISBN: 978-0-12-407785-0

**British Library Cataloguing-in-Publication Data**
A catalogue record for this book is available from the British Library

**Library of Congress Cataloging-in-Publication Data**
A catalog record for this book is available from the Library of Congress

For information on all Academic Press publications
visit our website at http://store.elsevier.com/

Typeset by TNQ Books and Journals
www.tnq.co.in

# Contents

## Section V
## Cord Blood Banking: A Current State of Affairs

### 15. Cord Blood Banking: Operational and Regulatory Aspects

*Cristina Navarrete*



### 16. Cord Blood Unit Selection for Unrelated Transplantation

*Andromachi Scaradavou*



### 17. Quality Management Systems Including Accreditation Standards

*Andreas Papassavas, Theofanis K. Chatzistamatiou, Efstathios Michalopoulos, Markella Serafetinidi, Vasiliki Gkioka, Elena Markogianni and Catherine Stavropoulos-Giokas*

# List of Contributors

**Julia Bosch** Institute for Transplantation Diagnostics and Cell Therapeutics, University of Duesseldorf Medical School, Duesseldorf, Germany

**Shijie Cai** Stem Cell Research Laboratory, Nuffield Division of Clinical Laboratory Sciences, Radcliffe Department of Medicine, University of Oxford, Oxford, UK; NHS Blood and Transplant, John Radcliffe Hospital, Oxford, UK; Weatherall Institute of Molecular Medicine, University of Oxford, UK

**Lee Carpenter** NHS Blood and Transplant and Radcliffe Department of Medicine, University of Oxford, Oxford, UK

**Keith M. Channon** Department of Cardiovascular Medicine, Radcliffe Department of Medicine, University of Oxford, UK

**Dominique Charron** Laboratoire "Jean Dausset," Immunology and Histocompatibility Hôpital Saint-Louis AP-HP, Université Paris Diderot, Paris, France

**Theofanis K. Chatzistamatiou** Hellenic Cord Blood Bank, Biomedical Research Foundation Academy of Athens, Athens, Greece

**Audrey Cras** Assistance Publique-Hôpitaux de Paris, Saint-Louis Hospital, Cell Therapy Unit, Cord Blood Bank and CIC-BT501, Paris, France; INSERM UMRS 1140, Paris Descartes, Faculté de Pharmacie, Paris, France

**Robert Danby** Department of Haematology, Oxford University Hospitals NHS Trust, Churchill Hospital, Oxford, UK; NHS Blood and Transplant, Oxford Centre, John Radcliffe Hospital, Oxford, UK

**Francesco Dazzi** Regenerative Medicine, Department of Haematology, King's College London, London, UK

**Amalia Dinou** Hellenic Cord Blood Bank, Biomedical Research Foundation of Academy of Athens, Athens, Greece

**Dominique Farge** Assistance Publique-Hôpitaux de Paris, Saint-Louis Hospital, Internal Medicine and Vascular Disease Unit, CIC-BT501, Paris, France

**Lydia Foeken** World Marrow Donor Association, WMDA Office, Leiden, The Netherlands

**Antonio Galleu** Regenerative Medicine, Department of Haematology, King's College London, London, UK

**Marietta Giannakou** MEP, Head of the Greek EPP Parliamentary Delegation, Former Minister of National Education and Religious Affairs, Former Minister of Health, Welfare and Social Security

**Vasiliki Gkioka** Hellenic Cord Blood Bank, Biomedical Research Foundation Academy of Athens (BRFAA), Greece; Evaluation Expert, Hellenic Transplant Organization, Athens, Greece

**Aspasia Goula** Organizational Culture in Health Services, Technological Educational Institute of Athens, Greece

**Gregory Katz** ESSEC Business School, Chair of Therapeutic Innovation; Fondation Générale de Santé, Paris, France

**Cheen P. Khoo** Stem Cell Research Laboratory, Nuffield Division of Clinical Laboratory Sciences, Radcliffe Department of Medicine, University of Oxford, Oxford, UK; NHS Blood and Transplant, John Radcliffe Hospital, Oxford, UK

**Gesine Kögler** Institute for Transplantation Diagnostics and Cell Therapeutics, University of Duesseldorf Medical School, Duesseldorf, Germany

**George Koutitsas** Process Analysis and Strategy Implementation Expert, National Insurance, Athens, Greece

**Joanne Kurtzberg** The Robertson Clinical and Translational Cell Therapy Program and Carolinas Cord Blood Bank, Duke University, Durham, NC USA

**Paul Leeson** Department of Cardiovascular Medicine, Radcliffe Department of Medicine, University of Oxford, UK

**Stefanie Liedtke** Institute for Transplantation Diagnostics and Cell Therapeutics, University of Dusseldorf Medical School, Dusseldorf, Germany

**Pascale Loiseau** Laboratoire "Jean Dausset," Immunology and Histocompatibility Hôpital Saint-Louis AP-HP, Université Paris Diderot, Paris, France

**Daniel Markeson** Stem Cell Research Laboratory, Nuffield Division of Clinical Laboratory Sciences, Radcliffe Department of Medicine, University of Oxford, Oxford, UK; NHS Blood and Transplant, John Radcliffe Hospital, Oxford, UK; Department of Plastic and Reconstructive Surgery, Stoke Mandeville Hospital, Aylesbury, UK; University College London Centre for Nanotechnology and Regenerative Medicine, Division of Surgery and Interventional Science, Royal Free Hospital, London, UK

**Elena Markogianni** Hellenic Cord Blood Bank, Biomedical Research Foundation Academy of Athens (BRFAA), Greece

**Emeline Masson** Laboratoire "Jean Dausset," Immunology and Histocompatibility Hôpital Saint-Louis AP-HP, Université Paris Diderot, Paris, France

**Efstathios Michalopoulos** Hellenic Cord Blood Bank, Biomedical Research Foundation Academy of Athens (BRFAA), Greece

**Anna Rita Migliaccio** Tisch Cancer Institute, Mount Sinai School of Medicine, New York, NY, USA

**Maria Mitrossili** Health Law of Technological Educational Institute of Athens, Greece; Institutional Technological Institute of Athens, Athens, Greece

**Cristina Navarrete** Histocompatibility and Immunogenetic Services and NHS-Cord Blood Bank, National Blood and Transplant (NHSBT), England, UK; Division of Infection and Immunity, University College London, London, UK

**Laura Newton** Stem Cell Research Laboratory, Nuffield Division of Clinical Laboratory Sciences, Radcliffe Department of Medicine, University of Oxford, Oxford, UK; NHS Blood and Transplant, John Radcliffe Hospital, Oxford, UK; Department of Cardiovascular Medicine, Radcliffe Department of Medicine, University of Oxford, UK

**Yannis Nikolados** Economists in Health Management, Technological Institute of Athens, Greece

**Amanda L. Olson** MD Anderson Center, Department of Stem Cell Transplantation and Cellular Therapy, University of Texas, Houston, Texas, USA

**Paul J. Orchard** Department of Pediatrics, Division of Blood and Marrow Transplantation, University of Minnesota, Minneapolis, Minnesota, USA

**Daniela Orsini** World Marrow Donor Association, WMDA Office, Leiden, The Netherlands

**Andreas Papassavas** Hellenic Cord Blood Bank, Biomedical Research Foundation Academy of Athens (BRFAA), Greece

**Thalia Papayannopoulou** Department of Medicine/ Hematology, University of Washington, Seattle, WA, USA

**George Pierrakos** Primary Health Management, Technological Educational Institute of Athens, Greece

**Sergio Querol** Barcelona Cord Blood Bank and Haematopoietic Progenitor Cell Unit, Banc Sang i Teixits, Barcelona, Spain

**Teja Falk Radke** Institute for Transplantation Diagnostics and Cell Therapeutics, University of Duesseldorf Medical School, Duesseldorf, Germany

**Paolo Rebulla** Foundation Ca' Granda Ospedale Maggiore Policlinico, Milano, Italy

**Vanderson Rocha** Department of Haematology, Oxford University Hospitals NHS Trust, Churchill Hospital, Oxford, UK; NHS Blood and Transplant, Oxford Centre, John Radcliffe Hospital, Oxford, UK; Eurocord, Hôpital Saint Louis APHP, University Paris VII IUH, Paris, France

**Marcos Sarris** Health and Sociology and Quality of Life, Technological Educational Institute of Athens, Greece

**Aurore Saudemont** Anthony Nolan Research Institute and University College London, London, UK

**Andromachi Scaradavou** National Cord Blood Program, New York Blood Center, New York, NY, USA; Department of Pediatrics, Memorial Sloan-Kettering Cancer Center, New York, NY, USA

**Markella Serafetinidi** Hellenic Cord Blood Bank, Biomedical Research Foundation Academy of Athens (BRFAA), Greece

**Elizabeth J. Shpall** MD Anderson Center, Department of Stem Cell Transplantation and Cellular Therapy, University of Texas, Houston, Texas, USA

**Angela R. Smith** Department of Pediatrics, Division of Blood and Marrow Transplantation, University of Minnesota, Minneapolis, Minnesota, USA

**Sotiris Soulis** Health Economics and Social Protection, Technological Educational Institute of Athens, Greece

**Stamatia Sourri** Stem Cell Research Laboratory, Nuffield Division of Clinical Laboratory Sciences, Radcliffe Department of Medicine, University of Oxford, Oxford, UK; NHS Blood and Transplant, John Radcliffe Hospital, Oxford, UK

**Catherine Stavropoulos-Giokas** Hellenic Cord Blood Bank, Biomedical Research Foundation Academy of Athens, Athens, Greece

**Jessica M. Sun** The Robertson Clinical and Translational Cell Therapy Program and Carolinas Cord Blood Bank, Duke University, Durham, NC USA

**LingYun Sun** Department of Immunology, The Affiliated Drum Tower Hospital of Nanjing University Medical School, Nanjing, China

**Antoine Toubert** Laboratoire d'Immunologie et Histocompatibilité, INSERM UMR1160, and Université Paris Diderot, Sorbonne Paris Cité, Institut Universitaire d'Hématologie, Hôpital Saint-Louis, Paris, France

**Dandan Wang** Department of Immunology, The Affiliated Drum Tower Hospital of Nanjing University Medical School, Nanjing, China

**Suzanne M. Watt** Stem Cell Research Laboratory, Nuffield Division of Clinical Laboratory Sciences, Radcliffe Department of Medicine, University of Oxford, Oxford, UK; NHS Blood and Transplant, John Radcliffe Hospital, Oxford, UK

**Youyi Zhang** Stem Cell Research Laboratory, Nuffield Division of Clinical Laboratory Sciences, Radcliffe Department of Medicine, University of Oxford, Oxford, UK; NHS Blood and Transplant, John Radcliffe Hospital, Oxford, UK

**Yong Zhao** Hackensack University Medical Center, Hackensack, NJ, USA

# Foreword

## TWENTY-FIVE YEARS OF CORD BLOOD TRANSPLANT

Since the first human cord blood transplant, performed in 1988, cord blood banks have been established worldwide for collection and cryopreservation of cord blood for allogeneic hematopoietic stem cell transplant. Umbilical cord blood (UCB) has now become a commonly used source of hematopoietic stem cells for allogeneic transplantation. Today, a global network of cord blood banks and transplant centers has been established for a common inventory with an estimated 600,000 UCB banked and an estimated 30,000 UCB units distributed worldwide for adults and children with severe hematological diseases. Several studies have shown that the number of cells is the most important factor for engraftment while some degree of HLA mismatches is acceptable. The absence of ethical concern, the unlimited supply of cells explains the increasing interest of using cord blood for stem cell therapy.

Much has been learned in a relatively short time on the properties of cord blood hematopoietic progenitors and their clinical application. Cord blood transplant needs to meet several new challenges. First, several methods of improvement of the speed of engraftment and decreasing transplant-related mortality are investigated such as the increase of donor pool to decrease the number of HLA mismatches or the use of double cord blood transplants. Other methods are currently investigated such as cord blood intrabone infusion, ex vivo expansion with cytokine cocktails or homing factors or addition of mesenchymal stromal cells. More interestingly, nonhematopoietic stem cells have been isolated from cord blood and placenta and could be used for the treatment of auto-immune diseases or for regenerative medicine.

**E. Gluckman MD, FRCP**
**Professor Emeritus of Hematology, Eurocord**
**Assistance publique des hôpitaux de Paris (APHP)**
**Institut universitaire d'Hématologie (IUH) Hospital**
**Saint Louis Paris, France**

Section I

# Introduction

Chapter 1

# Introduction to Cord Blood Stem Cells

**Dominique Charron[1], Catherine Stavropoulos-Giokas[2] and Cristina Navarrete[3]**

*[1]Laboratory "Jean Dausset," Immunology & Histocompatibility, Hopital Saint Louis, Paris, France; [2]Hellenic Cord Blood Bank, Biomedical Research Foundation Academy of Athens (BRFAA), Greece; [3]Histocompatibility and Immunogenetic Services and NHS-Cord Blood Bank, National Blood and Transplant (NHSBT), England, UK; Division of Infection and Immunity, University College London, London, UK*

Hematopoietic stem cell transplantation (HSCT) can be curative for selected malignant and nonmalignant diseases. In recent years, cord blood (CB) has become a standard alternative source of hematopoietic stem cells (HSC) to bone marrow (BM) or peripheral blood (PB) for allogeneic HSCT mainly in patients who lack a human leukocyte antigen (HLA)-matched donor. This transition of CB from a waste, as it was previously considered, to a valuable cell product has been relatively rapid, and it has been accompanied by the establishment of dedicated facilities for the safe storage of these CB units, i.e., cord blood banks (CBBs) and in the development of procedures, standards, and regulations to secure the safe storage of high-quality CB units, i.e., CB banking.

The efficacy of CB HSC for allogeneic transplantation has significantly increased over the past years and its use is nowadays widespread in many transplantation centers. Since the first CB transplantation was successfully performed in a child with Fanconi anemia with his HLA-identical sibling, the number of allogeneic unrelated and related transplants carried out for various hematological and nonhematological disorders has increased steadily. To perform these transplants, CBBs were established for the collection, cryopreservation, selection, and release of unrelated CB units for national and international exchanges. More than 25,000 unrelated CB transplants have been performed worldwide, provided by CBBs that have collected more than 800,000 CB units.

CB cells have proliferative advantage and decreased immune reactivity when compared to other sources of hematopoietic stem cells. These properties should give a clear advantage for engraftment and diminution of acute Graft-versus-Host Disease (aGvHD). However due to the reduced volume, total nucleated cells (TNC) and HSC numbers present in a CB unit compared to BM or PB, the engraftment potential of CB has been limited. In order to overcome this issue, clinical protocols for using two CB units have been successfully developed and implemented.

Several studies have shown that TNC dose and HLA matching are important factors for survival after transplantation. Donor search algorithms have now been developed indicating that the best units should contain more than $2 \times 10^7$ TNC and more than $2 \times 10^5$ $CD34^+$ cells per kilogram of the patient's body weight. If only HLA-A, -B, and DRB1 matching (6/6) are considered, the number of HLA mismatches should not be more than two. Additional advantages of banked CB compared to BM are the absence of risk to the donor, the rapid and direct availability of the cells, and the absence of infectious disease at birth.

CB also contains stem cells that have retained embryonic properties and they can be isolated, and, when cultured in appropriate conditions, they give rise to cell lines that can be used for tissue engineering and regeneration of nonhematopoietic organs or tissues.

In the context of regenerative medicine or tissue engineering, CB is also a suitable source of stem cells such as mesenchymal stem cells (MSCs), and of endothelial cell progenitors. These cells have the potential to differentiate into various cell lineages including hepatocytes, muscle, cardiac myoblasts, pancreatic islets, keratinocytes, and neuronal cells and blood vessels, respectively.

These cells could be used to replace damaged or degenerating tissues or to learn more about development and cellular signaling, in normal cells or tissues or in pathological conditions. Possible applications include vascularization or bone reconstruction, and many more. Also, these cells could be used to deliver missing gene products.

Although the current results are very promising, more research is needed in order for these cells to be used in clinical settings. Considering the availability of CB and the absence of ethical problems associated with their collection, banking, and manipulation, compared to other sources of immature cells, CB could probably become a valuable source of cells to be used in stem cell therapy.

In this book, we have tried to cover the main fields of interest regarding CB. The book starts with elements of the biology behind the product, and continues with a series of articles on current and future clinical applications covering both transplantation and the field of regenerative medicine. There is a section dedicated to the organization and function

**Cord Blood Stem Cells Medicine. http://dx.doi.org/10.1016/B978-0-12-407785-0.00001-3**

of the CB banking sector, the current state of regulatory affairs, and the challenges of the future. And last, but not least, the ethical, economical, societal, and public health impact of CB banking is presented.

The presence of relatively mature HSCs in human CB was demonstrated by *Knudtzon* in 1974. Ten years later, *Ogawa and colleagues* documented the presence of primitive HSCs in CB. The current knowledge on CB hematopoiesis, the biology of hematopoietic stem and progenitor cells present, and their characteristics that give CB its unique properties are exposed in Chapter 3, that discusses how understanding the cells opens the road to transplantation.

However, it was not until 1989 that experimental and clinical studies were published indicating that human CB could be used in a clinical setting. The first successful cord blood transplantation (CBT) reported in 1988 was made possible by close collaboration among three groups: *A.D. Auerbach* (Rockefeller University, New York, USA), *H.E. Broxmeyer* (Indiana University, Indianapolis, USA), and *E. Gluckman* (Saint-Louis Hospital, Paris, France). Based on the diagnostic test of Auerbach for Fanconi anemia and the basic work on stem cell biology of Broxmeyer, Gluckman performed the first CBT in Saint-Louis Hospital, on a 6-year-old boy from North Carolina with severe Fanconi's anemia, using cryopreserved CB of his HLA-identical younger sister who was unaffected by the disorder.

Since the initial HLA-matched sibling CB transplant was carried out, the substantial logistics and clinical advantages of CB as a source of HSCs for transplantation have become clear. The proliferative capacity of HSCs in CB is superior to that of cells in BM or PB from adults. Also, the immaturity of lymphocytes in CB dampens the risk of aGvHD which remains the main obstacle to the success of allogeneic transplantation of HSCs. An additional advantage of CB is the low viral (cytomegalovirus and Epstein-Barr virus) load of neonatal/fetal blood compared to adult blood.

The immune cells present in CB have unique properties that confer to CB its characteristics in the context of transplantation. The relative immaturity of these cells could explain the lower incidence of GvHD, but also opens roads for the improvement of the therapeutic potential of CB, by taking advantage of their inherent plasticity and flexibility. An overview of the mechanisms involved in the regulation and function of the immune component of CB is given in Chapter 4, as well as proposed strategies for expanding the clinical scope of CB from an immunological perspective.

The histocompatibility component is equally important. HLA is one of the decisive factors for selecting a CB unit and the impact of HLA matching on outcomes is critical. This affects the immunogenetic typing strategy for banking. In Chapter 6, the impact of both HLA and non-HLA immunogenetic factors is presented and discussed, toward a comprehensive immunogenetic assessment to establish the basis for a successful donor choice for a given patient.

Serious disadvantages of CB are the low number of HSCs compared with BM or mobilized PB, the increased risk of graft failure, the delayed hematopoietic engraftment, and the lack of donor lymphocyte transfusion for immune therapy. Additional possible disadvantages of the CB seem to be the difficulty in searching the large number of existing CBBs for matched grafts, the variation in the required handling, and, finally, uncertainty concerning aspects of post-thawing cell recovery and overall quality of CB among the CBBs.

A number of strategies have been proposed in order to overcome these limitations resulting in experimentation followed by trials. In Chapter 11, these strategies are presented: from administration of multiple units, to special pretransplantation regimens, passing from expansion and further manipulation of CB-derived cell subpopulations. The field of CBT is not static and novel approaches are being evaluated with the goal of optimizing transplantation outcomes.

Hematological malignant and nonmalignant diseases remain the main indications of CB, which is seen as an alternative to BM or peripheral blood stem cells (PBSC). Chapter 7 focuses on allogeneic related and unrelated transplantation in children and adults, giving a comprehensive view of the state of the field.

But other clinical conditions suitable for CBT have emerged. Chapter 8 is dedicated to the treatment of primary immune deficiencies and metabolic diseases by CBT. As these diseases affect children, CB, in some cases, in conjunction with *assisted reproductive technology* and preimplantation screening, is a precious graft source, especially in cases where the delay in engrafting is of paramount importance. Another application field is the use of CB in the treatment of autoimmune diseases, as described in Chapter 9. Following the successful use of HSCT for the treatment of severe autoimmune diseases, the use of CB with the added benefit of CB-derived MSCs with known immune-modulatory properties is very promising. The use of CB-derived immune-modulating T regulatory cell (Treg)s and MSCs in inflammatory disorders is examined in Chapter 10, and in Chapter 12, an immunomodulation model for treating diabetes with CB-derived cells is described.

Over those years much has been learned about the properties of CB hematopoietic progenitors and their clinical applications. But, CB is a complex tissue containing several subpopulations of nonhematopoietic cells, summarized in Chapter 2, that have unique functional characteristics and distinct differentiation capacity, and have their origins in different stages of fetal development. A whole chapter (Chapter 5) is dedicated to the endothelial cell compartment of CB, and their potential use as therapy. The nonhematopoietic stem cells that have been isolated from CB such as MSCs and endothelial cells, etc., can be grown and differentiated in various tissues and potentially in the near future, the CB cells will be used for the treatment of diseases such as diabetes, arthritis, burns, neurological disorders, and myocardial infarction.

This leads us away from CB transplantation and toward the realm of regenerative medicine. This is as yet an uncharted territory for now most of the work has been carried out using animal models and clinical trials have not yet yielded definitive results. CB, as described in Chapter 13, is being used as a source of cells for trials for neurological diseases and conditions, although routine administration (as in hematological diseases) is not for the near future. Another potential use is as a source of induced pluripotent stem cells (iPSCs). The biology and possible applications of iPSCs are presented in Chapter 14, as well as issues related to their eventual clinical use and ongoing trials.

The first efforts in CB banking were initiated in the laboratory of *Broxmeyer* at the Indiana University School of Medicine, where 7 of the first 10 units of CB collected for allogeneic transplantation use, were stored. These preliminary results led to the institution of CBBs. Besides many similarities, CB banking and traditional blood banking show a number of important differences.

The first CBBs were established in 1993 at the New York Blood Center (National CB program) by Pablo Rubinstein, in Milan (Milan CBB) by Girolamo Sirchia, and at the Bone Marrow Donor Center in Dusseldorf by Peter Wernet. These first three programs aimed at the implementation of large repositories of cryopreserved CB collected from healthy newborns. National regulatory agencies and transplant centers are aware of the need for international standards in order to promote quality throughout all phases of CB banking, with the production of high-quality CB units for transplantation. These standards, which include all practical aspects of CB banking, such as mother's informed consent, collection techniques, labeling and identification, infectious and genetic disease testing, HLA typing, methodology of cell processing, cryopreservation, transportation, and release, have been extensively published.

Two major models of CBBs that have emerged: those dedicated to the collection and distribution of CB units from unrelated donors for allogeneic transplantation, routinely called "public" CBBs; and those dedicated to the storage of CB for related and autologous transplantation, often referred as "private" banks. Chapter 20 discusses the value of the two models regarding the best interest of the newborn/donor and the factors influencing the parents' choice. Most of the clinical usage covered by this book concerns unrelated allogeneic banks, but one should not forget that the majority of cryopreserved CB units in the world are in fact stored for personal use. There is a whole industry centered on this, as is very thoroughly explained in Chapter 23: on the one hand publicly funded public banks and on the other for-profit commercial banks; both business models are analyzed for their efficacy, cost-effectiveness, and also their perspectives from a fiscal point of view.

A whole section of this book is dedicated to the regulatory and practical aspects of CB banking. Part of the attraction of CB as a source of HSCs for transplantation comes from the fact that it is a readily available and fully traceable product, of known quality, thoroughly tested and prepared according to stringent standards. In addition, CB collection poses no risk to the donor and provides a unique opportunity for recruiting donors from ethnic minorities. These donors, who are traditionally underrepresented in unrelated volunteer donor registries, can provide a higher frequency of rare haplotypes, thus reducing the time from initial donor search to final selection to a median time of 25–36 days, rather than months, typically required for an unrelated donor.

The whole process of banking is explained in successive chapters, starting with an overview of the CB banking process in Chapter 15 that takes the reader from the initial donor recruitment and the collection of CB after obtaining maternal consent, to the processing that takes place in dedicated laboratories, the testing and HLA typing to the subsequent selection and administration of a CB unit for transplantation. It continues with a clinician's point of view in Chapter 16: how the selection of the most suitable available unit for a given patient is made, and how the quality indices influence this choice. A large part of this chapter is dedicated to the importance of HLA matching and the influence a particular strategy regarding HLA compatibility can have on transplantation outcome.

The need for standards in order to ensure the quality of the administered product was recognized early on, with the first accreditations taking place in the 1990s. Today, accreditation standards are but a part of integrated quality management schemes that regiment all aspects of CBB operation and extend to the transplantation centers, as described in Chapter 17. In addition, legislators have set some general sets of rules for CB banking establishments (whatever their character: public or private) in the developed world, with developing emerging nations following, as CB is seen as a national resource that can in part ensure the well-being of the population. Chapter 18 is dedicated to the regulations governing those establishments, with emphasis on EU and North American legislation, and a general overview is given for the rest of the world.

Emphasis is also placed on the international cooperation and the existence of organizations like WMDA that regulate the distribution of CB units. Their operation is described in Chapter 19, and it encompasses all stages from the initial search from donor to the feedback from the end user.

CBBs are relatively new entities, and their role is most probably not fully developed. Regulatory and financial issues aside, the existence of CBB has some impact on society as a whole. Although CB appears to be conflict-free from ethical and moral viewpoints, the development of CBBs, both public and private, and their potential use in the broader field of regenerative medicine as opposed to the well-defined scope of HSCT, raises ethical dilemmas that will need to be contemplated in a rapidly advancing scientific context in a diverse but globalized environment. These are discussed in Chapter 22, while Chapter 24 discusses the place of CBB in Public Health Policy planning in the EU. Chapter 21 gives some insight into what could be the future of CBBs, with regenerative medicine applications, such as those previously described, being integrated in their operations.

The future of CBBs will not be limited to HSCT, but will include the fields of immunotherapy and regenerative medicine. Researchers and clinicians could take advantage of the whole spectrum of stem cells that make up CB, the cells' plasticity and scientific advances in the fields of induced differentiation (as with endothelial progenitors), pluripotency (CB as a source of iPSC), or immunomodulation. CBBs have the advantage of already satisfying a number of quality and regulatory requirements for clinical administration, as they already follow stringent protocols with an emphasis on safety and traceability, and already adhere to Good Manufacturing Process standards.

In this book, we have tried to give an up-to-date description of the field of CB in human medicine. We believe that we are at the beginning of an era when the concept of advanced therapies and regenerative medicine will finally be translated into everyday practice, and that CB could very well be at the center of this: this is reflected by the title of the book which emphasizes the importance CB could achieve in medicine.

Section II

# Cord Blood Cells Biology

Chapter 2

# Cord Blood Content

Gesine Kögler, Julia Bosch, Stefanie Liedtke and Teja Falk Radke
*Institute for Transplantation Diagnostics and Cell Therapeutics, University of Duesseldorf Medical School, Duesseldorf, Germany*

### Chapter Outline



## 1. BIOLOGICAL BACKGROUND OF CORD BLOOD CELLS—DEVELOPMENT OF HEMATOPOIETIC AND NONHEMATOPOIETIC CELLS

Cord blood (CB) is characterized by a unique richness in hematopoietic stem and progenitor cells, particularly those "early" cells which are detected in in vitro assays like the long-term culture-initiating cell (LTC-IC) assay and high proliferative potential-colony forming cell (HPP-CFC) assay or in vivo due to their potential to repopulate non-obese diabetic/severe combined immunodeficiency (NOD/SCID) mice (SRC—SCID repopulating cells). It has been extensively shown that hematopoietic stem cells (HSC) develop during embryogenesis and fetal life in a complex process involving multiple anatomic sites and niches (yolk sac, the aorta–gonad–mesonephros region, placenta, and fetal liver),[1] before they colonize the bone marrow (BM), as summarized recently.[2] As fetal and neonatal hematopoietic cells are markedly different from adult HSC with respect to their proliferative capacity, it was conceivable that different mechanisms and/or niches control engraftment and self-renewal of HSC during fetal and adult life. Since fetal blood is formed in close association with organs, the search for cell functions as niches similar to cell types present in adult BM environment (osteoblasts, endothelial cells, fibroblasts, reticular cells) was a logical consequence.

Although the HSC contained are currently the most relevant cells in CB with regard to clinical application, it also contains nonhematopoietic cell types which bear interesting properties that can potentially be utilized in regenerative medicine.

## 2. ENDOTHELIAL CELLS IN CB

Endothelial progenitor cells (EPCs) have been investigated as a potential source of cells for vascular repair. First described in 1997,[3] EPCs have been characterized by many investigators based on their morphology and surface antigen expression,[4] but frequently without stringent in vivo analysis of function.[5]

By the group of Mervin Yoder,[6] endothelial cord forming cells (ECFCs) in CB have been shown to be the only circulating cells that possess all the characteristics of an endothelial cell progenitor, including distinct functions. To isolate ECFCs, CB-derived mononuclear cells (MNC) or $CD34^+/CD45^-$ cells are plated on a collagen-coated surface and form adherent colonies with a cobblestone-like morphology between day 7 and 14.[6] ECFCs are rare cells, found at a concentration of about 0.05–0.2 cells/ml in adult

Cord Blood Stem Cells Medicine. http://dx.doi.org/10.1016/B978-0-12-407785-0.00002-5

peripheral blood. They are enriched in human umbilical CB, being found at a concentration of about 2–5 cells/ml. ECFC can be enriched from each CB sample (fresh or cryopreserved) applying the isolated CD34$^+$-subpopulation as a basis. ECFC progeny express the cell surface antigens CD31, CD105, CD144, CD146, von-Willebrand factor, and kinase insert domain receptor (KDR), but do not express the hematopoietic or monocyte/macrophage cell surface antigens CD14, CD45, or CD115.[6] Additionally, they are characterized by uptake of acetylated-low-density lipoprotein (AcLDL). Functionally ECFC progeny form tubes when plated alone and form de novo functionally active human blood vessels in vivo. One potential clinical use of ECFCs is in the treatment of patients with ischemia and defective wound healing due to impaired neoangiogenesis.[7] The authors state that the ability of implanted endothelial cells to form a vascular network when the host's angiogenic response is inhibited suggests that this strategy could be useful in treating patients with impaired wound healing. These and other reports suggest that ECFCs represent an excellent cell source for vascular engineering strategies. While there are not so many data available of the use of ECFCs in human clinical trials, the results with preclinical rodent studies provide some hope for patients who suffer from poor vascular function. Moreover, based on their growth kinetic, they are interesting candidates for tissue engineering in combination with MSC, induced pluripotent stem (iPS) cells, or mature tissue cells.

## 3. STROMAL CELLS IN CB AND CORD TISSUE AS COMPARED TO BM

The heterotopic transplantation of BM results in the formation of ectopic bone and marrow.[8] This "osteogenic potential" is associated with nonhematopoietic stromal cells coexisting with HSC in the BM.[9,10] Friedenstein and colleagues originally called these cells "osteogenic" or "stromal stem cells,"[10,11] in the following years the terms "mesenchymal stem cells," "mesenchymal stromal cells," or "skeletal stem cells" have been widely used in the literature.[12] In the present article, the BM-derived nonhematopoietic cells are referred to as "bone marrow mesenchymal/multipotent stromal cells" (BM MSC). Several groups proved the in vivo osteogenic potential—including the recruitment of hematopoietic cells of recipient origin—of BM MSC after transplantation on a hydroxyapatite scaffold.[13,14] After the original reports by Friedenstein, many other sources of MSC-like cells were described, for example adipose[15] or fetal tissue. In 2000, the first adherent cells (also termed "MSC") from CB (as summarized in Koegler et al. [16]) or in the umbilical cord tissue,[17,18] which revealed an immunophenotype (CD45$^-$, CD13$^+$, CD29$^+$, CD73$^+$, CD105$^+$) similar to BM MSC, however without any in vivo reconstitution studies. This was accompanied by many other publications, as summarized by Kogler et al.[18] In 2004, our group was able to detect cells in CB with a different proliferative potential, the so-called unrestricted somatic stromal cells (USSC),[19] and in the following years these data were confirmed by other groups (Figure 1).[20–22]

Several markers described in the literature for defined subtypes of MSC from BM, e.g., CD271, CD140b, STRO-1, GD2, or NG2 could not be correlated to functional subpopulations in CB.

In the publication of 2004,[19] we described USSC as a homogenous cell population with respect to their phenotype. During the following years, further detailed characterization in vitro and in vivo applying clonal cell population

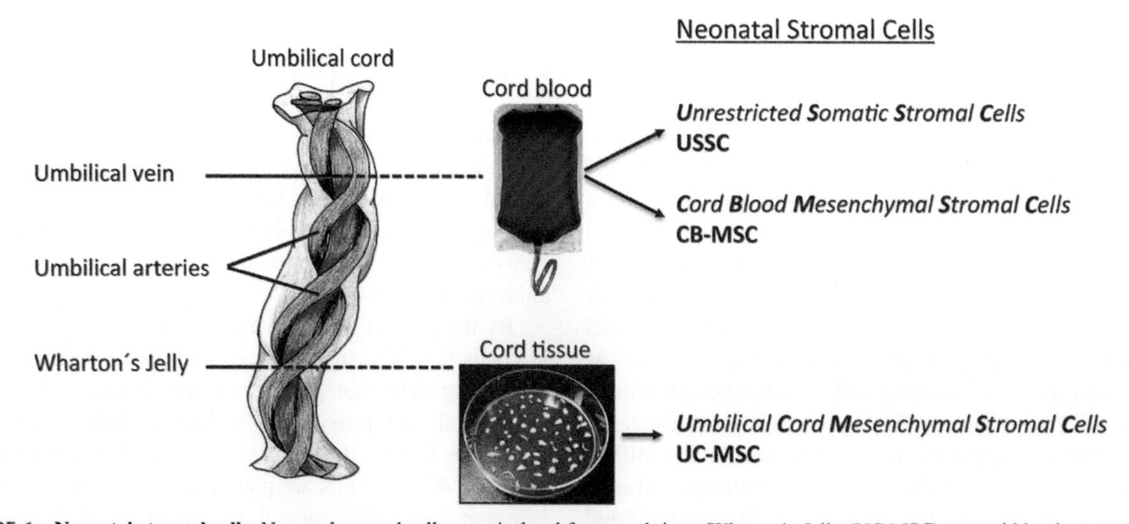


**FIGURE 1 Neonatal stromal cells.** Neonatal stromal cells were isolated from cord tissue/Wharton's Jelly (UC MSC) or cord blood, respectively. In cord blood, two distinct nonhematopoietic stromal cell populations were described so far: USSC and CB MSC. UC MSC, umbilical cord mesenchymal stromal cells; USSC, unrestricted somatic stromal cells; CB MSC, cord blood mesenchymal stromal cells.

isolated and expanded from CB, clearly revealed distinct cell populations.[23,24] We termed them according to the revisited MSC concept:[25] USSC and CB mesenchymal/multipotent stromal cells (CB MSC).[26] We could also define that clonal USSC and CB MSC lines differ most likely in their developmental origin reflected by a distinct homeobox (*HOX*) gene expression and expression of the delta-like 1 homolog (DLK-1), resulting in a completely different differentiation and regeneration in vivo applying specific models for neural, cardial, liver, and mesodermal (skeletal) regeneration. About 20 (out of 39) *HOX*-genes are expressed in CB MSC (*HOX*-positive), whereas native USSC (*HOX*-negative) reveal no *HOX*-gene expression.[27] In addition, USSC display a lineage-specific lack of the adipogenic differentiation potential along with the expression of DLK-1 (Figure 2).[23,24,28]

During recent years more than 300 cell lines were generated, characterized, and cryopreserved. Moreover, it was clearly documented that neonatal CB cells delivered at birth (>36 weeks of gestation) do not contain any OCT4A positive "embryonic-like" cells, and that the majority of data published by other groups were misleading due to artifacts based on OCT4A pseudogenes.[29] Due to the developmental status, it is unlikely that OCT4A is found in fetal, neonatal CB of 36 weeks gestation.[30] Moreover, more than 25,000 allogeneic CB transplantations have been performed to date without any germ cell tumor formation (including teratoma). In contrast to iPS cells (also from CB cells)[31] that form teratomas, the absence of tumor formation in clinically relevant models (nude mice) is one of the most important features indicating why native neonatal CB subpopulations are promising candidates in regenerative cell-based approaches.

Beside the cell populations described above, other cells with "embryonic characteristics" were claimed to be present in CB. These so-called very small embryonic-like (VSEL) cells were described as being pluripotent and positive for expression of OCT4.[32] As of today, when nearly every cell can be reprogrammed to a pluripotent state with a defined set of transcription factors resulting in a true "pluripotent nature," this puts the idea of pluripotency in CB to a break. In addition, recent reports doubt the existence of these cells.[33] This resulted in a report in Nature News, summarizing studies refuting the existence of VSEL cells in tissue, including CB.[34] Since there are many publications about the regenerative ability of CB cells, we will try to focus on the ones that, in addition to the ECFC already highlighted above, could have therapeutic applicability based on cell numbers generated as well as functional data available in vitro and in vivo.

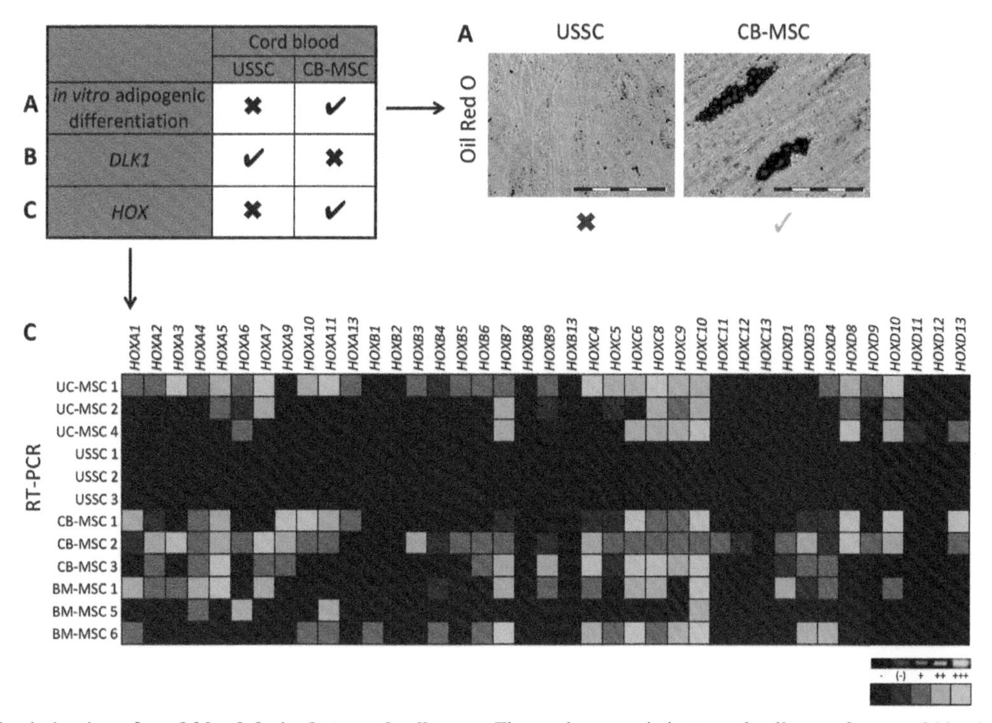


**FIGURE 2 Discrimination of cord blood-derived stromal cell types.** The nonhematopoietic stromal cell types from cord blood—USSC and CB MSC—can be distinguished on the basis of differentiation potential and gene expression. (A) In vitro adipogenic differentiation potential (21 days). Lipid vacuoles were visualized by Oil Red O staining. Scale bar: 100 µm. (B) RT-PCR analyses to evaluate the expression of *DLK-1* in USSC and CB MSC (Kluth et al. 2010, Stem Cells Dev). (C) RT-PCR of *HOX*-genes in the indicated cell types (black: absent (−); green: present, different expression levels are reflected by the color-intensity). UC MSC, umbilical cord mesenchymal stromal cells; USSC, unrestricted somatic stromal cells; CB MSC, cord blood mesenchymal stromal cells; BM MSC, bone marrow mesenchymal stromal cells; DLK-1, delta-like one homolog; HOX, homeobox.

## 4. ISOLATION, EXPANSION, AND CHARACTERIZATION OF CB-DERIVED ADHERENT CELLS FROM CB

Cultures of USSC and CB stromal cells are generated by the same method. Briefly, blood is collected from the umbilical cord vein directly after clamping, with informed consent of the mother. MNC are obtained by density-based Ficoll gradient separation and are subsequently cultured in plastic culture flasks containing serum-enriched medium supplemented with defined amounts of dexamethasone. As soon as colonies of adherently growing cells are observed, usually after 7–21 days in culture at 37 °C with 5% $CO_2$ and humidified atmosphere, the dexamethasone is omitted to avoid unwanted triggering of osteogenic differentiation.

Classification of the adherent cells into USSC and CB MSC can be assessed only after generation by determining the adipogenic differentiation potential, expression of *DLK-1* as well as *HOX*-gene expression. Over recent years, we initiated adherent cell cultures from 1009 CB samples, from which 40% ($n=394$) gave rise to an average of 1–11 colonies per individual blood sample. Due to their very low frequency in CB (1–100 per $1*10^{-8}$ MNC),[23,24,28] other fetal sources (cord) were also used,[35,36] however, resulting in stromal cells had a completely different functionality, as discussed below.

## 5. GENERATION OF CELL CLONES AND CLONAL POPULATIONS

The availability of clones as well as clonal populations is mandatory to characterize stromal populations derived from CB as compared to those derived from BM.

Colony-clonal populations were obtained during generation of cell lines by applying special cloning cylinders. In this case, cell lines were generated as described before and, if distinct, separate colonies were observed, a cloning cylinder was attached on a single colony and cells were trypsinated according to the standard protocol. While these cells could be regarded as clonal in being derived from one single developing colony, expansion of single cells is mandatory for true clonality. Therefore, cells of one colony or of already established cell lines, respectively, were plated at low density into six-well cell culture plates. Employing an AVISO CellCelector™, single cells were picked and transported to a defined destination well of a 96-well cell culture plate. For verification, pictures were taken before and after each picking process to document successful single cell selection. By subsequent cultivation, initially with preconditioned medium, clonal lines were established which then could be expanded under standard conditions. In the case of cells picked from primary colonies, the remaining cells were expanded as bulk culture and referred to as initial cell line.

## 6. USSC AND CB MSC

USSC and CB MSC exhibit a comparable proliferative potential (Figure 3(A)[36]), and can be distinguished on the basis of gene expression and differentiation potential (Figure 2(A)). Both cell populations can be expanded under conditions that conform to clinical requirements ("good manufacturing practice;" GMP).[37] Here, the target cell number to be reached is more than $1.5\times10^9$ cells within 30 cumulative population doublings (corresponding to passage 4–5).

Applying in vitro adipogenic differentiation assays, CB MSC, but not USSC, can differentiate into adipocytes (Figure 2(B)). Former results indicated a correlation of this absent adipogenic potential and the expression of *DLK-1* in USSC, since USSC but not CB MSC express *DLK-1*. Therefore, this marker can be applied to discriminate the CB stromal cells on a transcript but not protein level.[23]

Applying microarray- and PCR-analyses, the presence of *HOX*-genes was defined as a distinguishing feature between USSC and CB MSC besides the expression of *DLK-1*: USSC do not express *HOX*-genes, while CB MSC are *HOX*-positive (Figure 2(C)[27]).

In addition to the discriminating features presented in Figure 2, further differences between USSC and CB MSC were detected. USSC possess a higher hematopoiesis-supporting capacity in coculture experiments.[24] With respect to the immunophenotype, CB MSC reveal a stronger expression of CD146 (melanoma adhesion molecule, MCAM) than USSC. During flow cytometric analysis, no endothelial ($CD31^{negative}$), leukocytic ($CD45^{negative}$), epithelial ($CD326^{negative}$), or antigen-presenting ($HLA\text{-}DR^{negative}$) phenotype was detected in any cell type analyzed.[36]

## 7. GENERATION OF ADHERENT CELLS FROM THE WHARTON'S JELLY (CORD)

The umbilical cord consists of one vein and two arteries, which are embedded in a connective tissue, the so-called "Wharton's Jelly" (Figure 1). Many protocols to isolate stromal cells from Wharton's Jelly (often referred to as umbilical cord mesenchymal stromal cells, UC MSC), with or without an enzymatic digestion step, have been published over time. In our group, UC MSC were isolated as described in[36] by the outgrowth of stromal cells from umbilical cord tissue cut into small pieces and cultivated in standard culture medium. Commonly, in vitro differentiation assays were performed to prove the identity of the isolated cells as being MSC-like with the potential to give rise to cells of skeletal tissues (osteoblasts, chondroblasts, adipocytes).[38] In accordance with our results, further studies reported a restricted differentiation potential of so-called UC MSC concerning in vitro[39,40] and in vivo[41] assays. Compared to USSC and CB MSC from CB as well as to

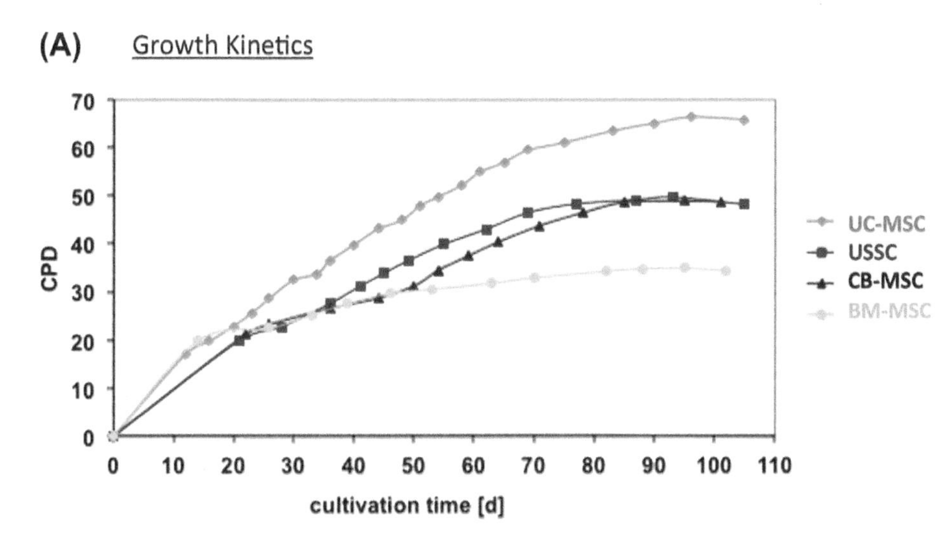


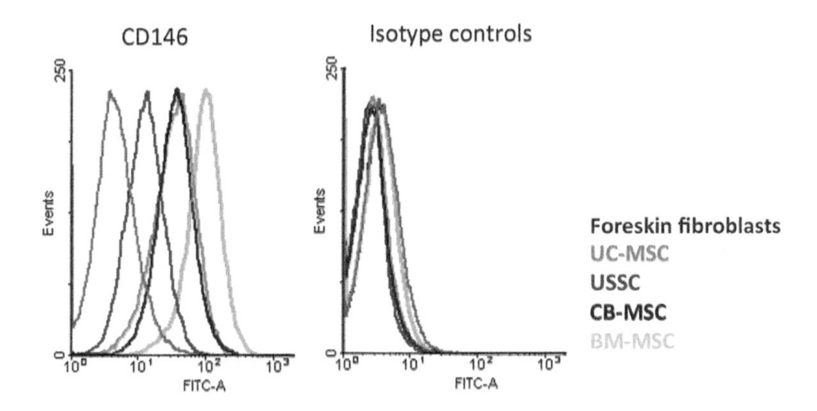


**FIGURE 3 Characterization: growth kinetics and immunophenotype.** (A) Representative growth kinetics. (B) Flow cytometric analysis of CD146 and the corresponding isotype controls. Foreskin fibroblasts were used as negative control. CPD, cumulative population doublings; UC MSC, umbilical cord mesenchymal stromal cells; USSC, unrestricted somatic stromal cells; CB MSC, cord blood mesenchymal stromal cells; BM MSC, bone marrow mesenchymal stromal cells.

BM MSC, these cells derived from the cord tissue demonstrated the highest proliferation potential (Figure 3(A)), but the lowest (or completely absent) differentiation potential toward the osteogenic, chondrogenic, and adipogenic lineage.[36] Therefore, it also has to be questioned whether the term mesenchymal or multipotent does apply at all to stromal cells from cord tissue.

## 8. GENE EXPRESSION PROFILES

Microarray gene expression analysis is a useful tool in comparing different cell types and provided further insight into the gene expression profile of CB-derived USSC and CB MSC in comparison to BM MSC.[35]

The expression of the majority of genes was shared by USSC, CB MSC, and BM MSC (Figure 4(A)). The expression pattern of USSC and CB MSC was more similar to each other than to BM MSC, which reflects the common CB origin.

These arrays also gave hints for differences in regard to differentiation capacity, and subsequent analysis of specific gene expressions was carried out. Special attention was paid to genes associated with the process of osteogenesis and in according quantitative PCR experiments, a stronger expression of bone sialoprotein (*BSP*), osterix (*OSX*, correctly *SP7*), bone morphogenetic protein 4 (*BMP4*) and osteocalcin (*OC*) in BM MSC compared to the CB-derived cell types was shown (Figure 4(B)[35]). Unlike BM MSC, the CB stromal cells lacked the typical bone signature.

Thus, USSC and CB MSC exhibit a more immature status than BM MSC with respect to the genetic control of bone formation.

## 9. CORRELATION OF *HOX*-GENE EXPRESSION AND REGENERATIVE POTENTIAL

The vertebrate skeleton consists of elements which vary regarding their embryonic origin. The craniofacial skeleton is derived from the cranial neural crest, the axial skeleton from the paraxial mesoderm, and the limb skeleton from the lateral plate mesoderm.[42] Depending on

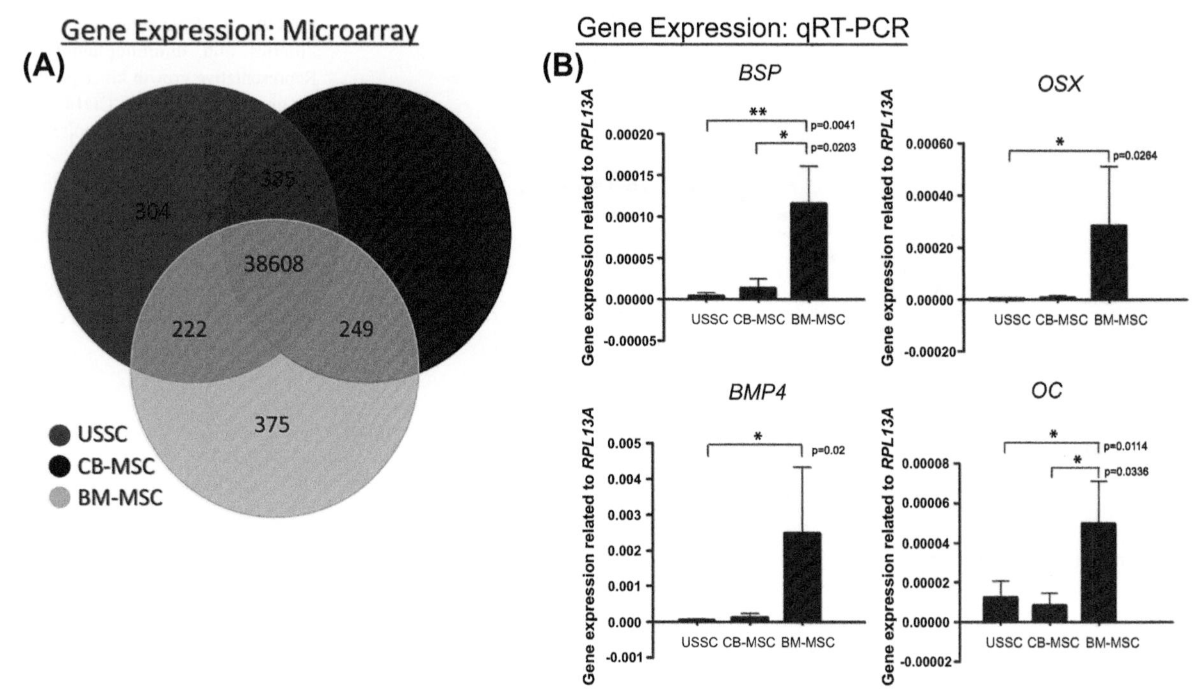


**FIGURE 4 Gene expression profile.** (A) The gene expression of USSC, CB MSC, and BM MSC (each in triplicate) was analyzed applying PrimeView™ Human Gene Expression Arrays. The Venn-diagram illustrates the common and unique expression of genes. The gene expression was filtered for each cell type: A probe set had to be expressed above the background (twentieth and hundredth percentile of the raw signal distribution) in at least two out of three replicates. This resulted in 39,519 transcripts for the "group USSC," 39,493 transcripts for CB MSC, and 39,454 transcripts for BM MSC which were compared. (B) Quantitative RT-PCR analyses of genes expressed differentially in USSC, CB MSC, and BM MSC. Illustrated are the arithmetic means and standard deviations of at least three different cell lines per cell type. $*p=0.01–0.05$, significant; $**p=0.001–0.01$, very significant (unpaired *t*-test). RPL13A was used as housekeeping gene. USSC, unrestricted somatic stromal cells; CB MSC, cord blood mesenchymal stromal cells; BM MSC, bone marrow mesenchymal stromal cells; BSP, bone sialoprotein; OSX (SP7), osterix; BMP4, bone morphogenetic protein 4; OC (BGLAP), osteocalcin; RPL13A, ribosomal protein L13A.

the embryonic origin, a different *HOX*-gene expression pattern can be detected. While most parts of the skeleton—including the corresponding tissue-derived progenitor cells—are *HOX*-positive, the craniofacial skeleton is *HOX*-negative.[43,44]

Leucht and coworkers performed transplantation experiments in mice and analyzed the influence of the *Hox*-gene expression.[44] As illustrated in Figure 5(A), skeletal progenitor cells derived from the mesodermal tibia (*Hox*$^{positive}$) were transplanted in defects of the ectodermal mandible (*Hox*$^{negative}$), which resulted in cartilage formation instead of bone regeneration. On the contrary, the transplantation of mandibular cells (*Hox*$^{negative}$) in a defect of the *Hox*-positive tibia resulted in effective bone repair (Figure 5(A)). The *Hox*-positive cells retained their *Hox*-gene expression after transplantation in a *Hox*-negative tissue. In contrast, *Hox*-gene expression was adapted in the previously *Hox*-negative cells following transplantation in a *Hox*-positive tissue. These results led to the conclusion, that the *Hox*-negative status of the mandibular cells and the potential to adapt a *Hox*-gene expression after transplantation may be regarded as beneficial concerning the regenerative potential.

As described above, CB-derived USSC lack the expression of *HOX*-genes, while CB MSC are *HOX*-positive (Figure 2(C)). As illustrated in Figure 5(B), we transferred the in vivo model described by Leucht et al. to an in vitro coculture model.[26] USSC were tagged using the green fluorescent protein (GFP) and cocultivated with *HOX*-positive cells (UC MSC, CB MSC, or BM MSC) for 5 days. After coculture and cell sorting, the previously *HOX*-negative USSC expressed *HOX*-genes (Figure 6(A)). Analogous to the *Hox*-negative mandibular cells, which started to express *Hox*-genes after transplantation in the tibial defect (Figure 5(A)), USSC were able to adapt a *HOX*-gene expression after cocultivation with *HOX*-positive stromal cells. Besides the *HOX*-gene expression, the adipogenic differentiation potential was affected by the direct coculture. USSC, which failed to differentiate into adipocytes, exhibited an adipogenic potential after coculture with CB MSC (Figure 6(B)). The direct coculture represents a simple way to modulate the cell fate decision of neonatal stromal cells. Due to their ability to adapt to the surrounding tissue, USSC are a promising cell type for the potential clinical application in the future, if the mechanisms are enlightened and proven in vivo.

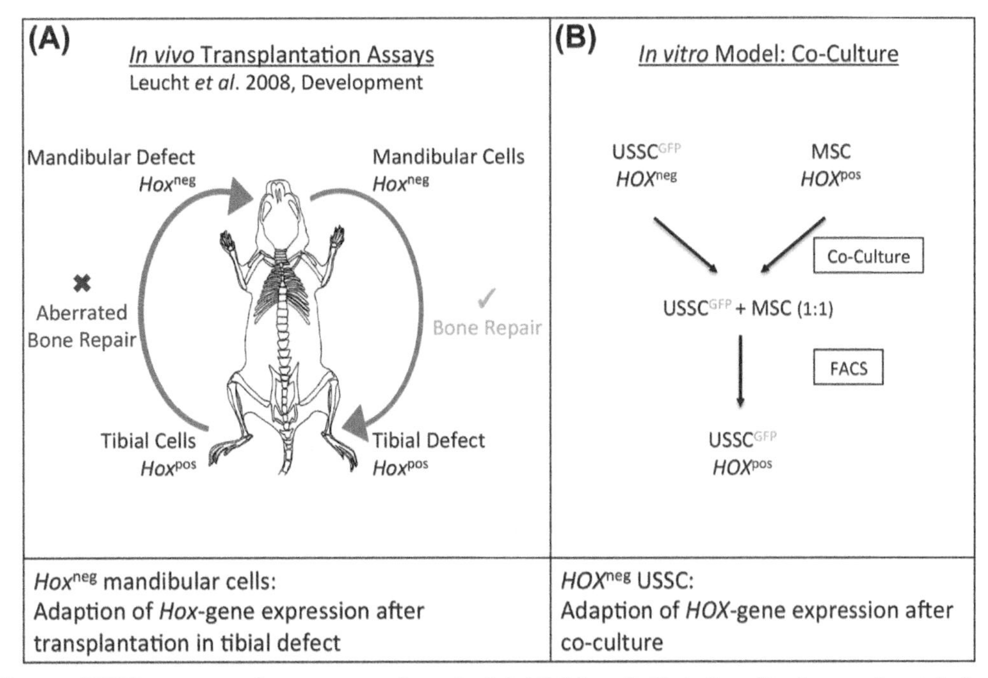


**FIGURE 5 Influence of *HOX*-gene expression on regenerative potential.** (A) Schematic illustration of in vivo experiments by Leucht et al. (2008, Development) analyzing the influence of *HOX*-gene expression of transplanted cells on bone repair. (B) Scheme of the in vitro coculture (5 days) of *HOX*-negative USSC with other *HOX*-positive cells. *HOX*, homeobox; USSC, unrestricted somatic stromal cells; MSC, mesenchymal stromal cells; GFP, green fluorescent protein; FACS, fluorescence-activated cell sorting.

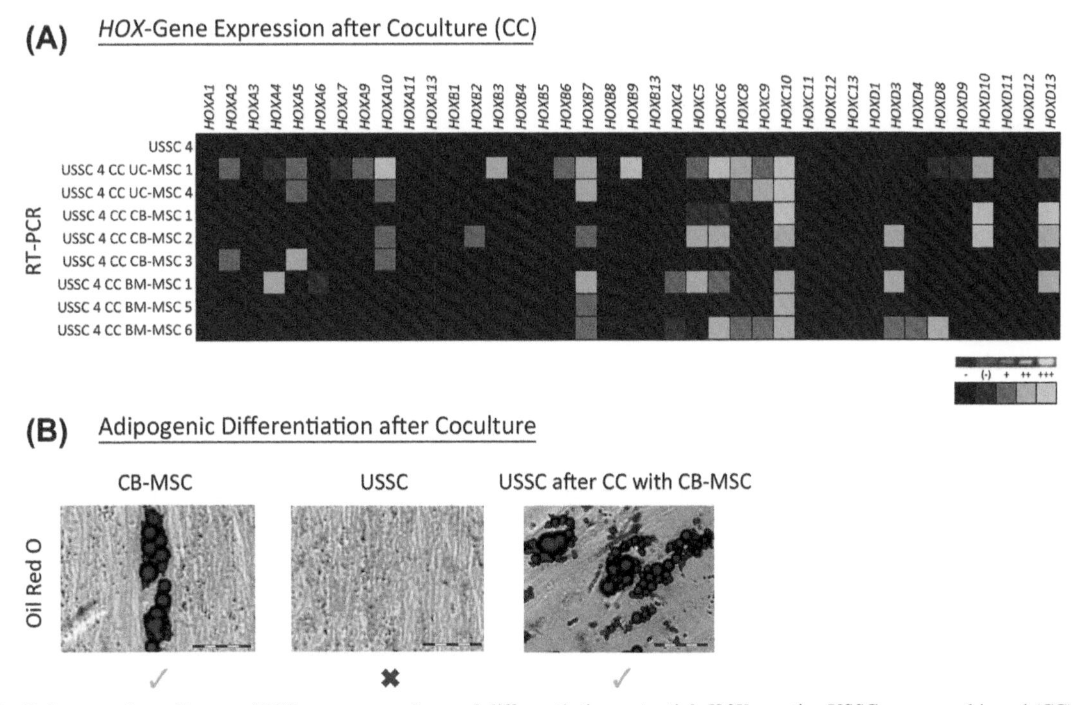


**FIGURE 6 Influence of coculture on *HOX*-gene expression and differentiation potential.** *HOX*-negative USSC were cocultivated (CC) with *HOX*-positive cells (UC MSC, CB MSC, or BM MSC) for 5 days, followed by GFP-mediated cell sorting. (A) RT-PCR of *HOX*-genes in the indicated cell types (black: absent (–), green: present, different expression levels are reflected by the color-intensity). (B) After FACS, USSC were cultivated for one further passage before adipogenic differentiation and Oil Red O staining (d21). Scale bar: 25 μm. *HOX*, homeobox; UC MSC, umbilical cord mesenchymal stromal cells; USSC, unrestricted somatic stromal cells; CB MSC, cord blood mesenchymal stromal cells; BM MSC, bone marrow mesenchymal stromal cells; GFP, green fluorescent protein; FACS, fluorescence-activated cell sorting.

## 10. BONE AND CARTILAGE FORMING POTENTIAL OF CORD BLOOD STROMAL CELLS

For the purpose of bone and cartilage regeneration (lost due to trauma, surgical resection of tumors, skeletal disorders, and aging), cell-based strategies are currently the gold standard of treatment. The use of freshly isolated CD146-positive BM MSC, in contrast to the extensively expanded counterpart, provides an important therapeutic tool for bone regeneration, although not for cartilage.

Bone is a highly vascularized connective tissue undergoing continuous remodeling and regeneration processes. The intrinsic regeneration potential is initiated in response to injury, as well as during normal skeletal development reflected by continuous remodeling throughout adult life.[45] Many bone and cartilage associated diseases require regeneration in large scale, e.g., large bone defects also known as "critical size" defects due to trauma, surgical resection of tumor, infection, or skeletal disorders. Especially for treatment of these defects, a cell-based strategy is the most promising approach as long as a sufficient number of cells can be supplied. Clinically, stromal cells can be used as cell suspension expanded by culture or simply as BM concentrate.[46] In this context, ex vivo expanded mesenchymal stromal cells (BM MSC) have demonstrated their ability to function as a tissue repair model in manifold therapeutic applications investigated in clinical trials.[47] However, the outcome of tissue repair is strongly associated with the applied cell concentration, which is lower in BM transplants compared to cultured cells. Moreover, large-scale cell amplification by ex vivo expansion harbors the risks of dilution of the relevant osteogenic clones, xenogenic incompatibility, and cellular transformation.[10,48] For de novo cartilage repair there are no established methods available, simply based on the fact that adult BM does not contain the early chondrogenic progenitors in sufficient amounts to regenerate large areas of defects. For all clinical applications one should choose the best characterized cells for a directed and specific tissue repair. As shown in Figure 4 and summarized in Bosch et al.,[35] Kluth et al.,[23] Liedtke et al.,[26] and many other publications in vitro and in vivo, fetal stromal cells (both USSC and CB MSC) have specific signatures for bone and cartilage formation.

## 11. WHY DO WE HAVE THESE PROGENITORS OR ELUSIVE CELLS IN CB?

The correct formation of the skeleton during embryogenesis and fetal development, and its preservation during adult life is essential and is maintained through the complementary activities of bone-forming osteoblasts and bone-resorbing osteoclasts.[49] The stability and strength of bones is accomplished by mineralization of the extracellular matrix, leading to a deposition of calcium hydroxyapatite.[50] The osteogenesis can be split into two different processes: intramembranous and endochondral ossification.[51] The intramembranous ossification is characterized by mesenchymal cells that condense and directly differentiate into osteoblasts and thereby deposit bone matrix. This process of bone formation is limited to certain parts of the skull as well as part of the clavicle. All other bones of the skeleton are formed by endochondral ossification. The formation of skeletal elements by endochondral ossification begins with the migration of undifferentiated mesenchymal cells to the zones that are destined to become bone. The undifferentiated cells condense, resulting in an increase in cell packing and forming of the cartilaginous anlagen. This process is regulated by mesenchymal–epithelial cell interactions. The next step, the aggregation of chondrogenic progenitor cells into precartilage condensations, is dependent on cell–cell and cell–matrix interactions.[51] The following transition from a chondrogenic progenitor cell to a chondrocyte is marked by a change in the extracellular matrix composition. The chondrocytes thereby acquire a rounded morphology and undergo hypertrophy (substantial increase in size). This chondrocyte hypertrophy triggers the initial osteoblast differentiation from perichondrial cells. Blood vessels start to invade the cartilage from the perichondrium and thereby transport osteoclast cells into the bone to degrade the existing cartilage matrix producing marrow cavity. Additionally, the blood vessels transport perichondrial cells to nascent BM, where they differentiate into osteoblasts. Many different transcription factors and regulatory signals are involved in the endochondral ossification, such as the transcription factor of the sex-determining region Y (SRY)-related high mobility group box, SOX9,[50] the Runt domain-containing transcription factor 2 (RUNX2),[52] Osterix (OSX),[50] BSP,[53] and parathyroid hormone-related protein (PTHLH).[52] All of the transcription factors are regulated by a variety of developmental signals, including Hedgehog (HH) proteins, NOTCH signaling, WNT signaling, BMP signaling, and fibroblast growth factor (FGF)-signaling.[50] It has been shown already that distinct populations can be defined, each of them representing progenitors of skeletal development during fetal life.[27,54] The transcription factor analysis of the subpopulation suggests that the stromal components in CB are elusive cells circulating from different stages of fetal development.

For bone- and cartilage-forming cells it can be concluded that CB contains natural progenitors, however with a different signature in vitro and a distinct in vivo regenerative capacity as compared to BM MSC.

## 12. USSC AND MSC FROM CB SUPPORT HEMATOPOIETIC CELLS

Another potential application of neonatal stromal cells is as a supportive mean for hematopoietic cells in general and for HSC in particular. Although technical procedures and protocols have been optimized since Eliane Gluckman's first transplantation in 1988,[55] the limited number of cells available from one single CB still is one of the major hurdles for transplantation purposes. With a harvested volume of usually between 50 and 150 ml, total nucleated cell (NC) counts and numbers of HSC rarely exceed $2 * 10^9$ or $1 * 10^7$, respectively. With requirements of at least $3.7 * 10^7$ cells per kilogram body weight for NC, and a desired amount of more than $1 * 10^5$/kg for HSC, the utilization in adults is highly restricted.

Accordingly, great efforts were made to find appropriate ways to overcome this problem and to provide cell counts sufficient even for heavy-weight patients. Among the ideas pursued, two basic principles can be distinguished: the expansion of the hematopoietic cells prior to transplantation in vitro, and the improvement of engraftment during or after transplantation in vivo.

Both approaches have their own advantages and disadvantages which have to be taken into consideration. For example, in vitro expansion not only requires GMP-grade equipment and reagents, but also requires development a certain time ahead of transplantation. Additionally, exhaustion of stem cells might occur and result in impaired capability of long-term reconstitution. On the other hand, manipulating engraftment in vivo is much more complicated and it goes without saying that any possible negative side effect on the patient has to be thoroughly excluded.

Double cord blood transplantation (DCBT), since being described in 2005 by the group of John E. Wagner,[56] doubtlessly has become a standard method in clinics. Here, the property of CB to be less restricted in regard to HLA-matching is utilized to successfully combine two individual CB units for treatment of a single patient. However, this still might not be sufficient for heavy-weight patients and/or patients with rare HLA-type.

Cytokine-driven expansion is performed in vitro by addition of specific hematopoietic cytokines. These cytokines, namely stem cell factor (SCF) as the most prominent one, lead to an extensive proliferation already at low concentration and multiple protocols for cytokine cocktails are described in literature. While this expansion results in increased numbers of hematopoietic stem and progenitor cells, the influence on the stemness of these cells has to be evaluated. For example, already short expansion in medium enriched with SCF, thrombopoietin (THPO), FMS-related tyrosine kinase 3 ligand (FLT3LG), and interleukin 6 (IL-6) leads to a strong expansion of cells isolated by fluorescence-activated cell sorting (FACS). On day 3, nearly all resulting cells are positive for surface expression of CD34, but analysis has already revealed distinct changes on protein level.[57] Since cytokines are very potent, fine tuning can be considered as most critical in order to prevent cells from exhaustion and loss of long-term repopulating cells. Novel protocols target the NOTCH-pathway and might help in finally achieving efficient expansion of early stem and progenitors cells,[58] but currently the clinical application is restricted to partial expansion only (either of half a CB unit or, in case of DCBT, of one complete unit). Another aspect might also be the manipulation of not the total cell number, but the engraftment itself. Here, the application of prostaglandin ex vivo prior to transplantation has recently been reported as resulting in a more rapid engraftment in a clinical trial.[59]

Finally, as a third option, nonhematopoietic stromal cells are used as a supportive tool, either in vitro or in vivo. Here, CB-derived stromal cells demonstrate a promising effect which also bears potential clinical importance. Therefore, the following section will comprise and discuss data from cocultivation as well as cotransplantation in an animal model.

Multipotent/mesenchymal stromal cells, such as BM MSC, CB MSC, or USSC, secrete various cytokines. Among these cytokines, many are known for their influence on hematopoietic cells, either as inducer of proliferation, as chemoattractor, or for differentiation. In the hematopoietic niche located in the adult BM, stromal cells are involved in maintaining the quiescence of HSC but they also are responsible for the mobilization and chemotaxis of HSC after injury (e.g., through secretion of stromal-derived factor 1, which strongly attracts HSC via the chemokine receptor CD184, also know as CXCR4).

Assuming that these cells do not only provide a cytokine composition which is well tuned to HSC, but also provide cell–cell contact, which might be an additional need, cocultivation of stromal cells and HSC in order to mimic the natural hematopoietic environment is a sound approach for in vitro expansion prior to transplantation.

Since stromal cells do grow adherently on plastic surfaces, as in culture flasks or well plates, they form a so-called feeder layer without the need of further manipulation. Though, to prevent these cells from overgrowing, they have to be stopped from proliferating in long-term expansion. This can be achieved either by treatment with mitomycin C or by irradiation. Subsequently, either the total fraction of MNC or isolated CD34-positive HSC from CB can be seeded on these layers in medium, e.g., Dulbecco's modified Eagle medium (DMEM) with 30% fetal calf serum (FCS).

Within a few days, while controls of HSC in FCS-enriched medium without stroma results in maintenance or even loss of cells, a definite expansion of the nonadherent cells can be observed on various feeder cells, such as USSC, CB MSC, or BM MSC.

However, flow cytometric analysis of these cells after staining against human CD34 and human CD45 (pan-leukocyte marker) will reveal that expression of CD34 is not detectable on the surface of all cells and that the percentage of CD34-positive cells decreases with days in culture. For estimation of HSC-expansion, this percentage has to be set off against the total cell counts. Further analysis comprises the assessment of colony-forming cells (CFC), which is a more functional test. In this assay, defined numbers of cells are seeded in growth factor-enriched semisolid media that induces strong proliferation and differentiation in HSC. This leads to the formation of clearly distinguishable erythroid (red) or myeloid (white) colonies within 14 days of culture and correlates with the presence of stem and progenitor cells in the sample (a complete scheme of coculture and respective analysis is depicted in Figure 7(A)).

For CB-derived stromal cells, it was already published in 2005 that these are capable of expanding hematopoietic cells in coculture in specific medium (Myelocult H5100) more than 100-fold within 4 weeks,[60] which was distinctly higher than on BM MSC. Interestingly, while total cell count constantly increased, expansion of CD34-positive cells and CFC reached a maximum by day 14 and decreased with further coculture. This might indicate that an exaggerated expansion here also leads to exhaustion of early stem and progenitor cells.

However, more recent data demonstrate that this ability to support HSC is much stronger in USSC than in CB MSC. Even in short time expansions for a maximum of 14 days and using DMEM with FCS as medium, which is less supportive to HSC than Myelocult H5100, expansion of total cells, CD34-positive cells as well as of CFC was clearly detectable on USSC-feeder. Under these conditions, cocultivation on CB MSC also resulted in expansion of total cell count, but only a maintenance of CD34-positive cells and CFC could be achieved[24] (see Figure 7(B)), similar to stromal cells derived from BM (data not published).

While cocultivation on neonatal stromal cells might provide a gentle expansion of HSC in vitro, application as supportive cells in vivo is another promising approach. Since stromal cells from CB can be generated, expanded, and cryopreserved under GMP-conform conditions,[37] they are potentially directly applicable, e.g., in transplantation.

First of all, it had to be verified that these cells do not bear the risk of intrinsic tumorigenicity. According to experiments in which CB- and BM-derived stromal cells were injected subcutaneously in the flank of nude mice, no tumor formation related to the human cells was demonstrated.[61]

In a further step to evaluate effects on engraftment and hematopoiesis, USSC cells were cotransplanted intravenously with CB-derived HSC, and isolated by magnetic-activated cell sorting (MACS) into immune-deficient mice (NOD/SCID) after sublethal irradiation. Additionally, a group of animals received stromal cells

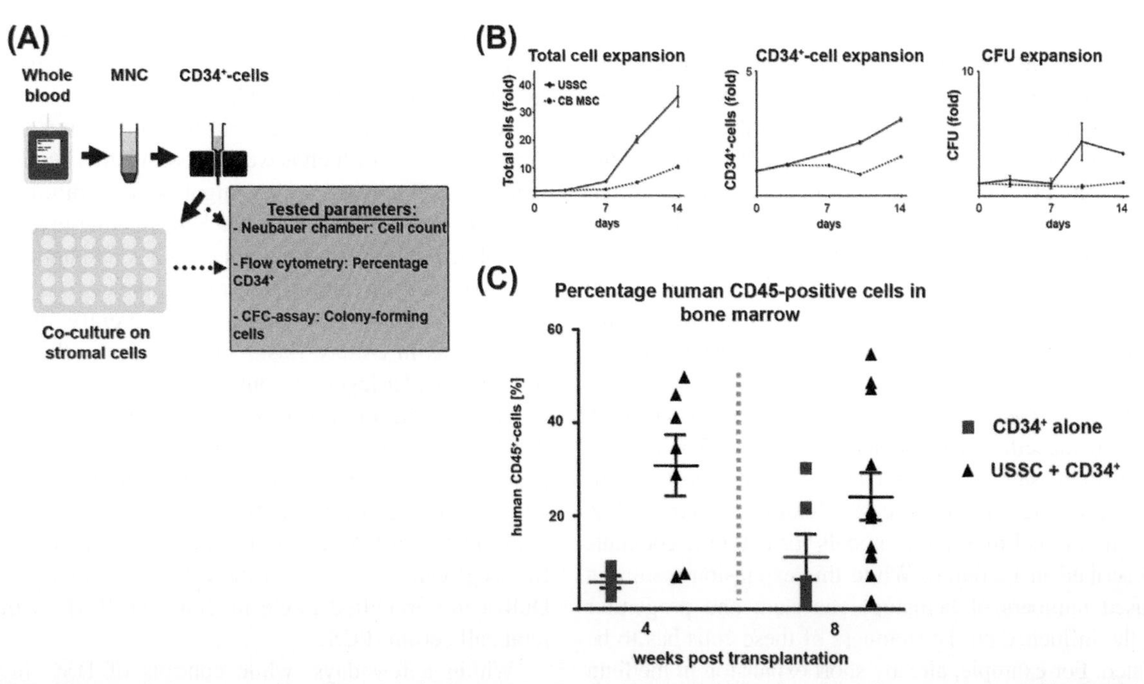


FIGURE 7 **Hematopoietic support of cord blood-derived stromal cells.** (A) Schematic illustration of coculture experiments with MACS-isolated CD34-positive cells on stromal feeder layers. (B) Under identical conditions, coculture of cord blood-derived HSC resulted in a higher expansion of total cells, CD34-positive stem/progenitor cells and colony-forming cells on USSC than on CB MSC. (C) In NOD/SCID-mice, cotransplantation of USSC resulted in higher percentages of human leukocytes in the bone marrow after 4–8 weeks. MNC, mononucleated cells; MACS, magnetic-activated cell sorting; CFC, colony forming cells; USSC, unrestricted somatic stromal cells; CB MSC, cord blood mesenchymal stromal cells.

only and another group HSC only. After 4 and 8 weeks, respectively, animals were sacrificed and analyzed. For assessment of reconstitution, occurrence of human hematopoietic cells in peripheral blood, spleen, and BM was analyzed by flow cytometry while solid organs (brain, heart, kidney, liver) were tested in immunohistochemistry by staining against human nuclei.

Albeit animals that received stromal cells only were completely negative for the presence of human cells in all organs tested and no difference could be observed in the percentage of engrafted human HSC between animals that were transplanted with CD34-positive cells alone or cotransplanted with stromal cells, the latter ones had a significantly higher percentage of human leukocytes (detected by expression of human CD45 in flow cytometry) in the BM.[61] While after 4 weeks cotransplantation resulted in a 5.2-fold higher amount, this difference was less marked after 8 weeks (2.1-fold higher in cotransplanted animals). This leads to the conclusion that probably a paracrine effect enhances the initial engraftment of the HSC and results in faster generation of more differentiated cells within the first period after transplantation. The impact observed was higher than that reported for other cotransplanted cells and also could be beneficial for human patients in regard to a faster reconstitution. Apart from this effect, CB-derived stromal cells are also reported to be more immunosuppressive than BM MSC and might help in avoidance of severe acute graft-versus-host-disease (GvHD). So far, clinical results are still rare but first reports are promising, including the combination of cotransplantation with the already established DBCT.[62]

Despite these clearly demonstrated effects on hematopoietic cells in vitro and in vivo, the mechanisms are yet not fully understood. However, it will at least in part be related to cytokines secreted by the stromal cell type utilized. For example, CB-derived stromal cells in general demonstrated at least a two-fold higher expression of SCF in quantitative real-time PCR than BM stromal cells, while vice versa, IGF-2 was expressed 10-fold more in BM MSC than in USSC or CB MSC. The findings that CB MSC do express a higher level of SCF, known as the most potent inducer of proliferation in hematopoietic stem and progenitor cells, is somewhat surprising since capacity for hematopoietic support in coculture expansion was comparable to that of BM MSC. Either the positive effect of SCF is counteracted by an inhibiting factor in CB MSC and BM MSC or enhanced in USSC. Analysis of additional cytokines revealed further differences between USSC and other stromal cells: IGFB-1 is expressed 10-fold higher than in CB MSC or BM MSC, while SDF-1 expression is five-fold lower (Figure 8(A)). Especially the lower expression of stromal-cell-derived factor 1 (SDF-1) might explain the differences in hematopoietic support, since this cytokine is known not only as a potent chemoattractant (in its secreted form[63]) for HSC, but also functions as an inducer of quiescence (if presented membrane-bound by stromal cells[64]).

Taken together, these results not only confirm that USSC are distinctly different from CB MSC or BM MSC, but also might indicate that USSC are of different origin than CB MSC. Chou and Lodish[65] reported a DLK-1-positive subpopulation in murine fetal liver cells with an increased capacity of hematopoietic support. With USSC being positive for DLK-1 on the transcript level, showing a hematopoietic support superior to comparable stromal cells and expressing some cytokines described by Chou and Lodish (Figure 8(B)), it can be speculated that these cells are descendants of the fetal liver while CB MSC are the progeny the of fetal BM cells.

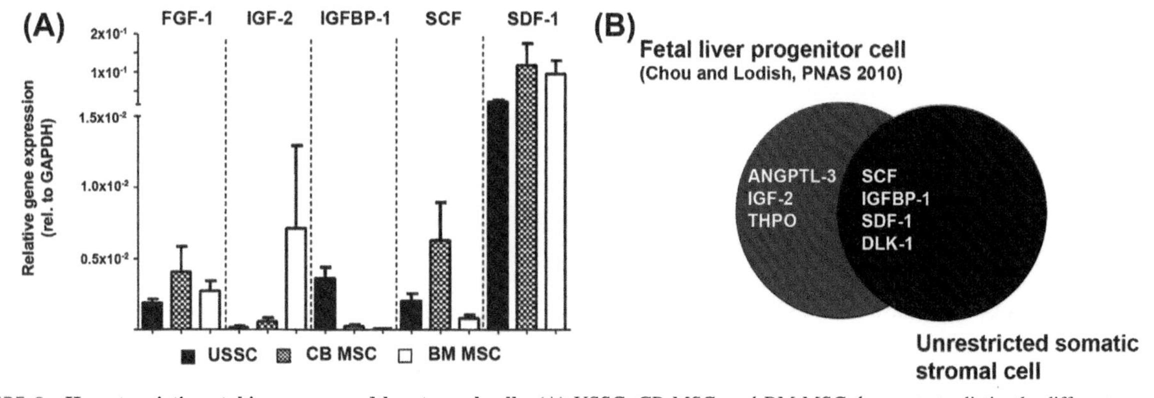


**FIGURE 8 Hematopoietic cytokines expressed by stromal cells.** (A) USSC, CB MSC, and BM MSC demonstrate distinctly different expression profiles for selected hematopoietic cytokines in real-time PCR analysis. (B) Comparing USSC with a subpopulation of murine fetal liver cell described by Chou and Lodish in 2010[65] reveals similarities with regard to expression of specific cytokines. GAPDH, glycerinaldehyde 3-phosphat dehydrogenase; FGF-1, fibroblast growth factor 1; IGF-2, insulin-like growth factor 2; IGFBP1, insulin-like growth factor-binding protein 1; SCF, stem cell factor; SDF-1, stromal cell-derived factor 1; ANGPTL-3, angiopoietin-like 3; THPO, thrombopoietin; USSC, unrestricted somatic stromal cells; CB MSC, cord blood mesenchymal stromal cells; BM MSC, bone marrow mesenchymal stromal cells.

## 13. LIVER REGENERATION AND POTENTIAL OF CB-DERIVED STEM CELLS TO UNDERGO HEPATIC DIFFERENTIATION

To date the clinical importance of cellular therapies for hepatic surgery is demonstrated by application of mature donor hepatocytes in certain disease forms. The marginal number of available donor tissues, and thus donor cells, determines the global restriction for therapeutic hepatocyte-based applications. Thus, appliance of differentiated stem cells could increase availability of hepatic cells for transplantation purposes.

The consideration of nonhepatic stem cells as sources for hepatocytes or hepatocyte-like cells in rodents rose after identification of donor cells in livers after stem cell transplantation.[66] Further studies revealed that a broad spectrum of extrahepatic stem cells repopulate rodent livers.[67] Nevertheless, it has been discussed controversially whether extrahepatic stem cells retain nonhepatic features after transplantation and whether the hepatic transformation is caused by cell fusion with hepatic cells or is committed intrinsically.[68] Other groups reported that transplanted CB stem cells express some liver markers but differ in many aspects from definitive hepatocytes.[69] A principal proof of the ability of extrahepatic stem cells to undergo hepatic differentiation is performed by in vitro differentiation analysis. Many scientists focused on differentiation of umbilical CB stem cells into the hepatic lineage. Different cell culture strategies were applied to induce differentiation. Mostly the authors described protocols reflecting embryonic development of the liver to achieve proper differentiation. For stromal cells from CB it could be shown to respond to signaling molecules of hepatic differentiation, followed by induction of hepatic gene expression, performance of hepatic functions, and in vivo liver repopulation capacity.[70]

First reports for USSC of hepatic differentiation could be reported after transplantation of USSC in a noninjury model, the preimmune fetal sheep model. USSC engrafted into host livers and revealed over 20% cells of total livers.[19] These cells could be characterized by human albumin and expression of hepatocyte paraffin 1 (HepPar1), further serum of born sheep showed human albumin content demonstrating the functionality of engrafted USSC to produce and release serum proteins.[19] In in vitro approaches it was reported that USSC differentiate into cells expressing liver markers like albumin, hepatocyte nuclear factor 4 α (HNF4α), and glycogen synthase 2 (GYS2) to different protocols[71] were applied. Ghodsizad et al.[72] could demonstrate the engraftment of USSC into mature sheep livers after portal injection. In this noninjury model over 80% of total cells around portal veins contributed to a human phenotype. Technically an adult sheep model was applied where portal vein cannulation was performed under systemic administration of heparin, while a port access sheet was placed into the portal vein. An angiography catheter was placed into the left portal vein and acute hepatic ischemia was induced by selective injection of micro beads. Subsequently, USSC were injected into the right portal vein. Sheep were sacrificed after 1 month and livers were analyzed by HepPar1 and human albumin staining of formalin-fixed sections. A majority of chimeric liver parenchymal cells showed HepPar1 and human albumin expression. In order to demonstrate hepatic-like phenotype resulting from differentiation and not from fusion events, single HepPar1 positive and negative cells were microdissected from 2-μm liver sections for single cell PCR analysis. Coamplification of species-specific genes, such as human VH1, human TCRVβ7.2, ovine VH7, and TCRCδ loci from single cells allowed selective detection of human DNA in HepPar1 positive and ovine DNA in HepPar1 negative cells, thereby demonstrating that USSC differentiation in vivo was not caused by cell fusion.

The differentiation capacity of USSC into endodermal tissue was further proven by in vitro assays, although the in vitro testing is always more "artificial" as compared to in vivo regenerative or injury models.

It could be demonstrated that induction of hepatic markers in USSC is possible upon treatment with growth factors, retinoic acid, and administration of different coculture assays. Hepatic-like identity of differentiated USSC was confirmed by functional assays for glycogen synthesis (periodic acid-Schiff staining) and albumin secretion (enzyme-linked immunosorbent assay; ELISA).[71,73] Assessment of the hepatic differentiation was also analyzed applying a novel three-stage protocol resembling embryonic developmental processes of hepatic endoderm. Hepatic preinduction was performed by Activin A and FGF4 resulting in enhanced expression of SOX17 and Forkhead box protein A2 (FOXA2) as demonstrated by real-time PCR and immunohistochemical analysis. Differentiation into mature hepatic cells was achieved sequentially by retinoic acid, FGF4, hepatocyte growth factor (HGF), epidermal growth factor (EGF), and oncostatin M (OSM) resulting in gene expression of GYS2, glucose 6-phoshatase (G6PC), fructose 1,6-bisphosphatase (FBP1), arginase1 (ARG1), and HNF4α after differentiation, thus indicating a more mature state. Functional testing specified the hepatic-like nature of differentiated USSC by albumin secretion, urea formation, and cytochrome-p450-3A4 (CYP3A4) enzyme activity.[74] In order to characterize the differentiated cells at a metabolic level, USSC were incubated with [1–13C] glucose, and neutralized perchloric acid extracts were analyzed by nuclear magnetic resonance spectroscopy. Corresponding to GYS2, G6PC, and FBP1 expression, formation of both glycogen and some gluconeogenetic activity could be observed providing evidence of a hepatocyte-like glucose metabolism in differentiated USSC.[74] Since USSC already express transcription factors of the early hepatic endoderm development, such as GATA binding protein 6 (GATA6), hematopoietically expressed homeobox gene (HEX), or prospero-related homeobox 1 (PROX), they are also an attractive source for reprogramming.[74]

Since USSC resemble in their expression status (*DLK-1*-positive, *HOX*-negative) fetal liver cells, their regenerative potential toward endoderm is logical biological evidence, however, the potential so far is much lower as compared to mature liver cells and reprogrammed cells (iPS cells) differentiated toward the endodermal lineage.

## 14. CARDIAL REGENERATION IN VIVO

Human heart infarction involves the dramatic loss of cardiomyocytes. Therefore investigators sought to identify endogenous cells or stem cells with the ability to differentiate into committed cardiomyocytes and to regenerate the myocardium.[75,76] Dozens of stem cell types have been reported to have cardiac potential, including pluripotent embryonic stem cells, iPS cells and adult progenitor cells from BM, peripheral blood, or intrinsic cardiac cells.[77] For donor cell types as BM MSC, MNC from BM, and endothelial cells from different sources already in clinical studies, the dominant in vivo effect may be neoangiogenesis and not cardiac specification. Moreover, despite some encouraging results from clinical trials, disagreement exists about the efficiency of the treatment, making this issue controversial in the field of stem cell therapy. Moreover for skeletal myoblasts, despite integration and survival, arrhythmia was found in clinical trials.

These obstacles provided the basis why researchers were interested in CB subpopulations, since CB exhibit great functional plasticity and can adapt in a new environmental niche to give rise to cell lineages for the new tissue site as described above for the adaptation of *HOX*-negative cells to *HOX*-positive tissue.[26] In two early reports, surgically implanted USSC into the infarcted heart in a porcine model revealed improvement of the regional and global left ventricular function, but only very few cells were detected in the myocardium after 4–8 weeks, and it could not be documented whether the effects were a result of de novo formation of cardiomyocytes together with a vascular restoration, or a paracrine effect.[20,78] USSC transplantation in a rat model resulted in functional improvement after myocardial infarction, however the observed functional effects could not be confirmed in two reports.[79,80] Besides minor variations in terms of isolation and cultivation of CB-derived adherent cells, the reason for the "negative" results remained unclear.[80] Mechanistic insights are required to evaluate the route of application of cells and the discrimination between differentiation and paracrine mechanisms. Applying a Wistar rat model with cyclosporin A for immunosuppression and an additional model of nude rats, Ding et al.[81] recently analyzed the fate of cells directly after coronary delivery. In detail, the differentiation of cells into vascular cells and cardiomyocytes was assessed. A major finding of the study was that about 80% of the initially infused USSC were retained in the heart directly after transplantation, however, the retained USSC underwent apoptosis so that the long-term engraftment was very low (0.13%). This small fraction adopted a cardiomyocytic phenotype (morphologically, positive for alpha-actinin and the mitochondrial human protein, hMITO as well as human nuclei). Some cells were incorporated into the vascular wall. The major observation of the study of Ding et al. is that the majority of USSC disappeared over time, independent of the applied model (also in nude rats lacking T cells). The reason for this observation was based on a high apoptosis activity analyzed by active-caspase 3 in the USSC. In addition, a substantial infiltration of CD11b positive cells into the myocardium 7 days after transplantation was noticed. In summary, the functional implications of the study clearly document that the number of surviving and differentiating USSC after transplantation was very small, so that reported beneficial effects observed in different models were unlikely to be a result of functional replacement of cardiomyocytes, but of paracrine factors stimulating the repair of resident cardiac progenitors in the heart. Moreover, the integration of USSC into the endothelium may support vasculogenic regeneration.

## 15. IN VITRO DIFFERENTIATION POTENTIAL TOWARD CARDIOMYOCYTES

In 2007, Nishiyama et al. described an MSC population in CB expressing GATA4 and Nkx2.5, cardiac actin as well as troponin T at default state, and termed these cells "cardiac progenitors." They had a very limited proliferation potential and were genetically manipulated for expansion.[82] In order to critically evaluate and confirm published data, eGFP-labeled USSC were cocultivated with neonatal rat cardiomyocytes. A significant number of USSC were stained positive for the cardiac markers alpha-actinin and cardiac troponin T, respectively, 3 and 7 days after cocultivation. A clear striation pattern of cardiac troponin T could be observed together with a spontaneous contraction of the cells. Gene expression profile of the isolated cells revealed the expression of GATA4 and the cardiac markers cTnT and cardiac actin, but only after cocultivation with neonatal cardiomyocytes.[81]

In summary, the application of USSC or MSC from CB in cardiac regeneration may involve both paracrine mechanisms, inflammatory mediators that support cardiomyogenesis as well as support of neovascularization. It will be of critical importance to understand the mechanisms at the cellular level ongoing in vivo. In vitro differentiation experiments can never reflect the in vivo situation, however, they can help to compare the distinct progenitor cells to the adult mature cardiomyocyte and differentiated embryonic or reprogrammed stem cells.

## 16. CB SUBPOPULATIONS FOR NEURONAL REGENERATION

During the past 10 years CB has created great interest as a valuable source for neural stem cells or scientifically more precisely for the support of neural regeneration.

Although studies including our own have shown how subsets of CB cells differentiate under defined conditions into neurons, astrocytes, and microglia in vitro by more or less "artificial methods and substances,"[83] which do not reflect the in vivo situation, it is as of today common knowledge that CB stem cells secrete trophic factors that initiate and maintain the process of repair toward neurons in vivo.[84] In the meantime this resulted in the first clinical trials in treatment of neurological disorders such as cerebral palsy in children. Here, application of autologous or allogeneic matched CB resulted in amelioration of the motor skills and cognitive functions.[85]

In 2001, Chen et al.[86] were the first to demonstrate that the infusion or intracerebral transplantation of CB stem cells into rats that had been stroke-induced by occlusion of the middle cerebral artery displayed beneficial effects. The underlying mechanisms of these observations have not yet been elucidated, though the most straightforward idea is that stem cells differentiate into mature cell types and simply replace the lost tissue. However, there is increasing evidence that transplanted cells may secrete neurotrophic or neuroprotective factors[87] that can counteract degeneration or promote regeneration. It was demonstrated that USSC isolated from human CB are strongly attracted by HGF that is secreted by ischemia-damaged brain tissue and by apoptotic neurons in vitro and in vivo.[87] In opposition, necrotic neurons do not secrete hepatocyte growth factor and therefore have no potential to initiate migration of USSC. In all paradigms used in this study the secretion of HGF by neural target tissue and the expression of the HGF receptor c-MET in USSC directly correlated to migrational potency of USSC but also MSC, indicating that the HGF/c–MET axis is the driving force for migration toward neuronal injury.

Besides stroke and cerebral palsy, the treatment of spinal cord injury with stem cells is in the focus of many researchers. The major problem in spinal cord injury is the breakdown of the blood–spinal cord barrier associated with invasion of inflammatory cells, the activation of the glia, and subsequent axonal degeneration.[84] Stem cell populations derived from BM or peripheral blood have been transplanted in animal models and in pilot clinical studies as also summarized. Reports on the functional recovery were frequently only based on a single behavioral test. Although improvement of sensory and motor activity was reported in some studies,[88] no recovery was observed in others. Preclinical studies with MSC isolated from BM or CB in rodent spinal cord injury suggested variable mechanisms underlying the observed effect as differentiation into oligodendroglia. Here, Schira et al. transplanted USSC into a rodent model of acute spinal cord injury and investigated their survival, migration, and neural differentiation potential as well their influence on axonal regrowth, lesion size, and protection from spinal tissue loss. Moreover three different locomotor tasks (open field Basso–Beattie–Bresnahan locomotion score, horizontal ladder walking test, and CatWalk gait analysis) were applied. In the report, immune-suppressed adult rats received a highly reproducible dorsal hemisection injury at thoracic level Th8. Immediately after hemisection USSC were transplanted close to the site of the injury. Two days after transplantation grafted cells were located at the injection site, and one week after transplantation in the lesion center but without revealing immunoreactivity for the axon marker neurofilament. Three weeks after transplantation, USSC were mainly confined to the injury site, but in close proximity to the grafted USSC, neurofilament positive host cells were present in the lesion center. Although the USSC itself did not differentiate toward neurons or glia cells, they reduced the tissue loss significantly. This leads to improved locomotor function 16 weeks after transplantation. In three different test systems the grafting of USSC into the traumatic spinal cord injury significantly reduced the lesion size and enhanced the amount of spared tissue, indicating a strong neuroprotective function of USSC similar to BM MSC. USSC release a wide range of cytokines including stromal cell-derived factor 1 (SDF-1),[60] which induces homing of HSC and neural stem cells in ischemic and injured brain, and HGF, which is a known survival factor of neural development. The different growth factors or the combination of several as in other models of tissue regeneration are likely to participate in the positive regeneration effects observed.

Although most studies using CB stem cells for the treatment of neurological disorders in animal models, but also already in the clinic are very promising, several questions such as the route of administration, the amount of cells and the mode of action, the observed side effects, and mainly the *scientific* mechanisms have to be addressed before stem cells from CB can be brought into the clinical arena as a therapeutic strategy to treat neurological disorders.

## 17. REPROGRAMMED SUBPOPULATIONS FROM CB

Since their first description by the group of Yamanaka in 2007[89] iPS including CB-derived cells have received major attention by the scientific CB community and two distinct fields of interest were established: the application in regenerative medicine to deliver specific cells or tissue on demand, and the use in diagnostic systems. While the original protocol described lentiviral insertion of genes for the transcription factors Oct4, Sox2, Klf4, and c-Myc, this method bears the potential risk of disrupting normal genes or of activating oncogenes in close proximity to the integration site. This can be avoided by using more sophisticated integration-free methods, such as appliance of episomal plasmids[90] Sendai-virus,[91] mRNA[92] or direct addition of the according proteins,[93] and small molecules. Although generation frequencies reported for different reprogramming systems vary, they are still generally low, ranging from 0.001% to 0.1%. Furthermore, especially the protooncogene c-Myc is seen critically, therefore multiple different combinations of factors, including L-Myc and Lin28, are currently proposed for generation, with only Oct4 and Sox2 being overexpressed in the most minimalist protocols.[94] In regard to cell origin, publications demonstrate that cells from CB in general, and in particular CD34-positive HSC, are a better choice than adult cells. Not only that generation frequencies for iPS are higher, these biologically younger cells also bear a lower risk of acquired genetic alterations. In addition, isolation and cryopreservation of these cells is well established, making CB cells an easily available source. Although, as a disadvantage, it has to be taken into consideration that innate genetic mutations of the donor might not be obvious directly at birth,[95] CB-derived cells are the most promising candidates for standardized generation of iPS. Nearly all cell subpopulations, including MSC, USSC, ECFC, and CD34-positive cells from CB were analyzed in reprogramming using lentivirus- and Sendai-virus-based protocols, respectively. Both approaches resulted in cells with embryonal stem cell-like morphology with no detectable difference in growth or expression of pluripotency markers. However, for standardization CB-derived cells, which have a constant frequency both in fresh and cryopreserved CB, are among the most interesting candidates. iPS are a promising tool for disease modeling, toxicological tests, and basic science, and further improvements of generation (integration-free), maintenance (serum-/feeder-free), and differentiation (new differentiation pathways, higher efficiency) are likely to further extend their applicability. Therefore, although iPS are already becoming a laboratory standard, it is very likely that it will still take some years until they are finally transferable into the clinic, since it is mandatory that the technology needs to be improved first before CB-IPS banking is routinely feasible.

## 18. CONCLUSION

CB contains valuable nonhematopoietic progenitor cells from different stages of fetal development circulating in the CB as "elusive" cells. In order to characterize them with regard to their true differentiation potential, for each cell population clonal cells and in vivo experimental design is required. The results summarized here clearly show that USSC are different from CB MSC. The difference exists both for *HOX*-genes, playing an important role in skeletal formation as well as for transcription factors detecting differences between these cell populations, but also in differences to BM MSC. Although USSC and MSC have no clear "bone signature" they are able to differentiate under appropriate conditions toward bone and cartilage. USSC share many markers with early liver development and are therefore able to reconstitute the liver upon injury. Although there were in vitro differentiation data available to show differentiation of CB-derived cells toward neurons (ectodermal tissue) and cardiomyocytes, all in vivo experiments clearly indicate that no transdifferentiation occurred. Though, as recently also described by Bianco et al. for cultivated mesenchymal cells from BM,[25,96] it has to be considered that the positive data published in different models are probably mainly related to trophic factors, immune modulation, or antiinflammatory effects. Based on the developmental advantage of CB subpopulations they might be ideal tools to analyze the fate of the distinct population in extensive preclinical models and define the mechanisms behind the improvements observed in beginning clinical trials.

## LIST OF ACRONYMS AND ABBREVIATIONS

**AcLDL** Acetylated-low density lipoprotein
**ARG1** Arginase1
**BM MSC** Bone marrow mesenchymal/multipotent stromal cells
**BMP4** Bone morphogenetic protein 4
**BSP** Bone sialoprotein
**c-MET** MET or MNNG HOS transforming gene
**CB** Cord blood
**CB MSC** Cord blood mesenchymal/multipotent stromal cells
**CB-SC** Cord blood stromal cells
**CD** Cluster of differentiation
**CFC** Colony-forming cell
**CPD** Cumulative population doubling
**CYP3A4** Cytochrome-p450-3A4
**DCBT** Double cord blood transplantation
**DLK-1** Delta-like 1 homolog
**DMEM** Dulbecco's modified Eagle medium
**ECFC** Endothelial cord forming cell
**EGF** Epidermal growth factor
**EPC** Endothelial progenitor cell
**FACS** Fluorescence-activated cell sorting
**FBP1** Fructose 1,6-biphosphatase
**FCS** Fetal calf serum
**FGF** Fibroblast growth factor
**FLT3LG** Fms-related tyrosine kinase 3 ligand
**FOXA2** Forkhead box protein A2
**G6PC** Glucose 6-phosphatase
**GATA4** GATA binding protein 4
**GATA6** GATA binding protein 6

**GFP** Green fluorescent protein
**GMP** Good manufacturing practice
**GvHD** Graft-versus-host-disease
**GYS2** Glycogen synthase 2
**HepPar1** Hepatocyte paraffin 1
**HEX** Hematopoietically expressed homeobox gene
**HGF** Hepatocyte growth factor
**HH** Hedgehog protein
**HNF4α** Hepatocyte nuclear factor 4 alpha
**HOX** Homeobox
**HPP-CFC** High proliferative potential-colony-forming cell
**HSC** Hematopoietic stem cells
**IGF** Insulin-like growth factor
**IGFBP** Insulin-like growth factor-binding protein
**IL-6** Interleukin 6
**iPS** Induced pluripotent cells
**KDR** Kinase insert domain receptor
**LTC-IC** Long-term culture-initiating cell
**MNC** Mononuclear cells
**MSC** Mesenchymal/multipotent stromal cells
**NC** Nucleated cells
**Nkx2.5** NK2 homeobox 5
**NOD/SCID** Nonobese diabetic/severe combined immunodeficiency
**OC** Osteocalcin
**OCT4A** Octamer binding transcription factor 4
**OSM** Oncostatin M
**OSX (SP7)** Osterix
**PCR** Polymerase chain reaction
**PROX** Prospero-related homeobox 1
**PTHLH** Parathyroid hormone-related protein
**RPL13A** Ribosomal protein L13A.
**RUNX2** Runt domain-containing transcription factor 2
**SCF** Stem cell factor
**SCID** Severe combined immunodeficiency
**SDF-1** Stromal cell-derived factor 1
**SOX9** Sex-determining region Y-related high mobility group box 9
**SOX17** SRY (sex-determining region Y)-box 17
**SRC** SCID-repopulating cells
**TCR** T-cell receptor
**THPO** Thrombopoietin
**UC MSC** Umbilical cord mesenchymal stromal cells
**USSC** Unrestricted somatic stromal cells
**VSEL** Very small embryonic-like cells

## REFERENCES


1. Cumano A, Ferraz JC, Klaine M, Di Santo JP, Godin I. Intraembryonic, but not yolk sac hematopoietic precursors, isolated before circulation, provide long-term multilineage reconstitution. *Immunity* 2001;**15**:477–85.
2. Ciriza J, Thompson H, Petrosian R, Manilay JO, Garcia-Ojeda ME. The migration of hematopoietic progenitors from the fetal liver to the fetal bone marrow: lessons learned and possible clinical applications. *Exp Hematol* 2013;**41**:411–23.
3. Asahara T, Murohara T, Sullivan A, Silver M, van der Zee R, Li T, et al. Isolation of putative progenitor endothelial cells for angiogenesis. *Science* 1997;**275**:964–7.
4. Hill JM, Zalos G, Halcox JP, Schenke WH, Waclawiw MA, Quyyumi AA, et al. Circulating endothelial progenitor cells, vascular function, and cardiovascular risk. *N Engl J Med* 2003;**348**:593–600.
5. Peichev M, Naiyer AJ, Pereira D, Zhu Z, Lane WJ, Williams M, et al. Expression of VEGFR-2 and AC133 by circulating human CD34(+) cells identifies a population of functional endothelial precursors. *Blood* 2000;**95**:952–8.
6. Ingram DA, Mead LE, Tanaka H, Meade V, Fenoglio A, Mortell K, et al. Identification of a novel hierarchy of endothelial progenitor cells using human peripheral and umbilical cord blood. *Blood* 2004;**104**:2752–60.
7. Shepherd BR, Enis DR, Wang F, Suarez Y, Pober JS, Schechner JS. Vascularization and engraftment of a human skin substitute using circulating progenitor cell-derived endothelial cells. *FASEB J* 2006;**20**:1739–41.
8. Tavassoli M, Crosby WH. Transplantation of marrow to extramedullary sites. *Science* 1968;**161**:54–6.
9. Friedenstein AJ. *Osteogenic stem cells in bone marrow*. New York: Elsevier; 1990.
10. Owen M, Friedenstein AJ. Stromal stem cells: marrow-derived osteogenic precursors. *Ciba Found Symp* 1988;**136**:42–60.
11. Friedenstein AJ, Chailakhyan RK, Gerasimov UV. Bone marrow osteogenic stem cells: in vitro cultivation and transplantation in diffusion chambers. *Cell Tissue Kinet* 1987;**20**:263–72.
12. Bianco P, Robey PG, Saggio I, Riminucci M. "Mesenchymal" stem cells in human bone marrow (skeletal stem cells): a critical discussion of their nature, identity, and significance in incurable skeletal disease. *Hum Gene Ther* 2010;**21**:1057–66.
13. Kuznetsov SA, Krebsbach PH, Satomura K, Kerr J, Riminucci M, Benayahu D, et al. Single-colony derived strains of human marrow stromal fibroblasts form bone after transplantation in vivo. *J Bone Min Res* 1997;**12**:1335–47.
14. Sacchetti B, Funari A, Michienzi S, Di Cesare S, Piersanti S, Saggio I, et al. Self-renewing osteoprogenitors in bone marrow sinusoids can organize a hematopoietic microenvironment. *Cell* 2007;**131**:324–36.
15. Gronthos S, Franklin DM, Leddy HA, Robey PG, Storms RW, Gimble JM. Surface protein characterization of human adipose tissue-derived stromal cells. *J Cell Physiol* 2001;**189**:54–63.
16. Kogler G, Critser P, Trapp T, Yoder M. Future of cord blood for non-oncology uses. *Bone Marrow Transpl* 2009;**44**:683–97.
17. Weiss ML, Medicetty S, Bledsoe AR, Rachakatla RS, Choi M, Merchav S, et al. Human umbilical cord matrix stem cells: preliminary characterization and effect of transplantation in a rodent model of Parkinson's disease. *Stem Cells* 2006;**24**:781–92.
18. Reinisch A, Strunk D. Isolation and animal serum free expansion of human umbilical cord derived mesenchymal stromal cells (MSCs) and endothelial colony forming progenitor cells (ECFCs). *J Vis Exp* 2009;32:pii.
19. Kogler G, Sensken S, Airey JA, Trapp T, Muschen M, Feldhahn N, et al. A new human somatic stem cell from placental cord blood with intrinsic pluripotent differentiation potential. *J Exp Med* 2004;**200**:123–35.
20. Kim BO, Tian H, Prasongsukarn K, Wu J, Angoulvant D, Wnendt S, et al. Cell transplantation improves ventricular function after a myocardial infarction: a preclinical study of human unrestricted somatic stem cells in a porcine model. *Circulation* 2005;**112**:I96–104.
21. Kern S, Eichler H, Stoeve J, Kluter H, Bieback K. Comparative analysis of mesenchymal stem cells from bone marrow, umbilical cord blood, or adipose tissue. *Stem Cells* 2006;**24**:1294–301.
22. Chang YJ, Shih DT, Tseng CP, Hsieh TB, Lee DC, Hwang SM. Disparate mesenchyme-lineage tendencies in mesenchymal stem cells from human bone marrow and umbilical cord blood. *Stem Cells* 2006;**24**:679–85.

23. Kluth SM, Buchheiser A, Houben AP, Geyh S, Krenz T, Radke TF, et al. DLK-1 as a marker to distinguish unrestricted somatic stem cells and mesenchymal stromal cells in cord blood. *Stem Cells Dev* 2010;**19**:1471–83.
24. Kluth SM, Radke TF, Kogler G. Increased haematopoietic supportive function of USSC from umbilical cord blood compared to CB MSC and possible role of DLK-1. *Stem Cells Int* 2013;**2013**:985285.
25. Bianco P, Cao X, Frenette PS, Mao JJ, Robey PG, Simmons PJ, et al. The meaning, the sense and the significance: translating the science of mesenchymal stem cells into medicine. *Nat Med* 2013;**19**:35–42.
26. Liedtke S, Freytag EM, Bosch J, Houben AP, Radke TF, Deenen R, et al. Neonatal mesenchymal-like cells adapt to surrounding cells. *Stem Cell Res* 2013;**11**:634–46.
27. Liedtke S, Buchheiser A, Bosch J, Bosse F, Kruse F, Zhao X, et al. The HOX Code as a "biological fingerprint" to distinguish functionally distinct stem cell populations derived from cord blood. *Stem Cell Res* 2010;**5**:40–50.
28. Kluth SM, Radke TF, Kogler G. Potential application of cord blood-derived stromal cells in cellular therapy and regenerative medicine. *J Blood Transfus* 2012:2012.
29. Liedtke S, Enczmann J, Waclawczyk S, Wernet P, Kogler G. Oct4 and its pseudogenes confuse stem cell research. *Cell Stem Cell* 2007;**1**:364–6.
30. Liedtke S, Stephan M, Kogler G. Oct4 expression revisited: potential pitfalls for data misinterpretation in stem cell research. *Biol Chem* 2008;**389**:845–50.
31. Zaehres H, Kogler G, Arauzo-Bravo MJ, Bleidissel M, Santourlidis S, Weinhold S, et al. Induction of pluripotency in human cord blood unrestricted somatic stem cells. *Exp Hematol* 2010;**38**:809–18. 818 e1-2.
32. Kucia M, Reca R, Campbell FR, Zuba-Surma E, Majka M, Ratajczak J, et al. A population of very small embryonic-like (VSEL) CXCR4(+) SSEA-1(+)Oct-4+ stem cells identified in adult bone marrow. *Leukemia* 2006;**20**:857–69.
33. Miyanishi M, Mori Y, Seita J, Chen JY, Karten S, Chan CKF, et al. Do pluripotent stem cells exist in adult mice as very small embryonic stem cells? *Stem Cell Rep* 2013;**1**:198–208.
34. Abbott A. Doubt cast over tiny stem cells. *Nature* 2013;**499**:390.
35. Bosch J, Houben AP, Hennicke T, Deenen R, Köhrer K, Liedtke S, et al. Comparing the gene expression profile of stromal cells from human cord blood and bone marrow: lack of the typical "bone" signature in cord blood cells. *Stem Cells Int* 2013;**2013**:631984.
36. Bosch J, Houben AP, Radke TF, Stapelkamp D, Bunemann E, Balan P, et al. Distinct differentiation potential of "MSC" derived from cord blood and umbilical cord: are cord-derived cells true mesenchymal stromal cells? *Stem Cells Dev* 2012;**21**:1977–88.
37. Aktas M, Buchheiser A, Houben A, Reimann V, Radke T, Jeltsch K, et al. Good manufacturing practice-grade production of unrestricted somatic stem cell from fresh cord blood. *Cytotherapy* 2010;**12**:338–48.
38. Dominici M, Le Blanc K, Mueller I, Slaper-Cortenbach I, Marini F, Krause D, et al. Minimal criteria for defining multipotent mesenchymal stromal cells. The International Society for Cellular Therapy position statement. *Cytotherapy* 2006;**8**:315–7.
39. Capelli C, Gotti E, Morigi M, Rota C, Weng L, Dazzi F, et al. Minimally manipulated whole human umbilical cord is a very rich source of clinical-grade human mesenchymal stromal cells expanded in human platelet lysate. *Cytotherapy* 2011;**13**:786–801.
40. Sudo K, Kanno M, Miharada K, Ogawa S, Hiroyama T, Saijo K, et al. Mesenchymal progenitors able to differentiate into osteogenic, chondrogenic, and/or adipogenic cells in vitro are present in most primary fibroblast-like cell populations. *Stem Cells* 2007;**25**:1610–7.
41. Kaltz N, Funari A, Hippauf S, Delorme B, Noel D, Riminucci M, et al. In vivo osteoprogenitor potency of human stromal cells from different tissues does not correlate with expression of POU5F1 or its pseudogenes. *Stem Cells* 2008;**26**:2419–24.
42. Olsen BR, Reginato AM, Wang W. Bone development. *Annu Rev Cell Dev Biol* 2000;**16**:191–220.
43. Creuzet S, Couly G, Vincent C, Le Douarin NM. Negative effect of Hox gene expression on the development of the neural crest-derived facial skeleton. *Development* 2002;**129**:4301–13.
44. Leucht P, Kim JB, Amasha R, James AW, Girod S, Helms JA. Embryonic origin and Hox status determine progenitor cell fate during adult bone regeneration. *Development* 2008;**135**:2845–54.
45. Dimitriou R, Jones E, McGonagle D, Giannoudis PV. Bone regeneration: current concepts and future directions. *BMC Med* 2011;**9**:66.
46. Krampera M, Marconi S, Pasini A, Galie M, Rigotti G, Mosna F, et al. Induction of neural-like differentiation in human mesenchymal stem cells derived from bone marrow, fat, spleen and thymus. *Bone* 2007;**40**:382–90.
47. Tuan RS, Boland G, Tuli R. Adult mesenchymal stem cells and cell-based tissue engineering. *Arthritis Res Ther* 2003;**5**:32–45.
48. Prockop DJ, Brenner M, Fibbe WE, Horwitz E, Le Blanc K, Phinney DG, et al. Defining the risks of mesenchymal stromal cell therapy. *Cytotherapy* 2010;**12**:576–8.
49. DeLise AM, Fischer L, Tuan RS. Cellular interactions and signaling in cartilage development. *Osteoarthritis Cartilage* 2000;**8**:309–34.
50. Long F. Building strong bones: molecular regulation of the osteoblast lineage. *Nat Rev Mol Cell Biol* 2011;**13**:27–38.
51. Goldring MB, Tsuchimochi K, Ijiri K. The control of chondrogenesis. *J Cell Biochem* 2006;**97**:33–44.
52. Adams SL, Cohen AJ, Lassova L. Integration of signaling pathways regulating chondrocyte differentiation during endochondral bone formation. *J Cell Physiol* 2007;**213**:635–41.
53. Ogbureke KU, Fisher LW. SIBLING expression patterns in duct epithelia reflect the degree of metabolic activity. *J Histochem Cytochem* 2007;**55**:403–9.
54. Buchheiser A, Houben AP, Bosch J, Marbach J, Liedtke S, Kogler G. Oxygen tension modifies the "stemness" of human cord blood-derived stem cells. *Cytotherapy* 2012;**14**:967–82.
55. Gluckman E, Broxmeyer HA, Auerbach AD, Friedman HS, Douglas GW, Devergie A, et al. Hematopoietic reconstitution in a patient with Fanconi's anemia by means of umbilical-cord blood from an HLA-identical sibling. *N Engl J Med* 1989;**321**:1174–8.
56. Barker JN, Weisdorf DJ, DeFor TE, Blazar BR, McGlave PB, Miller JS, et al. Transplantation of 2 partially HLA-matched umbilical cord blood units to enhance engraftment in adults with hematologic malignancy. *Blood* 2005;**105**:1343–7.
57. Falkenberg H, Radke TF, Kogler G, Stuhler K. Proteomic profiling of ex vivo expanded CD34-positive haematopoetic cells derived from umbilical cord blood. *Stem Cells Int* 2013;**2013**:245695.
58. Dahlberg A, Delaney C, Bernstein ID. Ex vivo expansion of human hematopoietic stem and progenitor cells. *Blood* 2011;**117**:6083–90.
59. Cutler C, Multani P, Robbins D, Kim HT, Le T, Hoggatt J, et al. Prostaglandin-modulated umbilical cord blood hematopoietic stem cell transplantation. *Blood* 2013;**122**(17):3074–81.
60. Kogler G, Radke TF, Lefort A, Sensken S, Fischer J, Sorg RV, et al. Cytokine production and hematopoiesis supporting activity of cord blood-derived unrestricted somatic stem cells. *Exp Hematol* 2005;**33**:573–83.
61. Jeltsch KS, Radke TF, Laufs S, Giordano FA, Allgayer H, Wenz F, et al. Unrestricted somatic stem cells: interaction with CD34(+) cells in vitro and in vivo, expression of homing genes and exclusion of tumorigenic potential. *Cytotherapy* 2010;**13**(3):357–65.

62. Lee SH, Lee MW, Yoo KH, Kim DS, Son MH, Sung KW, et al. Co-transplantation of third-party umbilical cord blood-derived MSCs promotes engraftment in children undergoing unrelated umbilical cord blood transplantation. *Bone Marrow Transpl* 2013;**48**:1040–5.
63. Mohle R, Bautz F, Rafii S, Moore MA, Brugger W, Kanz L. The chemokine receptor CXCR-4 is expressed on CD34+ hematopoietic progenitors and leukemic cells and mediates transendothelial migration induced by stromal cell-derived factor-1. *Blood* 1998;**91**:4523–30.
64. Nie Y, Han YC, Zou YR. CXCR4 is required for the quiescence of primitive hematopoietic cells. *J Exp Med* 2008;**205**:777–83.
65. Chou S, Lodish HF. Fetal liver hepatic progenitors are supportive stromal cells for hematopoietic stem cells. *Proc Natl Acad Sci USA* 2010;**107**:7799–804.
66. Petersen BE, Bowen WC, Patrene KD, Mars WM, Sullivan AK, Murase N, et al. Bone marrow as a potential source of hepatic oval cells. *Science* 1999;**284**:1168–70.
67. Ishikawa F, Drake CJ, Yang S, Fleming P, Minamiguchi H, Visconti RP, et al. Transplanted human cord blood cells give rise to hepatocytes in engrafted mice. *Ann NY Acad Sci* 2003;**996**:174–85.
68. Wang X, Willenbring H, Akkari Y, Torimaru Y, Foster M, Al-Dhalimy M, et al. Cell fusion is the principal source of bone-marrow-derived hepatocytes. *Nature* 2003;**422**:897–901.
69. Hengstler JG, Brulport M, Schormann W, Bauer A, Hermes M, Nussler AK, et al. Generation of human hepatocytes by stem cell technology: definition of the hepatocyte. *Expert Opin Drug Metab Toxicol* 2005;**1**:61–74.
70. Lee KD, Kuo TK, Whang-Peng J, Chung YF, Lin CT, Chou SH, et al. In vitro hepatic differentiation of human mesenchymal stem cells. *Hepatology* 2004;**40**:1275–84.
71. Sensken S, Waclawczyk S, Knaupp AS, Trapp T, Enczmann J, Wernet P, et al. In vitro differentiation of human cord blood-derived unrestricted somatic stem cells towards an endodermal pathway. *Cytotherapy* 2007;**9**:362–78.
72. Ghodsizad A, Fahy BN, Waclawczyk S, Liedtke S, Gonzalez Berjon JM, Barrios R, et al. Portal application of human unrestricted somatic stem cells to support hepatic regeneration after portal embolization and tumor surgery. *ASAIO J* 2012;**58**:255–61.
73. Kogler G, Sensken S, Wernet P. Comparative generation and characterization of pluripotent unrestricted somatic stem cells with mesenchymal stem cells from human cord blood. *Exp Hematol* 2006;**34**:1589–95.
74. Waclawczyk S, Buchheiser A, Flogel U, Radke TF, Kogler G. In vitro differentiation of unrestricted somatic stem cells into functional hepatic-like cells displaying a hepatocyte-like glucose metabolism. *J Cell Physiol* 2010;**225**:545–54.
75. Lee J, Terracciano CM. Cell therapy for cardiac repair. *Br Med Bull* 2010;**94**:65–80.
76. Wollert KC, Drexler H. Cell therapy for the treatment of coronary heart disease: a critical appraisal. *Nat Rev Cardiol* 2010;**7**:204–15.
77. Passier R, van Laake LW, Mummery CL. Stem-cell-based therapy and lessons from the heart. *Nature* 2008;**453**:322–9.
78. Ghodsizad A, Niehaus M, Kogler G, Martin U, Wernet P, Bara C, et al. Transplanted human cord blood-derived unrestricted somatic stem cells improve left-ventricular function and prevent left-ventricular dilation and scar formation after acute myocardial infarction. *Heart* 2009;**95**:27–35.
79. Moelker AD, Baks T, Wever KM, Spitskovsky D, Wielopolski PA, van Beusekom HM, et al. Intracoronary delivery of umbilical cord blood derived unrestricted somatic stem cells is not suitable to improve LV function after myocardial infarction in swine. *J Mol Cell Cardiol* 2007;**42**:735–45.
80. Rabald S, Marx G, Mix B, Stephani C, Kamprad M, Cross M, et al. Cord blood cell therapy alters LV remodeling and cytokine expression but does not improve heart function after myocardial infarction in rats. *Cell Physiol Biochem* 2008;**21**:395–408.
81. Ding Z, Burghoff S, Buchheiser A, Kogler G, Schrader J. Survival, integration, and differentiation of unrestricted somatic stem cells in the heart. *Cell Transpl* 2013;**22**:15–27.
82. Nishiyama N, Miyoshi S, Hida N, Uyama T, Okamoto K, Ikegami Y, et al. The significant cardiomyogenic potential of human umbilical cord blood-derived mesenchymal stem cells in vitro. *Stem Cells* 2007;**25**:2017–24.
83. Goodwin HS, Bicknese AR, Chien SN, Bogucki BD, Quinn CO, Wall DA. Multilineage differentiation activity by cells isolated from umbilical cord blood: expression of bone, fat, and neural markers. *Biol Blood Marrow Transpl* 2001;**7**:581–8.
84. Schira J, Gasis M, Estrada V, Hendricks M, Schmitz C, Trapp T, et al. Significant clinical, neuropathological and behavioural recovery from acute spinal cord trauma by transplantation of a well-defined somatic stem cell from human umbilical cord blood. *Brain* 2012;**135**:431–46.
85. Min K, Song J, Kang JY, Ko J, Ryu JS, Kang MS, et al. Umbilical cord blood therapy potentiated with erythropoietin for children with cerebral palsy: a double-blind, randomized, placebo-controlled trial. *Stem Cells* 2013;**31**:581–91.
86. Chen J, Sanberg PR, Li Y, Wang L, Lu M, Willing AE, et al. Intravenous administration of human umbilical cord blood reduces behavioral deficits after stroke in rats. *Stroke* 2001;**32**:2682–8.
87. Trapp T, Kogler G, El-Khattouti A, Sorg RV, Besselmann M, Focking M, et al. Hepatocyte growth factor/c-MET axis-mediated tropism of cord blood-derived unrestricted somatic stem cells for neuronal injury. *J Biol Chem* 2008;**283**:32244–53.
88. Kang KS, Kim SW, Oh YH, Yu JW, Kim KY, Park HK, et al. 37-year-old spinal cord-injured female patient, transplanted of multipotent stem cells from human UC blood, with improved sensory perception and mobility, both functionally and morphologically: a case study. *Cytotherapy* 2005;**7**:368–73.
89. Takahashi K, Tanabe K, Ohnuki M, Narita M, Ichisaka T, Tomoda K, et al. Induction of pluripotent stem cells from adult human fibroblasts by defined factors. *Cell* 2007;**131**:861–72.
90. Yu J, Hu K, Smuga-Otto K, Tian S, Stewart R, Slukvin II, et al. Human induced pluripotent stem cells free of vector and transgene sequences. *Science* 2009;**324**:797–801.
91. Fusaki N, Ban H, Nishiyama A, Saeki K, Hasegawa M. Efficient induction of transgene-free human pluripotent stem cells using a vector based on Sendai virus, an RNA virus that does not integrate into the host genome. *Proc Jpn Acad Ser B Phys Biol Sci* 2009;**85**:348–62.
92. Warren L, Manos PD, Ahfeldt T, Loh YH, Li H, Lau F, et al. Highly efficient reprogramming to pluripotency and directed differentiation of human cells with synthetic modified mRNA. *Cell Stem Cell* 2010;**7**:618–30.
93. Zhou H, Wu S, Joo JY, Zhu S, Han DW, Lin T, et al. Generation of induced pluripotent stem cells using recombinant proteins. *Cell Stem Cell* 2009;**4**:381–4.
94. Kim JB, Zaehres H, Wu G, Gentile L, Ko K, Sebastiano V, et al. Pluripotent stem cells induced from adult neural stem cells by reprogramming with two factors. *Nature* 2008;**454**:646–50.
95. Takahashi K, Yamanaka S. Induced pluripotent stem cells in medicine and biology. *Development* 2013;**140**:2457–61.
96. Bianco P. Reply to MSCs: science and trials. *Nat Med* 2013;**19**:813–4.

Chapter 3

# Cord Blood Hematopoiesis: The Road to Transplantation

Anna Rita Migliaccio[1] and Thalia Papayannopoulou[2]

[1]*Tisch Cancer Institute, Mount Sinai School of Medicine, New York, NY, USA;* [2]*Department of Medicine/Hematology, University of Washington, Seattle, WA, USA*

## Chapter Outline



## 1. USE OF PLACENTAL CORD BLOOD AS A SOURCE FOR HEMATOPOIETIC STEM/PROGENITOR CELLS TRANSPLANTATION: A HISTORICAL PERSPECTIVE

Nowadays, allogeneic hematopoietic stem cell (HSC) transplantation is an established life-saving procedure for numerous inherited and acquired blood disorders. However, only 30% of the patients who may benefit from an allogeneic stem cell transplant have a matched sibling donor. The remaining 70% of the patients must find a match in national and international registries of marrow donors. Although over the years the ability of these registries to provide suitable matches has greatly improved, the probability of finding a match for non-Caucasian patients, African–American, and other minorities, is much lower (20–45% vs 60%).[1] This moral imperative prompted investigations into the use of placental cord blood (PCB), a discarded resource, for transplantation.

PCB had been recognized as a rich source of hematopoietic stem/progenitor cells (HSPCs) by developmental biology studies performed in the 1980s.[2,3] Based on this recognition, proof-of-principle experiments established that as little as 10 μL of PCB is sufficient to protect mouse models from sublethal irradiation.[4] Proof-of-principle that PCB contains numbers of HSPCs sufficient to reconstitute hematopoiesis in man was obtained in 1986 in a pediatric patient with Fanconi anemia.[5] These experiments led to the hypothesis that PCB, if stored with rational criteria, may represent an alternative, potentially more potent, source of HSPCs for transplantation of those patients who cannot find a graft in bone marrow registries. This hypothesis was tested by developing good manufacturing practice conditions to store placental/umbilical cord blood for clinical use in 1995,[6] and by demonstrating that PCB stored with these criteria may be used for allogeneic transplants of patients who would not otherwise have a compatible graft in 1998.[7] Presently, PCB is considered an established alternative, and potentially superior, source of HSPCs for transplantation, and as of 2011, more than 25,000 PCB transplants have been performed worldwide.[1] Several national placental blood banking programs have been initiated and more than 500,000 cord blood units have been donated for public use.[1]

It is widely recognized that the number of hematopoietic cells present in a single PCB is sufficient to restore the marrow and to sustain long-term hematopoietic recovery in related and unrelated recipients. However, the number

Cord Blood Stem Cells Medicine. http://dx.doi.org/10.1016/B978-0-12-407785-0.00003-7

of white blood cells (WBCs) transplanted per kilogram of body weight (WBC dose) is strongly predictive of the speed of myeloid engraftment and represents the single parameter that mostly affects, besides the underling disease, the morbidity and mortality of PCB transplantation.[1,7] Since the number of WBC of a PCB is limited, the use of PCB as stem source is restricted to patients with reduced body weights (20–40kg), i.e., mostly pediatric.[1] This chapter will summarize the biological properties of HSPCs that established their usefulness for transplantation, strategies under investigation to overcome current barriers to the use of PCB for transplantation in adults, and possible future clinical applications of PCB in nonhematopoietic diseases.

## 2. CHARACTERIZATION OF PCB HSPCs

The cell ultimately responsible for hematopoietic reconstitution after transplantation is the HSC.[8,9] The gold standard for assessing stem cell function is transplantation in immunocompromised hosts. For human stem cells several immunodeficient xenotransplantation models have been developed as surrogate assays.[10,11] In these xenotransplantation assays, PCB has been observed to generate more hematopoietic cells than either adult human bone marrow or mobilized peripheral blood.[12]

The capacity of HSCs to generate progenitor cells of different lineages can be assessed very efficiently by their clonogenic potential in vitro to form colonies representing different lineages (colony-forming cells, CFC) under specific semisolid culture conditions.[8,9] Extensive studies in mice have established that the number of CFC in a given tissue is genetically determined and correlates with the number of stem cells present in the tissue, as determined by in vivo reconstitution assays.[13] In humans, the use of CFC as a surrogate assay for the HSC has been validated by clinical studies indicating that in autologous bone marrow transplants, the CFC dose (number of cells injected per kilogram of body weight) is positively correlated with early engraftment time.[14] Likewise, the analyses of the outcomes of transplantation with autologous mobilized peripheral blood, indicated that time to engraftment is also inversely correlated with the dose of cells positive for CD34, an antigen expressed on the surface of human stem/progenitor cells.[15–18] These observations, based on CFC or CD34+ cell determinations, provided a yardstick for quantitative comparisons between stem cells present in circulation during the course of human ontogenesis.

Fetal blood obtained under ultrasound guidance for diagnostic or therapeutic purposes was compared to neonatal and adult blood.[19,20] It was found that fetal or neonatal blood contained HSPCs similar in frequency to those present in adult bone marrow samples, and 100 times greater than those present in adult blood. In fact, CD34+ cells represent ~30% of the cells in the lymphocyte gate both from fetal blood and PCB (Table 1). In addition, the frequency of CD34+ cells lacking the CD38 antigen, a phenotype that defines primitive hematopoietic progenitors,[15] in fetal–neonatal blood is 1–2%, similar to that of CD34+/CD38– cells in adult bone marrow, but it is <0.01 in adult blood. In fetal blood and PCB, CFC are so numerous that they can be enumerated by culturing 1 µL of sample directly, while enumeration of CFC in adult blood usually requires enrichment in mononuclear cells by density separation (Table 1).

In addition to their presence in great numbers, hematopoietic progenitors present in fetal blood and PCB are considered more immature than those present in adult blood. This "immaturity" is represented by the size of the colonies they generate (a measure of the number of divisions

**TABLE 1** Frequency of Stem/Progenitor Cells (Defined by Flow Cytometry or by Colony Assay) in Fetal Blood, Cord Blood, and Adult Blood. Flow Cytometry Determination of the Frequency of Stem/Progenitor Cells in Adult Marrow is also Reported for Comparison

| | Fetal Blood | Cord Blood | Adult Blood | Adult Marrow |
|---|---|---|---|---|
| Flow cytometry (percent of the cells in the lymphocyte gate) | | | | |
| CD34+ | 30.7 | 31.0 | 0.01 | 18.9 |
| CD34+, CD38– | 1.04 | 1.81 | <0.01 | 0.60 |
| Colony-forming cells/microliter of blood | | | | |
| BFU-E | 0.5±0.1 | 2.7±0.1 | bd | nd |
| CFU-GM | 14.7±2.0 | 26.1±0.5 | bd | nd |
| CFU-GEMM | 3.5±0.8 | 12.2±0.2 | bd | nd |
| Total | 18.0±1.6 | 40.9±0.6 | bd | nd |

F, G-CSF, and EPO. bd = below detectable levels. nd = not done in this set of experiments. Colony growth was measured in cultures stimulated with SCF, IL-3, GM-CSF, G-CSF and EPO.

completed before their terminal maturation is activated); the presence within a single colony of cells of multiple lineages (an indication that their differentiation potential is not restricted); and the number of times their progeny will generate additional colonies upon replating (a measure of self-replication events allowed before commitment to maturation becomes irreversible) (Table 2). The majority of the progenitor cells detected in fetal blood and PCB give rise to colonies visible to the naked eye (>1 mm in diameter) containing more than 5000 cells of erythroid, megakaryocytic, myeloid, and monocytic lineages and upon replating, each of them generates at least 150 secondary multilineage colonies, the cells from which tertiary multilineage colonies can be generated upon replating (Table 2).[21,22] By contrast, the majority of the colonies generated by mononuclear cells from adult blood contain mainly erythroid or myeloid cells, an indication that they derive from more restricted progenitor cells and they can be replated only once.

In addition to the differences mentioned above, progenitor cells from fetal blood, PCB, or adult blood express other intrinsic differences. Fetal blood and PCB progenitor cells, in addition to their increased division number, proliferate more rapidly than adult cells: their average doubling time in culture is 20h compared to 32h of adult cells,[23] their progeny reach terminal maturation in vitro 6days earlier[23] and with fewer growth factors (1–2 growth factors vs the 3–4 required by adult cells)[24,25] (Table 2). Furthermore, when transplanted into irradiated adult recipients, fetal HSPCs expand faster than adult bone marrow cells.[26,27]

A key role in the duplication of the chromosomes is played by the telomere, a repeated DNA sequence located at the end of a chromosome, which guarantees the fidelity of DNA duplication.[28] Telomeres are duplicated by a specific enzyme complex including the telomerase enzyme. Since the elongation reaction catalyzed by this enzyme is partially inefficient, the length of the telomere decreases with each cell division.[29] It is not surprising then that the telomere length of CD34+ cells purified from fetal liver and PCB is longer than that of cells purified from adult marrow[30] (Table 2). Mice with targeted deletion of their telomerase gene (by homologous recombination) have telomereless chromosomes and experience normal life span only until the third generation.[31] By virtue of their longer telomeres, PCB CD34+ cells, even upon ex vivo expansion, likely remain active in the transplanted recipient longer than cells from adult bone marrow.

Furthermore, during ontogenesis fetal/PCB stem cells, beyond their aforementioned proliferative advantages, are distinguished by additional intrinsic properties manifested through differential expression of different sets of genes. Thus, selective expression of some genes (Sox 17, Ezh2, Hmga2[32–34]) is required for maintenance of fetal properties, whereas other genes (bmi1, Gfi1, Etv6, and Cebpα[35–38]) are important in adult stem cells but dispensable for fetal stage HSCs. In addition to these changes, there are differences in the expression program of downstream cells representing specific fetal lineages, most notably exemplified by hemoglobin switches and growth factor responsiveness in erythroid cells.[39–41]

It is of interest that the behavior of fetal stem/progenitor cells changes gradually postnatally (after the first 2–3 weeks in mice and during the first 2 years in humans). Although microenvironmental (ME) influences have been implicated to mediate these changes, there is also evidence for a cell autonomous control mechanism, as illustrated by gradual changes in expression profiles, mentioned above, in conjunction to responses from ME cues prevailing in adult environments.

## 3. STRATEGIES TO OVERCOME CURRENT LIMITATIONS IN PCB TRANSPLANTATION

The use of PCB for transplantation has specific advantages over the use of bone marrow or mobilized peripheral blood: it is easy to collect with no risk to the mother or the child; it is available immediately on demand; it may be received for use on average 25–36 days earlier than unrelated bone

**TABLE 2** Differences Among Progenitor Cells Present in Fetal, Cord, and Adult Blood

| | Fetal Cells | Neonatal Cells | Adult Cells | References |
|---|---|---|---|---|
| Duplication time in vitro | 20±2h | 26±2h | 32±2h | 23 |
| Differentiation time in vitro | 10 days | 14 days | 16 days | 23 |
| Differentiation program | Fetal Hb | In switching | Adult Hb | 39,40 |
| Proliferation potential | 4 replating | 3 replating | 1 replating | 21,22 |
| Response to single GF | 100% | 30–60% | 10–30% | 24,25 |
| Average telomere length | 12.8±0.3 | 12.3±0.4 | 8.4±0.3 | 29,30 |
| Presence in circulation | High | High | Low | - |

marrow donation; there is a tolerance 1–2 antigen mismatch without influencing graft-versus-malignancy effects; it has a lower risk of infection transmission; there is an increase in the likelihood of non-Caucasian patients finding a match.[1,42] However, PCB has one serious drawback: it contains a limited number of cells. Therefore, the results of the PCB transplants in adult patients are not as impressive as those obtained in pediatric transplantation programs. Strategies to overcome this PCB limitation by increasing the number and/or the engraftment performance of the cells present in the graft, or to influence engraftment by altering the conditioning regiment of the host, are continuously being explored. These strategies are summarized in Table 3 and discussed further below.

## 4. INCREASING THE HSPCs OF THE GRAFT

The undertaking of systematic collecting and storing of cord blood for transplantation was pioneered by the New York Blood Center in NY, USA, supported by NIH funding (National Heart, Blood and Lung Institute, Bethesda, WA, USA).[6] The first 6000 PCB units collected from 1992 to 1999 by this program were routinely characterized for their volume and WBC content. Selected units were also assayed for CFC (40.8% of the units stored) and CD34+ (5% of the units stored) content. PCB units collected under this program provided within a few years more than 550 PCB grafts in the US and 17 other countries.[42] The results of this significant clinical experience, the first of its type, were instrumental in establishing the use of PCB as a source of stem cells for transplantation. Of the 381 patients transplanted as of 10/31/1996, CFC counts were available on 130 units, most of which (67%) had a graft volume ≤60 ml. These data allowed comparison of predictions of time to engraftment obtained on the basis of WBC and CFC dose. By multivariable analyses, CFC dose was found to predict time to neutrophil engraftment better than WBC counts.[43,44] These data validated the importance of CFC level for engraftment and have been translated into three currently pursued strategies to improve the outcome of PCB transplantation by increasing the number of HSPCs present in the graft: (1a) by stem/progenitor cell expansion ex vivo; (1b) by transplantation of two PCB; and (1c) by collecting cells present in additional extraembryonic tissues.

### 4.1 HSPC Expansion Ex vivo

The recognition that PCB HSPCs are less likely than adult cells to decrease or lose their proliferative potential when expanded ex vivo with a cocktail of cytokines (Table 3) has motivated several attempts to manipulate PCB grafts ex vivo. However, because of the possibility that any expansion process may ultimately exhaust HSCs, a conservative study design has always been adopted. As a result all studies initiated up to now use a combination of unmanipulated PCB together with ex vivo expanded PCB. The ex vivo expanded PCB may be represented either by a portion of the same PCB or by a different PCB unit.

The first attempts used cocktails of cytokines that had proved themselves effective in rendering human HSPCs susceptible to retroviral infection, such as the combination of stem cell factor, FLT-3 ligand, and thrombopoietin.[15] More recent ex vivo expansion strategies are exploring the use of coculture with mesenchymal stem cells (MSCs) (either generated ex vivo from the bone marrow of a family member or off-the-shelf commercial cell lines), or cultures with activators of the Notch and PGE2 signaling pathways.[1]

Notch delivers paracrine signals among neighboring cells that promote cell growth during embryogenesis. HSPCs from PCB are still capable of responding to this signal. Stimulation of PCB with a cytokine cocktail composed of stem cell factor, FLT-3, thrombopoietin, and interleukin 6 in combination with a molecularly-engineered chimera of the extracellular domains of Notch ligand isoforms results in a 222-fold increase in the number of CD34+ cells and a 16-fold increase in repopulating cells in a mouse model.[45] These results inspired a clinical trial that has recruited so far 100 patients.[1] Preliminary results of this trial are very promising, as they indicate that this procedure makes time to neutrophil recovery 1 week shorter than that of double PCB transplants.

The use of PGE2 as a cell cycle regulator was suggested by earlier studies, which demonstrated that PGE2 enhances the formation of hematopoietic colonies from human progenitor cells.[46] This effect was rediscovered by mass screening that identified PGE2 in zebra fish as a potent stimulator of stem cell proliferation.[47] These findings, along with effects of PGE2 on stem/progenitor cell trafficking[48] and the results of preclinical studies indicating that PGE2 treatment enhances engraftment of human cord blood in xenotransplantation models,[49] formed the basis of a clinical trial currently in progress to evaluate the engraftment of double PCB units, one of which is treated with PGE2 shortly before infusion. The results presented so far with a small number of patients indicate that the PGE2-treated PCB prevails.[50]

### 4.2 Transplantation of Two PCB

The concept of considering the use of two unmanipulated PCB for the same transplant (double cord blood transplant) was primarily inspired by the known low immunogenicity of PCB. Double PCB transplants were pioneered at the University of Minnesota. The initial program involved 21 adults with hematological malignancies who received sequential infusion of two PCB after host conditioning. All patients engrafted neutrophils within 15–41 days. Presently, there have been at least 438 double PCB transplants performed worldwide (Table 3). These larger trials confirmed

**TABLE 3 Summary of the Present Clinical Experience with Cord Blood Transplantation and of Ongoing Clinical Trials Testing Strategies to Improve the Outcome of Cord Blood Transplantation**

| Source | No. of Cases (up to 2011) | Disease Treated | Advantages | Limitations | www.clinicaltrials.gov number |
|---|---|---|---|---|---|
| **Clinical Experience with Established Protocols** | | | | | |
| Single PCB | 25,000 | Benign hematological disorders and leukemias | Source of stem cells for patients who otherwise do not have a match | Cell dose: 2.5–5.0 × $10^7$ total nucleated cells/kg of body weight | Not applicable |
| | | | 1–2 antigen mismatch tolerance | This dose limits the use of single PCB to patients with a body weight of 20–40 kg (mostly children) | |
| Double PCB | 438 | Hematological malignancy | Increased speed of engraftment | Increased GVHD rates (grade II skin GVHD) | NCT00412360 Randomized children with leukemia with single versus double PCB |
| | | | Decreased relapse rates | One PCB prevails | |
| **Strategies to Improve the Stem/Progenitor Cells in the Graft** | | | | | |
| Coculture with MSC | <100 | Hematological malignancies | Early neutrophil and platelet recovery | Double PCB transplantation in which the nonexpanded unit prevails | NCT00498316 |
| Notch-mediated expansion | <100 | Hematological malignancies | Time to neutrophil recovery 1 week shorter than with double PCB | Unknown | NCT01701323 |
| Ex vivo PGE2 treatment | Unknown | Hematological malignancy | Prompt (17 days) neutrophil engraftment | Double, one unmanipulated and one PGE2-treated, PCB transplantation The PGE2-treated unit prevails | NCT00890500 |
| Collection of cells from other discarded sources | Recruiting | Hematological and nonhematological malignancy | Not available | Increased risk of bacterial contamination | NCT01586455 NCT00596999-Cell gene |
| **Strategies to Improve the Lodging Ability** | | | | | |
| Intrabone injection | 32 | Acute leukemia | Rapid recovery | Increased risk of infection | NCT00696046 |
| Copper chelator (tetraethylenepentamine) Sponsored by Gamida cell Ltd | Unknown | Hematological malignancies | Increased migration toward SDF1 and enhanced homing to bone marrow | Unknown: analyses of results on single PCB transplants in progress | NCT01221857 Trial open for double PCB transplants |
| Ex vivo treatment with fucosyltransferase-VI | Recruiting | Hematological malignancies | Not available | Study design: Double PCB transplant, one treated and one not treated | NCT01471067 |

*Continued*

**TABLE 3 Summary of the Present Clinical Experience with Cord Blood Transplantation and of Ongoing Clinical Trials Testing Strategies to Improve the Outcome of Cord Blood Transplantation—cont'd**

| | | | | | |
|---|---|---|---|---|---|
| Ex vivo treatment with sitagliptin, a CD26/dipeptidyl peptidase inhibitor | 24 | High-risk hematological malignancies | Encouraging but not yet conclusive | Study design: Systemic administration to patients receiving single PCB transplant | NCT00862719 |
| **Strategies to Affect the Microenvironment of the Recipient** | | | | | |
| Reduced conditioning regimen (12 different protocols) | <150 patients in total | Hematology malignancy and hematology benign diseases | Improve long-term survival | Unknown | Nonregistered |
| **Strategies to Improve Recovery of Immunological Functions of the Recipient** | | | | | |
| Cord blood-derived virus-specific cytotoxic T lymphocytes | Recruiting | Hematologic malignant and hematologic benign diseases | Not applicable | Study design: Administration of PCB-derived multivirus-specific T cells for prevention and treatment of EBV, CMV, and adenovirus | NCT01017705 |
| Cord blood-derived NK cells | Recruiting | Refractory AML patients | Endpoint: NK-cell expansion and function in vivo | Study design: Phase 2 trial using T-cell deleted double PCB recipients with posttransplantation IL-2 | NCT01464359 |
| Cord blood-derived Tregs | 23 | Various | Endpoint: Safety, reduced GVHD compared with historic control was observed | Study design: Injection of Treg purified from a third PCB and amplified in vitro to recipients of double PCB transplants | NCT00602693 |

AML, acute myeloid leukemia; EBV, Epstein Barr Virus; CMV, cytomegalovirus; GVHD, graft versus host disease; NK, Natural Killer; PCB, placental cord blood; MSC, mesenchymal stem cells.

that double PCB transplants have increased the speed of engraftment and decreased relapse rates compared to historical controls with single PCB transplants. It is of interest that one PCB, not necessarily the one with greater cell dose, prevails and the recipients may express single PCB hematopoiesis as early as 12 days after transplantation.[51] It was recently shown that effector CD8+ T cells play a critical role in the dominant unit that actively rejects the losing unit of PCB.[52] However, no factors have been identified that predict which unit will become dominant and further investigations in that direction are necessary.

In addition to early engraftment, recipients of double PCB have been consistently reported to experience low relapse rates.[1] This intriguing finding, the mechanism of which is not completely understood, will be analyzed in a randomized prospective trial which will evaluate the outcome of randomized children with leukemia transplanted with either a single or double PCB. Apart from the PGE2 trial described above, a trial with double PCB, one of which is expanded on MSCs, has been initiated. In this trial engraftment was prompt with neutrophil recovery in 15 days and platelet recovery in 40 days. Based on these preliminary results, a large multinational randomized trial has been planned that will compare the results of double unmanipulated PCB with those of double PCB in which one of the units will be expanded in coculture with MSC[1] (Table 3).

### 4.3 Collecting Cells Present in Additional Extraembryonic Tissues

The amount of blood lost during normal birth is on average 250 mL.[53] Since the average volume of a single PBC collection is 50–150 mL, a significant portion of blood and its content of HSPCs is still discarded. This realization has motivated the use of alternative maneuvers (such as perfusion of cord and placenta vessels with solutions containing proteases) that might increase the number of HSPCs collected. The placenta, however, is a contractile muscle the vessels of which collapse within minutes after birth. Therefore, perfusion of this organ is challenging and difficult to standardize into good manufacturing practice conditions. In addition, these procedures increase the number of cells collected by only 20%, but increase the probability of bacterial contamination by several fold.

Circulating HSPCs in fetal blood allow the colonization of new hematopoietic sites as they become available. In the fetus, the major hematopoietic organ is the liver but, as development of the skeleton progresses, new hematopoietic niches become available in the bones, the permanent hematopoietic site after birth.[54] Further, it has been recently discovered that, in addition to the liver, HSPCs are present in great numbers in proximity to the endothelium of blood vessels, especially those of the placenta which contain HSPCs in numbers far greater than expected on the basis of its blood volume.[55] Physical/enzymatic dissociation of the cellular elements of the placenta followed by physical enrichment of the stem/progenitor cells has been proposed as a procedure that might provide great numbers of HSPCs.[55] A clinical trial is currently open for recruitment that will evaluate the performance of cells collected from the placenta as a source for transplantation in hematological and nonhematological disorders.[1]

## 5. IMPROVING THE LODGING CAPACITY OF THE PCB GRAFT

### 5.1 Intrabone Injection

PCB units are generally infused intravenously and cells remain in circulation for only a few hours before reaching the marrow or other tissue sites. Although studies in mice have demonstrated that intrafemoral injection does not affect the stem cell dose that ensures radioprotection,[56] intrafemoral injection bypasses the homing to other tissue sites, especially lung, and may increase the homing of human HSPCs in xenotransplantation models.[57] Based on these considerations, a phase I/II study was conducted to evaluate the effect of intrabone transplantation of PCB in acute leukemia, reporting modest improvements in speed of neutrophil recovery compared to historical controls.[58]

### 5.2 In vitro or In vivo Treatments with Homing-enhancing Agents

One of the major pathways in stem cell mobilization and homing is the one mediated by the chemokine SDF1 (also known as CXCL12) produced by bone marrow stromal cells (including endothelial cells) interacting with the CXCR4 receptor expressed by HSPCs.[59,60] Down-modulation of CXCR4 expression induces HSPC mobilization, whereas increased levels of SDF1 in marrow improves HSPC homing. Other pathways include those with participation of integrin, selectin, or CD44 ligands.[61,62] These observations suggest that efforts to increase either the activity of SDF1 in the marrow of the recipient, or enhance the ability of PCB HSPCs to interact with ligands within bone marrow should improve homing/engraftment of donor cells.

One of the proteins identified by Dr Broxmeyer's group is dipeptidyl peptidase 4 (DDP4), a protein expressed both as a type II cell surface protein recognized by the CD26 antibody and in a soluble form.[63,64] DDP4 truncates proteins with a penultimate alanine, proline, or other selective amino acids at their aminoterminal, including several chemokines, colony-stimulating factors, and interleukins that regulate HSPC functions. DDP4 cleaves and inactivates SDF1 and inhibition of its activity in vivo with small peptides, such as diprotin A, enhances chemotaxis toward SDF1 and homing and engraftment of HSPCs.[64] Based on these results, a clinical trial was conducted to evaluate the effects of systemic administration of sitagliptin, a small inhibitor of DDP4 approved by the Food and Drug Administration for treatment of type 2 diabetes in patients with high-risk hematological malignancies transplanted with one PCB.[65] The results of the first 24 patients enrolled in the trial are indicative of enhanced time to neutrophil recovery, but must be confirmed by a larger trial.

Other studies in xenotransplantation models have demonstrated that engraftment of PCB CD34+ cells occurs more rapidly and is of greater magnitude if the cells are pretreated with an enzyme (fucosyltransferase-VI) that increases the level of fucosylation on specific membrane proteins.[66] These observations provided the rationale for a clinical trial open for recruitment that will evaluate hematopoietic recovery in patients receiving double PCB, one untreated and one pretreated with fucosyltransferase-VI.[1]

## 6. IMPROVING THE STEM CELL RECEPTIVITY OR IMMUNOLOGICAL STATUS OF THE HOST

### 6.1 Reduced Conditioning Regimen

The use of a reduced conditioning regimen has been shown by Kurzberg et al. to improve the outcome of PCB transplantation in children with hematological malignancy.[67] Based on this encouraging early result, at least 12 different protocols are investigating the effects of a reduced conditioning regimen on the outcome of transplantation in 40–60 year old adults with hematological malignancies[1] (Table 3).

### 6.2 Injection of PCB-derived Immune Cells

PCB transplantation is associated with a delayed recovery of the immune functions and a higher risk of morbidity and mortality due to opportunistic infections. In addition, although promising, the outcomes of double PCB are associated with increased expression of graft-versus-host disease (GVHD). To address these immunological concerns, clinical trials are in progress to assess outcomes of transplantation of PCB in association with administration of PCB-derived immune cells. The possibility of generating anti-EBV, CMV, and adenovirus-specific T cells ex vivo from PCB[68] has been exploited in a trial (NCT01017705) to analyze the use of PCB-derived T cells primed to deliver a multiviral response for prevention and treatment of viral infections after PCB transplantation. Another trial will analyze the efficacy of posttransplantation IL-2 administration to increase the cytotoxicity of PCB-derived natural killer cells and the frequency of remission. Tregs (regulatory T cells), a cell population that modulates immune functions, generated ex vivo from a third PCB, will be administered to patients receiving double PCB transplants in an attempt to decrease GVHD.[69]

## 7. ADDITIONAL CLINICAL USES OF CORD BLOOD UNDER INVESTIGATION

The original patent on PCB as a source of stem cell transplantation filed in 1991 listed as many as 98 possible different applications.[4] These included not only transplantation to cure congenital (35 applications) and acquired (20 applications) defects of the HSCs, but also to restore altered immunological functions in patients with solid tumors (12 applications), autoimmune diseases (5 applications), bacterial or fungal infections, and to treat congenital metabolic disorders (9 applications).

In addition to HSCs and to progenitor cells restricted toward all myeloid and lymphoid lineages, PCB contains stem and progenitor cells for endothelial cells (hemangioblasts and hemogenic endothelium), MSCs, and unrestricted somatic stem cells.[70] This recognition has been exploited and fueled the development of several for-profit PCB banks established worldwide aiming at dedicated PCB uses.[71]

The number of clinical trails that are exploring new applications of PCB is also exponentially increasing.[70,71] A search of ClinicalTrials.gov database, a USA registry of Government and privately supported clinical trials, with the keywords "cord blood stem cells" performed in 2011 generated over 120,000 hits. The majority of the hits referred to trials for hematological disorders, but 31 regarded applications for nonhematological disorders. In addition to transplantation to generate hematopoietic cells as a source of metabolites not produced in metabolic diseases (as mentioned in the original US patent), additional clinical trials are exploiting the presence of endothelial stem/progenitor cells in PCB to generate in vitro blood vessels for regenerative medicine in stroke and other conditions of vessel degeneration (see as an example[72]). The presence of MSC also suggests clinical applications for unmanipulated and in vitro modified PCB as tissue repair in hearing loss, Alzheimer's disease, and to prevent GVHD.[73,74] The presence of unrestricted somatic stem cells is encouraging the use of PCB possibly in combination with synthetic scaffold devices for repairing cartilage and bone injury, for osteoarthritis and systemic sclerosis.[75–77] Last but not least, PCB has been also suggested as a possible source to provide ex vivo expanded cells as a cell therapy for burns and wound healing.[78]

## 8. CONCLUDING REMARKS

The exciting biology of PCB cell repertoire and their functional potential is currently unfolding. Exploitation of this knowledge will certainly expand in the future the roster of clinical indications using PCB. As a renewable resource of primary human cells with superior functional potential, PCB has unique advantages over other sources of cells for potential uses in regenerative medicine, and all indications thus far suggest a productive journey in future therapeutic uses of PCB.

## REFERENCES


1. Oran B, Shpall E. Umbilical cord blood transplantation: a maturing technology. *Hematology* 2012:215–22.
2. Nakahata T, Ogawa M. Hemopoietic colony-forming cells in umbilical cord blood with extensive capability to generate mono- and multipotential hemopoietic progenitors. *J Clin Invest* 1982;**70**:1324–8.
3. Broxmeyer HE, Douglas GW, Hangoc G, Cooper S, Bard J, English D, et al. Human umbilical cord blood as a potential source of transplantable hematopoietic stem/progenitor cells. *Proc Natl Acad Sci USA* 1989;**86**:3828–32.
4. Boyse EA, Broxmeyer HE, Douglas GW. Preservation of fetal and neonatal hematopoietic stem and progenitor cells of the blood. U.S. Pat No 5,004,681 1991;15:14–16.
5. Gluckman E, Broxmeyer HA, Auerbach AD, Friedman HS, Douglas GW, Devergie A, et al. Hematopoietic reconstitution in a patient with Fanconi's anemia by means of umbilical-cord blood from an HLA-identical sibling. *N Engl J Med* 1989;**321**:1174–8.
6. Rubinstein P, Dobrila L, Rosenfield RE, Adamson JW, Migliaccio G, Migliaccio AR, et al. Processing and cryopreservation of placental/umbilical cord blood for unrelated bone marrow reconstitution. *Proc Natl Acad Sci USA* 1995;**92**:10119–22.
7. Rubinstein P, Carrier C, Scaradavou A, Kurtzberg J, Adamson J, Migliaccio AR, et al. Outcomes among 562 recipients of placental blood transplants from unrelated donors. *N Engl J Med* 1998;**339**:1565–77.
8. Metcalf D. The molecular control of cell division, differentiation, commitment and maturation in haemopoietic cells. *Nature* 1989;**339**:27–30.
9. Weissman IL, Shizuru JA. The origins of the identification and isolation of hematopoietic stem cells, and their capability to induce donor-specific transplantation tolerance and treat autoimmune diseases. *Blood* 2008;**112**:3543–53.
10. Dick JE. Human stem cell assays in immune-deficient mice. *Curr Opin Hematol* 1996;**3**:405–9.
11. Greiner DL, Hesselton RA, Shultz LD. SCID mouse models of human stem cell engraftment. *Stem Cells* 1998;**16**(3):166–77.
12. Wang JC, Doedens M, Dick JE. Primitive human hematopoietic cells are enriched in cord blood compared with adult bone marrow or mobilized peripheral blood as measured by the quantitative in vivo SCID-repopulating cell assay. *Blood* 1997;**89**(11):3919–24.
13. Phillips RL, Reinhart AJ, Van Zant G. Genetic control of murine hematopoietic stem cell pool sizes and cycling kinetics. *Proc Natl Acad Sci USA* 1992;**89**:11607–11.
14. Spitzer G, Verma DS, Fisher R, Zander A, Vellekoop L, Litam J, et al. The myeloid progenitor cell—its value in predicting hematopoietic recovery after autologous bone marrow transplantation. *Blood* 1980;**55**:317–23.
15. Dao MA, Shah AJ, Crooks GM, Nolta JA. Engraftment and retroviral marking of CD34+, and CD34+CD38− human hematopoietic progenitors assessed in immune-deficient mice. *Blood* 1998;**91**:1243–55.
16. Molineux G, Pojda Z, Hampson IN, Lord BI, Dexter TM. Transplantation potential of peripheral blood stem cells induced by granulocyte colony-stimulating factor. *Blood* 1990;**76**:2153–8.
17. Berenson RJ, Bensinger WI, Hill RS, Andrews RG, Garcia-Lopez J, Kalamasz DF, et al. Engraftment after infusion of CD34+ marrow cells in patients with breast cancer or neuroblastoma. *Blood* 1991;**77**:1717–22.
18. Schiller G, Vescio R, Freytes C, Spitzer G, Sahebi F, Lee M, et al. Transplantation of CD34+ peripheral blood progenitor cells after high-dose chemotherapy for patients with advanced multiple myeloma. *Blood* 1995;**86**:390–7.
19. Hohlfeld P, Forestier F, Kaplan C, Tissot JD, Faffos F. Fetal thrombocytopenia: a retrospective survey of 5194 fetal blood samplings. *Blood* 1994;**84**:1851–6.
20. Eddleman KA, Chervenak FA, George-Siegel P, Migliaccio G, Migliaccio AR. Circulating hematopoietic stem cell populations in human fetuses: Implications for fetal gene therapy and alterations with in utero red cell transfusion. *Fetal Diagn Ther* 1996;**11**:231–40.
21. Migliaccio G, Baiocchi M, Hamel N, Eddleman K, Migliaccio AR. Circulating progenitor cells in human ontogenesis: response to growth factors and replating potential. *J Hemother* 1996;**5**:161–70.
22. Lu L, Xiao M, Shen RN, Grigsby S, Broxmeyer HE. Enrichment, characterization and responsiveness of single primitive CD34+++ human umbilical cord blood hematopoietic progenitors with high proliferative and replating potential. *Blood* 1993;**81**:41–8.
23. Peschle C, Migliaccio AR, Migliaccio G, Ciccariello R, Lettieri F, Quattrin S, et al. Identification and characterization of three classes of erythroid progenitors in human fetal liver. *Blood* 1981;**58**:565–72.
24. Emerson SG, Sieff CA, Wang EA, Wong GG, Clark SC, Nathan DG. Purification of fetal hematopoietic progenitors and demonstration of recombinant multipotential colony-stimulating activity. *J Clin Inv* 1985;**76**:1286–90.
25. Ruggieri L, Heimfeld S, Broxmeyer HE. Cytokine-dependent ex vivo expansion of early subsets of CD34+ cord blood myeloid progenitors is enhanced by cord blood plasma, but expansion of the more mature subsets of progenitor is favored. *Blood Cells* 1994;**20**:436–54.
26. Micklem HS, Ford CE, Evans EP, Ogden DA, Papworth DS. Competitive in vivo proliferation of foetal and adult haematopoietic cells in lethally irradiated mice. *J Cell Physiol* 1972;**79**(2):293–8.
27. Harrison DE, Zhong RK, Jordan CT, Lemischka IR, Astle CM. Relative to adult marrow, fetal liver repopulates nearly five times more effectively long-term than short-term. *Exp Hematol* 1997;**25**(4):293–7.
28. Holt SE, Shay JW, Wright WE. Refining the telomere-telomerase hypothesis of aging and cancer. *Nat Biotech* 1996;**14**:836–9.
29. Vaziri H, Dragowska W, Allsopp RC, Thomas TE, Harley CB, Lansdorp PM. Evidence for a mitotic clock in human hematopoietic stem cells: loss of telomeric DNA with age. *Proc Natl Acad Sci USA* 1994;**91**:9857–60.
30. Lansdorp PM, Dragowska W, Mayani H. Ontogeny-related changes in proliferative potential of human hematopoietic cells. *J Exp Med* 1993;**178**:787–91.
31. Blasco MA, Lee H-W, Hande MP, Samper E, Lansddorp PM, DePinho RA, et al. Telomere shortening and tumor formation by mouse cells lacking telomerase RNA. *Cell* 1997;**91**:25–34.
32. Kim I, Saunders TL, Morrison SJ. Sox17 dependence distinguishes the transcriptional regulation of fetal from adult hematopoietic stem cells. *Cell* 2007;**130**(3):470–83.
33. Mochizuki-Kashio M, Mishima Y, Miyagi S, Negishi M, Saraya A, Konuma T, et al. Dependency on the polycomb gene Ezh2 distinguishes fetal from adult hematopoietic stem cells. *Blood* 2011;**118**(25):6553–61.
34. Copley MR, Babovic S, Benz C, Knapp DJ, Beer PA, Kent DG, et al. The Lin28b–let-7–Hmga2 axis determines the higher self-renewal potential of fetal haematopoietic stem cells. *Nat Cell Biol* 2013;**15**:916–25.
35. Park IK, Qian D, Kiel M, Becker MW, Pihalja M, Weissman IL, et al. Bmi-1 is required for maintenance of adult self-renewing haematopoietic stem cells. *Nature* 2003;**423**(6937):302–5.
36. Hock H, Hamblen MJ, Rooke HM, Schindler JW, Saleque S, Fujiwara Y, et al. Gfi-1 restricts proliferation and preserves functional integrity of haematopoietic stem cells. *Nature* 2004;**431**(7011):1002–7.

37. Hock H, Meade E, Medeiros S, Schindler JW, Valk PJ, Fujiwara Y, et al. Tel/Etv6 is an essential and selective regulator of adult hematopoietic stem cell survival. *Genes Dev* 2004;**18**(19):2336–41.
38. Ye M, Zhang H, Amabile G, Yang H, Staber PB, Zhang P, et al. C/EBPa controls acquisition and maintenance of adult haematopoietic stem cell quiescence. *Nat Cell Biol* 2013;**15**(4):385–94.
39. Kidoguchi K, Ogawa M, Karem JD, McNeil JS, Fitch MS. Hemoglobin biosynthesis in individual bursts in culture: studies of human umbilical cord blood. *Blood* 1979;**53**:519–22.
40. Stamatoyannopoulos G, Rosenblum BB, Papayannopoullou T, Brice M, Nakamoto B, Shepard TH. HbF and HbA production in erythroid cultures from human fetuses and neonates. *Blood* 1979;**54**:440–50.
41. Bowie MB, Kent DG, Copley MR, Eaves CJ. Steel factor responsiveness regulates the high self-renewal phenotype of fetal hematopoietic stem cells. *Blood* 2007;**109**:5043–8.
42. Rubinstein P, Rosenfield RE, Adamson JW, Stevens CE. Stored placental blood for unrelated bone marrow reconstitution. *Blood* 1993;**81**:1679–90.
43. Migliaccio AR, Adamson JW, Stevens CE, Dobrila NL, Carrier C, Rubinstein P. Cell dose and speed of engraftment in placental/umbilical cord blood transplantation: graft progenitor cell content is a better predictor than nucleated cell quantity. *Blood* 2000;**96**:2717–22.
44. Stevens CE, Gladstone J, Taylor P, Scaradavou A, Migliaccio AR, Visser J, et al. Placental/umbilical cord blood for unrelated bone marrow reconstitution: relevance of nucleated red blood cells. *Blood* 2002;**100**:2662–4.
45. Delaney C, Heimfeld S, Brashem-Stein C, Voorhies H, Manger RL, Bernstein ID. Notch-mediated expansion of human cord blood progenitor cells capable of rapid myeloid reconstitution. *Nat Med* 2010;**16**:232–6.
46. Pelus LM. Association between colony forming units-granulocyte macrophage expression of Ia-like (HLA-DR) antigen and control of granulocyte and macrophage production. A new role for prostaglandin E. *J Clin Invest* 1982;**70**:568–78.
47. North TE, Goessling W, Walkley CR, Lengerke C, Kopani KR, Lord AM, et al. Prostaglandin E2 regulates vertebrate haematopoietic stem cell homeostasis. *Nature* 2007;**447**:1007–11.
48. Hoggatt J, Mohammad KS, Singh P, Hoggatt AF, Chitteti BR, Speth JM, et al. Differential stem- and progenitor-cell trafficking by prostaglandin E2. *Nature* 2013;**495**:365–9.
49. Goessling W, Allen RS, Guan X, Jin P, Uchida N, Dovey M, et al. Prostaglandin E2 enhances human cord blood stem cell xenotransplants and shows long-term safety in preclinical nonhuman primate transplant models. *Cell Stem Cell* 2011;**8**:445–58.
50. Cuttler CS, Shoemaker D, Ballen KK, et al. FT1050 (16,16-dimethyl Prostaglandin E2)-enhanced umbilical cord blood accelerates hematopoietic engraftment after reduced intensity conditioning and double umbilical cord blood transplantation [abstract]. *Blood* 2011;**118**:653.
51. Scaradavou A, Brunstein CG, Eapen M, Le-Rademacher J, Barker JN, Chao N, et al. Double unit grafts successfully extend the application of umbilical cord blood transplantation in adults with acute leukemia. *Blood* 2013;**121**:752–8.
52. Gutman JA, Turtle CJ, Manley TJ, Heimfeld S, Bernstein ID, Riddell SR, et al. Single-unit dominance after double-unit umbilical cord blood transplantation coincides with a specific CD8+ T-cell response against the nonengrafted unit. *Blood* 2010;**115**(4):757–65.
53. Zhang J, Bowes Jr WA. Birth-weight-for-gestational-age patterns by race, sex, and parity in the United States population. *Obstet Gynecol* 1995;**86**:200–8.
54. Charbord P, Tavian M, Humeau L, Péault B. Early ontogeny of the human marrow from long bones: an immunohistochemical study of hematopoiesis and its microenvironment. *Blood* 1996;**87**:4109–19.
55. Dzierzak E, Robin C. Placenta as a source of hematopoietic stem cells. *Trends Mol Med* 2010;**16**:361–7.
56. van Bekkum DW. Intrafemoral and intravenous (retro-orbital) administration of syngeneic bone marrow. *Exp Hematol* 2011;**39**:413.
57. McKenzie JL, Gan OI, Doedens M, Dick JE. Human short-term repopulating stem cells are efficiently detected following intrafemoral transplantation into NOD/SCID recipients depleted of CD122+ cells. *Blood* 2005;**106**:1259–61.
58. Frassoni F, Gualandi F, Podestà M, Raiola AM, Ibatici A, Piaggio G, et al. Direct intrabone transplant of unrelated cord-blood cells in acute leukaemia: a phase I/II study. *Lancet Oncol* 2008;**9**:831–9.
59. Broxmeyer HE. Chemokines in hematopoiesis. *Curr Opin Hematol* 2008;**15**(1):49–58.
60. Aiuti A, Tavian M, Cipponi A, Ficara F, Zappone E, Hoxie J, et al. Expression of CXCR4, the receptor for stromal cell-derived factor-1 on fetal and adult human lympho-hematopoietic progenitors. *Eur J Immunol* 1999;**29**:1823–31.
61. Bonig H, Papayannopoulou T. Hematopoietic stem cell mobilization: updated conceptual renditions. *Leukemia* 2013;**27**(1):24–31.
62. Xia L, McDaniel JM, Yago T, Doeden A, McEver RP. Surface fucosylation of human cord blood cells augments biding to P-selectin and E-selectin and enhances engraftment in the bone marrow. *Blood* 2009;**113**:3091–6.
63. Broxmeyer HA. Enhancing engraftment of cord blood cells via insight into the biology of stem/progenitor cell function. *Ann N Y Acad Sci* 2012;**1266**:151–60.
64. Ou X, O-Leary HA, Broxmeyer HE. Implications of DPP4 modification of proteins that regulate stem/progenitor and more mature cell types. *Blood* 2013;**122**:161–9.
65. Farag SS, Srivastava S, Messina-Graham S, Schwartz J, Robertson MJ, Abonour R, et al. In vivo DPP-4 inhibition to enhance engraftment of single-unit cord blood transplants in adults with hematological malignancies. *Stem Cells Dev* 2013;**22**:1007–15.
66. Robinson SN, Simmons PJ, Thomas MW, Brouard N, Javni JA, Trilok S, et al. Ex vivo fucosylation improves human cord blood engraftment in NOD-SCID IL-2Rgamma(null) mice. *Exp Hematol* 2012;**40**:445–56.
67. Kurtzberg J, Prasad VK, Carter SL, Wagner JE, Baxter-Lowe LA, Wall D, et al. Results of the cord blood transplantation study (COBLT): clinical outcomes of unrelated umbilical cord blood transplantation in pediatric patients with hematological malignancies. *Blood* 2008;**112**:4318–27.
68. Hanley PJ, Cruz CR, Savoldo B, Leen AM, Stanojevic M, Khalil M, et al. Functionally active virus-specific T cells that target CMV, adenovirus and EBV can be expanded from naïve T-cell populations in cord blood and will target a range of viral epitopes. *Blood* 2009;**114**:1958–67.
69. Brunstein CG, Miller JS, Cao Q, McKenna DH, Hippen KL, Curtsinger J, et al. Infusion of ex vivo expanded T regulatory cells in adults transplanted with umbilical cord blood: safety profile and detection kinetics. *Blood* 2011;**117**:1061–70.
70. Pelosi E, Castelli G, Testa U. Human umbilical cord blood is a unique and safe source of stem cells suitable for treatment of hematological diseases and for regenerative medicine. *Blood Cell Mol Dis* 2012;**49**:20–8.
71. Ilic D, Miere C, Lazic E. Umbilical cord blood stem cells: clinical trials in non-hematological disorders. *Br Med Bull* 2012;**102**:43–57.

72. Roura S, Pujal JM, Bayes-Genis A. Umbilical cord blood for cardiovascular therapy: from promise to fact. *Ann N Y Acad Sci* 2012;**1254**:66–70.
73. Avanzini MA, Bernardo ME, Cometa AM, Perotti C, Zaffaroni N, Novara F, et al. Generation of mesenchymal stromal cells in the presence of platelet lysate: a phenotypic and functional comparison of umbilical cord blood- and bone marrow-derived progenitors. *Haematologica* 2009;**94**:1649–60.
74. Zhao Y. Stem cell educator therapy and induction of immune balance. *Curr Diab Rep* 2012;**12**:517–23.
75. Longo UG, Loppini M, Berton A, La Verde L, Khan WS, Denaro V. Stem cells from umbilical cord and placenta for musculoskeletal tissue engineering. *Curr Stem Cell Res Ther* 2012;**7**:272–81.
76. Nagano M, Kimura K, Yamashita T, Ohneda K, Nozawa D, Hamada H, et al. Hypoxia responsive mesenchymal stem cells derived from human umbilical cord blood are effective for bone repair. *Stem Cells Dev* 2010;**19**:1195–210.
77. Polini A, Pisignano D, Parodi M, Quarto R, Scaglione S. Osteoinduction of human mesenchymal stem cells by bioactive composite scaffolds without supplemental osteogenic growth factors. *PLoS One* 2011;**6**(10):e26211.
78. Branski LK, Gauglitz GG, Herndon DN, Jeschke MG. A review of gene and stem cell therapy in cutaneous wound healing. *Burns* 2009;**35**:171–80.

Chapter 4

# Immunobiology of Cord Blood Cells

**Sergio Querol[1], Aurore Saudemont[2] and Antoine Toubert[3]**

*[1]Barcelona Cord Blood Bank and Haematopoietic Progenitor Cell Unit, Banc Sang i Teixits, Barcelona, Spain; [2]Anthony Nolan Research Institute and University College London, London, UK; [3]Laboratoire d'Immunologie et Histocompatibilité, INSERM UMR1160, and Université Paris Diderot, Sorbonne Paris Cité, Institut Universitaire d'Hématologie, Hôpital Saint-Louis, Paris, France*

## Chapter Outline



## 1. INTRODUCTION

Cord blood (CB) is a well-established source of hematopoietic stem cells (HSC) for transplantation.[1,2] The key immunological advantages of CB include the tolerance to antigen mismatches that confers a decreased incidence of Graft-versus-Host Disease (GvHD) while maintaining the pursued antitumor effect.[2] In the past years, significant progresses have been made to improve the clinical outcome of CB transplantation (CBT), but delayed hematopoietic engraftment and increased treatment-related mortality remain obstacles to the widespread use of CB. Although CBT has mainly been used as a treatment of blood disorders, it has now been proposed that CB cell therapies could be used to support CBT, to treat nonhematopoietic diseases, or as a form of cellular regenerative therapy or immune modulation.[3]

This chapter will cover the definition of immune properties of CB cells and its consequences in CBT and cell therapy.

## 2. IMMUNE PROPERTIES OF CORD BLOOD CELLS

CB contains higher numbers of T cells, natural killer (NK) cells, and B cells in comparison to peripheral blood (PB) or to the bone marrow (BM), but fewer γδ T cells and NKT cells.[4–6] Moreover, most T lymphocytes present in CB are naïve, with CB being usually characterized by an increased number of CD45RA+ lymphocytes over CD45RO memory lymphocytes.[4,5] In addition, several studies have reported that overall CB mononuclear cells produce less cytokines than PB mononuclear cells.[6–8] This section presents the characteristics—phenotype and function—of each immune cell type present in CB in comparison to their adult counterparts.

### 2.1 Monocytes and Dendritic Cells

Monocytes are innate immune cells that can act as antigen presenting cells (APC), and are key in the production of inflammatory cytokines and in the generation of a specific immune response. Monocytes are characterized by their high expression of CD14, which has been reported to be similar between CB and PB monocytes. CB and PB exhibit similar frequencies of monocytes and monocyte subsets.[9] In addition, the expression of markers key in APC function such as CD11c, CD80/CD86, CD163, and HLA-DR was similar in CB and PB monocytes. Monocyte subsets in CB produced higher levels of IL-12 and TNF in response

Cord Blood Stem Cells Medicine. http://dx.doi.org/10.1016/B978-0-12-407785-0.00004-9

to peptidoglycan,[9] however lower production of cytokines, in particular TNF, by CB monocytes in response to other stimuli has also been reported.[10] CB and PB monocytes express similar levels of Toll-like receptors (TLR) but CB monocytes do not produce similar levels of TNF in response to TLR ligands.[11,12]

Dendritic cells (DC) are APC that play a key role in the induction of T-cell responses, in immunity after hematopoietic stem cell transplantation (HSCT), and in this context in the induction of GvHD. Absolute numbers of myeloid and lymphoid DC, two different types of circulating DC in humans, were reported to be lower in CB than in the BM but higher than in PB.[13] The frequency of lymphoid DC in CB is also much higher than in PB, with most DC in CB being lymphoid.[14] Impaired APC functions of CB DC, such as capture of antigens or stimulation of allogeneic T cells have been reported as compared to PB DC.[14,15] TLR expression is similar on CB and PB APC, however the responses to TLR ligands, in particular, production of cytokines such as TNF are different with weaker responses generated by CB DC.[16]

DC derived from CB monocytes had a similar phenotype as those derived from PB monocytes with, however, a lower expression of HLA-DR, CD40, and CD80.[17,18] Moreover, CB-derived DC exhibited impaired IL-12 secretion in response to stimuli that impacted on the capacity of T cells to produce cytokines, in particular IFN-γ.[17,19] However, another study reported CB-derived DC to be immature with lower capacity to produce cytokines such as IFN and TNF-α.[20] In addition, CB-derived DC expressed lower levels of CD1a, and MHC class II, and exhibited lower endocytosis capacity as well as lower ability to stimulate T cells in mixed lymphocyte reaction.[21]

## 2.2 NK Cells

NK cells are innate immune cells that can eliminate tumor cells or virus-infected cells without prior stimulation that are also key effectors of the Graft-versus-Leukemia (GvL) effect after HSCT. It has been reported that NK cells are more abundant in CB than in PB, constituting up to 33% of CB lymphocytes.[22] The two major populations of NK cells, $CD56^{dim}$ and $CD56^{bright}$ NK cells, exist in similar proportions in CB and PB.[23–25] Phenotypic and functional similarities and differences between PB and CB NK cells have been reported.[26] Some groups found that CB NK cells are mature,[23] and others that CB NK cells are phenotypically and functionally immature as compared to PB NK cells.[27,28]

CB NK cells are less cytotoxic than resting PB NK cells against K562 cells.[23,24,28] A lower expression of adhesion molecules and a higher expression of inhibitory receptors might account for the reduced cytotoxicity of CB NK cells.[24,28,29] Nevertheless, CB NK cell cytotoxicity can be enhanced after IL-2, IL-12, or IL-15 stimulation.[23,27,28] No difference in production of IFN-γ between CB and PB NK cells has been observed.[23,25,29] Moreover, CB NK cells seem to have a great potential to migrate to the BM, but not to lymph nodes as compared to PB NK cells, as suggested by the trafficking receptors they expressed.[28] In terms of repertoire acquisition and functional education, CB NK cells are functionally educated toward their cognate human leukocyte antigen (HLA) class I ligand.[30]

## 2.3 Unconventional Lymphoid T Cells

The so-called unconventional lymphoid T cells include NKT cells, γδ T cells, and mucosal-associated invariant T cells (MAIT). They play an increasing role at the interface between innate and adaptive immunity.

NKT cells are a rare subset of T cells with immune regulatory functions. The invariant NKT cell (iNKT) subset expresses the Vα24-Vβ11 TCR in humans. Low numbers of NKT cells have been found in CB as well as γδ T cells, however it is possible to expand efficiently both cell types in vitro.[31] Interestingly, CB NKT cells show an activated phenotype as compared to PB NKT cells,[32,33] highlighting the fact that NKT cells get stimulated before birth as opposed to other cell types in CB. Memory NKT cells were present at the same frequency in CB and PB, were polyclonal, and were poor cytokine producers in response to stimuli.[32] CB NKT cells showed a Th2-biased response as demonstrated by their phenotype and cytokine production.[34] In fact, another study describes that CB NKT cells have a preferential capacity to differentiate into IL-4+IFN-γ − NKT2 cells[35] while another group reported that NKT cells could produced both IL-4 and IFN-γ depending on the conditions used to expand these cells.[36]

MAIT express the semi-invariant TCR Vα7.2 in humans and are restricted by the MHC-Ib molecule MR1. As with conventional T cells, iNKT and MAIT are selected in the thymus in an MHC-dependent manner on CD1d and MR1, respectively. In the periphery, MAIT have a high ability to expand following exposure to antigenic stimuli, especially of bacterial origin,[37] and are therefore mostly of memory phenotype. In CB they are already functional while keeping a naïve phenotype as for other T cells.[38]

## 2.4 T Cells

As previously mentioned, CB contains the same percentage of CD3+ T cells as PB, however absolute numbers of T cells are increased in CB.[5,6] Notably, the ratio between CD4 and CD8 T cells is similar in CB as in PB. CB T cells are naïve, with most T cells being CD45RA+CD62L+ recent thymic emigrants.[39,40] CB contains low levels of memory or mature T cells that express CD45RO. Interestingly, it has been reported that a higher percentage of T cells from CB divide in response to stimuli, but that CB T cells maintain a CD45RA phenotype even in these conditions.[41] Recently, CB CD4+ T cells have been reported to have a higher calcium flux following CD3 engagement than their adult CD4+ naïve counterpart but similar to adult CD4+ recent thymic emigrants,[42] highlighting their specific properties compared to adult naïve T cells. CB CD4 T cells are immature and exhibit a CD45RA phenotype. Consistently, CD4+ CD25+ regulatory T (Treg) cells are mostly naïve, have a higher frequency in CB, and can suppress the proliferation of allogeneic effector cells after stimulation with cytokines and anti-CD3/anti-CD28 as efficiently as PB Treg cells.[43–45]

CB T cells have been reported to be less cytolytic.[46] This could be explained by the fact that CB lymphocytes express lower levels of perforin.[47] In addition, CB mononuclear cells have been shown to produce less cytokines than their adult counterparts[8,48] and to express lower levels of T helper (Th)1 over Th2 cytokines, which correlate with a lower expression of T-bet in CB T cells.[49] Moreover, in comparison to PB T cells, CB T cells express lower levels of NF-κB, STAT4, and NFATc2 (nuclear factor of activated T cells), transcription factors key in the regulation of cytokine expression.[49]

## 2.5 B Cells

CB B cells have been shown to be immature and naïve with an increased percentage of CD5+ B cells and B1 cells.[50,51] CD20+CD27+CD43+ B1 cells have been found in CB and it was demonstrated that these cells are functional in comparison to those in adult PB.[52] Moreover, CD34+ Pax-5+ B cell progenitors are also present in CB. Interestingly, it was shown that B cell progenitors in CB exhibit a different phenotype from those in the BM.[53] CB B cells express different membrane immunoglobulin isotypes; however, they have been shown to secrete low levels of IgG and IgA. Moreover, CB B cells were functional after culture with T-cell supernatants.[54,55] However, it has been reported that CD40 signaling in CB B cells is inefficient.[56]

# 3. IMMUNE RECONSTITUTION AFTER CBT

A main concern after CBT is the profound and long-lasting immunodeficiency,[57] since CB units have one to two log lower HSC absolute numbers compared to BM or PB as stem cell source. A potent immune reconstitution (IR) is essential to limit the infection risk and disease relapse. Reconstitution of the different lymphocyte populations (B, T, NK, NKT cells) and APC of myeloid origin (monocytes, macrophages, and DC) should be considered not only quantitatively but foremost qualitatively in terms of functional subsets. General parameters of IR have been recently reviewed extensively in adults and children;[58–61] therefore we will focus here on some aspects of CBT IR, including evaluation of T-cell diversity and recovery, immunological issues of double CBT, and the impact of Killer-cell immunoglobulin-like receptor (KIR)–ligand incompatibility in CBT.

## 3.1 T-cell IR after CBT: Thymic-independent versus Thymic-dependent Pathways

Immaturity and naïvety of CB T cells could explain the lower GvHD risk but also raises some concern about the quality of IR in CB transplanted patients. Evaluation of IR after HSCT has recently improved through the development of direct methods for analysis of T-cell diversity (Immunoscope or spectratyping and high-resolution TCR sequencing), ex vivo thymic function (TCR rearrangement excision circles), and viral-specific immune responses (HLA class I tetramers). Regeneration of the T-cell population proceeds normally on two different pathways.

The thymic-independent pathway includes transfer of graft-derived mature donor T cells followed by antigen-driven expansions, mainly specific for reactivating herpes viruses such as cytomegalovirus (CMV) and Epstein Barr virus (EBV). This provides the first wave of T-cell reconstitution after transplantation, which is helpful to control viral reactivations early after transplant and which will be limited in CBT due to the lack of memory T cells. Homeostatic expansion of CB naïve T cells is however possible, as shown by comparing CBT and T-cell-depleted transplants from unrelated PB stem cell[62] or HLA haploidentical settings.[63] In vivo T-cell depletion using thymoglobulin (ATG) has a profound detrimental effect on IR in CBT.[64]

The thymic-dependent pathway involves selection of graft-derived precursor cells, accounts for the more durable reconstitution of the T-cell compartment, and generates a diverse TCR repertoire. Thymopoiesis peaked approximately 6–12 months after HSCT[60] and is strongly impacted by ATG conditioning regimen.[57] In the first year post-transplant, thymic naïve CD31+CD45RA+CD4+ T cells are lower after CBT than after BM as HSC source.[65] In the long term, 2 years after transplant and beyond, analysis of T-cell IR after CBT compared to identical sibling BM transplant precisely matched for age and GvHD grade indicated that, despite the much lower number of CD34+ cells infused, recovery of T lymphocyte diversity was comparable, or even better, in CBT recipients.[66] An efficient recovery of T-cell diversity in CD4 and CD8 populations has been reported after double CBT by high-throughput sequencing techniques. At 6 months post-transplant, CD4+ T-cell diversity was about 28-fold higher compared to T-cell-depleted recipients.[67] An efficient thymic function may also be crucial for a durable antileukemic immune response in children after CBT.[63] Peculiar properties of lymphoid progenitors in CB could favor a strong thymic rebound and a durable long-term reconstitution of the TCR diversity. Indeed, a CD34+Lin-CD10+CD24− population of lymphoid progenitors with thymic seeding properties (thymic seeding progenitors) can be found in healthy adult donors in the periphery, but also at least at a five-fold higher frequency in CB samples.[68]

In the short term after transplant (first 3 months), specific immune responses against herpes viruses (EBV, CMV) measured by the frequency of antigen-specific CD8 T cells may be delayed and less efficient in CBT compared to BM transplant. Clearance of CMV viremia after double CBT appears dependent on the reconstitution of thymopoiesis[69] but also on in vivo expansion of CMV-specific CD4 and CD8 naïve T cells.[70] There is a high risk of EBV reactivation and development of post-transplant lymphoproliferative disorders in such a naïve immune environment. These patients need a close monitoring of EBV reactivation and of the EBV-specific T-cell immune responses in order to initiate early specific therapies such as rituximab (anti-CD20 monoclonal antibody)[71] and could also benefit from viral-specific adoptive cellular therapies (see below).

### 3.2 Double Cord Blood Transplantation

Double-unit CBT may increase the availability of this transplant procedure to adults, improve engraftment, and speed neutrophil recovery.[72] Single CB unit dominance is usually achieved within the first month of the double CB transplant. Engraftment kinetics of the various leukocyte subsets are variable after double CBT.[73] Single-donor dominance is dependent on the dose of CD3+, CD4+, and naïve CD3+CD8+ T cells,[74] in favor of an alloreactive graft-versus-graft effect. With regard to the HLA choice between the units and between units and the recipient, the best choice presently appears to match at the allelic level for HLA-A, -B, -DRB1 between CB units and the recipient but not unit to unit.[75] However, as in other HSCT settings, a complete HLA-matching at the allelic level for HLA-A, -B, -C, and -DRB1 between CB units and the recipient appears to be the best choice.[76] The naïvety of CB lymphocytes and the possible situation of a KIR-ligand mismatch in single- and even more frequently in double-unit CBT raises the issue of a potential antileukemic effect of KIR-ligand mismatches, very much in line with the GvL effect of such mismatch in other T-cell-depleted HSCT settings, especially haplo identical HSCT.[77] Indeed, data still remain controversial for CBT, a beneficial effect of KIR-ligand incompatibility with decreased relapse and improved overall survival having been reported by some[78] but not other authors.[79] Finally, double-unit CBT seems to be associated with an increased risk of severe infections and delayed IR while preserving a GvL activity.[80,81]

## 4. GENERATION OF IMMUNE COMPETENT CELLS FROM CORD BLOOD

In this section we will focus only on cellular immunotherapy approaches that use CB-derived lymphocytes. CB cell therapies have mainly consisted of the adoptive transfer of a well-defined cellular product to treat the slow and often incomplete IR observed after CBT with two main objectives: improving the immune control of pathogens without increasing the risk of alloreactivity; and enhancing cytotoxicity against hematological cancer.

### 4.1 Evidence of CB Immune Competent Cells Post-infusion

Clinical protocols tuning the immunosuppressive regimen required for CB engraftment and GvHD prophylaxis demonstrated the ability of CB lymphocytes to undergo maturation in vivo. It has recently been shown that omission of T-cell depletion in vivo promotes rapid expansion of naïve CD4+ CB lymphocytes and restores adaptive immunity within 2 months after CBT in children.[82] Notably, early T-cell expansion was thymic-independent with a rapid shift from a naïve to a central memory phenotype and early recovery of Treg cells. Viral infections were more frequent in those patients, however the infections resolved rapidly in most cases and virus-specific T-lymphocytes were detected within 2 months post-CBT. Acute GvHD was frequent but steroid responsive, and the incidence of chronic GvHD was low, further confirming the skewed immune response of CB toward tolerance. This study is the proof of concept that T cells present in the graft can expand in vivo and induce protective cellular immunity against infections. The authors suggested that this could be due to the specific immunological and ontological qualities of fetal-derived lymphocytes as mentioned above.

Donor lymphocyte infusions (DLIs) are often used to treat both viral infections and leukemia relapses after transplantation, but are associated with potentially life-threatening GvHD. In this context, are CB T cells a good DLI alternative in CBT? As previously mentioned, CB T cells are mostly naïve CD4 T cells.[6] Notably, infused T lymphocytes will need to meet antigens in secondary lymphoid organs in order to generate effector cells. Davis et al. also showed that expansion with IL-7 promotes the preservation of a broad polyclonal T-cell receptor repertoire and of a phenotype that favors lymph node homing.[83] IL-7 is present at high levels in the serum of patients who are lymphopenic post-transplantation,[84] and this has also been observed in adult recipients of CB. Falkenburg et al. already showed the competency of CB-derived lymphocytes having normal frequencies of cytotoxic and helper T-lymphocyte precursors against noninherited maternal antigens and noninherited paternal antigens.[85] Furthermore, Canto et al. analyzed the maturation of CB CD4+CD45RA+ cells indicating that Th effector cells generated from naïve CB cells were intrinsically as competent as naïve PB cells and respond to TCR-mediated stimulation.[86] Moreover, CB effector cells produce higher IL-10, but its inhibitory effect on proliferation may be partially compensated by the higher production of IL-2 and enhanced expression of CD25.[87] Altogether, Foxp3 expression, a characteristic of Treg cells, and the low level of IL-17 gene expression suggest that CB-derived CD4+ T cells may be an appropriate source for DLI. A first attempt to improve IR could be the post-engraftment infusion of donor CB lymphocytes, spare from the original infused graft in the absence of immunosuppression.

These data demonstrate the therapeutic ability of CB-derived cells to foster IR and eventually control opportunistic infections and relapse. The availability of a well-developed cellular immunotherapy technology based on selection, activation, expansion, and/or gene modification of primary cells may facilitate the engineering of different cellular products derived from CB for adoptive therapy. Ex vivo generation of immune cells could remedy the cell dose problem intrinsic in the limited amount of blood normally collected from the residual placenta. However, the relative immune (in)competence secondary to the naïve nature of CB immunity, a unique characteristic of newborns, can be overcome by protocols aiming at an ex vivo maturation of these cells. We present below different CB-derived cellular products for immunotherapy that have been proposed in the scientific literature.

## 4.2 Ex vivo Expansion of Naïve T Lymphocytes

CB T cells can be expanded ex vivo using different strategies. Robinson et al. showed that a combination of IL-2, IL-12, anti-CD3, and IL-7 significantly enhanced the proliferation, activation, maturation, and cytotoxic potential of CB T cells of both fresh and thawed CB mononuclear cells.[88] Following this finding, several authors have proposed the use of CB aliquots followed by ex vivo manipulation as DLI after CBT.[89] Notably, ex vivo manipulation may also include protocols aiming at TCR-transduction for cellular engineering. It is known that naïve lymphocytes isolated from PB provide superior in vivo survival and function. Therefore, the use of CB lymphocytes, which comprise mainly naïve cells, for transducing viral or cancer-specific TCR makes a lot of sense. There is growing evidence that comparable TCR expression can be achieved in adult and CB cells, but the latter expressed an earlier differentiation profile. Further antigen-driven stimulation skewed adult lymphocytes toward a late differentiation phenotype associated with immune exhaustion. In contrast, CB T cells retained a less-differentiated phenotype after antigen stimulation, remaining CD57-negative but were still capable of antigen-specific polyfunctional cytokine expression and cytotoxicity.[90] CB T cells also retained longer telomeres and in general possessed higher telomerase activity indicative of greater proliferative potential. CB lymphocytes therefore have properties indicating prolonged survival and effector function favorable to immunotherapy, especially in settings where donor lymphocytes are unavailable such as in solid organ and CBT.[90]

## 4.3 CB as Source of Treg Cells

CB is a superior source of naïve Treg cells compared to adult PB. CB harbors a subset of CD4+CD25 high Treg cells that require antigen stimulation to show expansion and become functional, suppressive Treg cells.[91] Initial studies have shown that CB-derived Tregs can be effectively expanded ex vivo using anti-CD3/CD28 beads with IL-2. In an animal model of allogeneic GvHD, ex vivo expanded CB-Treg cells significantly prevented immunoreactivity in vivo by modulating cytokine secretion and polarizing the Treg/Th17 balance toward Treg cells, which suggests the potential use of expanded CB Tregs as a therapeutic approach for this complication.[92] Yang et al. observed in the serum of the CB-Treg-treated mice that the production of transforming growth factor-β increased continuously, as opposed to IL-17, which decreased quickly. Consistent with the changes of cytokines, the percentage of mouse CD4+ forkhead box protein 3+ Treg cells increased, while that of Th17 cells decreased.[92]

This finding has been also shown in a humanized mouse model of alloreactivity. Interestingly, the authors' approach was to combine primary, nonexpanded Treg cells from different donors, as an attempt to increase the diversity of antigens to suppress, to increase the cell dose of the therapeutic product. In vitro, pooled freshly isolated CB Treg cells were as suppressive as adult PB-derived Treg cells. However, in a mouse model of human skin allodestruction, pooled CB Treg cells were more potent at suppressing alloresponses and prolonging skin survival compared with pooled PB-derived Treg cells. The higher survival capacity of CB Treg cells and their lower expression of HLA-ABC suggest that lower immunogenicity may account for their superior efficacy in vivo.[93] Other intrinsic characteristics of CB Tregs may also play a role in this enhanced suppressor activity.

CB Treg cells can be expanded in vitro. The feasibility and safety of expanded Treg infusion after double CBT in humans has been reported[94] (study registered at http://www.clinicaltrials.gov as NCT00602693).

## 4.4 CB-derived NK Cells

Recent clinical studies suggest that the adoptive transfer of donor NK cells may improve clinical outcome in hematological malignancies and in some solid tumors by direct antitumor effects as well as by reduction of GvHD. NK cells have also been shown to enhance transplant engraftment during allogeneic HSCT for hematological malignancies.[95] Based on these observations, different groups are pursuing the possibility of in vitro generation of functional NK cells, and attempt to define a convenient therapeutic cellular product to support CBT or to be used as co-adjuvant therapy for cancer. Some publications have already shown the ability to generate ex vivo functional NK cells from CB.[96] Ex vivo generation of immune competent CB-derived NK cells has been attempted from two different sources described below.

### *4.4.1 From CB Lymphopid Progenitor Cells*

CB CD34 selected cells can be differentiated in vitro toward the NK cell lineage for immunotherapeutic purposes. Stroma and cytokines support NK cell differentiation resulting in a phenotype consistent with $CD56^{bright}$ NK cells. For clinical-grade production, heparin-based methods could serve as an alternative to stroma.[97] The ex vivo differentiation system faithfully reproduces major steps of the differentiation of NK cells from their progenitors, and constitutes an excellent model to study NK cell differentiation as well as being valuable to generate large-scale NK cells for immunotherapy. There are evidences that these cells are functional in in vivo models. Some authors have demonstrated that these cells can actively migrate to the bone marrow, spleen, and liver early after infusion. Also, these cells can be directed to inflamed tissues upon expression of CXCR3 and CCR6 via the CXCR3/CXCL10-11 and CCR6/CCL20 axis. A low dose of IL-15 mediates efficient survival, expansion, and maturation of CB NK cells in vivo and findings in AML models showed that a single CB-NK cell infusion combined with supportive IL-15 administration efficiently inhibited growth of human leukemia cells implanted in the femur of mice, resulting in significant prolongation of mice survival.[98]

### *4.4.2 From Mature CB NK Cells*

Following isolation of CD56+CD3− CB NK cells, different methods to expand in long-term cultures these mature cells to yield high numbers of relatively pure cellular populations per donor have been described. As proposed by Kang et al., cells generated could be comparable to those previously utilized in clinical applications. Ex vivo-expanded CB NK cells exhibited a marked increase in antitumor cytolytic activity coinciding with the significantly increased expression of NKG2D, NKp46, and NKp44 activating receptors. Strong cytolytic activity was observed against a wide range of tumor cell lines in vitro. These CB-derived NK cells display a distinct microRNA expression profile, immunophenotype, and greater antitumor capacity in vitro compared to PB NK cells used in recent clinical trials.[99]

## 4.5 Adoptive Immunotherapy Using Highly Enriched Antigen-specific T Cells

Cellular immune functions are impaired after HSCT and solid organ transplantation or in cancer and autoimmune diseases treated with intensified immunosuppression. Thus, control of opportunistic pathogens is lost and severe infections break out. Defective cellular immunity can be restored upon endogenous IR or, if delayed, exogenous IR with pathogen-specific T lymphocytes selected or expanded from appropriate donors can be applied.[100] Methods for the identification of antigens, selection and expansion of specific T cells, and safer manipulation of cellular products have been applied with promising advances.[20] For CB, the challenge is that the majority of T lymphocytes have not been exposed to infections and therefore there are no possibilities to select specific populations from the memory compartment to generate cellular products. This implies that in CB, a pre-activation step is necessary. This step is still under scrutiny in many research laboratories. To give evidence of the progress of this approach, publications orientated to the in vitro generation of specific T cells targeting pathogen and leukemia are presented below.

### *4.5.1 In vitro Generation of Pathogen-specific T Cells from Cord Blood Mononuclear Cells*

Seminal advances come from the Bollard group in Houston.[101] In an effort to better mimic the in vivo priming conditions of naïve T cells, they established a method that used CB-derived DC transduced with an adenoviral vector (Ad5f35pp65) containing the immunodominant CMV antigen pp65, hence driving T-cell specificity toward CMV and adenovirus. For the first stimulation, the authors use these matured DC as well as CB-derived T cells in the presence of the cytokines IL-7, IL-12, and IL-15. At the second stimulation, EBV-transformed B cells or EBV-lymphoblastoid cell lines (LCL), which express both latent and lytic EBV antigens, were used instead. Ad5f35pp65-transduced EBV-LCL were used to stimulate T cells in the presence of IL-15 at the second stimulation. Subsequent stimulations used Ad5f35pp65-transduced EBV-LCL and IL-2. From $50 \times 10^6$ CB mononuclear cells, they were able to generate upward of $150 \times 10^6$ virus-specific T cells that lyse antigen-pulsed targets and release cytokines in response to antigenic stimulation. These cells were manufactured in a good manufacturing practice (GMP)-compliant manner using only the 20% fraction of a fractionated CB unit and have been translated for clinical use.

Leukemia-specific cytotoxic T lymphocytes (CTL) can be expanded from CB. Below, we present papers that explore two models to generate CTL clones specific against leukemic antigens, in particular Wilms' Tumor antigen 1 (WT1) and acute lymphoblastic leukemia (ALL) blasts.

WT1-CTL can be generated from healthy adult donors and from CB by stimulation with an overlapping pool of peptides derived from full-length WT1 and selecting antigen-specific cells based on the expression of CD137. The rapid expansion with anti-CD3 and IL-2 resulted in a 100- to 200-fold expansion. These CTL lysed WT1 expressing targets, including leukemia lines, in an HLA-restricted manner.[102] Furthermore, another group proved the ability of CB T lymphocytes to specifically recognize ALL xenogeneic blasts. CD34+-derived DC electroporated with total RNA from an ALL xenograft generate antileukemic CTL with specificity for the ALL xenograft while sparing autologous CB mononuclear cells. CD4+ T cells dominated the CD3+ T-cell compartment of the CTL, although CD8+ T cells accounted for an average of 30% of the CD3+ T cells present. Expansion of both CD4+ and CD8+ memory and terminal effector memory subsets from predominantly naïve cells was evident. NK cells accounted for an average of 13% of the final antitumor lymphoid cells produced. Blocking experiments confirmed that the CD8+ T-cell compartment was responsible for the antileukemic activity of the polyclonal CTL pool,[103] suggesting an important role of CD8 cells present in the CB grafts in immunological responses.

These promising results both suggest the ability to promote specific CTL clones against pathogens and cancer and open the door for developing ex vivo systems to produce more efficiently cells in clinical-grade conditions for cellular immunotherapy using banked CB units. This will help selection of HLA-specific clones using the increased number of CB units banked worldwide and eliminate requirement for adult donors as a cellular source.

## 5. CONCLUSION

The increasing knowledge of CB immunity has proven in the past years that CB should be considered as a reliable source of HSC with beneficial properties in terms of tolerance induction and maintenance of an antitumor effect. However, enhancement of immune recovery, especially pathogen-specific immunity, remains a challenge. Characterization of CB immune populations and CBT immune recovery will further benefit from the recent technological improvements in "-omics." CB immune cells and also CB lymphoid progenitors appear inherently flexible and plastic. CB cells are in a global immunosuppressive environment to prevent an aberrant immune activation during the period of intense bacterial colonization occurring at birth.[104] A better understanding of the environmental and epigenetic mechanisms driving CB plasticity could help in purposely designing cell therapy products derived from CB in CBT but also in broader clinical indications, such as autoimmunity and aging.

## LIST OF ABBREVIATIONS

**ALL** Acute lymphoblastic leukemia
**APC** Antigen presenting cells
**BM** Bone marrow
**CB** Cord blood
**CBT** CB transplantation
**CTL** Cytotoxic T lymphocytes
**CMV** Cytomegalovirus
**DC** Dendritic cells
**EBV** Epstein Barr virus
**GMP** Good Manufacturing practice
**GvHD** Graft-versus-Host Disease
**GvL** Graft-versus-Leukemia
**HSC** Hematopoietic stem cells
**HSCT** Hematopoietic stem cell transplantation
**HLA** Human leukocyte antigen
**LCL** Lymphoblastoid cell lines
**NK** Natural killer cells
**NKT** Natural killer T cells
**NFATc2** Nuclear factor of activated T cells
**PB** Peripheral blood
**Treg** Regulatory T cells
**Th** T helper
**TLR** Toll-like receptors
**WT1** Wilms' Tumor antigen 1

## REFERENCES


1. Ballen KK, Gluckman E, Broxmeyer HE. Umbilical cord blood transplantation: the first 25 years and beyond. *Blood* 2013;**122**(4):491–8.
2. Metheny L, Caimi P, de Lima M. Cord blood transplantation: can we make it better? *Front Oncol* 2013;**3**:238.
3. Iafolla MA, Tay J, Allan DS. Transplantation of umbilical cord blood-derived cells for novel indications in regenerative therapy or immune modulation: a scoping review of clinical studies. *Biol Blood Marrow Transpl* 2013;**20**(1):20–5.
4. Han P, Hodge G, Story C, Xu X. Phenotypic analysis of functional T-lymphocyte subtypes and natural killer cells in human cord blood: relevance to umbilical cord blood transplantation. *Br J Haematol* 1995;**89**(4):733–40.
5. Szabolcs P, Park KD, Reese M, Marti L, Broadwater G, Kurtzberg J. Coexistent naïve phenotype and higher cycling rate of cord blood T cells as compared to adult peripheral blood. *Exp Hematol* 2003;**31**(8): 708–14.
6. D'Arena G, Musto P, Cascavilla N, Di Giorgio G, Fusilli S, Zendoli F, et al. Flow cytometric characterization of human umbilical cord blood lymphocytes: immunophenotypic features. *Haematologica* 1998;**83**(3):197–203.
7. Krampera M, Tavecchia L, Benedetti F, Nadali G, Pizzolo G. Intracellular cytokine profile of cord blood T- and NK-cells and monocytes. *Haematologica* 2000;**85**(7):675–9.
8. Chalmers IM, Janossy G, Contreras M, Navarrete C. Intracellular cytokine profile of cord and adult blood lymphocytes. *Blood* 1998;**92**(1):11–8.
9. Sohlberg E, Saghafian-Hedengren S, Bremme K, Sverremark-Ekström E. Cord blood monocyte subsets are similar to adult and show potent peptidoglycan-stimulated cytokine responses. *Immunology* 2011;**133**(1):41–50.

10. Brichard B, Varis I, Latinne D, Deneys V, de Bruyere M, Leveugle P, et al. Intracellular cytokine profile of cord and adult blood monocytes. *Bone Marrow Transpl* 2001;**27**(10):1081–6.
11. Levy O, Zarember KA, Roy RM, Cywes C, Godowski PJ, Wessels MR. Selective impairment of TLR-mediated innate immunity in human newborns: neonatal blood plasma reduces monocyte TNF-alpha induction by bacterial lipopeptides, lipopolysaccharide, and imiquimod, but preserves the response to R-848. *J Immunol* 2004;**173**(7):4627–34.
12. Levy O. Innate immunity of the human newborn: distinct cytokine responses to LPS and other Toll-like receptor agonists. *J Endotoxin Res* 2005;**11**(2):113–6.
13. Szabolcs P, Park KD, Reese M, Marti L, Broadwater G, Kurtzberg J. Absolute values of dendritic cell subsets in bone marrow, cord blood, and peripheral blood enumerated by a novel method. *Stem Cells* 2003;**21**(3):296–303.
14. Borràs FE, Matthews NC, Lowdell MW, Navarrete CV. Identification of both myeloid CD11c+ and lymphoid CD11c- dendritic cell subsets in cord blood. *Br J Haematol* 2001;**113**(4):925–31.
15. Sorg RV, Kögler G, Wernet P. Identification of cord blood dendritic cells as an immature CD11c-population. *Blood* 1999;**93**(7):2302–7.
16. Drohan L, Harding JJ, Holm B, Cordoba-Tongson E, Dekker CL, Holmes T, et al. Selective developmental defects of cord blood antigen-presenting cell subsets. *Hum Immunol* 2004;**65**(11):1356–69.
17. Goriely S, Vincart B, Stordeur P, Vekemans J, Willems F, Goldman M, et al. Deficient IL-12(p35) gene expression by dendritic cells derived from neonatal monocytes. *J Immunol* 2001;**166**(3):2141–6.
18. Hunt DW, Huppertz HI, Jiang HJ, Petty RE. Studies of human cord blood dendritic cells: evidence for functional immaturity. *Blood* 1994;**84**(12):4333–43.
19. Langrish CL, Buddle JC, Thrasher AJ, Goldblatt D. Neonatal dendritic cells are intrinsically biased against Th-1 immune responses. *Clin Exp Immunol* 2002;**128**(1):118–23.
20. Encabo A, Solves P, Carbonell-Uberos F, Miñana MD. The functional immaturity of dendritic cells can be relevant to increased tolerance associated with cord blood transplantation. *Transfusion* 2007;**47**(2):272–9.
21. Liu E, Tu W, Law HK, Lau YL. Decreased yield, phenotypic expression and function of immature monocyte-derived dendritic cells in cord blood. *Br J Haematol* 2001;**113**(1):240–6.
22. Kotylo PK, Baenzinger JC, Yoder MC, Engle WA, Bolinger CD. Rapid analysis of lymphocyte subsets in cord blood. *Am J Clin pathology* 1990;**93**(2):263–6.
23. Dalle J-H, Menezes J, Wagner E, Blagdon M, Champagne J, Champagne M, et al. Characterization of cord blood natural killer cells: implications for transplantation and neonatal infections. *Pediatr Res* 2005;**57**(5):649–55.
24. Tanaka H, Kai S, Yamaguchi M, Misawa M, Fujimori Y, Yamamoto M, et al. Analysis of natural killer (NK) cell activity and adhesion molecules on NK cells from umbilical cord blood. *Eur J Haematol* 2003;**71**(1):29–38.
25. Luevano M, Daryouzeh M, Alnabhan R, Querol S, Khakoo S, Madrigal A, et al. The unique profile of cord blood natural killer cells balances incomplete maturation and effective killing function upon activation. *Hum Immunol* 2012;**73**(3):248–57.
26. Verneris MR, Miller JS. The phenotypic and functional characteristics of umbilical cord blood and peripheral blood natural killer cells. *Br J Haematol* 2009;**147**(2):185–91.
27. Gaddy J, Risdon G, Broxmeyer HE. Cord blood natural killer cells are functionally and phenotypically immature but readily respond to interleukin-2 and interleukin-12. *J Interferon & Cytokine Res* 1995;**15**(6):527–36.
28. Luevano M, Daryouzeh M, Alnabhan R, Querol S, Khakoo S, Madrigal A, et al. The unique profile of cord blood natural killer cells balances incomplete maturation and effective killing function upon activation. *Hum Immunol* 2012;**73**(3):248–57.
29. Wang Y, Xu H, Zheng X, Wei H, Sun R, Tian Z. High expression of NKG2A/CD94 and low expression of granzyme B are associated with reduced cord blood NK cell activity. *Cell Mol Immunol* 2007;**4**(5):377–82.
30. Schönberg K, Fischer JC, Kögler G, Uhrberg M. Neonatal NK-cell repertoires are functionally, but not structurally, biased toward recognition of self HLA class I. *Blood* 2011;**117**(19):5152–6.
31. Musha N, Yoshida Y, Sugahara S, Yamagiwa S, Koya T, Watanabe H, et al. Expansion of CD56+ NK T and gamma delta T cells from cord blood of human neonates. *Clin Exp Immunol* 1998;**113**(2):220–8.
32. D'Andrea A, Goux D, De Lalla C, Koezuka Y, Montagna D, Moretta A, et al. Neonatal invariant Valpha24+ NKT lymphocytes are activated memory cells. *Eur J Immunol* 2000;**30**(6):1544–50.
33. van Der Vliet HJ, Nishi N, de Gruijl TD, von Blomberg BM, van den Eertwegh AJ, Pinedo HM, et al. Human natural killer T cells acquire a memory-activated phenotype before birth. *Blood* 2000;**95**(7):2440–2.
34. Harner S, Noessner E, Nadas K, Leumann-Runge A, Schiemann M, Faber FL, et al. Cord blood Vα24-Vβ11 natural killer T cells display a Th2-chemokine receptor profile and cytokine responses. *PLoS One* 2011;**6**(1):e15714.
35. Kadowaki N, Antonenko S, Ho S, Rissoan MC, Soumelis V, Porcelli SA, et al. Distinct cytokine profiles of neonatal natural killer T cells after expansion with subsets of dendritic cells. *J Exp Med* 2001;**193**(10):1221–6.
36. Okada H, Nagamura-Inoue T, Mori Y, Takahashi TA. Expansion of Valpha24(+)Vbeta11(+) NKT cells from cord blood mononuclear cells using IL-15, IL-7 and Flt3-L depends on monocytes. *Eur J Immunol* 2006;**36**(1):236–44.
37. Le Bourhis L, Martin E, Péguillet I, Guihot A, Froux N, Coré M, et al. Antimicrobial activity of mucosal-associated invariant T cells. *Nat Immunol* 2010;**11**(8):701–8.
38. Gold MC, Eid T, Smyk-Pearson S, Eberling Y, Swarbrick GM, Langley SM, et al. Human thymic MR1-restricted MAIT cells are innate pathogen-reactive effectors that adapt following thymic egress. *Mucosal Immunol* 2013;**6**(1):35–44.
39. Ruiz-Hernandez RJA, Cabrera C, Noukwe F, deHaro J, Borras F, Blanco J, et al. Distribution of CD31 on CD4 T-Cells from cord blood, peripheral blood and tonsil at different stages of differentiation. *Open Immunol J* 2010;**3**:19–26.
40. Kimmig S, Przybylski GK, Schmidt CA, Laurisch K, Möwes B, Radbruch A, et al. Two subsets of naive T helper cells with distinct T cell receptor excision circle content in human adult peripheral blood. *J Exp Med* 2002;**195**(6):789–94.
41. Parmar S, Robinson SN, Komanduri K, St John L, Decker W, Xing D, et al. Ex vivo expanded umbilical cord blood T cells maintain naive phenotype and TCR diversity. *Cytotherapy* 2006;**8**(2):149–57.
42. Palin AC, Ramachandran V, Acharya S, Lewis DB. Human neonatal naive CD4+ T cells have enhanced activation-dependent signaling regulated by the microRNA miR-181a. *J Immunol* 2013;**190**(6):2682–91.

43. Takahata Y, Nomura A, Takada H, Ohga S, Furuno K, Hikino S, et al. CD25+CD4+ T cells in human cord blood: an immunoregulatory subset with naive phenotype and specific expression of forkhead box p3 (Foxp3) gene. *Exp Hematol* 2004;**32**(7):622–9.
44. Godfrey WR, Spoden DJ, Ge YG, Baker SR, Liu B, Levine BL, et al. Cord blood CD4(+)CD25(+)-derived T regulatory cell lines express FoxP3 protein and manifest potent suppressor function. *Blood* 2005;**105**(2):750–8.
45. Lee CC, Lin SJ, Cheng PJ, Kuo ML. The regulatory function of umbilical cord blood CD4(+) CD25(+) T cells stimulated with anti-CD3/anti-CD28 and exogenous interleukin (IL)-2 or IL-15. *Pediatr Allergy Immunol* 2009;**20**(7):624–32.
46. Risdon G, Gaddy J, Stehman FB, Broxmeyer HE. Proliferative and cytotoxic responses of human cord blood T lymphocytes following allogeneic stimulation. *Cell Immunol* 1994;**154**(1):14–24.
47. Berthou C, Legros-Maïda S, Soulié A, Wargnier A, Guillet J, Rabian C, et al. Cord blood T lymphocytes lack constitutive perforin expression in contrast to adult peripheral blood T lymphocytes. *Blood* 1995;**85**(6):1540–6.
48. Derniame S, Lee F, Domogala A, Madrigal A, Saudemont S. Unique effects of mycophenolate mofetil on cord blood T cells: implications for GVHD prophylaxis. *Transplantation* 2014;**97**(8):870–8.
49. Nitsche A, Zhang M, Clauss T, Siegert W, Brune K, Pahl A. Cytokine profiles of cord and adult blood leukocytes: differences in expression are due to differences in expression and activation of transcription factors. *BMC Immunol* 2007;**8**:18.
50. Durandy A, Thuillier L, Forveille M, Fischer A. Phenotypic and functional characteristics of human newborns' B lymphocytes. *J Immunol* 1990;**144**(1):60–5.
51. Paloczi K, Batai A, Gopcsa L, Ezsi R, Petranyi GG. Immunophenotypic characterisation of cord blood B-lymphocytes. *Bone Marrow Transpl* 1998;**22**(Suppl. 4):S89–91.
52. Griffin DO, Holodick NE, Rothstein TL. Human B1 cells in umbilical cord and adult peripheral blood express the novel phenotype CD20+ CD27+ CD43+ CD70-. *J Exp Med* 2011;**208**(1):67–80.
53. Sanz EA, Alvarez-Mon M, Martínez AC, de la Hera A. Human cord blood CD34+Pax-5+ B-cell progenitors: single-cell analyses of their gene expression profiles. *Blood* 2003;**101**(9):3424–30.
54. Tucci A, Mouzaki A, James H, Bonnefoy JY, Zubler RH. Are cord blood B cells functionally mature? *Clin Exp Immunol* 1991;**84**(3): 389–94.
55. Splawski JB, Jelinek DF, Lipsky PE. Delineation of the functional capacity of human neonatal lymphocytes. *J Clin Invest* 1991;**87**(2):545–53.
56. Ha YJ, Mun YC, Seong CM, Lee JR. Characterization of phenotypically distinct B-cell subsets and receptor-stimulated mitogen-activated protein kinase activation in human cord blood B cells. *J Leukoc Biol* 2008;**84**(6):1557–64.
57. Komanduri KV, St John LS, de Lima M, McMannis J, Rosinski S, McNiece I, et al. Delayed immune reconstitution after cord blood transplantation is characterized by impaired thymopoiesis and late memory T-cell skewing. *Blood* 2007;**110**(13):4543–51.
58. Mackall C, Fry T, Gress R, Peggs K, Storek J, Toubert A. Background to hematopoietic cell transplantation, including post transplant immune recovery. *Bone Marrow Transpl* 2009;**44**(8): 457–62.
59. Seggewiss R, Einsele H. Immune reconstitution after allogeneic transplantation and expanding options for immunomodulation: an update. *Blood* 2010;**115**(19):3861–8.
60. Toubert A, Glauzy S, Douay C, Clave E. Thymus and immune reconstitution after allogeneic hematopoietic stem cell transplantation in humans: never say never again. *Tissue Antigens* 2012;**79**(2):83–9.
61. Fuji S, Kapp M, Einsele H. Monitoring of pathogen-specific T-cell immune reconstitution after allogeneic hematopoietic stem cell transplantation. *Front Immunol* 2013;**4**:276.
62. Oshrine BR, Li Y, Teachey DT, Heimall J, Barrett DM, Bunin N. Immunologic recovery in children after alternative donor allogeneic transplantation for hematologic malignancies: comparison of recipients of partially T cell-depleted peripheral blood stem cells and umbilical cord blood. *Biol Blood Marrow Transpl* 2013;**19**(11):1581–9.
63. Clave E, Lisini D, Douay C, Giorgiani G, Busson M, Zecca M, et al. Thymic function recovery after unrelated donor cord blood or T-cell depleted HLA-haploidentical stem cell transplantation correlates with leukemia relapse. *Front Immunol* 2013;**4**:54.
64. Lindemans CA, Chiesa R, Amrolia PJ, Rao K, Nikolajeva O, de Wildt A, et al. The impact of thymoglobulin prior to pediatric unrelated umbilical cord blood transplantation on immune-reconstitution and clinical outcome. *Blood* 2014;**123**(1):126–32.
65. Charrier E, Cordeiro P, Brito RM, Mezziani S, Herblot S, Le Deist F, et al. Reconstitution of maturating and regulatory lymphocyte subsets after cord blood and BMT in children. *Bone Marrow Transpl* 2013;**48**(3):376–82.
66. Talvensaari K, Clave E, Douay C, Rabian C, Garderet L, Busson M, et al. A broad T-cell repertoire diversity and an efficient thymic function indicate a favorable long-term immune reconstitution after cord blood stem cell transplantation. *Blood* 2002;**99**(4):1458–64.
67. van Heijst JW, Ceberio I, Lipuma LB, Samilo DW, Wasilewski GD, Gonzales AM, et al. Quantitative assessment of T cell repertoire recovery after hematopoietic stem cell transplantation. *Nat Med* 2013;**19**(3):372–7.
68. Six EM, Bonhomme D, Monteiro M, Beldjord K, Jurkowska M, Cordier-Garcia C, et al. A human postnatal lymphoid progenitor capable of circulating and seeding the thymus. *J Exp Med* 2007;**201**(13):3085–93.
69. Brown JA, Stevenson K, Kim HT, Cutler C, Ballen K, McDonough S, et al. Clearance of CMV viremia and survival after double umbilical cord blood transplantation in adults depends on reconstitution of thymopoiesis. *Blood* 2010;**115**(20):4111–9.
70. McGoldrick SM, Bleakley ME, Guerrero A, Turtle CJ, Yamamoto TN, Pereira SE, et al. Cytomegalovirus-specific T cells are primed early after cord blood transplant but fail to control virus in vivo. *Blood* 2013;**121**(14):2796–803.
71. Clave E, Agbalika F, Bajzik V, Peffault de Latour R, Trillard M, Rabian C, et al. Epstein-Barr virus (EBV) reactivation in allogeneic stem-cell transplantation: relationship between viral load, EBV-specific T-cell reconstitution and rituximab therapy. *Transplantation* 2004;**77**(1):76–84.
72. Barker JN, Weisdorf DJ, DeFor TE, Blazar BR, McGlave PB, Miller JS, et al. Transplantation of 2 partially HLA-matched umbilical cord blood units to enhance engraftment in adults with hematologic malignancy. *Blood* 2005;**105**(3):1343–7.
73. Somers JA, Brand A, van Hensbergen Y, Mulder A, Oudshoorn M, Sintnicolaas K, et al. Double umbilical cord blood transplantation: a study of early engraftment kinetics in leukocyte subsets using HLA-specific monoclonal antibodies. *Biol Blood Marrow Transpl* 2013;**19**(2):266–73.
74. Milano F, Heimfeld S, Gooley T, Jinneman J, Nicoud I, Delaney C, et al. Correlation of infused CD3+CD8+ cells with single-donor dominance after double-unit cord blood transplantation. *Biol Blood Marrow Transpl* 2013;**19**(1):156–60.

75. Ponce DM, Gonzales A, Lubin M, Castro-Malaspina H, Giralt S, Goldberg JD, et al. Graft-versus-host disease after double-unit cord blood transplantation has unique features and an association with engrafting unit-to-recipient HLA match. *Biol Blood Marrow Transpl* 2013;**19**(6):904–11.
76. Eapen M, Klein JP, Ruggeri A, Spellman S, Lee SJ, Anasetti C, et al. Impact of allele-level HLA matching on outcomes after myeloablative single unit umbilical cord blood transplantation for hematologic malignancy. *Blood* 2014;**123**(1):133–40.
77. Velardi A, Ruggeri L, Mancusi A. Killer-cell immunoglobulin-like receptors reactivity and outcome of stem cell transplant. *Curr Opin Hematol* 2012;**19**(4):319–23.
78. Willemze R, Rodrigues CA, Labopin M, Sanz G, Michel G, Socié G, et al. KIR-ligand incompatibility in the graft-versus-host direction improves outcomes after umbilical cord blood transplantation for acute leukemia. *Leukemia* 2012;**23**(3):492–500.
79. Brunstein CG, Wagner JE, Weisdorf DJ, Cooley S, Noreen H, Barker JN, et al. Negative effect of KIR alloreactivity in recipients of umbilical cord blood transplant depends on transplantation conditioning intensity. *Blood* 2009;**113**(22):5628–34.
80. Jacobson CA, Turki AT, McDonough SM, Stevenson KE, Kim HT, Kao G, et al. Immune reconstitution after double umbilical cord blood stem cell transplantation: comparison with unrelated peripheral blood stem cell transplantation. *Biol Blood Marrow Transpl* 2012;**18**(4):565–74.
81. Ruggeri A, Peffault de Latour R, Carmagnat M, Clave E, Douay C, Larghero J, et al. Outcomes, infections, and immune reconstitution after double cord blood transplantation in patients with high-risk hematological diseases. *Transpl Infect Dis* 2011;**13**(5):456–65.
82. Chiesa R, Gilmour K, Qasim W, Adams S, Worth AJ, Zhan H, et al. Omission of in vivo T-cell depletion promotes rapid expansion of naïve CD4+ cord blood lymphocytes and restores adaptive immunity within 2 months after unrelated cord blood transplant. *Br J Haematol* 2012;**156**(5):656–66.
83. Davis CC, Marti LC, Sempowski GD, Jeyaraj DA, Szabolcs P. Interleukin-7 permits Th1/Tc1 maturation and promotes ex vivo expansion of cord blood T cells: a critical step toward adoptive immunotherapy after cord blood transplantation. *Cancer Res* 2010;**70**(13):5249–58.
84. Thiant S, Yakoub-Agha I, Magro L, Trauet J, Coiteux V, Jouet JP, et al. Plasma levels of IL-7 and IL-15 in the first month after myeloablative BMT are predictive biomarkers of both acute GVHD and relapse. *Bone Marrow Transpl* 2010;**45**(10):1546–52.
85. Falkenburg JH, Luxemburg-Heijs SA, Lim FT, Kanhai HH, Willemze R. Umbilical cord blood contains normal frequencies of cytotoxic T-lymphocyte precursors (ctlp) and helper T-lymphocyte precursors against noninherited maternal antigens and noninherited paternal antigens. *Ann Hematol* 1996;**72**(4):260–4.
86. Cantó E, Rodríguez-Sánchez JL, Vidal S. Naive CD4+ cells from cord blood can generate competent Th effector cells. *Transplantation* 2005;**80**(6):850–8.
87. Miyagawa Y, Kiyokawa N, Ochiai N, Imadome K, Horiuchi Y, Onda K, et al. Ex vivo expanded cord blood CD4 T lymphocytes exhibit a distinct expression profile of cytokine-related genes from those of peripheral blood origin. *Immunology* 2009;**128**(3):405–19.
88. Robinson KL, Ayello J, Hughes R, van de Ven C, Issitt L, Kurtzberg J, et al. Ex vivo expansion, maturation, and activation of umbilical cord blood-derived T lymphocytes with IL-2, IL-12, anti-CD3, and IL-7. Potential for adoptive cellular immunotherapy post-umbilical cord blood transplantation. *Exp Hematol* 2002;**30**(3):245–51.
89. Ayello J, van de Ven C, Fortino W, Wade-Harris C, Satwani P, Baxi L, et al. Characterization of cord blood natural killer and lymphokine activated killer lymphocytes following ex vivo cellular engineering. *Biol Blood Marrow Transplant* 2006;**12**(6):608–22.
90. Frumento G, Zheng Y, Aubert G, Raeiszadeh M, Lansdorp PM, Moss P, et al. Cord blood T cells retain early differentiation phenotype suitable for immunotherapy after TCR gene transfer to confer EBV specificity. *Am J Transpl* 2013;**13**(1):45–55.
91. Mayer E, Bannert C, Gruber S, Klunker S, Spittler A, Akdis CA, et al. Cord blood derived CD4+ CD25(high) T cells become functional regulatory T cells upon antigen encounter. *PLoS One* 2012;**7**(1):e29355.
92. Yang J, Fan H, Hao J, Ren Y, Chen L, Li G, et al. Amelioration of acute graft-versus-host disease by adoptive transfer of ex vivo expanded human cord blood CD4+CD25+ forkhead box protein 3+ regulatory T cells is associated with the polarization of Treg/Th17 balance in a mouse model. *Transfusion* 2012;**52**(6):1333–47.
93. Milward K, Issa F, Hester J, Figueroa-Tentori D, Madrigal A, Wood KJ. Multiple unit pooled umbilical cord blood is a viable source of therapeutic regulatory T cells. *Transplantation* 2013;**95**(1):85–93.
94. Brunstein CG, Miller JS, Cao Q, McKenna DH, Hippen KL, Curtsinger J, et al. Infusion of ex vivo expanded T regulatory cells in adults transplanted with umbilical cord blood: safety profile and detection kinetics. *Blood* 2011;**117**(3):1061–70.
95. Velardi A. Natural killer cell alloreactivity 10 years later. *Curr Opin Hematol* 2012;**19**(6):421–6.
96. Luevano M, Madrigal A, Saudemont A. Generation of natural killer cells from hematopoietic stem cells in vitro for immunotherapy. *Cell Mol Immunol* 2012;**9**(4):310–20.
97. Dezell SA, Ahn YO, Spanholtz J, Wang H, Weeres M, Jackson S, et al. Natural killer cell differentiation from hematopoietic stem cells: a comparative analysis of heparin- and stromal cell-supported methods. *Biol Blood Marrow Transpl* 2012;**18**(4):536–45.
98. Cany J, van der Waart AB, Tordoir M, Franssen GM, Hangalapura BN, de Vries J, et al. Natural killer cells generated from cord blood hematopoietic progenitor cells efficiently target bone marrow-residing human leukemia cells in NOD/SCID/IL2Rg(null) mice. *PLoS One* 2013;**8**(6):e64384.
99. Kang L, Voskinarian-Berse V, Law E, Reddin T, Bhatia M, Hariri A, et al. Characterization and ex vivo expansion of human placenta-derived natural killer cells for Cancer immunotherapy. *Front Immunol* 2013;**1**(4):101.
100. Fuji S, Kapp M, Grigoleit GU, Einsele H. Adoptive immunotherapy with virus-specific T cells. *Best Pract Res Clin Haematol* 2011;**24**(3):413–9.
101. Hanley PJ, Lam S, Shpall EJ, Bollard CM. Expanding cytotoxic T lymphocytes from umbilical cord blood that target cytomegalovirus, Epstein-Barr virus, and adenovirus. *J Vis Exp* 2012;**63**:e3627.
102. Krishnadas DK, Stamer MM, Dunham K, Bao L, Lucas KG. Wilms' tumor 1-specific cytotoxic T lymphocytes can be expanded from adult donors and cord blood. *Leuk Res* 2011;**35**(11):1520–6.
103. Cullup H, Hsu AK, Kassianos AJ, McDonald K, Radford KJ, Rice AM. CD34+ cord blood DC-induced antitumor lymphoid cells have efficacy in a murine xenograft model of human ALL. *J Immunother* 2011;**34**(4):362–71.
104. Elahi S, Ertelt JM, Kinder JM, Jiang TT, Zhang X, Xin L, et al. Immunosuppressive CD71+ erythroid cells compromise neonatal host defence against infection. *Nature* 2014;**504**(7478):158–62.

Chapter 5

# Cord and Cord Blood-derived Endothelial Cells

Suzanne M. Watt[1], Paul Leeson[2], Shijie Cai[1,3], Daniel Markeson[1,4,5], Cheen P. Khoo[1], Laura Newton[1,2], Youyi Zhang[1], Stamatia Sourri[1] and Keith M. Channon[2]

[1]*Stem Cell Research Laboratory, Nuffield Division of Clinical Laboratory Sciences, Radcliffe Department of Medicine, University of Oxford, Oxford, UK; NHS Blood and Transplant, John Radcliffe Hospital, Oxford, UK;* [2]*Department of Cardiovascular Medicine, Radcliffe Department of Medicine, University of Oxford, UK;* [3]*Weatherall Institute of Molecular Medicine, University of Oxford, UK;* [4]*Department of Plastic and Reconstructive Surgery, Stoke Mandeville Hospital, Aylesbury, UK;* [5]*University College London Centre for Nanotechnology and Regenerative Medicine, Division of Surgery and Interventional Science, Royal Free Hospital, London, UK*

## Chapter Outline



## 1. INTRODUCTION

The blood vascular network delivers nutrients, oxygen, factors, and cells to tissues and organs and removes toxic waste products, while the lymphatic vascular system collects immune cells, tissue-extravasated fluids, and macromolecules and returns them to the venous circulation. While the vasculature is fundamental to human survival, when dysregulated or damaged it plays a significant role in pathologies (e.g., cardiovascular disease, acute lung injury, renal disease, cancer, diabetic retinopathies, diabetes, preeclampsia), but, in the regenerative medicine setting, it is also crucial for most tissue repair and regeneration. The vasculature has therefore become a major therapeutic target, and predictions are that many millions of individuals will reap the benefit of research conducted in this area.

Blood vessels are formed by three main mechanisms: **vasculogenesis** (de novo generation from endothelial precursors or angioblasts); **angiogenesis** (remodeling and expansion of pre-existing vessels by intussusceptive angiogenesis or extant endothelial cell sprouting and proliferation); and **arteriogenesis** (the growth of collateral vessels from arterioles by endothelial cell proliferation and enhanced lumen formation downstream of a vascular obstruction).[1–3] The resulting arteries, veins, and smaller vessels differ in their complexity. For example, the vascular wall of human arteries and veins is composed of three layers, the lumen facing *tunica intima*, the *tunica media*, and the outer *tunica adventitia*, while the capillaries comprise endothelia ensheathed in or associated with, to varying degrees, pericytes/stromal cells as well as other cell types.[3,4] Furthermore, significant endothelial heterogeneity not only exists among the arterial, venous, and lymphatic macrovasculature,[5] but also between and within the different tissue and organ microvascular capillary beds.[6] Interestingly, embryonic stem cell-derived endothelial precursor cells have been shown to be educated to adopt the phenotype of the tissue in which they reside following transplantation, at least when examined in murine models.[6]

Cord Blood Stem Cells Medicine. http://dx.doi.org/10.1016/B978-0-12-407785-0.00005-0

During placentation and fetal life, vascular development is closely associated with hematopoietic development and there is evidence that human endothelial and hematopoietic cells differentiate in the placenta, yolk sac, and embryo either from a common bipotent hemangioblast progenitor cell, or that hematopoietic cells transiently bud off from hemogenic endothelium by the process of endothelial hematopoietic transition.[7–10] There are reports that human umbilical cord blood contains hemogenic endothelium.[11,12] Unfortunately, there is no single cell surface marker to identify endothelial stem cells or their progeny and, given their close association with hematopoietic cells during development, it is not surprising that cells within the endothelial lineage share key cell surface markers in common with hematopoietic cells. This lack of an easy phenotypic distinction between these cell types has also led to substantial confusion over the past one to two decades about the identity of progenitor cells belonging to the endothelial lineage compared to those proangiogenic cells, which belong to the hematopoietic lineage, circulate in the blood and promote blood vessel formation.

In 1963, Stump et al. demonstrated that "circulating endothelial cells" could repopulate dacron grafts after transplantation into the aorta of pigs.[13] Subsequent research by Asahara and colleagues led to the hypothesis that "endothelial progenitor cells (EPCs)" could circulate in the human blood and contribute to revascularization by becoming incorporated into active sites of blood vessel regeneration in damaged tissues (e.g., as assessed in rodent hind limb ischemic or bone marrow transplant models).[14–16] These studies ignited interest in the concept of postnatal vasculogenesis and subsequently led to the use of such cells in clinical trials or as diagnostic markers. However, more recent evidence, which has been reviewed in detail,[17–19] supports the view that these so-called "EPCs" comprise proangiogenic cells (PACs) or circulating angiogenic cells (CACs) of hematopoietic origin and not endothelial lineage cells.

A major step forward in identifying precursors for the endothelial lineage postnatally came from the identification of late outgrowth endothelial cells (late OECs). These could, for example, be detected when human peripheral blood cells (following bone marrow or mobilized peripheral blood allografting) were cultured on collagen 1-coated plates in endothelial growth medium for several weeks and they were also shown to be of both donor and recipient origin.[20] In 2004/2005, Ingram et al., using a similar culture technique, were instrumental in identifying a hierarchy of endothelial precursor cells in adult human peripheral and umbilical cord blood, as well as in the umbilical cord.[21,22] This endothelial lineage hierarchy was reminiscent of the hierarchical organization described, some years earlier, for clonogenic myeloid lineage cells where more primitive high proliferative potential colony-forming cells (HPP-CFCs) gave rise to more mature low proliferative potential (LPP-) CFCs and then smaller myeloid cell clusters.[23–25] Using clonogenic assays, Ingram et al. plated single early passage late OECs from human peripheral blood, umbilical cord blood, and umbilical cord into individual wells of 96-well collagen 1-coated plates and assessed the size and number of primary endothelial colonies or clusters developing over 14 days.[21,22] They classified the originating cells as high proliferative potential endothelial colony-forming cells (HPP-ECFC), low proliferative potential endothelial colony-forming cells (LPP-ECFC), endothelial clusters, and mature nondividing endothelial cells based on the cell numbers generated by day 14 (Figure 1).

It is now generally accepted that these ECFCs are synonymous with late OECs and that they are distinct from proangiogenic cells (PACs) or early outgrowth cells of hematopoietic origin (referred to as early EPCs). However, their relationship with endothelial stem cells remains to be defined. A discussion of the properties of ECFCs derived from umbilical cord, umbilical cord blood, and placenta will form the basis of this review, and, as appropriate, comparisons will be made with proangiogenic cells and ECFCs derived from other sources.

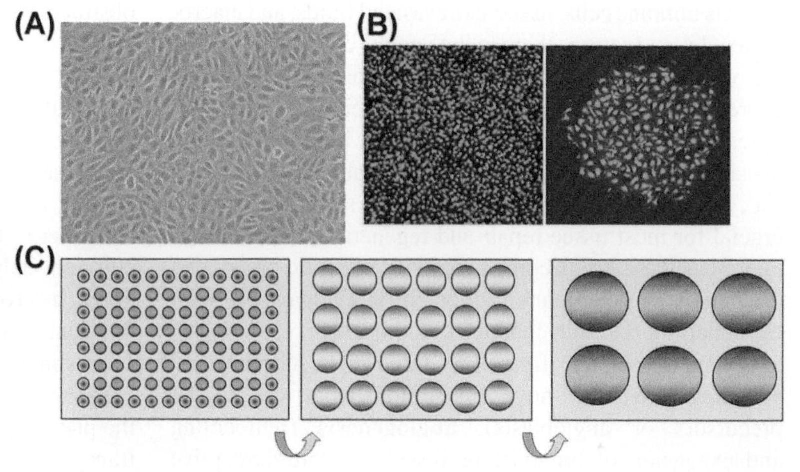


**FIGURE 1** Clonogenic Endothelial Colony-Forming Cell (ECFC) Assay. (A) A primary culture of human umbilical cord endothelial cells. (B) Single human umbilical cord endothelial cells transduced with eGFP were plated into 96-well plates (C left) and cultured for 14 days. Single colonies that formed and contained >2000 cells were derived from HPP-ECFCs (B left), while those with >50 and up to 2000 cells were derived from LPP-ECFC (B right). (C) Colonies were replated into 24-well plates to assess secondary colony formation. Secondary colonies that were 50% confluent or more after 7 days were replated into six-well plates to access the growth of tertiary colonies.

## 2. THE HUMAN ENDOTHELIAL COLONY-FORMING CELL HIERARCHY

Ingram, Yoder, and colleagues[21,22] were the first to describe, in human umbilical cord blood, adult peripheral blood and tissues, a hierarchy of lineage-restricted endothelial precursors based on their proliferative potential in vitro. In this hierarchy (Figure 1), these endothelial precursors gradually lose their proliferative capacity, with HPP-ECFC giving rise to LPP-ECFC which then generate endothelial clusters and eventually mature into nondividing endothelial cells.[21,22] HPP-ECFC were identified if they generated endothelial colonies of more than 2000 cells but retained their ability to form secondary and, in some cases, tertiary colonies upon replating in vitro. LPP-ECFC were defined as those endothelial precursors capable of forming colonies of greater than 50 cells and up to 2000 cells provided they did not form secondary colonies upon replating, while endothelial clusters contained 2–50 cells. Other studies have demonstrated that human umbilical cord HPP-ECFCs that generated larger primary colonies (>4000 cells) also generated more tertiary colonies and therefore generally possessed higher proliferative potentials than those forming smaller colonies (<4000 cells), while the progeny of ECFCs from the larger primary colonies preferentially contributed to vascular tubule formation in vitro in the presence of stromal cells.[26] In vivo, the early passage ECFCs from term human umbilical cord, umbilical cord blood, placenta, or adult peripheral blood could also form vascular structures and inosculate with murine vessels when implanted with stromal cells in a collagen–fibronectin matrix or in matrigel in vivo in murine immunodeficient models of vasculogenesis.[27–39] Notably, extended culture of ECFCs or late OECs decreased their vasculogenic potential,[32] while proangiogenic myeloid cells could promote vessel formation both in vitro and in vivo.[40,41] Furthermore, clonogenic murine nestin$^+$ stromal cell lines support not only hematopoiesis,[42] but also vasculogenesis/angiogenesis in vitro.[43]

Human ECFC-derived progeny from umbilical cord blood, umbilical cord, placenta, and adult peripheral blood has typical cobblestone morphology in vitro (Figure 1(A)) and express combinations of cell surface markers, which are not specific for endothelial lineage cells but together are consistent with an endothelial phenotype, e.g., CD31, CD144, CD146, and CD105, and lack CD45 and CD14 expression.[17–19,44] Expression of the CD34 surface marker was more evident on ECFC-derived cells at earlier passage in vitro and declined after serial passaging,[26,45] although early passage placental ECFC were reported to be CD34 negative.[27] These cells also take up acetylated low density lipoproteins (LDL) and express UEA lectin, characteristics which are not, however, exclusive to endothelial cells.[17–19] However, cell surface CD34 was generally significantly higher and persisted in culture for longer on umbilical cord blood than on adult peripheral blood ECFC-derived cells.[44] Human umbilical cord ECFCs when cultured in vitro could resist short exposures to acute (24 h) hypoxia, but not longer chronic (14 day) exposures to severe (1.5%) hypoxia.[26] They have also been shown to be more proliferative at higher (8% and 21% oxygen) rather than lower (<8% oxygen) oxygen tensions.[45]

## 3. DIFFERENCES BETWEEN HUMAN UMBILICAL CORD BLOOD ECFCs AND THOSE SOURCED FROM ADULT PERIPHERAL BLOOD, UMBILICAL CORD, AND PLACENTA

Differences were first noted between ECFCs derived from adult human peripheral blood and those found in umbilical cord blood. In relation to ECFC content, umbilical cord blood from healthy donors has been demonstrated to contain significantly more ECFCs than adult peripheral blood, and these have a higher proliferative potential, with population doubling times reported as being 2.5-fold higher in umbilical cord blood than in adult peripheral.[21,45] The estimated number of ECFCs in adult human peripheral blood donors is around $1.1 \pm 0.5$ ECFCs per $10^8$ mononuclear cells or 0.05–0.2 ECFCs per ml of blood, although this can vary with age and disease status.[21,45] This compares with around 400 proangiogenic cells per ml of normal human adult peripheral blood.[46,47] On average, reports suggest that there are approximately 10- to 50-fold more ECFCs in the human umbilical cord blood at term (e.g., various reports state $12 \pm 4$, $8.2 \pm 1.8$, $30.8 \pm 23.8$, $40.1 \pm 6.9$ per $10^8$ mononuclear cells) than in adult peripheral blood from normal blood donors.[21,44,48–50] Furthermore, the umbilical cord blood endothelial precursors from early passage endothelial outgrowth cultures were more proliferative and found to undergo at least 100 population doublings compared to 20–30 for adult peripheral blood and to contain enhanced telomerase activity.[21] This higher proliferative capacity has been confirmed by others. For example, Athanassopoulos et al.[44] showed that umbilical cord blood ECFCs gave rise to $1–2 \times 10^{13}$ endothelial cells over a 7-week period, in contrast to $10^9–2 \times 10^{11}$ endothelial cells from adult peripheral blood. Ingram et al.[21] also noted that some single human umbilical cord blood HPP-ECFC could generate as many as $10^{12}$ endothelial progeny over 12–13 weeks of culture and that, upon replating, approximately five-fold more umbilical cord blood ECFCs generated secondary colonies than peripheral blood ECFCs. Interestingly, there also appeared to be greater heterogeneity in ECFC content in adult human peripheral blood from different healthy Caucasoid blood donors than in umbilical cord blood from healthy Caucasoid donors at term.[44] Human umbilical cord and cord blood ECFC-derived cells could generate significantly more microvascular networks in vitro than similar cells from adult peripheral blood in a 14-day ECFC–stromal cell coculture assay, although the content of the latter was more variable than those sourced from umbilical cord or cord blood at birth.[44] Human umbilical cord and cord blood ECFC-derived cells could also form increased vessels in surrogate vasculogenesis models in vivo than such cells when sourced from adult peripheral blood.[51] Studies in Rhesus monkeys have demonstrated a similar hierarchy of ECFC, which expresses similar surface markers to those found in the human.[52] With increasing postnatal age, the HPP-ECFCs in these nonhuman primates declined in the circulation and lost their ability to form vessels in immunodeficient mice,[52] confirming similar observations in humans.

Further examples of differences between human umbilical cord and adult peripheral blood ECFCs include reports of the greater sensitivity of the latter to hydrogen peroxide induced oxidative stress than umbilical cord ECFCs,[53] although He et al.[54] have shown opposing findings. However, there are differences in their transcriptional profiles, with the former expressing less prothrombotic and proinflammatory genes[55] and differences in their miRNA profiles.[56] Human umbilical cord blood ECFCs have additionally been found to possess more robust angiogenic potentials (short-term proliferative and migratory) under physiological and ambient normoxic and ischemic conditions although the comparison was with human dermal microvascular endothelial cells rather than adult peripheral blood.[45]

Differences have been reported to exist between human umbilical cord blood, umbilical cord, and placental ECFCs. Studies on early passage (p3–4) human umbilical vein and aortic endothelial cell cultures revealed that these contained fewer HPP-ECFC than the umbilical cord blood itself,[22] although it is reported that higher numbers of ECFCs are found in umbilical arteries than the umbilical vein.[57] In contrast, the human placenta contained significantly more ECFCs than umbilical cord blood and the ECFCs isolated from the human placental chorionic villi at term generated significantly more vessels in vivo in NOD-Scid mice when transplanted with human dermal fibroblasts than those from umbilical cord blood.[27]

## 4. OBSTETRIC FACTORS AND ECFC CONTENT IN UMBILICAL CORD BLOOD AT TERM

Human umbilical cord blood has the advantage that it can be tissue typed using next generation sequencing to high resolution to improve graft matching, prescreened for viral biomarkers, assessed for hematopoietic stem/progenitor cell (HSPC) content (generally using CD34 positivity and clonogenic myeloid cell content), and processed and banked ready for transplantation.[58] Since the first umbilical cord blood transplant in 1988,[59] knowledge of the optimal selection criteria for umbilical cord blood units for HSPC transplantation has grown significantly. Key criteria depend on obstetric factors at the time of collection, the total nucleated and $CD34^+$ cell content, optimal HLA and blood group matching of the recipient and donor, and the quality of the processing, storage, and testing procedures.[58,60,61] Other quality measurements such as HLA typing to assess the donor's noninherited maternal HLA antigens also influence transplant success.[62] Typical obstetric factors that have been assessed in relation to nucleated cell and HSPC content in human umbilical cord blood include birth and placental weight, infant gender, parity of the mother, nucleated red cell content, low venous pH, prolonged first stage of labor, and Apgar scores[a]. Positive correlations have been observed between placental weight and infant weight at birth with umbilical cord blood volume, total nucleated and $CD34^+$ cell content, and total hematopoietic clonogenic myeloid cells, and it has been suggested that each 500 g increase in birth weight contributes to a 28% increase in $CD34^+$ cell counts.[61] Based on such parameters, a "cord blood Apgar" score has been developed and this is designed to predict umbilical cord blood HSPC engraftment posttransplant and hence provide the optimal umbilical cord blood unit for transplantation into an individual recipient.[63]

In an attempt to define obstetric factors that might affect the content of umbilical cord blood ECFCs at term in healthy deliveries and hence predict their potential for banking, Coldwell et al.[50] examined potential correlations between prenatal, perinatal, and postnatal factors recorded for the mother and infant and the ECFC content of human umbilical cord blood as assessed in day 14 mononuclear cell clonogenic cultures. Prenatal factors examined included maternal parity, gravidity, age, ethnic group, the infant's gender, and the gestational period. Perinatal features examined were the duration and stages of labor, delivery method, route of delivery, weight of placenta, birth weight, ethnic group of the infant, and Apgar scores at 1, 5, and 10 min. Postnatal factors included the umbilical cord blood volume, nucleated cell count, nucleated red blood cell concentration, and ECFC content. Notable demographic factors for 52 umbilical cord blood donors included mode of delivery (27% vaginal, 71% caesarean, and 2% caesarean after first stage labor), average maternal age of 29.3 years, infant gender (61% male, 39% female), ethnic origin (90% white caucasoid), neonatal weight ($3.43 \pm 0.47$ kg), placental weight ($704 \pm 144$ g), gestational age (37–41 weeks; $39 \pm 1$), and Apgar scores (3.3% with $Apgar_{1min}$ scores $<7$). Parity and

[a] The Apgar score was developed as a simple means to quickly assess the health of the newborn immediately after birth (1–10 min) and hence measures the baby's ability to adapt to the extrauterine environment.[58] The Apgar score ranges from 0 to 10 and assesses five criteria: Appearance, Pulse, Grimace, Activity (muscle tone), and Respiration, which are allocated a score of 0–2, and with normal total Apgar scores ranging from 7 to 10.[100]

gravidity were higher than the UK national averages. The only statistically significant predictor of ECFC frequency in umbilical cord blood at term was found to be placental weight and a formula was devised to predict the ECFC frequency from placental weight, where $E_{pred}$ is the predicted ECFC per cord blood unit and P is placental weight (g):

$$E_{pred} = \left(1.2421 + \frac{3.226 \times 10^3}{P^{2/3}}\right)^{10}$$

Notably, although there were no correlations with such obstetric factors as parity, mode of delivery, or infant gender, skewing of the donor population in these studies was toward uncomplicated pregnancies and caesarian deliveries at term from Caucasoid mothers, mostly with normal Apgar scores and gestational ages. The formula above might therefore be used with defined ethnic groups with healthy term deliveries to confirm its robustness.

## 5. ECFC CONTENT AND FUNCTION IN UMBILICAL CORD BLOOD BASED ON GESTATIONAL AGE

The significant numbers of ECFCs that circulate in the human umbilical cord blood and their decline in numbers, immaturity, and function in the circulation in adult life are reminiscent of HSPCs, which are found in higher numbers, are more proliferative, and are more immature in umbilical cord blood than in the adult circulation under normal homeostatic conditions, with the adult human peripheral blood containing principally committed erythroid progenitors.[64] Hypotheses for the circulation of restricted hematopoietic progenitor cells in the adult circulation are: (1) that this is a remnant of the circulating HSPC pool during fetal development; or (2) that these circulating progenitors have the opportunity to rapidly enter the bone marrow or relevant tissues in response to hematological stress, thereby allowing their rapid maturation and maximizing the delivery of oxygen to tissues when most needed. Eaves and colleagues have also provided evidence from murine studies for a developmental switch from a fetal to adult HSC type at a specific time shortly after birth.[65,66] Whether such a switch occurs with ECFCs has not been determined. One hypothesis for the higher content of circulating HPP-ECFCs at birth as opposed to adult peripheral blood is that circulating and highly proliferative ECFCs are readily available to enter the developing tissues and contribute to organogenesis (e.g., kidney, retina, lungs, bone) by enhancing microvascular development during late gestation and in early postnatal life, in response to rapid tissue and organ growth.

The importance of vascular development during the later stages of fetal development is exemplified as follows. Firstly, in premature infants who develop respiratory distress syndrome and subsequent bronchopulmonary dysplasia following treatment with supplemental oxygen and positive-pressure ventilation, both alveolarization and the development of the pulmonary microvasculature are disrupted.[67] Experimental evidence in surrogate models suggests that cellular therapies and biologics associated with neovascularization and derived from the umbilical cord blood/cord, placenta, or bone marrow (ECFCs, mesenchymal stem cells (MSCs), PACs) may provide treatment options in ameliorating these problems.[68,69] However, if autologous cell therapies are required, such cells, particularly if sourced from preterm infants, would need to be functionally effective. In this respect, there is an association of increased cardiovascular and other disease risk in adulthood with very low birth weight and evidence of impairment of ECFC function in the umbilical cord of low birth weight infants.[70–74] A second example where ECFCs potentially play a role in tissue development is in the ability of ECFCs derived from umbilical cord blood to enhance bone formation by circulating fetal osteogenic progenitor cells.[75]

An assessment of the content of ECFCs in umbilical cord blood during the third trimester of pregnancy has come from three main studies, although some data are conflicting. Firstly, to address whether ECFCs were present in the umbilical cord blood throughout the third trimester of pregnancy and whether ECFC proliferation is negatively affected by hyperoxia (effects which may contribute to poor vascular growth and bronchopulmonary dysplasia), Ingram and colleagues examined the ECFC content of preterm umbilical cord blood and the effects of hyperoxia on human umbilical cord blood ECFC function.[48,49] In the first studies, Javed et al.[48] compared the ECFC content in preterm human umbilical cord blood (24–37 week gestational age) to that of term umbilical cord blood and found approximately twice the number of ECFC at 37–40 weeks as at 32–36 weeks of gestation. Furthermore, around three-fold more ECFCs were found in the umbilical cord blood at term than at 24–31 weeks of gestation.[49] There were no differences reported between the ECFC content of umbilical cord blood based on mononuclear cell number at 24–28 weeks of gestation when compared to 29–32 weeks of gestation. Baker and colleagues subsequently compared the ECFC content and the proliferative potential of ECFCs in human umbilical cord blood from preterm (28–35 weeks gestational age; n = 26) and term (n = 24) infants.[49] In contrast to the studies from Javed et al.,[48] they detected around five-fold more ECFCs in these preterm as opposed to the term umbilical cord blood samples (51 ± 7 vs. 12 ± 4 ECFCs per $10^8$ mononuclear cells). Furthermore, they found that, while both sources of ECFCs formed vascular networks in matrigel, the preterm ECFCs were more proliferative, continued to proliferate once they had reached confluence in vitro, and their rate of proliferation decreased in response to hyperoxia (40% oxygen), when compared to term ECFCs. In these studies, the numbers of HPP-ECFC and LPP-ECFC were not individually assessed. In contrast to the studies of Baker et al.[49] but in agreement with those of Javed et al.,[48] Ligi et al.[70] demonstrated a significant increase in ECFCs over the third trimester of pregnancy. In agreement with studies of Coldwell et al.,[50] they found that ECFC content varied significantly in babies with increased birth weights (i.e., >2.7 kg and by around 10- to 30-fold, up to almost 60 ECFC per $10^8$ mononuclear cells plated).[70] However, by extending the culture period for ECFCs from 14 to 28 days, Ligi et al.[70] found an inverse correlation between the time taken for ECFCs to form endothelial colonies in vitro and infant birth weight. They further demonstrated that the low birth weight early passage preterm ECFC-derived cells had impaired proliferative and migratory function both in vitro and in vivo and had upregulated angiostatic gene profiles compared to those obtained from umbilical cord blood at term.[70] In particular, they demonstrated a reduction in VEGF and an increase in antiangiogenic sVEGFR-1 and PF4, but not thrombospondin-1, in the associated umbilical cord-derived serum.[71] No other obstetric factors that they examined (maternal age, parity, gravidity, mode of delivery, infant gender, and Apgar score) were found to predict the ECFC content in umbilical cord blood[70,71] in agreement with data from Coldwell et al.[50]

As a second example, ECFCs from umbilical cord blood may also contribute to bone formation. Liu et al.[75] demonstrated that healthy human umbilical cord blood ECFCs expressed higher levels of osteogenic and proangiogenic factors than did adult peripheral blood ECFCs, with secreted factors from the ECFCs promoting osteogenesis in vitro by fetally-derived MSCs. Taken together with studies from Roubelakis et al.[34] who demonstrated enhanced vasculogenesis of human umbilical cord ECFC-derived cells in the presence of second trimester MSCs and from canine and rodent models of enhanced bone repair when MSCs are coimplanted with ECFCs from different sources,[76–78] these results suggest that umbilical cord blood ECFCs may enhance skeletal as well as vascular development in the presence of MSCs in the perinatal period. In the event of bone injury in the early postnatal period, umbilical cord or cord blood and placental ECFCs and MSCs may prove effective in enhancing bone repair.

In conclusion, these and related studies suggest that the content of ECFCs in umbilical cord blood at term is individually variable, that their numbers may be inversely related to placental weight, and that umbilical cord blood ECFC functions are decreased when subjected to chronic severe hypoxia or hyperoxia or when obtained from preterm low birth weight infants. The research also demonstrates that it is not as easy to predict ECFC content in umbilical cord blood obtained at term from healthy infants based on obstetric factors as it is to predict the HSPC content, where a "cord blood Apgar score" to predict the likelihood of HSPC engraftment posttransplant has been developed. Finally, circulating and tissue resident ECFCs may contribute to rapid fetal tissue and organ growth in the third trimester of pregnancy and in early postnatal life. On the contrary, the effects of an adverse antenatal environment on vasculogenesis from ECFCs or their precursors may predispose to diseases or disorders affecting the vasculature postnatally and in adult life.

## 6. THE DIABETIC ENVIRONMENT AFFECTS UMBILICAL CORD BLOOD ECFCs

The incidence of diabetes (Type 1, Type 2, and gestational diabetes) is increasing significantly worldwide (www.diabetes.org.uk). Maternal diabetes is associated with morbidities in the fetus and neonate and the increased long-term risk of later cardiovascular and other diseases in these infants.[79] It has been reported that hyperglycemia is a primary causative agent for vascular disease in diabetics, and that this reduces vascular function and the growth of new blood vessels.[80] Additionally, it has been demonstrated that proangiogenic cells (otherwise termed "early EPCs") are reduced in the circulation of patients with diseases which affect the vasculature including diabetes, rheumatoid arthritis, preeclampsia, peripheral vascular disease, and coronary artery disease.[81]

The effects of hyperglycemia on endothelial precursor and proangiogenic cell function have been assessed in vitro and in vivo in surrogate models of human vasculogenesis. Chen et al.[82] showed that, when both PACs and ECFCs (late OEC) from healthy human blood donors were exposed in vitro to high glucose concentrations (10–25 mM), their numbers and proliferative abilities declined, both cell types showed increased senescence and the ECFCs were less migratory and formed fewer tubules in matrigel assays. They also provided evidence that hyperglycemic conditions impaired nitric oxide (NO) mechanisms rather than oxidative stress mechanisms in the cells by decreasing eNOS, FoxO1, and Akt phosphorylation and bioavailable NO. Building on these studies, Ingram et al.[82] examined both the ECFC content of human umbilical cord blood from diabetic and healthy mothers at 38–42 weeks of gestation and the effects of hyperglycemia on normal umbilical cord blood ECFC function in vitro. They found (albeit in a sample of four infants without hyperglycemia and born to diabetic mothers) almost a four-fold reduction in umbilical cord blood ECFCs compared to healthy controls ($13.2 \pm 2.6$ vs. $40.1 \pm 6.9$ ECFC colonies per $10^8$ mononuclear cells, respectively). The former ECFCs also showed a reduced proliferative response to FGF-2 and VEGF following serum starvation, and a 66% reduction in vascular tubule formation in matrigel in vitro. Transplantation of early passage human ECFCs in NOD-Scid mice revealed a two-fold reduction in the resulting perfused vasculature in the umbilical cord blood from diabetic pregnancies. These researchers also demonstrated that hyperglycemic conditions, particularly when mimicked with 10–15 mM dextrose and compared to normoglycemic conditions (5 mM dextrose), reduced the numbers of human term umbilical cord blood ECFC produced in vitro and their ability to form tubules in matrigel assays in vitro by inducing cellular senescence rather than apoptosis. Further reports have indicated a three-fold increase in oxidative DNA damage in ECFCs obtained from diabetic pregnancies.[53]

## 7. PROCESSING AND CRYOPRESERVATION AFFECT ECFC NUMBERS IN UMBILICAL CORD BLOOD

It is notable that the content of ECFCs has been reported to vary approximately 5- to 30-fold in human umbilical cord blood at term. While this may in part be due to obstetric or genetic factors or to environments which affect ECFC numbers, the quality of the collection, processing, and, if used, cryopreservation of umbilical cord blood prior to ECFC culture may also affect the ECFC content of human umbilical cord blood. The importance of such quality control measures is exemplified in the human HSPC transplant setting. Prior to HSPC transplantation and in accordance with FACT-Netcord accreditation standards, donated umbilical cord blood is collected, held at 22 ± 2°C for up to 48h prior to processing, then cryopreserved under controlled rate freezing conditions, and stored long-term below −150°C.[58] Our own practice at the UK NHS cord blood bank is to maintain umbilical cord blood units at 22 ± 2°C for up to 24 h before processing because of better viability of HSPCs postthawing and prior to transplant, and the success of transplant outcomes using such umbilical cord blood units.[83] More specifically, the impact on $CD34^+$ cell viability in human umbilical cord blood units (n = 28) was assessed on three consecutive days, both before and after cryopreservation, by flow cytometry using 7-aminoactinomycin-D (7-AAD) and Annexin-V methods. The mean viability of $CD34^+$ cells precryopreservation remained high (>92% ± 4%) over three days, whereas storage time significantly affected the viability of $CD34^+$ cells postcryopreservation, falling by approximately 5% per extra day in storage from 84% ± 6% on day 1, to 79 ± 7 ($p < 0.0057$) on day 2, and to 74% ± 8% ($p < 0.0001$) on day 3 for 7-AAD staining and by around 7% per extra day in storage from 77 ± 9 on day 1, to 70 ± 13 ($p < 0.0194$) on day 2, and to 63 ± 15 ($p < 0.0002$) on day 3 for Annexin-V. Several studies have shown that the functionality of $CD34^+$ cells can be reduced significantly after cryopreservation and that Annexin-V negative human $CD34^+$ cells correlated with NOD-Scid mouse hematopoietic repopulating ability, indicating as would be expected that apoptotic $CD34^+$ cells fail to engraft.[84] Scaradavou and colleagues[85] have also demonstrated that umbilical cord blood units with <75% $CD34^+$ viable cells postthaw are very unlikely to engraft in double umbilical cord blood transplants.

As cell viability and recovery after banking are critical for the subsequent use of cells as a cell therapeutic or in the research setting, studies have therefore also been conducted on the recovery of ECFCs from cryopreserved human umbilical cord blood units. Although Lin et al.[20] found that there were no statistically significant differences in the late OEC (ECFC) content between cryopreserved and fresh umbilical cord mononuclear cells, both Coldwell et al.[50] and Vanneaux et al.[86] recovered significantly fewer ECFCs from cryopreserved than from fresh umbilical cord blood units. Storage of umbilical cord blood at 22 ± 2°C for up to 24 h was preferable to storage at 4°C prior to freezing,[50] with storage for the former yielding four times more ECFCs than the latter and with the HPP-ECFC being less affected by cryopreservation than the total ECFC population. When fresh cord blood units were divided into two equal parts and analyzed for ECFC content before and after cryopreservation, approximately one-third of the cryopreserved umbilical cord units yielded ECFC compared to 100% of the fresh units and the total ECFC recovery after cryopreservation in those units, which contained ECFCs was around 50% lower than that found in the original fresh umbilical cord blood sample. Broxmeyer and colleagues[87] noted that human umbilical cord blood units which had been stored for over 20 years still contained ECFCs but with expected recoveries of 10–20% based on historical controls and with a loss of proliferative potential. Thus, the lower content of ECFC recovered from cryopreserved human umbilical cord blood units needs to be considered if autologous cord blood units are banked for later uses in regenerative medicine. Improved cryopreservation technologies are also needed for these cells, as ECFCs do not all survive cryopreservation in the 10% dimethylsulfoxide solution used for HSPCs. It is important to also note that, once ECFCs are cultured, cryopreservation does not appear to affect the recovery of the resulting endothelial lineage precursors.

## 8. THE PHENOTYPIC IDENTITY OF ECFCs

A key research question is whether cells within the ECFC lineage hierarchy can be distinguished from each other or from other cell types in blood or tissues. Since endothelial and hematopoietic precursors share cell surface markers in common, and as both cell types enhance vessel formation, the main thrust of the research to date has been to distinguish these precursor cells from one another. When sections of the umbilical cord are stained, it is relatively easy to identify endothelial cells with CD34 or CD31, vascular smooth muscle cells with smoothelin or calponin, and pericytes/mesenchymal stem/stromal cells in Wharton's jelly with CD10 (Figure 2). However, of these markers, CD34, CD31, and CD10 are shared with hematopoietic cells,[23] which in contrast to ECFC also express CD45.[17–19,41]

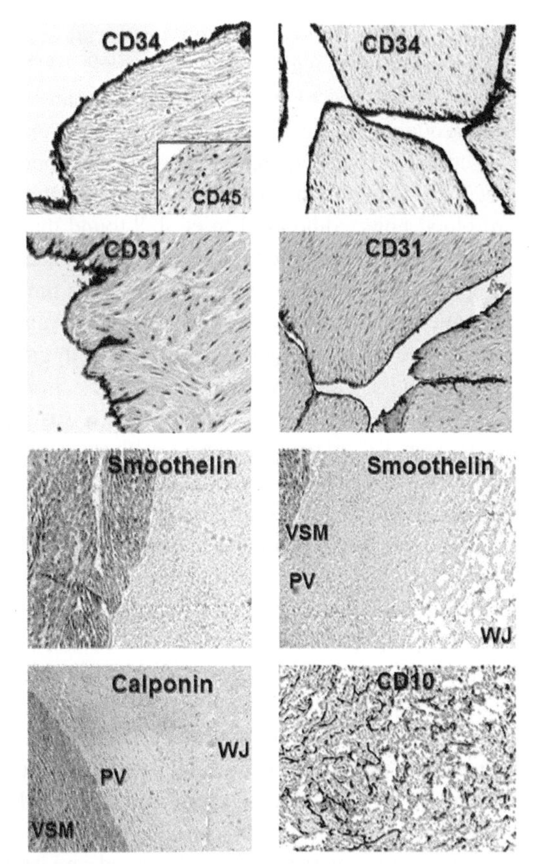


**FIGURE 2** Immunohistochemical Staining of Sections of Human Umbilical Cord. Paraffin-embedded sections were stained for the different markers indicated on each image. CD34 and CD31 were detected on the endothelium in the umbilical vein and arteries. No staining of these cells was detected with CD45 staining. Vascular smooth muscle (VSM) cells were positive for smoothelin and calponin, while the perivascular (PV) and Wharton's jelly (WJ) cells were negative. CD10 was positive on Wharton's jelly stromal cells only as indicated.

Because of their rarity, studies on the identity of human HSPCs sourced from human umbilical cord blood, granulocyte-colony stimulating factor (G-CSF) mobilized peripheral blood, and bone marrow were fraught with difficulty until the discovery of CD34 and CD133 cell surface markers allowed enrichment of these cells.[88,89] These markers identified cell subpopulations, which occurred at frequencies of around or less than 1%, contained HSPCs, formed clonogenic hematopoietic colonies in vitro, and could repopulate hematopoiesis in vivo in immunodeficient mice, rhesus monkeys, and the human after transplantation into the peripheral blood.[90] Subsequent studies[91–93] have allowed the enrichment of HSCs and their immediate progeny based on multiple markers, which include lineage negative selection, CD45 (the common leukocyteleukocyte antigen), CD34 positivity and then division into subsets based on such markers as CD133 positivity, CD110 (the thrombopoietin receptor), CD135 (the Flt-3 receptor), CD123 (the IL-7 receptor), CD49f (alpha 6 integrin), and CD90 (also known as Thy-1). A further marker used is vital staining for aldehyde dehydrogenase (ALDH) and segregation of progenitors into ALDH$^{hi}$ or ALDH$^{lo}$ cells.[94] From these studies, the HSCs were estimated to represent less than 1 in 10,000 mononuclear cells in human umbilical cord blood and yet these cells can reconstitute hematopoiesis over an individual's life span. The identity of the ECFCs or their precursors in umbilical cord blood has presented a further challenge as these cells also generally represent much less than 1 in 10,000 mononuclear cells. Although endothelial cells in different tissues possess different expression profiles either at the cell surface or transcriptome levels,[6] it is not currently anticipated that the circulating endothelial precursors that generate these endothelia possess such differences.

In developing ECFC enrichment procedures, the first studies defined the phenotype of endothelial lineage cells and the ECFCs contained therein after the primary culture (but low passage) of human ECFCs from umbilical cord, umbilical cord blood, placenta, and adult peripheral blood into endothelial cells. The cells generated were CD45$^-$, CD31$^+$, CD144$^+$, CD146$^+$, and CD105$^+$, but demonstrated variable expression of CD34 and lacked CD133.[17–19] These markers do not segregate HPP-ECFC from LPP-ECFC within endothelial cultures. It has, however, been possible to segregate ECFC with different proliferative abilities on the basis of their ALDH activity. Nagano et al.[95] have demonstrated in this way that early passage ALDH$^{lo}$ umbilical cord blood endothelial precursors were more proliferative in both normoxia and hypoxia (5% $O_2$), showed enhanced migration toward CXCL12, failed to form tubules in short-term matrigel assays, and showed enhanced function in vivo in an ischemic hind limb model when compared to ALDH$^{hi}$ cells, suggesting the latter were more mature.

Prior to culture, ECFCs have been enriched from fresh human umbilical cord blood or adult peripheral blood by immunomagnetic or flow sorting. These are based initially on CD34 or CD31 positivity and CD45 negativity, as this distinguishes the ECFC from PACs.[96–99] Using polychromatic flow cytometry, Estes and colleagues[98,99] defined ECFCs as being CD31$^+$CD34$^{bright}$CD45$^-$CD133$^-$CD14$^-$CD41a$^-$CD235a$^-$, proangiogenic cells as CD31$^+$CD34$^{bright}$CD45$^{dim}$CD133$^+$CD14$^-$CD41a$^-$CD235a$^-$, and nonangiogenic cells were CD31$^+$CD34$^{bright}$CD45$^{dim}$CD133$^-$CD14$^-$CD41a$^-$CD235a$^-$. The ECFC-containing subset was below the detection limit in the blood of healthy adult donors, but detectable in individuals undergoing treatment for cancers and those with vascular disease. Umbilical cord blood ECFCs also expressed CD146 and CD105.[99] Mund et al.[99] have also reported that flow cytometric isolation of ECFCs has a detrimental effect on their recovery and has recommended immunomagnetic selection of these cells. Our own studies suggest that a variety of factors including the antibody clone used, prolonged staining, the interval between collection of the umbilical cord blood and the magnetic bead isolation or flow cytometry, the temperature of the collection and of subsequent isolation all affect ECFC recoveries.

## 9. THE USE OF UMBILICAL CORD BLOOD OR PLACENTAL ECFCs AS A CELLULAR PRODUCT IN REGENERATIVE MEDICINE

There are a number of clinical scenarios alluded to earlier (e.g., lung injury, bone repair) in which ECFCs along with proangiogenic hematopoietic cells or MSCs sourced before or at birth may provide cellular products for promoting revascularization. However, the treatment of full thickness skin loss represents a challenging clinical entity and is used here as an exemplar of where umbilical cord-, cord blood-, or placental-derived ECFCs may provide a useful clinical application in the future. There are currently no tissue-engineered substrates that can fully replicate the epidermal and dermal in vivo niches to fulfill both aesthetic and functional demands. The current gold standard treatment of autologous skin grafting is inadequate because of poor textural durability, scarring, and associated contracture. This is particularly the case when split thickness grafts are used, even when in combination with one of the many proprietary dermal substitutes currently available for clinical use.

One option that is currently being explored that may improve the quality of existing dermal substitutes is to prevascularize them so that incorporation with the wound bed occurs more rapidly (reviewed in [41]). This would increase the trajectory of wound healing and, based on studies examining time-to-healing versus scarring, would improve the final aesthetic result. When considering a suitable endothelial cell source for this purpose, it is necessary to have cells that can be easily isolated and expanded rapidly in vitro prior to their clinical application, making ECFCs an excellent option. However, as stated above, HPP-ECFCs are more readily harvested from umbilical cord, cord blood, or the placenta than peripheral blood and are also found in much higher numbers in the former. This leaves two potential options that would enable clinicians to use such ECFCs for clinical applications: (1) isolate ECFCs from fresh umbilical cord, cord blood, or placenta and freeze them in a bank for use should the need arise at some time in the donor's future (as stated above, it would be unsatisfactory to freeze cells prior to culture as the prospect of isolating HPP-ECFCs from such a harvest would be significantly reduced); (2) use freshly isolated and expanded ECFCs from umbilical cord, cord blood, or placenta in an acute wound reconstruction scenario. In adults with large wounds, for example, due to trauma or a significant burn, option 1 would be a useful, although an expensive clinical application unless the cell bank is used to treat various other vascular diseases or to promote tissue regeneration where revascularization is a key. Option 2 may provide a more suitable model particularly where a congenital condition is identified before or at birth. The conditions where skin reconstruction is required in neonates such as giant congenital melanocytic nevus and aplasia cutis congenita have historically been managed with split skin grafts. Therefore, using recently isolated and expanded umbilical cord, cord blood, or placental ECFCs to produce a prevascularized scaffold layer underneath the graft to improve the quality of skin grafts applied to wounds in these patients maybe a useful novel application for these cells.

In view of the vasculogenic properties of ECFCs, human skin substitutes are just one of several potential clinical roles for these cells. ECFCs have been shown to contribute to vessel formation when implanted subcutaneously within collagen, fibrin, or matrigel matrices and also to improve perfusion in animal hind limb ischemia models or to promote bone repair, therefore patients with impaired vascular function may also benefit from umbilical cord-, cord blood-, or placental-derived ECFCs either in the acute postnatal period, or later in life using cryopreserved ECFCs.

## 10. CONCLUSIONS

Studies over the past one to two decades have demonstrated that both proangiogenic cells of hematopoietic origin and endothelial precursor cells circulate in the umbilical cord blood and peripheral blood. Although these share cell surface markers in common, they can be segregated from one another principally on the basis of differential cell surface expression of the leukocyte common antigen, CD45. A hierarchy of endothelial precursors was found to exist in the umbilical cord blood, adult peripheral blood, and in some tissues including the umbilical cord and placenta. Those endothelial precursors that proliferate significantly are termed ECFC and these may have high (HPP-ECFC) or low (LPP-ECFC) proliferative abilities. The proangiogenic cell content in the circulation is significantly higher than the ECFC content and the proliferative abilities of the ECFCs decline with age. Both proangiogenic hematopoietic cells and ECFC are involved in vessel formation. It has been hypothesized that, in response to tissue injury or expansion, the proangiogenic hematopoietic cells first create a proangiogenic environment and this permits or encourages the ECFC to form or repair the vasculature in late gestation and postnatally. The numbers of ECFCs in umbilical cord blood appear to increase during the third trimester of pregnancy and may contribute to perinatal organogenesis. Perinatally and in the early postnatal period, ECFC function can be compromised by adverse environments, such as hyperoxia, severe hypoxia, oxidative stress, and hyperglcemia, and it has been proposed that this can contribute to certain diseases that affect the vasculature in adult life. ECFCs have been shown to improve or contribute to revascularization in preclinical surrogate models, such as islet cell transplantation, bone repair, retinal and hind limb ischemia, and myocardial infarction, and to reduce bronchopulmonary dysplasia. They may also have a role in congenital wound repair. Whether there are sufficient ECFCs that can be harnessed from blood or tissues for vascular repair in the human is unclear, as the origin of the circulating ECFCs is not understood. However, cells with the potential to give rise to hemogenic endothelium have been reported in human umbilical cord blood. Furthermore, it is encouraging that fully functional human tracheal grafts, which are precolonized with autologous epithelial and chondrogenic cells ex vivo, can be fully revascularized by the recipient's own cells posttransplantation, and that it is now possible to reprogram cells with exogenous factors to enhance their vasculogenic potential.

## GLOSSARY

**Apgar Score** Developed by Dr Virgina Apgar as a backronym of her surname and used as a simple means to quickly assess the health of the newborn immediately after birth (1–10 min). Hence the Apgar Score measures the baby's ability to adapt to the extrauterine environment by assessing five criteria: Appearance, Pulse, Grimace, Activity (muscle tone), and Respiration, which are allocated a score of 0–2, and with normal total Apgar scores ranging from 7 to 10.

**CAC** Circulating angiogenic cell thought to be of hematopoietic origin and which promotes vessel formation. May be equivalent to a PAC.

**ECFC** A progenitor cell for the endothelial lineage that forms clonogenic colonies of >50 cells over 14 days of culture.

**EPC** A term used to describe endothelial progenitor cells, but more often describes proangiogenic hematopoietic cells, and thus there are recent suggestions that the use of this acronym should be discontinued.

**HPP-CFC** A progenitor cell for the myeloid lineage that forms large clonogenic colonies visible with the naked eye over 14 days of culture.

**HPP-ECFC** A progenitor cell for the endothelial lineage that forms clonogenic colonies of ≥500 cells over 14 days of culture.

**HSPC** Includes multipotent hematopoietic stem cells for the myeloid and lymphoid lineages and their committed progeny, the hematopoietic progenitor cells.

**LPP-CFCs** A progenitor cell for the myeloid lineage that forms clonogenic colonies of ≥50 cells over 14 days of culture, but is not visible with the naked eye.

**LPP-ECFCs** A progenitor cell for the endothelial lineage that forms clonogenic colonies of ≥50 and <500 cells over 14 days of culture.

**Obstetric Factors** Typical obstetric factors include birth and placental weight, infant gender, parity of the mother, nucleated red cell content of cord blood, low venous pH of cord blood, prolonged first stage of labor, and Apgar scores.

**PAC** Proangiogenic cell that promotes vessel formation.

## LIST OF ACRONYMS AND ABBREVIATIONS

**ALDH** Aldehyde Dehydrogenase
**ECFCs** Endothelial Colony-Forming Cells
**EPCs** Endothelial Progenitor Cells
**FGF-2** Fibroblast Growth Factor-2
**HLA** Human Leucocyte Antigen
**HPP-CFCs** High Proliferative Potential Colony-Forming Cells
**HPP-ECFCs** High Proliferative Potential Endothelial Colony-Forming Cells
**HSPC** Hematopoietic Stem Progenitor Cells
**LPP-CFCs** Low Proliferative Potential Colony-Forming Cells
**LPP-ECFCs** Low Proliferative Potential Endothelial Colony-Forming Cells
**NHS** National Health Service
**NO** Nitric Oxide
**NOD-Scid** Nonobese Diabetic–Severe Combined Immunodeficient
**OEC** Outgrowth Endothelial Cells
**PACs** Proangiogenic Cells
**PF4** Platelet Factor 4
**sVEGF** Soluble Vascular Endothelial Growth Factor Receptor-1
**UEA** Ulex Europaeus Lectin
**VEGF** Vascular Endothelial Growth Factor

## ACKNOWLEDGMENT

The authors wish to acknowledge the support of NHS Blood and Transplant, the National Institute of Health Research (NIHR) under its Programme Grants Schemes RP-PG-0310-1001 and RP-PG-0310-1003, the NIHR Oxford Biomedical Research Centre, the British Heart Foundation and Restore Burns and Wound Healing Trust. The views expressed in this publication are those of the authors and not necessarily those of the NHS, the NIHR or the Department of Health. We would like to thank Mrs Wendy Slack for expert assistance with sourcing and formatting the references.

## REFERENCES

1. Marcelo KL, Goldie LC, Hirschi KK. Regulation of endothelial cell differentiation and specification. *Circ Res* 2013;**112**:1272–87.
2. Potente M, Gerhardt H, Carmeliet P. Basic and therapeutic aspects of angiogenesis. *Cell* 2011;**146**:873–87.
3. Watt SM, Gullo F, van der Garde M, Markeson D, Camicia RP, Khoo CP, et al. The angiogenic properties of mesenchymal stem/stromal cells and their therapeutic potential. *Brit Med Bull* 2013;**108**:25–53.
4. Corselli M, Chen CW, Sun B, Yap S, Rubin JP, Peault B. The tunica adventitia of human arteries and veins as a source of mesenchymal stem cells. *Stem Cells Dev* 2012;**21**:1299–308.
5. Aird WC. Phenotypic heterogeneity of the endothelium: I. Structure, function, and mechanisms. *Circ Res* 2007;**100**:158–73.
6. Nolan DJ, Ginsberg M, Israely E, Palikuqi B, Poulos MG, James D, et al. Molecular signatures of tissue-specific microvascular endothelial cell heterogeneity in organ maintenance and regeneration. *Dev Cell* 2013;**26**:204–19.
7. Zambidis ET, Sinka L, Tavain M, Jokubaitis V, Park TS, Simmons P, et al. Emergence of human angiohematopoietic cells in normal development and from cultured embryonic stem cells. *Ann N Y Acad Sci* 2007;**1106**:223–32.
8. Dieterlen-Lèvre F, Jaffredo T. Decoding the hemogenic endothelium in mammals. *Cell Stem Cell* 2009;**4**:189–90.
9. Clarke RL, Yzaguirre AD, Yashiro-Ohtani Y, Bondue A, Blanpain C, Pear WS, et al. The expression of Sox17 identifies and regulates haemogenic endothelium. *Nat Cell Biol* 2013;**15**:502–10.
10. Nimmo R, Ciau-Uitz A, Ruiz-Herguido C, Soneji S, Bigas A, Patient R, et al. miR-142-3p controls the specification of definitive hemangioblasts during ontogeny. *Dev Cell* 2013;**26**(3):237–49.
11. Wu X, Lensch MW, Wylie-Sears J, Daley GQ, Bischoff J. Hemogenic endothelial progenitor cells isolated from human umbilical cord blood. *Stem Cells* 2007;**25**:2770–6.

12. Pelosi E, Castelli G, Martin-Padura I, Bordoni V, Santoro S, Conigliaro A, et al. Human haemato-endothelial precursors: cord blood CD34+ cells produce haemogenic endothelium. *PLoS One* 2012;**7**:e51109.
13. Stump MM, Jordan Jr GL, Debakey ME, Halpert B. Endothlium grown from circulating blood on isolated intravascular dacron hub. *Am J Pathol* 1963;**43**:361–7.
14. Asahara T, Murohara T, Sullivan A, Silver M, van der Zee R, Li T, et al. Isolation of putative progenitor endothelial cells for angiogenesis. *Science* 1997;**275**:964–7.
15. Takahashi T, Kalka C, Masuda H, Chen D, Silver M, Kearney M, et al. Ischemia- and cytokine-induced mobilization of bone marrow-derived endothelial progenitor cells for neovascularization. *Nat Med* 1999;**5**:434–8.
16. Kalka C, Masuda H, Takahashi T, Kalka-Moll WM, Silver M, Kearney M, et al. Transplantation of ex vivo expanded endothelial progenitor cells for therapeutic neovascularization. *Proc Natl Acad Sci USA* 2000;**97**:3422–7.
17. Watt SM, Athanassopoulos A, Harris AL, Tsaknakis G. Human endothelial stem/progenitor cells, angiogenic factors and vascular repair. *J R Soc Interface* 2010;**7**(Suppl 6):S731–51.
18. Yoder MC, Ingram DA. The definition of EPCs and other bone marrow cells contributing to neoangiogenesis and tumor growth: is there common ground for understanding the roles of numerous marrow-derived cells in the neoangiogenic process? *Biochim Biophys Acta* 2009;**1796**:50–4.
19. Basile DP, Yoder MC. Circulating and tissue resident endothelial progenitor cells. *J Cell Physiol* 2014;**229**(1):10–6.
20. Lin Y, Weisdorf DJ, Solovey A, Hebbel RP. Origins of circulating endothelial cells and endothelial outgrowth from blood. *J Clin Invest* 2000;**105**:71–7.
21. Ingram DA, Mead LE, Tanaka H, Meade V, Fenoglio A, Mortell K, et al. Identification of a novel hierarchy of endothelial progenitor cells using human peripheral and umbilical cord blood. *Blood* 2004;**104**:2752–60.
22. Ingram DA, Mead LE, Moore DB, Woodard W, Fenoglio A, Yoder MC. Vessel wall-derived endothelial cells rapidly proliferate because they contain a complete hierarchy of endothelial progenitor cells. *Blood* 2005;**105**:2783–6.
23. Bradley TR, Hodgson GS. Detection of primitive macrophage progenitor cells in mouse bone marrow. *Blood* 1979;**54**:1446–50.
24. Bertoncello I, Bradley TR, Hodgson GS, Dunlop JM. The resolution, enrichment, and organization of normal bone marrow high proliferative potential colony-forming cell subsets on the basis of rhodamine-123 fluorescence. *Exp Hematol* 1991;**19**:174–8.
25. Bertoncello I, Bradley TR, Watt SM. An improved negative immunomagnetic selection strategy for the purification of primitive hemopoietic cells from normal bone marrow. *Exp Hematol* 1991;**19**:95–100.
26. Zhang Y, Fisher N, Newey SE, Smythe J, Tatton L, Tsaknakis G, et al. The impact of proliferative potential of umbilical cord-derived endothelial progenitor cells and hypoxia on vascular tubule formation in vitro. *Stem Cells Dev* 2009;**18**:359–75.
27. Rapp BM, Saadatzedeh MR, Ofstein RH, Bhavsar JR, Tempel ZS, Moreno O, et al. Resident endothelial progenitor cells from human placenta have greater vasculogenic potential than circulating endothelial progenitor cells from umbilical cord blood. *Cell Med* 2011;**2**:85–96.
28. Yoon CH, Hur J, Park KW, Kim JH, Lee CS, Oh IY, et al. Synergistic neovascularization by mixed transplantation of early endothelial progenitor cells and late outgrowth endothelial cells: the role of angiogenic cytokines and matrix metalloproteinases. *Circulation* 2005;**112**:1618–27.
29. Au P, Daheron LM, Duda DG, Cohen KS, Tyrrell JA, Lanning RM, et al. Differential in vivo potential of endothelial progenitor cells from human umbilical cord blood and adult peripheral blood to form functional long-lasting vessels. *Blood* 2008;**111**:1302–5.
30. Au P, Tam J, Fukumura D, Jain RK. Bone marrow-derived mesenchymal stem cells facilitate engineering of long-lasting functional vasculature. *Blood* 2008;**111**:4551–8.
31. Reinisch A, Hofmann NA, Obenauf AC, Kashofer K, Rohde E, Schallmoser K, et al. Humanized large-scale expanded endothelial colony-forming cells function in vitro and in vivo. *Blood* 2009;**113**:6716–25.
32. Traktuev DO, Prater DN, Merfled-Clauss S, Sanjeevaiah AR, Saadtzadeh MR, Murphy M, et al. Robust functional vascular network formation in vivo by cooperation of adipose progenitor and endothelial cells. *Circ Res* 2009;**104**:1410–20.
33. Hendrickx B, Verdonck K, Van den Berge S, Dickens S, Eriksson E, Vranckx JJ, et al. Integration of blood outgrowth endothelial cells in dermal fibroblast sheets promotes full thickness wound healing. *Stem Cells* 2010;**28**:1165–77.
34. Roubelakis MG, Tsaknakis G, Pappa KI, Anagnou NP, Watt SM. Spindle shaped human mesenchymal stem/stromal cells from amniotic fluid promote neovascularization. *PLoS One* 2013;**8**:e54747.
35. Melero-Martin JM, Khan ZA, Picard A, Wu X, Paruchuri S, Bischoff J. In vivo vasculogenic potential of human blood-derived endothelial progenitor cells. *Blood* 2007;**109**:4761–8.
36. Melero-Martin JM, De Obaldia ME, Kang SY, Khan ZA, Yuan L, Oettgen P, et al. Engineering robust and functional vascular networks in vivo with human adult and cord blood-derived progenitor cells. *Circ Res* 2008;**103**:194–202.
37. Koike N, Fukumura D, Gralla O, Au P, Schechner JS, Jain RK. Tissue engineering: creation of long-lasting blood vessels. *Nature* 2004;**428**:138–9.
38. Merfeld-Clauss S, Gollahalli N, March KL, Traktuev DO. Adipose tissue progenitor cells directly interact with endothelial cells to induce vascular network formation. *Tissue Eng Part A* 2010;**16**:2953–66.
39. Melero-Martin JM, De Obaldia ME, Allen P, Dudley AC, Klagsbrun M, Bischoff J. Host myeloid cells are necessary for creating bioengineered human vascular networks in vivo. *Tissue Eng Part A* 2010;**16**:2457–66.
40. Wara AK, Croce K, Foo S, Sun X, Icli B, Tesmenitsky Y, et al. Bone marrow-derived CMPs and GMPs represent highly functional pro-angiogenic cells: implications for ischemic cardiovascular disease. *Blood* 2011;**118**:6461–4.
41. Markeson D, Pleat JM, Sharpe JR, Harris A, Seifalian AM, Watt SM, et al. Scarring, stem cells, scaffolds and skin repair. *J Tiss Eng Regen Med* 2014 [Epub ahead of print].
42. Zhou B, Tsaknakis G, Coldwell KE, Khoo CP, Roubelakis MG, Chang CH, et al. A novel function for the haemopoietic supportive murine bone marrow MS-5 mesenchymal stromal cell line in promoting human vasculogenesis and angiogenesis. *Br J Haematol* May 2012;**157**:299–311.
43. Carpenter L, Malladi R, Yang CT, French A, Pilkington KJ, Forsey RW, et al. Human induced pluripotent stem cells are capable of B-cell lymphopoiesis. *Blood* 2011;**117**(15):4008–11.
44. Athanassopoulos A, Tsaknakis G, Newey SE, Harris AL, Kean J, Tyler MP, et al. Microvessel networks in pre-formed in artificial clinical grade dermal substitutes in vitro using cells from haematopoietic tissues. *Burns* 2012;**38**:691–701.

45. Decaris ML, Lee CI, Yoder MC, Tarantal AF, Leach JK. Influence of the oxygen microenvironment on the proangiogenic potential of human endothelial colony forming cells. *Angiogenesis* 2009;**12**:303–11.
46. Lowndes SA, Adams A, Timms A, Fisher N, Smythe J, Watt SM, et al. Phase I study of copper-binding agent ATN-224 in patients with advanced solid tumors. *Clin Cancer Res* 2008;**14**:7526–34.
47. Smythe J, Fox A, Fisher N, Frith E, Harris AL, Watt SM. Measuring angiogenic cytokines, circulating endothelial cells, and endothelial progenitor cells in peripheral blood and cord blood: VEGF and CXCL12 correlate with the number of circulating endothelial progenitor cells in peripheral blood. *Tissue Eng Part C Methods* 2008;**14**:59–67.
48. Javed MJ, Mead LE, Prater D, Bessler WK, Foster D, Case J, et al. Endothelial colony forming cells and mesenchymal stem cells are enriched at different gestational ages in human umbilical cord blood. *Pediatr Res* 2008;**64**:68–73.
49. Baker CD, Ryan SL, Ingram DA, Seedorf GJ, Abman SH, Balasubramaniam V. Endothelial colony-forming cells from preterm infants are increased and more susceptible to hyperoxia. *Am J Respir Crit Care Med* 2009;**180**:454–61.
50. Coldwell KE, Lee SJ, Kean J, Khoo CP, Tsaknakis G, Smythe J, et al. Effects of obstetric factors and storage temperatures on the yield of endothelial colony forming cells from umbilical cord blood. *Angiogenesis* 2011;**14**:381–92.
51. Yoder MC, Mead LE, Prater D, Krier TR, Mroueh KN, Li F, et al. Redefining endothelial progenitor cells via clonal analysis and hematopoietic stem/progenitor cell principals. *Blood* 2007;**109**:1801–9.
52. Shelley WC, Leapley AC, Huang L, Critser PJ, Zeng P, Prater D, et al. Changes in the frequency and in vivo vessel-forming ability of rhesus monkey circulating endothelial colony-forming cells across the lifespan (birth to aged). *Pediatr Res* 2012;**71**:156–61.
53. Case J, Ingram DA, Haneline LS. Oxidative stress impairs endothelial progenitor cell function. *Antioxid Redox Signal* 2008;**10**: 1895–907.
54. He T, Peterson TE, Holmuhamedov EL, Terzic A, Caplice NM, Oberley LW, et al. Human endothelial progenitor cells tolerate oxidative stress due to intrinsically high expression of manganese superoxide dismutase. *Arterioscler Thromb Vasc Biol* 2004;**24**:2021–7.
55. Nuzzolo ER, Capodimonti S, Martini M, Iachininoto MG, Bianchi M, Cocomazzi A, et al. Adult and cord blood endothelial progenitor cells have different gene expression profiles and immunogenic potential. *Blood Transfus* 2014;**12**(Suppl. 1):S367–74.
56. Cheng CC, Lo HH, Huang TS, Cheng YC, Chang ST, Chang SJ, et al. Genetic module and miRNome trait analyses reflect the distinct biological features of endothelial progenitor cells from different anatomic locations. *BMC Genomics* 2012;**13**:447–60.
57. Sipos PI, Rens W, Schlecht H, Fan X, Wareing M, Hayward C, et al. Uterine vasculature remodeling in human pregnancy involves functional macrochimerism by endothelial colony forming cells of fetal origin. *Stem Cells* 2013;**31**:1363–70.
58. Watt SM. Umbilical cord blood stem cell banking. In: Moo-Young M, Butler M, Webb C, Moreira A, Grodzinski B, Cui ZF, et al., editors. *Comprehensive Biotechnology*. 2nd ed. Amsterdam, The Netherlands: Elsevier BV Press; 2011. p. 397–406.
59. Gluckman E, Broxmeyer HA, Auerbach AD, Friedman HS, Douglas GW, Devergie A, et al. Hematopoietic reconstitution in a patient with Fanconi's anaemia by means of umbilical cord blood from an HLA identical sibling. *N Eng J Med* 1989;**321**:1174–8.
60. Ramirez P, Wagner JE, DeFor TE, Blazar BR, Verneris MR, Miller JS, et al. Factors predicting single-unit predominance after double umbilical cord blood transplantation. *Bone Marrow Transpl* 2012;**47**:799–803.
61. Ballen KK, Koreth J, Chen YB, Dey BR, Spitzer TR. Selection of optimal alternative graft source: mismatched unrelated donor, umbilical cord blood, or haploidentical transplant. *Blood* 2012;**119**: 1972–80.
62. Rocha V, Spellman S, Zhang MJ, Ruggeri A, Purtill D, Brady C, et al. Effect of HLA-matching recipients to donor noninherited maternal antigens on outcomes after mismatched umbilical cord blood transplantation for hematologic malignancy. *Biol Blood Marrow Transpl* 2012;**18**:1890–6.
63. Page KM, Zhang L, Mendizabal A, Wease S, Carter S, Shoulars K, et al. The Cord Blood Apgar: a novel scoring system to optimize selection of banked cord blood grafts for transplantation (CME). *Transfusion* 2012;**52**:272–83.
64. Ho PJ, Hall GW, Watt S, West NC, Wimperis JW, Wood WG, et al. Unusually severe heterozygous beta-thalassemia: evidence for an interacting gene affecting globin translation. *Blood* 1998;**92**: 3428–35.
65. Bowie MB, McKnight KD, Kent DG, McCaffrey L, Hoodless PA, Eaves CJ. Identification of a new intrinsically timed developmental checkpoint that reprograms key hematopoietic stem cell properties. *J Clin Invest* 2006;**116**:2808–16.
66. Bowie MB, Kent DG, Dykstra B, McKnight KD, McCaffrey L, Hoodless PA, et al. Identification of a new intrinsically timed developmental checkpoint that reprograms key hematopoietic stem cell properties. *Proc Natl Acad Sci USA* 2007;**104**:5878–82.
67. Hayes Jr D, Feola DJ, Murphy BS, Shook LA, Ballard HO. Pathogenesis of bronchopulmonary dysplasia. *Respiration* 2010;**79**: 425–36.
68. O'Reilly M, Thébaud B. The promise of stem cells in bronchopulmonary dysplasia. *Semin Perinatol* 2013;**37**:79–84.
69. Balasubramaniam V, Ryan SL, Seedorf GJ, Roth EV, Heumann TR, Yoder MC, et al. Bone marrow-derived angiogenic cells restore lung alveolar and vascular structure after neonatal hyperoxia in infant mice. *Am J Physiol Lung Cell Mol Physiol* 2010;**298**:L315–23.
70. Ligi I, Simoncini S, Tellier E, Vassallo PF, Sabatier F, Guillet B, et al. A switch toward angiostatic gene expression impairs the angiogenic properties of endothelial progenitor cells in low birth weight preterm infants. *Blood* 2011;**118**:1699–709.
71. Ligi I, Simoncini S, Tellier E, Grandvuillemin I, Marcelli M, Bikfalvi A, et al. Altered angiogenesis in low birth weight individuals: a role for anti-angiogenic circulating factors. *J Matern Fetal Neonatal Med* 2014;**27**(3):233–8.
72. Barker DJ, Winter PD, Osmond C, Margetts B, Simmonds SJ. Weight in infancy and death from ischaemic heart disease. *Lancet* 1989;**2**:577–80.
73. Johansson S, Iliadou A, Bergvall N, Tuvemo T, Norman M, Cnattingius S. Risk of high blood pressure among young men increases with the degree of immaturity at birth. *Circulation* 2005;**112**:3430–6.
74. Kistner A, Jacobson L, Jacobson SH, Svensson E, Hellstrom A. Low gestational age associated with abnormal retinal vascularization and increased blood pressure in adult women. *Pediatr Res* 2002;**51**:675–80.
75. Liu Y, Teoh SH, Chong MS, Lee ES, Mattar CN, Randhawa NK, et al. Vasculogenic and osteogenesis-enhancing potential of human umbilical cord blood endothelial colony-forming cells. *Stem Cells* 2012;**30**:1911–24.

76. Usami K, Mizuno H, Okada K, Narita Y, Aoki M, Kondo T, et al. Composite implantation of mesenchymal stem cells with endothelial progenitor cells enhances tissue-engineered bone formation. *J Biomed Mater Res A* 2009;**90**:730–41.
77. Seebach C, Henrich D, Kähling C, Wilhelm K, Tami AE, Alini M, et al. Endothelial progenitor cells and mesenchymal stem cells seeded onto beta-TCP granules enhance early vascularization and bone healing in a critical-sized bone defect in rats. *Tissue Eng Part A* 2010;**16**(6):1961–70.
78. Chandrasekhar KS, Zhou H, Zeng P, Alge D, Li W, Finney BA, et al. Blood vessel wall-derived endothelial colony-forming cells enhance fracture repair and bone regeneration. *Calcif Tissue Int* 2011;**89**:347–57.
79. Ehrenthal DB, Catov JM. Importance of engaging obstetrician/gynecologists in cardiovascular disease prevention. *Curr Opin Cardiol* 2013;**28**:547–53.
80. Paneni F, Beckman JA, Creager MA, Cosentino F. Diabetes and vascular disease: pathophysiology, clinical consequences, and medical therapy: part I. *Eur Heart J* 2013;**34**:2436–43.
81. Richardson MR, Yoder MC. Endothelial progenitor cells: quo vadis? *J Mol Cell Cardiol* 2011;**50**:266–72.
82. Chen YH, Lin SJ, Lin FY, Wu TC, Tsao CR, Huang PH, et al. High glucose impairs early and late endothelial progenitor cells by modifying nitric oxide-related but not oxidative stress-mediated mechanisms. *Diabetes* June 2007;**56**(6):1559–68.
83. Guttridge MG, Soh TG, Belfield H, Sidders C, Watt SM. Storage time affects umbilical cord blood viability. *Transfus Med* 2014;**54**(5):1278–85.
84. Radke TF, Barbosa D, Duggleby RC, Saccardi R, Querol S, Kogler G. The assessment of parameters affecting the quality of cord blood by the applicance of the annexin V staining method and correlation with CFU assays. *Stem Cells Int* 2013;**2013**:823–912.
85. Scaradavou A, Smith KM, Hawke R, Schaible A, Abboud M, Kernan NA, et al. Cord blood units with low CD34+ cell viability have a low probability of engraftment after double unit transplantation. *Biol Blood Marrow Transpl* 2010;**16**:500–8.
86. Vanneaux V, El-Ayoubi F, Delmau C, Driancourt C, Lecourt S, Grelier A, et al. In vitro and in vivo analysis of endothelial progenitor cells from cryopreserved umbilical cord blood: are we ready for clinical application? *Cell Transpl* 2010;**19**:1143–55.
87. Broxmeyer HE, Lee MR, Hangoc G, Cooper S, Prasain N, Kim YJ, et al. Hematopoietic stem/progenitor cells, generation of induced pluripotent stem cells, and isolation of endothelial progenitors from 21- to 23.5-year cryopreserved cord blood. *Blood* 2011;**117**:4773–7.
88. Civin CI. CD34 stem cell stories and lessons from the CD34 wars: the Landsteiner Lecture 2009. *Transfusion* 2010;**50**:2046–56.
89. Miraglia S, Godfrey W, Yin AH, Atkins K, Warnke R, Holden JT, et al. A novel five-transmembrane hematopoietic stem cell antigen: isolation, characterization, and molecular cloning. *Blood* 1997;**90**:5013–21.
90. Goardon N, Marchi E, Atzberger A, Quek L, Schuh A, Soneji S, et al. Coexistence of LMPP-like and GMP-like leukemia stem cells in acute myeloid leukemia. *Cancer Cell* 2011;**19**:138–52.
91. Goergens A, Radtke S, Moellmann M, Cross M, Duering J, Hoen PA, et al. Revision of the human hematopoietic tree: granulocyte subtypes derive from distinct hematopoietic lineages. *Cell Reports* 2013;**3**:1539–51.
92. Laurenti E, Doulatov S, Zandi S, Plumb I, Chen J, April C, et al. The transcriptional architecture of early human hematopoiesis identifies multilevel control of lymphoid commitment. *Nat Immunol* 2013;**14**:756–63.
93. Larochelle A, Savona M, Wiggins M, Anderson S, Ichwan B, Keyvanfar K, et al. Human and rhesus macaque hematopoietic stem cells cannot be purified based only on SLAM family markers. *Blood* 2011;**117**:1550–4.
94. Hess DA, Meyerrose TE, Wirthlin L, Craft TP, Herrbrich PE, Creer MH, et al. Functional characterization of highly purified human hematopoietic repopulating cells isolated according to aldehyde dehydrogenase activity. *Blood* 2004;**104**:1648–55.
95. Nagano M, Yamashita T, Hamada H, Ohneda K, Kimura K, Nakagawa T, et al. Identification of functional endothelial progenitor cells suitable for the treatment of ischemic tissue using human umbilical cord blood. *Blood* 2007;**110**:151–60.
96. Case J, Mead LE, Bessler WK, Prater D, White HA, Saadatzadeh MR, et al. Human CD34+AC133+VEGFR-2+ cells are not endothelial progenitor cells but distinct, primitive hematopoietic progenitors. *Exp Hematol* 2007;**35**:1109–18.
97. Timmermans F, Van Hauwermeiren F, De SM, Raedt R, Plasschaert F, De Buyzere ML, et al. Endothelial outgrowth cells are not derived from CD133+ cells or CD45+ hematopoietic precursors. *Arterioscler Thromb Vasc Biol* 2007;**27**:1572–9.
98. Estes ML, Mund JA, Ingram DA, Case J. Identification of endothelial cells and progenitor cell subsets in human peripheral blood. *Curr Protoc Cytom* 2010;**52**:9.33.1–1.
99. Mund JA, Estes ML, Yoder MC, Ingram Jr DA, Case J. Flow cytometric identification and functional characterization of immature and mature circulating endothelial cells. *Arterioscler Thromb Vasc Biol* 2012;**32**:1045–53.
100. Apgar V. A proposal for a new method of evaluation of the newborn infant. *Curr Res Anesth Analg* 1953;**32**:260–7.

Chapter 6

# HLA and Immunogenetics in Cord Blood Transplantation

**Dominique Charron, Emeline Masson and Pascale Loiseau**
*Laboratoire "Jean Dausset," Immunology and Histocompatibility Hôpital Saint-Louis AP-HP, Université Paris Diderot, Paris, France*

## Chapter Outline



## 1. INTRODUCTION

Cord blood (CB) transplantation is a fast growing medical procedure that originated 25 years ago when the first umbilical cord blood transplantation was performed and reported.[1] CB was identified as a valid source of hematopoietic stem cells (HSCs), therefore with a real potential for reconstituting hematopoiesis in vivo. Cryopreservation of cord blood cells was achieved with no significant loss in their hematopoietic capacity. This led to the concept of storing and banking CB cells for deferred use, whether in an autologous or an allogeneic setting.

The first cord blood transplantation (CBT) was performed in a child suffering from Fanconi anemia, using a human leukocyte antigen (HLA) identical CB collected and cryopreserved at birth from a sibling.[1] The full hematopoietic restoration and the absence of Graft-versus-Host Disease (GvHD) were highly encouraging. This original patient is alive and doing well after 25 years with full hematological and immunological reconstitution.

Once the hematopoietic capacity of CBT was recognized, two concerns were immediately identified: the number of stem cells in the cord blood unit (CBU) and the immunogenetic barrier. The importance of immunogenetic differences between cord and recipient was addressed concurrently with the issues of allogeneic bone marrow transplantation (BMT). Even transplantation from an HLA-identical sibling could result in allogeneic-driven complication, which reveals that both HLA and non-major histocompatibility complex (MHC) immunogenetics are key elements to the success of HSC transplantation, no matter what the source of stem cells.

This chapter reviews the impact of immunogenetic factors, HLA and non-HLA, in the outcome of CBT. While early studies suggested that the requirement for HLA matching could be less stringent than in unrelated donor bone marrow transplantation (UDBMT), it is today evident that a full HLA matching remains an optimum according to most recent studies. This will naturally affect the HLA

Cord Blood Stem Cells Medicine. http://dx.doi.org/10.1016/B978-0-12-407785-0.00006-2

typing resolution level required for CB banking in the future. Clinical data will further help define the optimal strategy of immunogenetic typing for cord blood banking. On the other hand, the higher level of HLA mismatching observed in CBT compared to UDBMT cohorts points to the possibility that anti-HLA antibodies in the recipient could have an effect on engraftment, as evidenced in a series of recent reports. Finally, data on the role of non-HLA immunogenetics in CBT remains scarce and warrants further investigation.

## 2. STRUCTURE, FUNCTION, AND POLYMORPHISM OF HLA

HLA genes are the human MHC and code for molecules expressed on the surface of almost all nucleated cells. The HLA molecules control the immune response through recognition of "self and non-self" and therefore have a decisive role in the compatibility of tissue and organ transplantations. Immune responses against HLA incompatibility represent the major barrier to hematopoietic stem cell transplantation (HSCT).

The human MHC spans over 4 Mb of the short arm of chromosome 6 (6p21.3) and contains a large number of genes that are subdivided into three classes (class I, II, and III) on the basis of functional and structural differences. Two of these three classes (class I and II) code for the HLA molecules. The class I region encodes the classical HLA molecules HLA-A, -B, and -C. The class II region comprises the HLA-DR region (containing the DRA, DRB1, and depending on the haplotype DRB3, DRB4, or DRB5 genes), the HLA-DP region (containing the DPA1, DPB1 genes) and the HLA-DQ region (containing the DQA1, DQB1 genes).

The main function of HLA molecules is presenting the antigenic peptides to T lymphocytes, initiating the specific immune response. $CD4^+$ T cells recognize peptides presented by HLA class II, and $CD8^+$ T cells recognize peptides presented by HLA class I.

HLA molecules have similar structures, with three regions: extracellular; intramembranous; and short intracellular. The extracellular region is divided into domains, coded by the exons of HLA genes. The domains further from the cell membrane constitute the peptide-binding groove, interacting with T cell receptors. The class I molecules are heterodimers constitutively expressed on all nucleated cells, however at varying levels. They are composed of a polymorphic glycoprotein chain encoded by a class I gene that is noncovalently associated with and stabilized by a nonpolymorphic β2-microglobulin chain, encoded by a gene located on chromosome 15. Three domains characterize the extracellular region: alpha 1; alpha 2; and alpha 3 domains. The alpha 1 and 2 domains are the site of the peptide-binding groove and are encoded by the second and third exon of the gene. The high polymorphism of HLA class I is located in these two domains. The class II molecules are heterodimers composed of an alpha and a beta chain, coded by the A and B genes of the class II loci. HLA class II expression is restricted to cells of the immune system (antigen presenting cells). Class II molecules have a similar structure to class I; the peptide-binding groove is constituted by the first domain of alpha and beta chains encoded by the second exon of the corresponding genes, with most of the polymorphism located in this site. The class II beta chain is much more polymorphic than the alpha chain and therefore, current HLA typing is mainly performed on the beta chain (HLA-DRB1, HLA-DRB3, HLA-DRB4, HLA-DRB5, HLA-DQB1, and HLA-DPB1).

HLA class I and II genes are the most polymorphic regions of the human genome. More than 12,000 alleles (2946 HLA-A, 3693 HLA-B, 2466 HLA-C, 1582 DRB1, 509 DQB1, and 472 DPB1 alleles) were recognized in the last release 3.18.0 (2014-10) of the "IMGT/HLA Database" that provides a specialist database for sequences of the HLA and comprises the official sequences for the World Health Organization (WHO) Nomenclature Committee for factors of the HLA system (http://www.ebi.ac.uk/imgt/hla/). The number of known alleles is continuing to increase because of the use of sequence-based typing methods in many HLA laboratories. Most of the new alleles are detected only once and the new mutations are mainly concentrated in the exons encoding the antigen-presentation domains. However, to date, the sequencing of introns is not included in most settings of sequence-based typing, and extensive intron sequencing would probably allow the detection of even more new alleles with intronic mutations.

The set of HLA alleles inherited from one parent is referred to as a haplotype and is located on chromosome 6. On the haplotypes, certain alleles are in linkage disequilibrium, meaning that certain alleles occur together with a greater frequency than would be expected by chance. This is more frequently observed between closely located loci such as HLA-B and C genes or HLA-DRB1 and DQB1, but also between alleles of the same region (DPA1 and DPB1, DQA1 and DQB1). Two hot spots of recombination also exist in the HLA region, between HLA-DP and DQ, and HLA-A and B regions. Certain haplotypes are common in particular ethnic groups; in HSCT, the probability of identifying an HLA-matched unrelated donor is higher when patient and donor originate from the same ethnic group.

## 3. HLA TYPING TECHNIQUES

### 3.1 Serological Typing

HLA polymorphism was first explored by a serological method, microlymphocytotoxicity typing, which is based on the recognition of shared epitopes on the HLA molecules by specific sera or monoclonal antibodies.[2] It defines a low resolution typing (also called generic typing), and is still in use in many laboratories, especially for organ transplantation. This method is also useful to clarify the absence on the cell surface of some null alleles detected by molecular typing. Microlymphocytotoxicity defines 22 HLA-A, 36 HLA-B, 9 HLA-C, 14 HLA-DR, and 6 HLA-DQ specificities, but is completely inadequate to cover the diversity of the HLA system.

## 3.2 DNA-based Techniques[3,4]

The DNA-based techniques for HLA typing were introduced to alleviate the deficiencies of serological typing. They are based on the nucleotide sequence information of the polymorphic DNA segments, using polymerase chain reaction (PCR) technology. A number of HLA typing methods have been developed, mainly PCR-SSP (sequence-specific primers), reverse PCR-SSOP (sequence-specific oligonucleotide probes), or PCR-SBT (sequence-based typing). PCR is an "in vitro" method of nucleic acid synthesis by which a particular segment of DNA can be specifically replicated. It involves two primers that flank the sequence of the DNA fragment to be amplified. These primers hybridize to opposite strands of the target sequence and are oriented so that DNA synthesis by the enzyme Taq polymerase proceeds across the region between the primers. The use of cycles of heat denaturation, annealing of the primers to their complementary sequences, and extension of the annealed primers by DNA polymerase, results in an exponential accumulation of the specific target fragment, about $2n$ (where $n$ is the number of cycles of amplification).

Different levels of HLA typing resolution were defined: the *low resolution* or two-digit typing (e.g., HLA-A*02) provides a definition of a broad group of alleles that corresponds to a serologically defined antigen. The *intermediate resolution* typing defines a combination of specific alleles. The use of National Marrow Donor Program (NMDP) codes (http://bioinformatics.nmdp.org/HLA/hla-res-idx.html) can be helpful in this setting to describe allele ambiguities (A*68:02 or 68:08 or 68:09 or 68:18N = A*68UDD). The *high resolution* typing is defined as "identifying HLA alleles at a level of resolution which defines the first and second fields according to WHO Nomenclature, at least resolving all ambiguities resulting from polymorphisms located within exons 2 and 3 for HLA class I loci, and exon 2 for HLA class II loci, and all ambiguities that encompass a null allele" (standards of the European Federation for Immunogenetics). Finally, the *allelic* or four-digit typing (e.g., A*02:01) discriminates the individual alleles and resolves the tissue type to the allele level, with no ambiguity.

### 3.2.1 PCR-SSP Technique

The principle of PCR-SSP is the development of typing systems based on the extension of primers with 3′ ends that are either matched or mismatched with the target sequence. Completely matched primers will be efficiently used in the PCR reaction. However, primers with a 3′ single-base mismatched with the target sequence do not amplify efficiently. So, choosing primers in polymorphic regions allows specific amplification. Commercial SSP trays of HLA class I and II contain multiple pairs of PCR primers that are designed to anneal within DNA regions present in certain alleles or groups of alleles. If the corresponding alleles are present, their specific amplicons can be detected by gel electrophoresis. The number of primer pairs needed to perform typing is a function of the gene polymorphism and also the level of typing resolution required. While somewhat cumbersome as a method, SSP is still used for emergency HLA typing and to resolve ambiguities that result from PCR-SSO or -SBT methods.

### 3.2.2 Reverse Dot Blot PCR-SSOP Technique

One of the first widely employed methods for detecting HLA alleles was PCR-SSO typing. In this approach, the PCR primers were selected from a conserved region on the exon of a particular locus to give locus-specific amplification. To analyze the sequence polymorphisms in the amplified exons, oligonucleotide probes are hybridized under conditions where a probe complementary to the target genomic sequence would bind and a probe with a single destabilizing mismatch would not. A typical pattern of oligonucleotide probes hybridized to the amplified DNA is necessary to define an allele or a group of alleles. The resulting probe hybridization pattern is interpreted as an HLA genotype with appropriate genotyping software. In the reverse dot blot PCR-SSO technique, PCR product is labeled with biotinylated primers and hybridized to a panel of immobilized probes. This process makes SSO typing with large numbers of probes easy. In the early approaches, the oligonucleotide probes were blotted onto a solid media, which was often a nylon membrane. This concept of labeled amplicons and immobilized probes was later extended from nylon membranes to beads, as in the widely used flow cytometry Luminex HLA typing system, and to microarray assays. These latter media, which allow the fixation of many probes, are well adapted to the increasing need of probes in relation to the increasing number of HLA alleles. The reverse dot blot PCR-SSO technique is well adapted to large series of HLA typing and is able to give low or intermediate resolution typing according to the number of probes used.

### 3.2.3 PCR-SBT Typing

The great advantage of SBT is that it allows a complete determination of the exon sequences and not only of the polymorphic regions, as do PCR-SSO and SSP techniques. It is the only technique that directly detects the nucleotide sequence of an allele, thus allowing exact assignment. The technique by Sanger that uses dye terminator chemistry is most commonly used for routine HLA typing. DNA sequencing developed by Sanger takes advantage of using 2′,3′ dideoxynucleotides as substrates such that when it is incorporated at the 3′ end of the growing chain, chain elongation is stopped selectively at A, C, G, T, because the chain lacks a 3′ hydroxyl group. Four different dyes are used to identify the A, C, G, and T extension reaction, and each dye emits light at a different wavelength when excited by a laser. Amplification and direct sequencing by PCR-SBT of the relevant exons constitute one approach to high-resolution typing. However, just as for SSO typing, ambiguities resulting from *cis/trans* assignment of base calls in heterozygous samples represent a substantial source of ambiguities. Complementary typing using PCR-SSP might be necessary to reach high resolution typing. Another way to deal with this problem is the practice of workflows involving a preliminary typing by SSO or SSP, followed by the separate amplification of the two alleles of a heterozygote and the sequencing of the separated alleles. However, given the large and continually increasing number of alleles in the HLA sequence database, dealing with the ambiguity of most HLA typing methods remains a critical challenge.

### 3.2.4 Next-generation Sequencing

One technical approach to the issue of ambiguity is to harness the power of next-generation sequencing methods to achieve very high resolution and very high-throughput HLA typing. Three next-generation sequencing systems exist and the resolution of HLA genotyping ambiguity is facilitated by two properties of all next-generation sequencing systems: (1) the "clonal sequencing of DNA molecule," which allows setting the phase of linked polymorphisms within the amplicon; and (2) the "massively parallel sequencing," which allows determination of a very large number of sequences in a single run and, thus, the capability of providing many exonic and intronic sequences per gene.

In the next few years, new sequencing techniques will undoubtedly allow the sequencing of HLA class I and class II genes at an allelic resolution in a faster, more automated, and more cost-effective way.

## 4. CBT RESULTS

Since the first report of a successful CBT in 1988, great interest has developed in the use of CB as an alternative stem cell source to treat cancer and genetic diseases. The parameters that affect clinical outcome have also been intensely studied in this clinical setting. As CBT moved into ever more clinical settings (from sibling HLA-matched donors to unrelated mismatched donors, from children to adult patients, from single to double CBT, from myeloablative to reduced intensity conditioning regimens), the clinical and biological characteristics that affect CBT have been more defined.

### 4.1 Impact of HLA in Myeloablative Single Unrelated Unit CBT for Malignant Hematological Diseases in Children and Adults

Very rapidly after the first unrelated single unit CBT in 1996, retrospective studies showed that unrelated CBT (UCBT) is an acceptable alternative to HLA unrelated BMT. UCBT is able to reconstitute hematopoiesis, but with a delayed recovery, is not associated with a higher incidence of GvHD despite increased HLA disparities, and does not result with a higher risk of relapse[5] compared to UDBMT. A recent retrospective analysis of the data concerning 1525 adult patients with acute leukemia transplanted between 2002 and 2006 with CB ($n = 165$), PBSC ($n = 888$), or bone marrow ($n = 472$), revealed that the leukemia-free survival after CBT with 4–6/6 HLA match was comparable with that after 8/8 and 7/8 allele-matched unrelated donor transplantation. However, transplant-related mortality (TRM) was higher after UCBT than after 8/8 allele-matched unrelated donor transplantation. Grade 2–4 acute and chronic GvHD were lower in CB compared with 10/10 allele-matched unrelated donor transplantation, while the incidence of chronic, but not acute GvHD, was lower after CB than after 8/8 allele-matched BMT. These data support the use of UCB for malignant diseases when a fully matched donor is unavailable, and when a transplant is needed urgently.[6]

In earlier series, two factors were recognized as important for hematopoietic recovery: cell dose measured as total nucleated cells (TNC); and HLA-matching (defined as matched or mismatched based on HLA-A and -B low resolution and HLA-DRB1 high resolution typing). In 1997, the Eurocord group first described the association of TNC dose and HLA-match with neutrophil and platelet recovery and survival in 143 patients, mostly children, receiving a related or unrelated CBT. A closer HLA-match was associated with a better engraftment and survival.[7] These results have since been confirmed in a series of 562 children and adults who received unrelated CBT.[8] The Minnesota group also showed that the number of HLA disparities (6/6 and 5/6 vs 4/6 or more) was associated with better engraftment and lower transplant related mortality. In addition, 1-year mortality was increased two times after CBT for patients receiving a 2 or 3/6 HLA graft. These two early studies clearly highlighted that HLA-matching and cell dose are crucial factors for improving the outcome of CBT, although they predominantly involved children.

Other studies reported the outcome and risk factors of single unit unrelated CBT in adults and demonstrated the feasibility of CBT with delayed hematopoietic engraftment.[9–11] The impact of cell dose in these studies was also found to be associated with engraftment or survival in different cohorts after myeloablative conditioning regimen,[12,13] but the number of HLA disparities was not associated with any outcome. This observation was probably related to the small number of patients reported or to the advanced disease status of a large portion of the studied patients (53% in Arcese et al. study 2006).[12,13] The same phenomenon exists in UDBMT: the impact of HLA-mismatches does not appear in patients with advanced disease despite a strong impact observed in patients with early disease and the same observation could also exist in UCBT.

Two large more recent studies have examined cell dose and HLA in conjunction with the outcome of single unit CBT for malignant diseases after myeloablative conditioning regimen. The COBLT study (191 pediatric patients with hematologic malignancies that had at least a 3/6 HLA match CB) found that both HLA and cell dose played a role in engraftment: cell dose was positively associated with neutrophil and platelet recovery, whereas better HLA-matching was associated with neutrophil but not platelet engraftment. This study also found that disparate HLA matching was significantly associated with grade III–IV acute GvHD (5 or 6/6 matched vs 4/6 or less). Survival was impacted by the total precryopreserved TNC dose (>2.5 TNC × $10^7$/kg) but not by HLA-matching. Even if HLA-matching decreases GvHD, overall survival may not be affected because of competing contributions of GvHD and graft-versus-leukemia.[14]

A cohort study of the New York Blood Center's CBT (1061 patients who received single-unit myeloablative CBT for leukemia and myelodysplasia) found the impact of HLA-matching on neutrophil and platelet engraftment, aGvHD, TRM, treatment failure, and overall mortality. The study of the combined impact of TNC dose and HLA-match demonstrated that HLA match can compensate for low TNC dose: the best outcome for neutrophil and platelet engraftment, aGvHD, TRM, treatment failure, and overall mortality was associated with the transplantation of 0-MM units ("6/6" HLA-matched), regardless of the TNC dose; the next best survival outcomes (TRM, treatment failure, and overall mortality) were observed in recipients of 1-MM with a TNC dose of 2.5 × $10^7$/kg or greater, or 2-MM units with a TNC dose of 5.0 × $10^7$/kg or greater. Importantly, there was no difference in survival outcomes between 1-MM units that provided a TNC dose of 2.5–4.9 × $10^7$/kg and 2-MM units that provided a dose of 5.0 × $10^7$/kg or greater. However, the extent to which better HLA-match can compensate for low TNC dose has limits. No advantage was detected in TRM for 1-MM units over 2-MM units when the TNC dose was less than 2.5 × $10^7$/kg.[15] A more recent Japanese study confirmed the effect of HLA disparities on mortality in 498 children, but not in 1880 adults, who received a single CBT.[16]

## 4.2 Impact of HLA in Reduced Intensity Conditioning Regimen and Double Unrelated Unit CBT for Malignant Hematological Diseases

CBT is often used as a "last chance" transplant due to high TRM and low engraftment rates. Early efforts to reduce TRM and low engraftment focused on reducing the conditioning regimen toxicity and increasing the CB cell dose. Reduced intensity conditioning (RIC) showed reduced early mortality and allowed treatment of older patients.

To abrogate the obstacle of cell dose in adult patients, Barker and colleagues showed that two partially HLA-matched CB units could be infused into a patient and that this infusion enhanced engraftment in adults with hematologic malignancy. Only one CB unit eventually provided predominant hematopoietic reconstitution.[17] After RIC regimen, rapid and complete donor chimerism has also been achieved in adult recipients of CBT.[18] Following the use of RIC regimen and double CB transplants, the number of adult patients transplanted with CB has increased. Brunstein and colleagues demonstrated the effectiveness of double CBT (dCBT) after myeloablative conditioning regimen in 128 dCBT. The authors showed that leukemia-free survival after dCBT is comparable to that observed after matched-related and matched-unrelated transplantation, and that the higher risk of TRM is counterbalanced by lower relapse rates.[19] They suggested that for patients without an available HLA-matched donor, the use of two partially HLA-matched UCB units is a suitable alternative. MacMillan and coworkers retrospectively compared the outcomes of single versus double CBT. Although cGvHD rates were similar (17% vs 18%), grade II–IV aGvHD rates were greater in the double CBT group (58% vs 39%, $p < 0.01$)[20] but they did not find an effect of HLA matching of the engrafting unit on aGvHD. A higher risk of grade II–IV aGvHD associated with less relapse was also observed by Veineris et al. in recipients receiving dCBT.[21] The study of 110 patients with hematological diseases treated mostly by dCBT ($n = 93$) after a nonmyeloablative regimen showed that neutrophil recovery was high and achieved in 92%, but that neither TNC, $CD34^+$, $CD3^+$ cell counts, nucleated cell viability, nor HLA-matching were predictive of engraftment.[22] When the effect of HLA-match on engraftment after dCBT was examined, single-unit dominance was observed in almost all double CBT outcomes.[23] A high incidence of sustained donor engraftment was observed (93%) with a median time to neutrophil recovery of 23 days. Engraftment was associated with TNC dose and $CD34^+$ cells infused and also with the viability of the $CD34^+$ cells post thaw, but not with HLA disparity even at high resolution typing.[23] Furthermore, after dCBT with a myelo- or nonmyeloablative conditioning regimen, grade III to IV aGvHD incidence was found to be lower if the engrafting unit HLA-A, -B, -DRB1 allele match was >4/6 to the recipient, whereas engrafting unit infused TNC and unit-to-unit HLA-match were not significant.[24]

The Eurocord registry in collaboration with the French Society for blood and marrow transplantation and cellular therapy (SFGM-TC) addressed the feasibility of RIC using a single CBT for hematological malignancies ($n$ = 176) and studied a cohort of 155 single or double CBT with a homogenous RIC regimen ($n$ = 155). The results suggest that TNC, $CD34^+$ cell dose, and the number of HLA disparities continue to play a critical role in engraftment and risk of TRM after CBT in the RIC setting.[25] Several studies compared single versus dCBT in adult patients but results are somehow contradictory, a controversy far from being resolved. Indeed, the influence of HLA-matching and TNC on engraftment as well as other transplantation outcomes after dCBT or RIC regimen should be retackled in further studies using larger and more homogeneous series of patients. This would allow the drawing of definitive conclusions on the impact of HLA matching in dCBT as well as in RIC setting. Although double unit grafts have been widely adopted as an easy strategy to increase graft cell doses in unrelated CBT, there is still little information to guide transplant centers in the selection of the graft. Nevertheless, based on the importance of TNC dose and HLA-match demonstrated in large analyses of single-unit CBT[15] as well as a recent analysis of 84 dCBT recipients,[23] current opinion with regard to CB selection recommends screening for grafts with 4–6 HLA-matches and $>2.0 \times 10^7$ cells/kg for each unit of a double-unit graft. This approach increases the likelihood of engraftment by infusing 2 units with at least an adequate cell dose in each unit. This recommendation may need to be revised in the future when much larger numbers of dCBT recipients will be available for analysis.[26]

## 4.3 Impact of HLA in CBT for Nonmalignant Diseases

Nonmalignant diseases comprise acquired aplastic anemia and inborn diseases including hemoglobinopathies, hereditary metabolic disorders, and primary immunodeficiencies.

### 4.3.1 CBT in Patients with Hemoglobinopathies and Inherited Bone Marrow Failures

In view of the low incidence of GvHD associated with the procedure, allogeneic CBT is particularly appealing for patients with inherited and primary immunodeficiency disorders. Several studies showed that survival outcomes after matched related CBT were comparable to matched related BMT with a lower risk of GvHD, and demonstrated that related donor CBT is a safe and effective option for patients with hemoglobinopathies.[27–30] Based on this, Locatelli and colleagues recommended collection and freezing of CB units in families in which a child is affected with genetic or hematological disease.[28] The same observation was reported for inherited bone marrow failures in a Eurocord/EBMT (European Society for Blood and Marrow Transplantation) study. In patients with hereditary bone marrow failure syndromes, related CBT was associated with excellent outcomes (cumulative incidence of neutrophil and platelet recovery at 60 days is 95% and 90%, respectively, and a low rate of acute grade III-IV and chronic GvHD of 5% and 10%, respectively).[31]

CBT using unrelated HLA mismatched units were performed in small cohorts or in single cases with encouraging results; this suggests that CBT should be considered in patients if they lack a suitable matched-related donor.[32,33] The outcome of HLA-mismatched unrelated CBT after a myeloablative ($n$ = 30) or RIC ($n$ = 14) conditioning regimen in 44 patients suffering from inherited bone marrow failure has been recently reported. The risk of graft failure and chronic GvHD in this cohort was high (50% and 53% respectively).[31] In order to overcome the risk of rejection, the authors recommended a selection of the CB units similar to that for aplastic anemia: high number of nucleated cells infused (>4 × 107 CNT/kg) and no more than one HLA mismatch. The results of CBT performed in 88 patients with primary immunodeficiency patients (SCID ($n$ = 40), Wiskott-Aldrich syndrome ($n$ = 23), chronic granulomatous disease ($n$ = 7), severe congenital neutropenia ($n$ = 5), and other immunodeficiencies ($n$ = 13)) were reported and confirmed the negative impact of two or more HLA mismatches.[34]

Altogether these studies emphasize the importance of both HLA matching and TNC dose, but the use of unrelated CB for patients with inherited bone marrow failure still requires larger studies and better consideration of the conditioning regimens.

### 4.3.2 CBT in Patients with Severe Acquired Aplastic Anemia

Recently, Eurocord/EBMT collaboration reported results of 71 UCBT performed for severe acquired aplastic anemia (SAA) using a single or double CB, with an RIC in 68% of patients. This study highlights the fundamental role of cell dose for both engraftment and overall survival (OS) in patients with SAA undergoing CBT. Higher prefreezing TNC dose ($>3.9 \times 10^7$/kg) significantly increases engraftment and overall survival.[35] The importance of TNC dose infused for engraftment and survival was already shown by Minck et al.[36]

Eurocord also studied a cohort of 279 patients with nonmalignant diseases who received a single UCBT between 1994 and 2005. A total of 40% of the patients suffered from SAA, 36% from primary immunodeficiency and 24% from hereditary metabolic disorders. Engraftment, GvHD, TRM, and OS were influenced by HLA mismatching. OS (CI of at 100 months = 49%) was influenced by cell dose and by the number of HLA mismatches. The group of patients who received a CBT with $<3.5 \times 10^7$ TNC/kg and a 2–3

HLA-mismatched transplant had <10% survival. Increasing cell dose partially abrogated the effect of HLA mismatches, and there was no statistical difference between the groups receiving >3.5 × $10^7$ TNC/kg with a 0–1, 2-, or 3-HLA-mismatched CBT.[37] Thus, for nonmalignant disorders, HLA matching is critical and patients should receive a higher cell dose to obtain engraftment than patients with a malignant disease; for patients who need a 4/6 HLA graft, a CB unit containing more than 4–5 × $10^7$/kg at collection should be targeted. It has been suggested that, for each HLA disparity, the TNC dose should be increased by an increment of 1.5 × $10^7$/kg, however this recommendation has not yet been validated.[37]

# 5. OTHER HLA CRITERIA FOR CBU SELECTION

## 5.1 HLA Mismatch Direction

Stevens examined the relationship between direction of HLA mismatch and transplantation outcomes in 1202 recipients of single CB units. Patients with mismatch in the "GvH direction only" had lower TRM (HR = 0.5, P = 0.062), overall mortality (HR = 0.5, P = 0.019), and treatment failure (HR = 0.5, P = 0.016), resulting in outcomes similar to those of matched CB grafts. In contrast, mismatches in the "rejection direction only" had slower engraftment, higher graft failure, and higher relapse rates (HR = 2.4, P = 0.010). Based on these findings, the authors recommended that transplant centers give priority to "GvH direction only" mismatched units over other mismatches and to avoid selecting "rejection direction only" mismatches, if possible.[38] However, the same study realized with 2977 Japanese patients and with 1565 patients from the Eurocord registry did not confirm the above findings.[39,40] The lack of clear evidence does not recommend changing the current practice for CBU selection.

## 5.2 HLA High Resolution Typing

Many studies stressed the importance of both HLA matching (based on HLA-A, -B at low resolution and -DRB1 at high resolution) and TNC infused; however, questions arise: which is more important and would a higher HLA matching improve CBT outcome? Various studies addressed this question. For instance, a study on 122 UCBT performed by Kogler et al. revealed the large number of actual HLA disparities when analyzed with HLA-A, -B, -C, -DRB1, -DQB1 allelic subtyping (5 cases with 0 mismatch and 12 with 1 mismatch); the authors conclude that there is no benefit from additional high resolution typing for HLA-A, -B, -C, -DRB1, -DQB1 locus for outcomes after UCBT in this limited and heterogeneous group of patients.[41] However, the number of matched or with 1-mismatch unit CBT was probably too small to allow detecting the effect of HLA disparity. Kurtzberg et al. showed in 191 children suffering from hematologic malignancy that high resolution matching for HLA-A, -B, -DRB1 locus increased the risk of severe acute GvHD if the donor–recipient pairs were matched for fewer than 5 of 6 alleles.[14] Nonetheless, in a larger cohort of 803 patients transplanted with a single UCB for leukemia and myelodysplastic syndrome, Eapen et al. found that a mismatch at HLA-C locus increased the risk of TRM in HLA-A, -B, -DRB1 matched transplants, and that the risk of TRM was higher with mismatches at two, three or four loci compared with matched units. The risk of TRM was not significantly different between matched and one mismatch units.[42] More recently, in a larger cohort of 1568 single UCBT for hematologic malignancy, Eapen M and colleagues also reassessed the overall effect of HLA disparity based on high resolution typing for HLA-A, -B, -C, and -DRB1 loci on nonrelapse mortality. Only 7% of units were allele matched at HLA-A, -B, -C, -DRB1; 15% were mismatched at one, 26% at two, 30% at three, 16% at four, and 5% at five alleles. NRM was higher with units mismatched at more than one allele compared to HLA-matched units ($p$ <0.001). The observed effects were independent of cell dose and patient age. These data supported allele-level HLA-matching in the selection of single UCB units.[43] Accordingly, matching for HLA-C and for HLA class I alleles should be included in the CB unit selection when possible, in order to minimize the mortality risks.

## 5.3 (HLA) Noninherited Maternal Antigen

During pregnancy, bidirectional transplacental trafficking of maternal cells into the fetus exposes him or her to maternal cells expressing both inherited maternal antigens and noninherited maternal antigens (NIMA), resulting in the development of NIMA specific responses. Two recent studies evaluated the impact of fetal exposure to NIMA on the outcome of unrelated CBT. The first one included 1121 patients with hematologic malignancies who received single unit CBT and demonstrated a survival advantage to choosing CBU in which there was a match between the patient and the NIMA of the unit.[44] The second study compared 48 NIMA-matched to 116 non-NIMA-matched pairs and confirmed the survival advantage of the NIMA-matched CBT (55% vs 38% respectively).[45] These observations suggested that in utero exposure of the fetus to NIMA antigens induces regulator T cells to those maternal antigens. Consequently, when faced with the choice of multiple HLA-mismatched UCB containing adequate cell doses, selecting an NIMA-matched CB may improve survival after mismatched CBT.

However, in a mouse model, it was shown that exposure to NIMA antigens could lead to tolerance induction or sensitization,[46–48] and that the difference of reactivity was influenced by the amount of maternal microchimerism and the timing of exposure. An in vitro test could predict the reactivity to NIMA antigens by the level of IFN-γ producing cells.[46] Importantly, however, in human, the graft-versus-leukemia effect is not abrogated, as demonstrated by the low rate of relapse observed in patients transplanted for myeloid leukemia with NIMA matched transplants.[44] This might be explained in part by the sensitization of CB cells against NIMA antigens. In fact, regulatory T cells could coexist with cytotoxic T cells in CB similar to that which has been shown for the minor histocompatibility antigens HA-1 in healthy individuals.[49] Another possibility is that during pregnancy, the maternal T cells have been exposed and sensitized against inherited paternal antigens (IPA) expressed on the fetal cells that enter the maternal circulation. Some of these maternal T cells can cross the placenta and enter the fetus where they can persist for a long time. When infused with the CBU, these cells might reduce the risk of relapse and possibly increase the risk of GvHD in recipients bearing the IPA antigens. This hypothesis was supported by the fact that patients who share HLA antigens with their graft's IPAs had lower risk of relapse in CB transplant than those without shared antigens.[50]

## 5.4 Anti-HLA Antibodies

Anti-HLA antibodies may occur due to the allo-immunization against HLA through blood transfusions, pregnancy, and organ transplantations, but also in some unexposed individuals.[51,52] In patients with hematologic diseases, anti-HLA antibodies may be more frequently detected because of the frequent use of transfusion therapy.[51] In a series of 294 patients, Ruggeri et al. estimated the incidence of preformed anti-HLA antibodies to be 23% among individuals being considered for CBT.[53] The impact of donor specific antibody (DSA) on outcome has been extensively described in solid organ transplantation,[54] in which HLA matching is less respected than in HSCT. In a case control study of unrelated donor recipients, Spellman et al. reported pretransplant DSA to be associated with graft failure and higher mortality.[55] In the CBT setting, three series reporting respectively 386,[56] 73,[57] and 294 patients[53] receiving single or double CBT showed an increased risk of graft failure and lower survival for patients with DSA. Only one report of 126 double UCBT showed no association between the presence of DSA and transplant outcome.[58] Moreover, the association between graft failure and antibody response proves stronger in case of intense antibody response.[53,57] It is therefore preferable to consider screening and identification of specific anti-HLA antibodies when selecting a CBU. Units against which there are very intense antibody responses should be particularly avoided. DSA search is an effective tool to be included in the algorithm for the best CBU selection.

## 5.5 Non-HLA Immunogenetics

### 5.5.1 ABO Blood Group

ABO major incompatibility was found to be associated with a worse outcome in HSCT from related and unrelated donors. In CBT, a recent study did not observe an impact of ABO incompatibility on aGvHD or cGvHD in recipients of single and dCBT, both in univariate and multivariate analysis.[59] However, this has not been confirmed by other studies; ABO major incompatibility has been associated with decreased survival and disease-free survival rates in adults with hematological malignancies,[13,60] and with higher TRM after RIC UCBT.[25] Therefore, when several CBU are available, the use of a unit that is ABO compatible should be taken into consideration.

### 5.5.2 Killer Cell Immunoglogulin-like Receptor

Natural killer (NK) cells are part of the innate immune system and are involved in viral immunity and cancer surveillance. The physiology of NK cells is finely regulated, to control cytotoxicity and cytokine production. NK cells express on their surface inhibitory killer cell immunoglobulin-like receptors (KIR) that recognize allotypic determinants (KIR ligand) shared by certain HLA alleles. KIR 2DL1 recognizes HLA-C alleles coding for a Lysine residue in position 80 (group C2 HLA-C alleles) and KIR2DL2 or KIR2DL3 recognize HLA-C alleles coding for an asparagine residue in position 80 (group C1 HLA-C alleles). KIR 3DL1 is the receptor for HLA-B alleles sharing Bw4 supertypic specificity and KIR3DL2 was shown to function as a receptor for HLA-A3/-A11. NK cells expressing inhibitory KIR receptors that do not recognize ligands on the target cells are released from HLA inhibition and mediate cytotoxicity and cytokine production, leading to target destruction. In allogeneic HSCT, NK alloreactivity is determined by the specificity of the KIR receptors on the donor cells for recipient HLA class I. Some donors have a subset of NK cells that do not express inhibitory KIRs recognizing their cognate HLA class I ligand on recipient cells. This corresponds to a KIR-ligand mismatch situation, leading to the risk of allogeneic reaction. Studies in haploidentical transplants showed that KIR ligand incompatibilities in the GvH direction were associated with decreased incidence of relapse and an improved disease-free and overall survival.[61] However, several studies concerning HSCT with unrelated donors did not reveal any impact. In UCBT for acute leukemia, KIR ligand incompatibilities were associated with a decreased incidence of relapse and an improved disease free and overall survival.[62] However, Burnstein and coworkers did not confirm this effect and rather observed a significant association between KIR-ligand incompatibilities and increased risk of grade III–IV GvHD and death.[63] A recent study showed that KIR-ligand incompatibility was not associated with relapse reduction after double umbilical cord blood transplantation.[64] Currently, the impact of KIR-ligand incompatibilities remains unclear. Other series of patients and functional studies analyzing the impact of KIR-ligand matching are needed before this factor can be included in the algorithm of CBU selection.

**Recommendations for Cord Blood Unit (CBU) Selection in 2014. Three Criteria are Major for the Choice of CBU: Human Leukocyte Antigen (HLA) Matching, Total Nucleated Cell Dose at Freezing, and Patient HLA Antibodies**

**Human Leukocyte Antigen (HLA) Match**

- Compatibility 6/6, 5/6, or 4/6
  - Antigenic HLA-A and B, allelic DRB1
- Prefer HLA-A and B mismatches to DRB1
- For *nonmalignant diseases*:
  - Matching is particularly critical and DRB1 mismatch should be avoided
- *Double cord blood transplantation* (*CBT*):
  - 4/6 minimum between the two units

**Total Nucleated Cell (TNC) Dose at Freezing**

|  |  |
|---|---|
| Compatibility 6/6 or 5/6 | • Malignant diseases: ≥3 × 10e7 TNC/kg<br>• Nonmalignant diseases: ≥3.5 × 10e7 TNC/kg |
| Compatibility 4/6 | • Malignant diseases: ≥3.5 × 10e7 TNC/kg<br>• Nonmalignant diseases: ≥4 × 10e7 TNC/kg |

➠ If these cell doses are not met with one unit, a *double CBT* should be performed
≥2 × 10e7 TNC/kg for each unit

**Anti-HLA Antibodies**

Exploration of anti-HLA immunization by a sensitive method for all recipients
➠ Units against which there are a very intense antibody response should be particularly avoided

**Other Criteria for CB Selection**

When faced with multiple cord blood units (CBU) that fit the above criteria, the following criteria should be considered:

- *HLA high resolution typing*: Matching for HLA-C (-DQB1) and for HLA class I alleles should be included in the CBU selection HLA mismatch in the "GvH direction only" should be preferred.
- *Dose of CD34+ cells* ≥1.5 × $10^5$/kg
- *ABO matching*: The use of ABO compatible units should be taken in consideration
- *Noninherited maternal antigen match*: Select an NIMA-matched CBU?

### 5.5.3 Cytokine Polymorphisms and Minor Histocompatibility Antigens

In HLA-identical BMT, certain recipient cytokine gene polymorphisms and minor histocompatibility antigen (mHag) differences influence the occurrence and severity of acute GvHD. Little was done concerning the impact of cytokine polymorphisms and mHag in CBT. Nevertheless, the impact of TNF-α and IL-10 polymorphism genotypes and of H-Y, HA1, and CD31 mHag incompatibilities on acute GvHD occurrence was analyzed in a cohort of 115 unrelated mismatched CBT,[65] but no impact was detected. However, Mommas et al. observed pre-existing and ex vivo-generated specific T cells directed against the maternal mHag HA1 in CB, indicating that CB T cells can elicit cytotoxic T cells against maternal mHag which can contribute to the in vivo graft-versus-leukemia activity after transplantation.[66] The absence of clinical impact observed in the studied CBT cohort might be explained in part by the fact that the HLA mismatch effects are very strong and could mask the effect of cytokine polymorphisms or mHag incompatibility.

## 6. CONCLUSION

The genetic risks associated with HLA loci and allele histoincompatibility represent a central and critical issue in HSCT.[67,68] While originally considered as a less stringent requirement in CBT compared to BMT, all recent data point out the benefit of a full HLA matching in this context. The newer data provide a rationale to implement comprehensive pretransplant genetic assessment in order to lower the risks of GvHD, infections, graft failure, relapse, TRM, and overall survival. System biology approaches based on data from experimental, clinical epidemiology, and modeling will undoubtedly bring a forward looking strategy in this field in order to establish the tailored and personalized treatment unique to each patient.[69]

## LIST OF ABBREVIATIONS

**aGvHD** Acute graft versus host disease
**BMT** Bone marrow transplantation
**CB** Cord blood
**CBT** Cord blood transplantation
**CBU** Cord blood unit
**cGvHD** Chronic graft versus host disease
**dCBT** Double cord blood transplantation
**DSA** Donor specific antibodies
**EFI** European Federation for Immunogenetics
**GvHD** Graft versus host disease
**HLA** Human leukocyte antigen
**HR** Hazards ratio
**HSC** Hematopoietic stem cell
**HSCT** Hematopoietic stem cell transplantation
**IMA** Inherited maternal antigen
**IPA** Inherited paternal antigen
**KIR** Killer cell immunoglogulin-like receptor
**mHag** Minor histocompatibility antigen
**MHC** Major histocompatibility complex
**NIMA** Noninherited maternal antigen
**OS** Overall survival

**PBSC** Peripheral blood stem cell
**PCR** Polymerase chain reaction
**RIC** Reduced intensity conditioning
**SAA** Severe acquired aplastic anemia
**SBT** Sequence-based typing
**SSO** Sequence-specific oligonucleoprobe
**SSP** Sequence-specific primer
**TNC** Total nucleated cell
**TRM** Transplant related mortality
**UCBT** Unrelated cord blood transplantation
**UDBMT** Unrelated donor bone marrow transplantation

## REFERENCES


1. Gluckman E, Broxmeyer HA, Auerbach AD, Friedman HS, Douglas GW, Devergie A, et al. Hematopoietic reconstitution in a patient with Fanconi's anemia by means of umbilical-cord blood from an HLA-identical sibling. *N Engl J Med* 1989;**321**(17):1174–8.
2. Terasaki PI, McClelland JD. Microdroplet assay of human serum cytotoxins. *Nature* 1964;**204**:998–1000.
3. Bontadini A. HLA techniques: typing and antibody detection in the laboratory of immunogenetics. *Methods* 2012;**56**(4):471–6.
4. Erlich H. HLA DNA typing: past, present, and future. *Tissue Antigens* 2012;**80**(1):1–11.
5. Rocha V, Cornish J, Sievers EL, Filipovich A, Locatelli F, Peters C, et al. Comparison of outcomes of unrelated bone marrow and umbilical cord blood transplants in children with acute leukemia. *Blood* 2001;**97**(10):2962–71.
6. Eapen M, Rocha V, Sanz G, Scaradavou A, Zhang MJ, Arcese W, et al. Effect of graft source on unrelated donor haemopoietic stem-cell transplantation in adults with acute leukaemia: a retrospective analysis. *Lancet Oncol* 2010;**11**(7):653–60.
7. Gluckman E, Rocha V, Boyer-Chammard A, Locatelli F, Arcese W, Pasquini R, et al. Outcome of cord-blood transplantation from related and unrelated donors. Eurocord Transplant Group and the European Blood and Marrow Transplantation Group. *N Engl J Med* 1997;**337**(6):373–81.
8. Rubinstein P, Carrier C, Scaradavou A, Kurtzberg J, Adamson J, Migliaccio AR, et al. Outcomes among 562 recipients of placental-blood transplants from unrelated donors. *N Engl J Med* 1998;**339**(22):1565–77.
9. Laughlin MJ, Eapen M, Rubinstein P, Wagner JE, Zhang MJ, Champlin RE, et al. Outcomes after transplantation of cord blood or bone marrow from unrelated donors in adults with leukemia. *N Engl J Med* 2004;**351**(22):2265–75.
10. Ooi J, Iseki T, Takahashi S, Tomonari A, Takasugi K, Shimohakamada Y, et al. Unrelated cord blood transplantation for adult patients with de novo acute myeloid leukemia. *Blood* 2004;**103**(2):489–91.
11. Rocha V, Labopin M, Sanz G, Arcese W, Schwerdtfeger R, Bosi A, et al. Transplants of umbilical-cord blood or bone marrow from unrelated donors in adults with acute leukemia. *N Engl J Med* 2004;**351**(22):2276–85.
12. Laughlin MJ, Barker J, Bambach B, Koc ON, Rizzieri DA, Wagner JE, et al. Hematopoietic engraftment and survival in adult recipients of umbilical-cord blood from unrelated donors. *N Engl J Med* 2001;**344**(24):1815–22.
13. Arcese W, Rocha V, Labopin M, Sanz G, Iori AP, de Lima M, et al. Unrelated cord blood transplants in adults with hematologic malignancies. *Haematologica* 2006;**91**(2):223–30.
14. Kurtzberg J, Prasad VK, Carter SL, Wagner JE, Baxter-Lowe LA, Wall D, et al. Results of the cord blood transplantation study (COBLT): clinical outcomes of unrelated donor umbilical cord blood transplantation in pediatric patients with hematologic malignancies. *Blood* 2008;**112**(10):4318–27.
15. Barker JN, Scaradavou A, Stevens CE. Combined effect of total nucleated cell dose and HLA match on transplantation outcome in 1061 cord blood recipients with hematologic malignancies. *Blood* 2010;**115**(9):1843–9
16. Atsuta Y, Kanda J, Takanashi M, Morishima Y, Taniguchi S, Takahashi S, et al. Different effects of HLA disparity on transplant outcomes after single-unit cord blood transplantation between pediatric and adult patients with leukemia. *Haematologica* 2013;**98**(5):814–22.
17. Barker JN, Weisdorf DJ, DeFor TE, Blazar BR, McGlave PB, Miller JS, et al. Transplantation of 2 partially HLA-matched umbilical cord blood units to enhance engraftment in adults with hematologic malignancy. *Blood* 2005;**105**(3):1343–7.
18. Barker JN, Weisdorf DJ, DeFor TE, Blazar BR, Miller JS, Wagner JE. Rapid and complete donor chimerism in adult recipients of unrelated donor umbilical cord blood transplantation after reduced-intensity conditioning. *Blood* 2003;**102**(5):1915–9.
19. Brunstein CG, Gutman JA, Weisdorf DJ, Woolfrey AE, Defor TE, Gooley TA, et al. Allogeneic hematopoietic cell transplantation for hematologic malignancy: relative risks and benefits of double umbilical cord blood. *Blood* 2010;**116**(22):4693–9.
20. MacMillan ML, Weisdorf DJ, Brunstein CG, Cao Q, DeFor TE, Verneris MR, et al. Acute graft-versus-host disease after unrelated donor umbilical cord blood transplantation: analysis of risk factors. *Blood* 2009;**113**(11):2410–5.
21. Verneris MR, Brunstein CG, Barker J, MacMillan ML, DeFor T, McKenna DH, et al. Relapse risk after umbilical cord blood transplantation: enhanced graft-versus-leukemia effect in recipients of 2 units. *Blood* 2009;**114**(19):4293–9.
22. Brunstein CG, Barker JN, Weisdorf DJ, DeFor TE, Miller JS, Blazar BR, et al. Umbilical cord blood transplantation after nonmyeloablative conditioning: impact on transplantation outcomes in 110 adults with hematologic disease. *Blood* 2007;**110**(8):3064–70.
23. Avery S, Shi W, Lubin M, Gonzales AM, Heller G, Castro-Malaspina H, et al. Influence of infused cell dose and HLA match on engraftment after double-unit cord blood allografts. *Blood* 2011;**117**(12):3277–85. quiz 3478.
24. Ponce DM, Gonzales A, Lubin M, Castro-Malaspina H, Giralt S, Goldberg JD, et al. Graft-versus-host disease after double-unit cord blood transplantation has unique features and an association with engrafting unit-to-recipient HLA match. *Biol Blood Marrow Transpl* 2013;**19**(6):904–11.
25. Rocha V, Mohty M, Gluckman E, Rio B. Reduced-intensity conditioning regimens before unrelated cord blood transplantation in adults with acute leukaemia and other haematological malignancies. *Curr Opin Oncol* 2009;**21**(Suppl. 1):S31–4.
26. Barker JN, Byam C, Scaradavou A. How I treat: the selection and acquisition of unrelated cord blood grafts. *Blood* 2011;**117**(8):2332–9.
27. Brichard B, Vermylen C, Ninane J, Cornu G. Persistence of fetal hemoglobin production after successful transplantation of cord blood stem cells in a patient with sickle cell anemia. *J Pediatr* 1996;**128**(2):241–3.
28. Locatelli F, Rocha V, Reed W, Bernaudin F, Ertem M, Grafakos S, et al. Related umbilical cord blood transplantation in patients with thalassemia and sickle cell disease. *Blood* 2003;**101**(6):2137–43.

29. Fang J, Huang S, Chen C, Zhou D, Li CK, Li Y, et al. Umbilical cord blood transplantation in Chinese children with beta-thalassemia. *J Pediatr Hematol Oncol* 2004;**26**(3):185–9.
30. Walters MC, Quirolo L, Trachtenberg ET, Edwards S, Hale L, Lee J, et al. Sibling donor cord blood transplantation for thalassemia major: experience of the Sibling Donor Cord Blood Program. *Ann N Y Acad Sci.* 2005;**1054**:206–13.
31. Bizzetto R, Bonfim C, Rocha V, Socié G, Locatelli F, Chan K, et al. Outcomes after related and unrelated umbilical cord blood transplantation for hereditary bone marrow failure syndromes other than Fanconi anemia. *Haematologica* 2011;**96**(1):134–41.
32. Tan PH, Hwang WY, Goh YT, Tan PL, Koh LP, Tan CH, et al. Unrelated peripheral blood and cord blood hematopoietic stem cell transplants for thalassemia major. *Am J Hematol* 2004;**75**(4): 209–12.
33. Gluckman E, Rocha V, Ionescu I, Bierings M, Harris RE, Wagner J, et al. Results of unrelated cord blood transplant in fanconi anemia patients: risk factor analysis for engraftment and survival. *Biol Blood Marrow Transpl* 2007;**13**(9):1073–82.
34. Morio T, Atsuta Y, Tomizawa D, Nagamura-Inoue T, Kato K, Ariga T, et al. Outcome of unrelated umbilical cord blood transplantation in 88 patients with primary immunodeficiency in Japan. *Br J Haematol* 2011;**154**(3):363–72.
35. Peffault de Latour R, Rocha V, Socié G. Cord blood transplantation in aplastic anemia. *Bone Marrow Transpl* 2013;**48**(2):201–2.
36. Min CK, Kim DW, Lee JW, Han CW, Min WS, Kim CC. Hematopoietic stem cell transplantation for high-risk adult patients with severe aplastic anemia; reduction of graft failure by enhancing stem cell dose. *Haematologica* 2001;**86**(3):303–10.
37. Rocha V, Gluckman E. Improving outcomes of cord blood transplantation: HLA matching, cell dose and other graft- and transplantation-related factors. *Br J Haematol* 2009;**147**(2):262–74.
38. Stevens CE, Carrier C, Carpenter C, Sung D, Scaradavou A. HLA mismatch direction in cord blood transplantation: impact on outcome and implications for cord blood unit selection. *Blood* 2011;**118**(14):3969–78.
39. Cunha R, Loiseau P, Ruggeri A, Sanz G, Michel G, Paolaiori A, et al. Impact of HLA mismatch direction on outcomes after umbilical cord blood transplantation for hematological malignant disorders: a retrospective Eurocord-EBMT analysis. *Bone Marrow Transpl* 2014;**49**(1):24–9.
40. Kanda J, Atsuta Y, Wake A, Ichinohe T, Takanashi M, Morishima Y, et al. Impact of the direction of HLA mismatch on transplantation outcomes in single unrelated cord blood transplantation. *Biol Blood Marrow Transpl* 2013;**19**(2):247–54.
41. Kogler G, Enczmann J, Rocha V, Gluckman E, Wernet P. High-resolution HLA typing by sequencing for HLA-A, -B, -C, -DR, -DQ in 122 unrelated cord blood/patient pair transplants hardly improves long-term clinical outcome. *Bone Marrow Transpl* 2005;**36**(12):1033–41.
42. Eapen M, Klein JP, Sanz GF, Spellman S, Ruggeri A, Anasetti C, et al. Effect of donor-recipient HLA matching at HLA A, B, C, and DRB1 on outcomes after umbilical-cord blood transplantation for leukaemia and myelodysplastic syndrome: a retrospective analysis. *Lancet Oncol* 2011;**12**(13):1214–21.
43. Eapen M, Klein JP, Ruggeri A, Spellman S, Lee SJ, Anasetti C, et al. Impact of allele-level HLA matching on outcomes after myeloablative single unit umbilical cord blood transplantation for hematologic malignancy. *Blood* 2014;**123**(1):133–40.
44. van Rood JJ, Stevens CE, Smits J, Carrier C, Carpenter C, Scaradavou A. Reexposure of cord blood to noninherited maternal HLA antigens improves transplant outcome in hematological malignancies. *Proc Natl Acad Sci USA* 2009;**106**(47):19952–7.
45. Rocha V, Spellman S, Zhang MJ, Ruggeri A, Purtill D, Brady C, et al. Effect of HLA-matching recipients to donor noninherited maternal antigens on outcomes after mismatched umbilical cord blood transplantation for hematologic malignancy. *Biol Blood Marrow Transpl* 2012;**18**(12):1890–6.
46. Araki M, Hirayama M, Azuma E, Kumamoto T, Iwamoto S, Toyoda H, et al. Prediction of reactivity to noninherited maternal antigen in MHC-mismatched, minor histocompatibility antigen-matched stem cell transplantation in a mouse model. *J Immunol* 2010;**185**(12):7739–45.
47. Andrassy J, Kusaka S, Jankowska-Gan E, Torrealba JR, Haynes LD, Marthaler BR, et al. Tolerance to noninherited maternal MHC antigens in mice. *J Immunol* 2003;**171**(10):5554–61.
48. Matsuoka K, Ichinohe T, Hashimoto D, Asakura S, Tanimoto M, Teshima T. Fetal tolerance to maternal antigens improves the outcome of allogeneic bone marrow transplantation by a $CD4^+$ $CD25^+$ T-cell-dependent mechanism. *Blood* 2006;**107**(1):404–9.
49. van Halteren AG, Jankowska-Gan E, Joosten A, Blokland E, Pool J, Brand A, et al. Naturally acquired tolerance and sensitization to minor histocompatibility antigens in healthy family members. *Blood* 2009;**114**(11):2263–72.
50. van Rood JJ, Scaradavou A, Stevens CE. Indirect evidence that maternal microchimerism in cord blood mediates a graft-versus-leukemia effect in cord blood transplantation. *Proc Natl Acad Sci USA* 2012;**109**(7):2509–14.
51. Schonewille H, Haak HL, van Zijl AM. Alloimmunization after blood transfusion in patients with hematologic and oncologic diseases. *Transfusion* 1999;**39**(7):763–71.
52. Idica A, Sasaki N, Hardy S, Terasaki P. Unexpected frequencies of HLA antibody specificities in the sera of pre-transplant kidney patients. *Clin Transpl* 2006:161–70.
53. Ruggeri A, Rocha V, Masson E, Labopin M, Cunha R, Absi L, et al. Impact of donor-specific anti-HLA antibodies on graft failure and survival after reduced intensity conditioning-unrelated cord blood transplantation: a Eurocord, Societe Francophone d'Histocompatibilite et d'Immunogenetique (SFHI) and Societe Francaise de Greffe de Moelle et de Therapie Cellulaire (SFGM-TC) analysis. *Haematologica* 2013;**98**(7):1154–60.
54. Patel AM, Pancoska C, Mulgaonkar S, Weng FL. Renal transplantation in patients with pre-transplant donor-specific antibodies and negative flow cytometry cross matches. *Am J Transpl* 2007;**7**(10):2371–7.
55. Spellman S, Bray R, Rosen-Bronson S, Haagenson M, Klein J, Flesch S, et al. The detection of donor-directed, HLA-specific alloantibodies in recipients of unrelated hematopoietic cell transplantation is predictive of graft failure. *Blood* 2010;**115**(13):2704–8.
56. Takanashi M, Atsuta Y, Fujiwara K, Kodo H, Kai S, Sato H, et al. The impact of anti-HLA antibodies on unrelated cord blood transplantations. *Blood* 2010;**116**(15):2839–46.
57. Cutler C, Kim HT, Sun L, Sese D, Glotzbecker B, Armand P, et al. Donor-specific anti-HLA antibodies predict outcome in double umbilical cord blood transplantation. *Blood* 2011;**118**(25):6691–7.
58. Brunstein CG, Noreen H, DeFor TE, Maurer D, Miller JS, Wagner JE. Anti-HLA antibodies in double umbilical cord blood transplantation. *Biol Blood Marrow Transpl* 2011;**17**(11):1704–8.

59. Romee R, Weisdorf DJ, Brunstein C, Wagner JE, Cao Q, Blazar BR, et al. Impact of ABO-mismatch on risk of GVHD after umbilical cord blood transplantation. *Bone Marrow Transpl* 2013;**48**(8):1046–9.
60. Michallet M, Le QH, Mohty M, Prébet T, Nicolini F, Boiron JM, et al. Predictive factors for outcomes after reduced intensity conditioning hematopoietic stem cell transplantation for hematological malignancies: a 10-year retrospective analysis from the Societe Francaise de Greffe de Moelle et de Therapie Cellulaire. *Exp Hematol* 2008;**36**(5):535–44.
61. Ruggeri L, Capanni M, Urbani E, Perruccio K, Shlomchik WD, Tosti A, et al. Effectiveness of donor natural killer cell alloreactivity in mismatched hematopoietic transplants. *Science* 2002;**295**(5562):2097–100.
62. Willemze R, Rodrigues CA, Labopin M, Sanz G, Michel G, Socié G, et al. KIR-ligand incompatibility in the graft-versus-host direction improves outcomes after umbilical cord blood transplantation for acute leukemia. *Leukemia* 2009;**23**(3):492–500.
63. Brunstein CG, Wagner JE, Weisdorf DJ, Cooley S, Noreen H, Barker JN, et al. Negative effect of KIR alloreactivity in recipients of umbilical cord blood transplant depends on transplantation conditioning intensity. *Blood* 2009;**113**(22):5628–34.
64. Garfall A, Kim HT, Sun L, Ho VT, Armand P, Koreth J, et al. KIR ligand incompatibility is not associated with relapse reduction after double umbilical cord blood transplantation. *Bone Marrow Transpl* 2013;**48**(7):1000–2.
65. Kögler G, Middleton PG, Wilke M, Rocha V, Esendam B, Enczmann J, et al. Recipient cytokine genotypes for TNF-alpha and IL-10 and the minor histocompatibility antigens HY and CD31 codon 125 are not associated with occurrence or severity of acute GVHD in unrelated cord blood transplantation: a retrospective analysis. *Transplantation* 2002;**74**(8):1167–75.
66. Mommaas B, Stegehuis-Kamp JA, van Halteren AG, Kester M, Enczmann J, Wernet P, et al. Cord blood comprises antigen-experienced T cells specific for maternal minor histocompatibility antigen HA-1. *Blood* 2005;**105**(4):1823–7.
67. Charron D. Immunogenetics today: HLA, MHC and much more. *Curr Opin Immunol* 2005;**17**(5):493–7.
68. Charron D, Petersdorf E. The HLA system in hematopoietic stem cell transplantation. In: Socie G, Blazar BR, editors. *Immune biology of allogeneic hematopoietic stem cell transplantation*. Elsevier ed; 2013. p. 19–32.
69. Charron D. HLA, immunogenetics, pharmacogenetics and personalized medicine. *Vox Sang* 2011;**100**(1):163–6.

Section III

# Cord Blood Cells for Clinical Use

Chapter 7

# Clinical Use of Umbilical Cord Blood Cells

Robert Danby[1,2] and Vanderson Rocha[1,2,3]
[1]*Department of Haematology, Oxford University Hospitals NHS Trust, Churchill Hospital, Oxford, UK;* [2]*NHS Blood and Transplant, Oxford Centre, John Radcliffe Hospital, Oxford, UK;* [3]*Eurocord, Hôpital Saint Louis APHP, University Paris VII IUH, Paris, France*

## Chapter Outline



## 1. INTRODUCTION

Allogeneic hematopoietic stem cell (HSC) transplantation is the transfer of HSC from a healthy donor into an immunosuppressed host, allowing the formation of new donor hematopoiesis and reestablishing functional immunity. Over the last 50 years, allogeneic HSC transplantation has become an established curative therapy for the treatment of many malignant and nonmalignant disorders, particularly acute leukemia and lymphoproliferative conditions. Worldwide, there are over 25,000 allogeneic HSC transplants (HSCTs) performed each year. In 2012, 15,351 allogeneic HSCTs were reported to the European Blood and Marrow Transplantation (EBMT) group, performed across 48 countries (mainly European) in over 650 transplant centers.[1] Of these, 14,165 were first transplants with 71% ($n$=10,080) treating leukemia, 15% ($n$=2182) lymphoproliferative disorders, and 12% ($n$=1734) nonmalignant disorders.

Over the last 30 years, there have been many improvements in clinical HSC transplantation. One of the most significant advancements has been the extended use of "alternative donors." For many years, a human leukocyte antigen (HLA)-matched sibling was the only donor routinely used. However, only 25–30% of patients who need an allogeneic HSCT will have a suitable HLA-identical sibling, with only a one in four chance of any individual sibling being HLA-matched. For the remainder, the preferred strategy is to search for an unrelated HLA-matched volunteer donor through international donor registries. However, to extend the possibility of HSC transplantation to even more patients, single HLA-mismatched unrelated donors, cord blood (CB) donors, and full-haplotype mismatched family members have increasingly been used. Of the 15,531 allogeneic HSCT reported to the 2012 EBMT survey, 38% ($n$=5806) of donors were HLA-identical siblings, 0.3% ($n$=46) syngeneic twins, 49% ($n$=7530)

Cord Blood Stem Cells Medicine. http://dx.doi.org/10.1016/B978-0-12-407785-0.00007-4

**TABLE 1 Comparison of Alternative Donor Sources, Advantages, and Disadvantages of Using Single HLA-Mismatched Unrelated Donors, Cord Blood, and Haploidentical Family Members**

| | Mismatched Unrelated | Cord Blood | Haploidentical |
|---|---|---|---|
| HLA matching | 9–10/10 (HLA-A, -B, -C, DRB1 ± DQB1) at allelic level | 4–6/6 (HLA-A and -B (antigen); HLA-DRB1 (allele) | 50% |
| Rare haplotypes available | 2–10% | 20% | N/A |
| Donor availability | 3–6 months | 1 month | Immediate |
| Donor attrition | 20–30% | 1–2% | <1% |
| Risk to donor | Low | None | Low |
| Cell dose | High | Low | High |
| Post-transplant immunotherapy | Yes | No | Yes (limited dose) |
| Cost | Low/medium | High | Low |
| Clinical advantages | Good engraftment; low relapse | Low GvHD | Good engraftment |
| Clinical disadvantages | Increased GvHD | Graft failure; delayed myeloid recovery; delayed immune reconstitution; increased infections | Delayed immune recovery; increased infections; increased relapse (T-cell depletion/RIC) |

Abbreviations: GvHD, Graft-versus-Host Disease; HLA, human leukocyte antigen; RIC, reduced intensity conditioning.
Adapted from Rocha and Locatelli (2008).[11]

unrelated volunteers, 4% ($n=694$) unrelated CB, and 8% ($n=1217$) other family members.[1] Nowadays, an alternative HSC donor can be found for virtually all patients and many retrospective studies have shown that both CB donors and haploidentical family donors are suitable alternatives to HLA-matched or -mismatched unrelated donors with acceptable overall outcomes.[2,3] The decision on whether to employ an HLA-mismatched unrelated volunteer, an unrelated CB unit, or a haploidentical relative depends on patient-, disease-, and transplant-related factors. The advantages and limitations of each of these strategies (HLA-mismatched donor, CB, and haploidentical relative) are summarized in Table 1.

## 2. CLINICAL CB TRANSPLANTATION

Umbilical CB transplantation has extended the possibility of performing allogeneic HSC transplantation to patients that otherwise lack a suitable HLA-matched donor. In the 2012 EBMT survey, there were 758 allogeneic CB transplants performed with 8% ($n=58$) from HLA-identical sibling donors, 92% ($n=694$) from unrelated donors, and 0.8% ($n=6$) from other family members.[1] Due to the relative immaturity of placental cells, CB lymphocytes have lower alloreactivity compared to conventional bone marrow (BM) or peripheral blood (PB) lymphocytes.[4] As such, HLA mismatches between CB and recipient are better tolerated, with a lower incidence of Graft-versus-Host Disease (GvHD) and less-stringent HLA-matching requirements compared to BM and peripheral blood stem cell (PBSC) transplants.[5–7] Traditionally, only HLA matching at HLA-A and HLA-B (low resolution; antigen), and HLA-DRB1 (high resolution; allele) have been used, with mismatches at one or two HLA-loci usually being tolerated if sufficient cell doses are transplanted.[8–10] Suitable CB units for transplantation can, therefore, be found for the vast majority of patients, including those with infrequent haplotypes and/or from ethnic minority groups that are underrepresented in international volunteer donor registries. Cryopreserved CB units also have many logistical advantages including: being immediately available; avoiding long delays to transplantation; lack of donor attrition with the ability to process and store the donor cells long-term; and having no associated risks to the donor.[11]

Today, clinical CB transplantation includes:

1. Use of CB from an HLA-identical sibling donor (related CB transplantation); administered as a single CB unit or in combination with BM cells from the same donor.
2. Use of cryopreserved ("banked") CB units from unrelated donors; administered as single (one donor) or double CB units (two different donors).
3. Investigational use of unrelated CB injected directly into bone, expanded in vitro, or in combination with third-party donor cells (haploidentical or mesenchymal cells).
4. Use of autologous CB cells; remains controversial with little scientific data to support its routine use in clinical practice (*this will not be discussed in this chapter*).

## 3. CLINICAL USE OF RELATED CB CELLS FOR ALLOGENEIC TRANSPLANTATION

In 1988, Gluckman et al. reported the first allogeneic HSCT using CB as the source of HSC for hematopoietic reconstitution in a 5-year-old boy with Fanconi anemia (FA).[12] After his initial diagnosis, his mother became pregnant again with a girl who was known to be HLA-identical to her brother but not affected by the same disorder. The CB cells were collected at delivery, cryopreserved, and later transplanted to her brother. The transplant was successful and the patient remains alive and well today. Since this first report, many cases of CB transplantation from HLA-identical siblings have been reported. In Europe alone, around 700 cases have been performed, although the number of related CB transplants does not appear to be increasing year-on-year (V. Rocha, personal information). Most related CB transplant recipients are children, and the vast majority of the CB units are HLA-identical. Around 150 of these children also received BM cells from the same donor in addition to CB, due to the relatively low number of CB cells available. In the 519 related CB transplants with outcome data available, the cumulative incidence of neutrophil engraftment by day 60 was 91±3% with a median of 22 days (range, 12–80 days).[13] The number of infused $CD34^+$ cells was associated with the cumulative incidence of neutrophil recovery, being 95±2% if the infused $CD34^+$ dose was $>1.4\times10^5$/kg, but only 90±3% if the $CD34^+$ dose was lower ($P=0.02$). The cumulative incidence of acute and chronic GvHD at 100 days and 4 years was 12±3% and 13±2%, respectively. The 4-year cumulative incidence of nonrelapse mortality (NRM) for malignant diseases was 8±2%. The 4-year overall survival (OS) for all patients was 75±2%, 91±3% for patients with nonmalignant conditions, and 56±4% for patients with malignant diseases.

In 2000, Eurocord in collaboration with the International Blood and Marrow Transplantation Registry (IBMTR) compared 113 HLA-identical sibling CB transplants (1990–1997) to 2052 HLA-identical sibling bone marrow transplants (BMTs) performed over the same period.[14] The median number of total nucleated cells (TNC) infused in the CB transplants was $0.47\times10^8$/kg compared to $3.5\times10^8$/kg for the BM transplants ($P<0.001$). This comparative retrospective analysis demonstrated inferior neutrophil recovery (day 60) for CB transplantation compared to BM (89% (95% CI, 82–94%) vs 98% (95% CI, 97–99%), respectively; $P=0.02$). However, CB transplantation was associated with a lower risk of acute GvHD (100 days: 14% (95% CI, 8–22%) vs 24% (95% CI, 22–26%), respectively; $P=0.02$) and chronic GvHD (3 years: 6% (95% CI, 2–13%) vs 15% (95% CI, 13–17%), respectively; $P=0.02$). This lower incidence of GvHD seen with CB transplantation probably reflects the relative immunological immaturity of CB. T cells within CB are antigen naïve, being less responsive to allogeneic stimulation and produce lower levels of effector cytokines compared to activated T cells from adult blood.[4] CB dendritic cells also have lower antigen-presenting activity and reduced expression of co-stimulatory molecules (CD80, CD86).[15] In addition, CB contains more regulatory T cells (Tregs), with greater potential for expansion and increased suppressive function compared to adult Tregs.[16] OS at 3 years was similar after CB transplantation and BMTs for both malignant (46% (95% CI, 31–62%) vs 55% (95% CI, 52–58%), respectively; $P=0.69$) and nonmalignant (86% (95% CI, 75–94%) vs 84% (95% CI, 81–87%), respectively; $P=0.82$) disorders.

### 3.1 Related CB Transplantation for Malignant Disorders

Many studies have reported outcomes after related CB transplants for patients with malignant diseases.[14,17] In 2010, Eurocord and the EBMT group reported risk factor analysis and long-term outcomes of 147 HLA-identical sibling CB transplant recipients with hematological malignancies.[18] The CB transplants were performed from 1990 to 2008 with a median follow-up of 6.7 years (range, 7 months to 18 years). All donors were younger siblings and the patient's diagnosis pre-dated the donor's birth. Of the 142 CB transplants with further information available, the mother was pregnant at the time of diagnosis in 60 cases, and the CB unit was stored or used immediately after birth. In 82 cases, the CB donor was conceived after diagnosis, although it is unknown whether any of these pregnancies were conceived with the intention of providing an HLA-matched donor. The median age of the patients was 5 years (range, 1–32 years), and acute leukemia was the most common diagnosis ($n=109$; 74%). The CB grafts contained a median of $4.1\times10^7$/kg (95% CI, $1.20–7.45\times10^7$/kg) TNC after thawing. The cumulative incidence of neutrophil recovery at day 60 was 90±3%. Higher infused TNC dose ($>4.1\times10^7$/kg) (HR 1.72 (95% CI, 1.20–2.45); $P=0.003$) and use of methotrexate as GvHD prophylaxis (HR 0.48 (95% CI, 0.31–0.73); $P<0.001$) were independent predictors of neutrophil recovery. The cumulative incidence of acute and chronic GvHD was 12±3% and 10±2% at day 100 and 2 years, respectively. The 5-year cumulative incidence of NRM, relapse, disease-free survival (DFS), and OS were 9±2%, 47±4%, 44±4%, and 55±4%, respectively. In subgroup analysis of patients with acute leukemia, 5-year DFS were 57±9%, 46±7%, 31±13%, and 21±11% if transplanted in first complete remission (CR1), second CR (CR2), third CR (CR3), and refractory disease, respectively. In multivariate analysis, later transplants (performed after 2000) ($P=0.003$), a higher infused TNC dose ($>4.1\times10^7$/kg) ($P=0.02$), and transplanting early/intermediate disease ($P=0.04$) were associated with improved DFS. Related CB transplantation, therefore, provides a suitable alternative to BM or PBSC transplants in children with malignant diseases and, importantly, avoids any risk to the sibling donor. However, cell dose may become a limiting factor in larger children.

## 3.2 Related CB Transplantation for Nonmalignant Disorders

The majority of related CB transplants are performed in children with nonmalignant genetic disorders, with hemoglobinopathies (e.g., thalassemia major (TM) or sickle cell disease (SCD)) and BM failure syndromes (e.g., FA) being the most frequent indications. Several years ago, the Eurocord group retrospectively reviewed the results of 44 CB transplants performed for thalassemia ($n$=33) or SCD ($n$=11).[19] The median age of the recipients was 5 years (range, 1–20 years) and the median number of TNC infused was $4.0\times10^7$/kg (range, $1.2–10\times10^7$/kg). Eight patients had graft failure (1 SCD and 7 thalassemia). The cumulative incidence of neutrophil recovery by day 60 was 89%, with a median time of 23 days (range, 12–60 days). The estimated probabilities of acute GvHD (grade II–IV) and limited chronic GvHD were 11% and 6%, respectively. Thirty-six of the 44 children remained disease-free, with a median follow-up of 24 months (range, 4–76 months), and no patient died after transplantation during the follow-up period. The 2-year DFS was 79% and 90% in the thalassemia and SCD groups, respectively. Of note, the use of methotrexate for GvHD prophylaxis was again associated with a greater risk of treatment failure (HR 6.6 (95% CI, 1.47–25.86); $P$=0.01). More recently, Eurocord in collaboration with the EBMT group retrospectively compared outcomes in a larger cohort of 485 patients with TM or SCD who were treated with an HLA-identical sibling CB transplant ($n$=96) or BMT ($n$=389).[20] In keeping with the earlier studies, CB transplants had inferior neutrophil recovery compared to BMT, with a cumulative incidence at day 60 of 90±4% and 92±1%, respectively ($P$=0.01). The median time to neutrophil recovery was also slower in the CB transplant recipients, with a median time of 23 days (range, 9–606 days) compared to 19 days (range, 8–56 days) in the BMT group ($P$=0.002). CB transplant recipients had less acute GvHD (10±3% vs 21±2%; $P$=0.04), and no CB transplant patients experienced extensive chronic GvHD. With a median follow-up of 70 months, the 6-year OS after CB transplant and BMT was very similar at 95±1% and 97±2%, respectively ($P$=0.92). The 6-year DFS was 86±2% and 80±5% in TM patients after BMT and CB transplant, respectively, while the corresponding DFS in SCD patients was 92±2% and 90±5%, respectively. However, in multivariate analysis, DFS was not statistically different between CB transplant and BMT recipients. Within the CB transplants only ($n$=96), use of methotrexate (HR 3.81 (95% CI, 1.40–10.87); $P$=0.004) and later transplants (performed after 1999) (HR 0.33 (95% CI, 0.12–0.89); $P$=0.02) were independent predictors of DFS. In summary, related CB transplantation therefore offers excellent outcomes for patients with TM and SCD, and may even be preferable to BMT because of the lower incidence of GvHD.

## 3.3 Directed Related Donor CB Banking

In view of the encouraging clinical outcomes of HLA-identical sibling CB transplantation, there has been increased interest in the use of directed donor CB banking. If the mother of a child diagnosed with a malignant or nonmalignant disorder that is potentially curable by HSC becomes pregnant, efforts should be made to cryopreserve the CB from the new sibling. If HSC transplantation becomes necessary and the new sibling is a suitable donor, preemptive collection of the CB has, therefore, avoided the risks and discomfort associated with a BM harvest and may be associated with lower rates of GvHD. A number of studies have reported efforts to systematically identify, collect, and store CB units for family-directed use.[21,22] Over a 10-year period, NHS Blood and Transplant for England and Wales collected 268 directed CB units; 87% ($n$=235) were collected for an existing sibling that may require an HSCT and 13% ($n$=35) were collected because there was a pre-existing family history of serious inherited disease.[21] Sixty-five (28%) of the CB units collected for existing siblings were HLA matched, but only 13 of these were used for transplantation. None of the CB units collected with a preexisting family history of inherited disease were used. In a similar report, only one of 48 CB units collected at delivery from a woman with an existing child that may have required HSC transplantation was used.[22] Therefore, while directed CB banking may have an important role, it is associated with a relatively low probability of usage and careful consideration needs to be given to which units are stored long-term. Furthermore, directed CB banking raises important ethical questions, particularly as to whether any pregnancy should ever be conceived with the aim of providing an HLA-matched sibling donor for a child with a pre-existing disease. With the increased development and use of private cord banks, every effort must be made to ensure these transplants and cord banks always meet the current international clinical and ethical standards required for CB transplantation.

# 4. CLINICAL USE OF UNRELATED CB CELLS FOR ALLOGENEIC TRANSPLANTATION

Following the promising results with related CB transplantation and the lower incidence of GvHD, many centers proposed the use of unrelated CB and the formation of volunteer unrelated CB banks. In 1991, the first public CB bank was established at the New York Blood Center (NYBC). Shortly afterward in 1993, the first unrelated CB transplant was performed in a 4-year-old boy with acute lymphoblastic leukemia (ALL).[23] To date, over 130 public CB banks across the world have now collected over 600,000 volunteer, altruistic CB units. Basic HLA typing and clinical data for each unit are stored on searchable international registry databases, such as the as Bone Marrow Donor World Wide and NetCord Foundation, allowing transplant centers to identify and locate potentially suitable CB units for transplantation. As a result of these initiatives, over 30,000 unrelated CB transplants have now been performed worldwide.

Initial progress in unrelated CB transplantation was made with the improved understanding of minimum cell dose and HLA-matching requirements for successful outcomes. CB units, on average, contain only 10% of the number of CD34+ HSC/progenitor cells compared to BM grafts, and only 5% compared to PBSC grafts. From the seminal publications on unrelated CB transplantation in the late 1990s, it was recognized that low CB cell doses (TNC and/or CD34+ cells) were associated with an increased risk of graft failure, significant delays in engraftment, and increased risk of early transplant-related mortality (TRM), mainly due to infection.[17,24] Inferior transplant outcomes were also associated with a higher number of HLA mismatches between CB unit and recipient. This new understanding, therefore, directly led to improvements in collection and use of CB units with higher cell doses and less HLA disparity. Combined with improvements in supportive care, several studies showed outcomes of unrelated CB transplantation comparable to that of HSCT using conventional donors.[6,25] Furthermore, with increased use of CB transplantation in adults and the need to increase the infused CB cell dose in larger patients, the use of double CB transplants was pioneered.[26] The use of two CB units, each from a different unrelated donor, produced a significant reduction in the risk of graft failure and opened up the possibility of HSCT with CB donors for all patients.

## 4.1 Unrelated CB Transplantation in Children

Many published studies have demonstrated that unrelated CB transplantation in children is associated with sustained myeloid engraftment, a low incidence of GvHD, and comparable overall outcomes to using conventional HSCT donors.[5,9,27] Results from the first prospective phase 2 multicenter trial of unrelated CB transplantation (COBLT) for children (<18 years) with hematological malignancies were published in 2008.[27] Unrelated CB transplants were performed in 191 children with a median age of 7.7 years (range, 0.9–17.9 years). Most patients were transplanted for acute leukemia ($n$=161; 84%). The median collected TNC and CD34+ dose was $5.1\times10^7$/kg (range, $1.5–23.7\times10^7$/kg) and $1.9\times10^5$/kg (range, $0.0–25.3\times10^5$/kg), respectively. CB units and recipients were HLA-matched for 5–6/6 HLA-loci (HLA-A and -B (antigen); HLA-DRB1 (allele)) in 39% of transplants ($n$=75), with the reminder having two or three HLA disparities. The cumulative incidence of neutrophil recovery (day 42), acute GvHD (day 100), chronic GvHD (2 years), and relapse (2 years) was 80% (95% CI, 75–85%), 20% (95% CI, 14–26%), 21% (95% CI, 15–28%), and 20% (95% CI, 15–26%), respectively. In multivariate analysis, greater HLA matching ($P$=0.04) and higher TNC dose ($P$=0.04) were independently associated with improved neutrophil recovery. Two-year OS was 50% (95% CI, 42–57%). In multivariate analysis, cytomegalovirus (CMV) serostatus ($P$<0.01), ABO matching ($P$=0.02), recipient gender ($P$<0.01), and TNC dose ($P$=0.04) were independent predictors for OS. This prospective trial, therefore, reinforced the use of CB donors in children with malignant diseases. Many retrospective series of children receiving unrelated CB transplantation for specific diseases have also been published, including ALL,[9] acute myeloid leukemia (AML),[28] myelodysplastic syndromes (MDS),[29] hemoglobinopathies,[30] Hurler syndrome,[31] FA,[32] and primary immunodeficiency.[33]

### 4.1.1 Malignant Hematological Disorders in Children

*Acute lymphoblastic leukemia*: Although the outcomes in children treated for ALL with chemotherapy alone remain very good, approximately 20–30% will require an allogeneic HSCT.[34] Eurocord retrospectively reviewed the results of 532 unrelated CB transplants in children with ALL (Eurocord, unpublished data). B-cell precursor ALL was the most frequent phenotype. Patients were transplanted in CR1 ($n$=186; 35%), CR2 ($n$=238; 45%), or CR3/advanced disease ($n$=108; 20%). The median age was 6.8 years. Most transplants used single CB units, with a median infused TNC dose of $4\times10^7$/kg, and were mismatched at one or two HLA-loci. The cumulative incidence of neutrophil recovery, acute GvHD, and TRM were 82%, 27%, and 21%, respectively. In multivariate analysis, infused TNC >$4\times10^7$/kg and disease status (CR1) at transplantation were associated with improved neutrophil recovery. The 2-year cumulative incidence of relapse was 37%, with disease status and use of total body irradiation (TBI) being independent predictors of relapse. The 2-year leukemia-free survival (LFS) was 38% (49% for CR1, 42% for CR2, 10% CR3/advanced disease). In 2012, Eurocord also analyzed the outcomes of 170 unrelated CB transplants in children with ALL who had minimal residual disease (MRD) monitoring available prior to transplantation.[35] All children, median age 6.5 years (range, <1–17 years), were transplanted in CR using a myeloablative conditioning (MAC) regimen. At 4 years post-transplant, the cumulative incidence of relapse was 30±3%. In multivariate analysis, positive MRD before transplant was an independent predictor of relapse (HR 2.2 (95% CI, 1.2–3.9); $P$=0.001). LFS was also improved in patients transplanted with negative MRD (54% vs 29%, $P=0.003$) with MRD also being an independent predictor of LFS (HR 0.5 (95% CI, 0.3–0.8); $P=0.003$). MRD assessment prior to CB transplantation may therefore identify children at higher risk of relapse after CB transplantation and, as such, further approaches to reduce their risk of relapse post-transplant should be investigated.

*Acute myeloid leukemia*: In 2003, Eurocord reviewed the results of 95 CB transplants performed in children with AML.[28] As with ALL, most transplants used single CB units, with a median collected TNC dose of $5.2\times10^7$/kg, and with one or two HLA disparities. The median age of the children was 4.8 years (range, <1–15 years). The cumulative incidence of neutrophil recovery at day 60 was $78\pm4\%$ and the incidence of acute GvHD at day 100 was $35\pm5\%$. The 2-year incidence of relapse was $29\pm5\%$ with 2-year LFS being $59\pm11\%$ if transplanted in CR1, $50\pm8\%$ for CR2, and $21\pm9\%$ for relapsed disease. In multivariate analysis, disease status and major ABO incompatibility were independently associated with LFS and OS. Following these findings, Eurocord analyzed the impact of cytogenetic and/or molecular markers on the outcomes of 390 CB transplants in children with AML (Eurocord, unpublished data). The median infused TNC and CD34 cell doses were $4.9\times10^7$/kg and $1.9\times10^5$/kg, respectively, and the majority (81%) of CB grafts had one or two HLA disparities with the recipient. MAC regimens were used for 87% of the transplants. The cumulative incidence of day 60 neutrophil recovery was 85% and was significantly associated with the infused TNC dose and CD34 cell dose. At day 100, the cumulative incidence of acute GvHD (grade II–IV) was 34%. At 2 years, LFS was 63% if transplanted in CR1, 43% for CR2, and 22% for more advanced disease. In subgroup analysis of patients transplanted in CR2, favorable disease status based upon prognostic cytogenetic/molecular markers ($P=0.005$) and a previous CR longer than 7 months ($P=0.03$) were associated with improved LFS. In summary, the results of CB transplantation for childhood AML remain encouraging when no HLA-identical donors are available. Furthermore, use of cytogenetic/molecular markers may be able to identify children in CR with high risk of relapse post-CB transplantation, and thus, may benefit from novel interventions to further improve outcomes.

*MDS and juvenile myelomonocytic leukemia*: MDS are a rare group of diseases in children but HSC transplantation remains the only curative treatment. In a collaborative study between European Working Group on childhood MDS, the Center for International Blood and Marrow Transplant Research (CIBMTR), and the Eurocord-EBMT Group, the outcomes of 70 children with MDS (refractory cytopenia ($n=33$); refractory anemia with excess blasts ($n=28$); refractory anemia with excess blasts in transformation ($n=9$)), who received a CB transplantation were retrospectively examined.[29] The median age of the children at transplantation was 7 years (range, 1–18 years) and all received MAC regimens. The collected TNC dose was $5.75\times10^7$/kg (range, $1–28\times10^7$/kg) and 38 transplants (54%) were HLA-matched at ≥5/6 HLA-loci. The day 60 cumulative incidence of neutrophil recovery was 76% (95% CI, 64–74%). A TNC dose $>6\times10^7$/kg (HR 0.55 (95% CI, 0.33–0.93); $P=0.02$), CB units matched ≥5/6 HLA-loci (HR 0.47 (95% CI, 0.25–0.90); $P=0.02$), use of TBI conditioning regimens (HR 0.47 (95% CI, 0.25–0.85); $P=0.01$), and presence of monosomy 7 (HR 0.58 (95% CI, 0.33–0.99); $P=0.045$) all favored improved engraftment. The 3-year DFS was 39% (95% CI, 33–45%). CB transplants performed before 2001 (HR 2.38 (95% CI, 1.14–5.00); $P=0.02$) and karyotypes other than monosomy 7 (HR 2.04 (95% CI, 1.11–3.70); $P=0.02$) were independent predictors for higher risk of treatment failure. In a separate analysis, Eurocord analyzed the outcomes of 110 children with juvenile myelomonocytic leukemia, receiving a single-unit CB transplant (1995–2010).[36] The median TNC dose infused was $7.1\times10^7$/kg (range, $1.7–27.6\times10^7$/kg) and 59% ($n=65$) of CB units had 0–1 HLA mismatches. The median follow-up was 44 months (range, 3–169 months). The cumulative incidence of neutrophil recovery by day 60 was $82\pm4\%$, with a median time of 25 days (range, 10–60 days). The cumulative incidence acute GvHD (grade II–IV) at 100 days was $41\pm4\%$. At 5 years, the cumulative incidence of relapse was $33\pm5\%$, with age at diagnosis >1.4 years independently associated with increased relapse (HR 2.8 (95% CI, 1.4–7.0); $P=0.004$). The 5-year DFS and OS were $44\pm5\%$ and $52\pm5\%$, respectively. In multivariate analysis, factors associated with better DFS were age at diagnosis >1.4 years ($P=0.005$), 0–1 HLA-matched CB units ($P=0.009$), and the absence of monosomy 7 ($P=0.02$). In summary, these data demonstrate that unrelated CB transplantation provides reasonable outcomes, in the absence of a suitable HLA-matched sibling donor, for the relatively small group of children with MDS.

### 4.1.2 Nonmalignant Hematological Disorders in Children

*Congenital BM failure syndromes*: Several years ago, the Eurocord group reported the outcomes of CB transplantation in 93 patients, mainly children, with FA.[32] The median age of the recipients was 8.6 years (range, 1–45 years). The median TNC dose collected and infused were $5.9\times10^7$/kg (range, $0.8–24.0\times10^7$/kg) and $4.9\times10^7$/kg (range, $1–19.2\times10^7$/kg), respectively. Half the CB units ($n=37$) were matched with the recipient at 5–6/6 HLA-loci. The 60-day cumulative incidence of neutrophil recovery was $60\pm5\%$. In multivariate analysis, neutrophil recovery was improved by using Fludarabine as part of the conditioning regimen ($P=0.05$) and an infused TNC dose $\geq4.9\times10^7$/kg ($P=0.03$). The cumulative incidence of acute GvHD (grade II–IV) and chronic GvHD was 32±5% and 16±4% respectively. The 3-year OS was 40±5% and was improved if the recipient's CMV serology was negative (HR 2.82 (95% CI, 1.45–5.59); $P<0.001$); the infused TNC dose was ≥4.9×$10^7$/kg (HR 1.75 (95% CI, 0.99–3.16); $P=0.05$); and if the conditioning regimen contained Fludarabine (HR 1.79 (95% CI, 1.02–3.13); $P=0.04$). Eurocord also reviewed the outcomes of 44 CB transplants given to patients with congenital BM failure syndromes other than FA (Diamond Blackman ($n=18$); congenital amegakaryocytic thrombocytopenia ($n=13$); dyskeratosis congenital ($n=6$); severe congenital neutropenia ($n=15$); Shwachman–Diamond syndrome ($n=1$); unclassified ($n=1$)).[37] The median infused TNC dose was 6.1×$10^7$/kg (range, 0.3–18×$10^7$/kg) with three CB transplants using double-unit grafts. The day 60 cumulative incidence of neutrophil recovery was 55% with 17 patients having primary graft failure. The 100-day cumulative incidence of acute GvHD was 24% and the 2-year cumulative incidence of chronic GvHD was 53%. The 2-year OS was 61% (95% CI, 47–75%) and was improved with age under 5 years (RR 0.29 (range, 0.1–0.8); $P=0.02$), and an infused TNC dose ≥6.1×$10^7$/kg (RR 0.31 (range, 0.11–0.86); $P=0.03$). Both studies showed reasonable results, although there was a high incidence of primary graft failure. Therefore, in this clinical setting, it is recommend that selected CB units should have high cell doses (infused TNC >4×$10^7$/kg) and no more than one HLA disparity. In addition, due to the high probability that these patients will have received multiple transfusions prior to transplantation, it is recommended that they are tested for the presence of donor-specific anti-HLA antibodies (DSA).

*Metabolic disorders*: CB transplantation has become an attractive option for the treatment-inherited metabolic disorders, such as Hurler syndrome, due to the potential for cure and the low rates of GvHD. In 2013, Eurocord retrospectively reviewed the outcomes after HSCT (CB ($n=116$), unrelated donor ($n=105$), HLA-matched sibling ($n=37$)) in 258 children with Hurler syndrome.[31] The median age of the children was 16.7 months (range, 2.1–228 months). In the CB group, the median TNC and $CD34^+$ cell dose infused were 8.8×$10^7$/kg (range, 1.2–32×$10^7$/kg) and 3.0×$10^5$/kg (range, 0.2–105×$10^5$/kg), respectively. The 5-year event-free survival (EFS) and OS for all transplants were 63±3% and 74±3%, respectively. When analyzing the outcomes according to the donor, the 5-year EFS was 81% for HLA-matched sibling donors and HLA-matched (6/6) CB transplants; 68% and 57% for one- and two-antigen mismatched CB transplants, respectively; and 66% and 41% for matched and mismatched unrelated transplants, respectively. In multivariate analysis, the age at transplantation (<16.7 months) and donor source were independent predictors of EFS. Compared with HLA-matched sibling donors and CB transplants, CB transplants mismatched at two HLA-loci (HR 2.5 (95% CI, 1.09–5.87); $P=0.03$) and HLA-mismatched unrelated transplants (HR 2.7 (95% CI, 1.30–5.54); $P=0.007$) had significantly lower EFS. Matched (10/10) unrelated donors or 5/6 UCB (when combined) showed a nonsignificant trend ($P=0.07$) toward lower EFS. Based upon these observations, CB transplantation remains a suitable option for the treatment of Hurler syndrome, although the use of CB units with two or more HLA mismatches is not recommended due to the inferior outcomes.

*Primary immunodeficiency*: CB transplantation has been successfully used in the treatment of children with primary immunodeficiency. In 2012, Eurocord retrospectively compared the outcomes of 74 CB transplants to 175 mismatched related donor (MMRD) transplants in patients with severe combined immunodeficiency (SCID) or Omenn syndrome.[33] In the CB transplant cohort, 67% ($n=50$) of patients received grafts with 0 or 1 HLA mismatches. The 5-year OS was 57±6% for CB transplants and 62±4% MMRD transplants, respectively ($P=0.68$). For the CB transplants, the 5-year OS was higher in patients with SCID receiving HLA-matched (6/6) (76%, $n=21$) or single HLA-mismatched (5/6) (62%, $n=29$) CB grafts compared to 4/6 HLA-mismatched CB units (35%, $n=24$). These data demonstrate that CB transplantation can be considered for severe immunodeficiencies, although use of CB units with ≥2 HLA disparities is not recommended.

*Hemoglobinopathies*: Unrelated CB transplantation and use of alternative donors have been used in the treatment of TM and SCD. However, the use of unrelated CB transplantation remains controversial, as some studies have shown high rates of graft rejection.[38] In 2012, the Eurocord–CIBMTR collaboration published a retrospective analysis of 51 unrelated CB transplants for children with TM ($n=35$) or SCD ($n=16$).[30] Twenty-five CB units had 0–1 HLA disparities, with the remainder mismatched at 2–3 HLA-loci. The median TNC dose infused was 5×$10^7$/kg (range, 1.1–23×$10^7$/kg). The day 60 cumulative incidence of engraftment was 63±9% with CB units containing >5×$10^7$/kg, and only 32±8% with lower cell doses. Twenty-seven patients had primary graft failure (20 with thalassemia and 7 with SCD), which was the main cause of treatment failure. The cumulative incidence of acute GvHD at day 100 was 23±2%. OS and DFS were 62±9% and 21±7% for the TM patients and 94±6% and 50±9% for the SCD patients, respectively. In multivariate analysis, engraftment (HR 2.2 (95% CI, 0.96–3.6); $P=0.05$) and DFS (HR 0.4 (95% CI, 0.2–0.8); $P=0.01$) were improved with a higher infused TNC dose (>5×$10^7$/kg). Therefore, these data rehighlight the relatively high rates of primary graft failure with unrelated CB transplantation for hemoglobinopathies. It is, therefore, recommended that if transplanting children with hemoglobinopathies, only CB units with high cell doses should be considered.

### 4.1.3 Outcomes of CB Transplantation Compared to Other Graft Sources in Children

As the number of registered volunteer donors and/or number of cryopreserved CB units continues to rise, for many children, the search process may identify multiple donor options. Therefore, to aid the clinician in selecting the most appropriate donor, several retrospective studies have attempted to directly compare the outcomes of CB transplantation with unrelated BM transplantation in children with acute leukemia.[5,39] Overall, recipients of CB transplantation were generally transplanted sooner compared to children given an unrelated BMT. In CB transplantation, neutrophil and platelet recovery were delayed, acute GvHD was decreased, but OS was not significantly different compared to BMTs. The Eurocord group reported higher TRM with CB transplants (HR 2.13 (95% CI, 1.20–3.76); $P<0.01$), mainly due to infection-related mortality as a result of delayed engraftment. However, all patients in the Eurocord series were transplanted before 1998 when CB transplantation was still in its infancy. A subsequent *meta*-analysis was published in 2007.[40] In total, the outcomes of 161 pediatric CB transplants were compared to the 316 pediatric unrelated BMT. In this analysis, the incidence of acute GvHD did not differ although the incidence of chronic GvHD was lower with CB transplantation. Furthermore, there was no significant difference in 2-year OS between the two transplant methods. Of note, all of the earlier studies analyzed HLA-matched BMT using low-resolution typing for HLA-A and -B, and high-resolution typing for HLA-DRB1. However, current selection of unrelated BMT donors is now based upon high-resolution typing (allelic) at HLA class I (HLA-A, -B, and -C) and class II (DRB1 ± DQB1). Therefore, on behalf of the CIBMTR and the New York Cord Blood program, Eapen et al. (2007) reviewed the outcomes in 885 children with acute leukemia transplanted using unrelated CB ($n=503$) or unrelated BM ($n=282$).[9] CB units and recipients were HLA-matched ($n=35$) or HLA-mismatched for one ($n=201$) or two HLA-loci ($n=267$) (antigen level for HLA-A, -B; allelic level for HLA-DRB1). BM grafts and recipients were matched at the allele level for HLA-A, -B, -C, and -DRB1 ($n=116$) or mismatched at one ($n=44$) or two alleles ($n=122$). Overall, CB recipients were younger, more likely be non-Caucasian, have relapsed disease, receive an HLA-mismatched graft, and have a female donor. Compared to BMT (HLA-matched and -mismatched), the cumulative incidence of neutrophil recovery at day 42 was significantly lower after a mismatched CB transplant ($P<0.0001$). Matched CB transplants also showed a nonsignificant trend toward lower neutrophil recovery ($P=0.06$). TRM was higher in transplants using two-antigen HLA-mismatched CB (any cell dose) or using one-antigen HLA-mismatched CB grafts with low cell doses (TNC $<3\times10^7$/kg) compared to HLA-matched BMT. The 5-year LFS was 38% using HLA-matched BM; 37% using HLA-mismatched BM; 60% using HLA-matched CB; 36% using single mismatched CB with TNC $<3\times10^7$/kg; 45% using single mismatched CB with TNC $>3\times10^7$/kg; and 33% using two HLA-matched CB. Overall, these data support the use of unrelated CB transplantation and demonstrate it is an acceptable alternative to unrelated HLA-matched BM transplantation in children. However, the choice of whether to use an unrelated BM donor or CB must take other factors into consideration including the urgency of the transplant, the cell dose available, and the degree of HLA matching.

## 4.2 Unrelated CB Transplantation in Adults

Initially, the use of CB transplantation was largely restricted to children, mainly because of the lower cell doses (per kilogram recipient body weight) available to adults and higher TRM. However, over the last 10 years, unrelated CB transplantation has increasingly been used in adults. To date, more than 10,000 unrelated CB transplants have been performed in Europe and reported to Eurocord, with almost 50% of these performed in adults (V. Rocha, personal information). The number of adults receiving a CB transplant has increased since 2004, and from 2006 the number of adults has surpassed the number of children being transplanted with CB. This increase has been largely driven by improved collection and selection of CB units; new developments in CB transplantation including use of double CB transplantation (i.e., using two CB units from different unrelated donors) and RIC regimens; more clearly defined indications; and greater transplant center experience. As such, an increasing number of studies are now demonstrating similar outcomes in adults when using CB compared to HLA-matched or -mismatched unrelated donors.[41,42] In contrast to children, the majority of CB transplants in adults are performed for malignant disorders using double CB transplantation.

### 4.2.1 Malignant Hematological Disorders in Adults

*Acute lymphoblastic leukemia*: Many of the larger studies of unrelated CB transplantation in adults with acute leukemia combine the results for ALL and AML together. However, smaller single-center studies have shown promising results in adult CB transplantation for ALL.[43] A Japanese study retrospectively reported outcomes after CB transplantation in 256 adults with ALL.[44] The median age of the patients was 40 years (range, 16–74 years), with 39% of the patients having Philadelphia positive (Ph+) disease. All transplants used a single CB unit with median infused TNC dose of $2.50\times10^7$/kg (range, 1.51–$5.00\times10^7$/kg). The 2-year DFS and OS rates were 36% (95% CI, 33–39%) and 42% (95% CI, 39–45%), respectively. In multivariate analysis, younger patients (<51 years) (HR 1.9 (95% CI, 1.3–2.8); $P=0.001$), disease remission at transplantation (HR 2.2 (95% CI, 1.5–3.2); $P<0.0001$), no acute GvHD (grade III–IV) (HR 2.0 (95% CI, 1.2–3.2); $P=0.006$), and the presence of chronic GvHD (HR 2.4 (95% CI, 1.1–5.1); $P=0.02$) were independently associated with improved OS. More recently, Eurocord and the Acute Leukemia Working Party (ALWP) of EBMT performed a larger retrospective survey on the outcomes after CB transplantation for 421 adults with ALL.[45] B-ALL was the most common phenotype ($n=271$; 65%) and 46% ($n=195$) patients were in CR1, 32% ($n=136$) in CR2, and 22% ($n=90$) had advanced disease at transplant. The median age of the patients was 32 years (range, 18–76 years). Single CB grafts were used in 59% ($n=248$) and the median-collected TNC dose was $4.0\times10^7$/kg (range, 1.4–$9.4\times10^7$/kg). The majority of CB transplants (61%) used CB units with ≥2 HLA disparities and an MAC regimen (74%). The cumulative incidence of day 60 neutrophil recovery was 78±2%. The cumulative incidence of acute GvHD (100 days) and chronic GvHD were 33±2% and 26±2%, respectively. The 2-year NRM was 42±2% with younger age (<35 years), CR at transplantation, and RIC regimens associated with lower NRM. The incidence of relapse at 2 years was 28±2%, with CR at transplantation and MAC associated with lower relapse risk. The estimated 2-year LFS was 39% for patients in CR1, 31% for CR2, and 8% for advanced disease. In multivariate analysis, age ≥35 years (HR 1.3 (95% CI, 1.1–1.7); $P=0.03$), advanced disease (HR 2.8 (95% CI, 2.2–3.7); $P<0.0001$), and use of MAC (HR 1.4 (95% CI, 1.1–1.9); $P=0.03$) were associated with inferior LFS. In summary, unrelated CB transplantation in adult patients with ALL is an important option for those without an HLA-identical sibling. However, further studies will be required to determine the optimal conditioning regimens to reduce NRM without increasing relapse in this high-risk population. Furthermore, it remains to be determined whether post-CB transplantation adjuvant therapies, e.g., tyrosine kinase inhibitors, further improve outcomes.

*Acute myeloid leukemia*: CB transplantation is a potential curative treatment for many adult patients with AML that otherwise lack a suitable HLA-matched donor.[46] In 2013, Eurocord retrospectively reviewed the results of 604 CB transplants performed in adults (median age 41 years) with AML between 2000 and 2011 (Eurocord, unpublished data). At transplantation, 38% ($n=229$) were in CR1, 38% ($n=228$) in CR2–3, and the remaining 24% ($n=147$) had advanced disease. Of those with further data evaluable at diagnosis ($n=339$), 31% were classified as high risk using cytogenetic and molecular prognostic markers. Double CB grafts were used in 40% of the transplants ($n=243$) with a median infused TNC and CD34$^+$ dose of $3.1\times10^7$/kg and $1.2\times10^5$/kg, respectively. Thirty-nine percent of CB units were mismatched at 0–1 HLA-loci with the remaining 61% having 2–3 HLA disparities. RIC regimens were used in approximately half (49%) of these transplants. The cumulative incidence of neutrophil recovery, acute GvHD (II–IV), and 1-year TRM were 80%, 26%, and 21%, respectively. The cumulative incidence of relapse at 2 years was 38%; being 27% for patients in CR1, 29% in CR2–3, and 56% for advanced disease. On subgroup analysis of conditioning regimen, 2-year LFS was 50% for patients in CR1, 27% for CR2–3, and 17% for advanced disease when using MAC. For those CB transplants using RIC, the corresponding 2-year LFS were 35%, 44%, and 18%, respectively. Although longer follow-up will be required (median follow-up only 13 months), this large series provides useful data on the overall outcomes of CB transplantation in adults with AML. In particular, it highlights the need to carefully consider the conditioning regimen, taking into account disease status and age of the patients.

*Myelodysplastic syndromes*: Compared to acute leukemia, the recent outcomes of CB transplantation in adult patients with MDS are generally less well reported. However, in 2011, Eurocord and EBMT reported the outcomes in 108 adults who received CB transplants for MDS ($n=69$) or secondary AML ($n=39$).[47] The median age was 43 years (range, 18–72 years). Seventy-seven patients (71%) received single CB units and 57 (53%) patients were given MAC. The cumulative incidence of neutrophil recovery was 78±4% at 60 days with a median time of 23 days (range, 6–51 days). The 2-year NRM was 49±5% and was significantly higher after MAC (HR 2.38 (95% CI, 1.32–4.17); $P=0.009$). Two-year DFS and OS were 30±5% and 34±5%, respectively. Patients with high-risk disease (blasts >5% and/or IPSS≥int-2) had significantly worse DFS (HR 0.57 (95% CI, 0.32–0.99); $P=0.047$).

*Lymphoid malignancies*: The role of CB transplantation for patients with lymphoma or chronic lymphoid leukemia (CLL) also appears promising. Several years ago, Eurocord and the Lymphoma Working Party of EBMT evaluated 104 unrelated CB transplants in adults (median 41 years, range, 16–65 years) with lymphoid malignancies.[48] Seventy-eight (75%) patients received a single CB transplant and 64 (62%) received an RIC regimen. Progression-free survival (PFS) and OS at 1-year were 40% and 48%, respectively. PFS was improved in those with chemosensitive disease (HR 0.54 (95% CI, 0.31–0.93); $P=0.03$), who received higher cell dose (≥$2\times10^7$/kg) (HR 0.49 (95% CI, 0.29–0.84); $P=0.009$) and following use of low-dose TBI (HR 0.40 (95% CI, 0.23–0.69); $P=0.001$).

### 4.2.2 Single Unrelated CB Transplants in Adults

As single CB transplantation was increasingly being used to treat adults, it became necessary to understand how the outcomes compared to conventional BMT, particularly considering the lower cell doses available in CB. Three large retrospective studies directly compared the results of single CB transplantation with related or unrelated BMT in adults.[6,25,49] All three studies demonstrated that CB transplantation was associated with delayed neutrophil and platelet recovery and a lower incidence of acute or chronic GvHD compared to BM transplantation. However, the results for TRM and DFS were more discrepant. In the Eurocord study of 98 CB transplants and 584 HLA-matched BMT (1998–2002), TRM (HR 1.13 (95% CI, 0.78–1.64); $P=0.50$) and LFS (HR 0.95 (95% CI, 0.72–1.25); $P=0.70$) were not significantly different between the groups.[6] At a similar time, an IBMTR and NCBP study demonstrated in 150 CB transplants and 450 BMT (1996–2001), that TRM was higher (HR 1.89 (95% CI, 1.45–2.48); $P<0.001$) and DFS lower (HR 1.48 (95% CI, 1.18–1.86); $P=0.001$) in CB transplants compared to HLA-matched BM.[25] Of note, there were no significant differences when comparing CB with HLA-mismatched BM. In contrast, a Japanese analysis of 68 CB transplants with 45 BMT (1996–2003) observed lower TRM (HR 0.32 (95% CI, 0.12–0.86); $P=0.02$) and improved DFS (HR 0.27 (95% CI, 0.14–0.51); $P<0.01$) when comparing CB with BM transplantation.[49] These studies highlight the difficulties of comparing different nonrandomized studies, particularly with differing combinations of HLA-matched and -mismatched CB and BM transplants. Unsurprisingly, an overall *meta*-analysis of these studies, comparing 316 adults undergoing CB transplantation with 996 adults undergoing unrelated BMT, demonstrated delayed neutrophil recovery in CB transplantation but no significant difference in TRM and DFS.[40]

One significant limitation of the three studies above, particularly in relation to today's clinical practice, is that the BMT donors were HLA typed using low-resolution methods for HLA-A and -B and high-resolution typing for HLA-DRB1. However, selection of an unrelated donor is now based on high-resolution (allelic) typing for class I (HLA-A, -B, and -C) and class II (HLA-DRB1 ± DQB1). Therefore, in 2010, Eurocord in collaboration with the CIBMTR compared the outcomes of 165 unrelated CB transplants with 1360 unrelated donor HSCT (BM ($n=472$) and PBSC ($n=888$)) in adult patients (>16 years) with acute leukemia (2002–2006).[50] CB transplants were matched at HLA-A and -B at the antigen level and HLA-DRB1 at allele level ($n=10$) or mismatched for one or two antigens ($n=155$). Only CB grafts meeting the current recommended selection criteria for choosing CB units (≤2/6 HLA disparities and TNC dose at freezing $>2.5\times10^7$/kg) were included in this analysis. PBSC and BM grafts were matched at high resolution (allele) for HLA-A, -B, -C, and DRB1 ($n=632$ and $n=332$, respectively) or mismatched at one HLA-locus ($n=256$ and $n=140$, respectively). Day 42 neutrophil recovery was inferior after CB transplantation (80%) compared to PBSC (96%) and BM (93%) ($P<0.0001$). Acute GvHD (grade II–IV) was lower in CB transplants compared to HLA-matched PBSC (HR 0.57 (95% CI, 0.42–0.77); $P<0.01$) but not when compared to HLA-matched BM (HR 0.78 (95% CI, 0.56–1.08); $P=0.13$). However, chronic GvHD was lower in CB transplants compared to both HLA-matched PBSC and BM (HR 0.38 (95% CI, 0.27–0.53); $P<0.01$ and HR 0.63 (95% CI, 0.44–0.90); $P=0.01$, respectively). TRM was higher after CB transplantation compared to allele-matched PBSC (HR 1.62 (95% CI, 1.18–2.23); $P<0.01$) or BM (HR 1.69 (95% CI, 1.19–2.39); $P<0.01$). However, importantly, LFS after CB transplantation was not significantly different to either HLA-matched or single HLA-mismatched PBSC or BM transplantation. These findings were supported by a similar Japanese analysis of 351 CB transplants with 1028 BMT (1996–2005) which also observed similar DFS (HR 0.97 (95% CI, 0.92–1.35); $P=0.75$) when comparing CB transplants with single HLA-mismatched BM (HLA-DRB1).[51] These results, therefore, support the use of CB transplantation when an HLA-matched donor is unavailable. Furthermore, CB transplantation may be considered when an urgent HSCT is required and there may be a significant delay in identifying a suitable HLA-matched donor.

### 4.2.3 Double Unrelated CB Transplants in Adults

To overcome the issue of low infused CB cell dose in larger children and adults, in 2001, the Minneapolis group demonstrated that transplanting two partially HLA-matched CB units from different donors was feasible.[52] Although two CB units are infused, one unit normally wins the immunological battle to provide long-term donor engraftment. However, despite the fact that most double CB transplant recipients are heavier than patients receiving a single unit, the cumulative incidence of myeloid recovery does not differ between the two groups, suggesting an initial response from the non-engrafting unit. For malignant disease, there may also be an additional immunological benefit of using two CB units. In analysis of 177 patients receiving a myeloablative CB transplant for acute leukemia (ALL ($n=88$) and AML ($n=89$)), the cumulative incidence of relapse was significantly lower in those patients (CR1/2 only) receiving two CB units (RR 0.5 (95% CI, 0.2–1.0); $P=0.04$).[53] It was, therefore, proposed that using two CB units might have a greater early Graft-versus-Leukemia (GvL) effect. However, DFS was not significantly different between single and double CB transplant recipients, at 40% (95% CI, 30–51%) and 51% (95% CI, 41–62%), respectively ($P=0.35$), possibly due to the limited sample size. Shortly afterward, Eurocord and the ALWP of EBMT reviewed the outcomes after single CB (TNC $>2.5\times10^7$/kg) or double CB transplantation in adult patients with acute leukemia in remission (AML and ALL) (Eurocord, unpublished data). Using double CB grafts was associated with a reduced risk relapse and improved LFS, but only in patients transplanted in CR1 using an RIC regimen. Eurocord has also reported similar findings in CB transplantation for lymphoid malignancies.[48] In a retrospective review of CB transplant outcomes in 104 adults receiving a transplant for lymphoid malignancy (NHL ($n$=61), Hodgkin's lymphoma ($n$=29), and CLL ($n$=14)), the risk of relapse or progression at 1 year was significantly lower in double CB transplants compared to single CB transplants (13% vs 38%, $P$=0.009). This remained significant in multivariate analysis (RR 0.28 (95% CI, 0.09–0.87); $P$=0.03).

Double CB transplantation has led to a significant increase in the number of adult patients receiving CB transplants and the results from many retrospective studies support the efficacy and safety of the procedure.[26] In 2011, Eurocord reported the outcomes of 35 double CB transplants in recipients with high-risk hematological diseases. The median infused TNC and CD34$^+$ doses were $4\times10^7$/kg (range, 1.8–$9.7\times10^7$/kg) and $3\times10^5$/kg (range, 0.5–$7.5\times10^5$/kg), respectively. The cumulative incidence of neutrophil recovery at day 60 was 72±8% with a median time of 25 days (range, 11–42 days). The predicted 2-year OS was 48±8%.[54] Likewise, in a long-term SGGM-TC follow-up study of 136 double CB transplants in patients with hematological malignancies, the cumulative incidence of neutrophil engraftment by day 60 was 91% (95% CI, 89–94%), with 2-year TRM 27% (95% CI, 23–31%).[55] The 3-year OS and PFS were 41% (95% CI, 34–51%) and 35% (95% CI, 24–44%), respectively. More recently, Eurocord and EBMT compared the results of 239 single ($n$=156) and double ($n$=83) CB transplants in adults with acute leukemia in CR1 using MAC regimens.[56] The cumulative incidence of neutrophil engraftment at day 60 was similar in single and double CB transplants (82±6% vs 90±6%, respectively), although a TNC dose >$3.2\times10^7$/kg was independently associated with higher neutrophil engraftment (HR 0.63 (95% CI, 0.36–0.86); $P$=0.01). Double CB transplantation had higher rates of acute GvHD (grade II–IV) at day 100 but there was no difference in the 2-year risk of relapse in this study. LFS was 43±3% at 2 years post-transplant. Interestingly, in subgroup analysis, single CB transplants using non-TBF (Thiotepa/Busulfan/Fludarabine) conditioning regimens were associated with inferior LFS compared to double CB transplants (HR 1.6 (95% CI, 1.03–2.49); $P$=0.03). However, no difference in LFS was observed between single CB transplants using TBF conditioning and double CB transplants ($P$=0.98). These important findings, therefore, suggest that the overall outcomes of single CB transplantation, when using sufficient cell doses (TNC >$2.5\times10^7$/kg) and specific MAC regimens (e.g., TBF), may have similar results to double CB transplantation in adult patients with hematological malignancy. Further studies will be required to confirm these observations.

To compare the results of double CB transplantation to using other donor sources, Brunstein et al. (2010) published a retrospective study of 536 patients with malignant disease (AML ($n$=211), ALL ($n$=236), CML ($n$=70), and MDS ($n$=19)) transplanted with an HLA allele-matched (8/8) related donor (MRD; $n$=204) or unrelated donor (MUD; $n$=152), single allele-mismatched unrelated donor (MMUD; $n$=52) or a double CB graft ($n$=128).[41] The majority of patients were adults, with a median age of 25 years (range, 10–46 years), and all received an MAC regimen with cyclophosphamide and TBI. For the CB transplants, the median TNC dose was $4.0\times10^7$/kg (range, 2–$30\times10^7$/kg) and the majority (61%) had ≥2 HLA mismatches with the recipient in one or both of the CB units transplanted. Double CB transplantation was associated with a significantly lower cumulative incidence of neutrophil recovery, acute GvHD (grades II–IV), and chronic GvHD compared to other donor sources. The risk of NRM at 2 years was higher after double CB transplantation (34% (95% CI, 25–42%)) compared to MRD (24% (95% CI, 17–39%)) and MUD transplants (14% (95% CI, 9–20%)). Conversely, the 5-year risk of relapse was lower following double CB transplantation (CB 15% (95% CI, 9–22%); MRD 43% (95% CI, 35–52%); MUD 37% (95% CI, 29–46%); and MMUD 35% (95% CI, 21–48%)). Overall, the 5-year LFS was comparable for all groups (CB 51% (95% CI, 41–59%); MRD 33% (95% CI, 26–41%); MUD 48% (95% CI, 40–56%; and MMUD 38% (95% CI, 25–51%)). Therefore, in the absence of an HLA-matched sibling, LFS after double CB transplantation appears comparable to the results when using a MUD and provides a suitable alternative in adult patients requiring an HSCT.

As the use of double CB transplantation in adults has continued to increase, there has been a corresponding expansion in the use of RIC regimens to allow older adults to benefit from this procedure. In 2007, the Minnesota group reported the results for 110 adult patients receiving a CB transplant for hematological diseases using a non-myeloablative regimen consisting of single fraction of TBI (200 cGy), Fludarabine, and cyclophosphamide (TCF).[57] The median age of the recipients was 51 years (range, 17–69 years). The majority of transplants ($n$=93; 85%) used two CB units with a median infused TNC and CD34$^+$ dose of $3.7\times10^7$/kg (range, 1.1–$5.3\times10^7$/kg) and $4.7\times10^5$/kg (range, 0.7–$18.8\times10^5$/kg), respectively. Of note, the cell doses were similar between the single and double CB transplants (TNC dose 3.3 vs $3.7\times10^7$/kg ($P$=0.2) and CD34 dose 3.8 vs $4.9\times10^5$/kg, ($P$=0.60), respectively). Neutrophil recovery occurred in 92% of transplants with a median time of 12 days (range, 0–32 days). TRM, DFS, and OS at 3 years were 26% (95% CI, 18–34%), 38% (95% CI, 28–48%), and 45% (95% CI, 34–56%), respectively. The Société Française de Greffe de Moelle-Thérapie Cellulaire (SFGM-TC) in collaboration with Eurocord later reviewed the outcomes of 155 RIC CB transplants (Eurocord, unpublished data). The cumulative incidence of neutrophil engraftment by day 60 was 80±3% with higher CD34 cell dose (>$1.2\times10^5$/kg) (HR 1.51; $P=0.04$) and greater HLA matching (0–1 HLA disparities) (HR 1.5; $P=0.05$) associated with improved neutrophil recovery. The 18-month cumulative incidence of TRM was 18±3% with an estimated OS and DFS of 62±5% and 51±4%, respectively. These studies, therefore, demonstrated that RIC regimens have acceptable outcomes following CB transplantation and have become the basis for many CB transplants worldwide, particularly in older patients. However, as with MAC regimens, CB cell dose and HLA matching still appear to be critical factors for the overall results, particularly engraftment.

Several groups have demonstrated that RIC CB transplants for hematological malignancy have comparable outcomes to RIC transplants using conventional HSC donors. In 2012, Brunstein et al. analyzed the outcomes of 160 double CB transplants with 414 PBSC transplants using unrelated donors.[42] All patients were adults (>18 years) with acute leukemia (ALL ($n=50$), AML ($n=523$)). Four groups were evaluated: CB transplants receiving TBI/Cyclophosphamide/Fludarabine conditioning (CB-TCF) ($n=120$); CB transplants receiving other RIC conditioning (CB-other) ($n=40$); HLA-matched (8/8) PBSC transplants ($n=313$); and single HLA-mismatched (7/8) PBSC transplants ($n=111$). Compared with HLA-matched PBSC transplants, TRM and overall mortality were similar after double CB transplants using TCF conditioning (RR 0.72; $P=0.72$ and RR 0.93; $P=0.60$, respectively) but higher after double CB transplants using other RIC regimens (HR 2.70; $P=0.0001$ and HR 1.79; $P=0.004$, respectively). However, compared with single HLA-mismatched PBSC transplants, TRM, but not overall mortality, was lower after double CB transplants using TCF conditioning (RR 0.57; $P=0.04$ and RR 0.87; $P=0.41$, respectively). Two-year OS were 37% (95% CI, 28–48%), 19% (95% CI, 4–34%), 44% (95% CI, 38–50%), and 37% (95% CI, 27–46%) for the CB-TCF, CB-other, PBSC HLA-matched, and PBSC HLA-mismatched transplants, respectively. Therefore, double CB transplants using a TCF RIC regimen appears to provide comparable outcomes to using an HLA-mismatched unrelated PBSC donor in adult patients with acute leukemia. However, care must be taken as not all RIC regimens produce the same results. A similar comparison of donor source (matched sibling, unrelated, or CB) in RIC HSCT for older patients (>50 years) with AML in CR was recently published.[58] The 3-year cumulative incidence of TRM was 18% (95% CI, 10–28%), 14% (95% CI, 5–28%), and 24% (95% CI, 15–34%) with sibling, unrelated, and CB transplantation, respectively ($P=0.22$). The corresponding 3-year LFS was 48% (95% CI, 38–62%), 57% (95% CI, 42–78%), and 33% (95% CI, 23–45%), respectively ($P=0.009$). However, in multivariate analysis, compared to sibling HSCT, use of unrelated donors (HR 1.26 (95% CI, 0.58–2.73); $P=0.56$) or CB grafts (HR 1.23 (95% CI, 0.72–2.08); $P=0.45$) was not associated with a significant difference in LFS. Therefore, these data support the continued use of RIC-unrelated CB transplantation for the treatment of older patients with AML with comparable outcomes to using conventional donor sources.

## 5. SELECTION OF CB UNITS FOR TRANSPLANTATION

### 5.1 Cell Dose and HLA Matching

The two most important considerations when selecting CB units for HSCT are cell dose and the HLA matching between the recipient and CB unit. Many retrospective studies have found them to be critical independent factors for engraftment and overall outcome, including survival.[8–10] Eurocord analyzed the interaction between cell dose and HLA disparity in 550 single-unit CB transplants (1994–2001) in recipients with malignant disorders.[8] The 60-day cumulative incidence of neutrophil engraftment for all patients was 74% (95% CI, 70–78%), with an incidence of 83% for those using HLA-matched (6/6) CB grafts compared to 53% when using CB units with at least three HLA mismatches. In multivariable analysis, the number of HLA disparities (HR 0.79 (95% CI, 0.68–0.91); $P=0.001$) and the TNC dose at freezing (≥$4\times10^7$/kg) (HR 1.004 (95% CI, 1.001–1.006); $P<0.0001$) were independently associated with neutrophil recovery. Eapen et al. (2007) also analyzed the interaction between cell dose and HLA, comparing outcomes of 503 CB transplants and 282 unrelated BMT in children with acute leukemia.[9] The probability of neutrophil recovery by day 42 was similar after unrelated BMT- or HLA-matched (6/6) CB transplants. For the single HLA-mismatched (5/6) CB transplants, a cutoff for the collected TNC was observed with higher cell doses (>$3.0\times10^7$/kg) resulting in a higher probability of engraftment. However, a dose effect was not seen in CB transplants mismatched at two HLA-loci (4/6), suggesting that cell dose may not be able to overcome the increasing adverse impact of HLA mismatching in the setting of two or more HLA disparities. Later, Barker et al. (2010) analyzed the interaction between the collected TNC dose and HLA matching in a retrospective analysis of 1061 CB transplants.[10] All patients received single-unit MAC CB transplants for the treatment of acute leukemia or myelodysplasia. TNC dose was significantly associated with neutrophil and platelet engraftment in a dose-responsive manner. Using a reference TNC dose of 2.5–$4.9\times10^7$/kg, the HR for neutrophil engraftment was 0.7 (95% CI, 0.6–0.8) ($P<0.001$) for TNC 0.7–$2.4\times10^7$/kg; 1.2 (95% CI, 1.0–1.5) ($P<0.001$) for TNC 5.0–$9.9\times10^7$/kg; and 1.8 (95% CI, 1.3–2.5) ($P<0.001$) for TNC >$10.0\times10^7$/kg. Likewise, the degree of HLA matching was also associated with engraftment. Using a reference group of single HLA-mismatched transplants (5/6), the HR for neutrophil engraftment was 1.8 (95% CI, 1.3–2.5) ($P<0.001$) for matched CB units (6/6); 1.0 (95% CI, 0.9–1.20) ($P=0.90$) for two HLA mismatches; and 0.8 (95% CI, 0.6–1.1) ($P=0.16$) for three HLA mismatches. Analyzing the interaction between TNC dose and HLA matching (reference group was CBU with a single HLA mismatch and TNC $2.5–5.0\times10^7$/kg) on engraftment, TRM and overall mortality demonstrated that the best outcomes were achieved in matched CB transplants irrespective of TNC dose. CB transplants using CB with a single HLA mismatch and TNC $>2.5\times10^7$/kg or two HLA mismatches and TNC $>5.0\times10^7$/kg/kg were next. Using CB units with two HLA mismatches and TNC $>5.0\times10^7$/kg had faster engraftment than CBU with a single HLA mismatch and TNC dose $>2.5\times10^7$/kg, although there was no significant difference in survival. Transplants using CBU with two HLA mismatches and TNC $2.5–5.0\times10^7$/kg had higher mortality, followed by CBU with 1–2 HLA mismatches and TNC $<2.5\times10^7$/kg or >2 HLA mismatches irrespective of dose. In a more recent analysis of 1658 single CB transplants for hematological malignancies using high-resolution typing at HLA-A, -B, -C, and -DRB1, using CB units with TNC $<3.0\times10^7$/kg was associated with significantly higher NRM, independent of HLA matching.[59] However, compared to CB units containing TNC $>3.0\times10^7$/kg, further increases in cell dose were not associated with any further improvement in NRM.

In all these analyses, the interaction between cell dose and HLA matching was studied in patients with malignant disorders. However, in nonmalignant conditions, such as hemoglobinopathies, the required CB cell dose and the interactions with HLA matching may be somewhat different. This is because patients with nonmalignant conditions can have a full hematopoiesis and/or are unlikely to have received chemotherapy or immunosuppression before transplant conditioning. In addition, they have often received multiple previous transfusions, increasing the risk of developing anti-HLA antibodies and, thus, increasing the risk of nonengraftment and mortality.

In 2009, Eurocord published recommendations to guide clinicians when selecting a CB unit for transplantation, taking into account the impact of diagnosis, cell dose, and HLA incompatibilities.[7] For malignant diseases, Eurocord currently recommends a minimum target TNC dose of $3.0–3.5\times10^7$/kg at collection or $2.5–3.0\times10^7$ kg at infusion. If this cell dose cannot be achieved with a single CB unit, double CB transplantation should be considered. Well-matched CB units (6/6 for HLA-A and -B antigen; HLA-DRB1 allele) should ideally be selected, although one- or two-antigen mismatched units (5/6 or 4/6 HLA-matched) can be used if HLA-matched CB units are not available. Recently, Eapen et al. (2014) also demonstrated that HLA-allele mismatches, when using high-resolution HLA typing at HLA-A, -B, -C, and -DRB1, have an important impact on NRM.[59] In this study, 42% of the CB units with 2-antigen mismatches (4/6 HLA matched) had ≥4 allele mismatches and were associated with higher NRM. It is therefore recommended that additional high-resolution HLA typing should be performed if possible, especially for CB units matched at only 4/6 HLA-loci. When considering locus-specific effects, single allele mismatches at HLA-A, -C, or -DRB1 were associated with a three-fold increase in NRM. An isolated allele mismatch at HLA-B was better tolerated, although the likelihood of identifying such a unit is relatively low because HLA-B and -C loci are in linkage disequilibrium. Increasing cell dose may attenuate the effect of HLA mismatching, but not for CB grafts with ≥3 of 6 or ≥4 of 8 HLA incompatibilities and, therefore, selection of these units is not recommended for routine use. Patients with a nonmalignant disease should receive a higher TNC dose to obtain engraftment which should not be $<4.0\times10^7$/kg at collection or $<3.5\times10^7$/kg at infusion. HLA matching also has an important role in engraftment, GvHD, TRM, and survival when treating nonmalignant conditions. The effects of HLA mismatches are partially reduced by increasing the cell dose. However, CB units containing ≥2 HLA incompatibilities with a TNC dose of $<3.5\times10^7$ TNC/kg should generally be avoided. Experience of double CB transplantation for nonmalignant disorders remains too limited to allow routine recommendation in this setting. A summary of the current CB selection criteria is shown in Table 2.

## 5.2 Other Considerations in Selecting a CB Unit

In addition to cell dose and HLA matching, there are a number of other factors that need to be considered when selecting a CB unit for transplantation.

### 5.2.1 *Markers of Cell Content*

It remains to be determined which phenotypic marker is the best measure of cell content in CB grafts and how these relate to clinical outcomes following transplantation. The majority of studies have reported the TNC dose at freezing and/or after thawing. However, when selecting units based upon prefreeze counts, it should be noted that there is a median loss of approximately 20% of nucleated cells after thawing. $CD34^+$ cell dose may be a better phenotypic marker since it more accurately reflects HSC content and the loss of CD34 cells after thawing may be less pronounced. However, lack of previous standardization of $CD34^+$ quantification could mean that older CD34 counts are less reliable. Another factor that remains to be resolved is the association of TNC and/or $CD34^+$ cell viability with clinical outcomes, since the techniques used for measuring viability have varied between different transplant centers

**TABLE 2 Recommendations for Selection of Cord Blood Units in Allogeneic Transplantation, Proposed Selection Criteria for Choosing Suitable Cord Blood Units for Allogeneic Hematopoietic Stem Cell Transplantation**

1. Initial selection of cord blood units (CBU) should consider:
   a. HLA typing of the recipient and CBU
   b. CBU cell dose
   c. Patient diagnosis
   d. Avoid CBU that matches the specificity of any anti-HLA antibodies within the recipient.

2. Recommendations for HLA matching
   HLA matching is currently based upon low-resolution (antigen) typing for HLA-A and -B, and high-resolution typing (allele) for HLA-DRB1. If many CBUs are potentially available, consider allele typing at HLA-A, -B, -C, and -DRB1*.
   a. Select an HLA-matched (6/6 or 8/8) CBU.
   b. When an HLA-matched CBU is unavailable; select a unit matched at 4/6 or 5/6 HLA-loci. Avoid CBU with HLA-DRB1 mismatches. If many CBUs matched for 4/6 HLA-loci are available, avoid units with ≥4 allele mismatches when performing allele typing of HLA-A, -B, -C, and -DRB1.
   c. CBU matched for ≤3/6 HLA-loci are not recommended.

*N.B. The impact of HLA allele typing in CB transplantation has only been analyzed for patients transplanted with single CB units for malignancies using a myeloablative conditioning regimen.

*Other considerations*:

- Selection of CB units based on direction of HLA mismatches is not recommended.

3. Recommendations for cell dose

**Malignant disorders**

Nucleated cell dose: TNC dose ≥3.0–3.5 × $10^7$/kg (at freezing); ≥2.5–3.0 × $10^7$/kg (after thawing)

CD34+ cell dose: CD34 1.0–1.7 × $10^5$/kg (at freezing); 1.0–1.2 × $10^5$/kg (after thawing)

*Other considerations*:

- If the TNC dose infused is 1.0–2.0 × $10^7$/kg, CD34+ cell dose and/or CFU-GM should be taken into consideration.
- In double CB transplantation, the TNC dose should be ≥1.5 × $10^7$/kg at freezing for at least one unit.

**Nonmalignant disorders**

Nucleated cell dose: TNC dose ≥4.0 × $10^7$/kg (at freezing); ≥3.5 × $10^7$/kg (after thawing)
CD34+ cell dose: CD34+ dose >1.7 × $10^5$/kg (at freezing or after thawing)

*Other considerations*:

- For patients with BM failure syndromes (aplastic anemia or congenital bone marrow failure) or hemoglobinopathies, the TNC dose at freezing should be ≥5 × $10^7$/kg

**TABLE 2 Recommendations for Selection of Cord Blood Units in Allogeneic Transplantation, Proposed Selection Criteria for Choosing Suitable Cord Blood Units for Allogeneic Hematopoietic Stem Cell Transplantation—cont'd**

4. Other selection criteria to consider (*if many CBU available*)
   a. Used accredited cord blood banks (to ensure safety and reliability)
   b. ABO compatibility (avoid major ABO mismatch)
   c. NIMA (consider NIMA-matched CBU; as part of clinical trial)

*N.B. selection of CBU based on KIR-ligand compatibility and direction of HLA mismatch is not recommended*

Abbreviations: CB, cord blood; CBU, cord blood unit; HLA, human leukocyte antigen; KIR, killer cell immunoglobulin receptor; NIMA, noninherited maternal antigen; TNC, total nucleated cell.

and over time. In double CB transplantation, however, viable CD34+ cell dose has been associated with the engrafted unit.[60] While the TNC or CD34+ dose remains useful, these markers may not necessarily correlate with functional HSC. Colony-forming unit, granulocyte-macrophage (CFU-GM) in CB grafts may have better HSC potential but few studies have documented a definite correlation with outcomes. Therefore, due to current technical aspects, the use of this cell marker is difficult to apply for routine selection of CB units. However, where possible, the results of CFU-GM after thawing should be collected to aid future studies and to assist the treating physicians in cases of CB graft failure.

### 5.2.2 Influence of HLA-C Matching

While HLA matching at antigen level (low or intermediate resolution) for HLA-A and -B and allele-level matching for HLA-DRB1 continues to be the current standard for CB unit selection, some studies have analyzed the impact of additional matching for HLA-C. In 2011, Eurocord in collaboration with CIBMTR retrospectively analyzed the impact of additional matching for HLA-C (antigen level) in 803 single CB transplants performed for AML ($n$=727) or MDS ($n$=76).[61] Day-28 neutrophil recovery was significantly lower for transplants mismatched at three/four HLA-loci (matched 70% (95% CI 57–79%); one mismatch 64% (95% CI 55–71%); two mismatches 64% (95% CI 57–69%); three mismatches 54% (95% CI 48–60%); four mismatches 44% (95% CI 32–55%)). More specifically, mismatching at HLA-DRB1 in the presence of mismatches at any other two HLA-loci and mismatching at HLA-A in the presence of mismatches at three or four HLA-loci were associated with inferior neutrophil engraftment. TRM was higher when CB units were mismatched at two ($n$=259; HR 3.27 (95% CI, 1.42–7.54); $P$=0.006), three ($n$=253; HR 3.34 (95% CI, 1.45–7.71); $P$=0.005), or four loci ($n$=75; HR 3.51

(95% CI, 1.44–8.58); $P=0.006$) compared to matched units ($n=69$; HR 1.00). In addition, TRM using CB units mismatched at HLA-C was greater compared to fully matched CB units (8/8) (HR 3.97 (95% CI, 1.27–12.40); $P=0.02$). TRM was also higher after CB transplantation with a single mismatch at HLA-A, -B, or -DRB1 and mismatched at HLA-C ($n=234$) compared with CB transplants matched at HLA-C with a single mismatch at HLA-A, -B, or -DRB1 (HR 1.70 (95% CI, 1.06–2.74); $P=0.03$). Additional matching at HLA-C was therefore recommended.

### 5.2.3 Allelic Typing

In unrelated BMT, current donor selection is based upon high-resolution typing (allelic) at HLA class I (HLA-A, -B, and -C) and class II (DRB1 ± DQB1) loci. Allelic disparities at these HLA-loci are associated with inferior transplant outcomes.[62] In light of these findings, several retrospective analyses have evaluated the impact of HLA-allele mismatches in CB transplantation. In 2005, Koegler et al. demonstrated significant changes in HLA-matching status in 122 CB transplants when reassigning them using allelic typing for HLA-A, -B, and -DRB1.[63] Only 9 of the 16 originally reported as HLA-matched (6/6) remained fully matched when using the new allelic level HLA typing. When additional typing at HLA-C and HLA-DQB1 were included, only 14% were matched at 9–10/10 alleles, 63% were matched for 6–8/10 alleles, and the remaining 23% were more extensively mismatched. In 2008, the Cord Blood Transplant Study (COBLT) retrospectively analyzed 179 pediatric CB transplants using high-resolution typing at HLA-A, -B, and -DRB1.[27] Using high-resolution typing, 9% of transplants were HLA-matched (6/6), 65% had one or two HLA-allele mismatches, and 26% were allele mismatched at >2 HLA-loci. This contrasts to 9% (6/6), 89% (4–5/6), and 3% (3/6) when matching using conventional approaches (low level at HLA-A and -B; high level at HLA-DRB1). HLA matching (5–6/6) by high-resolution typing was associated with a lower incidence of severe acute GvHD ($P=0.02$) but had no effect on neutrophil or platelet engraftment. Although there was a trend toward improved survival for patients given grafts matched at 6/6 by high-resolution typing, the study was not sufficiently powered to reach statistical significance. Recently, in a much larger CIBMTR/Eurocord analysis, the outcomes of 1658 MAC single CB transplants for hematological malignancies were analyzed using high-resolution typing at HLA-A, -B, -C, and -DRB1.[59] Day-28 neutrophil recovery was significantly reduced in CB transplants mismatched at ≥3 alleles compared to fully-matched CB (10/10) (OR 0.56 (95% CI, 0.36–0.88), $P=0.01$; OR 0.55 (95% CI, 0.34–0.88), $P=0.01$; OR 0.45 (95% CI, 0.25–0.82), $P=0.009$ for three, four, and five allele mismatches, respectively). Neutrophil recovery was also reduced in CB transplants mismatched at ≥3 alleles compared to CB transplants mismatched at one or two alleles (8–9/10) (three to four mismatches: OR 0.69 (95% CI, 0.55–0.86), $P=0.001$; five mismatches: OR 0.56 (95% CI, 0.35–0.89), $P=0.01$). NRM was also significantly associated with the number of high-resolution mismatches. Single allele mismatches at HLA-A, -C, or -DRB1 were associated with increased NRM (HR 3.05 (95% CI, 1.52–6.14), $P=0.02$; HR 3.04 (95% CI, 1.28–7.20), $P=0.01$; HR 2.93 (95% CI, 1.38–6.25), $P=0.005$, respectively). Based upon these findings, it has been proposed that the best HLA-allele match should be selected, although mismatches at one or two alleles are acceptable. However, CB transplants with mismatches at three or more alleles should be used carefully due to the increased risk of graft failure and NRM. Further studies will need to be performed to confirm these findings and to assess the impact of high-resolution HLA matching on CB donor availability.

### 5.2.4 Direction of HLA Mismatches

Despite the increasing number of cryopreserved CB units available for transplantation, the majority of CB transplants still use a unit mismatched for at least one HLA antigen/allele. If the recipient is homozygous at an HLA-locus but the CB donor is heterozygous at the same site (only one antigen/allele matching the recipient), there is a mismatch in the Host-versus-Graft (HvG) direction (risk of rejection). Conversely, if the CB donor is homozygous but the recipient heterozygous at the same HLA-locus (only one antigen/allele matching the donor), the mismatch is in the Graft-versus-Host (GvH) direction. When a mismatched antigen/allele is present in the recipient and donor, the mismatch is bidirectional.

In the setting of an HLA mismatch between the CB and recipient, there remains debate as to whether direction of HLA mismatch should be considered when selecting the unit. In a retrospective analysis of HLA-mismatch direction in 1202 single-unit CB transplants, 890 transplants had bidirectional HLA mismatches, 58 GvH mismatches only, 40 HvG mismatches only, and 145 had other combinations.[64] Recipients of HvG mismatches only had a nonsignificant trend toward lower engraftment compared to those with a single bidirectional mismatch (HR 0.7 (95% CI, 0.4–1.1); $P=0.1$). Conversely, those with no HLA mismatches or GvH mismatches only had improved engraftment (HR 1.5 (95% CI, 1.1–2.0), $P=0.006$; HR 1.6 (95% CI, 1.2–2.2), $P=0.003$, respectively). In subgroup analysis of those with malignant disorders, recipients of CB units with HLA mismatches in the GvH direction only had lower overall mortality (HR 0.5 (95% CI, 0.3–0.9); $P=0.02$). In a Japanese study of 2977 patients with leukemia or MDS receiving a single-unit CB transplant, recipients of CB units with HLA mismatches in the GvH direction only showed a nonsignificant trend toward improved neutrophil and platelet recovery compared to single bidirectional mismatched transplants ($P=0.08$ and $P=0.05$, respectively). However, the presence of HLA mismatches in the GvH direction only or the HvG direction only was not associated with a significant difference in overall mortality compared to single bidirectional mismatched transplants. This observation was later supported by a recent Eurocord analysis of 1565 single CB transplants in which the direction of HLA mismatch had no significant effect on overall mortality or survival.[65] Therefore, although possibly influencing engraftment, the published data do currently support the routine selection of CB units based upon the direction of HLA mismatch.

### 5.2.5 Anti-HLA Antibodies

DSA may have important implications for engraftment in CB transplantation. Since the majority of CB transplants use HLA-mismatched units, the presence of anti-HLA antibodies in the recipient, specific against the mismatched antigen/allele in the cord, should be considered. In a Japanese study, 386 MAC single-unit CB transplants were retrospectively analyzed for the presence of anti-HLA antibodies pretransplant.[66] Of the 89 testing positive, 20 had specificity against an HLA antigen within the CB. The cumulative incidence of neutrophil recovery was 83% (95% CI, 79–87%) for the antibody-negative group, 73% (95% CI, 61–82%) for the anti-HLA antibody positive group, and 32% (95% CI, 13–53%) for the DSA-positive group. In a multivariate analysis, patients with DSA had significantly lower neutrophil and platelet recovery compared to the DSA-negative group (RR 0.23 (95% CI, 0.09–0.56), $P=0.001$; RR 0.31 (95% CI, 0.12–0.81), $P=0.02$, respectively).

In double CB transplantation, the impact of DSA shows differing results. Retrospective analysis of 73 double CB transplants showed an increased risk of graft failure (5.5% vs 18.2% vs 57.1% for none, single, and dual DSA positivity; $P=0.0001$) and lower 3-year OS (0.0% vs 45.0%, $P=0.04$) for patients with positive DSA.[67] However, a similar report showed no association between the presence of DSA and transplant outcomes in 126 double CB transplant recipients, although lower thresholds were used to detect DSA.[68] More recently, in a Eurocord analysis of 294 RIC unrelated CB transplants (60% double CB transplants), 5% of recipients had DSA. Day-60 neutrophil engraftment (44% vs 81%; $P=0.006$) and 1-year TRM (46% vs 32%; $P=0.06$) were inferior in the presence of DSA. Therefore, it is now recommended that recipients should be screened pretransplant for the presence of anti-HLA antibodies and only CB units that do not match the specificity of any anti-HLA antibodies in the recipient should be selected.

### 5.2.6 Killer-immunoglobulin Receptor Ligand Matching

In HSCT, natural killer (NK)-cell alloreactivity stems from mismatches between the inhibitory receptors of self-MHC class I molecules (Killer cell Immunoglobulin-like Receptors (KIR)) on NK cells and MHC class I antigens. In haploidentical and HLA-mismatched unrelated HSCT, donor KIR-ligand incompatibility in the GvHD direction has been associated with reduced relapse and improved LFS.[69] Eurocord, therefore, assessed KIR-compatibility in the outcomes of 218 patients with acute leukemia in CR (AML ($n=94$) and ALL ($n=124$)) who had received a single-unit unrelated CB transplant.[70] KIR-ligand incompatible CB transplants in the GvH direction had lower relapse (HR 0.53 (95% CI, 0.3–0.99); $P=0.05$) and improved OS (HR 2.0 (95% CI, 1.2–3.2); $P=0.004$). These findings were more marked in patients with AML, with 2-year relapse of 5% versus 36% ($P=0.005$) and 2-year LFS of 73% versus 38% ($P=0.01$) in CB transplants with or without KIR-ligand mismatches, respectively. However, in 2009, the Minnesota group published contrasting results from an analysis of KIR-ligand matching in 257 patients receiving CB transplantation for malignant diseases.[71] In the MAC cohort, there was no significant effect observed with KIR mismatching on TRM, relapse, or survival. Furthermore, in the RIC cohort, KIR-ligand mismatching between the engrafted unit and the recipient resulted in significantly higher rates of acute GvHD (grade III–IV) (42% (95% CI, 27–59%) vs 13% (95% CI, 5–21%); $P<0.01$) and TRM (27% (95% CI, 12–42%) vs 12% (95% CI, 5–19%); $P=0.03$), and inferior survival (32% (95% CI, 15–59%) vs 52% (95% CI, 47–67%); $P=0.03$). In keeping with these findings, Tanaka et al. (2013) also found no association between KIR-ligand incompatibility in the GvH direction and the incidence of GvHD, relapse, NRM, or OS in 643 T-replete single CB transplants (no antithymocyte globulin (ATG)).[72] Similarly, in the setting of double CB transplantation, analysis of 80 transplants did not find any difference in engraftment, relapse, PFS, or OS between groups receiving KIR-ligand-compatible or -incompatible CB units.[73] Therefore, current data do not support the routine use of KIR matching in the selection of CB units. However, further studies are warranted.

### 5.2.7 Noninherited Maternal Antigens

Fetal exposure to noninherited maternal antigens (NIMA) may promote lasting tolerance in transplant recipients. A report from the NYBC retrospectively evaluated the impact on NIMA matching on 1059 single CB transplants mismatched for one or two HLA antigens. Of these patients, 79 recipients had a mismatched antigen that was identical to a donor NIMA.[74] NIMA-matched transplants were associated with higher neutrophil recovery (RR 1.3 (95% CI, 1.01–1.7); $P=0.04$), lower TRM (RR 0.7 (95% CI, 0.5–0.97); $P=0.03$), and overall mortality rates (RR 0.7 (95% CI, 0.5–0.97); $P=0.03$). A Eurocord-CIBMTR study also showed lower TRM (RR 0.48 (95% CI, 0.23–1.01); $P=0.05$) and overall mortality (RR 0.61 (95% CI, 0.38–0.98); $P=0.04$) after

162 CB transplants when selecting a NIMA-matched CB unit.[75] However, the frequency of NIMA matching in these studies was below 10% and therefore may not be practical for routine use. CB banks are increasingly capturing HLA data from the mother of the CB donor to enable larger studies and, after other selection criteria have been met, it may become possible to select a CB unit based on NIMA matching.

### *5.2.8 ABO Compatibility*

Major ABO incompatibility between the recipient and CB unit has previously been associated with reduced OS and DFS (RR 1.55 (95% CI, 1.05–2.29); $P=0.03$) in unrelated single-unit CB transplants in adults with hematological malignancies.[76] However, in a more recent analysis of 503 consecutive CB transplants (single and double) from the University of Minnesota, the presence of an ABO mismatch did not significantly influence the rates of GvHD, TRM, DFS, or OS.[77] Similarly, in retrospective analysis of 191 adults with malignant disease receiving an MAC single-unit CB transplant, CB–recipient ABO mismatches were not associated with any significant difference in GvHD, TRM, or overall mortality. Therefore, although ABO matching between the CB and recipient may be considered if there are many well-matched CB units available, this should not be to the detriment of more important selection criteria such as cell dose and HLA matching.

### *5.2.9 CB Bank Procedures, Quality, and Accreditation*

The techniques used to collect, process, and store CB can significantly influence the quality and size (TNC and CD34$^{+}$ cell count) of the final CB unit. Immediately after delivery, CB is collected by sterile puncture and drainage of the umbilical cord vein. Improved yields may be obtained by additional collection from the placental vessels and/or placental perfusion, although it remains to be determined whether such techniques are practical for routine CB collection without increasing contamination from maternal cells.[78] Validated and standardized operating procedures for processing CB and cryopreservation are necessary to maximize cell recovery and ensure reliability between different CB banks. Immediate processing and/or storage of CB at 4 °C produce higher post-thaw recoveries and greater viability than initial storage at room temperature.[79] Use of modern automated systems for red cell depletion and volume reduction has also improved CB processing, although there is still an associated cell loss with these procedures. Cryopreservation, thawing, and washing CB also cause a further 20% cell loss.[80] Therefore, minimizing CB processing and improving good manufacturing compliant methods to enhance cell recovery could increase the number of cells available for infusion. Of note, a common concern raised during the selection of CB is the influence of storage duration on the CB unit and the subsequent impact on transplant outcomes. Some CB banks in the United States and Europe were established in the early 1990s and, therefore, some of the earlier units will have been stored for over 20 years. Although there are no formally published data on the impact of CB storage time on outcomes, Eurocord analyzed storage time in 1351 unrelated MAC CB transplants using a single CB unit. The median time of CB storage was 2.3 years (range, 0.3–14 years). CB storage time did not significantly affect neutrophil recovery or OS in this analysis (Eurocord, unpublished data).

As the number of unrelated CB banks across the world increases, so too does the number of CB units available for transplantation. Furthermore, with the increasing number of CB transplants being performed, improved selection criteria, and greater international cooperation, there are an increasing number of CB units being exchanged internationally. As such, many transplant centers and national regulatory authorities became aware of the need for international standards for CB collection, processing, testing, banking, selection, and release. Founded in 1998, NetCord is the international CB banking arm of Eurocord, aiming to promote high-quality CB banking and clinical use of CB for allogeneic HSCT. Through its online virtual office (www.netcord.org), CB units from member banks are made available for unrelated donor transplantation. Approximately 25 CB banks, mostly European, are full or associate NetCord members accounting for approximately 50% of worldwide CB units available. In 2013, NetCord and the Foundation for the Accreditation of Cellular Therapy (FACT) published the fifth edition of the International Standards for Cord Blood Collection, Banking, and Release for Administration.[81] The major objective of these standards is "to promote quality medical practices, laboratory processes, and banking to achieve consistent production of high quality placental and umbilical cord blood units for administration."[82] These standards cover donor management; collection, processing, testing of CB; cryopreservation, and storage of CB cells; listing, search, selection, reservation of CB units; and release and distribution to clinical transplant centers. To be compliant with these standards, CB banks must use validated methods, equipment, and reagents, and must maintain a documented Quality Management Program. The accreditation process includes submission of written documentation and on-site inspection. NetCord-FACT-accredited CB banks are reinspected every 3 years.

## 6. NEW STRATEGIES TO IMPROVE OUTCOMES AFTER CB TRANSPLANTATION

The main concern when using CB cells for allogeneic HSCT continues to be the low infused cell dose, particularly when treating adult patients or nonmalignant conditions. This translates into an increased risk of graft failure (10–20%), delayed hematopoietic engraftment, and delayed immune reconstitution.[6,9,25,54] Therefore, as well as using double CB transplants and RIC regimens (both discussed above), many other experimental approaches are currently being investigated to increase the infused HSC dose and/or improve engraftment (Table 3).

## 6.1 Expansion of CB

Ex vivo expansion of CB has successfully been used to increase the number of HSC for long-term engraftment, as well as enhancing the number of committed progenitors to attenuate the initial aplastic period. The expanded CB unit can then be given alone or in combination with a second unmanipulated unit. Although the expanded unit improves early hematopoietic recovery, it is the unmanipulated unit that usually provides long-term engraftment.[83] CB expansion has developed using three main techniques: (1) liquid culture; (2) culture with supporting mesenchymal stromal cells; and (3) continuous perfusion. In liquid culture, isolated $CD34^+$ or $CD133^+$ HSC are expanded in the presence of growth factors including stem cell factor (SCF), thrombopoietin (TPO), granulocyte colony stimulating factor (GCSF), and/or fms-like tyrosine kinase 3 ligand (FLT-3-L). Early feasibility studies demonstrated that $CD34^+$ cells could be successfully isolated from a fraction of the CB unit, expanded in liquid culture and then reinfused with the remainder in the original CB unit following an MAC regimen.[84] Modifications to this approach, by the addition of the copper chelator TEPA or immobilized Notch ligand delta-1, were later tested in phase I/II clinical trials.[85,86] The median fold expansions for TNCs were 219 (range, 2–260) and 562 (range, 146–1496) for the two methods, respectively. Both trials demonstrated successful engraftment in 9 out of 10 patients, although in the later study, there was a predominance for donor $CD33^+$ and $CD14^+$ cell engraftment from the expanded unit. In coculture systems, mesenchymal stromal cells provide a supporting hematopoietic microenvironment for HSC proliferation. In 2012, de Lima et al. published the results of 24 double-unit CB transplants in which in one unit was expanded ex vivo with mesenchymal stem cells.[83] TNCs and $CD34^+$ cells expanded by a median of 12.2 and 30.1 times, respectively, and the median TNC dose infused was $8.34 \times 10^7$/kg; higher than in their conventional unmanipulated double CB transplants. Neutrophil engraftment was achieved in 23 of the 24 patients at a median of 15 days (range, 9–42 days). This compared favorably to 80 CIBMTR historical controls that received unmanipulated double CB transplants (median 24 days (range, 12–52 days)). Finally, in the continuous perfusion system, isolated HSC are continually supplied with fresh culture media and gaseous exchange. In a phase I study, the median fold increase in TNC of 2.4 (range, 1.0–8.5) was achieved.[87] After infusion of the expanded cells with the remainder of the original CB unit, 21 of the 26 patients attained neutrophil engraftment with a median time of 22 days (range, 13–40 days). In summary, several CB expansion protocols appear promising, although it remains to be determined if these strategies will actually improve clinical outcomes following CB transplantation. Furthermore, it needs to be established whether the increase in committed progenitors is at the expanse of long-term HSC. There are many prospective clinical trials of CB expansion currently in progress, including commercial expansion of CB using MPC (Mesoblast) and/or Nicord®.

**TABLE 3 Current Strategies to Improve Cord Blood Transplantation**

1. Increasing cell dose
   a. Improved CB collection, processing, freezing, and thawing
   b. Double CB transplantation
   c. Ex vivo expansion of CB
   d. Infusion of CB with BM from same donor
   e. Infusion of CB with third-party donor cells
2. Improved delivery and homing cord blood HSC to bone marrow
   a. Direct intrabone infusion of CB
   b. Inhibition of CD26 peptidase
   c. Ex vivo fucosylation of CB HSC/HPC
3. Improved selection of cord blood units
   a. Enhanced HLA matching (HLA-C; high-resolution HLA-A, -B, -C, -DRB1)
   b. Detection of donor-specific HLA antibodies
   c. KIR matching
   d. NIMA matching
   e. ABO matching
4. Modification of transplant conditioning
   a. Reduced intensity conditioning
   b. Modification of Graft-versus-Host Disease prophylaxis
5. Post-transplant use of growth factors/cytokines
   a. Granulocyte colony stimulating factor (GCSF)/stem cell factor (SCF)
   b. Thrombopoietin peptide mimetic (Romiplostim) or agonists (Eltrombopag)
6. Infusion of accessory cells
   a. Mesenchymal stem cells
   b. Regulatory T cells

Abbreviations: BM, bone marrow; CB, cord blood; HLA, human leukocyte antigen; HPC, hematopoietic progenitor cells; HSC, hematopoietic stem cells.

## 6.2 Intrabone Injection

After intravenously injecting CB into the recipient, the cells pass through the peripheral circulation to the BM microvasculature. From here, a highly regulated process of cell adhesion and migration allows homing of the HSC and hematopoietic progenitor cells (HPC) to the BM niche where they engraft, proliferate and differentiate into normal hematopoiesis. However, from animal models, it has been estimated that only around 10% of the HSC/HPC actually make it to the BM niche, the rest being sequestrated in the lungs, liver, and spleen.[88] Therefore, to overcome the problem of sequestration within other organs, direct intrabone infusion of CB has been tested. In a Eurocord analysis of single-unit intrabone CB transplantation ($n=87$) with double-unit intravenous CB transplantation ($n=149$), intrabone infusion was associated with improved neutrophil engraftment at day 30 (76% vs 62%; $P=0.01$) and improved platelet engraftment by day 180 (74% vs 64%; $P=0.003$).[89] In multivariate analysis, intrabone CB transplantation had improved neutrophil (HR 1.5 (95% CI, 1.04–2.17); $P=0.03$) and platelet recovery (HR 1.97 (95% CI, 1.35–2.29); $P=0.004$) compared to intravenous CB transplantation. Intrabone infusion was also associated with a lower incidence of acute GvHD and showed a trend toward improved DFS. Larger phase II clinical trials of intrabone infusion of CB cells for hematological malignancies are currently in progress.

## 6.3 Enhance Homing of CB HSC Cells

Homing, migration, and engraftment of HSC to the BM niche are highly dependent upon the chemo-attractant stromal cell-derived factor-1 (SDF-1) (chemokine (C-X-C motif) ligand 12 (CXCL12)). SDF-1 levels, produced by BM endothelium, increase following conditioning regimens and the infused HSC follow the SDF-1 gradient toward the BM.[90] SDF-1 binds to its receptor, CXCR4 (fusin/CD184), on the surface of HSC, activating a series of cytoskeletal changes required for adhesion and migration of cells across the endothelium.[91] In mouse models, inhibition of the membrane-bound extracellular peptidase dipeptidyl peptidase-4 (CD26), which cleaves SDF-1, enhanced long-term engraftment in CB $CD34^+$ cells.[92] Fucosylation (the addition of a fructose) of CB HSC is also required for interaction with the cell adhesion molecules (P- and E-selectin) expressed in the BM microvasculature. In mouse models, treatment of CB HSC with guanosine diphosphate fucose and alpha 1-3 fucosyl transferase VI improved adhesion and rolling of the cells on P- and E-selectin and improved HSC engraftment.[93] These preclinical biological observations provide important areas for future clinical development of CB transplantation. As such, multicenter phase II trials of the CD26 peptidase inhibitor (Sitagliptin) and CB fucosylation are currently in progress.

## 6.4 Use of Growth Factors

Use of recombinant growth factors in vivo may hasten neutrophil and platelet recovery, although relatively little published data exist. In 2004, Gluckman et al. published the results from 550 CB transplants given to patients with hematological malignancies.[8] GCSF was administered to 60% of these transplants and was independently associated with improved neutrophil recovery (HR 1.66 (95% CI, 1.34–2.05); $P<0.0001$). There has also been recent interest in use of the TPO peptide mimetic (Romiplostim) and/or the nonpeptide small molecule TPO receptor (*c-Mpl*) agonist (Eltrombopag) in CB transplantation. In mouse models, Eltrombopag increased the expansion of human CB $CD34^+$, $CD45^+$, and $CD41^+$ cells with an associated increase in platelets and white cells.[94] Consequently, there are now several early phase trials of Eltrombopag in CB transplantation currently recruiting.

## 6.5 Co-infusion of CB with Third-party Cells

To overcome the issue of the low cell doses in CB, several groups have used co-infusion of cells from third-party haploidentical family members. While the haploidentical graft supports early cell recovery, it is the CB unit that usually provides long-term engraftment. In 2010, the outcomes from 55 combined CB/haploidentical transplants for high-risk myeloproliferative and lymphoproliferative disorders were published.[95] The median TNC and $CD34^+$ cell dose from the CB were $2.39\times10^7$/kg (range, 1.14–$4.30\times10^7$/kg) and $1.1\times10^5$/kg (range, 0.35–$3.7\times10^5$/kg), respectively. This was combined with positively selected third-party haploidentical $CD34^+$ and/or $CD133^+$ cells, median dose $2.4\times10^6$/kg (range, 1.05–$3.34\times10^6$/kg). The maximum cumulative incidence of neutrophil recovery was 96% (95% CI, 91–100%) with a median time of 10 days (range, 9–36 days). The cumulative incidence of full CB chimerism was 91% (95% CI, 84–99%) at the median time of 44 days (range, 11–186 days). In a similar report, 45 patients were transplanted using an unrelated CB graft and $CD34^+$ selected cells from a haploidentical family member following an RIC regimen (Fludarabine, Melphalan, and ATG).[96] The cumulative incidence of neutrophil engraftment at day 50 was 95% (95% CI, 87–100%) with a median time of 11 days. The CB unit accounted for medians of 10%, 78%, and 95% of PB cells at day 30, 100, and 180, respectively. The cumulative incidence of acute and chronic GvHD was 25% (95% CI, 11–39%) and 6%, respectively, with 1-year TRM 28% (95% CI, 13–43%), relapse 30% (95% CI, 14–44%), and OS 55% (95% CI, 39–71%). Interestingly, in contrast to these two studies, Chen et al. have recently reported results in T-replete combined CB/haploidentical transplants.[97] In a prospective study, 50 patients with hematological malignancy received MAC followed by combined CB/haploidentical grafts. Forty-eight patients engrafted within 20 days with a median time to neutrophil recovery of 13 days (range, 11–20 days). However, in this setting, all surviving patients achieved sustained haploidentical engraftment. Therefore, in summary, combining CB and haploidentical grafts improves early cell recovery, although which graft eventually provides long-term hematopoiesis appears to be dependent on the T-cell depletion of the haploidentical graft. Ongoing phase II/III studies comparing double CB transplant with combined haploidentical/CB transplantation in hematological disease are in progress.

## 7. CONCLUSION

CB transplantation has made significant progress over the last 25 years and now provides a suitable alternative to conventional HSC transplantation when HLA-matched sibling or unrelated donors are not available. With adequate CB cell doses (e.g., TNC $\geq 3.0 \times 10^7$/kg at collection; TNC $\geq 2.5 \times 10^7$/kg at infusion) and appropriate HLA matching ($\leq 2$ HLA mismatches at HLA-A and -B (antigen) and HLA-DRB1 (allele)), the rates of engraftment and TRM continue to improve, the incidence of GvHD remains low, and DFS and OS now appear comparable to using other donor sources. In children, use of related or unrelated single CB units provides very good results in the treatment of both malignant and nonmalignant conditions. However, due to the high rates of graft failure with nonmalignant disorders, higher cell doses and/or more stringent HLA matching are recommended. In adults, where the main transplant indication is malignant disease, use of unrelated double CB transplantation and RIC regimens has increased the applicability of CB transplantation to older patients, with a reduction in failed engraftment and TRM. As such, DFS now appears similar to using HLA-matched or -mismatched unrelated donors. However, care must be taken as not all conditioning regimens have the same outcomes. As our understanding of the biology of these diseases improves and use of prognostic cytogenetic and molecular makers becomes more common, further improvements may be made by identifying those patients at high risk of relapse following CB transplantation.

While the overall results of CB transplantation appear promising, further improvements can be made. With the increased availability of high-quality cryopreserved CB units, selection of CB units with higher cell doses and/or more closely HLA matched may be possible. Furthermore, use of additional selection criteria such as matching at HLA-C, high-resolution (allele) matching, and NIMA matching may improve outcomes, although this should be at the expense of donor availability for those patients with limited donor options. Although use of ex vivo-expanded CB has not significantly increased over the last few years, other developments such as intrabone injection, use of growth factors (GSCF and/Romiplostim/Eltrombopag), and combined infusion of third party or accessory cells all appear promising.

Finally, an important question that remains unanswered is how will CB transplants compare to other alternative donor sources over the next 5–10 years? Over the last few years, the number of CB transplants being performed each year appears to have plateaued. In fact, in the 2012 EBMT survey, the number of CB transplants fell from 833 in 2011 to 758 in 2012.[1] In contrast, the number of haploidentical donor transplants increased by 24% to 1217. As new methods of haploidentical HSC transplantation have developed and outcomes improved, the use of haploidentical donors has become more popular. Haploidentical donors are almost always available and donor costs are considerably cheaper compared to CB transplantation. However, haploidentical transplantation is not without its problems, including delayed immune reconstitution and high risk of relapse when using RIC regimens and T-cell depletion. For these reasons, it is anticipated that for selected patients, CB transplantation will continue to have a crucial role in the management of hematological disorders for the foreseeable future.

## LIST OF ACRONYMS AND ABBREVIATIONS

**ALL** Acute lymphoblastic leukemia
**ALWP** Acute Leukemia Working Party
**AML** Acute myeloid leukemia
**BM** Bone marrow
**BMT** Bone marrow transplantation
**CB** Cord blood
**CI** Cumulative incidence
**CIBMTR** Center for International Blood and Marrow Transplant Research
**CR** Complete remission
**DFS** Disease-free survival
**EBMT** European Blood and Marrow Transplant
**EFS** Event-free survival
**GvHD** Graft-versus-Host Disease
**GvL** Graft-versus-Leukemia
**HLA** Human leukocyte antigen
**HR** Hazard ratio
**HSC** Hematopoietic stem cells
**IBMTR** International Blood and Marrow Transplantation Registry
**IPSS** International prognostic scoring system
**LFS** Leukemia-free survival
**NRM** Nonrelapse mortality
**MAC** Myeloablative conditioning
**MDS** Myelodysplastic syndromes
**MMUD** Mismatched unrelated donor
**MRD** Matched related donor
**MTX** Methotrexate
**MUD** Matched unrelated donor
**NHL** Non-Hodgkin's lymphoma
**OS** Overall survival
**PB** Peripheral blood
**PBPC** Peripheral blood progenitor cells
**PBSC** Peripheral blood stem cells
**RIC** Reduced intensity conditioning
**RR** Relative risk
**SCD** Sickle cell disease
**SFGM-TC** Société Française de Greffe de Moelle-Thérapie Cellulaire

**TBI** Total body irradiation
**TM** Thalassemia major
**TNC** Total nucleated cells
**TRM** Transplant-related mortality
**UCB** Umbilical cord blood

## REFERENCES


1. Passweg JR, Baldomero H, Peters C, Gaspar HB, Cesaro S, Dreger P, et al. European Society for B, Marrow Transplantation E. Hematopoietic SCT in Europe: data and trends in 2012 with special consideration of pediatric transplantation. *Bone Marrow Transplant* 2014;**49**(6):744–50.
2. Ruggeri A, Ciceri F, Gluckman E, Labopin M, Rocha V. Eurocord, Acute Leukemia Working Party of the European B, Marrow Transplant G. Alternative donors hematopoietic stem cells transplantation for adults with acute myeloid leukemia: umbilical cord blood or haploidentical donors? *Best Pract Res Clin Haematol* 2010;**23**(2):207–16.
3. Brunstein CG, Fuchs EJ, Carter SL, Karanes C, Costa LJ, Wu J, et al. Blood, Marrow Transplant Clinical Trials N. Alternative donor transplantation after reduced intensity conditioning: results of parallel phase 2 trials using partially HLA-mismatched related bone marrow or unrelated double umbilical cord blood grafts. *Blood* 2011;**118**(2): 282–8.
4. Chen L, Cohen AC, Lewis DB. Impaired allogeneic activation and T-helper 1 differentiation of human cord blood naive CD4 T cells. Biology of blood and marrow transplantation. *J Am Soc Blood Marrow Transplant* 2006;**12**(2):160–71.
5. Rocha V, Cornish J, Sievers EL, Filipovich A, Locatelli F, Peters C, et al. Comparison of outcomes of unrelated bone marrow and umbilical cord blood transplants in children with acute leukemia. *Blood* 2001;**97**(10):2962–71.
6. Rocha V, Labopin M, Sanz G, Arcese W, Schwerdtfeger R, Bosi A, et al. Acute Leukemia Working Party of European B, Marrow Transplant G, Eurocord-Netcord R. Transplants of umbilical-cord blood or bone marrow from unrelated donors in adults with acute leukemia. *N Engl J Med* 2004;**351**(22):2276–85.
7. Rocha V, Gluckman E. Eurocord-Netcord R, European B, Marrow Transplant G. Improving outcomes of cord blood transplantation: HLA matching, cell dose and other graft- and transplantation-related factors. *Br J Haematol* 2009;**147**(2):262–74.
8. Gluckman E, Rocha V, Arcese W, Michel G, Sanz G, Chan KW, et al. Factors associated with outcomes of unrelated cord blood transplant: guidelines for donor choice. *Exp Hematol* 2004;**32**(4):397–407.
9. Eapen M, Rubinstein P, Zhang MJ, Stevens C, Kurtzberg J, Scaradavou A, et al. Outcomes of transplantation of unrelated donor umbilical cord blood and bone marrow in children with acute leukaemia: a comparison study. *Lancet* 2007;**369**(9577):1947–54.
10. Barker JN, Scaradavou A, Stevens CE. Combined effect of total nucleated cell dose and HLA match on transplantation outcome in 1061 cord blood recipients with hematologic malignancies. *Blood* 2010;**115**(9):1843–9.
11. Rocha V, Locatelli F. Searching for alternative hematopoietic stem cell donors for pediatric patients. *Bone Marrow Transplant* 2008; **41**(2):207–14.
12. Gluckman E, Broxmeyer HA, Auerbach AD, Friedman HS, Douglas GW, Devergie A, et al. Hematopoietic reconstitution in a patient with Fanconi's anemia by means of umbilical-cord blood from an HLA-identical sibling. *N Engl J Med* 1989;**321**(17):1174–8.
13. Gluckman E, Ruggeri A, Rocha V, Baudoux E, Boo M, Kurtzberg J, et al. National Marrow Donor P. Family-directed umbilical cord blood banking. *Haematologica* 2011;**96**(11):1700–7.
14. Rocha V, Wagner Jr JE, Sobocinski KA, Klein JP, Zhang MJ, Horowitz MM, et al. Graft-versus-host disease in children who have received a cord-blood or bone marrow transplant from an HLA-identical sibling. Eurocord and International Bone Marrow Transplant Registry Working Committee on Alternative Donor and Stem Cell Sources. *N Engl J Med* 2000;**342**(25):1846–54.
15. Yamazaki S, Inaba K, Tarbell KV, Steinman RM. Dendritic cells expand antigen-specific Foxp3$^+$ CD25$^+$ CD4$^+$ regulatory T cells including suppressors of alloreactivity. *Immunol Rev* 2006;**212**: 314–29.
16. Kim YJ, Broxmeyer HE. Immune regulatory cells in umbilical cord blood and their potential roles in transplantation tolerance. *Crit Rev Oncol Hematol* 2011;**79**(2):112–26.
17. Gluckman E, Rocha V, Boyer-Chammard A, Locatelli F, Arcese W, Pasquini R, et al. Outcome of cord-blood transplantation from related and unrelated donors. Eurocord Transplant Group and the European Blood and Marrow Transplantation Group. *N Engl J Med* 1997;**337**(6):373–81.
18. Herr AL, Kabbara N, Bonfim CM, Teira P, Locatelli F, Tiedemann K, et al. Long-term follow-up and factors influencing outcomes after related HLA-identical cord blood transplantation for patients with malignancies: an analysis on behalf of Eurocord-EBMT. *Blood* 2010;**116**(11):1849–56.
19. Locatelli F, Rocha V, Reed W, Bernaudin F, Ertem M, Grafakos S, et al. Related umbilical cord blood transplantation in patients with thalassemia and sickle cell disease. *Blood* 2003;**101**(6):2137–43.
20. Locatelli F, Kabbara N, Ruggeri A, Ghavamzadeh A, Roberts I, Li CK, et al. Eurocord, European B, Marrow Transplantation G. Outcome of patients with hemoglobinopathies given either cord blood or bone marrow transplantation from an HLA-identical sibling. *Blood* 2013;**122**(6):1072–8.
21. Smythe J, Armitage S, McDonald D, Pamphilon D, Guttridge M, Brown J, et al. Directed sibling cord blood banking for transplantation: the 10-year experience in the national blood service in England. *Stem Cells* 2007;**25**(8):2087–93.
22. Goussetis E, Peristeri I, Kitra V, Papassavas AC, Theodosaki M, Petrakou E, et al. Low usage rate of banked sibling cord blood units in hematopoietic stem cell transplantation for children with hematological malignancies: implications for directed cord blood banking policies. *Blood Cells Mol Dis* 2011;**46**(2):177–81.
23. Kurtzberg J, Laughlin M, Graham ML, Smith C, Olson JF, Halperin EC, et al. Placental blood as a source of hematopoietic stem cells for transplantation into unrelated recipients. *N Engl J Med* 1996; **335**(3):157–66.
24. Rubinstein P, Carrier C, Scaradavou A, Kurtzberg J, Adamson J, Migliaccio AR, et al. Outcomes among 562 recipients of placental-blood transplants from unrelated donors. *N Engl J Med* 1998;**339**(22): 1565–77.
25. Laughlin MJ, Eapen M, Rubinstein P, Wagner JE, Zhang MJ, Champlin RE, et al. Outcomes after transplantation of cord blood or bone marrow from unrelated donors in adults with leukemia. *N Engl J Med* 2004;**351**(22):2265–75.
26. Barker JN, Weisdorf DJ, DeFor TE, Blazar BR, McGlave PB, Miller JS, et al. Transplantation of 2 partially HLA-matched umbilical cord blood units to enhance engraftment in adults with hematologic malignancy. *Blood* 2005;**105**(3):1343–7.

27. Kurtzberg J, Prasad VK, Carter SL, Wagner JE, Baxter-Lowe LA, Wall D, et al. Results of the Cord Blood Transplantation Study (COBLT): clinical outcomes of unrelated donor umbilical cord blood transplantation in pediatric patients with hematologic malignancies. *Blood* 2008;**112**(10):4318–27.
28. Michel G, Rocha V, Chevret S, Arcese W, Chan KW, Filipovich A, et al. Unrelated cord blood transplantation for childhood acute myeloid leukemia: a Eurocord Group analysis. *Blood* 2003;**102**(13):4290–7.
29. Madureira AB, Eapen M, Locatelli F, Teira P, Zhang MJ, Davies SM, et al. Marrow Transplant G, Center of International B, Marrow Transplant R, European Working Group on childhood MDS. Analysis of risk factors influencing outcome in children with myelodysplastic syndrome after unrelated cord blood transplantation. *Leukemia* 2011;**25**(3):449–54.
30. Ruggeri A, Eapen M, Scaravadou A, Cairo MS, Bhatia M, Kurtzberg J, et al. Umbilical cord blood transplantation for children with thalassemia and sickle cell disease. *Biol Blood Marrow Transplant* 2011;**17**(9):1375–82.
31. Boelens JJ, Aldenhoven M, Purtill D, Ruggeri A, Defor T, Wynn R, et al. Outcomes of transplantation using various hematopoietic cell sources in children with Hurler syndrome after myeloablative conditioning. *Blood* 2013;**121**(19):3981–7.
32. Gluckman E, Rocha V, Ionescu I, Bierings M, Harris RE, Wagner J, et al. Results of unrelated cord blood transplant in fanconi anemia patients: risk factor analysis for engraftment and survival. Biology of blood and marrow transplantation. *J Am Soc Blood Marrow Transplant* 2007;**13**(9):1073–82.
33. Fernandes JF, Rocha V, Labopin M, Neven B, Moshous D, Gennery AR, et al. Transplantation in patients with SCID: mismatched related stem cells or unrelated cord blood? *Blood* 2012;**119**(12):2949–55.
34. Mitchell C, Richards S, Harrison CJ, Eden T. Long-term follow-up of the United Kingdom medical research council protocols for childhood acute lymphoblastic leukaemia, 1980–2001. *Leukemia* 2010;**24**(2):406–18.
35. Ruggeri A, Michel G, Dalle JH, Caniglia M, Locatelli F, Campos A, et al. Impact of pretransplant minimal residual disease after cord blood transplantation for childhood acute lymphoblastic leukemia in remission: an Eurocord, PDWP-EBMT analysis. *Leukemia* 2012;**26**(12):2455–61.
36. Locatelli F, Crotta A, Ruggeri A, Eapen M, Wagner JE, Macmillan ML, et al. Analysis of risk factors influencing outcomes after cord blood transplantation in children with juvenile myelomonocytic leukemia: a EUROCORD, EBMT, EWOG-MDS, CIBMTR study. *Blood* 2013;**122**(12):2135–41.
37. Bizzetto R, Bonfim C, Rocha V, Socie G, Locatelli F, Chan K, et al. Outcomes after related and unrelated umbilical cord blood transplantation for hereditary bone marrow failure syndromes other than Fanconi anemia. *Haematologica* 2011;**96**(1):134–41.
38. Kamani NR, Walters MC, Carter S, Aquino V, Brochstein JA, Chaudhury S, et al. Unrelated donor cord blood transplantation for children with severe sickle cell disease: results of one cohort from the phase II study from the Blood and Marrow Transplant Clinical Trials Network (BMT CTN). Biology of blood and marrow transplantation. *J Am Soc Blood Marrow Transplant* 2012;**18**(8): 1265–72.
39. Barker JN, Davies SM, DeFor T, Ramsay NK, Weisdorf DJ, Wagner JE. Survival after transplantation of unrelated donor umbilical cord blood is comparable to that of human leukocyte antigen-matched unrelated donor bone marrow: results of a matched-pair analysis. *Blood* 2001;**97**(10):2957–61.
40. Hwang WY, Samuel M, Tan D, Koh LP, Lim W, Linn YC. A meta-analysis of unrelated donor umbilical cord blood transplantation versus unrelated donor bone marrow transplantation in adult and pediatric patients. *Biol Blood Marrow Transplant J Am Soc Blood Marrow Transplant* 2007;**13**(4):444–53.
41. Brunstein CG, Gutman JA, Weisdorf DJ, Woolfrey AE, Defor TE, Gooley TA, et al. Allogeneic hematopoietic cell transplantation for hematologic malignancy: relative risks and benefits of double umbilical cord blood. *Blood* 2010;**116**(22):4693–9.
42. Brunstein CG, Eapen M, Ahn KW, Appelbaum FR, Ballen KK, Champlin RE, et al. Reduced-intensity conditioning transplantation in acute leukemia: the effect of source of unrelated donor stem cells on outcomes. *Blood* 2012;**119**(23):5591–8.
43. Bachanova V, Verneris MR, DeFor T, Brunstein CG, Weisdorf DJ. Prolonged survival in adults with acute lymphoblastic leukemia after reduced-intensity conditioning with cord blood or sibling donor transplantation. *Blood* 2009;**113**(13):2902–5.
44. Matsumura T, Kami M, Yamaguchi T, Yuji K, Kusumi E, Taniguchi S, et al. Allogeneic cord blood transplantation for adult acute lymphoblastic leukemia: retrospective survey involving 256 patients in Japan. *Leukemia* 2012;**26**(7):1482–6.
45. Tucunduva L, Ruggeri A, Sanz G, Furst S, Socie G, Michallet M, et al. Risk factors for outcomes after unrelated cord blood transplantation for adults with acute lymphoblastic leukemia: a report on behalf of Eurocord and the Acute Leukemia Working Party of the European Group for Blood and Marrow Transplantation. *Bone Marrow Transplant* 2014;**49**(7):887–94.
46. Majhail NS, Brunstein CG, Shanley R, Sandhu K, McClune B, Oran B, et al. Reduced-intensity hematopoietic cell transplantation in older patients with AML/MDS: umbilical cord blood is a feasible option for patients without HLA-matched sibling donors. *Bone Marrow Transplant* 2012;**47**(4):494–8.
47. Robin M, Sanz GF, Ionescu I, Rio B, Sirvent A, Renaud M, et al. Unrelated cord blood transplantation in adults with myelodysplasia or secondary acute myeloblastic leukemia: a survey on behalf of Eurocord and CLWP of EBMT. *Leukemia* 2011;**25**(1):75–81.
48. Rodrigues CA, Sanz G, Brunstein CG, Sanz J, Wagner JE, Renaud M, et al. Analysis of risk factors for outcomes after unrelated cord blood transplantation in adults with lymphoid malignancies: a study by the Eurocord-Netcord and lymphoma working party of the European group for blood and marrow transplantation. *J Clin Oncol* 2009;**27**(2):256–63.
49. Takahashi S, Iseki T, Ooi J, Tomonari A, Takasugi K, Shimohakamada Y, et al. Single-institute comparative analysis of unrelated bone marrow transplantation and cord blood transplantation for adult patients with hematologic malignancies. *Blood* 2004;**104**(12): 3813–20.
50. Eapen M, Rocha V, Sanz G, Scaradavou A, Zhang MJ, Arcese W, et al. Effect of graft source on unrelated donor haemopoietic stem-cell transplantation in adults with acute leukaemia: a retrospective analysis. *Lancet Oncol* 2010;**11**(7):653–60.
51. Atsuta Y, Morishima Y, Suzuki R, Nagamura-Inoue T, Taniguchi S, Takahashi S, et al. Comparison of unrelated cord blood transplantation and HLA-mismatched unrelated bone marrow transplantation for adults with leukemia. Biology of blood and marrow transplantation. *J Am Soc Blood Marrow Transplant* 2012;**18**(5):780–7.
52. Barker JN, Weisdorf DJ, Wagner JE. Creation of a double chimera after the transplantation of umbilical-cord blood from two partially matched unrelated donors. *N Engl J Med* 2001;**344**(24):1870–1.

53. Verneris MR, Brunstein CG, Barker J, MacMillan ML, DeFor T, McKenna DH, et al. Relapse risk after umbilical cord blood transplantation: enhanced graft-versus-leukemia effect in recipients of 2 units. *Blood* 2009;**114**(19):4293–9.
54. Ruggeri A, Peffault de Latour R, Carmagnat M, Clave E, Douay C, Larghero J, et al. Outcomes, infections, and immune reconstitution after double cord blood transplantation in patients with high-risk hematological diseases. *Transplant Infect Dis Off J Transplant Soc* 2011;**13**(5):456–65.
55. Wallet HL, Sobh M, Morisset S, Robin M, Fegueux N, Furst S, et al. Double umbilical cord blood transplantation for hematological malignancies: a long-term analysis from the SFGM-TC registry. *Exp Hematol* 2013;**41**(11):924–33.
56. Ruggeri A, Sanz G, Bittencourt H, Sanz J, Rambaldi A, Volt F, et al. Comparison of outcomes after single or double cord blood transplantation in adults with acute leukemia using different types of myeloablative conditioning regimen, a retrospective study on behalf of Eurocord and the Acute Leukemia Working Party of EBMT. *Leukemia* 2014;**28**(4):779–86.
57. Brunstein CG, Barker JN, Weisdorf DJ, DeFor TE, Miller JS, Blazar BR, et al. Umbilical cord blood transplantation after nonmyeloablative conditioning: impact on transplantation outcomes in 110 adults with hematologic disease. *Blood* 2007;**110**(8):3064–70.
58. Peffault de Latour R, Brunstein CG, Porcher R, Chevallier P, Robin M, Warlick E, et al. Similar overall survival using sibling, unrelated donor, and cord blood grafts after reduced-intensity conditioning for older patients with acute myelogenous leukemia. *Biol Blood Marrow Transplant* 2013;**19**(9):1355–60.
59. Eapen M, Klein JP, Ruggeri A, Spellman S, Lee SJ, Anasetti C, et al. Impact of allele-level HLA matching on outcomes after myeloablative single unit umbilical cord blood transplantation for hematologic malignancy. *Blood* 2014;**123**(1):133–40.
60. Barker JN, Abboud M, Rice RD, Hawke R, Schaible A, Heller G, et al. A "no-wash" albumin-dextran dilution strategy for cord blood unit thaw: high rate of engraftment and a low incidence of serious infusion reactions. *Biol Blood Marrow Transplant* 2009;**15**(12): 1596–602.
61. Eapen M, Klein JP, Sanz GF, Spellman S, Ruggeri A, Anasetti C, et al. Effect of donor-recipient HLA matching at HLA A, B, C, and DRB1 on outcomes after umbilical-cord blood transplantation for leukaemia and myelodysplastic syndrome: a retrospective analysis. *Lancet Oncol* 2011;**12**(13):1214–21.
62. Lee SJ, Klein J, Haagenson M, Baxter-Lowe LA, Confer DL, Eapen M, et al. High-resolution donor-recipient HLA matching contributes to the success of unrelated donor marrow transplantation. *Blood* 2007;**110**(13):4576–83.
63. Kogler G, Enczmann J, Rocha V, Gluckman E, Wernet P. High-resolution HLA typing by sequencing for HLA-A, -B, -C, -DR, -DQ in 122 unrelated cord blood/patient pair transplants hardly improves long-term clinical outcome. *Bone Marrow Transplant* 2005;**36**(12):1033–41.
64. Stevens CE, Carrier C, Carpenter C, Sung D, Scaradavou A. HLA mismatch direction in cord blood transplantation: impact on outcome and implications for cord blood unit selection. *Blood* 2011;**118**(14): 3969–78.
65. Cunha R, Loiseau P, Ruggeri A, Sanz G, Michel G, Paolaiori A, et al. Impact of HLA mismatch direction on outcomes after umbilical cord blood transplantation for hematological malignant disorders: a retrospective Eurocord-EBMT analysis. *Bone Marrow Transplant* 2014;**49**(1):24–9.
66. Takanashi M, Atsuta Y, Fujiwara K, Kodo H, Kai S, Sato H, et al. The impact of anti-HLA antibodies on unrelated cord blood transplantations. *Blood* 2010;**116**(15):2839–46.
67. Cutler C, Kim HT, Sun L, Sese D, Glotzbecker B, Armand P, et al. Donor-specific anti-HLA antibodies predict outcome in double umbilical cord blood transplantation. *Blood* 2011;**118**(25): 6691–7.
68. Brunstein CG, Noreen H, DeFor TE, Maurer D, Miller JS, Wagner JE. Anti-HLA antibodies in double umbilical cord blood transplantation. Biology of blood and marrow transplantation. *J Am Soc Blood Marrow Transplant* 2011;**17**(11):1704–8.
69. Ruggeri L, Mancusi A, Burchielli E, Capanni M, Carotti A, Aloisi T, et al. NK cell alloreactivity and allogeneic hematopoietic stem cell transplantation. *Blood Cells Mol Dis* 2008;**40**(1):84–90.
70. Willemze R, Rodrigues CA, Labopin M, Sanz G, Michel G, Socie G, et al. Acute Leukaemia Working Party of the E. KIR-ligand incompatibility in the graft-versus-host direction improves outcomes after umbilical cord blood transplantation for acute leukemia. *Leukemia* 2009;**23**(3):492–500.
71. Brunstein CG, Wagner JE, Weisdorf DJ, Cooley S, Noreen H, Barker JN, et al. Negative effect of KIR alloreactivity in recipients of umbilical cord blood transplant depends on transplantation conditioning intensity. *Blood* 2009;**113**(22):5628–34.
72. Tanaka J, Morishima Y, Takahashi Y, Yabe T, Oba K, Takahashi S, et al. Effects of KIR ligand incompatibility on clinical outcomes of umbilical cord blood transplantation without ATG for acute leukemia in complete remission. *Blood Cancer J* 2013;**3**:e164.
73. Garfall A, Kim HT, Sun L, Ho VT, Armand P, Koreth J, et al. KIR ligand incompatibility is not associated with relapse reduction after double umbilical cord blood transplantation. *Bone Marrow Transplant* 2013;**48**(7):1000–2.
74. van Rood JJ, Stevens CE, Smits J, Carrier C, Carpenter C, Scaradavou A. Reexposure of cord blood to noninherited maternal HLA antigens improves transplant outcome in hematological malignancies. *Proc Natl Acad Sci USA* 2009;**106**(47):19952–7.
75. Rocha V, Spellman S, Zhang MJ, Ruggeri A, Purtill D, Brady C, et al. Effect of HLA-matching recipients to donor noninherited maternal antigens on outcomes after mismatched umbilical cord blood transplantation for hematologic malignancy. *Biol Blood Marrow Transplant* 2012;**18**(12):1890–6.
76. Arcese W, Rocha V, Labopin M, Sanz G, Iori AP, de Lima M, et al. Eurocord-Netcord Transplant G. Unrelated cord blood transplants in adults with hematologic malignancies. *Haematologica* 2006;**91**(2):223–30.
77. Romee R, Weisdorf DJ, Brunstein C, Wagner JE, Cao Q, Blazar BR, et al. Impact of ABO-mismatch on risk of GVHD after umbilical cord blood transplantation. *Bone Marrow Transplant* 2013;**48**(8):1046–9.
78. Broxmeyer HE, Cooper S, Hass DM, Hathaway JK, Stehman FB, Hangoc G. Experimental basis of cord blood transplantation. *Bone Marrow Transplant* 2009;**44**(10):627–33.
79. Louis I, Wagner E, Dieng MM, Morin H, Champagne MA, Haddad E. Impact of storage temperature and processing delays on cord blood quality: discrepancy between functional in vitro and in vivo assays. *Transfusion* 2012;**52**(11):2401–5.
80. McManus MP, Wang L, Calder C, Manes B, Evans M, Bruce K, et al. Comparison of pre-cryopreserved and post-thaw-and-wash-nucleated cell count on major outcomes following unrelated cord blood transplant in children. *Pediatr Transplant* 2012;**16**(5):438–42.

81. FACT-NetCord. *Fifth edition NetCord-fact international standards for cord blood collection, banking, and release for administration.* The Foundation for the Accreditation of Cellular Therapy University of Nebraska Medical Center; 2013.
82. Cord Blood Bank Standards. http://www.factwebsite.org/cbstandards/; 2014 [accessed 17.08.14].
83. de Lima M, McNiece I, Robinson SN, Munsell M, Eapen M, Horowitz M, et al. Cord-blood engraftment with ex vivo mesenchymal-cell coculture. *N Engl J Med* 2012;**367**(24):2305–15.
84. Shpall EJ, Quinones R, Giller R, Zeng C, Baron AE, Jones RB, et al. Transplantation of ex vivo expanded cord blood. Biology of blood and marrow transplantation. *Biol Blood Marrow Transplant* 2002;**8**(7):368–76.
85. Delaney C, Heimfeld S, Brashem-Stein C, Voorhies H, Manger RL, Bernstein ID. Notch-mediated expansion of human cord blood progenitor cells capable of rapid myeloid reconstitution. *Nat Med* 2010;**16**(2):232–6.
86. de Lima M, McMannis J, Gee A, Komanduri K, Couriel D, Andersson BS, et al. Transplantation of ex vivo expanded cord blood cells using the copper chelator tetraethylenepentamine: a phase I/II clinical trial. *Bone Marrow Transplant* 2008;**41**(9):771–8.
87. Jaroscak J, Goltry K, Smith A, Waters-Pick B, Martin PL, Driscoll TA, et al. Augmentation of umbilical cord blood (UCB) transplantation with ex vivo-expanded UCB cells: results of a phase 1 trial using the AastromReplicell System. *Blood* 2003;**101**(12): 5061–7.
88. van Hennik PB, de Koning AE, Ploemacher RE. Seeding efficiency of primitive human hematopoietic cells in nonobese diabetic/severe combined immune deficiency mice: implications for stem cell frequency assessment. *Blood* 1999;**94**(9):3055–61.
89. Rocha V, Labopin M, Ruggeri A, Podesta M, Gallamini A, Bonifazi F, et al. Unrelated cord blood transplantation: outcomes after single-unit intrabone injection compared with double-unit intravenous injection in patients with hematological malignancies. *Transplantation* 2013;**95**(10):1284–91.
90. Ponomaryov T, Peled A, Petit I, Taichman RS, Habler L, Sandbank J, et al. Induction of the chemokine stromal-derived factor-1 following DNA damage improves human stem cell function. *J Clin Invest* 2000;**106**(11):1331–9.
91. Peled A, Kollet O, Ponomaryov T, Petit I, Franitza S, Grabovsky V, et al. The chemokine SDF-1 activates the integrins LFA-1, VLA-4, and VLA-5 on immature human CD34(+) cells: role in transendothelial/stromal migration and engraftment of NOD/SCID mice. *Blood* 2000;**95**(11):3289–96.
92. Christopherson 2nd KW, Paganessi LA, Napier S, Porecha NK. CD26 inhibition on $CD34^+$ or lineage-human umbilical cord blood donor hematopoietic stem cells/hematopoietic progenitor cells improves long-term engraftment into NOD/SCID/Beta2null immunodeficient mice. *Stem Cells Dev* 2007;**16**(3):355–60.
93. Sahin AO, Buitenhuis M. Molecular mechanisms underlying adhesion and migration of hematopoietic stem cells. *Cell Adhesion Migr* 2012;**6**(1):39–48.
94. Sun H, Tsai Y, Nowak I, Liesveld J, Chen Y. Eltrombopag, a thrombopoietin receptor agonist, enhances human umbilical cord blood hematopoietic stem/primitive progenitor cell expansion and promotes multi-lineage hematopoiesis. *Stem Cell Res* 2012;**9**(2):77–86.
95. Sebrango A, Vicuna I, de Laiglesia A, Millan I, Bautista G, Martin-Donaire T, et al. Haematopoietic transplants combining a single unrelated cord blood unit and mobilized haematopoietic stem cells from an adult HLA-mismatched third party donor. Comparable results to transplants from HLA-identical related donors in adults with acute leukaemia and myelodysplastic syndromes. *Best Pract Res Clin Haematol* 2010;**23**(2):259–74.
96. Liu H, Rich ES, Godley L, Odenike O, Joseph L, Marino S, et al. Reduced-intensity conditioning with combined haploidentical and cord blood transplantation results in rapid engraftment, low GVHD, and durable remissions. *Blood* 2011;**118**(24):6438–45.
97. Chen J, Wang RX, Chen F, Sun AN, Qiu HY, Jin ZM, et al. Combination of a haploidentical SCT with an unrelated cord blood unit: a single-arm prospective study. *Bone Marrow Transplant* 2014;**49**(2):206–11.

Chapter 8

# Immunodeficiencies and Metabolic Diseases

**Paul J. Orchard and Angela R. Smith**
*Department of Pediatrics, Division of Blood and Marrow Transplantation, University of Minnesota, Minneapolis, Minnesota, USA*

## Chapter Outline



## 1. IMMUNODEFICIENCIES

### 1.1 Background

Primary immune deficiencies (PID) are rare diseases, but cause significant morbidity and typically lead to early mortality. The first patient successfully treated by hematopoietic stem cell transplantation (HSCT) was an infant with SCID who received bone marrow from his human leukocyte antigen (HLA)-identical sister in 1968.[1] Since then, HSCT has become the standard of care for treatment of children with severe PID. Transplantation has the capacity to cure the majority of patients with PID, and although a stem cell graft from an unaffected matched related donor (MRD) is preferred, that option exists for only a minority of patients, leaving alternative donor stem cell sources as the choice for the majority of patients. Mismatched-related donor (MMRD), matched-unrelated donor (MURD), and umbilical cord blood (UCB) have all been used, and outcomes vary by disease and stem cell source. The optimal stem cell source for PID patients with no available MRD is not clear.[2] UCB is an attractive source for many reasons, including rapid access to the donor unit, allowing for: (1) expedient transplants in patients at very high risk for life-threatening infection; (2) a lower risk of latent viral transmission in immunologically naïve recipients; (3) a lower risk of Graft-versus-Host Disease (GvHD) despite HLA mismatch; (4) a higher likelihood of finding an acceptable HLA-matched unit in patients who are of ethnic minorities and/or the product of consanguinity; and (5) at least hypothetically the potential that UCB stem cells may have greater self-renewing capacity than those obtained from an adult donor.[3] Published studies examining the use of UCB in PID are scarce, but its use is becoming more common. PID encompasses a variety of diseases, all of which are extremely rare. In this chapter, we focus on the more common disorders in this group treated with HSCT, specifically severe combined immune deficiency (SCID) and Wiskott–Aldrich syndrome (WAS).

### 1.2 Severe Combined Immune Deficiency

SCID is a rare genetic disease characterized by a deficiency of T- and B-lymphocyte function that leads to recurrent infections and early death in affected children. The SCID phenotype can result from mutations in multiple genes that encode components of the immune system, including cytokine receptor chains or signaling molecules and genes needed for antigen receptor development.[4] Except for those with SCID due to adenosine deaminase (ADA) deficiency,

Cord Blood Stem Cells Medicine. http://dx.doi.org/10.1016/B978-0-12-407785-0.00008-6

for which enzyme replacement therapy exists, the only proven curative therapy for children with SCID is allogeneic HSCT. Historically, the majority of SCID patients have active infection at the time of diagnosis, which negatively affects the success of HSCT. SCID can be identified at birth through quantification of T-cell receptor excision circles (TRECs) on dried blood spots used for newborn screening. Low or absent TRECs may be due to inadequate thymic production or excessive loss of T cells.[5–7] Newborn screening utilizing the TREC assay has recently been initiated in several states and will be helpful in diagnosing and treating these patients early. Regardless of the method of diagnosis, rapid transition to HSCT is imperative due to the extreme risk of life-threatening infection. In fact, studies have shown that the best outcomes are achieved if HSCT is performed in the first few months of life.[8] Because SCID is a genetic disease, less than 25% of affected children will have a healthy MRD available, so alternative stem cell sources are frequently utilized. MMRD are often considered because of immediate donor availability. While this approach is often successful in attaining T cell chimerism even without conditioning therapy,[8,9] most patients lack full B cell and myeloid donor engraftment, and some investigators have reported issues with lack of long-term immune reconstitution, graft failure (early or late), and inferior survival for SCID patients undergoing MMRD transplantation.[10–12] MURD stem cells are another potential alternative for transplantation in patients with SCID, but historically, it was thought to be an inadequate option for patients with SCID given the time lag required to obtain URD bone marrow and the concern regarding the inability of identifying a donor in the registries. Continued expansion of URD registries has improved our ability to identify donors, and this option should be explored in situations where it is feasible.

Similar to MMRD, UCB is an appealing alternative donor source because the units are stored and readily available, obviating any significant time delay following diagnosis. Several recent reports describe encouraging outcomes after UCB transplantation for variants of SCID.[13–19] The largest report of the role of UCBT for SCID to date is a series from the Japan Cord Blood Bank Network that included 40 subjects with SCID. The cumulative incidence of neutrophil engraftment at 100 days was 77%, grade II–IV aGvHD was 28%, and cGvHD was 13%. The 2-year overall survival (OS) in the SCID cohort was estimated at 71%.[17] Smaller series have shown similar outcomes, with nearly 100% engraftment and an overall survival of 63–75%.[13,14] The most common reported causes of death in all three of these series were GvHD and infection.[13,14,17] These results compare favorably with those seen after MURD transplant. Additionally, a recent retrospective review comparing UCB and MMRD showed that the two groups did not differ significantly in terms of 5-year overall survival, despite a higher incidence of chronic GvHD in UCB recipients.[20] Most reported UCB transplants have been done with pretransplant conditioning chemotherapy, but engraftment and immune reconstitution have been observed without the use of conditioning prior to UCB infusion as well.[13,18] Though most achieve donor T cell engraftment after HSCT for SCID, B cell engraftment and function are much more variable. Up to 60% of patients receiving unconditioned T cell-depleted MRD HSCT continue to require immune globulin replacement following HSCT. The use of pretransplant conditioning may improve the likelihood of B cell chimerism and engraftment, but even with conditioning there are a significant percentage of patients who require immune globulin replacement.[21] Though the number of UCBT reported in the literature is small comparatively, there appears to be a higher likelihood of B cell engraftment and function after UCBT, as only 20% of patients require intravenous immune globulin long term.[22] The reason for this is unclear, and larger studies that can control for molecular type of SCID and the use of pretransplant conditioning are needed to confirm this finding. The optimal alternative stem cell source for children with SCID remains controversial and the choice made appears to be based more on center bias rather than comparative data, but these data demonstrate that UCB is a safe and effective stem cell source for transplant in SCID and should be further investigated.

## 1.3 Wiskott–Aldrich Syndrome

WAS is a rare X-linked immune deficiency that affects between 1 and 10 males per million live births and is characterized by eczema, thrombocytopenia and recurrent sinopulmonary infections. Prior to 1968, when the first successful HSCT for WAS was performed,[23] supportive therapy, including transfusions and splenectomy, was the only option for these patients. As might be expected, the first transplants for WAS were performed using MRD grafts. Long-term follow-up of these patients demonstrated that when transplanted early with both myeloablative (MA) and immunosuppressive therapy, the majority of patients survived with normal platelet counts and immune function.[24] However, early reports cautioned against using alternative donors as very few patients receiving MMRD or MURD marrow survived, and splenectomy alone provided an acceptable survival and quality of life particularly for those >5 years of age.[25] As transplant practices and supportive care measures improved over the years, the utility of alternative donor transplant for children with WAS has been reevaluated. Several studies have delineated the effectiveness of MURD transplant for WAS,[26–28] but with few exceptions,[29–31] most reports of MMRD and haploidentical transplant for WAS have been discouraging with high rates of graft rejection and fatal Epstein Barr virus-associated post-transplant lymphoproliferative disease.[26,27,32] There have been few studies focused on the use of UCB in WAS. Several published case reports have shown encouraging results with sustained engraftment, low rates of GvHD, and excellent overall survival.[33–35] Kobayashi et al. reported a large case series of UCB transplants for WAS. In their cohort of 57 patients, 15 patients received UCB grafts. All received conditioning, though the regimen used varied (busulfan/cytoxan, n=33; radiation containing, n=14; other n=10). The 5-year OS and failure-free survival (defined as survival with treatment response) were 80% (±10.3) and 71.4% (±12.1), respectively. This was similar to that seen in the MRD and MURD groups (81.8%/64.3% and 80%/75.2%, respectively).[27] More recently, Morio et al. reported 23 patients with WAS in their report of 88 patients undergoing UCBT for PID. The cumulative incidence of GvHD was significantly higher in those with WAS compared to other diseases in the cohort (50% vs. 20%, p=0.021), but despite this the 5-year OS in WAS patients was still 82%.[17] These results are encouraging and indicate that more effort should be directed at further delineating the role of UCB transplant in children with WAS. Despite early reports to the contrary, URD transplant has clearly proved to be an effective curative treatment for young children with WAS who do not have a suitable MRD available. Outcomes after UCB appear to be very comparable to URD transplant, and due to the significantly reduced risk of GvHD and rapid availability, many centers now prefer UCB as the donor stem cell source for boys with WAS. Head-to-head outcome comparisons between MURD bone marrow and UCB in larger numbers of patients, however, are needed.

## 1.4 Other Immune Deficiencies

There are multiple case reports in the literature of UCBT for other PID such as hemophagocytic lymphohistiocytosis, IPEX, chronic granulomatous disease (CGD), severe congenital neutropenia, Shwachman–Diamond syndrome, X-linked lymphoproliferative disease, and others.[3] The majority of these patients received pretransplant conditioning therapy and, in general, results have been encouraging. One must be cautious, however, of reporting bias, as failures are less likely to be reported.

## 1.5 Summary and Future Directions

UCB transplantation is a viable alternative donor option for patients with PID and should be considered when an MRD is not immediately available. As discussed previously, UCB has several advantages in this population of patients and the expanding donor pool of UCB units makes it highly likely that a suitably matched UCB unit will be identifiable for nearly any PID patient requiring HSCT. This is true even for ethnic minority populations where the identification of a suitably matched URD has often been difficult. Additionally, UCB also offers the opportunity to couple modern molecular genetic diagnostic techniques, specifically preimplantation genetic diagnosis,[36] with genetically based HLA typing. This technique allows HLA typing and disease screening in single cells from early-stage embryos. Embryos that are unaffected by the particular disease and HLA matched to an affected sibling then can be selected for implantation. Subsequently, if a successful term pregnancy occurs, the UCB can be collected at birth and utilized for transplantation of an affected sibling as well as providing the high-risk couple with a child unaffected by the disease.[37,38]

The use of UCB for PID does have its challenges, however. In one large review, factors associated with overall mortality after UCBT for PID included infection and HLA disparity.[17] Infection is almost universally identified in patients with PID at some point prior to HSCT. Several studies have shown that active infection at the time of UCBT negatively affects outcomes and is associated with poor survival.[12,17,39] Therefore, controlling existing infection prior to UCBT is critically important. This is difficult given the nature of the underlying disease, but newborn screening should allow for earlier diagnosis prior to the onset of infection, at least for children with SCID. Additionally, though less-stringent HLA-matching requirements are one advantage of using UCB, higher degrees of HLA disparity (≥2 antigen mismatches) appear to negatively impact survival after UCBT in PID patients.[17] Other challenges with UCB that are not specific to PID include lack of availability of the utilized donor for subsequent infusions of hematopoietic stem cells or T cells, a reduced stem cell dose, the risk of transmitting unidentified genetic disease and a lack of viral-specific cytotoxic T cells within the UCB.[3] Strategies to overcome these disadvantages, including ex vivo expansion of UCB stem cells, expansion of viral-specific T cells, and co-infusion of UCB-derived T regulatory cells are being developed and will likely be applicable to PID patients as well. While there are very little data available that have addressed the issue of immune reconstitution after UCBT in PID, one small series has reported engraftment of donor T, B, and NK cells, and the recovery of T cell function by 60–100 days and NK cell function by 180 days.[40] Concern still exists related to the incidence of infectious complications in recipients of UCB, and in particular, for those patients who have pre-existing infectious complications already present at the time of HSCT. The future development of adoptive immunotherapy techniques utilizing UCB may help to overcome this obstacle.

It is clear from the literature that children with PID do not always require fully MA conditioning, and some with SCID may not need any conditioning at all to achieve adequate donor engraftment and cure of their disease. Based on this, reduced intensity conditioning (RIC) is often considered in an attempt to limit both acute and long-term complications, but there has been no prospective head-to-head comparison of RIC versus MA conditioning in UCBT for PID. In their retrospective review, Morio et al. showed that an RIC approach was associated with a significantly higher 5-year overall survival when compared to an MA approach; outcomes in relationship to donor status were not addressed.[17] It is impossible to conclude based on this that RIC is superior for all forms of PID and in all situations, but it may be beneficial for some types of PID and should be given consideration. Future studies aimed at defining adequate RIC regimens with UCB in PID will further expand the use of UCB in these diseases.

Much progress has been made in gene therapy in the last few years, and may prove an alternative for the treatment of several PIDs.[41] The current gene therapy approach is based on ex vivo transfer of the transgene into autologous HSCs using viral vectors followed by transplantation of the modified HSCs back to the patient with or without conditioning. Gene therapy is being utilized in several different PIDs, including ADA–SCID, X-linked SCID, WAS, and CGD. Outcomes vary depending on the disease being treated, but they have been encouraging overall. Insertional mutagenesis leading to malignant transformation has been the major success-limiting adverse event. To address this, retroviral vectors are being increasingly replaced by self-inactivating vectors which have shown similar efficacy but a reduced tendency for harmful mutagenesis. Studies using these vectors are ongoing. In patients with these diseases who lack an HLA-identical donor for allogeneic HCT, gene therapy should be given consideration.

## 2. METABOLIC DISEASES

### 2.1 Background—Lysosomal Storage Disease

The existence of the lysosome was first described by Christian de Duve in 1955.[42] Dr. de Duve, who also described the peroxisome, shared the Nobel Prize in 1974 "for their discoveries concerning the structural and functional organization of the cell."[43] The lysosome is an organelle with a low internal pH containing a variety of hydrolytic enzymes; there are over 50 lysosomal storage diseases (LSD) that have been described related to dysfunction of these enzymes.[44] These disorders are single gene defects, the vast majority of which are autosomal recessive in inheritance. While rare when considered individually, the cumulative incidence of these disorders has been estimated to be approximately 1 in 7000 births.[45] Loss-of-function of these enzymes leads to accumulation of substrate, the specifics of which are characteristic of the enzyme in question. In 1968, Dr. Neufeld's laboratory reported the accumulation of storage material observed in fibroblasts derived from patients with Hurler and Hunter syndrome could be reversed when cultured with fibroblasts from a normal individual.[46] Evidence was also provided that media obtained from normal cells could markedly reduce the amount of accumulated storage material in affected cells. This principle of "cross-correction" was fundamentally important in the consideration of allogeneic HSCT as a treatment for LSD. This internalization of lysosomal enzymes when provided into the environment was shown to be inhibited in the presence of mannose 6-phosphate, and the receptors responsible for this binding were quickly regenerated even in the absence of protein synthesis.[47] Based on these findings, it was postulated that the use of allogeneic hematopoietic stem cell transplantation could prove useful therapy for the lysosomal storage diseases (LSD), as donor-derived cells may provide a continuous source of enzyme to a recipient suffering from these disorders.

### 2.2 Early Experience in the Use of Transplantation for Lysosomal Disorders

As mentioned, hematopoietic stem cell transplantation was explored following the documentation of the potential for cellular cross-correction. The first description of the use of allogeneic transplantation for a lysosomal storage disorder was reported by Hobbs, describing the outcome of a patient transplanted for Hurler syndrome.[48] In this patient, they demonstrated improvement in the excretion of urinary glycosaminoglycans (GAG), a decrease in organomegaly and achievement of donor levels of α-L-iduronidase, the defective enzyme in MPSI. In 1984, a patient with Maroteaux-Lamy (MPSVI) was described following transplantation using a matched related donor with normal enzyme levels.[49] Post transplant, enzyme levels in leukocytes were shown to be normal, but in liver levels were only 16% of normal. Importantly, this demonstrated that while enzyme levels may be "normal" following transplant, this testing is almost exclusively reflective of levels in blood cells, but is not necessarily reflective of enzyme delivered to other sites such as bone, the heart valves, the cornea or the central nervous system (CNS). Since that time, the utilization of HSCT has been explored in other mucopolysaccharide disorders, including Hunter syndrome (MPSII),[50,51] Sanfilippo syndrome (MPSIII),[50,52] Maroteaux-Lamy syndrome (MPSVI),[49,52–54] and Sly syndrome (MPS VII).[52,55] In addition, transplantation has been used for other lysosomal diseases including I-cell disease,[52,56] fucosidosis,[57] α-mannosidosis,[52,53,58] aspartylglucosaminuria,[52,59] Farber's disease,[52,60] Gaucher's disease,[52,61,62] GM1 gangliosidosis,[63] Tay-Sachs disease and Sandhoff's disease,[64] Niemann-Pick A/B disease,[52,65,66] Wolman disease,[52,67] and the inherited leukodystrophies including globoid leukodystrophy (GLD, or Krabbe disease),[52,68–73] and metachromatic leukodystrophy (MLD).[52,73–78]

### 2.3 Obstacles in Establishing Response of Lysosomal Storage Disease to Transplantation

Assessment of outcomes in patients with LSD is difficult for a number of reasons. (1) These diseases are rare, and relatively few patients are treated in individual centers. While multi-institutional analyses using existing registries are useful, standardized data collection within registries is limited to transplant-related parameters such as the type of preparative regimen, the rate of neutrophil engraftment, incidence and severity of GvHD, and overall survival. In contrast, registry data are not well suited to capture disease-specific and functional outcomes. To develop a better understanding of the benefits and shortcomings of transplantation for individual diseases, this information is critical. For instance, the techniques and timing of imaging (hip X-rays, MRI's, etc.) are not standardized from center to center. In addition, there is to date no centralized database for storing images, which would facilitate review by orthopedic surgeons or radiologists with extensive experience in these disorders. (2) Also, for many inherited metabolic disorders, significant variation in phenotype exists, with a continuum from severe to attenuated forms. As might be expected, patients with more severe disease often have more rapid progression, and are therefore less responsive to intervention. On this basis, clearly defining the severity of disease in an individual patient is important in understanding the anticipated natural history, and how any intervention may impact expected outcomes. For a given individual, predicting the phenotype is often difficult, and while the genotype may be useful in some disorders, for many there is not an established phenotype–genotype correlation. (3) Within each phenotype, the diagnosis may be established at varying points in the course of disease progression. The state of the disease at the time of transplantation is therefore a confounding factor in the analysis of outcomes. In summary, based on the above concerns, the development of large, well-standardized multi-institutional collaborative efforts is key, and will be needed to achieve a better understanding of the relative benefits of transplantation.

## 2.4 Response of Hurler Syndrome to HSCT

To date, more than 500 patients with Hurler syndrome have undergone transplantation, making it the lysosomal disease for which there is the greatest experience with regard to the role of HSCT.[79,80] There are many clinical manifestations of Hurler syndrome, and it has become clear that while some respond well to transplantation, other disease-associated complications are less amenable to HSCT. Patients with Hurler syndrome commonly develop upper airway obstruction, corneal clouding, hearing difficulties, cardiac dysfunction and valvular changes, developmental deterioration, and extensive skeletal abnormalities termed dysostosis multiplex.[81,82] While the accumulation of storage material in the soft tissues is relatively responsive to transplantation, other complications, such as valvular and orthopedic changes are less successfully treated by HSCT.[79,83,84] In particular, the disease-related orthopedic issues, including hand,[83,85] knee,[83,86] hip,[87,88] back, and cervical spine[89,90] remain a significant cause of morbidity despite transplantation. Importantly, transplant has been shown to help preserve cognitive function.[50,91,92] This is thought to be due to the engraftment of donor microglial cells within the brain, as microglia are hematopoietically derived.[93,94] This is the presumed mechanism for any CNS effects of transplantation.

In 2003, recombinant enzyme (Aldurazyme) was FDA approved for the treatment of MPSI, as it has been shown to decrease urinary GAG, increase the forced vital capacity, and improve the distance in a 6 min walk.[95] However, the intravenously administed Aldurazyme does not cross the blood–brain barrier, and on this basis would not be expected to mediate beneficial changes in the CNS.[95–96] Patients with the attenuated forms of MPSI (Hurler–Scheie and Scheie) do not experience the cognitive deterioration observed in patients with Hurler syndrome. This has led to the current recommendation of proceeding to transplantation in young patients with Hurler syndrome to preserve cognition, while patients with attenuated disease are treated with enzyme therapy.[79,80,97] The use of UCB for Hurler syndrome has expanded significantly. Engraftment and chimerism rates utilizing cord blood grafts have been encouraging, and it has been suggested that it may be superior to that of marrow or peripheral blood.[79,83,92] This remains to be definitively demonstrated, as historical comparisons may differ in terms of donor HLA compatibility, use of targeted busulfan or supportive care.[83] There are data that expediency in proceeding to transplantation is important, both in regards to survival as well as disease-related outcomes.[79,83–85,98] On this basis, it is clear that there is an advantage in the use of cord blood as a graft source. As the median age at transplant in a recent large, multi-institutional analysis was 16.7 months,[79] these patients are sufficiently small in size that cell dose is not a limitation.

## 2.5 Other MPS Disorders

It is generally accepted that allogeneic transplantation is the standard of care for Hurler syndrome. However, for other MPS disorders, the role of transplantation is less clear. In MPSII (Hunter) and MPSIII (Sanfilippo), the neurologic deterioration has not been as responsive to intervention with transplant.[50–53] The majority of reported patients transplanted with these disorders have had clinically evident neurologic changes at the time of transplantation. Identification of patients very early in life, prior to the development of developmental deterioration, could potentially alter this experience. However, currently there is no clear role for the utilization of HSCT as therapy for patients with Hunter syndrome or Sanfilippo with symptomatic neurologic disease. Patients with Morquio syndrome (MPS IV) have significant musculoskeletal changes related to their disease.[99] To date there is no convincing evidence that HSCT has an impact on the course of the disease. In MPS VI (Maroteaux–Lamy syndrome), outcomes are encouraging with transplantation.[54] The availability of enzyme replacement therapy, however, has altered the role of HSCT for MPS VI. As there is no neurologic aspect to this disorder, it is not necessary to deliver enzyme to the CNS. As a result, few MPS VI patients are now referred for transplantation due to its increased risk in comparison to ERT. Whether transplantation and ERT for MPS VI are equally efficacious as therapy has not been addressed to date. MPS VII is a very rare condition, and there is limited experience in the use of transplantation as an intervention. However, the available information suggests that there may be a role for transplantation.[52,55] I-cell disease has a number of clinical manifestations similar to the MPS disorders. Rather than a deficiency of a lysosomal enzyme, the defect is in a phosphotransferase necessary for establishing the mannose- 6-phosphate signal required to achieve localization of enzymes to the lysosome.[100] It is to date not clear that transplantation alters the course of I-cell disease. However, if HSCT is shown to be of benefit, the mechanism would likely not be through the cross-correction of a single deficient enzyme as occurs in other MPS diseases, as there is no evidence for cross-correction of the missing gene product in I-cell disease. While it is possible that a number of enzymes with the mannose- 6-phosphate signal could be provided by the donor-derived graft, this remains to be demonstrated.

## 2.6 Indications of HSCT for Lysosomal Diseases of the CNS

Due to the availability of ERT for diseases such as Gaucher and Wolman disease, there will likely be few transplants performed in the future for these disorders. Similarly, ERT is being developed for α-mannosidosis,[101] which will likely limit the use of transplantation for this diagnosis. However, there are other lysosomal disorders with a neurologic phenotype that currently cannot be effectively treated with intravenous enzyme therapy. These include Tay–Sachs disease, Sandhoff, and GM1 gangliosidosis. While these diseases have attenuated forms, the majority of patients with these diseases are diagnosed in the first 12–18 months of life, and these diseases are rapidly progressive. There are no compelling data that symptomatic infants with these diseases benefit from transplantation.[64,102] It is possible that there may be responses in attenuated disease,[103] but more information is required. Other lysosomal diseases associated with neurologic deterioration, such as Batten and Niemann–Pick A, do not appear to respond to HSCT.[52,104] However, Niemann–Pick B, which is caused by the same enzyme deficiency as Niemann–Pick A (sphingomyelinase), has shown improvement in organomegaly and pulmonary disease, as well as clearance of bone marrow disease.[65,66] MLD is a demyelinating disease of both the central and peripheral nervous system, typically causing motor deficits in patients before 2 years of age.[75,105] The late-infantile form of the disease is rapidly progressive, and outcomes following intervention with HSCT have been poor.[50,53,75,78,106] For patients with attenuated MLD, including the juvenile form, reports suggest a beneficial role for transplantation.[50,78] Whether very early intervention with transplantation in asymptomatic patients with a late-infantile phenotype will result in acceptable outcomes is unclear; it appears it may change the natural history of the disease, but more information is necessary.[78] Globoid cell leukodystrophy, or Krabbe disease, also has a variable phenotype but commonly becomes apparent in the first year of life. These patients rapidly deteriorate due to central and peripheral demyelination. There is no role for transplantation in symptomatic infants with GLD due to rapid progression and poor outcomes.[53,68,69] However, there has been a great deal of interest in the potential of cord blood transplantation within the first few weeks of life. This approach clearly alters the course of the disease, but outcomes are variable.[71,107,108] This can only be achieved through the identification of patients soon after birth due to a prior family history, or through newborn screening. Screening has been performed for Krabbe in New York since 2006,[109] although debate remains about the usefulness of this approach.[110] Nevertheless, the ability to identify patients very early in life to minimize progression is an important development in the field, and will rely almost exclusively on UCB as a stem cell source, unless a matched, related donor is available that is not a carrier. As carriers are assumed to have decreased levels of enzyme, they would not be considered optimal donors, as the amount of enzyme delivered would be expected to be approximately 50% of that which could be achieved with a donor that is not a carrier. This restriction would also apply to haploidentical transplantation using parental donors.

## 2.7 Other Inherited Metabolic Diseases

As opposed to the lysosomal diseases associated with white matter changes, adrenoleukodystrophy (ALD) is an X-linked, peroxisomal disorder. In approximately 40% of boys with ALD, an acute neuroinflammatory condition develops that is rapidly progressive and lethal. While there have been numerous reports of the ability of transplantation to stabilize disease,[77,111–113] the physiologic means by which this is accomplished remains to be determined. In some cases, patients are identified early in the course of the disease, and there is sufficient time to explore unrelated donors. However, in more advanced patients expediency is of great importance, as rapid neurologic dysfunction is observed and is not recoverable. In this population, UCB grafts are used extensively. Osteopetrosis (OP) is an inherited disorder of bone metabolism for which allogeneic transplantation has been demonstrated to be effective.[114–116] Deficient osteoclast function leads to alteration in the architecture of the bone, with a markedly decreased marrow space and insufficient hematopoiesis. Untreated, patients with recessive OP commonly die in the first decade of pancytopenia.[117]

As osteoclasts are derived from the monocytic lineage, successful allogeneic transplantation has resulted in remodeling of bone, which can appear normal by X-ray.[114–116] However, engraftment may be difficult to achieve, likely due to the abnormal bone structure. In addition, peri-transplant mortality has been high, with complications including hypercalcemia, veno-occlusive disease, pulmonary and airway issues, and pulmonary hypertension.[118–120] Due to the increase in mortality associated with HSCT, there have been attempts to explore reduced intensity regimens for OP. In one report, the use of UCB grafts for OP with a reduced intensity approach was not successful in achieving stable chimerism.[121] On this basis, at the University of Minnesota we have chosen to utilize an unrelated donor graft for patients with OP, and in circumstances where these are the only option, a fully ablative regimen is utilized. A large, multi-institutional review of these issues will be important. It should be acknowledged that the genotypes associated with recessive OP are varied, and in rare cases the defect may not be intrinsic to the osteoclast. In these situations transplantation would not be expected to be beneficial. Therefore determining the genotype is of great importance.

## 2.8 Changes in the Landscape in the Treatment of Inherited, Metabolic, and Storage Diseases

As previously mentioned, UCB transplantation has distinct advantages as a graft source in the treatment of these disorders. Proceeding quickly to HSCT is often of great importance, as there is often a narrow window of time in which to achieve optimal outcomes. With the availability of enzyme replacement for the lysosomal diseases, there are diagnoses for which transplantation has been replaced by ERT, such as Maroteaux–Lamy syndrome. Nevertheless, transplantation remains the only means of long-term delivery of enzyme to the brain, and therefore remains the standard of care for diseases such as Hurler. This is likely to continue for the near future. Still, transplantation remains insufficient as therapy for many of the disease-related complications of Hurler, most notably the orthopedic aspects of the disorder. There are preliminary biochemical and clinical data that higher enzyme levels after HSCT can result in better disease correction.[122] However, if full chimerism is achieved with a noncarrier donor, this may be the best correction possible using transplantation as an intervention. Assuming that there is limited delivery of enzyme to tissues such as bone, and that additional enzyme could result in more complete correction, added benefit could potentially be achieved with ERT after transplant. This combination therapy approach is being explored. In addition, enzyme therapy prior to transplantation has been used to reduce substrate accumulation, with the goal of potentially decreasing morbidity and mortality.[123,124] A significant improvement in overall survival has not yet been apparent in a large population of Hurler patients,[79] but could possibly be demonstrated in a higher risk population.[98] The evolving nature of neonatal screening will also affect the use of HSCT for these diseases. Newborn testing for Hurler syndrome, as well as inherited lysosomal leukodystrophies (ALD, MLD, and GLD) will likely become increasingly utilized over the next 5–10 years. For disorders with rapid progression requiring urgent transplantation such as GLD, and potentially MLD and others, UCB will almost certainly be the preferred graft source. In addition, gene therapy is currently being explored for several of these diseases, including MLD and ALD. It seems clear that these techniques will be further explored in these disorders to test an autologous, gene corrected approach as an alternative to allogeneic transplantation. In addition, higher levels of lysosomal gene product may be achievable, which could lead to better correction of these diseases. An extensive description of the various potential benefits and challenges of gene therapy for these disorders is, however, beyond the scope of this discussion.

## 2.9 Unanswered Questions in the Use of Transplantation for Metabolic Disorders

Why transplantation has been shown be of benefit for disorders such as MPS I, and not MPS II and MPS III, is unclear. It has been accepted that donor-derived microglia are responsible for delivering enzyme to the central nervous system, and it would seem likely that with an identical preparative regimen and graft source, microglial engraftment would be comparable across these disorders. However, it is possible that the delivery of a larger amount of enzyme is required to achieve relative correction in MPS II and MPS III than for Hurler. Is there variation in binding and internalization of one enzyme as opposed to another within the CNS? Are the kinetics of turnover within the lysosome different? What is the contribution of neuroinflammation in one disease vs. another? A related and important question is how to best achieve microglial engraftment. It is possible that modification of the preparative regimen may alter the proportion of donor-derived cells in the brain, or the time required for them to engraft. In addition, it is possible that UCB may achieve more complete microglial engraftment than marrow. This has not been demonstrated, but would be important to explore. As outcomes with lysosomal leukodystrophies remain variable, enhanced delivery of enzyme to the CNS could enhance function and quality of life. In addition, assuming neonatal transplantation is increasingly used in these diseases, what will prove to be the optimal conditioning regimen for an infant several weeks old? Is a fully ablative approach required, or would a less intensive regimen be sufficient? Finally, as other therapies are developed such as chaperone therapy and small molecules,[72] will combinations with transplantation prove to provide the optimal outcomes? Collaborative multi-institutional studies will be best suited to provide these insights.

## REFERENCES

1. Gatti R, Meuwissen H, Allen H, Hong R, Good R. Immunological reconstitution of sex-linked lymphopenic immunological deficiency. *Lancet* 1968;**2**:1366–9.
2. Dvorak CC, Cowan MJ. Hematopoietic stem cell transplantation for primary immunodeficiency disease. *Bone Marrow Transpl* 2008;**41**:119–26.
3. Gennery A, Cant A. Cord blood stem cell transplantation in primary immune deficiencies. *Curr Opin Allergy Clin Immunol* 2007;**7**: 528–34.
4. Buckley RH. The multiple causes of human SCID. *J Clin Invest* 2004;**114**:1409–11.
5. Puck JM. Neonatal screening for severe combined immune deficiency. *Curr Opin Allergy Clin Immunol* 2007;**7**:522–7.
6. Puck JM, SCID Newborn Screening Working Group. Population-based newborn screening for severe combined immunodeficiency: steps toward implementation. *J Allergy Clin Immunol* 2007;**120**:760–8.
7. Puck JM, Routes J, Filipovich AH, Sullivan K. Expert commentary: practical issues in newborn screening for severe combined immune deficiency (SCID). *J Clin Immunol* 2012;**32**:36–8.
8. Buckley RH, Schiff SE, Schiff RI, Markert L, Williams LW, Roberts JL, et al. Hematopoietic stem-cell transplantation for the treatment of severe combined immunodeficiency. *N Engl J Med* 1999;**340**:508–16.
9. Buckley RH, Schiff SE, Sampson HA, Schiff RI, Markert ML, Knutsen AP, et al. Development of immunity in human severe primary T cell deficiency following haploidentical bone marrow stem cell transplantation. *J Immunol* 1986;**136**:2398–407.
10. Friedrich W, Hönig M, Müller S. Long-term follow-up in patients with severe combined immunodeficiency treated by bone marrow transplantation. *Immunol Res* 2007;**38**:165–73.
11. Grunebaum E, Mazzolari E, Porta F, Dallera D, Atkinson A, Reid B, et al. Bone marrow transplantation for severe combined immune deficiency. *JAMA* 2006;**295**:508–18.
12. Antoine C, Müller S, Cant A, Cavazzana-Calvo M, Veys P, Vossen J, et al. Long-term survival and transplantation of haemopoietic stem cells for immunodeficiencies: report of the European experience 1968–99. *Lancet* 2003;**361**:553–60.
13. Bhattacharya A, Slatter MA, Chapman CE, Barge D, Jackson A, Flood TJ, et al. Single centre experience of umbilical cord stem cell transplantation for primary immunodeficiency. *Bone Marrow Transpl* 2005;**36**:295–9.
14. Díaz de Heredia C, Ortega JJ, Díaz MA, Olivé T, Badell I, González-Vicent M, et al. Unrelated cord blood transplantation for severe combined immunodeficiency and other primary immunodeficiencies. *Bone Marrow Transpl* 2008;**41**:627–33.
15. Frangoul H, Wang L, Harrell FE Jr, Manes B, Calder C, Domm J. Unrelated umbilical cord blood transplantation in children with immune deficiency: results of a multicenter study. *Bone Marrow Transpl* 2010;**45**:283–8.
16. Knutsen A, Wall D. Umbilical cord blood transplantation in severe T-cell immunodeficiency disorders: two-year experience. *J Clin Immunol* 2000;**20**:466–76.
17. Morio T, Atsuta Y, Tomizawa D, Nagamura-Inoue T, Kato K, Ariga T, et al. Outcome of unrelated umbilical cord blood transplantation in 88 patients with primary immunodeficiency in Japan. *Br J Haematol* 2011;**154**:363–72.
18. Toren A, Nagler A, Amariglio N, Neumann Y, Golan H, Bilori B, et al. Successful human umbilical cord blood stem cell transplantation without conditioning in severe combined immune deficiency. *Bone Marrow Transpl* 1999;**23**:405–8.
19. Tsuji Y, Imai K, Kajiwara M, Aoki Y, Isoda T, Tomizawa D, et al. Hematopoietic stem cell transplantation for 30 patients with primary immunodeficiency diseases: 20 years experience of a single team. *Bone Marrow Transpl* 2006;**37**:469–77.
20. Fernandes JF, Rocha V, Labopin M, Neven B, Moshous D, Gennery AR, et al. Transplantation in patients with SCID: mismatched related stem cells or unrelated cord blood? *Blood* 2012;**119**:2949–55.
21. Buckley RH. B-cell function in severe combined immunodeficiency after stem cell or gene therapy: a review. *J Allergy Clin Immunol* 2010;**125**:790–7.
22. Chan WY, Roberts RL, Moore TB, Stiehm ER. Cord blood transplants for SCID: better B-cell engraftment? *J Pediatr Hematol Oncol* 2013;**35**:e14–18.
23. Bach F, Albertini R, Joo P, Anderson J, Bortin M. Bone-marrow transplantation in a patient with the Wiskott-Aldrich syndrome. *Lancet* 1968;**2**:1364–6.
24. Rimm I, Rappeport J. Bone marrow transplantation for the Wiskott-Aldrich syndrome. Long-term follow-up. *Transplantation* 1990;**50**:617–20.
25. Mullen C, Anderson K, Blaese R. Splenectomy and/or bone marrow transplantation in the management of the Wiskott-Aldrich syndrome: long-term follow-up of 62 cases. *Blood* 1993;**82**:2961–6.
26. Filipovich AH, Stone JV, Tomany SC, Ireland M, Kollman C, Pelz CJ, et al. Impact of donor type on outcome of bone marrow transplantation for Wiskott-Aldrich syndrome: collaborative study of the International Bone Marrow Transplant Registry and the National Marrow Donor Program. *Blood* 2001;**97**:1598–603.
27. Kobayashi R, Ariga T, Nonoyama S, Kanegane H, Tsuchiya S, Morio T, et al. Outcome in patients with Wiskott-Aldrich syndrome following stem cell transplantation: an analysis of 57 patients in Japan. *Br J Haematol* 2006;**135**:362–6.
28. Pai SY, DeMartiis D, Forino C, Cavagnini S, Lanfranchi A, Giliani S, et al. Stem cell transplantation for the Wiskott-Aldrich syndrome: a single-center experience confirms efficacy of matched unrelated donor transplantation. *Bone Marrow Transpl* 2006;**38**:671–9.
29. Fischer A, Friedrich W, Fasth A, Blanche S, Le Deist F, Girault D, et al. Reduction of graft failure by a monoclonal antibody (anti-LFA-1 CD11a) after HLA nonidentical bone marrow transplantation in children with immunodeficiencies, osteopetrosis, and Fanconi's anemia: a European group for immunodeficiency/European group for bone marrow transplantation report. *Blood* 1991;**77**:249–56.
30. Inagaki J, Park Y, Kishimoto T, Yoshioka A. Successful unmanipulated haploidentical bone marrow transplantation from an HLA 2-locus-mismatched mother for Wiskott-Aldrich syndrome after unrelated cord blood stem cell transplantation. *J Pediatr Hematol Oncol* 2005;**27**:229–31.
31. Rumelhart S, Trigg M, Horowitz S, Hong R. Monoclonal antibody T-cell-depleted HLA-haploidentical bone marrow transplantation for Wiskott-Aldrich syndrome. *Blood* 1990;**75**:1031–5.

32. Brochstein JA, Gillio AP, Ruggiero M, Kernan NA, Emanuel D, Laver J, et al. Marrow transplantation from human leukocyte antigen-identical or haploidentical donors for correction of Wiskott- Aldrich syndrome. *J Pediatr* 1991;**119**:907–12.
33. Kaneko M, Watanabe T, Watanabe H, Kimura M, Suzuya H, Okamoto Y, et al. Successful unrelated cord blood transplantation in an infant with Wiskott-Aldrich syndrome following recurrent cytomegalovirus disease. *Int J Hematol* 2003;**78**:457–60.
34. Knutsen A, Steffen M, Wassmer K, Wall D. Umbilical cord blood transplantation in Wiskott Aldrich syndrome. *J Pediatr* 2003;**142**:519–23.
35. Slatter MA, Bhattacharya A, Flood TJ, Abinun M, Cant AJ, Gennery AR. Use of two unrelated umbilical cord stem cell units in stem cell transplantation for Wiskott-Aldrich syndrome. *Pediatr Blood Cancer* 2006;**47**:332–4.
36. Kuliev A, Rechitsky S, Tur-Kaspa I, Verlinsky Y. Preimplantation genetics: improving access to stem cell therapy. *Ann N Y Acad Sci* 2005;**1054**:223–7.
37. Verlinsky Y, Rechitsky S, Sharapova T, Morris R, Taranissi M, Kuliev A. Preimplantation HLA testing. *JAMA* 2004;**291**: 2079–85.
38. Verlinsky Y, Rechitsky S, Sharapova T, Laziuk K, Barsky I, Verlinsky O, et al. Preimplantation diagnosis for immunodeficiencies. *Reprod Biomed Online* 2007;**14**:214–23.
39. Cuvelier GD, Schultz KR, Davis J, Hirschfeld AF, Junker AK, Tan R, et al. Optimizing outcomes of hematopoietic stem cell transplantation for severe combined immunodeficiency. *Clin Immunol* 2009;**131**:179–88.
40. Knutsen A, Wall D. Kinetics of T-cell development of umbilical cord blood transplantation in severe T-cell immunodeficiency disorders. *J Allergy Clin Immunol* 1999;**103**:823–32.
41. Mukherjee S, Thrasher AJ. Gene therapy for PIDs: progress, pitfalls and prospects. *Gene* 2013;**525**:174–81.
42. De Duve C, Pressman BC, Gianetto R, Wattiaux R, Appelmans F. Tissue fractionation studies. 6. Intracellular distribution patterns of enzymes in rat-liver tissue. *Biochem J* 1955;**60**:604–17.
43. Blobel G. Christian de Duve (1917–2013). *Nature* 2013;**498**:300.
44. Fuller M, Tucker JN, Lang DL, Dean CJ, Fietz MJ, Meikle PJ, et al. Screening patients referred to a metabolic clinic for lysosomal storage disorders. *J Med Genet* 2011;**48**:422–5.
45. Meikle PJ, Hopwood JJ, Clague AE, Carey WF. Prevalence of lysosomal storage disorders. *J Am Med Assoc* 1999;**281**:249–54.
46. Fratantoni JC, Hall CW, Neufeld EF. Hurler and Hunter syndromes: mutual correction of the defect in cultured fibroblasts. *Science* 1968;**162**:570–2.
47. Rome LH, Weissmann B, Neufeld EF. Direct demonstration of binding of a lysosomal enzyme, alpha-L-iduronidase, to receptors on cultured fibroblasts. *Proc Natl Acad Sci U S A* 1979;**76**:2331–4.
48. Hobbs JR, Hugh-Jones K, Barrett AJ, Byrom N, Chambers D, Henry K, et al. Reversal of clinical features of Hurler's disease and biochemical improvement after treatment by bone-marrow transplantation. *Lancet* 1981;**2**:709–12.
49. Krivit W, Pierpont ME, Ayaz K, Tsai M, Ramsay NK, Kersey JH, et al. Bone-marrow transplantation in the Maroteaux-Lamy syndrome (mucopolysaccharidosis type VI). Biochemical and clinical status 24 months after transplantation. *N Engl J Med* 1984;**311**:1606–11.
50. Shapiro EG, Lockman LA, Balthazor M, Krivit W. Neuropsychological outcomes of several storage diseases with and without bone marrow transplantation. *J Inherit Metab Dis* 1995;**18**:413–29.
51. Vellodi A, Young E, Cooper A, Lidchi V, Winchester B, Wraith JE. Long-term follow-up following bone marrow transplantation for Hunter disease. *J Inherit Metab Dis* 1999;**22**:638–48.
52. Krivit W. Allogeneic stem cell transplantation for the treatment of lysosomal and peroxisomal metabolic diseases. *Springer Semin Immunopathol* 2004a;**26**:119–32.
53. Peters C, Steward CG, National Marrow Donor P, International Bone Marrow Transplant R, Party Working. on Inborn Errors, E.B.M.T.G. Hematopoietic cell transplantation for inherited metabolic diseases: an overview of outcomes and practice guidelines. *Bone Marrow Transpl* 2003;**31**:229–39.
54. Turbeville S, Nicely H, Rizzo JD, Pedersen TL, Orchard PJ, Horwitz ME, et al. Clinical outcomes following hematopoietic stem cell transplantation for the treatment of mucopolysaccharidosis VI. *Mol Genet Metab* 2011;**102**:111–5.
55. Yamada Y, Kato K, Sukegawa K, Tomatsu S, Fukuda S, Emura S, et al. Treatment of MPS VII (Sly disease) by allogeneic BMT in a female with homozygous A619V mutation. *Bone Marrow Transpl* 1998;**21**:629–34.
56. Grewal S, Shapiro E, Braunlin E, Charnas L, Krivit W, Orchard P, et al. Continued neurocognitive development and prevention of cardiopulmonary complications after successful BMT for I-cell disease: a long-term follow-up report. *Bone Marrow Transpl* 2003;**32**:957–60.
57. Vellodi A, Cragg H, Winchester B, Young E, Young J, Downie CJ, et al. Allogeneic bone marrow transplantation for fucosidosis. *Bone Marrow Transpl* 1995;**15**:153–8.
58. Mynarek M, Tolar J, Albert MH, Escolar ML, Boelens JJ, Cowan MJ, et al. Allogeneic hematopoietic SCT for alpha-mannosidosis: an analysis of 17 patients. *Bone Marrow Transpl* 2012;**47**:352–9.
59. Ringdén O, Remberger M, Svahn BM, Barkholt L, Mattsson J, Aschan J, et al. Allogeneic hematopoietic stem cell transplantation for inherited disorders: experience in a single center. *Transplantation* 2006;**81**:718–25.
60. Vormoor J, Ehlert K, Groll AH, Koch HG, Frosch M, Roth J. Successful hematopoietic stem cell transplantation in Farber disease. *J Pediatr* 2004;**144**(1):132–4.
61. Tsai P, Lipton JM, Sahdev I, Najfeld V, Rankin LR, Slyper AH, et al. Allogenic bone marrow transplantation in severe Gaucher disease. *Pediatric Res* 1992;**31**:503–7.
62. Goker-Alpan O, Wiggs EA, Eblan MJ, Benko W, Ziegler SG, Sidransky E, et al. Cognitive outcome in treated patients with chronic neuronopathic Gaucher disease. *J Pediatr* 2008;**153**:89–94.
63. Boomkamp SD, Butters TD. Glycosphingolipid disorders of the brain. *Sub-cell Biochem* 2008;**49**:441–67.
64. Jacobs JF, Willemsen MA, Groot-Loonen JJ, Wevers RA, Hoogerbrugge PM. Allogeneic BMT followed by substrate reduction therapy in a child with subacute Tay-Sachs disease. *Bone Marrow Transpl* 2005;**36**:925–6.
65. Vellodi A, Hobbs JR, O'Donnell NM, Coulter BS, Hugh-Jones K. Treatment of Niemann-Pick disease type B by allogeneic bone marrow transplantation. *Br Med J Clin Res Ed* 1987;**295**:1375–6.
66. Shah AJ, Kapoor N, Crooks GM, Parkman R, Weinberg KI, Wilson K, et al. Successful hematopoietic stem cell transplantation for Niemann-Pick disease type B. *Pediatrics* 2005;**116**:1022–5.
67. Tolar J, Petryk A, Khan K, Bjoraker KJ, Jessurun J, Dolan M, et al. Long-term metabolic, endocrine, and neuropsychological outcome of hematopoietic cell transplantation for Wolman disease. *Bone Marrow Transpl* 2009;**43**:21–7.

68. Escolar ML, Poe MD, Provenzale JM, Richards KC, Allison J, Wood S, et al. Transplantation of umbilical-cord blood in babies with infantile Krabbe's disease. *N Engl J Med* 2005;**352**: 2069–81.
69. Escolar ML, Poe MD, Martin HR, Kurtzberg J. A staging system for infantile Krabbe disease to predict outcome after unrelated umbilical cord blood transplantation. *Pediatrics* 2006;**118**:e879–89.
70. Lim ZY, Ho AY, Abrahams S, Fensom A, Aldouri M, Pagliuca A, et al. Sustained neurological improvement following reduced-intensity conditioning allogeneic haematopoietic stem cell transplantation for late-onset Krabbe disease. *Bone Marrow Transpl* 2008;**41**(9):831–2.
71. Duffner PK, Caviness VS Jr, Erbe RW, Patterson MC, Schultz KR, Wenger DA, et al. The long-term outcomes of presymptomatic infants transplanted for Krabbe disease: report of the workshop held on July 11 and 12. New York: Holiday Valley; 2008. *Genet Med* 2009;11(6):450–454.
72. Schiffmann R. Therapeutic approaches for neuronopathic lysosomal storage disorders. *J Inherited Metab Dis* 2010;**33**:373–9.
73. Kohlschutter A. Lysosomal leukodystrophies: krabbe disease and metachromatic leukodystrophy. *Handb Clin Neurol* 2013;**113**:1611–8.
74. Krivit W. In: Zimran A, editor. *Glycolipid Storage Disorders*. Abingdon, UK: Adis Communications; 2004. p. 91–100.
75. Gieselmann V, Krageloh-Mann I. Metachromatic leukodystrophy–an update. *Neuropediatrics* 2010;**41**:1–6.
76. Miyake N, Miyake K, Karlsson S, Shimada T. Successful treatment of metachromatic leukodystrophy using bone marrow transplantation of HoxB4 overexpressing cells. *Mol Ther* 2010;**18**:1373–8.
77. Orchard PJ, Tolar J. Transplant outcomes in leukodystrophies. *Semin Hematol* 2010;**47**:70–8.
78. Martin HR, Poe MD, Provenzale JM, Kurtzberg J, Mendizabal A, Escolar ML. Neurodevelopmental outcomes of umbilical cord blood transplantation in metachromatic leukodystrophy. *Biol Blood Marrow Transpl* 2013;**19**:616–24.
79. Boelens JJ, Aldenhoven M, Purtill D, Ruggeri A, Defor T, Wynn R, et al. Outcomes of transplantation using various hematopoietic cell sources in children with Hurler syndrome after myeloablative conditioning. *Blood* 2013;**121**:3981–7.
80. Orchard P, Boelens JJ, Raymond G. Multi-institutional assessments of transplantation for metabolic disorders. *Biol Blood Marrow Transpl* 2013;**19**:S58–63.
81. Malone BN, Whitley CB, Duvall AJ, Belani K, Sibley RK, Ramsay NK, et al. Resolution of obstructive sleep apnea in Hurler syndrome after bone marrow transplantation. *Int J Pediatric Otorhinolaryngology* 1988;**15**:23–31.
82. Peters C, Shapiro EG, Krivit W. Hurler syndrome: past, present, and future. *J Pediatr* 1998;**133**:7–9.
83. Aldenhoven M, Boelens JJ, de Koning TJ. The clinical outcome of Hurler syndrome after stem cell transplantation. *Biol Blood Marrow Transpl* 2008;**14**:485–98.
84. Polgreen LE, Tolar J, Plog M, Himes JH, Orchard PJ, Whitley CB, et al. Growth and endocrine function in patients with Hurler syndrome after hematopoietic stem cell transplantation. *Bone Marrow Transpl* 2008;**41**:1005–11.
85. Khanna G, Van Heest AE, Agel J, Bjoraker K, Grewal S, Abel S, et al. Analysis of factors affecting development of carpal tunnel syndrome in patients with Hurler syndrome after hematopoietic cell transplantation. *Bone Marrow Transpl* 2007;**39**:331–4.
86. Odunusi E, Peters C, Krivit W, Ogilvie J. Genu valgum deformity in Hurler syndrome after hematopoietic stem cell transplantation: correction by surgical intervention. *J Pediatric Orthop* 1999;**19**:270–4.
87. Weisstein JS, Delgado E, Steinbach LS, Hart K, Packman S. Musculoskeletal manifestations of Hurler syndrome: long-term follow-up after bone marrow transplantation. *J Pediatric Orthop* 2004;**24**:97–101.
88. Thawrani DP, Walker K, Polgreen LE, Tolar J, Orchard PJ. Hip dysplasia in patients with hurler syndrome (Mucopolysaccharidosis type 1H). *J Pediatric Orthop* 2013;**33**(6):635–43.
89. Hite SH, Peters C, Krivit W. Correction of odontoid dysplasia following bone-marrow transplantation and engraftment (in Hurler syndrome MPS 1H). *Pediatr Radiol* 2000;**30**:464–70.
90. Malm G, Gustafsson B, Berglund G, Lindstrom M, Naess K, Borgstrom B, von Dobeln U, Ringden O. Outcome in six children with mucopolysaccharidosis type IH, Hurler syndrome, after haematopoietic stem cell transplantation (HSCT). *Acta Paediatr* 2008;**97**(8):1108–12.
91. Peters C, Balthazor M, Shapiro EG, King RJ, Kollman C, Hegland JD, et al. Outcome of unrelated donor bone marrow transplantation in 40 children with Hurler syndrome. *Blood* 1996;**87**:4894–902.
92. Staba SL, Escolar ML, Poe M, Kim Y, Martin PL, Szabolcs P, et al. Cord-blood transplants from unrelated donors in patients with Hurler's syndrome. *N Engl J Med* 2004;**350**:1960–9.
93. Krivit W, Sung JH, Shapiro EG, Lockman LA. Microglia: the effector cell for reconstitution of the central nervous system following bone marrow transplantation for lysosomal and peroxisomal storage diseases. *Cell Transplant* 1995;**4**(4):385–92.
94. Priller J. Robert Feulgen Prize Lecture. Grenzganger: adult bone marrow cells populate the brain. *Histochem Cell Biol* 2003;**120**(2):85–91.
95. Miebach E. Enzyme replacement therapy in mucopolysaccharidosis type I. *Acta Paediatr Suppl* 2005;**94**(447):58–60. Discussion 57.
96. Wraith EJ, Hopwood JJ, Fuller M, Meikle PJ, Brooks DA. Laronidase treatment of mucopolysaccharidosis I. *BioDrugs: Clinical Immunotherapeutics, Biopharmaceuticals and Gene Therapy* 2005;**19**(1):1–7.
97. de Ru MH, Boelens JJ, Das AM, Jones SA, van der Lee JH, Mahlaoui N, et al. Enzyme replacement therapy and/or hematopoietic stem cell transplantation at diagnosis in patients with mucopolysaccharidosis type I: results of a European consensus procedure. *Orphanet Journal of Rare Diseases* 2011;**6**:55.
98. Orchard PJ, Milla C, Braunlin E, Defor T, Bjoraker K, Blazar BR, et al. Pre-transplant risk factors affecting outcome in Hurler syndrome. *Bone Marrow Transplant* 2009;**45**(7):1239–46.
99. Tomatsu S, Montano AM, Oikawa H, Smith M, Barrera L, Chinen Y, Thacker MM, et al. Mucopolysaccharidosis type IVA (Morquio A disease): clinical review and current. *Biotechnology* 2011;**12**(6):931–45.
100. Tiede S, Storch S, Lubke T, Henrissat B, Bargal R, Raas-Rothschild A, et al. Mucolipidosis II is caused by mutations in GNPTA encoding the alpha/beta GlcNAc-1-phosphotransferase. *Nat Med* 2005;**11**(10):1109–12.
101. Borgwardt L, Dali CI, Fogh J, Mansson JE, Olsen KJ, et al. Enzyme replacement therapy for alpha-mannosidosis: 12 months follow-up of a single centre, randomised, multiple dose study. *J Inherit Metab Dis* 2013;**36**(6):1015–24.
102. Bley AE, Giannikopoulos OA, Hayden D, Kubilus K, Tifft CJ, Eichler FS. Natural history of infantile G(M2) gangliosidosis. *Pediatrics* 2011;**128**(5):e1233–41.
103. Neudorfer O, Kolodny EH. Late-onset Tay-Sachs disease. *The Israel Medical Association Journal* 2004;**6**(2):107–11.

104. Lake BD, Steward CG, Oakhill A, Wilson J, Perham TG. Bone marrow transplantation in late infantile Batten disease and juvenile Batten disease. *Neuropediatrics* 1997;**28**((1):80–1.
105. Maria BL, Deidrick KM, Moser H, Naidu S. Leukodystrophies: pathogenesis, diagnosis, strategies, therapies, and future research directions. *J Child Neurol* 2003;**18**(9):578–90.
106. Biffi A, Lucchini G, Rovelli A, Sessa M. Metachromatic leukodystrophy: an overview of current and prospective treatments. *Bone Marrow Transplant* 2008;**42**(Suppl 2):S2–6.
107. Yagasaki H, Kato M, Ishige M, Shichino H, Chin M, Mugishima H. Successful cord blood transplantation in a 42-day-old boy with infantile Krabbe disease. *Int J Hematol* 2011;**93**(4):566–8.
108. Sharp ME, Laule C, Nantel S, Madler B, Aul RB, Yip S, et al. Stem cell transplantation for adult-onset Krabbe disease: report of a case. *JIMD Reports* 2013;**10**:57–9.
109. Duffner PK, Caggana M, Orsini JJ, Wenger DA, Patterson MC, Crosley CJ, et al. Newborn screening for Krabbe disease: the New York State model. *Pediatr Neurol* 2009;**40**(4):245–52. Discussion 53–55.
110. Lantos JD. Dangerous and expensive screening and treatment for rare childhood diseases: the case of Krabbe disease. *Developmental Disabilities Research Reviews* 2011;**17**(1):15–8.
111. Aubourg P, Blanche S, Jambaque I, Rocchiccioli F, Kalifa G, Naud-Saudreau C, et al. Reversal of early neurologic and neuroradiologic manifestations of X-linked adrenoleukodystrophy by bone marrow transplantation. *N Engl J Med* 1990;**322**(26):1860–6.
112. Peters C, Charnas LR, Tan Y, Ziegler RS, Shapiro EG, DeFor T, et al. Cerebral X-linked adrenoleukodystrophy: the international hematopoietic cell transplantation experience from 1982 to 1999. *Blood* 2004;**104**(3):881–8.
113. Miller WP, Rothman SM, Nascene D, Kivisto T, DeFor TE, Ziegler RS, et al. Outcomes after allogeneic hematopoietic cell transplantation for childhood cerebral adrenoleukodystrophy: the largest single-institution cohort report. *Blood* 2011;**118**(7): 1971–8.
114. Tolar J, Teitelbaum SL, Orchard PJ. Osteopetrosis. *N Engl J Med* 2004;**351**(27):2839–49.
115. Eapen M, Davies SM, Ramsay NK, Orchard PJ. Hematopoietic stem cell transplantation for infantile osteopetrosis. *Bone Marrow Transplant* 1998;**22**(10):941–6.
116. Coccia PF, Krivit W, Cervenka J, Clawson C, Kersey JH, Kim TH, et al. Successful bone-marrow transplantation for infantile malignant osteopetrosis. *N Engl J Med* 1980;**302**(13):701–8.
117. Shapiro F. Osteopetrosis. Current clinical considerations. *Clin Orthop Relat Res* 1993:34–44.
118. Steward CG, Pellier I, Mahajan A, Ashworth MT, Stuart AG, Fasth A, et al. Severe pulmonary hypertension: a frequent complication of stem cell transplantation for malignant infantile osteopetrosis. *Br J Haematol* 2004;**124**:63–71.
119. Corbacioglu S. Honig M. Lahr G. Stohr S. Berry G. Friedrich W, et al. Stem cell transplantation in children with infantile osteopetrosis is associated with a high incidence of VOD, which could be prevented with defibrotide. *Bone Marrow Transpl* 2006;**38**:547–53.
120. Martinez C, Polgreen LE, DeFor TE, Kivisto T, Petryk A, Tolar J, et al. Characterization and management of hypercalcemia following transplantation for osteopetrosis. *Bone Marrow Transplant* 2010;**45**(5):939–44.
121. Tolar J, Bonfim C, Grewal S, Orchard P. Engraftment and survival following hematopoietic stem cell transplantation for osteopetrosis using a reduced intensity conditioning regimen. *Bone Marrow Transpl* 2006;**38**:783–7.
122. Wynn RF, Wraith JE, Mercer J, O'Meara A, Tylee K, Thornley M, et al. Improved metabolic correction in patients with lysosomal storage disease treated with hematopoietic stem cell transplant compared with enzyme replacement therapy. *J Pediatr* 2009;**154**:609–11.
123. Cox-Brinkman J, Boelens JJ, Wraith JE, O'Meara A, Veys P, Wijburg FA, et al. Haematopoietic cell transplantation (HCT) in combination with enzyme replacement therapy (ERT) in patients with Hurler syndrome. *Bone Marrow Transpl* 2006;**38**:17–21.
124. Tolar J, Orchard PJ. alpha-L-iduronidase therapy for mucopolysaccharidosis type I. Biologics 2008;2:743–751.

Chapter 9

# Cord Blood Cells and Autoimmune Diseases

**LingYun Sun[1], Audrey Cras[2,3], Dandan Wang[1] and Dominique Farge[4]**

*[1]Department of Immunology, The Affiliated Drum Tower Hospital of Nanjing University Medical School, Nanjing, China; [2]Assistance Publique-Hôpitaux de Paris, Saint-Louis Hospital, Cell Therapy Unit, Cord Blood Bank and CIC-BT501, Paris, France; [3]INSERM UMRS 1140, Paris Descartes, Faculté de Pharmacie, Paris, France; [4]Assistance Publique-Hôpitaux de Paris, Saint-Louis Hospital, Internal Medicine and Vascular Disease Unit, CIC-BT501, Paris, France*

## Chapter Outline



## 1. INTRODUCTION

Auto immune diseases (AD) are a group of heterogeneous conditions, affecting 5–8% of the population. They were traditionally classified as "organ specific AD," such as autoimmune thyroiditis and diabetes mellitus, where the consequences of organ failure can be improved by a replacement opotherapy or an organ transplant, and as "diffuse or systemic AD," including Systemic Lupus Erythematosus (LED), Vasculitis, Rheumatoid arthritis (RA), Juvenile Immune Arthritis (JIA), Scleroderma (SSc), Multiple Sclerosis (MS), and others (neurological AD, rheumatological AD, inflammatory bowel diseases, immune cytopenia) where treatment is more difficult. In most cases conventional immunosuppressive therapy allows control of the original AD (SLE, RA, Vasculitis), but definitive cure is rarely achieved and life-long immunosuppression is required. Chronic immunosuppression in severe inflammatory or in refractory AD patients is associated with a high treatment-related morbidity and significant disease and treatment-related mortality. In this context, new therapeutic approaches are warranted.

Over the past two decades, based on innovative animal and clinical data, more than 3000 patients worldwide have been treated by hematopoietic stem cell (HSC) transplantation for an autoimmune disease (AD) alone and reported in the respective European Bone Marrow Transplant Association (EBMT), Center for International Blood and Marrow Research (CIBMTR), and Asian registries, with impressive clinical results, never previously observed in severe AD with any other therapies. In parallel, three major innovations enriched our current therapeutic approach and opened new perspectives in the field of use of cord blood and in the pathogenesis of autoimmunity. First was the discovery of bone marrow (BM)

Cord Blood Stem Cells Medicine. http://dx.doi.org/10.1016/B978-0-12-407785-0.00009-8

stromal cells or mesenchymal stem cells (MSCs), which can also be obtained from many other human tissues including umbilical cord. The promising immunomodulation and immunosuppressive properties of MSCs, as well as their regenerative potential, have been extensively studied in animals, and clinical applications are emerging in humans.[1] Second and importantly, after the first successful umbilical cord blood (UCB) transplantation for Fanconi's anemia in 1989,[2] UCB has been increasingly used as a source of cells for HSCs to treat patients with nonhematopoietic diseases. Last but not least, progress in the identification of the genetic background of each type of AD,[3] and in the respective differences between autoimmunity and autoinflammation,[4] shed new light on the pathophysiology of autoimmunity. These successive innovations over the past 25 years constitute the backbone of current use of cord blood-derived cells for immunomodulation and treatment of AD.

## 2. NEW INSIGHTS IN THE PATHOGENESIS OF ADs

### 2.1 Autoimmunity and Loss of Tolerance

Although there is not a single model of AD, conceptually they all share some similarities.[5] Indeed, ADs are viewed as a polyclonal activation of the immune system with a defect of B- or T-lymphocyte selection and altered lymphocytic reactions to autoantigen components,[6] although it is rare to identify a single antigenic epitope. The native immune system and its tissue environment play an important role to determine if exposure to a given antigen will induce an immune response or tolerance or anergy. The role of the genes coding for the major histocompatibility system molecules, but also of many other genes, is important in the regulation of the immune response, but the genetic background of AD does not explain all of the observed phenomena during the loss of tolerance.[3,7] The exact nature of the T-cells necessary to cause an autoimmune reaction remains unknown. Self-reactive T-cells appear to escape thymic deletion and persist in the peripheral blood, where they can be activated and induce the autoimmune process. Consequently, activation of various subtypes of regulatory T-cells occurs, particularly the suppressive T CD4+CD25+ cells, whose role is important before, during, and after the onset of AD. Alteration of these regulatory mechanisms over time contributes to the clinical expression and development of AD. Presence of organ-specific antibodies may precede AD clinical expression and organ damage. Use of corticosteroids, immunosuppressive drugs (antimetabolites, calcineurin, mTOR inhibitors), antilymphocyte poly- or monoclonal antibodies, or biological drugs suppresses or modulates the activation of immune response in order to control the AD.

### 2.2 Autoimmunity or Autoinflammation

Over the years, identification of the key mutations associated with all types of immune system perturbations and better knowledge of the genetic background of both common and rare ADs have led to revision of the traditional concept of autoimmunity.[3,4] The elucidation of the mechanisms associated with self-directed tissue inflammation independently of T- or B-cell abnormalities has transformed our understanding of immunological diseases, and McGonagle has recently underlined the importance of the autoinflammatory component of each AD as such: "the boundaries for AD, are set by mutations associated with the monogenic autoimmune diseases (APS-1 with *AIRE*/AIRE, ALPS with *FAS*/FAS, IPEX with *FOXP3*/FOPXP3...) with an increased propensity towards an adaptative response and the presence of autoantibodies. On the other hand, the boundaries of auto-inflammation are defined by mutations in cells or molecules involved in innate immune responses at disease-prone sites (e.g., Familial Mediterranean Fever with *MEFV*/pyrin, TRAPS with *TNFRSFIA*/TNFR1...), where disease expression cannot be explained by autoimmune mechanisms."[4] Most of the classical ADs are polygenic diseases with a predominant autoimmune component (SLE, T1D, autoimmune thyroiditis) background, whereas other polygenic ADs have a predominant autoinflammatory (Crohn's disease for instance) component. Therefore, optimal treatment of AD should take in account this continuum between autoimmunity and autoinflammation, and innovative therapies should first consider if the AD is purely autoinflammatory or autoimmune, or in most cases, a combination of autoinflammatory and autoimmune mechanisms that variably interact in the AD phenotypic expression.

### 2.3 ADs: Stem Cell Disorders

All immune system cells derive from HSC. The direct relationship between AD and the hematopoietic system was evidenced by the pioneering work of Ikehara in 1985, who first demonstrated that ADs originate from defects in the HSCs.[8] Thereafter, extensive experimental data from genetically prone and immunized animal models of AD treated with allogeneic, syngeneic, autologous, and pseudoautologous bone marrow transplantation (BMT) showed that allogeneic BMT (but not syngeneic or autologous) could be used to treat AD-prone mice.[9] Conversely, the AD transfer was possible in normal mice after allograft from a mouse with lupus nephritis, showing that AD was in fact a stem cell disorder. Animal models of AD can be divided into two categories. The hereditary and spontaneous AD models, such as murine BSB lupus and nonobese diabetic (NOD) mice, where autoimmune manifestations affect the majority of the animals of a susceptible line, with a strong genetic predisposition transported by the HSC and manifested by anomalies of thymic development and/or function of lymphocytes B-, T- or antigen presenting cells (APCs), macrophages. Other experimental models, such as arthritis adjuvant (AA) and experimental acute encephalomyelitis (EAE),[10] use active immunization by exposure to a foreign antigen to induce the AD. In AA, mediated by cytotoxic T-cells reacting against a normal component of the rat articular cartilage, if the allogeneic transplant, but also autologous or syngeneic graft, is performed before immunization, the AA onset and its evolution are unchanged. Conversely, if the transplants are performed after immunization, it accelerates healing and prevents relapse of AA.[11] In these models, intense conditioning before BMT allows deep immunoablation, which removes pathogenic autoreactive T-lymphocytes. Thereafter, during immune reconstitution, everything happens as if a tolerance of antigen-induced disease appeared with the reappearance of new T-lymphocytes and de novo repertoire. In agreement with this theory, it is important that after transplantation, the antigen is accessible to APC and presented to T-cells under development to induce tolerance. The frequency of relapses after BMT appears to decrease if the BMT is made earlier during the AD evolution or after high-intensity conditioning. The effectiveness of the allogeneic BMT can be explained by a reduction of self-reactive lymphocytes during conditioning and eradication of the residual immune cells by a mechanism of graft-versus-autoimmunity (GvA), secondary to the action of the immune system cells derived from normal donor like the graft-versus-leukemia (GvL) effect.[12,13] The effectiveness of the autologous BMT could result from a similar removal of the T-autoreactive cells during conditioning, followed by induction of a mechanism of self-tolerance by a phenomenon of rehabilitation of lymphocytes derived from HSC.[14] Analysis of immune reconstitution after autologous transplantation has confirmed this hypothesis in mouse models of EAE and murine lupus.[15] Extrapolation of these models would suggest that only allogeneic transplants can be followed by success in the treatment of the human AD with a strong genetic component. These experimental data were confirmed by several clinical reports in humans with complete clinical response of ADs to BMT performed for other conventional indications. The reciprocal, passive transfer of an AD to the receiver from an allogeneic donor was also observed.[16]

## 3. HEMATOPOIETIC STEM CELL TRANSPLANTATION FOR THE TREATMENT OF AD

### 3.1 Autologous Hematopoietic Stem Cell Transplantation and Reset of Tolerance

Based on experimental and clinical data, consensus indications for the use of BM-derived or peripheral hematopoietic stem cell transplantation (HSCT) to treat severe ADs were first published in 1997[17] and recently updated in 2012.[18] Indeed, over the past 15 years, use of HSCT has moved from the lab to the clinics, and AD patients can be considered for treatment by HSCT when matching the following criteria: (1) diagnosed with an AD severe enough to have an increased risk of mortality or advanced and irreversible disability; (2) the ADs must be unresponsive to conventional treatments; and (3) HSCT should be undertaken before irreversible organ damage so that significant clinical benefit can be achieved.

Most of the procedures were performed using autologous HSCT, and in the EBMT database with more than 1500 patients registered today (source EBMT registry), the most commonly transplanted diseases are multiple sclerosis (MS), scleroderma (SSc), Crohn's disease, and systemic lupus erythematosus (SLE), coming from over 215 transplant centers in 30 countries. Prevalence of female sex and young age reflects the natural distribution of the diseases. Long-lasting responses were obtained in all disease categories with an overall adjusted transplant-related mortality (TRM) being $7 \pm 3\%$ at 3 years, directly related to the type of AD disease (SSc and SLE have a higher risk), the year of transplant with a learning curve, and the intensity of conditioning (higher-intensity conditioning showed a higher risk of TRM but lower probability of disease progression). The promising results from phase I–II studies led to three major randomized phase III studies, whose results were presented as abstracts and are about to be published for systemic sclerosis (ASTIS), MS (ASTIMS), and Crohn's disease (ASTIC). Significant benefit was obtained in patients undergoing autologous BMT as compared to their respective controls, at the price of early toxicity due to the poor condition of patients at the time of transplant and also to the risk of HSCT, which remains an aggressive procedure, underlying the need for careful patient selection. The introduction of new biotherapies for inflammatory arthritis since 2002 resulted in a drop in activity in HSCT for rheumatoid arthritis (RA), whereas sustained prolonged remission was obtained for patients with severe or rapidly progressive SSc, SLE, MS, inflammatory bowel disease, and others. In this setting, analysis of the regenerating adaptive immune system showed normalization of the restricted T-cell repertoire, with sustained shifts in T- and B-cell subpopulations from memory to naïve cell dominance, supportive of thymic reprocessing and re-education of the reconstituting immune system.[19,20] In addition, restoration of normal or raised levels of CD4+ regulatory T-cells with disappearance of circulating plasmablasts was reported in juvenile immune arthritis,[21] and unusual CD8+ FoxP3+ regulatory T-cell subsets, capable of inhibiting the pathogenic T-cell response to autoepitopes in nucleosomes, were shown in SLE[22,23] following autologous HSCT. This has never been shown previously after the use of conventional immunosuppressive therapies. Such clinical and immunological results separated the nonspecific immunosuppressive changes, observed both in blood and in tissue after cytotoxic therapy[19,24] from immune re-educative changes supporting immune tolerance.[25] Therefore, for the first time in treating AD, the interruption of the vicious circle of autoimmunity allowed the emergence of normal regulatory mechanisms and eradication of the last autoreactive T-cell, which is one of the proposed mechanisms for using HSC in the treatment of AD.

## 3.2 Allogeneic HSCT

Allogeneic HSCT has been used infrequently in patients for the treatment of AD, despite the theoretical attraction of immune replacement and the graft-versus-autoimmune effect. The TRM risks far outweigh the risks of severe ADs, and therefore, allogeneic HSCT for AD are rarely justified, unless an underlying hematological malignancy is coexisting with the AD as recommended originally in 1997.[17] Today, allogeneic HSCT has mostly been used in the context of immune cytopenia, predominantly in the pediatric setting, with encouraging results in terms of relapse-free outcome. Patients who received an allogeneic HSCT showed a sustained response in 33% of the cases reported to the EBMT database.[26]

In exceptional circumstances, allogeneic HSCT may be considered for patients with refractory ITP, AIHA, and Evans syndrome. Well-matched unrelated allogeneic HSCT should be limited in the pediatric setting, but matched sibling allogeneic HSCT may be considered in patients under 50 years of age with refractory cytopenias. BM or UCB is recommended as a graft source for allogeneic HSCT in autoimmune cytopenia. If no human leukocyte antigen (HLA)-matched sibling donor is available, or if an adult is over 50 years of age, autologous HSCT is recommended. In other ADs, experience is limited to individual case reports in SSc, SLE, vasculitis, and RA, with no recommendations other than enrollment in a clinical trial. As the risks of syngeneic HSCT approximate to those of autologous HSCT, it may be considered as an alternative in those rare patients who have an identical twin, providing donor welfare is given high priority.

# 4. USE OF UCB-DERIVED CELLS AND CORD BLOOD MSCs FOR TREATING AD

The use of HSCs to induce tolerance by replacing (allogeneic) or resetting (autologous) immune responses in patients with ADs and the preclinical evidence for a GvA effect in replacement of a dysfunctional immune system by allogeneic HSCT are now supported by clinical evidence obtained in the past 20 years. Today, more than 3000 patients worldwide (EBMT, CIBMTR, Asian registry) have received a BM for an AD alone. In the meantime, discovery and identification of BM stromal cells or MSCs within the BM content and their therapeutic properties has led us and others to use MSCs derived from various tissues to treat AD patients. In the meantime, use of UCB transplantation increased for patients with nonmalignant hematopoietic diseases (Fanconi anemia, hereditary hemoglobinopathies), and several groups showed the possibility of deriving MSCs not only from cord blood but also from the umbilical cord matrix—Wharton's jelly (UCB-MSCs).[27]

## 4.1 MSCs Identification and Properties

MSCs can be derived from various tissues such as bone, cartilage, muscle, ligament, tendon, periodontal ligament, dental pulp, adipose tissue and cord blood, fetal liver, skin, gut, brain, or kidney.[28] MSCs were originally identified by Friedenstein in 1976 as a fibroblast-like cellular population capable of generating osteogenic precursors.[29] Caplan introduced the name of MSCs,[30] which was then changed after consensus into multipotent mesenchymal stromal cells due to their capacity to differentiate toward mesodermal, as well as endodermal and ectodermal, lineages.[31] Compared to other stem cell sources, such as HSCs, MSCs appear as promising source for overcoming autoimmunity due to their ability to differentiate toward various lineages in addition to their immunosuppressive capacities.[32] Although no specific membrane marker has yet been definitely identified on MSCs, several phenotypical characteristics have allowed their identification. MSCs express several cell surface antigens, such as CD73, CD90, CD105, CD146, and CD200,[33] as well as various integrins and adhesion molecules. As the MSCs are a nonhematopoietic cell line, they do not have hematopoietic markers such as CD34, CD14, and CD45.[34] Adult human MSCs show intermediate levels of MHC class I molecules on their cell surface and have no detectable levels of MHC class II. In addition, MSCs differentiated into adipose, bone, and cartilage cells express HLA class I but no HLA class II, which allows their transplantation across major histocompatibility complex barriers.[35] Due to their low immunogenicity, MSCs represent an appropriate stem cell source for allogeneic transplantation, irrespective of HLA compatibility. They can also synthesize trophic mediators such as growth factors and cytokines (M-CSF, IL-6, IL-11, IL-15, SCF, VEGF) involved in hematopoiesis regulation, cell signaling, and modulation of the immune response.[36] As shown in vitro and in vivo, MSCs modulate the immunologic activity of different cellular populations, the most important being their inhibitory effect on T-cell proliferation and dendritic cell differentiation, which are key factors for activating autoimmune disorders. MSCs are effective in inhibiting proliferation of CD4 and CD8 T-cells, as well as memory and naïve T-cells.[37] This mechanism may necessitate an initial cell contact phase, as well as several specific mediators, produced by MSCs such as TGF-β,

PGE2, and indoleamine-2,3-dioxygenase (IDO) (IDO is induced by IFN-γ, catalyzes the conversion of tryptophan to kynurenine, and inhibits T-cell responses by tryptophan depletion).[38] The ability reported in humans, rodents, and primates to suppress T-cell responses to mitogenic and antigenic triggering is explained by a complex mechanism of induction of "division arrest anergy," responsible for maintaining T-lymphocytes in a quiescent state. Thus, the MSCs determine the inhibition of cyclin D2 expression arresting cells in the G0/G1 phase of the cell cycle. MSCs also stimulate the production of regulatory T-cells, which inhibit lymphocyte proliferation in allogeneic transplantation.[39] BM-MSCs obtained from healthy human donors can indirectly reduce T-cell activation by inhibiting dendritic cell differentiation (mainly DC type I) from monocytes.[40] In addition, MSCs inhibit B-cell proliferation and activation in a dose-dependent manner and modulate their differentiation, antibody production, and chemotactic abilities.[41] Due to their supportive function for HSC in the BM niche, their selective activity on the cell cycle and their immunomodulatory effects, BM-derived MSCs have already been used in several phase I–II, and very few phase III clinical trials for the treatment of acute graft-versus-host disease following allogeneic transplantation for leukemia or hematological malignancies,[42] and a few cases report for AD such as systemic sclerosis,[43] MS,[44] and Crohn's disease.[45] The important differentiation capacity of MSCs has made them a useful therapeutic method in orthopedics by increasing new dense bone formation and total bone mineral content in osteogenesis imperfecta,[46] providing early bone regeneration in osteonecrosis of the femoral head and repair of large bone defects.[47] The MSCs capacity to secrete growth factors and enzymes, such as arylsulfatase A and α-L-iduronidase, which are deficient in metachromatic leukodystrophy and Hurler's disease, allows them, after in vitro expansion and intravenous administration, to enhance enzyme production and improve symptomatology in these inborn metabolism errors.[48] MSCs have also been administered as autologous transcoronary transplants in human infarcted myocardium, alone or in association with endothelial progenitors, with satisfactory results regarding myocardial contractility improvement.[49]

## 4.2 Isolation and Characterization of MSCs from UCB

UCB is used clinically as a cell source for HSCT for both malignant and nonmalignant diseases. Beyond this, umbilical cord and cord blood contained different stem or progenitor cells such as unrestricted somatic stem cells, very small embryonic-like stem cells, MSCs, and epithelial and endothelial progenitor cells. Because of the rarity of MSCs in the BM, where they represent 1 in 10,000 nucleated cells, UCB is a practical and promising source of MSC, which has been extensively explored, notably by Sun and coworkers in China. Isolation and characterization of MSCs from UCB still need to be evaluated and are controversial.[50] Indeed, depending on the variability between each donor as well as the methods of isolation and the contamination by a large number of cells such as fibroblastic cells, dendritic cells, adherent monocytes, macrophages, and osteoclastic cells arising within the cultures, neither study provided sufficient evidence to fulfill the qualifying criteria for MSCs. MSCs designation requires the following minimal criteria: (1) plastic-adherent in vitro expansion; (2) absence of hematopoietic surface markers CD14, CD11b, CD19, CD34, CD45, and HLA-DR, and presence of surface markers CD73, CD90, and CD105; and (3) in vitro differentiation into adipocytes, chondroblasts, and osteoblasts. MSCs can be isolated from the Ficoll layer of UCB and form an adherent population that allows their separation from hematopoietic progenitor cells by differential adherence selection. MSCs were then amplified in FGF-supplemented cell culture medium.[51–54] McElreavey and colleagues first reported the possibility of isolating MSCs from Wharton's jelly.[55] Several techniques have been reported to dissect Wharton's jelly mechanically and/or digest it enzymatically with collagenase and trypsin to culture homogeneous MSC populations.[56–58] A single piece of 5–10 $mm^3$ Wharton's jelly has the potential to yield as many as one billion MSCs in 30 days, thus showing that umbilical cords are a reliable and easily accessible source of MSCs.[59] Therefore, umbilical cord-derived MSCs represent a novel source of MSCs with higher accessibility and fewer ethical constraints than BM-derived MSCs.[50] UCB can be cryopreserved and stored for years without significant loss of viability, making them very attractive tools with widespread public and private biobanking worldwide.

The immunophenotypic as well as the multilineage differentiation properties of UCB-MSCs are similar to those reported for BM-MSCs. Importantly, the expression profiles of proteins and the cytokines in BM- and UCB-MSCs are very similar.[60,61] UCB-MSCs display a variety of in vitro immunomodulatory and anti-inflammatory capabilities.[62,63] UCB-MSCs have immunosuppressive action on lymphocyte proliferation in MLR by alloantigen and mitogens such as phytohemagglutinin and reduce the level of proinflammatory cytokines such as interferon-γ and tumor necrosis factor-α. Evidence has demonstrated that UCB-MSCs can not only suppress the function of mature dendritic cells, but also increase the portion of T-cells related to immune regulation. This regulation of immune response by MSCs is mediated by soluble factors and cell-to-cell contact mechanisms. Furthermore, like BM-MSCs, UCB-MSCs did not express major histocompatibility complex class II molecules and co-stimulatory molecules such as CD40, CD40 ligand, CD80, and CD86, which are involved in T-cell activation response for transplant rejection. Therefore, immunogenic phenotypes of UCB-MSCs can retain low immunogenicity under certain biological conditions that provide advantages for their use in allogeneic settings.

## 4.3 Use of UCB-derived MSC for AD in Preclinical Model Studies

### 4.3.1 Cord Blood and EAE

EAE animal models were generally and widely induced by recombinant myelin oligodendrocyte glycoprotein, with clinical manifestations and immune disorders similar to human MS. Given that autologous BM-MSCs from MS patients displayed a similar characteristic compared to that from healthy controls,[64] now most investigators use patients BM-derived MSCs for preclinical and clinical studies.[65,66] Recently Liu et al. showed that human umbilical cord-derived MSCs (UC-MSCs) injected intravenously could restore behavioral functions and attenuate the histopathological deficits of EAE mice, which were mediated by suppression of perivascular immune cell infiltrations and reduction in both demyelination and axonal injury in the spinal cord.[67] These data suggested that human umbilical cord-derived stem cells may be used as an alternative option in treating human MS.

### 4.3.2 Cord Blood and Experimental RA

RA is a chronic and systemic disease that primarily attacks synovial joints, leading to articular destruction and functional disability. Since both autologous BM and synovial-derived MSCs from RA patients demonstrated many functional defects, including impaired clonogenic and proliferative potential, differential expression of genes related to cell adhesion and cycle progression.[68] Therefore, allogeneic MSCs were widely used to treat RA. Liu et al. showed that human UC-MSCs intraperitoneal infusion attenuated the development of collagen-induced arthritis (CIA) in vivo, which may result from the modulation of regulatory T-cells and inflammatory cytokines like TNF-$\alpha$, monocyte chemotactic protein 1 (MCP-1), and IL-6.[70] Another study, however, showed that intraarticular injection of human UC-MSCs had no benefit in CIA mice, but accelerated the progression of arthritis. On the other hand, the combination of UC-MSCs and TNF inhibitor showed reduced disease activity in CIA mice. This was explained as the inhibition of MSCs activity by elevated TNF-$\alpha$ in vivo, because on exposure of TNF-$\alpha$, stem cells significantly decreased the expression of CD90, HLA-G, and the levels of IL-10 in vitro and in vivo.[71]

For complete Freund's adjuvant-induced RA in rat models, Greish et al. compared the treatment effects among human cord blood derived-MSCs, hematopoietic stem cells (CD34+), and methotrexate (MTX) groups. The results showed that both MSCs and HSCs treatment groups improved the signs of overall arthritis compared to the MTX group, with the MSCs treatment group having an enhanced efficacy. The effect was most likely through the modulation of cytokine expression.[72]

### 4.3.3 Cord Blood and Genetically Prone Lupus

SLE is an AD characterized by multiorgan involvements. Both Fas mutated MRL/lpr mice and NZB/W F1 mice are widely used as genetically prone lupus models, which demonstrated progressive nephritis, elevated serum autoimmune antibodies, and immune abnormalities. Dr Sun's team had firstly showed that human UC-MSCs alleviated lupus nephritis in MRL/lpr mice in a dose-dependent manner. Both single and multiple treatments with UC-MSCs were able to decrease the levels of 24-h proteinuria, serum creatinine, anti-double stranded DNA (dsDNA) antibody, and the extent of renal injury such as crescent formation. The further mechanism studies showed that UC-MSCs treatment inhibited renal expression of MCP-1 and high-mobility group box 1 expression, while it upregulated Foxp3+regulatory T-cells. Moreover, carboxyfluorescein diacetate succinimidyl ester-labeled UC-MSCs could be found in the lungs and kidneys postinfusion.[73] To NZB/W F1 mice, Chang et al. showed that human umbilical cord blood-derived MSCs (HUCB-MSCs) transplantation significantly delayed the development of proteinuria, decreased anti-dsDNA, alleviated renal injury, and prolonged the life span. Further mechanism studies showed that the treatment effect was mediated by inhibiting lymphocytes, inducing polarization of Th2 cytokines and inhibition of proinflammatory cytokine production rather than direct engraftment and differentiation into renal tissue.[74]

### 4.3.4 Cord Blood and Genetically Prone Diabetes

Type 1 diabetes (T1D) is characterized by T-cell-mediated autoimmune destruction of pancreatic $\beta$-cells. The pathophysiology of T1D in both humans and genetically prone NOD mice appears to be largely related to an innate defect in the immune system, culminating in a loss of self-tolerance and destruction of insulin-producing $\beta$-cells. While insulin replacement represents the current therapy, its metabolic control remains difficult. Pancreatic or islet transplantation can provide exogenous insulin independence but is limited by its intrinsic complications and the scarcity of organ donors. Stem cell therapy, based on the generation of insulin-producing cells (IPCs), represents an attractive possibility.[75] HUCB stem cell is a unique type of stem cell in cord blood.[76] Islet-like clusters can be obtained from HUCB cells at the end of a four-stage differentiation protocol. The cell clusters showed insulin and other pancreatic $\beta$-cell-related genes (PDX1, Hlxb9, Nkx2.2, Nkx6.1, and GLUT2), and released insulin and C-peptide in response to physiological glucose concentrations in vitro.[69] In addition to differentiation, HUCB stem cells also functioned as an immune modulator to control the immune responses.[68] In vivo experiments showed that human cord blood-derived T-cell-depleted mononuclear cell transplantation can generate IPCs in newborn NOD/*scid*/b2m$^{null}$ mice, as evidenced by the presence of human insulin at the RNA level and human chromosome-containing insulin-positive cells in situ.[78] NOD mice treated with HUCB cells significantly lowered their blood glucose levels and increased their life span, as well as reducing insulitis, in a dose-dependent manner.[79] Thus, administration of HUCB cells protects the islets from insulitis in NOD mice.

In addition to the effect on insulitis, HUCB-derived MSCs (HUCB-MSCs) effectively prevented diabetic renal injury NOD mice models. CM-DiI-labeled HUCB-MSCs confirmed a few engraftments of infused HUCB-MSCs in diabetic kidneys. The further mechanism studies showed that HUCB-MSCs-conditioned media inhibited TGF-β1-induced extracellular matrix upregulation and epithelial-to-mesenchymal transition in NRK-52E cells in a concentration-dependent manner,[80] which may partially explain the protective role on renal injury.

## 4.4 Clinical Applications of UCB-derived MSC and Ongoing Developments

### 4.4.1 Multiple Sclerosis

MS is the most frequent chronic inflammatory demyelinating disease, with a prevalence of 1 in 700 adults. MS may be categorized into relapsing-remitting (RR-MS), secondary progressive (SP-MS), primary progressive (PP-MS), and a rapidly evolving malignant (or Marburg) form. There is no curable treatment for this potentially disabling disease at the moment. Various immunomodulators, such as glatiramer acetate and beta interferon, and more recently, the oral sphingosine-1-phosphate receptor agonist fingolimod, are used as first-line treatments and were shown to be considerably more effective than beta interferon in delaying the progression of disability. Second-line treatments are mitoxantrone and the monoclonal antibody natalizumab. Nonetheless, subsets of nonresponders are recognized, necessitating maintenance of or frequent long-term immunosuppression. In this context, MS is the most frequent diagnosis for which HSCT has been used, and the majority of patients have SP-MS. The ADWP-updated guidelines recommend autologous HSCT for MS patients with RR-MS, with high inflammatory activity rapidly deteriorating despite the use of at least one or more lines of treatment. Patients with rapidly deteriorating "malignant" MS are also suitable candidates. Although the majority of MS patients reported in the literature have SP-MS, autologous HSCT is appropriate in this phase only when some inflammatory activity (clinically or at MRI) is still present. Except for "malignant" forms, patients who have lost the ability to walk should not be treated with HSCT. BEAM plus anti-T-cell serotherapy is recommended as conditioning.

Current therapeutic interventions for MS essentially modulate the immune system and reduce the inflammation by general and nonspecific mechanisms, but have little effect on the neurodegenerative component of the disease. Treatment strategies to prevent tissue damage or augment repair are needed. MSCs therapies probably exert their neuroprotective effects by secreting many soluble factors with immunomodulatory and trophic properties that might influence central nervous system inflammation and/or endogenous remyelination.[65,70] Most clinical studies have used autologous BM-MSCs.[64,81–83] A preliminary study on the application of UCB-MSCs on an MS patient demonstrated that this treatment is feasible and effective.[44] Recently, a case report[84] presented treatment of aggressive MS by multiple allogenic human UCB-MSCs and autologous BM-MSCs over a 4-year period with no significant adverse events.[84]

### 4.4.2 Systemic Lupus Erythematosus

SLE is a heterogeneous chronic AD with a prevalence of 40–50 per 100,000, predominantly affecting females (>85%). SLE is an inflammatory disease with protean manifestations ranging from relatively minor skin and joint symptoms to severe life-threatening major organ involvement such as nephritis and neuropsychiatric complications.[75] Conventional immunosuppressive or immunomodulatory therapy, such as glucocorticoids, cyclophosphamide (CYC), and mycophenolate mofetil (MMF), can control disease in most, but not all, lupus patients. There is a subset of patients whose disease does not respond or relapses despite continuing chemotherapy, and their prognosis remains poor. In addition, progressive immunosuppressive therapy may lead to the development of serious infection, cumulative drug toxicity, and an increased risk of cardiovascular disease and malignancy.[87] The outcome of active severe SLE due to kidney, lung, heart, or brain involvement has improved in adults and children with early diagnosis and new treatment with immunosuppressive agents combined with overall tighter control of blood pressure and infections. Response rates to standard first therapy with the classical NIH regimen[88,89] or the Euro-Lupus regimen[90] vary according to extent of visceral involvement, ethnic origin, and socioeconomic profile, but even with modern treatments, around 5–15% of patients with SLE evolve toward end-stage disease, and 10–15% die within 10 years. In severe SLE patients, refractory to conventional immunosuppressive therapies, autologous HSCT has been shown to achieve sustained clinical remissions in around half of patients, with qualitative immunological changes not seen with other forms of therapy.[91,92] In this high-risk population, TRM has been significant in multicenter (as opposed to single-center) settings and highlights the need for careful patient selection and recognition of the intrinsic immune suppression and other risks associated with advanced SLE, as well as the need for further clinical studies. In the revised guidelines, autologous HSCT is recommended in SLE ideally in the context of a multicenter clinical trial, but may be considered as treatment for carefully selected subpopulations of SLE patients early in their disease course. Such patients should have reliably predicting poor-prognostic factors.

UC-MSCs have been transplanted for severe lupus patients who were not responsive to conventional therapies. The 4-year follow-up demonstrated that about 50% patients acquired clinical remission after transplantation, although in 23% relapses occurred.[93] MSCs infusion induced disease remission for lupus nephritis,[94] diffuse alveolar hemorrhage,[95] and refractory cytopenia.[96] The multicenter clinical study showed that 32.5% patients achieved major clinical response (MCR, 13/40) and 27.5% patients achieved partial clinical response (PCR, 11/40) during 12-month follow-up. However, 7 out of 40 (17.5%) patients experienced disease relapse after 6-month follow-up after a prior clinical response, which indicated a repeated MSCs treatment after 6 months.[97] Now it is important to design a controlled study to further see the clinical efficacy between allogeneic MSCT and conventional immunosuppressive therapies, like CYC and MMF, or whether MSCT combined with immunosuppressive drug treatment is more effective than drugs alone.

### *4.4.3 Primary Sjogren's Syndrome*

Primary Sjogren's syndrome (pSS) is a chronic, systemic autoimmune disorder characterized by inflammation of exocrine glands and functional impairment of the salivary and lacrimal glands. It has extraglandular organ involvement, including lung (interstitial pneumonitis), renal (interstitial nephritis), peripheral and central nervous system manifestations, vasculitis of skin and other organs, and increased frequency of lymphoma. For visceral involvement, corticosteroids are used widely. Cytotoxic drugs such as azathioprine and MTX are also used. Other drugs such as leflunomide, ciclosporin A, CYC, and even biological agents are used for organ injury for pSS patients. However, there were also some patients who were not responsive to conventional therapies. Dr Sun's group has used UC-MSCs infusion for drug-resistant pSS patients and followed up for 12 months[98]. All 11 patients showed improvements in symptoms of xerostomia and/or xerophthalmia after MSCT. For the other 13 patients who showed severe systemic comorbidities, platelet counts, refractory hemolytic anemia, and autoimmune hepatitis improved after treatment. This clinical study demonstrated that umbilical cord-derived MSCT could be used as an alternative therapeutic option for refractory pSS patients.

## 5. CONCLUSION

A recent survey from all published literature on clinical use of UCB for nonhematological indications yielded a total of 691 publications with a total of 20 published articles describing the treatment of 317 patients and 12 using immunomodulatory therapy for AD conditions.[99] Most studies were performed in China, and used unprocessed bulk cells from UCB that were cryopreserved in accordance with standard UCB banking. Interestingly, MSCs were expanded successfully from 30% to 60% of UCB units, although newer approaches of collecting cells from Wharton's jelly or from placenta itself enhance the yield of MSC expansion from UCB. The majority of patients received HLA-compatible cord blood cells or third-party MSCs expanded in vitro. Past and future developments in the field of autoimmunity and cord blood have shown and will confirm that according to Ikehara's first hypothesis AD is a hematopoietic stem cell disorder.

## REFERENCES


1. Wang S, Qu X, Zhao RC. Clinical applications of mesenchymal stem cells. *J Hematol Oncol* 2012;**5**:19
2. Gluckman E, Broxmeyer HA, Auerbach AD, Friedman HS, Douglas GW, Devergie A, et al. Hematopoietic reconstitution in a patient with Fanconi's anemia by means of umbilical-cord blood from an HLA-identical sibling. *N Engl J Med* 1989;**321**(17):1174–8.
3. Rioux JD, Abbas AK. Paths to understanding the genetic basis of autoimmune disease. *Nature* 2005;**435**(7042):584–9.
4. McGonagle D, McDermott MF. A proposed classification of the immunological diseases. *PLoS Med* 2006;**3**(8):e297.
5. Davidson A, Diamond B. Autoimmune diseases. *N Engl J Med* 2001; **345**(5):340–50.
6. Burnet F. *The clonal selection theory of acquired immunity*. UK: Cambridge University Press; 1959.
7. Matzinger P. Tolerance, danger, and the extended family. *Annu Rev Immunol* 1994;**12**:991–1045.
8. Ikehara S, Good RA, Nakamura T, Sekita K, Inoue S, Oo MM, et al. Rationale for bone marrow transplantation in the treatment of autoimmune diseases. *Proc Natl Acad Sci USA* 1985;**82**(8):2483–7.
9. van Bekkum DW. Experimental basis of hematopoietic stem cell transplantation for treatment of autoimmune diseases. *J Leukoc Biol* 2002;**72**(4):609–20.
10. Karussis DM, Slavin S, Lehmann D, Mizrachi-Koll R, Abramsky O, Ben-Nun A. Prevention of experimental autoimmune encephalomyelitis and induction of tolerance with acute immunosuppression followed by syngeneic bone marrow transplantation. *J Immunol* 1992;**148**(6):1693–8.
11. van Bekkum DW, Bohre EP, Houben PF, Knaan-Shanzer S. Regression of adjuvant-induced arthritis in rats following bone marrow transplantation. *Proc Natl Acad Sci USA* 1989;**86**(24):10090–4.
12. Hinterberger W, Hinterberger-Fischer M, Marmont A. Clinically demonstrable anti-autoimmunity mediated by allogeneic immune cells favorably affects outcome after stem cell transplantation in human autoimmune diseases. *Bone Marrow Transpl* 2002;**30**(11):753–9.

13. Marmont AM, Gualandi F, Van Lint MT, Bacigalupo A. Refractory Evans' syndrome treated with allogeneic SCT followed by DLI. Demonstration of a graft-versus-autoimmunity effect. *Bone Marrow Transpl* 2003;**31**(5):399–402.
14. Ikehara S. Treatment of autoimmune diseases by hematopoietic stem cell transplantation. *Exp Hematol* 2001;**29**(6):661–9.
15. Abrahamsson S, Muraro PA. Immune re-education following autologous hematopoietic stem cell transplantation. *Autoimmunity* 2008;**41**(8):577–84.
16. Hough RE, Snowden JA, Wulffraat NM. Haemopoietic stem cell transplantation in autoimmune diseases: a European perspective. *Br J Haematol* 2005;**128**(4):432–59.
17. Tyndall A, Gratwohl A. Blood and marrow stem cell transplants in auto-immune disease: a consensus report written on behalf of the European League against Rheumatism (EULAR) and the European Group for Blood and Marrow Transplantation (EBMT). *Bone Marrow Transpl* 1997;**19**(7):643–5.
18. Snowden JA, Saccardi R, Allez M, Ardizzone S, Arnold R, Cervera R, et al. Haematopoietic SCT in severe autoimmune diseases: updated guidelines of the European Group for Blood and Marrow Transplantation. *Bone Marrow Transpl* 2012;47(6):770–90.
19. Farge D, Henegar C, Carmagnat M, Daneshpouy M, Marjanovic Z, Rabian C, et al. Analysis of immune reconstitution after autologous bone marrow transplantation in systemic sclerosis. *Arthritis Rheum* 2005;**52**(5):1555–63.
20. Muraro PA, Douek DC, Packer A, Chung K, Guenaga FJ, Cassiani-Ingoni R, et al. Thymic output generates a new and diverse TCR repertoire after autologous stem cell transplantation in multiple sclerosis patients. *J Exp Med* 2005;**201**(5):805–16.
21. de Kleer I, Vastert B, Klein M, Teklenburg G, Arkesteijn G, Yung GP, et al. Autologous stem cell transplantation for autoimmunity induces immunologic self-tolerance by reprogramming autoreactive T cells and restoring the CD4+CD25+ immune regulatory network. *Blood* 2006;**107**(4):1696–702.
22. Alexander T, Thiel A, Rosen O, Massenkeil G, Sattler A, Kohler S, et al. Depletion of autoreactive immunologic memory followed by autologous hematopoietic stem cell transplantation in patients with refractory SLE induces long-term remission through de novo generation of a juvenile and tolerant immune system. *Blood* 2009;**113**(1):214–23.
23. Zhang L, Bertucci AM, Ramsey-Goldman R, Burt RK, Datta SK. Regulatory T cell (Treg) subsets return in patients with refractory lupus following stem cell transplantation, and TGF-beta-producing CD8+ Treg cells are associated with immunological remission of lupus. *J Immunol* 2009;**183**(10):6346–58.
24. Bingham S, Veale D, Fearon U, Isaacs JD, Morgan G, Emery P, et al. High-dose cyclophosphamide with stem cell rescue for severe rheumatoid arthritis: short-term efficacy correlates with reduction of macroscopic and histologic synovitis. *Arthritis Rheum* 2002;**46**(3):837–9..
25. Muraro PA, Douek DC. Renewing the T cell repertoire to arrest autoimmune aggression. *Trends Immunol* 2006;**27**(2):61–7.
26. Rabusin M, Snowden JA, Veys P, Quartier P, Dalle JH, Dhooge C, et al. Long-term outcomes of hematopoietic stem cell transplantation for severe treatment-resistant autoimmune cytopenia in children. *Biol Blood Marrow Transpl* 2013;**19**(4):666–9
27. El Omar R, Beroud J, Stoltz JF, Menu P, Velot E, Decot V. Umbilical cord mesenchymal stem cells: the new Gold standard for mesenchymal stem cell-based therapies? *Tissue Eng Part B Rev* 2014;**20**(5):523–44
28. Chamberlain G, Fox J, Ashton B, Middleton J. Concise review: mesenchymal stem cells: their phenotype, differentiation capacity, immunological features, and potential for homing. *Stem Cells* 2007;**25**(11):2739–49.
29. Friedenstein AJ, Gorskaja JF, Kulagina NN. Fibroblast precursors in normal and irradiated mouse hematopoietic organs. *Exp Hematol* 1976;**4**(5):267–74.
30. Caplan AI. Mesenchymal stem cells. *J Orthop Res* 1991;**9**(5):641–50.
31. Horwitz EM, Le Blanc K, Dominici M, Mueller I, Slaper-Cortenbach I, Marini FC, et al. Clarification of the nomenclature for MSC: the International Society for Cellular Therapy position statement. *Cytotherapy* 2005;**7**(5):393–5.
32. Dazzi F, Horwood NJ. Potential of mesenchymal stem cell therapy. *Curr Opin Oncol* 2007;**19**(6):650–5.
33. Delorme B, Ringe J, Gallay N, Le Vern Y, Kerboeuf D, Jorgensen C, et al. Specific plasma membrane protein phenotype of culture-amplified and native human bone marrow mesenchymal stem cells. *Blood* 2008;**111**(5):2631–5.
34. Silva Jr WA, Covas DT, Panepucci RA, Proto-Siqueira R, Siufi JL, Zanette DL. The profile of gene expression of human marrow mesenchymal stem cells. *Stem Cells* 2003;**21**(6):661–9.
35. Le Blanc K, Tammik C, Rosendahl K, Zetterberg E, Ringdén O. HLA expression and immunologic properties of differentiated and undifferentiated mesenchymal stem cells. *Exp Hematol* 2003;**31**(10):890–6.
36. Caplan AI, Dennis JE. Mesenchymal stem cells as trophic mediators. *J Cell Biochem* 2006;**98**(5):1076–84.
37. Di Nicola M, Carlo-Stella C, Magni M, Milanesi M, Longoni PD, Matteucci P, et al. Human bone marrow stromal cells suppress T-lymphocyte proliferation induced by cellular or nonspecific mitogenic stimuli. *Blood* 2002;**99**(10):3838–43.
38. Meisel R, Zibert A, Laryea M, Göbel U, Däubener W, Dilloo D. Human bone marrow stromal cells inhibit allogeneic T-cell responses by indoleamine 2,3-dioxygenase-mediated tryptophan degradation. *Blood* 2004;**103**(12):4619–21.
39. Maccario R, Podestà M, Moretta A, Cometa A, Comoli P, Montagna D, et al. Interaction of human mesenchymal stem cells with cells involved in alloantigen-specific immune response favors the differentiation of CD4+ T-cell subsets expressing a regulatory/suppressive phenotype. *Haematologica* 2005;**90**(4):516–25.
40. Le Blanc K, Ringden O. Immunomodulation by mesenchymal stem cells and clinical experience. *J Intern Med* 2007;**262**(5):509–25.
41. Asari S, Itakura S, Ferreri K, Liu CP, Kuroda Y, Kandeel F, et al. Mesenchymal stem cells suppress B-cell terminal differentiation. *Exp Hematol* 2009;**37**(5):604–15.
42. Le Blanc K, Rasmusson I, Sundberg B, Götherström C, Hassan M, Uzunel M, et al. Treatment of severe acute graft-versus-host disease with third party haploidentical mesenchymal stem cells. *Lancet* 2004;**363**(9419):1439–41.
43. Christopeit M, Schendel M, Föll J, Müller LP, Keysser G, Behre G. Marked improvement of severe progressive systemic sclerosis after transplantation of mesenchymal stem cells from an allogeneic haploidentical-related donor mediated by ligation of CD137L. *Leukemia* 2008;**22**(5):1062–4.
44. Liang J, Zhang H, Hua B, Wang H, Wang J, Han Z, et al. Allogeneic mesenchymal stem cells transplantation in treatment of multiple sclerosis. *Mult Scler* 2009;**15**(5):644–6.
45. Duijvestein M, Vos AC, Roelofs H, Wildenberg ME, Wendrich BB, Verspaget HW, et al. Autologous bone marrow-derived mesenchymal stromal cell treatment for refractory luminal Crohn's disease: results of a phase I study. *Gut* 2010;**59**(12):1662–9

46. Horwitz EM, Gordon PL, Koo WK, Marx JC, Neel MD, McNall RY, et al. Isolated allogeneic bone marrow-derived mesenchymal cells engraft and stimulate growth in children with osteogenesis imperfecta: Implications for cell therapy of bone. *Proc Natl Acad Sci USA* 2002;**99**(13):8932–7.
47. Quarto R, Mastrogiacomo M, Cancedda R, Kutepov SM, Mukhachev V, Lavroukov A, et al. Repair of large bone defects with the use of autologous bone marrow stromal cells. *N Engl J Med* 2001;**344**(5):385–6.
48. Koç ON, Day J, Nieder M, Gerson SL, Lazarus HM, Krivit W. Allogeneic mesenchymal stem cell infusion for treatment of metachromatic leukodystrophy (MLD) and Hurler syndrome (MPS-IH). *Bone Marrow Transpl* 2002;**30**(4):215–22.
49. Katritsis DG, Sotiropoulou PA, Karvouni E, Karabinos I, Korovesis S, Perez SA, et al. Transcoronary transplantation of autologous mesenchymal stem cells and endothelial progenitors into infarcted human myocardium. *Catheter Cardiovasc Interv* 2005;**65**(3):321–9.
50. El Omar R, Beroud J, Stoltz JF, Menu P, Velot E, Decot V. Umbilical cord mesenchymal stem cells: the new gold standard for mesenchymal stem cell-based therapies? *Tissue Eng Part B Rev* 2014 Oct;**20**(5):523–44.
51. Malgieri A, Kantzari E, Patrizi MP, Gambardella S. Bone marrow and umbilical cord blood human mesenchymal stem cells: state of the art. *Int J Clin Exp Med* 2010;**3**(4):248–69.
52. Erices A, Conget P, Minguell JJ. Mesenchymal progenitor cells in human umbilical cord blood. *Br J Haematol* 2000;**109**(1):235–42.
53. Lee OK, Kuo TK, Chen WM, Lee KD, Hsieh SL, Chen TH. Isolation of multipotent mesenchymal stem cells from umbilical cord blood. *Blood* 2004;**103**(5):1669–75.
54. Mareschi K, Biasin E, Piacibello W, Aglietta M, Madon E, Fagioli F. Isolation of human mesenchymal stem cells: bone marrow versus umbilical cord blood. *Haematologica* 2001;**86**(10):1099–100.
55. McElreavey KD, Irvine AI, Ennis KT, McLean WH. Isolation, culture and characterisation of fibroblast-like cells derived from the Wharton's jelly portion of human umbilical cord. *Biochem Soc Trans* 1991;**19**(1):29S.
56. Friedman R, Betancur M, Boissel L, Tuncer H, Cetrulo C, Klingemann H. Umbilical cord mesenchymal stem cells: adjuvants for human cell transplantation. *Biol Blood Marrow Transpl* 2007;**13**(12):1477–86.
57. Lu LL, Liu YJ, Yang SG, Zhao QJ, Wang X, Gong W, et al. Isolation and characterization of human umbilical cord mesenchymal stem cells with hematopoiesis-supportive function and other potentials. *Haematologica* 2006;**91**(8):1017–26.
58. Romanov YA, Svintsitskaya VA, Smirnov VN. Searching for alternative sources of postnatal human mesenchymal stem cells: candidate MSC-like cells from umbilical cord. *Stem Cells* 2003;**21**(1):105–10.
59. Forraz N, McGuckin CP. The umbilical cord: a rich and ethical stem cell source to advance regenerative medicine. *Cell Prolif* 2011;**44**(Suppl. 1):60–9.
60. Feldmann Jr RE, Bieback K, Maurer MH, Kalenka A, Bürgers HF, Gross B, et al. Stem cell proteomes: a profile of human mesenchymal stem cells derived from umbilical cord blood. *Electrophoresis* 2005;**26**(14):2749–58.
61. Liu CH, Hwang SM. Cytokine interactions in mesenchymal stem cells from cord blood. *Cytokine* 2005;**32**(6):270–9.
62. Oh W, Kim DS, Yang YS, Lee JK. Immunological properties of umbilical cord blood-derived mesenchymal stromal cells. *Cell Immunol* 2008;**251**(2):116–23.
63. Wang M, Yang Y, Yang D, Luo F, Liang W, Guo S, et al. The immunomodulatory activity of human umbilical cord blood-derived mesenchymal stem cells in vitro. *Immunology* 2009;**126**(2): 220–32.
64. Mallam E, Kemp K, Wilkins A, Rice C, Scolding N. Characterization of in vitro expanded bone marrow-derived mesenchymal stem cells from patients with multiple sclerosis. *Mult Scler* 2010;**16**(8):909–18.
65. Bonab MM, Sahraian MA, Aghsaie A, Karvigh SA, Hosseinian SM, Nikbin B, et al. Autologous mesenchymal stem cell therapy in progressive multiple sclerosis: an open label study. *Curr Stem Cell Res Ther* 2012;**7**(6):407–14.
66. Cohen JA. Mesenchymal stem cell transplantation in multiple sclerosis. *J Neurol Sci* 2013;**333**(1–2):43–9
67. Liu R, Zhang Z, Lu Z, Borlongan C, Pan J, Chen J, et al. Human umbilical cord stem cells ameliorate experimental autoimmune encephalomyelitis by regulating immunoinflammation and remyelination. *Stem Cells Dev* 2013;**22**(7):1053–62.
68. Kastrinaki MC, Sidiropoulos P, Roche S, Ringe J, Lehmann S, Kritikos H, et al. Functional, molecular and proteomic characterisation of bone marrow mesenchymal stem cells in rheumatoid arthritis. *Ann Rheum Dis* 2008;**67**(6):741–9.
69. Jones E, Churchman SM, English A, Buch MH, Horner EA, Burgoyne CH, et al. Mesenchymal stem cells in rheumatoid synovium: enumeration and functional assessment in relation to synovial inflammation level. *Ann Rheum Dis* 2010;**69**(2):450–7.
70. Liu Y, Mu R, Wang S, Long L, Liu X, Li R, et al. Therapeutic potential of human umbilical cord mesenchymal stem cells in the treatment of rheumatoid arthritis. *Arthritis Res Ther* 2010;**12**(6):R210.
71. Wu CC, Wu TC, Liu FL, Sytwu HK, Chang DM. TNF-alpha inhibitor reverse the effects of human umbilical cord-derived stem cells on experimental arthritis by increasing immunosuppression. *Cell Immunol* 2012;**273**(1):30–40.
72. Greish S, Abogresha N, Abdel-Hady Z, Zakaria E, Ghaly M, Hefny M. Human umbilical cord mesenchymal stem cells as treatment of adjuvant rheumatoid arthritis in a rat model. *World J Stem Cells* 2012;**4**(10):101–9.
73. Gu Z, Akiyama K, Ma X, Zhang H, Feng X, Yao G, et al. Transplantation of umbilical cord mesenchymal stem cells alleviates lupus nephritis in MRL/lpr mice. *Lupus* 2010;**19**(13):1502–14.
74. Chang JW, Hung SP, Wu HH, Wu WM, Yang AH, Tsai HL, et al. Therapeutic effects of umbilical cord blood-derived mesenchymal stem cell transplantation in experimental lupus nephritis. *Cell Transplant* 2011;**20**(2):245–57.
75. Vija L, Farge D, Gautier JF, Vexiau P, Dumitrache C, Bourgarit A, et al. Mesenchymal stem cells: stem cell therapy perspectives for type 1 diabetes. *Diabetes Metab* 2009;**35**(2):85–93.
76. Zhao Y, Mazzone T. Human cord blood stem cells and the journey to a cure for type 1 diabetes. *Autoimmun Rev* 2010;**10**(2):103–7.
77. Chao KC, Chao KF, Fu YS, Liu SH. Islet-like clusters derived from mesenchymal stem cells in Wharton's jelly of the human umbilical cord for transplantation to control type 1 diabetes. *PLoS One* 2008;**3**(1):e1451.
78. Yoshida S, Ishikawa F, Kawano N, Shimoda K, Nagafuchi S, Shimoda S, et al. Human cord blood-derived cells generate insulin-producing cells in vivo. *Stem Cells* 2005;**23**(9):1409–16.
79. Ende N, Chen R, Reddi AS. Effect of human umbilical cord blood cells on glycemia and insulitis in type 1 diabetic mice. *Biochem Biophys Res Commun* 2004;**325**(3):665–9.

80. Park JH, Hwang I, Hwang SH, Han H, Ha H. Human umbilical cord blood-derived mesenchymal stem cells prevent diabetic renal injury through paracrine action. *Diabetes Res Clin Pract* 2012;**98**(3):465–73.
81. Auletta JJ, Bartholomew AM, Maziarz RT, Deans RJ, Miller RH, Lazarus HM, et al. The potential of mesenchymal stromal cells as a novel cellular therapy for multiple sclerosis. *Immunotherapy* 2012;**4**(5):529–47.
82. Connick P, Kolappan M, Crawley C, Webber DJ, Patani R, Michell AW, et al. Autologous mesenchymal stem cells for the treatment of secondary progressive multiple sclerosis: an open-label phase 2a proof-of-concept study. *Lancet Neurol* 2012;**11**(2):150–6.
83. Karussis D, Karageorgiou C, Vaknin-Dembinsky A, Gowda-Kurkalli B, Gomori JM, Kassis I, et al. Safety and immunological effects of mesenchymal stem cell transplantation in patients with multiple sclerosis and amyotrophic lateral sclerosis. *Arch Neurol* 2010;**67**(10):1187–94.
84. Yamout B, Hourani R, Salti H, Barada W, El-Hajj T, Al-Kutoubi A, et al. Bone marrow mesenchymal stem cell transplantation in patients with multiple sclerosis: a pilot study. *J Neuroimmunol* 2010;**227**(1–2):185–9.
85. Hou ZL, Liu Y, Mao XH, Wei CY, Meng MY, Liu YH, et al. Transplantation of umbilical cord and bone marrow-derived mesenchymal stem cells in a patient with relapsing-remitting multiple sclerosis. *Cell Adh Migr* 2013;**7**(5):404–7.
86. Rahman A, Isenberg DA. Systemic lupus erythematosus. *N Engl J Med* 2008;**358**(9):929–39.
87. Bernatsky S, Boivin JF, Joseph L, Manzi S, Ginzler E, Gladman DD, et al. Mortality in systemic lupus erythematosus. *Arthritis Rheum* 2006;**54**(8):2550–7.
88. Contreras G, Pardo V, Leclercq B, Lenz O, Tozman E, O'Nan P, et al. Sequential therapies for proliferative lupus nephritis. *N Engl J Med* 2004;**350**(10):971–80.
89. Illei GG, Austin HA, Crane M, Collins L, Gourley MF, Yarboro CH, et al. Combination therapy with pulse cyclophosphamide plus pulse methylprednisolone improves long-term renal outcome without adding toxicity in patients with lupus nephritis. *Ann Intern Med* 2001;**135**(4):248–57.
90. Houssiau FA, Vasconcelos C, D'Cruz D, Sebastiani GD, Garrido Ed, Ede R, Danieli MG, et al. Immunosuppressive therapy in lupus nephritis: the Euro-Lupus Nephritis trial, a randomized trial of low-dose versus high-dose intravenous cyclophosphamide. *Arthritis Rheum* 2002;**46**(8):2121–31.
91. Burt RK, Traynor A, Statkute L, Barr WG, Rosa R, Schroeder J, et al. Nonmyeloablative hematopoietic stem cell transplantation for systemic lupus erythematosus. *Jama* 2006;**295**(5):527–35.
92. Jayne D, Tyndall A. Autologous stem cell transplantation for systemic lupus erythematosus. *Lupus* 2004;**13**(5):359–65.
93. Wang D, Zhang H, Liang J, Li X, Feng X, Wang H, et al. Allogeneic mesenchymal stem cell transplantation in severe and refractory systemic lupus erythematosus: 4 years of experience. *Cell Transpl* 2013;**22**(12):2267–77.
94. Sun L, Wang D, Liang J, Zhang H, Feng X, Wang H, et al. Umbilical cord mesenchymal stem cell transplantation in severe and refractory systemic lupus erythematosus. *Arthritis Rheum* 2010;**62**(8):2467–75.
95. Shi D, Wang D, Li X, Zhang H, Che N, Lu Z, et al. Allogeneic transplantation of umbilical cord-derived mesenchymal stem cells for diffuse alveolar hemorrhage in systemic lupus erythematosus. *Clin Rheumatol* 2012;**31**(5):841–6.
96. Li X, Wang D, Liang J, Zhang H, Sun L. Mesenchymal SCT ameliorates refractory cytopenia in patients with systemic lupus erythematosus. *Bone Marrow Transplant* 2013;**48**(4):544–50.
97. Wang D, Li J, Zhang Y, Zhang M, Chen J, Li X, et al. Umbilical cord mesenchymal stem cell transplantation in active and refractory systemic lupus erythematosus: a multicenter clinical study. *Arthritis Res Ther* 2014;**16**(2):R79.
98. Xu J, Wang D, Liu D, Fan Z, Zhang H, Liu O, et al. Allogeneic mesenchymal stem cell treatment alleviates experimental and clinical Sjogren syndrome. *Blood* 2012;**120**(15):3142–51.
99. Iafolla MA, Tay J, Allan DS. Transplantation of umbilical cord blood-derived cells for novel indications in regenerative therapy or immune modulation: a scoping review of clinical studies. *Biol Blood Marrow Transpl* 2014;**20**(1):20–5.

Chapter 10

# Umbilical Cord as a Source of Immunomodulatory Reagents

Antonio Galleu and Francesco Dazzi
*Regenerative Medicine, Department of Haematology, King's College London, London, UK*

## Chapter Outline



One of the main successes of cellular therapies in the last decade has been the generation of immunomodulating reagents for clinical use. Umbilical cord can be used to source such cell populations. In this chapter we will concentrate on describing the features of and the clinical experience with *regulatory T-cells* and *mesenchymal stromal cells*. Although relevant immunomodulating activity can also be ascribed to hematopoietic stem cells, that is the subject of another chapter of this book. Similarly, other populations have been isolated and demonstrated to exhibit immunosuppressive activity but, probably because of their small concentration and the difficulty in their expansion, they have not been developed any further.

## 1. REGULATORY T-CELLS

### 1.1 Definition and Properties

The concept of a regulatory subpopulation of T-cells that actively suppresses immune responses was first introduced in the 1970s.[1] Since then, the ontogeny, differentiation, and functions of these cells have been extensively studied and identified as playing a crucial role in the induction and maintenance of self-tolerance and immune homeostasis.

Regulatory T-cells (Treg) were first distinguished from conventional T-cells because of their expression of both the interleukin-2 (IL-2) receptor α-chain CD25$^+$ T-cells[2,3] and the transcription factor forkhead box P3 (FOXP3)[4–7] within CD4$^+$ T-cells. It is now widely appreciated that human CD4$^+$FOXP3$^+$ T-cells are highly heterogeneous,[8] and the expression of both CD25 and FOXP3 does not uniformly correlate with Treg suppressive activity. Single cell analysis using transcriptomics and *FOXP3* methylation profiling has convincingly shown[9–11] the existence of two main populations of CD4$^+$FOXP3$^+$ Treg cells in humans: natural Treg (nTreg, CD45RA$^+$) and activated Treg-like cells (CD45RO$^+$).[11] Natural Treg are potently suppressive in vitro and, once activated in vivo, they proliferate and convert into CD45RO$^+$ effector Treg (effTreg).[11] Conversely, activated Treg-like cells are generated in the periphery from conventional T-cells and are characterized by nonregulatory functions and production of proinflammatory cytokines.[11] While nTreg and effTreg cells are thymus-derived, activated Treg-like cells derive from conventional T-cells and develop in the periphery under specific microenvironmental conditions. Very controversial is the ontogeny of activated Treg-like cells, particularly whether this population is part of adaptive or induced Treg (iTreg) cells.[12] How and why these cells can develop and maintain their suppressive activity under different *stimuli* is largely unknown.

**Cord Blood Stem Cells Medicine. http://dx.doi.org/10.1016/B978-0-12-407785-0.00010-4**

Treg cells are able to suppress the activation, proliferation, and effector functions of virtually any subset of immune cells, including CD4+ and CD8+ T-cells, natural killer (NK) and NKT-cells, B-cells, and antigen-presenting cells (APCs).[8] As for conventional T-cells, Treg cell development is also subject to the selection of those recognizing self-MHC, but the selection of their T-cell receptor (TCR) repertoire mirrors the rules for conventional T-cells. In fact, since their primary function is to control T-cells reactive against self-antigens, thymus-derived Treg cells are positively selected for their ability to recognize such self-antigens and competitively engage with potentially dangerous conventional T-cells. Although TCR engagement is crucial for their activation, the suppressive function seems to be delivered in a cognate-independent manner.[8] Their pivotal role in suppressing aberrant and excessive immune responses potentially harmful for the host has been widely confirmed in vivo in both animal models and humans.[2,13–17]

## 1.2 Treg Cells as a Therapeutic Tool

Several studies have indicated that quantitative or qualitative abnormalities of Treg cells contribute to the pathogenesis of immune-mediated diseases.[18] Reconstituting the failing component is therefore a plausible approach.

While a potential approach is to induce in vivo expansion of patients' Treg cells with the use of sirolimus or low-dose IL-2,[19–22] the adoptive transfer of Treg has attracted considerable interest and attention. There are currently over 150 registered clinical trials (http://www.clinicaltrials.gov) as both observational and interventional clinical studies recruiting patients to assess the efficacy/feasibility of Treg therapy in several diseases (April 2014).

Despite encouraging preclinical data, the path to translate Treg into clinical applications has unveiled some limiting factors. The main hurdle to Treg isolation is the lack of specific cell surface markers able to define Treg with stable suppressive activity in vivo after clinical-grade expansion in vitro. Although transiently expressed in activated Tconv,[7,23] FOXP3 remains the most important functional marker of Treg in human T-cells. However, because of its intracellular expression, it cannot be used to sort viable cells. Thus, different markers have been proposed, but the results published so far are inconclusive in regard to specificity.[8] CD25 expression remains at the moment the most pragmatic approach, although the CD25+ T-cell population contains a mixture of suppressive and effector cells.[8] However, in the last few years, increasing studies have identified a specific epigenetic status of the Treg-specific demethylation region (TSDR) related to the suppressive functions of Treg.[24,25] Furthermore, it has been demonstrated that only some of these specific CpG hypomethylation patterns are key regulators of the stability of *FOXP3* expression and Treg signature genes.[26] Although currently limited to mouse models, the study of these TSDR profiles could become the milestone for the selection and preparation of functionally stable Treg for clinical use.

Independently of the marker used, cell isolation has been achieved by means of automated magnetic bead-activated cell sorting (MACS) and fluorescence-activated cell sorting (FACS).[27,28] While both methods comply with GMP protocols, the main limitation of the first method is the high degree of contaminating non-Treg cells. Conversely, FACS-based methods allow a higher purity of the selected population (>98%),[29] but the available technology is less suitable for GMP isolation processes, thus restraining their use to highly specialized centers.

Another important issue is Treg cells survival and stability of their regulatory functions after in vitro expansion and following in vivo infusion. It has been shown that Treg cell numbers decrease following a transient postinfusion increment.[30,31] A further issue is the choice of when to treat. While most preclinical studies convincingly showed the efficacy of Treg infusions in preventing disease development,[32,33] only few have documented efficacy in reverting active disease.[34–36] The reason for this can be ascribed to an intrinsic resistance of the disease to Treg therapies. However, Treg could be converted into cells able to produce proinflammatory cytokines[37–39] or turn into effector T-cells in vivo[40–42] when exposed to inflammatory microenvironments.

### 1.2.1 Treg Cells in UC

In humans, Treg cells appear in the thymus as early as 13 weeks of gestation and, concomitant to other T-cells, their egress from the thymus occurs rapidly after their generation, being detectable in the periphery from 14 weeks of gestation.[43] After a first increase over the second trimester of gestation, their number starts to gradually decrease to finally reach a sort of plateau at birth as their proportion appears to be inversely correlated with gestational age.[44] No significant difference in the frequency of CD4+FOXP3+ T-cells between neonates and adults has been found.[45] Consistent with this, the proportion of CD4+CD25+ cells in adult peripheral blood (APB) and umbilical cord blood (UCB) also seems to be comparable. However, while UCB Treg cells present mainly a naïve phenotype, APB Treg cells are largely effTreg,[9–11,46] which reflects the naïvety of UCB Treg cells. In fact, the ratio of nTreg to effTreg cells gradually declines with age.[47,48]

Although the suppressive activity of UCB Treg has been well documented in many studies both in vitro and in vivo in xenogeneic models,[9–11,47,49] other studies have not confirmed these initial observations.[46,50,51] The reasons for these conflicting results remain still unclear and they may be related to differences in the isolation procedures used and the degree of stimulation during Treg preparation. In fact, UCB nTreg acquired full and potent suppressive activity both in vitro and in vivo upon stimulation in vitro, also when an initial functional impairment was observed.[46,51–53] This has practical implications on the use of these cells for therapeutic purposes whereby in vitro expansion—a necessity to generate a sufficient yield—induces nTreg cells to acquire a more mature phenotype.[51,54]

Several studies have demonstrated the feasibility of isolating Treg from UCB[9–11,49,54–56] and have led to one of the first clinical trials.[30] Furthermore, the Treg cells obtained from UCB exhibit particular features that make UCB a better source than APB. While APB Treg cells are a very heterogeneous population with a poor separation between the different subsets of cells with different intensity of CD25 within the $CD4^+$ population, UCB Treg cells appear more uniform. Therefore, UCB Treg cells are more readily purified, and the population obtained can be obtained with a single-step magnetic bead isolation protocol,[30,49,55,56] in contrast to the two-step protocols currently used for Treg isolation from APB.[57]

In their study, Miyara et al. demonstrated that nTreg and effTreg differed not only in the mutually exclusive expression of CD45RA and CD45RO, but also in the expression intensity of CD25. While effTreg showed a $CD25^{high}$ phenotype, nTreg were $CD25^{dim}$. Activated Treg-like cells were $CD25^{dim}$ and they lack suppressive functions.[11] The distribution of the cells within the $CD25^+$ population is crucial for an efficient selection of purified Treg cells because the density of marker expression cannot be distinguished easily by MACS, and contamination with $CD25^{dim}$ cannot be avoided. Thus, a higher degree of purity can be obtained from UCB, since $CD4^+CD25^+$ T-cells from UCB are more homogeneous, mainly comprising of $CD25^{dim}$ nTreg cells.[9–11] Conversely, contamination with $CD25^{dim}$ deeply affects the suppressive functions of Treg derived from APB, as only a very small proportion of the $CD4^+CD25^+$ cells in ABP consist of effTreg or nTreg cells, while the main part of $CD4^+CD25^+$ cells comprise $CD25^{dim}$ activated Treg-like cells.[10,11,54] Certainly, the differences in purity between the two sources can be reduced with the use of FACS-based techniques, as suggested by the findings that more potent suppressor cell line generation was obtained from APB in a feasibility study.[58] However, the opportunity to isolate an enriched population of nTreg seems to represent a potential advantage over effTreg. Many studies have demonstrated that a naïve phenotype is associated with significantly enhanced proliferation potential[59] and most importantly, with the highest capacity to maintain FOXP3 expression following expansion in vitro.[11,58] While effTreg cells undergo apoptosis during proliferation, nTreg cells seem to be more resistant to apoptosis.[11] Furthermore, effTreg cells seem to be more prone to $CD8^+$ T-cell killing in vitro due to a higher expression of HLA-ABC molecules,[49] as suggested by their rapid decrease after injection in vivo.[60]

It is noteworthy that the preponderance of nTreg in clinical-grade products could better impact the efficacy of the treatment in particular pathological conditions. In an allogeneic HSC transplantation (HSCT) setting, it has been recently observed that nTreg cells were more frequent in patients who did not develop acute graft-versus-host disease (aGvHD) in comparison with those who developed the disease, despite the similar frequency of the total Treg population between the two cohorts.[45] Furthermore, only Treg cells expressing CD62L (a chemokine important for the homing in secondary lymphoid organs) were able to prevent the onset of the disease in a mouse model of GvHD.[61,62] Interestingly, CD62L is one of the markers preferentially expressed on nTreg cells and it is lost after acquisition of an effector phenotype.[63]

Our lack of knowledge regarding an ideal marker able to clearly define pure Treg populations still represents one of the main pitfalls in the field, and further studies are needed in the future to decipher the complexity of these cells. However, UCB is a very promising source for Treg also in the context of these limitations, thanks to the features we have already discussed.

An important caveat to the use of UCB as a source of Treg is the low absolute number of cells that can be obtained from a unit of UCB. In order to overcome this important limitation, Milward et al. have recently investigated an alternative approach by using multiple allogeneic units of UCB pooled together to achieve therapeutic doses without any preliminary expansion in vitro.[49] Interestingly, they provided evidence that the magnitude of suppression of Treg cells both in vitro and in vivo was not hampered by the pooling process. Although promising, more extensive studies are needed to confirm the safety and the efficacy of this approach, and the necessity for a large in vitro expansion still remains one of the major limitations for the extensive use of Treg in clinical practice. In more detail, two relevant problems have to be addressed: (1) the balance between the efficiency of the expansion protocol and its adherence to GMP compliant procedures; and (2) the possibility that Treg cells may lose their suppressive potency, or alternatively acquire inflammatory properties, upon in vitro stimulation.

It has been demonstrated that UCB Treg cells, like their APB counterpart, can be driven to proliferate in vitro with anti-CD3/CD28 stimulation in the presence of recombinant IL-2 and they seem to retain their suppressive functions in vivo when adoptively infused in some disease animal models.[52,64] Although several variations to the aforementioned protocol have been proposed,[28] the majority of these protocols can be currently used only in experimental settings because of the reduced number of GMP-compliant reagents available. Relevant to this point is the use of third cells as adjuvant for the proliferation of Treg. It has been demonstrated that UCB Treg cells can be expanded up to

1250-fold after supplement to the culture of artificially modified APC expressing OX40 and 4-1BB, two members of the tumor necrosis factor receptor family that provide signals for Treg expansion and survival.[53] Most importantly, the presence of APC pulsed with specific antigens (Ag) is required for the generation and expansion of Ag-specific Treg. The possibility that different subsets of Treg (Ag-specific Treg vs polyclonal expanded counterpart) can impact the final outcome of the treatment has been evaluated in the last few years. Although adoptive transfer of ex vivo polyclonally expanded Treg promotes tolerance to allogeneic pancreatic islet grafts[65] or autoimmune diseases,[66,67] several studies suggest that alloantigen or Ag-specific Treg can deliver better protection against tissue rejection,[27] autoimmune diseases, and type 1 diabetes (T1D).[33] However, some important hurdles related to the production of such cells should be considered. Protocols for the production of Ag-specific Treg are very complex, expensive, and the tendency for the outgrowth of autoreactive Treg after in vitro expansion still remains an important drawback.[68,69] Thereby, the current available technology and the standards required for the expansion of clinical-grade cells narrow the use of third cells (such as APC) as stimulators for the expansion of Treg only in the context of highly specialized institutions and for controlled experimental settings only.

To further complicate the issues regarding the best expansion protocol, it is becoming clear that either the source of Treg or the isolation protocol used may affect the final yield of Treg. While UCB and APB Treg cells showed similar proliferative capacity when expanded using two cycles of polyclonal stimulation separated by two days of rest,[52] UCB was a superior source for Treg expansion when a single expansion step was used.[70,71]

During in vitro expansion, Treg cells can lose their regulatory functions.[72–74] Despite this limitation, it is important to stress that UCB Treg cells mainly comprise nTreg cells. This subset maintains a higher stability of FOXP3 expression[63,75] and degree of TSDR demethylation levels,[11,72] and is more resistant to apoptosis[11,63] upon in vitro stimulation, in comparison to effTreg cells (mainly present in APB). Moreover, although they rapidly convert their phenotype in CD45RO$^+$ after CD3/CD28 stimulation, their genetic signature seems to retain a more undifferentiated program, since they are characterized by high expression of genes involved in cell proliferation, chromatin modification, and modulation of gene expression. Conversely, effTreg cells from APB are mostly characterized by an upregulation of genes involved in effector functions.[70] Furthermore, as already discussed, UCB Treg cells are easily isolated with a lower likelihood of contamination of Tconv, thus limiting the possibility of an outgrowth of contaminating Tconv due to their shorter period to enter S-phase after stimulation.[76] However, rapamycin, a well-known immunosuppressant, has been used as a supplement in the expansion culture due to its ability to prevent the outgrowth of contaminating nonregulatory cells.[77] Nonetheless, the use of rapamycin represents a double-edged sword, since a reduction of the overall Treg proliferation has been observed, thus requiring a longer time for Treg expansion[65] with a higher risk of lack of suppressive activity.[75]

It has been shown that Tconv can be easily converted into Treg upon stimulation of a variety of molecules in vitro such as TGF-b,[78] retinoic acids,[79] and rapamycin.[79] Furthermore, in vitro iTreg cells with potent in vitro suppressive activity can be developed from UCB CD34$^+$ cells when in coculture with Notch ligand, Delta-like 1 expressing OP9 stromal cells.[80] Since expansion protocols of in vitro iTreg are very efficient and there is large availability of Tconv or CD34$^+$ cells as starting cells for manipulation, iTreg cells have been proposed as possible alternatives to nTreg for clinical uses.[81] With this aim, UCB Tconv cells comprise a richer fraction of T-cells with a naïve phenotype, with a reduced capacity to produce activating cytokines compared to their counterparts obtained from APB.[82,83]

Although very intriguing, the use of in vitro iTreg as a surrogate of ex vivo freshly isolated or ex vivo expanded nTreg remains controversial. The main question to be addressed focuses on the stability of the regulatory functions of these cells in the absence of the signals controlling the shift toward the regulatory phenotype and the possibility of a subsequent conversion toward the native inflammatory phenotype.[24,84] Furthermore, the ability of in vitro iTreg cells to keep their regulatory functions after injection in vivo has yet to be substantiated. It has been shown that such cells possessed regulatory activity in vivo in different mouse models;[85–91] however, these data are still conflicting[92,93] and need further confirmation. As aforementioned, it is becoming clear that nTreg cells and in vitro iTreg cells are distinct populations with different molecular, genetic signatures, and epigenetic status of the *FOXP3* promoter gene,[24,25,90] and only the hypomethylation patterns at the TSDR described in nTreg seem to be linked to a more stable FOXP3 expression, irrespective of the surrounding microenvironment or the proliferation status of cells.[26] Although limited to mouse models, these studies could provide significant insights into the mechanisms controlling Treg heterogeneity and could be crucial in unveiling the best method for selecting and preparing functionally stable Treg cells, irrespective of their "natural" or "induced" ontogeny.

### 1.2.2 Treg Cell Therapy in Transplantation

Solid organ transplantation is a widely used treatment for end-stage failure of several organs. In spite of remarkable ameliorations in the management of the graft survival in the short term, the outcome of transplanted patients and grafts still remains unsatisfactory because of chronic rejection and the toxicity associated with long-term immunosuppressive treatments.[94] Graft rejection is mediated by recipient alloreactive T-cells that recognize donor alloantigens presented by donor APC (direct pathway) and by recipient APC (indirect pathway). Since Treg cells employ a mainly indirect pathway of allorecognition for their immunoregulatory properties[95–98] and the observation that the indirect pathway seems to be the major driver of chronic allograft rejection both in mice and in humans,[99–102] it is conceivable that the adoptive transfer of Treg can be successfully used for preventing chronic allograft rejection, thus allowing the reduction of immunosuppressive drugs. Supported by the promising results obtained in preclinical models, three clinical trials are now underway for the evaluation of the safety and tolerability of the adoptive therapy with polyclonal in vitro expanded Treg in children or adult kidney-transplanted patients (trials numbers NCT01446484 and NCT02091232, respectively), and in adult liver-transplanted patients (trial number NCT01624077). While polyclonal Treg cells will be used in the first trial, the last two trials have chosen to infuse Treg cells after exposure to alloantigens (alloreactive Treg cells). We have already discussed the importance of Ag-specific Treg cells. However, whether this superiority can be applied to humans remains unknown, since all the available data come from animal models. Certainly, although primary end points of these studies are safety and feasibility, their results may help in shedding light on this important issue, thus paving the way for a better design of future clinical trials.

### 1.2.3 Treg Cell Therapy for GvHD

GvHD is currently one of the major and life-threatening complications of allogeneic HSCT, thereby limiting the use of this important procedure.[103,104] Because of its high morbidity and mortality, GvHD has provided an ideal setting in which to test Treg cells. The initial evidence of their therapeutic activity in mouse models[105] was subsequently confirmed in two clinical trials.[30,106] In a phase I dose-escalating trial, Brunstein et al. treated 23 patients undergoing double-cord blood transplantation with in vitro expanded UCB Treg cells, which were then infused on the day of the transplant and in 13 patients, also at day +15.[30] A reduced frequency of grade II–IV aGvHD was observed in comparison to historical controls. A role for Treg in GvHD prophylaxis was further supported by Di Ianni et al.[106] who treated 28 patients undergoing HLA-haploidentical HSCT with APB freshly isolated Treg cells in the absence of any GvHD prophylaxis and showed a significant reduction in the onset and severity of aGvHD. They also reported a faster immune reconstitution compared to historical controls, but these data were not confirmed in Brunstein's study.[30] Notably, Treg infusions were safe and well tolerated, with no dismal effects on infection, relapse, or early mortality rates in both studies. However, although the initial data did not report any impact on the incidence of opportunistic infections, more recently an increase of viral reactivation was observed in the Treg-treated group compared with historical controls when opportunistic infections were evaluated by calculating the infection density.[107] This parameter, taking into account multiple infections in an individual patient within a specific time range, produces a better picture of patient susceptibility to infections. The authors demonstrated that Treg patients were more prone to opportunistic infections only during the period when infused Treg cells were demonstrated in circulation.[107]

Among the most important information that can be obtained from these studies is that high doses (up to $3 \times 10^6$/kg cells) and multiple infusions were well tolerated. The level of Treg in the circulation seemed to survive transiently in the recipient, with a longer survival of freshly expanded compared to cryopreserved Treg cells (up to 14 days and 4 days, respectively). Notably, the mean purity of the cells, assessed by the positivity for FOXP3, was not higher than 70% in both studies and the targeted Treg dose was achieved only in 74% of cases in the study of Brunstein et al.[30] Taken together, these data clearly show that Treg can be obtained from UCB and they are promising in preventing GvHD. Although partly related to the poor condition of the patient cohort treated by Ianni et al. and the severity of the transplantation regimen used, the disappointing overall survival (50%) reported in that study[27] and the increase in infection rate in Treg-treated patients[107] open new interrogatives regarding the precautions and the need for tailored measures of supportive care for Treg-treated patients.

### 1.2.4 Treg Cell Therapy for Autoimmune Diseases

It is widely accepted that numerous immune-based abnormalities of the T-cell system play a major role in the development of systemic autoimmune diseases, and several studies have reported quantitative and qualitative defects of Treg in such diseases.[32] Several lines of evidence show the efficacy of Treg infusion in ameliorating systemic autoimmune ailments in preclinical models.[32] The promising results described in the previous paragraph suggest that Treg-based therapy will soon be considered as a therapeutic option also in autoimmune diseases.

It is commonly accepted that the most important pathogenetic event in the development of T1D is the T-cell-mediated destruction of beta islets in the pancreas, and that Treg impairment contributes to the development of the disease.[108] Therefore, the manipulation of Treg cell homeostasis offers an opportunity to enhance their suppressive potency and/or increase their number in vivo.[109] Extensive data have showed that the adoptive transfer of Treg can be efficacious in preventing and delaying the onset and progression of T1D in NOD mice,[66,110,111] with some studies reporting the efficacy of Treg in reversing ongoing T1D.[66,112] The first phase I clinical trial was conducted on a cohort of 10 recently diagnosed T1D children and showed safety and tolerability. Treg cells were generated from ex vivo expanded autologous $CD4^+CD25^{high}CD127^-$sorted cells[113] and infused in escalating doses. Although a prolonged survival of the pancreatic isles was claimed, the small number of patients enrolled does not allow any conclusion regarding efficacy to be drawn.

## 2. MESENCHYMAL STROMAL CELLS

The acronym for mesenchymal stromal cells (MSC) is widely used and generally accepted in scientific and clinical practice, but it is subject to a number of controversial challenges related to its real significance. MSC preparations consist of a highly heterogeneous population of fibroblast-like cells selected and expanded in vitro as plastic-adherent. There is ample evidence that a large proportion of these cell preparations acquire features of osteoblastic, adipocytic, and chondrocytic lineage under certain in vitro conditions,[114] but the validity and actual significance of this progenitor activity remains ambiguous. They lack expression of hematopoietic and endothelial markers (CD45, CD34, CD14 or CD11b, CD79a or CD19, and HLA-DR, CD31) while expressing a few nonspecific molecules (CD105, CD73, and CD90) involved in cell-to-cell contact.[115] Despite new sets of markers being recently proposed, the characterization of MSC still remains elusive,[116] thus making the comparison between studies difficult. Notably, populations of cells meeting these criteria vary extremely in terms of morphology, stage of differentiation, proliferation rate, and functional characteristics, as clearly demonstrated by early investigations.[117–119] Whether this complexity merely originates from culture manipulations or reflects the innate heterogeneity of in vivo repertoire of MSC subsets has not yet been clarified.[120]

First isolated from the bone marrow (BMMSC),[119] MSC can be generated from several tissues,[121–123] including UC.[124] Although BMMSC have been the most used in clinical studies and still remain the basis of comparison for MSC generated from all the other tissues, other sources are being sought. For logistical and proprietary reasons, adipose and UC tissues are probably the most attractive sources. Here, we will provide an overview of the current knowledge in the characterization, function, and preliminary clinical experience with UC-derived MSC (UCMSC).

### 2.1 UC-derived MSC

Sourcing MSC from UC tissues has major logistical advantages. First, it does not require the relatively invasive BM harvest procedure, which is associated with some—albeit minimal—risks for the donor. Secondly, provision is at the moment straightforward, as UC is usually discarded. Furthermore, the amount of MSC that can be derived is remarkably higher when compared to BM. Finally, cord tissue fragments can be successfully frozen and stored for a prolonged period of time before isolation of MSC.[125,126]

Although MSC can be isolated from virtually any component of UC—cord blood,[124,127,128] mucous proteoglycan-rich matrix (Wharton's jelly), and the vasculature[129–131]— Wharton's jelly certainly provides the most efficient source.[132]

UCMSC appear to have a higher proliferative capacity and life span in culture than MSC from adult tissues.[133,134] Differences in the differentiation potential between UCMSC and BMMSC are still controversial, probably because of the poor insight provided by these assays.[133,135,136] At transcriptome analysis, MSC from the two sources display different gene expression profiles,[137] with a dominant osteogenic profile in BMMSC and a prevalent expression of IL-1 and TNF alpha in UCMSC.[138,139] Despite their equivalent differentiation potential and comparable proliferative properties,[140] the gene expression profiles of UCMSC obtained from UCB or Wharton's jelly are different.[141] Such heterogeneity seems to be a constant feature of MSC preparations, regardless of their tissue of origin. Different culture conditions such as seeding density, duration of in vitro expansion, and culture supplements affect surface protein expression, clonogenicity, capacity to differentiate, and immunomodulatory functions.[142–147]

It is not known whether the heterogeneity in MSC clinical preparations has an impact on the clinical use of the cells. It is possible that different tissues harbor different subsets with varying properties. Alternatively, the different fractions obtained from distinct tissues might respond differently to different *stimuli*, thus explaining why MSC of different tissue origin respond differently when exposed to a specific in vitro stimulation. This latter possibility is suggested by the recent observation that MSC from fetal lung, fetal bone marrow, adult adipose tissue, or adult bone marrow migrate in response to different stimuli. In this study, the tissue of origin, rather than donor age, determined the migratory pattern of MSC.[148] Therefore, a more precise understanding of the mechanisms behind MSC therapeutic activity will be crucial to choose the best source or subset for the treatment of specific conditions.

### 2.2 MSC as a Therapeutic Tool and MSC Immunobiology

Like Treg cells, MSC mediate potent immunosuppressive and immunoregulatory effects on virtually any type of adaptive and innate immune responses. MSC nonselectively and nonspecifically suppress $CD4^+$ and $CD8^+$ T-lymphocytes independently of whether they are naïve or antigen-experienced and their effect is not MHC-restricted. They are able to modulate the activity of NK-cells and inhibit B-cell terminal differentiation and dendritic cell maturation and functions. Notably, the immunosuppressive activity of MSC can be mediated by both autologous and allogeneic MSC.[149] A fundamental concept underpinning MSC immunobiology is their plasticity. MSC are not constitutively inhibitory, but they deliver immunosuppressive functions only after being exposed to an inflammatory environment that "licenses" these properties.[114] Although MSC immunosuppressive function is elicited in the presence of some typical inflammatory cytokines such as interferon-γ (IFN-γ),[150,151] interleukin-1β, and tumor necrosis factor-α,[152] not all types of inflammation have the same effect. Indeed, MSC can acquire antigen-presenting functions in the presence of particular concentrations of IFN-γ.[114,153,154] Moreover, the immunosuppressive properties of MSC are impaired by Toll-like receptor 4 stimulation.[155]

UCMSC show potent immunosuppressive properties in many in vitro assays with a magnitude similar[136,156] or higher than those of BMMSC.[157] Whether these in vitro observations may mirror a difference in the clinical potency of MSC from different sources is still unknown. There is now consensus that, at least in vitro, the secretion of soluble factors is a fundamental means by which immunomodulation is delivered. Probably the most important mechanisms involve essential amino acid metabolism through the engagement of different molecules such as indoleamine-2,3-dioxygenase,[158] transforming growth factor beta 1, hepatocyte growth factor, prostaglandin E2 (PGE2), and soluble human leukocyte antigen G (HLA-G).[159,160] Interestingly, these pathways are shared by other cell types like macrophages, MDCS, fibrocytes, and even epithelial cells in response to viral infections. Therefore, it is plausible that different tissue sources and/or different cell subsets may not necessarily impact on the immunomodulating activities but for a better response to the environmental cues. In this context, the use of MSC from UC could provide a step forward not only because of the ease by which the cells are isolated and expanded, but also because younger and/or more immature cells are likely to display a wider and prompter repertoire of effector molecules. By way of an example, UCMSC do not express TLR-4[161]—the stimulation of which impairs MSC immunosuppressive function—thus making them particularly useful in the treatment of patients with Gram-negative sepsis.[162]

A crucial issue regarding the use of MSC as therapeutic agents relates to their postinfusion fate. It has been widely documented that most of the injected MSC are first trapped in lungs, with very few cells reaching the liver and the spleen afterward, and only traces capable of reaching the remaining organs thereafter.[163–166] Whether the migration to the sites of tissue injury or inflammation could impact the therapeutic efficacy of MSC is still unknown.[167] Apart from the large cellular size,[168,169] the repertoire of adhesion molecules expressed by MSC is likely to play a crucial role in lung retention. Consistently, MSC expressing higher levels of podocalyxin-like protein and α6-integrin (CD49f) aggregated less in vitro when incubated in suspension at room temperature. These MSC are less likely to produce lethal pulmonary emboli and display a more efficient engraftment in injured heart with a longer in vivo survival.[144] Furthermore, the expression of higher levels of CD49f in UCMSC than BMMSC from adults has been recently associated to a faster clearance from the lung,[170] thus suggesting a potentially more efficient migratory profile.

MSC have been shown to express HLA and costimulatory molecules (CD40, CD80, or CD86) at very low levels, and this is consistent with their ability to escape from the recognition of alloreactive lymphocytes in vitro.[171] Furthermore, the therapeutic efficacy of third-party MSC across the MHC barrier both in animal models[172,173] and in humans[172,174–176] has been interpreted as a proof-of-principle of MSC immune privilege. Despite these observations, MSC can be induced to express MHC class II in the presence of low dose of IFN-γ[154] and act as antigen-presenting cells rather than being immunosuppressive.[153] Furthermore, it has been also observed that allogeneic MSC can be rejected after infusion with the consequent generation of memory T-cells.[177–179] However, the therapeutic activity of MSC was conserved also after rejection.[178] This observation, supported by the absence of any correlation between treatment response and MSC engraftment[180] and the similar therapeutic efficacy between third-party, haploidentical, or MHC-compatible MSC,[181] suggests a dispensable role of the "immunoprivileged" status of MSC for the delivery of therapeutic activity. However, it is conceivable that MSC survival could play a decisive role in some clinical settings, as inferred by the observation that murine MSC encapsulated in alginate beads persist longer and improve the survival of mice in an experimental model of GvHD[182] as compared to control intravenously infused MSC. In the absence of conclusive evidence supporting the superiority of autologous versus allogeneic MSC, the logistic of preparing off-the-shelf reagents is far more attractive than the choice of autologous cells and is therefore the preferred selection in most studies. While third-party MSC can be stocked and be available on demand, the autologous product is not feasible for acute indications because of the long period of time required to generate a sufficient therapeutic dose. Notably, in aging individuals MSC might be intrinsically impaired either for their ability to expand and/or in particular functions and therefore would be limited for therapeutic applications.[143,183,184]

Furthermore, some diseases are also associated with qualitative alterations of MSC, and it cannot be excluded that such alterations could hinder their therapeutic efficacy or facilitate the relapse of the primary disease.[185–187]

### 2.2.1 MSC for HSC and Solid Organ Transplantation

Supported by the observation that MSC are able to produce soluble signals and cytokines that support hematopoiesis,[188,189] Koc et al. first reported that autologous BMMSC coinfused at the time of HSCT had a positive impact on hematopoietic recovery.[190] This seminal finding has prompted further studies that confirmed the ability of MSC in supporting hematopoietic engraftment.[191–197] Such properties may provide an advantage when considering that pretransplant conditioning regimens could lead to damage of the hematopoietic niche, potentially delaying hematopoietic recovery. Consistent with this, MSC can prevent graft failure.[195,198] These initial studies performed using BMMSC have been confirmed with MSC sourced from UC.[130,125] Furthermore, their efficacy has been recently demonstrated in humans. In their study, Wu et al. cotransplanted in vitro-expanded UCMSC in 50 people with refractory/relapsed hematologic malignancy undergoing haplo-HSCT with myeloablative conditioning. They observed a median time to neutrophil and platelet engraftment of 12 and 14 days posttransplant, respectively, and 66% estimated a 2-year disease/progression-free survival rate.[199] The same group has recently shown a reduced risk of graft failure in a cohort of 21 patients with severe aplastic anemia treated with the administration of UCMSC at the same time of haplo-HSCT without T-cell depletion.[200]

Although all these findings seem very promising, discordant results have also been published, with no improvement in the kinetics of HSC engraftment after MSC infusion.[201,202] These data suggest that, while MSC infusions are safe, the efficacy of the treatment needs to be further confirmed. One of the factors to be taken into account is the underlying mechanisms of disease, whereby the therapeutic effect of MSC in the improvement of engraftment may prominently be related to the capacity of MSC to reset the hemopoietic niche through an immunomodulating activity. Furthermore, concomitant therapies or different conditioning regimens at the moment of MSC administration may grossly affect therapeutic efficacy by changing the "licensing" cues in the patient inflammatory microenvironment.

One of the first observations supporting the use of third-party MSC as a therapeutic immunomodulating tool was the observation that the infusion of BMMSC prolonged the survival of skin allografts in baboons.[203] Since then[204] a fairly large number of patients undergoing solid organ transplantation have been treated with MSC, showing a protective effect against graft rejection. In the only randomized clinical trial performed, 159 patients receiving living donor kidney transplant were enrolled to assess the role of MSC treatment as a replacement for the standard anti-IL-2 receptor antibody induction therapy. Fifty-three and fifty-two patients received MSC in association with standard or reduced dose of calcineurin inhibitors (CNIs), while fifty-one patients in the control group were treated with anti-IL-2 receptor antibody plus standard dose of CNIs. The use of MSC (irrespective to the CNIs dose) resulted in a lower 1-year incidence of acute rejection, a faster recovery of renal graft function during the first month posttransplant, and a better estimated renal function at 1 year.[205]

While most of these studies showed that MSC infusions seem to promote graft survival, only few data have been published regarding the effect of MSC treatment in halting an ongoing allograft rejection,[206] and the use of MSC in this clinical setting needs further confirmation with larger studies.

### 2.2.2 UCMSC and GvHD Therapy

For ethical reasons, the first studies testing MSC-based treatments were carried out in severe inflammatory conditions like GvHD. Several studies have been published regarding the use of MSC for the treatment of steroid-resistant aGvHD. The results of the first clinical studies have been extremely encouraging with overall response rates up to 60% and 70% in adults and children, respectively.[181,207] Most importantly, patients who experienced a complete response to MSC treatment showed a higher overall survival when compared to patients with partial or no response.[181,208,209]

The number of BMMSC infused in the clinical studies reported to date has ranged from $4 \times 10^5$ to $1 \times 10^7$ per kg body weight,[181,194] with no significant correlation between the dose of MSC received and the clinical outcome. One, two or more infusions have been administered with no obvious pattern in respect to the outcome. For example, some patients responded only to the second infusion, whereas others failed to respond after multiple infusions.[181] Although most of these studies used BMMSC, it is becoming evident that also UCMSC can represent a valid and effective alternative.[157,210]

Despite the fairly large overall total number of patients treated with MSC among the several studies, the majority of each study was a pilot or simply a case report.[202,207,211–221] The larger published studies[181,210,222] were phase II trials without randomization and placebo control groups. The only phase III trials were two commercially driven studies carried out in 2009. While in one study BMMSC were tested as first line in association with steroids in 192 patients with aGvHD, in the second BMMSC were tested in 260 patients with steroid-resistant aGvHD. The results—unfortunately never published but in abstract form—did not show any statistically significant superiority on survival in the MSC group, but the study design and in particular the criteria for patient recruitment were poor. Therefore, no definitive conclusions can be drawn in terms of efficacy of MSC therapy and demonstration of a beneficial effect from MSC in a large placebo-controlled trial is still needed.

However, the evidence that MSC can be effective (disease remission) in such a severe disease is compelling and with at least 50% response rate as per our and other experiences.[223] In order to improve response rates and durability of this treatment modality, the next important step is to identify the conditions associated with therapeutic efficacy. Important information will be gathered by a better standardization of MSC preparations and by identifying the best timing for intervention. Currently, there are no promising tools to develop the first aspect because no significant correlation has ever been made between MSC donor/preparations and clinical responses[181] (Galleu et al., MS in preparation). Furthermore, MSC from different tissues seem to be similarly effective at least in preclinical models.[181,210,215,216,222] The only factor that seems to impact on therapeutic efficacy is MSC aging in cultures: early culture passages exhibiting a superior outcome than late ones.[223]

Focusing on the inflammatory environmental cues present at the time of MSC infusion appears to be a more productive strategy. The importance of the "licensing" step is not only supported by in vitro data as already discussed, but seems to be confirmed by the observation that, in contrast to the general efficacy of MSC for active GvHD, the prophylactic approach at time of transplantation, before any sign of GvHD, does not appear to reduce the development of acute or chronic GvHD.[191] Despite the very limited experience, the lack of efficacy in chronic GvHD is consistent with this notion.[224,225]

### *2.2.3 UCMSC and Other Disorders Characterized by Immune-dysregulation*

Thanks to their immunomodulating properties and the aforementioned activity in hampering GvHD, MSC are now being considered for treatment of other inflammatory diseases. Several studies have been performed in the last few years and the results published so far seem to confirm the potential of this therapeutic option in diseases such as Crohn's disease, systemic lupus erythematosus (SLE), multiple sclerosis, rheumatoid arthritis, or T1D.[226,227] The use of UCMSC in this field is quite recent and most of the promising data come from preclinical models;[228–230] thus the reproducibility and clinical significance of these findings remain to be confirmed. However, the first clinical trial using UCMSC for the treatment of SLE was published in 2010. In their study, Sun et al. treated 16 patients with severe and treatment-resistant SLE and reported a significant reduction of disease activity in all patients with improvement of renal function, increase in peripheral Treg cells, and a reestablished balance between Th1- and Th2-related cytokines.[231]

## 3. CONCLUSIONS

UC tissues offer a great opportunity for immune-based regenerative medicine because of the enormous logistic advantages. The efficient provision of immunomodulatory reagents, like MSC and Treg cells, makes UC an attractive tissue to both commercial and academic health-care enterprises. In this chapter we have provided an overview of the advances achieved in the field of inflammatory disorders by the clinical use of Treg and MSC. These strategies have been demonstrated to exhibit a very safe profile. By way of an example, a recent and extensive meta-analysis of more than 1000 patients treated worldwide with MSC for different conditions has extensively documented the lack of significant toxicity.[232] However, much work is still required to improve clinical results, including the identification of the best protocols to isolate and expand the cells, the development of a predictive potency assay, and finally the key question of patient stratification. A structured approach to tackle these questions will pave the way to maximize modalities to harness UC tissues and deliver novel and effective therapeutic tools. Inflammatory disorders are a huge financial burden on society, and the optimization of these new tools will have a profound impact on health care and quality of life.

## REFERENCES

1. Gershon RK, Kondo K. Cell interactions in the induction of tolerance: the role of thymic lymphocytes. *Immunology* 1970;**18**:723–37.
2. Sakaguchi S, Sakaguchi N, Asano M, Itoh M, Toda M. Immunologic self-tolerance maintained by activated T cells expressing IL-2 receptor alpha-chains (CD25). Breakdown of a single mechanism of self-tolerance causes various autoimmune diseases. *J Immunol* 1995;**155**:1151–64.
3. Baecher-Allan C, Brown JA, Freeman GJ, Hafler DA. CD4+CD25high regulatory cells in human peripheral blood. *J Immunol* 2001;**167**:1245–53.
4. Hori S, Nomura T, Sakaguchi S. Control of regulatory T cell development by the transcription factor Foxp3. *Science* 2003;**299**: 1057–61.
5. Fontenot JD, Gavin MA, Rudensky AY. Foxp3 programs the development and function of $CD4^+CD25^+$ regulatory T cells. *Nat Immunol* 2003;**4**:330–6.
6. Khattri R, Cox T, Yasayko SA, Ramsdell F. An essential role for Scurfin in $CD4^+CD25^+$ T regulatory cells. *Nat Immunol* 2003;**4**: 337–42.
7. Roncador G, Brown PJ, Maestre L, et al. Analysis of FOXP3 protein expression in human $CD4^+CD25^+$ regulatory T cells at the single-cell level. *Eur J Immunol* 2005;**35**:1681–91.
8. Sakaguchi S, Miyara M, Costantino CM, Hafler DA. FOXP3+ regulatory T cells in the human immune system. *Nat Rev Immunol* 2010;**10**:490–500.
9. Fritzsching B, Oberle N, Pauly E, et al. Naive regulatory T cells: a novel subpopulation defined by resistance toward CD95L-mediated cell death. *Blood* 2006;**108**:3371–8.
10. Seddiki N, Santner-Nanan B, Tangye SG, et al. Persistence of naive $CD45RA^+$ regulatory T cells in adult life. *Blood* 2006;**107**:2830–8.
11. Miyara M, Yoshioka Y, Kitoh A, et al. Functional delineation and differentiation dynamics of human $CD4^+$ T cells expressing the FoxP3 transcription factor. *Immunity* 2009;**30**:899–911.
12. Bilate AM, Lafaille JJ. Induced $CD4^+Foxp3^+$ regulatory T cells in immune tolerance. *Annu Rev Immunol* 2012;**30**:733–58.

13. Itoh M, Takahashi T, Sakaguchi N, et al. Thymus and autoimmunity: production of CD25+CD4+ naturally anergic and suppressive T cells as a key function of the thymus in maintaining immunologic self-tolerance. *J Immunol* 1999;**162**:5317–26.
14. Brunkow ME, Jeffery EW, Hjerrild KA, et al. Disruption of a new forkhead/winged-helix protein, scurfin, results in the fatal lymphoproliferative disorder of the scurfy mouse. *Nat Genet* 2001;**27**:68–73.
15. Chatila TA, Blaeser F, Ho N, et al. JM2, encoding a fork head-related protein, is mutated in X-linked autoimmunity-allergic disregulation syndrome. *J Clin Invest* 2000;**106**:R75–81.
16. Wildin RS, Ramsdell F, Peake J, et al. X-linked neonatal diabetes mellitus, enteropathy and endocrinopathy syndrome is the human equivalent of mouse scurfy. *Nat Genet* 2001;**27**:18–20.
17. Bennett CL, Christie J, Ramsdell F, et al. The immune dysregulation, polyendocrinopathy, enteropathy, X-linked syndrome (IPEX) is caused by mutations of FOXP3. *Nat Genet* 2001;**27**:20–1.
18. Hirota K, Hashimoto M, Yoshitomi H, et al. T cell self-reactivity forms a cytokine milieu for spontaneous development of IL-17+ Th cells that cause autoimmune arthritis. *J Exp Med* 2007;**204**:41–7.
19. Bestard O, Cruzado JM, Rama I, et al. Presence of FoxP3+ regulatory T Cells predicts outcome of subclinical rejection of renal allografts. *J Am Soc Nephrol* 2008;**19**:2020–6.
20. Lopez M, Clarkson MR, Albin M, Sayegh MH, Najafian N. A novel mechanism of action for anti-thymocyte globulin: induction of CD4+CD25+Foxp3+ regulatory T cells. *J Am Soc Nephrol* 2006;**17**:2844–53.
21. Koreth J, Matsuoka K, Kim HT, et al. Interleukin-2 and regulatory T cells in graft-versus-host disease. *N Engl J Med* 2011;**365**:2055–66.
22. Kennedy-Nasser AA, Ku S, Castillo-Caro P, et al. Ultra low-dose IL-2 for GVHD prophylaxis after allogeneic hematopoietic stem cell transplantation mediates expansion of regulatory T cells without diminishing antiviral and antileukemic activity. *Clin Cancer Res* 2014;**20**:2215–25.
23. Walker MR, Kasprowicz DJ, Gersuk VH, et al. Induction of FoxP3 and acquisition of T regulatory activity by stimulated human CD4+CD25− T cells. *J Clin Invest* 2003;**112**:1437–43.
24. Floess S, Freyer J, Siewert C, et al. Epigenetic control of the foxp3 locus in regulatory T cells. *PLoS Biol* 2007;**5**:e38.
25. Polansky JK, Kretschmer K, Freyer J, et al. DNA methylation controls Foxp3 gene expression. *Eur J Immunol* 2008;**38**:1654–63.
26. Ohkura N, Hamaguchi M, Morikawa H, et al. T cell receptor stimulation-induced epigenetic changes and Foxp3 expression are independent and complementary events required for Treg cell development. *Immunity* 2012;**37**:785–99.
27. Safinia N, Leech J, Hernandez-Fuentes M, Lechler R, Lombardi G. Promoting transplantation tolerance; adoptive regulatory T cell therapy. *Clin Exp Immunol* 2013;**172**:158–68.
28. Kim YJ, Broxmeyer HE. Immune regulatory cells in umbilical cord blood and their potential roles in transplantation tolerance. *Crit Rev Oncol Hematol* 2011;**79**:112–26.
29. Putnam AL, Brusko TM, Lee MR, et al. Expansion of human regulatory T-cells from patients with type 1 diabetes. *Diabetes* 2009;**58**:652–62.
30. Brunstein CG, Miller JS, Cao Q, et al. Infusion of ex vivo expanded T regulatory cells in adults transplanted with umbilical cord blood: safety profile and detection kinetics. *Blood* 2011;**117**:1061–70.
31. Marek-Trzonkowska N, Mysliwiec M, Dobyszuk A, et al. Therapy of type 1 diabetes with CD4CD25CD127-regulatory T cells prolongs survival of pancreatic islets - results of one year follow-up. *Clin Immunol* 2014;**153**(1):23–30.
32. Miyara M, Gorochov G, Ehrenstein M, Musset L, Sakaguchi S, Amoura Z. Human FoxP3+ regulatory T cells in systemic autoimmune diseases. *Autoimmun Rev* 2011;**10**:744–55.
33. Tang Q, Bluestone JA. Regulatory T-cell physiology and application to treat autoimmunity. *Immunol Rev* 2006;**212**:217–37.
34. Zhang X, Koldzic DN, Izikson L, et al. IL-10 is involved in the suppression of experimental autoimmune encephalomyelitis by CD25+CD4+ regulatory T cells. *Int Immunol* 2004;**16**:249–56.
35. Kohm AP, Carpentier PA, Anger HA, Miller SD. Cutting edge: CD4+CD25+ regulatory T cells suppress antigen-specific autoreactive immune responses and central nervous system inflammation during active experimental autoimmune encephalomyelitis. *J Immunol* 2002;**169**:4712–6.
36. Mottet C, Uhlig HH, Powrie F. Cutting edge: cure of colitis by CD4+CD25+ regulatory T cells. *J Immunol* 2003;**170**:3939–43.
37. Oldenhove G, Bouladoux N, Wohlfert EA, et al. Decrease of Foxp3+ Treg cell number and acquisition of effector cell phenotype during lethal infection. *Immunity* 2009;**31**:772–86.
38. Beriou G, Costantino CM, Ashley CW, et al. IL-17-producing human peripheral regulatory T cells retain suppressive function. *Blood* 2009;**113**:4240–9.
39. Koenen HJ, Smeets RL, Vink PM, van Rijssen E, Boots AM, Joosten I. Human CD25highFoxp3pos regulatory T cells differentiate into IL-17-producing cells. *Blood* 2008;**112**:2340–52.
40. Billiard F, Litvinova E, Saadoun D, et al. Regulatory and effector T cell activation levels are prime determinants of in vivo immune regulation. *J Immunol* 2006;**177**:2167–74.
41. Venigalla RK, Tretter T, Krienke S, et al. Reduced CD4+,CD25− T cell sensitivity to the suppressive function of CD4+,CD25high,CD127−/low regulatory T cells in patients with active systemic lupus erythematosus. *Arthritis Rheum* 2008;**58**:2120–30.
42. Ryba H, Zorena, Myseliwiec, Myseliwska. CD4+Foxp3+ regulatory T lymphocytes expressing CD62L in patients with long-standing diabetes type 1. *Central Eur J Immunol* 2009;**34**:4.
43. Darrasse-Jeze G, Marodon G, Salomon BL, Catala M, Klatzmann D. Ontogeny of CD4+CD25+ regulatory/suppressor T cells in human fetuses. *Blood* 2005;**105**:4715–21.
44. Takahata Y, Nomura A, Takada H, et al. CD25+CD4+ T cells in human cord blood: an immunoregulatory subset with naive phenotype and specific expression of forkhead box p3 (Foxp3) gene. *Exp Hematol* 2004;**32**:622–9.
45. Dong S, Maiella S, Xhaard A, et al. Multiparameter single-cell profiling of human CD4+FOXP3+ regulatory T-cell populations in homeostatic conditions and during graft-versus-host disease. *Blood* 2013;**122**:1802–12.
46. Thornton CA, Upham JW, Wikstrom ME, et al. Functional maturation of CD4+CD25+CTLA4+CD45RA+ T regulatory cells in human neonatal T cell responses to environmental antigens/allergens. *J Immunol* 2004;**173**:3084–92.
47. Valmori D, Merlo A, Souleimanian NE, Hesdorffer CS, Ayyoub M. A peripheral circulating compartment of natural naive CD4 Tregs. *J Clin Invest* 2005;**115**:1953–62.
48. Booth NJ, McQuaid AJ, Sobande T, et al. Different proliferative potential and migratory characteristics of human CD4+ regulatory T cells that express either CD45RA or CD45RO. *J Immunol* 2010;**184**:4317–26.
49. Milward K, Issa F, Hester J, Figueroa-Tentori D, Madrigal A, Wood KJ. Multiple unit pooled umbilical cord blood is a viable source of therapeutic regulatory T cells. *Transplantation* 2013;**95**:85–93.

50. Wing K, Lindgren S, Kollberg G, et al. CD4 T cell activation by myelin oligodendrocyte glycoprotein is suppressed by adult but not cord blood CD25$^+$ T cells. *Eur J Immunol* 2003;**33**:579–87.
51. Fujimaki W, Takahashi N, Ohnuma K, et al. Comparative study of regulatory T cell function of human CD25CD4 T cells from thymocytes, cord blood, and adult peripheral blood. *Clin Dev Immunol* 2008;**2008**:305859.
52. Fan H, Yang J, Hao J, et al. Comparative study of regulatory T cells expanded ex vivo from cord blood and adult peripheral blood. *Immunology* 2012;**136**:218–30.
53. Hippen KL, Harker-Murray P, Porter SB, et al. Umbilical cord blood regulatory T-cell expansion and functional effects of tumor necrosis factor receptor family members OX40 and 4-1BB expressed on artificial antigen-presenting cells. *Blood* 2008;**112**:2847–57.
54. Godfrey WR, Spoden DJ, Ge YG, et al. Cord blood CD4(+)CD25(+)-derived T regulatory cell lines express FoxP3 protein and manifest potent suppressor function. *Blood* 2005;**105**:750–8.
55. Bresatz S, Sadlon T, Millard D, Zola H, Barry SC. Isolation, propagation and characterization of cord blood derived CD4$^+$ CD25$^+$ regulatory T cells. *J Immunol Methods* 2007;**327**:53–62.
56. Figueroa-Tentori D, Querol S, Dodi IA, Madrigal A, Duggleby R. High purity and yield of natural Tregs from cord blood using a single step selection method. *J Immunol Methods* 2008;**339**:228–35.
57. Di Ianni M, Del Papa B, Zei T, et al. T regulatory cell separation for clinical application. *Transfus Apher Sci* 2012;**47**:213–6.
58. Hoffmann P, Eder R, Kunz-Schughart LA, Andreesen R, Edinger M. Large-scale in vitro expansion of polyclonal human CD4(+) CD25high regulatory T cells. *Blood* 2004;**104**:895–903.
59. Duhen T, Duhen R, Lanzavecchia A, Sallusto F, Campbell DJ. Functionally distinct subsets of human FOXP3$^+$ Treg cells that phenotypically mirror effector Th cells. *Blood* 2012;**119**: 4430–40.
60. Vukmanovic-Stejic M, Zhang Y, Cook JE, et al. Human CD4$^+$ CD25hi Foxp3$^+$ regulatory T cells are derived by rapid turnover of memory populations in vivo. *J Clin Invest* 2006;**116**:2423–33.
61. Ermann J, Hoffmann P, Edinger M, et al. Only the CD62L$^+$ subpopulation of CD4$^+$CD25$^+$ regulatory T cells protects from lethal acute GVHD. *Blood* 2005;**105**:2220–6.
62. Taylor PA, Panoskaltsis-Mortari A, Swedin JM, et al. L-Selectin(hi) but not the L-selectin(lo) CD4$^+$25$^+$ T-regulatory cells are potent inhibitors of GVHD and BM graft rejection. *Blood* 2004;**104**: 3804–12.
63. Hoffmann P, Eder R, Boeld TJ, et al. Only the CD45RA$^+$ subpopulation of CD4$^+$CD25high T cells gives rise to homogeneous regulatory T-cell lines upon in vitro expansion. *Blood* 2006;**108**:4260–7.
64. Parmar S, Liu X, Tung SS, et al. Third-party umbilical cord blood-derived regulatory T cells prevent xenogenic graft-versus-host disease. *Cytotherapy* 2014;**16**:90–100.
65. Battaglia M, Stabilini A, Roncarolo MG. Rapamycin selectively expands CD4$^+$CD25$^+$FoxP3$^+$ regulatory T cells. *Blood* 2005;**105**:4743–8.
66. Tang Q, Henriksen KJ, Bi M, et al. In vitro-expanded antigen-specific regulatory T cells suppress autoimmune diabetes. *J Exp Med* 2004;**199**:1455–65.
67. Tung KS, Setiady YY, Samy ET, Lewis J, Teuscher C. Autoimmune ovarian disease in day 3-thymectomized mice: the neonatal time window, antigen specificity of disease suppression, and genetic control. *Curr Top Microbiol Immunol* 2005;**293**:209–47.
68. Jiang S, Tsang J, Game DS, Stevenson S, Lombardi G, Lechler RI. Generation and expansion of human CD4$^+$ CD25$^+$ regulatory T cells with indirect allospecificity: potential reagents to promote donor-specific transplantation tolerance. *Transplantation* 2006;**82**:1738–43.
69. Tsang J, Jiang S, Tanriver Y, Leung E, Lombardi G, Lechler RI. In-vitro generation and characterisation of murine CD4$^+$CD25$^+$ regulatory T cells with indirect allospecificity. *Int Immunopharmacol* 2006;**6**:1883–8.
70. Torelli GF, Maggio R, Peragine N, et al. Functional analysis and gene expression profile of umbilical cord blood regulatory T cells. *Ann Hematol* 2012;**91**:155–61.
71. Lin SJ, Lu CH, Yan DC, Lee PT, Hsiao HS, Kuo ML. Expansion of regulatory T cells from umbilical cord blood and adult peripheral blood CD4CD25 T cells. *Immunol Res* 2014;**60**(1):105–11.
72. Hoffmann P, Boeld TJ, Eder R, et al. Loss of FOXP3 expression in natural human CD4$^+$CD25$^+$ regulatory T cells upon repetitive in vitro stimulation. *Eur J Immunol* 2009;**39**:1088–97.
73. Walker MR, Carson BD, Nepom GT, Ziegler SF, Buckner JH. De novo generation of antigen-specific CD4$^+$CD25$^+$ regulatory T cells from human CD4$^+$CD25$^-$ cells. *Proc Natl Acad Sci USA* 2005;**102**:4103–8.
74. Tran DQ, Ramsey H, Shevach EM. Induction of FOXP3 expression in naive human CD4$^+$FOXP3 T cells by T-cell receptor stimulation is transforming growth factor-beta dependent but does not confer a regulatory phenotype. *Blood* 2007;**110**:2983–90.
75. Marek N, Bieniaszewska M, Krzystyniak A, et al. The time is crucial for ex vivo expansion of T regulatory cells for therapy. *Cell Transpl* 2011;**20**:1747–58.
76. June CH, Blazar BR. Clinical application of expanded CD4$^+$25$^+$ cells. *Semin Immunol* 2006;**18**:78–88.
77. Basu S, Golovina T, Mikheeva T, June CH, Riley JL. Cutting edge: foxp3-mediated induction of pim 2 allows human T regulatory cells to preferentially expand in rapamycin. *J Immunol* 2008;**180**:5794–8.
78. Rao PE, Petrone AL, Ponath PD. Differentiation and expansion of T cells with regulatory function from human peripheral lymphocytes by stimulation in the presence of TGF-{β}. *J Immunol* 2005;**174**: 1446–55.
79. Battaglia M, Stabilini A, Migliavacca B, Horejs-Hoeck J, Kaupper T, Roncarolo MG. Rapamycin promotes expansion of functional CD4$^+$CD25$^+$FOXP3$^+$ regulatory T cells of both healthy subjects and type 1 diabetic patients. *J Immunol* 2006;**177**:8338–47.
80. Hutton JF, Gargett T, Sadlon TJ, et al. Development of CD4$^+$CD25$^+$FoxP3$^+$ regulatory T cells from cord blood hematopoietic progenitor cells. *J Leukoc Biol* 2009;**85**:445–51.
81. Curotto de Lafaille MA, Lafaille JJ. Natural and adaptive foxp3$^+$ regulatory T cells: more of the same or a division of labor? *Immunity* 2009;**30**:626–35.
82. Chang M, Suen Y, Lee SM, et al. Transforming growth factor-beta 1, macrophage inflammatory protein-1 alpha, and interleukin-8 gene expression is lower in stimulated human neonatal compared with adult mononuclear cells. *Blood* 1994;**84**:118–24.
83. Lee SM, Suen Y, Chang L, et al. Decreased interleukin-12 (IL-12) from activated cord versus adult peripheral blood mononuclear cells and upregulation of interferon-gamma, natural killer, and lymphokine-activated killer activity by IL-12 in cord blood mononuclear cells. *Blood* 1996;**88**:945–54.
84. Ohkura N, Kitagawa Y, Sakaguchi S. Development and maintenance of regulatory T cells. *Immunity* 2013;**38**:414–23.

85. Chen W, Jin W, Hardegen N, et al. Conversion of peripheral $CD4^+CD25^-$ naive T cells to $CD4^+CD25^+$ regulatory T cells by TGF-beta induction of transcription factor Foxp3. *J Exp Med* 2003;**198**:1875–86.
86. Huter EN, Stummvoll GH, DiPaolo RJ, Glass DD, Shevach EM. Cutting edge: antigen-specific TGF beta-induced regulatory T cells suppress Th17-mediated autoimmune disease. *J Immunol* 2008;**181**:8209–13.
87. Huter EN, Punkosdy GA, Glass DD, Cheng LI, Ward JM, Shevach EM. TGF-beta-induced $Foxp3^+$ regulatory T cells rescue scurfy mice. *Eur J Immunol* 2008;**38**:1814–21.
88. Hippen KL, Merkel SC, Schirm DK, et al. Generation and large-scale expansion of human inducible regulatory T cells that suppress graft-versus-host disease. *Am J Transpl* 2011;**11**:1148–57.
89. Mucida D, Kutchukhidze N, Erazo A, Russo M, Lafaille JJ, Curotto de Lafaille MA. Oral tolerance in the absence of naturally occurring Tregs. *J Clin Invest* 2005;**115**:1923–33.
90. Haribhai D, Lin W, Edwards B, et al. A central role for induced regulatory T cells in tolerance induction in experimental colitis. *J Immunol* 2009;**182**:3461–8.
91. Haribhai D, Williams JB, Jia S, et al. A requisite role for induced regulatory T cells in tolerance based on expanding antigen receptor diversity. *Immunity* 2011;**35**:109–22.
92. Hill JA, Feuerer M, Tash K, et al. Foxp3 transcription-factor-dependent and -independent regulation of the regulatory T cell transcriptional signature. *Immunity* 2007;**27**:786–800.
93. Selvaraj RK, Geiger TL. A kinetic and dynamic analysis of Foxp3 induced in T cells by TGF-beta. *J Immunol* 2007;**179**:11. p following 1390.
94. Meier-Kriesche HU, Schold JD, Kaplan B. Long-term renal allograft survival: have we made significant progress or is it time to rethink our analytic and therapeutic strategies? *Am J Transpl* 2004;**4**:1289–95.
95. Wise MP, Bemelman F, Cobbold SP, Waldmann H. Linked suppression of skin graft rejection can operate through indirect recognition. *J Immunol* 1998;**161**:5813–6.
96. Yamada A, Chandraker A, Laufer TM, Gerth AJ, Sayegh MH, Auchincloss Jr H. Recipient MHC class II expression is required to achieve long-term survival of murine cardiac allografts after costimulatory blockade. *J Immunol* 2001;**167**:5522–6.
97. Hara M, Kingsley CI, Niimi M, et al. IL-10 is required for regulatory T cells to mediate tolerance to alloantigens in vivo. *J Immunol* 2001;**166**:3789–96.
98. Spadafora-Ferreira M, Caldas C, Fae KC, et al. $CD4^+CD25^+Foxp3^+$ indirect alloreactive T cells from renal transplant patients suppress both the direct and indirect pathways of allorecognition. *Scand J Immunol* 2007;**66**:352–61.
99. Ensminger SM, Spriewald BM, Witzke O, et al. Indirect allorecognition can play an important role in the development of transplant arteriosclerosis. *Transplantation* 2002;**73**:279–86.
100. Hornick PI, Mason PD, Baker RJ, et al. Significant frequencies of T cells with indirect anti-donor specificity in heart graft recipients with chronic rejection. *Circulation* 2000;**101**:2405–10.
101. Vella JP, Spadafora-Ferreira M, Murphy B, et al. Indirect allorecognition of major histocompatibility complex allopeptides in human renal transplant recipients with chronic graft dysfunction. *Transplantation* 1997;**64**:795–800.
102. Ciubotariu R, Liu Z, Colovai AI, et al. Persistent allopeptide reactivity and epitope spreading in chronic rejection of organ allografts. *J Clin Invest* 1998;**101**:398–405.
103. Ferrara JL, Levine JE, Reddy P, Holler E. Graft-versus-host disease. *Lancet* 2009;**373**:1550–61.
104. Welniak LA, Blazar BR, Murphy WJ. Immunobiology of allogeneic hematopoietic stem cell transplantation. *Annu Rev Immunol* 2007;**25**:139–70.
105. Edinger M, Hoffmann P. Regulatory T cells in stem cell transplantation: strategies and first clinical experiences. *Curr Opin Immunol* 2011;**23**:679–84.
106. Di Ianni M, Falzetti F, Carotti A, et al. Tregs prevent GVHD and promote immune reconstitution in HLA-haploidentical transplantation. *Blood* 2011;**117**:3921–8.
107. Brunstein CG, Blazar BR, Miller JS, et al. Adoptive transfer of umbilical cord blood-derived regulatory T cells and early viral reactivation. *Biol Blood Marrow Transpl* 2013;**19**:1271–3.
108. Bluestone JA, Tang Q, Sedwick CE. T regulatory cells in autoimmune diabetes: past challenges, future prospects. *J Clin Immunol* 2008;**28**:677–84.
109. Marek-Trzonkowska N, Mysliwec M, Siebert J, Trzonkowski P. Clinical application of regulatory T cells in type 1 diabetes. *Pediatr Diabetes* 2013;**14**:322–32.
110. Salomon B, Lenschow DJ, Rhee L, et al. B7/CD28 costimulation is essential for the homeostasis of the $CD4^+CD25^+$ immunoregulatory T cells that control autoimmune diabetes. *Immunity* 2000;**12**: 431–40.
111. Masteller EL, Warner MR, Tang Q, Tarbell KV, McDevitt H, Bluestone JA. Expansion of functional endogenous antigen-specific $CD4^+CD25^+$ regulatory T cells from nonobese diabetic mice. *J Immunol* 2005;**175**:3053–9.
112. Tarbell KV, Petit L, Zuo X, et al. Dendritic cell-expanded, islet-specific $CD4^+$ $CD25^+$ $CD62L^+$ regulatory T cells restore normoglycemia in diabetic NOD mice. *J Exp Med* 2007;**204**:191–201.
113. Marek-Trzonkowska N, Mysliwiec M, Dobyszuk A, et al. Administration of $CD4^+CD25^{high}CD127^-$ regulatory T cells preserves beta-cell function in type 1 diabetes in children. *Diabetes Care* 2012;**35**:1817–20.
114. Marigo I, Dazzi F. The immunomodulatory properties of mesenchymal stem cells. *Semin Immunopathol* 2011;**33**:593–602.
115. Dominici M, Le Blanc K, Mueller I, et al. Minimal criteria for defining multipotent mesenchymal stromal cells. The international society for cellular therapy position statement. *Cytotherapy* 2006;**8**:315–7.
116. Sivasubramaniyan K, Lehnen D, Ghazanfari R, et al. Phenotypic and functional heterogeneity of human bone marrow- and amnion-derived MSC subsets. *Ann NY Acad Sci* 2012;**1266**:94–106.
117. Mets T, Verdonk G. In vitro aging of human bone marrow derived stromal cells. *Mech Ageing Dev* 1981;**16**:81–9.
118. Phinney DG, Kopen G, Righter W, Webster S, Tremain N, Prockop DJ. Donor variation in the growth properties and osteogenic potential of human marrow stromal cells. *J Cell Biochem* 1999;**75**:424–36.
119. Pittenger MF, Mackay AM, Beck SC, et al. Multilineage potential of adult human mesenchymal stem cells. *Science* 1999;**284**:143–7.
120. Pevsner-Fischer M, Levin S, Zipori D. The origins of mesenchymal stromal cell heterogeneity. *Stem Cell Rev* 2011;**7**:560–8.
121. Gronthos S, Mankani M, Brahim J, Robey PG, Shi S. Postnatal human dental pulp stem cells (DPSCs) in vitro and in vivo. *Proc Natl Acad Sci USA* 2000;**97**:13625–30.
122. Young HE, Steele TA, Bray RA, et al. Human reserve pluripotent mesenchymal stem cells are present in the connective tissues of skeletal muscle and dermis derived from fetal, adult, and geriatric donors. *Anat Rec* 2001;**264**:51–62.

123. Zannettino AC, Paton S, Arthur A, et al. Multipotential human adipose-derived stromal stem cells exhibit a perivascular phenotype in vitro and in vivo. *J Cell Physiol* 2008;**214**:413–21.
124. Erices A, Conget P, Minguell JJ. Mesenchymal progenitor cells in human umbilical cord blood. *Br J Haematol* 2000;**109**:235–42.
125. Friedman R, Betancur M, Boissel L, Tuncer H, Cetrulo C, Klingemann H. Umbilical cord mesenchymal stem cells: adjuvants for human cell transplantation. *Biol Blood Marrow Transpl* 2007;**13**:1477–86.
126. Roy S, Arora S, Kumari P, Ta M. A simple and serum-free protocol for cryopreservation of human umbilical cord as source of Wharton's jelly mesenchymal stem cells. *Cryobiology* 2014;**68**:467–72.
127. Tondreau T, Meuleman N, Delforge A, et al. Mesenchymal stem cells derived from CD133-positive cells in mobilized peripheral blood and cord blood: proliferation, Oct4 expression, and plasticity. *Stem Cells* 2005;**23**:1105–12.
128. Bieback K, Kluter H. Mesenchymal stromal cells from umbilical cord blood. *Curr Stem Cell Res Ther* 2007;**2**:310–23.
129. Romanov YA, Svintsitskaya VA, Smirnov VN. Searching for alternative sources of postnatal human mesenchymal stem cells: candidate MSC-like cells from umbilical cord. *Stem Cells* 2003;**21**:105–10.
130. Lu LL, Liu YJ, Yang SG, et al. Isolation and characterization of human umbilical cord mesenchymal stem cells with hematopoiesis-supportive function and other potentials. *Haematologica* 2006;**91**:1017–26.
131. Sarugaser R, Lickorish D, Baksh D, Hosseini MM, Davies JE. Human umbilical cord perivascular (HUCPV) cells: a source of mesenchymal progenitors. *Stem Cells* 2005;**23**:220–9.
132. Secco M, Zucconi E, Vieira NM, et al. Multipotent stem cells from umbilical cord: cord is richer than blood! *Stem Cells* 2008; **26**:146–50.
133. Kern S, Eichler H, Stoeve J, Kluter H, Bieback K. Comparative analysis of mesenchymal stem cells from bone marrow, umbilical cord blood, or adipose tissue. *Stem Cells* 2006;**24**:1294–301.
134. Rebelatto CK, Aguiar AM, Moretao MP, et al. Dissimilar differentiation of mesenchymal stem cells from bone marrow, umbilical cord blood, and adipose tissue. *Exp Biol Med (Maywood)* 2008;**233**:901–13.
135. Hsieh JY, Fu YS, Chang SJ, Tsuang YH, Wang HW. Functional module analysis reveals differential osteogenic and stemness potentials in human mesenchymal stem cells from bone marrow and Wharton's jelly of umbilical cord. *Stem Cells Dev* 2010;**19**:1895–910.
136. Manochantr S, Up Y, Kheolamai P, et al. Immunosuppressive properties of mesenchymal stromal cells derived from amnion, placenta, Wharton's jelly and umbilical cord. *Intern Med J* 2013;**43**:430–9.
137. Nekanti U, Rao VB, Bahirvani AG, Jan M, Totey S, Ta M. Long-term expansion and pluripotent marker array analysis of Wharton's jelly-derived mesenchymal stem cells. *Stem Cells Dev* 2010;**19**:117–30.
138. Panepucci RA, Siufi JL, Silva Jr WA, et al. Comparison of gene expression of umbilical cord vein and bone marrow-derived mesenchymal stem cells. *Stem Cells* 2004;**22**:1263–78.
139. Flynn A, Barry F, O'Brien T. UC blood-derived mesenchymal stromal cells: an overview. *Cytotherapy* 2007;**9**:717–26.
140. Boissel L, Tuncer HH, Betancur M, Wolfberg A, Klingemann H. Umbilical cord mesenchymal stem cells increase expansion of cord blood natural killer cells. *Biol Blood Marrow Transpl* 2008;**14**:1031–8.
141. Secco M, Moreira YB, Zucconi E, et al. Gene expression profile of mesenchymal stem cells from paired umbilical cord units: cord is different from blood. *Stem Cell Rev* 2009;**5**:387–401.
142. Colter DC, Class R, DiGirolamo CM, Prockop DJ. Rapid expansion of recycling stem cells in cultures of plastic-adherent cells from human bone marrow. *Proc Natl Acad Sci USA* 2000;**97**: 3213–8.
143. Colter DC, Sekiya I, Prockop DJ. Identification of a subpopulation of rapidly self-renewing and multipotential adult stem cells in colonies of human marrow stromal cells. *Proc Natl Acad Sci USA* 2001;**98**:7841–5.
144. Lee RH, Seo MJ, Pulin AA, Gregory CA, Ylostalo J, Prockop DJ. The CD34-like protein PODXL and alpha6-integrin (CD49f) identify early progenitor MSCs with increased clonogenicity and migration to infarcted heart in mice. *Blood* 2009;**113**:816–26.
145. Lange C, Cakiroglu F, Spiess AN, Cappallo-Obermann H, Dierlamm J, Zander AR. Accelerated and safe expansion of human mesenchymal stromal cells in animal serum-free medium for transplantation and regenerative medicine. *J Cell Physiol* 2007;**213**:18–26.
146. Abdelrazik H, Spaggiari GM, Chiossone L, Moretta L. Mesenchymal stem cells expanded in human platelet lysate display a decreased inhibitory capacity on T- and NK-cell proliferation and function. *Eur J Immunol* 2011;**41**:3281–90.
147. Bieback K, Kinzebach S, Karagianni M. Translating research into clinical scale manufacturing of mesenchymal stromal cells. *Stem Cells Int* 2011;**2010**:193519.
148. Maijenburg MW, Noort WA, Kleijer M, et al. Cell cycle and tissue of origin contribute to the migratory behaviour of human fetal and adult mesenchymal stromal cells. *Br J Haematol* 2010;**148**: 428–40.
149. Marigo I, Dazzi F. The immunomodulatory properties of mesenchymal stem cells. *Semin Immunopathol* 2011;**33**(6):593–602.
150. Krampera M, Cosmi L, Angeli R, et al. Role for interferon-gamma in the immunomodulatory activity of human bone marrow mesenchymal stem cells. *Stem Cells* 2006;**24**:386–98.
151. Polchert D, Sobinsky J, Douglas G, et al. IFN-gamma activation of mesenchymal stem cells for treatment and prevention of graft versus host disease. *Eur J Immunol* 2008;**38**:1745–55.
152. Ren G, Zhang L, Zhao X, et al. Mesenchymal stem cell-mediated immunosuppression occurs via concerted action of chemokines and nitric oxide. *Cell Stem Cell* 2008;**2**:141–50.
153. Stagg J, Pommey S, Eliopoulos N, Galipeau J. Interferon-gamma-stimulated marrow stromal cells: a new type of nonhematopoietic antigen-presenting cell. *Blood* 2006;**107**:2570–7.
154. Chan WK, Lau AS, Li JC, Law HK, Lau YL, Chan GC. MHC expression kinetics and immunogenicity of mesenchymal stromal cells after short-term IFN-gamma challenge. *Exp Hematol* 2008;**36**:1545–55.
155. Liotta F, Angeli R, Cosmi L, et al. Toll-like receptors 3 and 4 are expressed by human bone marrow-derived mesenchymal stem cells and can inhibit their T-cell modulatory activity by impairing Notch signaling. *Stem Cells* 2008;**26**:279–89.
156. Yoo KH, Jang IK, Lee MW, et al. Comparison of immunomodulatory properties of mesenchymal stem cells derived from adult human tissues. *Cell Immunol* 2009;**259**:150–6.
157. Wu KH, Chan CK, Tsai C, et al. Effective treatment of severe steroid-resistant acute graft-versus-host disease with umbilical cord-derived mesenchymal stem cells. *Transplantation* 2011;**91**:1412–6.
158. Meisel R, Zibert A, Laryea M, Gobel U, Daubener W, Dilloo D. Human bone marrow stromal cells inhibit allogeneic T-cell responses by indoleamine 2,3-dioxygenase-mediated tryptophan degradation. *Blood* 2004;**103**:4619–21.

159. Selmani Z, Naji A, Zidi I, et al. Human leukocyte antigen-G5 secretion by human mesenchymal stem cells is required to suppress T lymphocyte and natural killer function and to induce $CD4^{+}CD25^{high}FOXP3^{+}$ regulatory T cells. *Stem Cells* 2008;**26**:212–22.
160. Bouffi C, Bony C, Courties G, Jorgensen C, Noel D. IL-6-dependent PGE2 secretion by mesenchymal stem cells inhibits local inflammation in experimental arthritis. *PLoS One* 2010;**5**:e14247.
161. Raicevic G, Najar M, Stamatopoulos B, et al. The source of human mesenchymal stromal cells influences their TLR profile as well as their functional properties. *Cell Immunol* 2011;**270**:207–16.
162. Nemeth K, Leelahavanichkul A, Yuen PS, et al. Bone marrow stromal cells attenuate sepsis via prostaglandin E(2)-dependent reprogramming of host macrophages to increase their interleukin-10 production. *Nat Med* 2009;**15**:42–9.
163. Gholamrezanezhad A, Mirpour S, Bagheri M, et al. In vivo tracking of 111In-oxine labeled mesenchymal stem cells following infusion in patients with advanced cirrhosis. *Nucl Med Biol* 2011;**38**:961–7.
164. Kidd S, Spaeth E, Dembinski JL, et al. Direct evidence of mesenchymal stem cell tropism for tumor and wounding microenvironments using in vivo bioluminescent imaging. *Stem Cells* 2009;**27**:2614–23.
165. Ruster B, Gottig S, Ludwig RJ, et al. Mesenchymal stem cells display coordinated rolling and adhesion behavior on endothelial cells. *Blood* 2006;**108**:3938–44.
166. Harting MT, Jimenez F, Xue H, et al. Intravenous mesenchymal stem cell therapy for traumatic brain injury. *J Neurosurg* 2009;**110**:1189–97.
167. Ankrum JA, Ong JF, Karp JM. Mesenchymal stem cells: immune evasive, not immune privileged. *Nat Biotechnol* 2014;**32**:252–60.
168. Schrepfer S, Deuse T, Reichenspurner H, Fischbein MP, Robbins RC, Pelletier MP. Stem cell transplantation: the lung barrier. *Transpl Proc* 2007;**39**:573–6.
169. Gao J, Dennis JE, Muzic RF, Lundberg M, Caplan AI. The dynamic in vivo distribution of bone marrow-derived mesenchymal stem cells after infusion. *Cells Tissues Organs* 2001;**169**:12–20.
170. Nystedt J, Anderson H, Tikkanen J, et al. Cell surface structures influence lung clearance rate of systemically infused mesenchymal stromal cells. *Stem Cells* 2013;**31**:317–26.
171. Le Blanc K, Tammik L, Sundberg B, Haynesworth SE, Ringden O. Mesenchymal stem cells inhibit and stimulate mixed lymphocyte cultures and mitogenic responses independently of the major histocompatibility complex. *Scand J Immunol* 2003;**57**:11–20.
172. Horwitz EM, Gordon PL, Koo WK, et al. Isolated allogeneic bone marrow-derived mesenchymal cells engraft and stimulate growth in children with osteogenesis imperfecta: Implications for cell therapy of bone. *Proc Natl Acad Sci USA* 2002;**99**:8932–7.
173. Arinzeh TL, Peter SJ, Archambault MP, et al. Allogeneic mesenchymal stem cells regenerate bone in a critical-sized canine segmental defect. *J Bone Jt Surg Am* 2003;**85-A**:1927–35.
174. Koc ON, Day J, Nieder M, Gerson SL, Lazarus HM, Krivit W. Allogeneic mesenchymal stem cell infusion for treatment of metachromatic leukodystrophy (MLD) and Hurler syndrome (MPS-IH). *Bone Marrow Transpl* 2002;**30**:215–22.
175. Fouillard L, Bensidhoum M, Bories D, et al. Engraftment of allogeneic mesenchymal stem cells in the bone marrow of a patient with severe idiopathic aplastic anemia improves stroma. *Leukemia* 2003;**17**:474–6.
176. Cahill RA, Jones OY, Klemperer M, et al. Replacement of recipient stromal/mesenchymal cells after bone marrow transplantation using bone fragments and cultured osteoblast-like cells. *Biol Blood Marrow Transpl* 2004;**10**:709–17.
177. Eliopoulos N, Stagg J, Lejeune L, Pommey S, Galipeau J. Allogeneic marrow stromal cells are immune rejected by MHC class I- and class II-mismatched recipient mice. *Blood* 2005;**106**:4057–65.
178. Nauta AJ, Westerhuis G, Kruisselbrink AB, Lurvink EG, Willemze R, Fibbe WE. Donor-derived mesenchymal stem cells are immunogenic in an allogeneic host and stimulate donor graft rejection in a nonmyeloablative setting. *Blood* 2006;**108**:2114–20.
179. Zangi L, Margalit R, Reich-Zeliger S, et al. Direct imaging of immune rejection and memory induction by allogeneic mesenchymal stromal cells. *Stem Cells* 2009;**27**:2865–74.
180. von Bahr L, Batsis I, Moll G, et al. Analysis of tissues following mesenchymal stromal cell therapy in humans indicates limited long-term engraftment and no ectopic tissue formation. *Stem Cells* 2012;**30**:1575–8.
181. Le Blanc K, Frassoni F, Ball L, et al. Mesenchymal stem cells for treatment of steroid-resistant, severe, acute graft-versus-host disease: a phase II study. *Lancet* 2008;**371**:1579–86.
182. Zanotti L, Sarukhan A, Dander E, et al. Encapsulated mesenchymal stem cells for in vivo immunomodulation. *Leukemia* 2013;**27**:500–3.
183. Bertram H, Mayer H, Schliephake H. Effect of donor characteristics, technique of harvesting and in vitro processing on culturing of human marrow stroma cells for tissue engineered growth of bone. *Clin Oral Implants Res* 2005;**16**:524–31.
184. Stolzing A, Jones E, McGonagle D, Scutt A. Age-related changes in human bone marrow-derived mesenchymal stem cells: consequences for cell therapies. *Mech Ageing Dev* 2008;**129**:163–73.
185. Medyouf H, Mossner M, Jann JC, Nolte F, Raffel S, Herrmann C, et al. Myelodysplastic cells in patients reprogram mesenchymal stromal cells to establish a transplantable stem cell niche disease unit. *Cell Stem Cell* 2014;**14**(6):824–37.
186. Blau O, Baldus CD, Hofmann WK, et al. Mesenchymal stromal cells of myelodysplastic syndrome and acute myeloid leukemia patients have distinct genetic abnormalities compared with leukemic blasts. *Blood* 2011;**118**:5583–92.
187. Avanzini MA, Bernardo ME, Novara F, Mantelli M, Poletto V, Villani L, et al. Functional and genetic aberrations of in vitro-cultured marrow-derived mesenchymal stromal cells of patients with classical Philadelphia-negative myeloproliferative neoplasms. *Leukemia* 2014;**28**(8):1742–5.
188. Anklesaria P, Kase K, Glowacki J, et al. Engraftment of a clonal bone marrow stromal cell line in vivo stimulates hematopoietic recovery from total body irradiation. *Proc Natl Acad Sci USA* 1987;**84**:7681–5.
189. Majumdar MK, Thiede MA, Mosca JD, Moorman M, Gerson SL. Phenotypic and functional comparison of cultures of marrow-derived mesenchymal stem cells (MSCs) and stromal cells. *J Cell Physiol* 1998;**176**:57–66.
190. Koc ON, Gerson SL, Cooper BW, et al. Rapid hematopoietic recovery after coinfusion of autologous-blood stem cells and culture-expanded marrow mesenchymal stem cells in advanced breast cancer patients receiving high-dose chemotherapy. *J Clin Oncol* 2000;**18**:307–16.
191. Lazarus HM, Koc ON, Devine SM, et al. Cotransplantation of HLA-identical sibling culture-expanded mesenchymal stem cells and hematopoietic stem cells in hematologic malignancy patients. *Biol Blood Marrow Transpl* 2005;**11**:389–98.
192. Almeida-Porada G, Porada CD, Tran N, Zanjani ED. Cotransplantation of human stromal cell progenitors into preimmune fetal sheep results in early appearance of human donor cells in circulation and

boosts cell levels in bone marrow at later time points after transplantation. *Blood* 2000;**95**:3620–7.

193. Noort WA, Kruisselbrink AB, in't Anker PS, et al. Mesenchymal stem cells promote engraftment of human umbilical cord blood-derived CD34(+) cells in NOD/SCID mice. *Exp Hematol* 2002;**30**:870–8.
194. Macmillan ML, Blazar BR, DeFor TE, Wagner JE. Transplantation of ex-vivo culture-expanded parental haploidentical mesenchymal stem cells to promote engraftment in pediatric recipients of unrelated donor umbilical cord blood: results of a phase I-II clinical trial. *Bone Marrow Transpl* 2009;**43**:447–54.
195. Le Blanc K, Samuelsson H, Gustafsson B, et al. Transplantation of mesenchymal stem cells to enhance engraftment of hematopoietic stem cells. *Leukemia* 2007;**21**:1733–8.
196. Poloni A, Leoni P, Buscemi L, et al. Engraftment capacity of mesenchymal cells following hematopoietic stem cell transplantation in patients receiving reduced-intensity conditioning regimen. *Leukemia* 2006;**20**:329–35.
197. Guo M, Sun Z, Sun QY, et al. A modified haploidentical nonmyeloablative transplantation without T cell depletion for high-risk acute leukemia: successful engraftment and mild GVHD. *Biol Blood Marrow Transpl* 2009;**15**:930–7.
198. Ball LM, Bernardo ME, Roelofs H, et al. Cotransplantation of ex vivo expanded mesenchymal stem cells accelerates lymphocyte recovery and may reduce the risk of graft failure in haploidentical hematopoietic stem-cell transplantation. *Blood* 2007;**110**: 2764–7.
199. Wu Y, Wang Z, Cao Y, et al. Cotransplantation of haploidentical hematopoietic and umbilical cord mesenchymal stem cells with a myeloablative regimen for refractory/relapsed hematologic malignancy. *Ann Hematol* 2013;**92**:1675–84.
200. Wu Y, Cao Y, Li X, et al. Cotransplantation of haploidentical hematopoietic and umbilical cord mesenchymal stem cells for severe aplastic anemia: successful engraftment and mild GVHD. *Stem Cell Res* 2014;**12**:132–8.
201. Bernardo ME, Ball LM, Cometa AM, et al. Co-infusion of ex vivo-expanded, parental MSCs prevents life-threatening acute GVHD, but does not reduce the risk of graft failure in pediatric patients undergoing allogeneic umbilical cord blood transplantation. *Bone Marrow Transpl* 2011;**46**:200–7.
202. Gonzalo-Daganzo R, Regidor C, Martin-Donaire T, et al. Results of a pilot study on the use of third-party donor mesenchymal stromal cells in cord blood transplantation in adults. *Cytotherapy* 2009;**11**:278–88.
203. Bartholomew A, Sturgeon C, Siatskas M, et al. Mesenchymal stem cells suppress lymphocyte proliferation in vitro and prolong skin graft survival in vivo. *Exp Hematol* 2002;**30**:42–8.
204. Pileggi A, Xu X, Tan J, Ricordi C. Mesenchymal stromal (stem) cells to improve solid organ transplant outcome: lessons from the initial clinical trials. *Curr Opin Organ Transpl* 2013;**18**:672–81.
205. Tan J, Wu W, Xu X, et al. Induction therapy with autologous mesenchymal stem cells in living-related kidney transplants: a randomized controlled trial. *JAMA* 2012;**307**:1169–77.
206. Reinders ME, de Fijter JW, Roelofs H, et al. Autologous bone marrow-derived mesenchymal stromal cells for the treatment of allograft rejection after renal transplantation: results of a phase I study. *Stem Cells Transl Med* 2013;**2**:107–11.
207. Prasad VK, Lucas KG, Kleiner GI, et al. Efficacy and safety of ex vivo cultured adult human mesenchymal stem cells (Prochymal) in pediatric patients with severe refractory acute graft-versus-host disease in a compassionate use study. *Biol Blood Marrow Transpl* 2011;**17**:534–41.
208. Ball LM, Bernardo ME, Roelofs H, et al. Multiple infusions of mesenchymal stromal cells induce sustained remission in children with steroid-refractory, grade III-IV acute graft-versus-host disease. *Br J Haematol* 2013;**163**:501–9.
209. Kurtzberg J, Prockop S, Teira P, et al. Allogeneic human mesenchymal stem cell therapy (remestemcel-L, Prochymal) as a rescue agent for severe refractory acute graft-versus-host disease in pediatric patients. *Biol Blood Marrow Transpl* 2014;**20**:229–35.
210. Chen GH, Yang T, Tian H, et al. Clinical study of umbilical cord-derived mesenchymal stem cells for treatment of nineteen patients with steroid-resistant severe acute graft-versus-host disease. *Zhonghua Xue Ye Xue Za Zhi* 2012;**33**:303–6.
211. Arima N, Nakamura F, Fukunaga A, et al. Single intra-arterial injection of mesenchymal stromal cells for treatment of steroid-refractory acute graft-versus-host disease: a pilot study. *Cytotherapy* 2010;**12**:265–8.
212. Ringden O, Uzunel M, Rasmusson I, et al. Mesenchymal stem cells for treatment of therapy-resistant graft-versus-host disease. *Transplantation* 2006;**81**:1390–7.
213. Dander E, Lucchini G, Vinci P, et al. Mesenchymal stromal cells for the treatment of graft-versus-host disease: understanding the in vivo biological effect through patient immune monitoring. *Leukemia* 2012;**26**:1681–4.
214. Muller I, Kordowich S, Holzwarth C, et al. Application of multipotent mesenchymal stromal cells in pediatric patients following allogeneic stem cell transplantation. *Blood Cells Mol Dis* 2008;**40**:25–32.
215. Fang B, Song Y, Lin Q, et al. Human adipose tissue-derived mesenchymal stromal cells as salvage therapy for treatment of severe refractory acute graft-versus-host disease in two children. *Pediatr Transpl* 2007;**11**:814–7.
216. Fang B, Song Y, Liao L, Zhang Y, Zhao RC. Favorable response to human adipose tissue-derived mesenchymal stem cells in steroid-refractory acute graft-versus-host disease. *Transpl Proc* 2007;**39**:3358–62.
217. von Bonin M, Stolzel F, Goedecke A, et al. Treatment of refractory acute GVHD with third-party MSC expanded in platelet lysate-containing medium. *Bone Marrow Transpl* 2009;**43**:245–51.
218. Lucchini G, Introna M, Dander E, et al. Platelet-lysate-expanded mesenchymal stromal cells as a salvage therapy for severe resistant graft-versus-host disease in a pediatric population. *Biol Blood Marrow Transpl* 2010;**16**:1293–301.
219. Lucchini G, Dander E, Pavan F, et al. Mesenchymal stromal cells do not increase the risk of viral reactivation nor the severity of viral events in recipients of allogeneic stem cell transplantation. *Stem Cells Int* 2012;**2012**:690236.
220. Perez-Simon JA, Lopez-Villar O, Andreu EJ, et al. Mesenchymal stem cells expanded in vitro with human serum for the treatment of acute and chronic graft-versus-host disease: results of a phase I/II clinical trial. *Haematologica* 2011;**96**:1072–6.
221. Herrmann R, Sturm M, Shaw K, et al. Mesenchymal stromal cell therapy for steroid-refractory acute and chronic graft versus host disease: a phase 1 study. *Int J Hematol* 2012;**95**:182–8.
222. Kebriaei P, Isola L, Bahceci E, et al. Adult human mesenchymal stem cells added to corticosteroid therapy for the treatment of acute graft-versus-host disease. *Biol Blood Marrow Transpl* 2009;**15**:804–11.
223. von Bahr L, Sundberg B, Lonnies L, et al. Long-term complications, immunologic effects, and role of passage for outcome in

mesenchymal stromal cell therapy. *Biol Blood Marrow Transpl* 2012;**18**:557–64.

224. Weng JY, Du X, Geng SX, et al. Mesenchymal stem cell as salvage treatment for refractory chronic GVHD. *Bone Marrow Transpl* 2010;**45**:1732–40.
225. Zhou H, Guo M, Bian C, et al. Efficacy of bone marrow-derived mesenchymal stem cells in the treatment of sclerodermatous chronic graft-versus-host disease: clinical report. *Biol Blood Marrow Transpl* 2010;**16**:403–12.
226. Figueroa FE, Carrion F, Villanueva S, Khoury M. Mesenchymal stem cell treatment for autoimmune diseases: a critical review. *Biol Res* 2012;**45**:269–77.
227. Chhabra P, Brayman KL. Stem cell therapy to cure type 1 diabetes: from hype to hope. *Stem Cells Transl Med* 2013;**2**:328–36.
228. Anzalone R, Lo Iacono M, Loria T, et al. Wharton's jelly mesenchymal stem cells as candidates for beta cells regeneration: extending the differentiative and immunomodulatory benefits of adult mesenchymal stem cells for the treatment of type 1 diabetes. *Stem Cell Rev* 2011;**7**:342–63.
229. Chang JW, Hung SP, Wu HH, et al. Therapeutic effects of umbilical cord blood-derived mesenchymal stem cell transplantation in experimental lupus nephritis. *Cell Transpl* 2011;**20**:245–57.
230. Liu Y, Mu R, Wang S, et al. Therapeutic potential of human umbilical cord mesenchymal stem cells in the treatment of rheumatoid arthritis. *Arthritis Res Ther* 2010;**12**:R210.
231. Sun L, Wang D, Liang J, et al. Umbilical cord mesenchymal stem cell transplantation in severe and refractory systemic lupus erythematosus. *Arthritis Rheum* 2010;**62**:2467–75.
232. Lalu MM, McIntyre L, Pugliese C, et al. Safety of cell therapy with mesenchymal stromal cells (safecell): a systematic review and meta-analysis of clinical trials. *PLoS One* 2012;**7**:e47559.

Chapter 11

# Cord Blood Cells for Clinical Use: Expansion and Manipulation

Amanda L. Olson and Elizabeth J. Shpall
*MD Anderson Center, Department of Stem Cell Transplantation and Cellular Therapy, University of Texas, Houston, Texas, USA*

### Chapter Outline



## 1. INTRODUCTION

Approximately 30% of patients who present for allogeneic stem cell transplant for treatment of underlying disease will have a matched sibling donor. The National Marrow Donor Program (NMDP) began connecting patients with unrelated donors in 1987 with a registry of 10,000 volunteers. Currently, this has grown such that the NMDP and its cooperative international registries have 16 million unrelated volunteer donors.[1] It is estimated that through the NMDP and its cooperative registries approximately 60% of Caucasian patients, but only 20–45% of African-American and other non-European minority patients, will be able to find a suitably matched unrelated donor (MUD) and proceed to transplant.[2] Therefore, there remain an estimated 5000 patients per year who are candidates for alternative donor transplants. The options for stem cell source for these remaining patients include mismatched-related (often haploidentical) donors, umbilical cord blood (CB) or mismatched-unrelated donors (MMUD).

CB has emerged as an alternative source of hematopoietic stem cells (HSCs)[3] since the first human CB transplantation (CBT) was performed in 1988[4] to successfully treat a child with Fanconi anemia. The field has grown rapidly by expanding on the success in pediatric CBT pioneered by Drs Kurtzberg, Gluckman, Wagner, Broxmeyer, and others.[5,6] By 2011, more than 25,000 CBTs have been performed worldwide and more than 600,000 CB units have been collected by the network of public CB banks that have been established around the world.[7] CB is a rapidly available stem cell source and adequately matched grafts can be identified for diverse non-European patient populations.[2]

The many advantages of CB compared with the use of bone marrow (BM) or mobilized peripheral blood progenitor cells (PBPCs), which make it a desirable source of hematopoietic stem cells, include the relative ease of procurement and lack of donor attrition, as well as the ability to process, cryopreserve, and bank the donor CB cells long term.[8] Other advantages include ease of collection with little to no risk to the mother or newborn, prompt availability with patients receiving CBT in a median of 25–36 days earlier than those receiving unrelated BM,[9] low risk of infection transmission, decreased stringency of human leukocyte antigen (HLA)-matching requirements given the immunologic naivety of the CB cells and the relatively lower risk of graft-versus-host disease (GVHD) as compared to BM and peripheral blood stem cell (PBSC) grafts with preserved graft-versus-malignancy effects. In contrast to BM or PBPCs that generally require a high degree of HLA match between donor and patient,[10] the requirement

Cord Blood Stem Cells Medicine. http://dx.doi.org/10.1016/B978-0-12-407785-0.00011-6

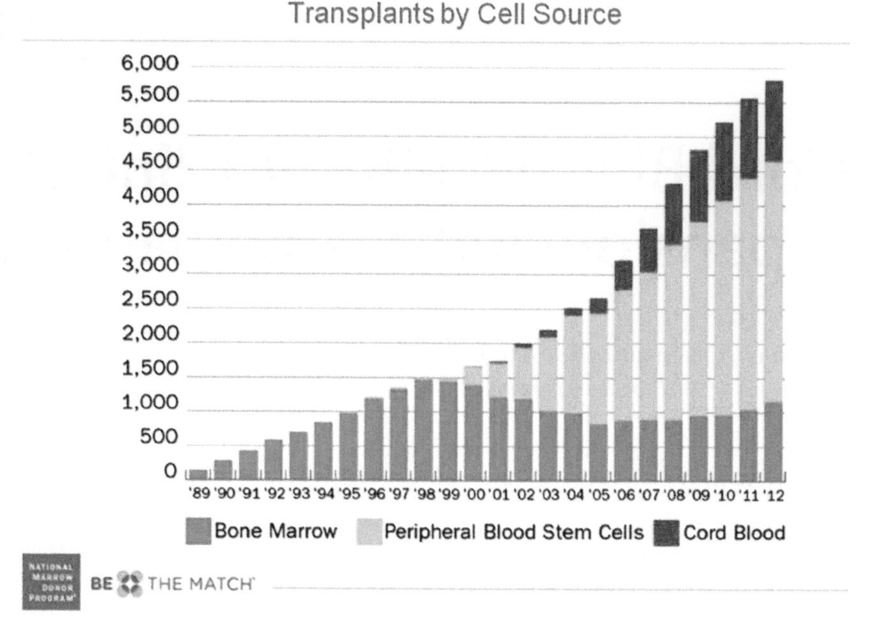


**FIGURE 1** In 2012, the NMDP facilitated nearly 1200 cord blood transplants, which represents 20% of the total number of NMDP transplants in that year.

for HLA match is less stringent with CBT. The reduced incidence of GVHD with partially HLA-mismatched CB is likely due to the lower numbers of T cells and the relatively immunologically naïve status of the lymphocytes in units of CB.[11] The tolerance of a partial HLA-mismatch increases donor access to HCT particularly for non-European minority and mixed ethnicity patients for whom it is often difficult to obtain a suitably matched related donor (MRD) or unrelated donor, as discussed above.

The first decade of CBT was important in defining the total nucleated cell (TNC) and CD34+ cell dose thresholds required for acceptable clinical outcomes, in extending the use of a CB graft from related to unrelated recipients and extending the use of CB graft to adult patients with hematologic malignancies. The lower cell dose (as compared to BM and PBSC grafts) was quickly identified as a critical limitation of CBT, particularly in larger pediatric and adult patients. CBT recipients receive, on average, 1/10th the number of CD34+ stem cells/progenitor cells compared to recipients of conventional BM grafts, and 1/20th of that received from a PBSC graft. This results in a significant delay in time to neutrophil and platelet engraftment and immune reconstitution, and also results in an increased risk of graft failure and early transplant-related mortality (TRM) in recipients of CBT.

The second decade of CBT was marked by improved outcomes. Outcomes were particularly improved in adult recipients of CBT as knowledge of the critical cell dose threshold requirements led to improved CB collection, cryopreservation, and banking of units, and therefore the availability of units with higher cell doses. Improvements in supportive care and optimization of conditioning regimen also contributed to the improved outcomes. Initial reports were published showing outcomes for recipients of CBT comparable with conventional donors.[12–14] Finally, the use of double cord blood transplant (dCBT) was pioneered and importantly demonstrated a significant reduction in the risk of graft failure. This extended the access to HCT, with CB donors for almost all patients in whom a suitable adult donor could not be identified.[15] CB is currently an accepted source of allogeneic hematopoietic stem cells. Figure 1 presents the growth in the number of marrow, PBPC, and CB transplants facilitated by the NMDP worldwide in pediatric and adult patients over time. Unfortunately, as will be discussed below, dCBT has not resulted in faster neutrophil recovery or immune reconstitution, with TNC and CD34+ cell dose remaining one of the major limitations associated with the use of CBT. The establishment of dCBT has, however, extended access to patients in which a single unit would not have been acceptable. Critically, dCBT has allowed for significant progress in the area of CB graft engineering, particularly in ex vivo expansion, as the double unit platform allows for manipulation of one unit as well as the ability to track both units in vivo.

## 2. CB TRANSPLANTATION IN PEDIATRIC PATIENTS

Although several studies have demonstrated a benefit of CBT in children with hematological malignancies,[16–21] the first prospective, multicenter trial of CBT in 191 pediatric patients with hematologic malignancies was reported by Kurtzberg et al. on behalf of the Cord Blood Transplantation Study (COBLT).[22] Despite 77% of the study recipients having high-risk disease, the overall survival (OS) of the study population was 57.3% at 1 year posttransplant. These results compared favorably with those published from registry and single center data showing disease-free survival (DFS) of 50–60% in early stage and 10–30% with more advanced and active disease. A landmark study was reported by Eapen et al.[23] on behalf of

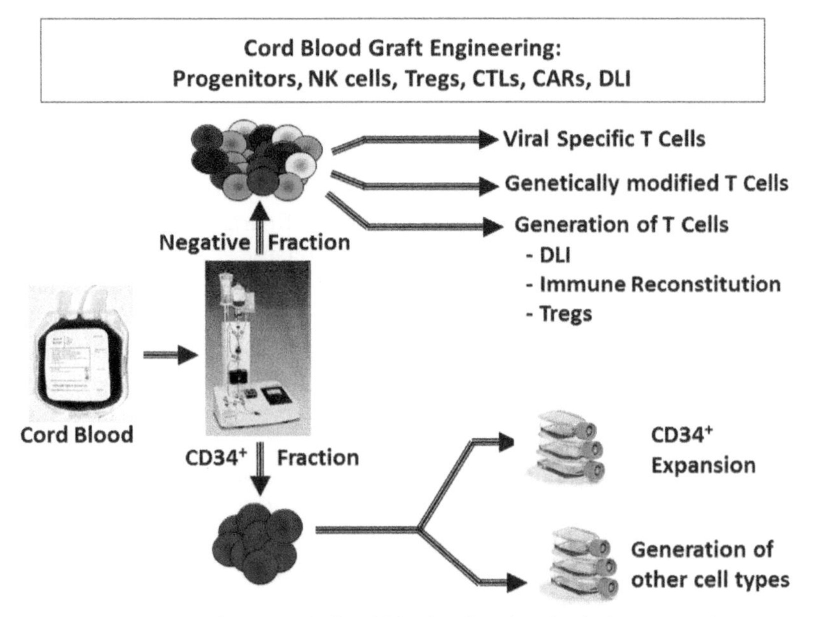


FIGURE 2 The potential of cord blood graft engineering is demonstrated.

The Center for International Blood and Marrow Transplant Research (CIMBTR), comparing the outcomes of 503 children (age<16 years) with acute leukemia who were transplanted with 4–6/6 HLA-matched CB with outcomes of 282 7–8/8 HLA-MUD BM recipients. In this study, CB compared favorably to the "gold standard" of 8/8 allele matched unrelated BM, thus supporting the use of HLA-matched or -mismatched CB in children with high-risk acute leukemia without MRD. Notably, the recipients of one antigen-mismatched CB units with a lower cell dose engrafted at a rate similar to recipients of two antigen-mismatched units, whereas one antigen-mismatched CB units with a higher cell dose had superior engraftment. This indicates that cell dose partially compensated for the degree of HLA mismatch.

The use of CBT has been evaluated for the treatment of metabolic diseases, hemoglobinopathies, and immune deficiencies in children.[24–28] Kurtzberg et al. have pioneered the use of CBT with preliminary very encouraging results in children with inherited metabolic disorders,[24,29] including Krabbe's disease and Hurler's syndrome. The Center for International Blood and Marrow Transplant Research (CIBMTR) registry studies have also demonstrated acceptable outcomes following CBT in patients with severe combined immune deficiency when compared with other donor sources.[30]

## 3. CB TRANSPLANTATION IN ADULT PATIENTS

The first large series of adult recipients of CBT was reported by Laughlin et al.[31] in 2001. This analysis of 68 heavily pretreated patients with advanced hematologic malignancies demonstrated the feasibility of performing CBTs following myeloablative conditioning (MAC) in adult patients, with an event-free survival (EFS) of 26%. Similar to that which was described in the pediatric series, a higher cryopreserved nucleated cell dose ($\geq 2.4 \times 10^7$/kg) was associated with faster rate of and higher probability of neutrophil recovery. Additionally, a higher cryopreserved CD34+ cell dose ($\geq 1.2 \times 10^5$/kg) was associated with a better EFS rate. Neither patient age nor HLA matching appeared to influence EFS in this study. Cornetta et al.[32] subsequently reported a 30% 6-month survival rate for the COBLT prospective study involving 34 adult patients (median age 34) who received MAC for advanced malignancies. Since that time results following CBT have improved. Two large registry studies compared disease outcomes for adults after CB or unrelated BM transplantation with MAC. Rocha et al. observed similar DFS, TRM, and relapse incidence, despite delay of engraftment in CB recipients with acute leukemia who received MAC when compared to results in recipients of MUD transplants.[12] Laughlin et al. observed higher TRM and shorter DFS with CB as compared to 6/6 HLA-MUD while the DFS was similar when compared to 5/6 HLA-MUD.[13] It is important to note that none of these studies compared CB with PBPC transplants.

Eurocord and the CIBMTR recently performed a study comparing unrelated BM (n=472) or PBPC (n=888) to CB (n=165) HCT adults with acute leukemia.[33] In this study, allele-level HLA typing was performed for adult donors at A, B, C, and DRB1 loci. Eight of eight and seven of eight matched donors were included. All CB units were HLA-typed at the antigen level for the A and B loci, with allele-level typing for DRB1. CB units matched at four of six, five of six, or six of six loci were included. Patients who received CB received a single unit containing a minimum of $2.5 \times 10^7$ TNC per kilogram body weight at cryopreservation. Multiple regression analyses revealed higher TRM but lower relapse and GVHD rates in patients who were transplanted with CB, which resulted in comparable DFS rates when compared to the other stem cell sources (Figure 2).

The results of this study and others confirmed that CBT is feasible in adults when a CB unit can be identified with an adequate number of cells per kilogram, and thus CB should be considered a graft option for allogeneic HCT in patients lacking an HLA-matched donor.

## 4. DOUBLE CB TRANSPLANTATION

CBT in adults was initially limited by the relatively lower number of progenitor cells present in a single CB unit per kilogram of body weight, which resulted in delayed hematopoietic recovery and an increased rate of engraftment. The search, for the majority of average-sized adults, will not yield a single CB unit containing the recommended nucleated cell dose of $2.5\times10^{7}$/kg.[34] To overcome the cell dose limitation, Barker and the University of Minnesota investigators pioneered the use of dCBT. These researchers explored the practice of sequentially infusing two CB units, rather than one unit, following MAC therapy.[15,35] They initially reported the safety and feasibility of dCBT in 21 adults with hematological malignancies after MAC HCT.[15] The outcomes were encouraging with all patients engrafting neutrophils at a median of 23 (range, 15–41) days posttransplant. By day 21 in the majority of the patients, only one of the two CB units infused was detected on molecular analysis. This unit would go on to be responsible for long-term hematopoiesis in those patients. In other studies, dominance of the engrafting or "winning" unit as early as day 12 has also been reported.[36,37] Ramirez et al. demonstrated that in the myeloablative setting, CD3+ cell dose was the driving factor associated with unit dominance, but in the nonmyeloablative setting, CD3+ cell dose and HLA match were independent factors associated with unit predominance.[38] To investigate dCBT biology, Eldjerou et al. established in vitro and murine models using cells from 39 patient grafts. Mononuclear cells and CD34+ cells from each unit both alone and in dCBT combination were assessed for multilineage engraftment in NOD/SCID/IL2R-$\gamma^{null}$ mice. MNC dCBT demonstrated single-unit dominance that correlated with human clinical engraftment in 18 of 21 cases ($P<0.001$). Unit dominance and clinical correlation were lost with CD34 selected dCBT, but add-back of CD34− cells restored unit dominance with the dominant unit correlating with human engraftment. The authors postulate that unit dominance is an in vivo phenomenon that is likely associated with a graft-versus-graft immune interaction probably mediated by CD34− cells.[39] However, the biological mechanisms responsible for single-donor predominance after dCBT remain incompletely understood and are under investigation by many groups.

Another limitation of dCBT is the higher incidence of acute GVHD. MacMillan et al.[40] reported a higher rate of grade II-IV acute GVHD in recipients of dCBT (58%, n=185) compared to recipients of single CB units (39%, n=80). This was determined to be due to an increased rate of grade II skin GVHD. The rates of grade III-IV GVHD were the same for both groups; however, the 1-year TRM was significantly lower after dCBT in comparison to single CBT (24% vs 39%).

In a study of 104 adult patients with lymphoid malignancies, Rodrigues et al.[41] reported a significantly lower risk of relapse at 1 year after dCBT in comparison to single CBT (at 13% vs 38%, respectively). A study prospectively comparing single versus double CBT (assignment of single versus double was based on the cell dose of the largest unit) was recently conducted in adult patients and confirmed previous reports of a lower relapse rate after dCBT versus single CBT (30.4% vs 59.3%).[42] The possible enhancement of graft versus malignancy may be due to greater alloreactivity when two CB units are infused, but this finding requires confirmation and further mechanistic studies, both of which are underway.

The University of Minnesota in collaboration with the Fred Hutchinson Cancer Center[43] reported the risks and benefits of dCBT (n=128) relative to those observed after transplantations with MRD (n=204), MUD (n=152), or 1-antigen-MMUD (n=52) after MAC in leukemia patients. The risk of relapse was significantly lower in recipients of dCBT (15%) compared with MRD (43%), MUD (37%), and MMUD (35%) transplant recipients. However, TRM was higher for dCBT (34%) compared with MRD (24%) and MUD (14%) transplant recipients. DFS after dCBT was comparable with DFS following MRD and MUD transplantation, supporting the use of two partially HLA-matched CB units when donor search does not reveal an HLA-matched donor. Whether dCBT is preferable to single CBT, when the cell dose in one unit is acceptable, is not known. The Blood and Marrow Transplant Clinical Trials Network (BMT-CTN) 0501 study (Clinical Trial # NCT00412360) randomized 224 children with acute leukemia to single versus double CB for MAC HCT. The results demonstrated no difference in day 42 neutrophil engraftment (89% in recipients of single unit CBT vs 86% in recipients of dCBT), no difference in 2 year posttransplant relapse rate (13% vs 14%), and no difference in DFS (68% vs 64%) between the two arms.[44] However, the authors conclude that dCBT remains useful in the instance where an adequately sized single unit is not available, such as is often the case in adult patients.

Double CBT is now commonly used in the adult setting to achieve an adequate cell dose and has dramatically increased access to HCT for adult and larger pediatric patients who do not have an adequately sized single CB unit available. However, despite doubling the cell dose with this approach, the median time to neutrophil recovery has not been significantly impacted, with a median time to neutrophil recovery of 26 days after a MAC regimen.[43] TRM was higher in dCBT recipients when compared to recipients of matched and

MMUD, with the majority of the TRM occurring within the first 100 days posttransplant. Analysis of the risk factors for TRM among dCBT recipients revealed a higher likelihood of TRM in patients with delayed myeloid recovery (time to ANC>500/μL) if the recovery was ≥26 days which is the median time to engraftment in dCBT recipients. Thus, the significant delay in neutrophil recovery that is observed in CBT recipients remains a critical barrier to successful outcomes in the CBT setting which could be overcome by clinically feasible ex vivo expansion of CB hematopoietic stem and progenitor cells and enhanced CB homing, as discussed below.

## 5. CB TRANSPLANTATION AFTER REDUCED INTENSITY REGIMENS

The development of reduced intensity conditioning (RIC) regimens was particularly important in extending HCT transplantation to adults. Barker et al. reported that a regimen containing fludarabine, cyclophosphamide, and low-dose total body irradiation (TBI)[45] was well tolerated with rapid neutrophil recovery, a sustained donor engraftment rate of 94%, and a low incidence of TRM. Additionally, Ballen et al. used an RIC regimen of fludarabine, melphalan, and rabbit antithymocyte globulin and reported a 1-year DFS of 67%.[46] Rocha et al., on behalf of the Eurocord Registry, also reported similar results using an RIC regimen of fludarabine, endoxan, and TBI with a 1-year TRM of 24% and DFS of 50%.[47] Multiple studies have subsequently been conducted with results that support the use of RIC CB transplant in patients who would not be able to tolerate more intensive preparative regimens.[35,42,46–54]

CIBMTR compared the outcomes in adult patients with acute leukemia who were transplanted with dCBT (n=161) and 8/8 (n=313) or 7/8 HLA-matched (n=111) PBPCs after RIC regimens between 2000 and 2009.[54] This analysis demonstrated comparable outcomes following dCBT if patients underwent conditioning therapy with TBI 200 cGy, cyclophosphamide, and fludarabine (TCF). Higher TRM and lower OS and DFS were observed in recipients of dCBT treated with alternative regimens. The authors note that TRM was not significantly different after dCBT with TCF and PBPC transplantations despite the slower neutrophil recovery with dCBT. Further, it was noted that the risk of relapse was not different between stem cell sources after RIC HCT, this is in contrast to the lower relapse incidence with dCBT in patients with acute leukemia following an MAC.[43] These results, along with single center studies supporting the use of RIC dCBT demonstrate that this approach should be considered a strategy for broadening the application of transplant therapy to patients with hematologic malignancies who might have been previously excluded based on age, co-morbidities, and the absence of an HLA-MUD.

BMT-CTN completed two parallel phase 2 trials studying RIC alternative donor HCT in order to study the reproducibility and the wider applicability of the above described results after RIC dCBT and compare them with HLA-haploidentical related donor marrow.[55] Fifty patients were treated in each study. All patients received an RIC conditioning regimen of fludarabine, cyclophosphamide, and low-dose TBI. TRM was significantly higher following dCBT (24% for CB vs 7% for haplo). However, the relapse rate was significantly higher following haplo-HCT (31% for CB vs 45% for haplo). The 1-year DFS was comparable at 46% for CBT and 48% for haplo-HCT. It is notable that the outcomes were analyzed at one year posttransplant and some speculate that with longer follow-up the differences in relapse following haplo-HCT may affect the OS.

An alternative to high-dose MAC or nonmyeloablative conditioning regimens has been proposed by Ponce et al.[56] This reduced intensity but still myeloablative regimen, evaluated in 30 adult patients with acute leukemia or myelodysplasia, consisted of cyclophosphamide 50 mg/kg, fludarabine 150 mg/m$^2$, thiotepa 10 mg/kg, and 400 cGy of total body irradiation with cyclosporine-A/mycophenolate mofetil immunosuppression. Ninety-seven percent of patients engrafted by a median of 26 days and platelet engraftment was 93% by day 180. TRM was 20% at day 180, relapse was 11% at 2 years posttransplant, and 2-year DFS was 60%. This regimen represents a potential alternative to high-dose MAC regimens in younger adult patients.

## 6. CB GRAFT MANIPULATION

Ex vivo expansion of CB stem cells and progenitor cells to enhance CB engraftment is an area of CB graft engineering that currently is under clinical investigation. As a means of overcoming the issues of delayed immune reconstitution in CBT recipients, the generation of immunotherapy from CB grafts is under active investigation both preclinically and now in the clinic with the generation of multivirus-specific cytotoxic T lymphocytes for treatment of viral infections. As relapse remains a significant obstacle to overcome for patients undergoing HCT for hematologic malignancy, the ex vivo expansion of CB-derived T cells that are genetically modified to express CD19 or CD20 CARs to prevent relapse and the ex vivo generation of increased numbers of CB-derived NK cells as a means of better disease control will be discussed. In the future it may be possible to manipulate a single unit of CB based on the individual's greatest risk factors, with generation of specific cell types as clinically appropriate (Figure 2).

## 7. CB STEM CELL AND PROGENITOR CELL EXPANSION TO ENHANCE ENGRAFTMENT

The low total cell and stem cell dose provided by a single or double CB graft results in a significant delay in hematopoietic recovery and increased risk of primary graft failure, especially in adult patients. There are now multiple strategies under clinical investigation aimed at overcoming this hurdle in the use of CB grafts. These strategies focus primarily on methods to increase the cell dose of a CB graft. Strategies include the transplantation of two CB units as discussed above,[15] ex vivo expansion of CB units,[57] direct intra-BM injection of CB cells,[58] coinfusion of a CB unit with a haploidentical T cell-depleted graft, the systemic addition of mesenchymal stem cells,[59] and the use of agents to enhance the homing of CB to the marrow,[60] all of which will be discussed below.

As a way to shorten time to neutrophil engraftment and to reduce the rate of graft failure, ex vivo expansion of CB-derived hematopoietic stem cells and progenitor cells is being studied by a number of investigators. Delaney et al. have published extremely promising results[61] using an engineered form of the Notch ligand Delta1 for the ex vivo generation of increased numbers of CB CD34+ stem and progenitor cells with the goal of reducing the time to engraftment. Preliminary results published in 2010 demonstrated both the safety and clinical feasibility of this approach and a significant decrease in the time to neutrophil recovery. Updated data (Delaney, personal communication) from this ongoing study, now with 17 patients, shows a median time to neutrophil recovery (ANC≥500) of 11days compared to 25days in a concurrent cohort of patients (n=36) treated with the same conditioning regimen and a double CB graft. The expanded cell graft in this study contributed almost exclusively to initial myeloid engraftment observed at one week. This demonstrates the enhanced capacity of the expanded cells to provide rapid myeloid recovery. Furthermore, all but two of the evaluable subjects had neutrophil engraftment prior to day 21, independent of whether the expanded cell graft persisted in vivo. The unit that was expanded ex vivo underwent positive selection for CD34+ cells prior to initiation of culture and the negative fraction from this unit was not infused at the time of transplant.

Using a strategy to culture CB cells with mesenchymal stem cells (MSCs)[62] Robinson et al. reported a 10- to 20-fold increase in TNC and a 16- to 40-fold increase in CD34+ cells. Based on these results, a clinical trial was initiated to evaluate the ex vivo coculture of CB mononuclear cells with either third party haploidentical family member marrow-derived MSCs or off-the-shelf Stro3+ MSCs from mesoblast (Clinical Trial # NCT00498316). The results with either source of MSCs were similar, with engraftment of neutrophils in 15days and platelets in 42days. In the majority of patients, long-term engraftment was provided by the unexpanded unit. Rates of engraftment were compared in 31 adults with hematologic malignancies who underwent dCBT with one of the units containing CB expanded ex vivo with allogeneic mesenchymal stem cells and 80 historical controls who received 2 units of unmanipulated CB. The median time to engraftment of neutrophils was significantly reduced in recipients of the dCBT with an expanded unit (a median of 15days) versus historical controls (a median of 24days). The median time of platelet engraftment was also significantly reduced at a median of 42days (median of 49days in historical controls).[63] These results provide the rationale for the prospective multinational randomized trial currently underway comparing unmanipulated dCBT with dCBT where one of the units will be expanded in MSC cocultures.

Unlike the previous two studies that utilize a double CB transplant platform in which one unit is manipulated and one ex vivo expanded, a trial in which a fraction of a single CB unit was ex vivo expanded using growth factors in conjunction with a copper chelator TEPA[64] was sponsored by Gamida Cell Ltd. Preliminary analysis revealed faster engraftment and improved survival compared to historical control recipients of single CBT reported to the international registries.[57] A more definitive analysis of this data is in progress.

In other preclinical studies by this group, CB-derived CD34+ cells cultured ex vivo with growth factors (stem cell factor, thrombopoietin, interleukin-6, and FMS-like tyrosine kinase 3) and nicotinamide (pyridine-3-carboxamide) displayed increased migration toward SDF-1 and enhanced homing to BM compared with untreated CB.[65] A multicenter pilot clinical trial led by Horwitz et al. is in progress to evaluate this strategy in the myeloablative dCBT setting (ClinicalTrial # NCT01221857). Preliminary analysis has revealed rapid engraftment (at a median of 10days for neutrophils and a median of 30days for platelets), with sustained engraftment from the expanded unit in the majority of patients evaluated to date (M. Horowitz, personal communication).

## 8. IMPROVING CB HOMING TO BM

While delayed and failed engraftment following CBT are postulated to be due to low TNC[66,67] and low CD34+ CB cell[68] doses as discussed above, there is also evidence that CD34+ cells found in CB may also have a defect in homing to the BM at transplant.[69,70] Despite being a critical part of the engraftment process, little is known about the mechanisms that regulate homing of infused stem cells to the marrow. A promising strategy to improve engraftment is to correct the decreased fucosylation of CB cell surface molecules which is thought to be impairing homing of CB-derived progenitor cells to the BM.[69] Using an NOD-SCID IL-2R$^{null}$ (NSG) mouse model,[71] the impact of fucosylation on the rate and magnitude of engraftment of human CB CD34+ cells was evaluated. Sublethally irradiated NSG mice received either FT-VI-treated, or untreated CB CD34+ cells intravenously and human engraftment was followed in PB by performing serial retro-orbital bleeds. Samples of PB were analyzed by flow cytometry to reveal the presence of human (hu)CD45+ cells. The recipients of the FT-VI-treated CB CD34+ cells showed a more rapid and greater magnitude of human engraftment than mice receiving the same number of untreated human CD34+ cells.[72] Long-term human engraftment was also followed. At 8 weeks after transplant, the proportion of huCD45+ cells in the PB of mice receiving FT-VI-treated CB CD34+ cells was greater than that found in the blood of mice receiving untreated CB CD34+ cells. These data suggest that while FT-VI-treatment increased both the rate and the magnitude of CB CD34+ engraftment, it did not compromise long-term engraftment. A clinical trial was recently opened where recipients will receive dCBT where one of the CB units will be fucosylated prior to infusion (Clinical Trial # NCT01471067). In future trials combining CB expansion with fucosylation may produce maximally rapid hematopoietic recovery in the patients.

## 9. PROSTAGLANDIN AND HOMING

Prostaglandin E2 ($PGE_2$) has also been shown to enhance hematopoietic stem cell homing, survival, and proliferation.[73] Cutler et al. are currently conducting a clinical trial to evaluate the ex vivo treatment of one of two CB units with $dmPGE_2$ prior to infusion (Clinical Trial # NCT00890500). After optimizing the procedure it appears that engraftment of neutrophils is prompt (at a median of 17 days), with long-term dominance of the $dmPGE_2$-modulated unit documented in the majority of recipients.[74]

Broxmeyer's group has shown that inhibition of the CD34+ CB cell surface protein CD26/dipeptidylpeptidase (DPPIV) enhanced engraftment in sublethally irradiated NOD/SCID mice.[75,76] This group is currently conducting a pilot clinical trial of systemic sitagliptin, a United States Food and Drug Administration approved DPPIV inhibitor for diabetes, to evaluate the potential for enhancing engraftment with the use of this drug in recipients of a single CBT (Clinical Trial # NCT00862719).

## 10. CB IMMUNE CELLS TO IMPROVE OUTCOME

Several obstacles remain that are associated with CBT. Importantly, CBT is associated with delayed immune reconstitution and a higher risk of morbidity and mortality due to viral and other infections posttransplant. GVHD remains a problem particularly in the dCBT setting.[40] Relapse is also of concern particularly in high-risk patients with persistent disease prior to transplant.[77] To address these major obstacles to widespread use of CBT, there are a number of exciting clinical studies in progress evaluating several different CB-derived immune cells in hopes of improving survival.

## 11. EXPANDING MULTIVIRUS-SPECIFIC CYTOTOXIC T LYMPHOCYTES FROM CB

CBT is associated with significant morbidity and mortality from viral infections including cytomegalovirus (CMV), Epstein Barr virus (EBV), and adenovirus (AD).[78] The susceptibility to viral infection is due to the antigen-inexperienced nature of CB, containing a higher percentage of naïve T cells compared to adult donor peripheral blood (PB). Virus-specific memory T cells confer protection against viral infections and reactivation, and even against relapse. Deficiencies associated with current pharmacological antiviral agents have increased the BM transplant community's interest in an immunotherapeutic approach to viral disorders. Adoptive transfer of T cells by donor lymphocyte infusions has been used to treat viral infection after allogeneic HSCT, but is also associated with the risk of GVHD. In contrast, peripheral blood-derived virus-specific cytotoxic T lymphocytes (CTL) directed to CMV, EBV, and AD can rapidly reconstitute antiviral immunity post-HSCT without GVHD.[79–82] Hanley et al.[83] developed a strategy using AD5f35pp65-transduced CB-derived antigen presenting cells (APC) comprised of dendritic cells and EBV-transformed lymphoblastoid cell lines, to generate large numbers of autologous CB-derived T cells specific for CMV, AD, and EBV. Based on this approach, a clinical trial led by Bollard et al. at Baylor College of Medicine sought to evaluate whether this strategy could be applied to recipients of CBT. This group is testing the use of CB-derived multivirus-specific T cells targeting EBV, CMV, and AD for the prevention and treatment of viral infections after CBT (Clinical trial # NCT01017705). Obstacles that had first to be circumvented with this approach included the naïvety of the CB-derived T cells and the limited cell numbers available for manipulation in the CB graft.[83] Because of these challenges, only a few investigators have attempted to generate CMV and EBV antigen-specific T cells from CB in proof-of-principle studies.[84,85] However, neither approach has been used clinically. Bollard et al. therefore set out to develop a GMP compliant methodology for direct translation anticipating that the ability to generate virus-specific CTL from CB for adoptive transfer recognizing a broad spectrum of epitopes recognized by both CD4+ and CD8+ T cells would minimize the risk of viral escape and maximize therapeutic benefit to CBT recipients at risk of severe viral disease.[83] Bollard et al. have opened a clinical trial using CB-derived multivirus-specific T cells for the prevention and treatment of viral infection after CBT (Clinical Trial # NCT01017705). To date 10 patients have received single CBT using the 80% fraction of a fractionated CB unit. All patients engrafted neutrophils and platelets at <30 or <60 days post CBT respectively. Multivirus-specific CTLs were generated from the remaining 20% fraction of the CB graft and have been infused to seven patients. Patients received the CTL at a median of 83 days post CBT (range 63–146 days). No infusion-related toxicities or GVHD were observed in any of the patients. With the exception of two patients, none developed a viral infection and all remain free of CMV, EBV, and AD infection/reactivation 2 months to 2 years after CBT. One patient developed CMV reactivation early post CBT and had virus clearance following two infusions of CTL. This patient was also positive for AD antigen in his stool, which resolved without additional therapy and he remains asymptomatic and virus free more than 2 years following CBT. Another patient had detectable EBV DNA in the peripheral blood that was controlled without antiviral therapy. To determine the persistence of the CB CTL these investigators used deep T cell receptor sequencing by Adaptive Biotechnologies. T cell clones present in the infused CTL but not present before CTL infusion were detected up to 1 year after CBT in all patients tested. Hence, administration of CB-derived virus-specific CTL to patients after CBT has so far been safe and can facilitate reconstitution of virus-specific T cells and control viral reactivation and infection in vivo.

## 12. CB-DERIVED NATURAL KILLER (NK) CELLS

NK cells are innate immune cells that recognize "non-self" by the absence of class I molecules and inhibitory receptors. Although the proportion of NK cells in CB is similar to that of adult peripheral blood, the NK cells found in CB are immature in phenotype and function. Resting CB NK cells have been reported to have significantly less cytotoxicity compared to peripheral blood NK cells,[86] but following cytokine stimulation the cytotoxicity of CB NK cells can be rapidly increased to levels that are comparable to peripheral blood NK cells.[87] Several groups have developed methods to isolate NK cells from peripheral blood of healthy donors.[88] However, CB-derived products are restricted by the finite number of cells available for expansion. This limitation has led to the development of NK cell expansion methods in vitro. Researchers have explored the use of cytokines, artificial antigen presenting cells, and the G-REX gas permeable device. After expansion, CB NK cells were noted to have cytolytic function in vitro against multiple hematologic tumor targets. However, whether ex vivo-expanded CB NK cells will be more efficacious in the clinical setting still remains to be tested. A Phase II trial using T cell-depleted dCBT with posttransplant interleukin-2 is being conducted by the University of Minnesota investigators in refractory AML patients with an aim to investigate the NK cell expansion and function in vivo (Clinical Trial # NCT01464359). Trials are also underway with the goal of evaluating dCBT in which a portion of one unit is used for expanding NK cells (Clinical Trial # NCT01619761) and expanded CB NK cells as an adjunct to an autologous stem cell transplant for patients with myeloma (Clinical Trial # NCT01729091).

## 13. CB-DERIVED REGULATORY T CELLS

Regulatory T cells (Tregs) are a subset of CD4+ T cells that coexpress CD25 (interleukin-2Rα chain) and high levels of Foxp3,[89] and are dependent on interleukin-2. They represent a novel CB cell-based approach for potentially reducing the risk of GVHD.[90] Brunstein et al. expanded Tregs obtained from a third CB unit and infused them into 23 patients undergoing dCBT.[91] No severe Treg-related acute toxicities were observed, and accrual to the study continues with refinements in the Treg generation procedure.

## 14. REDIRECTING SPECIFICITY OF CB-DERIVED T CELLS TO LEUKEMIA ANTIGENS

Targeting malignancies using exogenous receptors (chimeric antigen receptors or CARs) expressed on ex vivo-expanded T cells is an evolving therapy for both hematologic malignancies and solid tumors.[92] In recent years this approach has been translated to the CB setting by expanding CB-derived T cells that are genetically modified to express CD19 or CD20 CARs.[93,94] A clinical trial evaluating infusion of CB T cells that are generated to express CD19 using the sleeping beauty system[95] at day 49 (±7 days) posttransplant is in progress (Clinical Trial # NCT01362452).

## 15. CONCLUSION

CB is used increasingly as a source of allogeneic hematopoietic support for patients who need a transplant and do not have access to an HLA-matched donor. To overcome the limitation of low cell doses in single CB units, dCBT has been adopted for many patients with outcomes comparable to other donor sources. There are currently new strategies under development to improve engraftment with ex vivo expansion and homing, and to enhance immune reconstitution with the infusion of CB-derived NK cells and CTLs with antiviral and antileukemic specificities. Tregs are being evaluated to reduce the incidence of GVHD. Despite the many advantages of CB as a source of allogeneic stem cells, the use of CB immune cells for clinical use is still in its early stages. While the infusion of CB-derived multivirus-specific

T cells shows promise in the single CBT setting, this technology has not yet been applied in the dCBT setting. Further, the use of CB-derived leukemia-specific T cells and NK cells will require extensive testing in the clinical setting to enable researchers to evaluate and optimize the successes and failures of these approaches. Nevertheless, CB-derived immune-based therapies ultimately have the potential to greatly improve the outcomes for our patients. Prospective, multicenter clinical trials are needed to determine the efficacy of these promising technologies, both independently and eventually in combination.

## REFERENCES


1. Ballen KK, King RJ, Chitphakdithai P, Bolan CD, Agura E, Hartzman RJ, et al. The national marrow donor program 20 years of unrelated donor hematopoietic cell transplantation. *Biol Blood Marrow Transpl* 2008;**14**(Suppl. 9):2–7.
2. Barker JN, Byam CE, Kernan NA, Lee SS, Hawke R, Doshi KA, et al. Availability of cord blood extends allogeneic hematopoietic stem cell transplant access to racial and ethnic minorities. *Biol Blood Marrow Transpl: J Am Soc Blood Marrow Transpl* 2010;**16**(11):1541–8.
3. Broxmeyer HE, Douglas GW, Hangoc G, Cooper S, Bard J, English D, et al. Human umbilical cord blood as a potential source of transplantable hematopoietic stem/progenitor cells. *Proc Natl Acad Sci U S A* 1989;**86**(10):3828–32.
4. Gluckman E, Broxmeyer HA, Auerbach AD, Friedman HS, Douglas GW, Devergie A, et al. Hematopoietic reconstitution in a patient with Fanconi's anemia by means of umbilical-cord blood from an HLA-identical sibling. *N Engl J Med* 1989;**321**(17):1174–8.
5. Kurtzberg J, Laughlin M, Graham ML, Smith C, Olson JF, Halperin EC, et al. Placental blood as a source of hematopoietic stem cells for transplantation into unrelated recipients. *N Engl J Med* 1996;**335**(3):157–66.
6. Wagner JE, Rosenthal J, Sweetman R, Shu XO, Davies SM, Ramsay NK, et al. Successful transplantation of HLA-matched and HLA-mismatched umbilical cord blood from unrelated donors: analysis of engraftment and acute graft-versus-host disease. *Blood* 1996;**88**(3):795–802.
7. Gluckman E, Ruggeri A, Volt F, Cunha R, Boudjedir K, Rocha V. Milestones in umbilical cord blood transplantation. *Br J Haematol* 2011;**154**(4):441–7.
8. Broxmeyer HE, Lee MR, Hangoc G, Cooper S, Prasain N, Kim YJ, et al. Hematopoietic stem/progenitor cells, generation of induced pluripotent stem cells, and isolation of endothelial progenitors from 21- to 23.5-year cryopreserved cord blood. *Blood* 2011;**117**(18):4773–7.
9. Barker JN, Krepski TP, DeFor TE, Davies SM, Wagner JE, Weisdorf DJ. Searching for unrelated donor hematopoietic stem cells: availability and speed of umbilical cord blood versus bone marrow. *Biol Blood Marrow Transpl* 2002;**8**(5):257–60.
10. Petersdorf EW, Gooley TA, Anasettis C, Martin PJ, Smith AG, Mickelson EM, et al. Optimizing outcome after unrelated marrow transplantation by comprehensive matching of HLA class I and II alleles in the donor and recipient. *Blood* 1998;**92**(10):3515–20.
11. Gauderet L, Dulphy N, Douay C, Chalumeau N, Schaeffer V, Zilber MT, et al. The umbilical cord blood alphabeta T-cell repertoire: characteristics of a polyclonal and naive but completely formed repertoire. *Blood* 1998;**91**(1):340–6.
12. Rocha V, Labopin M, Sanz G, Arcese W, Schwerdtfefer R, Bosi A, et al. Transplants of umbilical-cord blood or bone marrow from unrelated donors in adults with acute leukemia. *N Engl J Med* 2004;**351**(22):2276–85.
13. Laughlin MJ, Eapen M, Rubinstein P, Wagner JE, Zhang MJ, Champlin RE, et al. Outcomes after transplantation of cord blood or bone marrow from unrelated donors in adults with leukemia. *N Engl J Med* 2004;**351**(22):2265–75.
14. Takajasjo S, Iseki T, Ooi J, Tomonari A, Takasugi K, Shimohakamada Y, et al. Single-institute comparative analysis of unrelated bone marrow transplantation and cord blood transplantation for adult patients with hematologic malignancies. *Blood* 2004;**104**(12):3813–20.
15. Barker JN, Weisdorf DJ, DeFor TE, Blazar BR, McGlave PB, Miller JS, et al. Transplantation of 2 partially HLA-matched umbilical cord blood units to enhance engraftment in adults with hematologic malignancy. *Blood* 2005;**105**(3):1343–7.
16. Hu JY, Wang S, Zhu JG, Zhou GH, Sun QB. Expression of B7 costimulation molecules by colorectal cancer cells reducestumorigenicity and induces anti-tumor immunity. *World J Gastroenterol* 1999;**5**(2):147–51.
17. Locatelli F, Rocha V, Chastang C, Arcese W, Michel G, Abecasis M, et al. Factors associated with outcome after cord blood transplantation in children with acute leukemia. Eurocord-Cord Blood Transplant Group. *Blood* 1999;**93**(11):3662–71.
18. Michel G, Rocha V, Chevret S, Arcese W, Chan KW, Filipovich A, et al. Unrelated cord blood transplantation for childhood acute myeloid leukemia: a Eurocord Group analysis. *Blood* 2003;**102**(13):4290–7.
19. Ohnuma K, Isoyama K, Ikuta K, Toyoda Y, Nakamura J, Nakajima F, et al. Cord blood transplantation from HLA-mismatched unrelated donors as a treatment for children with haematological malignancies. *Br J Haematol* 2001;**112**(4):981–7.
20. Gluckman E, Rocha V. Cord blood transplantation for children with acute leukaemia: a Eurocord registry analysis. *Blood Cells Mol Dis* 2004;**33**(3):271–3.
21. Wall DA, Carter SL, Kernan NA, Kapoor N, Kamani NR, Brochstein JA, et al. Busulfan/melphalan/antithymocyte globulin followed by unrelated donor cord blood transplantation for treatment of infant leukemia and leukemia in young children: the Cord Blood Transplantation study (COBLT) experience. *Biol Blood Marrow Transpl* 2005;**11**(8):637–46.
22. Kurtzberg J, Prasad VK, Carter SL, Wagner JE, Baxter-Lowe LA, Wall D, et al. Results of the Cord Blood Transplantation study (COBLT): clinical outcomes of unrelated donor umbilical cord blood transplantation in pediatric patients with hematologic malignancies. *Blood* 2008;**112**(10):4318–27.
23. Eapen M, Rubinstein P, Zhang MJ, Stevens C, Kurtzberg J, Scaradavou A, et al. Outcomes of transplantation of unrelated donor umbilical cord blood and bone marrow in children with acute leukaemia: a comparison study. *Lancet* 2007;**369**(9577):1947–54.
24. Escolar ML, Poe MD, Provenzale JM, Richards KC, Allison J, Wood S, et al. Transplantation of umbilical-cord blood in babies with infantile Krabbe's disease. *N Engl J Med* 2005;**352**(20):2069–81.
25. Staba SL, Escolar ML, Poe M, Kim Y, Martin PL, Szabolcs P, et al. Cord-blood transplants from unrelated donors in patients with Hurler's syndrome. *N Engl J Med* 2004;**350**(19):1960–9.
26. Kamani NR, Walters MC, Carter S, Aquino V, Brochstein JA, Chaudhury S, et al. Unrelated donor cord blood transplantation for children with severe sickle cell disease: results of one cohort from the phase II study from the blood and marrow transplant clinical trials network (BMT CTN). *Biol Blood Marrow Transpl* Aug 2012;**18**(8):1265–72.

27. Adamkiewicz TV, Szabolcs P, Haight A, Baker KS, Staba S, Kedar A, et al. Unrelated cord blood transplantation in children with sickle cell disease: review of four-center experience. *Pediatr Transpl* 2007;**11**(6):641–4.
28. Knutsen AP, Wall DA. Umbilical cord blood transplantation in severe T-cell immunodeficiency disorders: two-year experience. *J Clin Immunol* 2000;**20**(6):466–76.
29. Prasad VK, Mendizabal A, Parikh SH, Szabolcs P, Driscoll TA, Page K, et al. Unrelated donor umbilical cord blood transplantation for inherited metabolic disorders in 159 pediatric patients from a single center: influence of cellular composition of the graft on transplantation outcomes. *Blood* 2008;**112**(7):2979–89.
30. Fernandes JF, Rocha V, Labopin M, Neven B, Moshous D, Gennery AR, et al. Transplantation in patients with SCID: mismatched related stem cells or unrelated cord blood? *Blood* 2012;**119**(12):2949–55.
31. Laughlin MJ, Barker J, Bambach B, Koc ON, Rizzieri DA, Wagner JE, et al. Hematopoietic engraftment and survival in adult recipients of umbilical-cord blood from unrelated donors. *N Engl J Med* 2001;**344**(24):1815–22.
32. Cornetta K, Laughlin M, Carter S, Wall D, Weinthal J, Delaney C, et al. Umbilical cord blood transplantation in adults: results of the prospective Cord Blood Transplantation (COBLT). *Biol Blood Marrow Transpl: J Am Soc Blood Marrow Transplant* 2005;**11**(2):149–60.
33. Eapen M, Rocha V, Sanz G, Scaradavou A, Zhang MJ, Arcese W, Sirvent A, et al. Effect of graft source on unrelated donor haemopoietic stem-cell transplantation in adults with acute leukaemia: a retrospective analysis. *Lancet Oncol* 2010;**11**(7):653–60.
34. Rocha V, Gluckman E. Improving outcomes of cord blood transplantation: HLA matching, cell dose and other graft- and transplantation-related factors. *Br J Haematol* 2009;**147**(2):262–74.
35. Brunstein CG, Barker JN, Weisdorf DJ, DeFor TE, Miller JS, Blazar BR, et al. Umbilical cord blood transplantation after nonmyeloablative conditioning: impact on transplantation outcomes in 110 adults with hematologic disease. *Blood* 2007;**110**(8):3064–70.
36. Delaney C, Gutman JA, Appelbaum FR. Cord blood transplantation for haematological malignancies: conditioning regimens, double cord transplant and infectious complications. *Br J Haematol* 2009;**147**(2):207–16.
37. Gutman JA, Turtle CJ, Manley TJ, Heimfeld S, Bernstein ID, Riddell SR, et al. Single-unit dominance after double-unit umbilical cord blood transplantation coincides with a specific CD8+ T-cell response against the nonengrafted unit. *Blood* 2010;**115**(4):757–65.
38. Ramirez P, Wagner JE, DeFor TE, Blazar BR, Verneris MR, Miller JS, et al. Factors predicting single-unit predominance after double umbilical cord blood transplantation. *Bone Marrow Transpl* 2012;**47**(6):799–803.
39. Eldjerou LK, Chaudhury S, Baisre-de Leon A, He M, Arcila ME, Heller G, et al. An in vivo model of double-unit cord blood transplantation that correlates with clinical engraftment. *Blood* 2010;**116**(19):3999–4006.
40. MacMillan ML, Weisdorf DJ, Brunstein CG, Cao Q, DeFor TE, Verneris MR, et al. Acute graft-versus-host disease after unrelated donor umbilical cord blood transplantation: analysis of risk factors. *Blood* 2009;**113**(11):2410–5.
41. Rodrigues CA, Sanz G, Brunstein CG, Sanz J, Wagner JE, Renaud M, et al. Analysis of risk factors for outcomes after unrelated cord blood transplantation in adults with lymphoid malignancies: a study by the Eurocord-Netcord and lymphoma working party of the European group for blood and marrow transplantation. *J Clin Oncol* 2009;**27**(2):256–63.
42. Kindwall-Keller TL, Hegerfeldt Y, Meyerson HJ, Margevicius S, Fu P, van Heeckeren W, et al. Prospective study of one- versus two-unit umbilical cord blood transplantation following reduced intensity conditioning in adults with hematological malignancies. *Bone Marrow Transpl* Jul 2012;**47**(7):924–33.
43. Brunstein CG, Gutman JA, Weisdorf DJ, Woolfrey AE, Defor TE, Gooley TA, et al. Allogeneic hematopoietic cell transplantation for hematologic malignancy: relative risks and benefits of double umbilical cord blood. *Blood* 2010;**116**(22):4693–9.
44. Wagner JE, Eapen M, Carter SL, Haut PR, Peres E, Schultz KR, Thompson J, et al. No survival advantage after double umbilical cord blood (UCB) compared to single UCB transplant in children with hematologic malignancy: results of the Blood and Marrow Transplant Clinical Trials Network (BMT CTN 0501) randomized trial, *2012 ASH Annual Meeting*. Abstract 359. Presented December 10, 2012.
45. Barker JN, Weisdorf DJ, DeFor TE, Blazar BR, Miller JS, Wagner JE. Rapid and complete donor chimerism in adult recipients of unrelated donor umbilical cord blood transplantation after reduced-intensity conditioning. *Blood* 2003;**102**(5):1915–9.
46. Ballen KK, Spitzer TR, Yeap BY, McAfee S, Dey BR, Attar E, et al. Double unrelated reduced-intensity umbilical cord blood transplantation in adults. *Biol Blood Marrow Transpl* 2007;**13**(1):82–9.
47. Rocha V, Mohty M, Gluckman E, Rio B. Reduced-intensity conditioning regimens before unrelated cord blood transplantation in adults with acute leukaemia and other haematological malignancies. *Curr Opin Oncol* 2009;**21**(Suppl 1):S31–4.
48. Yuji K, Miyakoshi S, Kato D, Miura Y, Myojo T, Murashige N, et al. Reduced-intensity unrelated cord blood transplantation for patients with advanced malignant lymphoma. *Biol Blood Marrow Transpl* 2005;**11**(4):314–8.
49. Miyakoshi S, Kami M, Tanimoto T, Yamaguchi T, Narimatsu H, Kusumi E, Matsumura T, et al. Tacrolimus as prophylaxis for acute graft-versus-host disease in reduced intensity cord blood transplantation for adult patients with advanced hematologic diseases. *Transplantation* 2007;**84**(3):316–22.
50. Miyakoshi S, Yuji K, Kami M, Kusumi E, Kishi Y, Kobayashi K, et al. Successful engraftment after reduced-intensity umbilical cord blood transplantation for adult patients with advanced hematological diseases. *Clin Cancer Res* 2004;**10**(11):3586–92.
51. Bradley MB, Satwani P, Baldinger L, Morris E, Van de Ven C, Del Toro G, et al. Reduced intensity allogeneic umbilical cord blood transplantation in children and adolescent recipients with malignant and non-malignant diseases. *Bone Marrow Transpl* 2007;**40**(7):621–31.
52. Cutler C, Stevenson K, Kim HT, Brown J, McDonough S, Herrera M, et al. Double umbilical cord blood transplantation with reduced intensity conditioning and sirolimus-based GVHD prophylaxis. *Bone Marrow Transpl* 2011;**46**(5):659–67.
53. Uchida N, Wake A, Takagi S, Yamamoto H, Kato D, Matsuhashi Y, et al. Umbilical cord blood transplantation after reduced-intensity conditioning for elderly patients with hematologic diseases. *Biol Blood Marrow Transpl* 2008;**14**(5):583–90.
54. Brunstein CG, Eapen M, Ahn KW, Appelbaum FR, Ballen KK, Champlin RE, et al. Reduced intensity conditioning transplantation in acute leukemia: the effect of source of unrelated donor stem cells on outcomes. *Blood* Jun 7 2012;**119**(23):5591–8.
55. Brunstein CG, Fuchs EJ, Carter SL, Karanes C, Costa LJ, Wu J, et al. Alternative donor transplantation after reduced intensity conditioning: results of parallel phase 2 trials using partially HLA-mismatched related bone marrow or unrelated double umbilical cord blood grafts. *Blood* 2011;**118**(2):282–8.

56. Ponce DM, Sauter CS, Devlin SM, Lubin M, Gonzales AMR, Kernan NA, et al. A novel reduced-intensity conditioning regimen induces a high incidence of sustained donor-derived neutrophil and platelet engraftment after double-unit cord blood transplantation. *Biol Blood Marrow Transpl: J Am Soc Blood Marrow Transpl* 2013;**19**(5):799–803.
57. de Lima M, McMannis J, Gee A, Komanduri K, Couriel D, Andersson BS, et al. Transplantation of ex vivo expanded cord blood cells using the copper chelator tetraethylenepentamine: a phase I/II clinical trial. *Bone Marrow Transpl* 2008;**41**(9):771–8.
58. Frassoni F, Gualandi F, Podestà M, Raiola AM, Ibatici A, Piaggio G, et al. Direct intrabone transplant of unrelated cord-blood cells in acute leukaemia: a phase I/II study. *Lancet Oncol* 2008;**9**(9):831–9.
59. Macmillan ML, Blazar BR, DeFor TE, Wagner JE. Transplantation of ex-vivo culture-expanded parental haploidentical mesenchymal stem cells to promote engraftment in pediatric recipients of unrelated donor umbilical cord blood: results of a phase I-II clinical trial. *Bone Marrow Transpl* 2009;**43**(6):447–54.
60. Robinson SN, Simmons PJ, Thomas MW, Brouard N, Javni JA, Trilok S, et al. Ex vivo fucosylation improves human cord blood engraftment in NOD-SCID IL-2Rgamma(null) mice. *Exp Hematol* Jun 2012;**40**(6):445–56.
61. Delaney C, Heimfeld S, Brashem-Stein C, Voorhies H, Manger RL, Bernstein ID. Notch-mediated expansion of human cord blood progenitor cells capable of rapid myeloid reconstitution. *Nat Med* 2010;**16**(2):232–6.
62. Robinson SN, Ng J, Niu T, Yang H, McMannis JD, Karandish S, et al. Superior ex vivo cord blood expansion following co-culture with bone marrow-derived mesenchymal stem cells. *Bone Marrow Transpl* 2006;**37**(4):359–66.
63. de Lima M, McNiece I, Robinson SN, Munsell M, Eapen M, Horowitz M, et al. Cord-blood engraftment with ex vivo mesenchymal-cell coculture. *N Engl J Med* 2012;**367**(24):2305–15.
64. Peled T, Landau E, Mandel J, Glukhman E, Goudsmid N, Nagler A, et al. Linear polyamine copper chelator tetraethylenepentamine augments long-term ex vivo expansion of cord blood-derived CD34+ cells and increases their engraftment potential in NOD/SCID mice. *Exp Hematol* 2004;**32**(6):547–55.
65. Peled T, Shoham H, Aschengrau D, Yackoubov D, Frei G, Rosenheimer GN, et al. Nicotinamide, a SIRT1 inhibitor, inhibits differentiation and facilitates expansion of hematopoietic progenitor cells with enhanced bone marrow homing and engraftment. *Exp Hematol* 2012;**40**(4):342–55. e1.
66. Rubinstein P, Carrier C, Scaradavou A, Kurtzberg J, Adamson J, Migliaccio AR, et al. Outcomes among 562 recipients of placental-blood transplants from unrelated donors. *N Engl J Med* 1998;**339**(22):1565–77.
67. Gluckman E, Rocha V, Arcese W, Michel G, Sanz G, Chan KW, et al. Factors associated with outcomes of unrelated cord blood transplant: guidelines for donor choice. *Exp Hematol* 2004;**32**(4):397–407.
68. Wagner JE, Barker JN, DeFor TE, Barker KS, Blazar BR, Eide C, et al. Transplantation of unrelated donor umbilical cord blood in 102 patients with malignant and nonmalignant diseases: influence of CD34 cell dose and HLA disparity on treatment-related mortality and survival. *Blood* 2002;**100**(5):1611–8.
69. Xia L, McDaniel JM, Yago T, Doeden A, McEver RP. Surface fucosylation of human cord blood cells augments binding to P-selectin and E-selectin and enhances engraftment in bone marrow. *Blood* 2004;**104**(10):3091–6.
70. Hidalgo A, Frenette PS. Enforced fucosylation of neonatal CD34+ cells generates selectin ligands that enhance the initial interactions with microvessels but not homing to bone marrow. *Blood* 2005;**105**(2):567–75.
71. Shultz LD, Lyons BL, Burzenski LM, Gott B, Chen X, Chaleff S, et al. Human lymphoid and myeloid cell development in NOD/LtSz-scid IL2R gamma null mice engrafted with mobilized human hemopoietic stem cells. *J Immunol* 2005;**174**(10):6477–89.
72. Robinson SN, Simmons PJ, Thomas MW, Brouard N, Javni JA, Trilok S, et al. Ex vivo fucosylation improves human cord blood engraftment in NOD-SCID IL-2Rgamma(null) mice. *Exp Hematol* 2012;**40**(6):445–56.
73. Hoggatt J, Singh P, Sampath J, Pelus LM. Prostaglandin E2 enhances hematopoietic stem cell homing, survival, and proliferation. *Blood* 2009;**113**(22):5444–55.
74. Oran B, Dolan M, Cao Q, Brunstein C, Warlick E, Weisdorf D. Monosomal karyotype provides better prognostic prediction after allogeneic stem cell transplantation in patients with acute myelogenous leukemia. *Biol Blood Marrow Transpl* 2011;**17**(3):356–64.
75. Christopherson 2nd KW, Hangoc G, Broxmeyer HE. Cell surface peptidase CD26/dipeptidylpeptidase IV regulates CXCL12/stromal cell-derived factor-1 alpha-mediated chemotaxis of human cord blood CD34+ progenitor cells. *J Immunol* 2002;**169**(12):7000–8.
76. Christopherson 2nd KW, Hangoc G, Mantel CR, Broxmeyer HE. Modulation of hematopoietic stem cell homing and engraftment by CD26. *Science* 2004;**305**(5686):1000–3.
77. Oran B, Wagner JE, DeFor TE, Weisdorf DJ, Brunstein CG. Effect of conditioning regimen intensity on acute myeloid leukemia outcomes after umbilical cord blood transplantation. *Biol Blood Marrow Transpl: J Am Soc Blood Marrow Transpl* 2011;**17**(9):1327–34.
78. Sauter C, Abboud M, Jia X, Heller G, Gonzales A, Lubin M, et al. Serious infection risk and immune recovery after double-unit cord blood transplantation without antithymocyte globulin. *Biol Blood Marrow Transpl: J Am Soc Blood Marrow Transpl* 2011;**17**(10):1460–71.
79. Leen AM, Myers GD, Sili U, Huls MH, Weiss H, Leung KS, et al. Monoculture-derived T lymphocytes specific for multiple viruses expand and produce clinically relevant effects in immunocompromised individuals. *Nat Med* 2006;**12**(10):1160–6.
80. Doubrovina E, Oflaz-Sozmen B, Prockop SE, Kernan NA, Abramson S, Teruya-Feldstein J, et al. Adoptive immunotherapy with unselected or EBV-specific T cells for biopsy-proven EBV+ lymphomas after allogeneic hematopoietic cell transplantation. *Blood* 2012;**119**(11):2644–56.
81. Heslop HE, Slobod KS, Pule MA, Hale GA, Rousseau A, Smith CA, et al. Long-term outcome of EBV-specific T-cell infusions to prevent or treat EBV-related lymphoproliferative disease in transplant recipients. *Blood* 2010;**115**(5):925–35.
82. Peggs KS, Thomson K, Samuel E, Dyer G, Armoogum J, Chakraverty R, et al. Directly selected cytomegalovirus-reactive donor T cells confer rapid and safe systemic reconstitution of virus-specific immunity following stem cell transplantation. *Clin Infect Dis: Official Publ Infect Dis Soc Am* 2011;**52**(1):49–57.
83. Hanley PJ, Cruz CR, Savoldo B, Leen AM, Stanojevic M, Khalil M, et al. Functionally active virus-specific T cells that target CMV, adenovirus, and EBV can be expanded from naive T-cell populations in cord blood and will target a range of viral epitopes. *Blood* 2009;**114**(9):1958–67.
84. Park KD, Marti L, Kurtzberg J, Szabolcs P. In vitro priming and expansion of cytomegalovirus-specific Th1 and Tc1 T cells from naive cord blood lymphocytes. *Blood* 2006;**108**(5):1770–3.
85. Sun Q, Burton RL, Pollok KE, Emanuel DJ, Lucas KG. CD4(+) Epstein-Barr virus-specific cytotoxic T-lymphocytes from human umbilical cord blood. *Cell Immunol* 1999;**195**(2):81–8.
86. Verneris MR, Miller JS. The phenotypic and functional characteristics of umbilical cord blood and peripheral blood natural killer cells. *Br J Haematol* 2009;**147**(2):185–91.

87. Xing D, Ramsay AG, Gribben JG, Decker WK, Burks JK, Munsell M, et al. Cord blood natural killer cells exhibit impaired lytic immunological synapse formation that is reversed with IL-2 exvivo expansion. *J Immunother Hagerst Md: 1997* 2010;**33**(7):684–96.
88. Miller JS, Soignier Y, Panoskaltsis-Mortari A, McNearney SA, Yun GH, Fautsch SK, et al. Successful adoptive transfer and in vivo expansion of human haploidentical NK cells in patients with cancer. *Blood* 2005;**105**(8):3051–7.
89. Godfrey WR, Spoden DJ, Ge YG, Baker SR, Liu B, Levine BL, et al. Cord blood CD4(+)CD25(+)-derived T regulatory cell lines express FoxP3 protein and manifest potent suppressor function. *Blood* 2005;**105**(2):750–8.
90. Parmar S, Robinson SN, Komanduri K, St John L, Decker W, Xing D, et al. Ex vivo expanded umbilical cord blood T cells maintain naive phenotype and TCR diversity. *Cytotherapy* 2006;**8**(2):149–57.
91. Brunstein CG, Miller JS, Cao Q, McKenna DH, Hippen KL, Curtsinger J, et al. Infusion of ex vivo expanded T regulatory cells in adults transplanted with umbilical cord blood: safety profile and detection kinetics. *Blood* 2011;**117**(3):1061–70.
92. Kohn DB, Dotti G, Brentjens R, Savoldo B, Jensen M, Cooper LJ, et al. CARs on track in the clinic. *Mol Ther: J Am Soc Gene Ther* 2011;**19**(3):432–8.
93. Serrano LM, Pfeiffer T, Olivares S, Numbenjapon T, Bennitt J, Kim D, et al. Differentiation of naive cord-blood T cells into CD19-specific cytolytic effectors for posttransplantation adoptive immunotherapy. *Blood* 2006;**107**(7):2643–52.
94. Micklethwaite KP, Savoldo B, Hanley PJ, Leen AL, Demmler-Harrison GJ, Cooper L, et al. Derivation of human T lymphocytes from cord blood and peripheral blood with antiviral and anti-leukemic specificity from a single culture as protection against infection and relapse after stem cell transplantation. *Blood* 2010;**115**(13):2695–703.
95. Torikai H, Reik A, Liu PQ, Zhou Y, Zhang L, Maiti S, et al. A foundation for universal T-cell based immunotherapy: T cells engineered to express a CD19-specific chimeric-antigen-receptor and eliminate expression of endogenous TCR. *Blood* 2012;**119**(24):5697–705.

Chapter 12

# Cord Blood Stem Cells for Clinical Use: Diabetes and Cord Blood

**Yong Zhao**
*Hackensack University Medical Center, Hackensack, NJ, USA*

**Chapter Outline**



## 1. DIABETES AND GLOBAL CHALLENGES

Diabetes is a major global health issue, with prevalence rates exceeding 12.1% of the population in India, 11.6% in China, and 8.3% in the United States.[1–3] A new study reveals that current prevalence of prediabetes in China was 50.1% of population (493.4 million).[3] According to a report from the American Diabetes Association (Philadelphia), the total number of Americans living with diabetes will increase 64% by 2025, and diabetes-related Medicare expenditures will increase by 72% to $514 billion/year in the United States. Moreover, diabetes and its associated complications (e.g., cardiovascular diseases, stroke, kidney failure, and poor circulation) markedly decrease the quality of life, limiting the regular activity and productivity of individuals with the disease and creating significant economic and social burdens.[4] Thus, it is a top priority to find a cure for diabetes. Clinically, there are two major types of diabetes such as type 1 diabetes (T1D) caused by autoimmune destruction of islet β cells and type 2 diabetes (T2D). Insulin resistance is the hallmark of T2D. It is widely accepted that the inability of pancreatic β cells to function in compensating for insulin resistance leads to the onset of clinical diabetes. Persistent metabolic stresses including glucotoxicity, lipotoxicity, chronic metabolic inflammation, oxidative stress, and endoplasmic reticulum stress, cause progressive dysfunction of islet β cells and finally lead to the cellular death and absolute shortage of islet β cells in long-standing T2D subjects.[5] This chapter will focus on the treatment of diabetes by using cord blood-derived multipotent stem cells (CB-SC), designated Stem Cell Educator Therapy.

### 1.1 Type 1 Diabetes, Pathophysiology, and Challenges in Clinical Management

Type 1 diabetes (T1D) is a T cell-mediated autoimmune disease that reduces the population of pancreatic islet β cells and thereby limits insulin production and glucose homeostasis. Millions of individuals worldwide have T1D, and the number of individuals with diagnosed or undiagnosed T1D is increasing annually. While daily insulin injections offer some

Cord Blood Stem Cells Medicine. http://dx.doi.org/10.1016/B978-0-12-407785-0.00012-8

control over blood sugar levels and may delay the onset of chronic diseases initiated by glucose dysregulation, insulin supplementation is not a cure. It does not halt the persistent autoimmune response, nor can it reliably prevent devastating complications such as neuronal and cardiovascular diseases, blindness, and kidney failure. A true cure has proven elusive despite intensive research pressure over the past 25 years, and the failure of several recent clinical trials that were based on preliminary success in animal models[6–9] further highlights the challenges we face in conquering this disease. Ideally, therapeutic approaches to treating or curing T1D should address many or all of the underlying causes of autoimmunity in T1D. Unfortunately, the etiology of T1D remains largely unknown in humans. Possible triggers for autoimmunity in T1D include genetic, epigenetic, physical, social, and environmental factors. These factors may act independently or jointly to initiate or potentiate the development of autoimmunity.[10,11] As is expected in conditions with multiple contributing factors, T1D-related dysfunction in the immune system has been traced to dysfunctions in multiple cell types and targets including T cells, B cells, regulatory T cells (Tregs), monocytes/macrophages, dendritic cells (DCs), natural killer (NK) cells, and natural killer T (NKT) cells.[12] Due to the polyclonal nature of T1D-related autoimmune responses and the global challenges of immune regulation in T1D patients, therapies and trials that only target one or a few components of the autoimmune response are likely to fail, just as recent trials involving anti-CD3 antibody for T cells and anti-glutamic acid decarboxylase 65 (GAD65) vaccination have failed.[7–9] Successful therapies will likely restore immune balance and peripheral tolerance by addressing changes in multiple targets within the immune system.

## 1.2 Type 2 Diabetes, Pathophysiology, and Challenges in Clinical Management

Type 2 diabetes (T2D) is the most common type of diabetes, with prevalence rates exceeding 12.1% of the population in India, 9.7% in China, and 8.3% in the United States.[1,2] The incidence of T2D is increasing worldwide due to the popularization of a Western lifestyle characterized by overnutrition and limited exercise. Diabetes-associated complications (e.g., cardiovascular diseases, stroke, blindness, kidney failure, and emotional stress) markedly decrease quality of life of T2D patients, limiting the productivity of individuals with the disease and creating significant economic and social burdens. Thus, finding a cure for T2D is a top priority. In adults T2D has traditionally been characterized by elevated fasting blood glucose and an abnormal glucose tolerance test without evidence of autoimmune destruction of pancreatic islet β cells. However, evidence collected over the past decade indicates that the etiology of T2D includes an autoimmune component that initiates an inflammation affecting pancreatic islet β cells,[13–18] which provides new insight into the mechanism and potential treatment of insulin resistance. These findings suggest T2D may be a candidate for some of the therapies in development for T1D, including the use of stem cell-based regeneration of pancreatic islet β cells and immune modulation by adult multipotent stem cells derived from cord blood or bone marrow.[10,19–21] While stem cell transplantation research faces many technical and ethical barriers, the use of cord blood stem cells to modulate immune response may provide a more universally acceptable and feasible approach.

### *1.2.1 Pathophysiology of Metabolic Inflammation and Insulin Resistance in Type 2 Diabetes*

Insulin, a hormone produced by pancreatic islet β cells, plays a key role in regulating cell metabolism, growth, differentiation, survival, and homeostasis through receptors expressed in all tissue cells. The neuro-endocrine network controls insulin regulation from synthesis to release to uptake and action in peripheral tissues. Obesity and lack of exercise are associated with increased risk of insulin resistance, and recent evidence indicates this increased risk is due at least in part to adipocyte-mediated immune dysfunction and inflammation that may affect insulin regulation and uptake. Inflammatory cytokines derived from adipocytes and macrophages promote the development of insulin resistance in T2D through c-Jun N-terminal protein kinase (JNK) and/or IkappaB kinase beta/Nuclear factor-kappaB (IKKβ/NF-κB) pathways, including changes in the levels of tumor necrosis factor-α (TNFα), interleukin-1 (IL-1), IL-6, IL-17, monocyte chemoattractant protein-1 (MCP-1), resistin, plasminogen activator inhibitor-1 (PAI-1), and others.[18,22–24] Despite the complexity and multifactorial nature of T2D, metabolic inflammation is the most common step leading to insulin resistance in the disease (Figure 1). Although this relationship between insulin resistance and inflammation is a relatively recent finding, anti-inflammation therapy is rapidly gaining acceptance as an approach for the treatment of insulin-resistance in patients with T2D.[15,18,25–27]

### *1.2.2 Metabolic Abnormalities Cause Immune Dysfunctions in T2D*

The human immune system does not normally recognize single glucose and/or lipid molecules as antigens unless they occur in glycolipids or lipoproteins capable of stimulating an immune response. Like other cells, cells of the immune system rely on insulin signaling to utilize glucose and/lipid as regular fuels for energy to perform their normal functions. Overnutrition-related hyperglycemia and/or hyperlipidemia cause chronic toxicity to multiple body systems including the immune system, and interfere with the normal response to insulin. The resulting oxidative stress, mitochondrial dysfunction, and endoplasmic reticulum (ER) stress impair

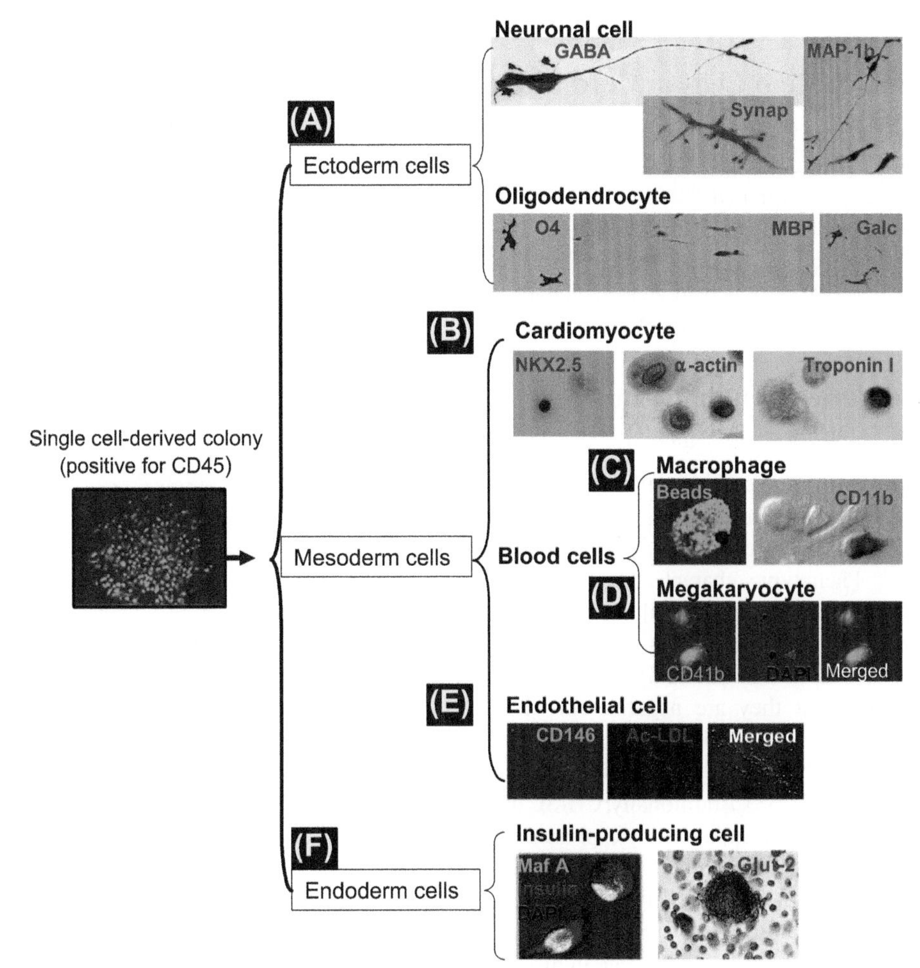


**FIGURE 1 Human cord blood-derived multipotent stem cells (CB-SC) give rise to multiple lineages in response to different physiological growth factors and inducers.** CB-SC were isolated from human umbilical cord blood using Ficoll-Hypaque ($\gamma = 1.077$) and plated in non-tissue culture-treated Petri dishes in RPMI 1640 medium supplemented with 7% fetal bovine serum, with formation of a single cell-derived colony positive for leukocyte common antigen CD45 (left panel). (A) Differentiation to neuronal cell and oligodendrocyte. CB-SC were treated with 100 ng/ml NGF for 10–14 days and characterized with lineage markers including γ-aminobutyric acid (GABA), microtubule-associated protein (MAP) 1B, synaptophysin (Synap), sulfatide O4, myelin basic protein (MBP), and galactocerebroside (Galc). (B) Differentiation to cardiomyocytes. CB-SC were treated with a chemical 3 μM 5-aza-2′ deoxycytidine for 24 h, followed by testing with cardiomyocyte markers including nuclear transcription factor Nkx2.5, cardiomyocyte-specific α-actin, and troponin I. (C) Differentiation to macrophages. CB-SC were treated with 50 ng/ml M-CSF for 7–10 days and then characterized with phagocytosis of fluorescence beads and surface marker CD11b/Mac-1. (D) Differentiation to megakaryocytes. CB-SC were treated with 10 ng/ml TPO for 10–14 days and then characterized with specific marker CD41b and polyploidy nuclear (red arrow). (E) Differentiation to endothelial cells. CB-SC were treated with 50 ng/ml VEGF for 10–14 days, and then characterized with specific marker CD146 and incorporation of the acetylated low density lipoprotein (Ac_LDL). (F) Differentiation to insulin-producing cells. CB-SC were treated with 10 nM exendin-4 + 25 mM glucose for 5–8 days and characterized with markers for β cell marker insulin and Glut2. Isotype-matched IgG served as negative controls for every experiment. *Refs 35, 10.*

intracellular homeostasis and can alter both innate and adaptive immune responses and promote inflammation.[16,28–30]

Healthy pancreatic islets are protected by a specialized basement membrane that serves as a barrier against immune cells and assists in maintaining homeostasis.[31,32] Studies using a humanized immune-mediated diabetic mouse model demonstrate that immune cells cannot cross the basement membrane into pancreatic islets unless triggered by antigen-presenting cells (APCs).[33] Donath and colleagues[28] found infiltrated macrophages inside pancreatic islets of T2D subjects. These macrophages may present islet β cell antigens to T cells and initiate the autoimmune responses, which is supported by evidence of autoimmune response in T2D similar to the response observed in T1D.[13,14] About 10% of subjects with T2D are diagnosed with "latent autoimmune diabetes in adults" following a positive test for at least one of the known T1D-related autoantibodies (e.g., islet cell antibodies, anti-protein tyrosine phosphatase-like protein IA2, anti-insulin, and GAD65).[17,34] In addition to these humoral autoimmune responses, Ismail and colleagues reported that some T2D patients who test negative for islet autoantibodies have T cells responsive to islet proteins in the peripheral blood.[17] Thus, it appears that autoimmune responses contribute to the pathogenesis of T2D in some, if not all, patients.

# 2. STEM CELLS IN CORD BLOOD

## 2.1 Introduction of Stem Cells in Cord Blood

Human cord blood contains several types of stem cells including hematopoietic stem cells (HSC), multipotent stem cells that have been designated CB-SC,[35] mesenchymal stem cells (MSC), endothelial progenitor cells (EPC), and monocyte-derived stem cells.[36] Phenotypic characterizations of HSC, MSC, and EPC have been reviewed recently.[37] Here, we focus on the novel type of stem cells, CB-SC.

## 2.2 Cord Blood-derived Multipotent Stem Cells

CB-SC are a unique type of stem cell identified from human cord blood,[10,35] which are different from other types of stem cells including HSC, MSC (Table 1), EPC, and monocyte-derived stem cells.[36] Phenotypic characterization demonstrates that CB-SC display embryonic cell markers (e.g., transcription factors OCT-4 and Nanog, stage-specific embryonic antigen (SSEA)-3, and SSEA-4) and leukocyte common antigen CD45, but they are negative for blood cell lineage markers (e.g., CD1a, CD3, CD4, CD8, CD11b, CD11c, CD13, CD14, CD19, CD20, CD34, CD41a, CD41b, CD83, CD90, CD105, and CD133). Additionally, CB-SC display very low immunogenicity, as indicated by expression of a very low level of major histocompatibility complex (MHC) antigens and failure to stimulate the proliferation of allogeneic lymphocytes.[35,38] They can give rise to three embryonic layer-derived cells in the presence of different inducers[10] (Figure 1). More specifically, CB-SC tightly adhere to culture dishes with a large rounded morphology and are resistant to common detaching methods (trypsin/EDTA), making it easy to collect suspended lymphocytes and separate with CB-SC after ex-vivo coculture.[35,38,39] Thus, during Stem Cell Educator Therapy, only the CB-SC-educated autologous lymphocytes are returned to the subjects.[40]

**TABLE 1** Phenotypic Comparison between CB-SCs and MSCs

| | CB-SCs | MSCs |
|---|---|---|
| Major source | Human cord blood | Human bone marrow, adipose tissue, placenta, umbilical cord |
| Morphology | Round | Spindle or fibroblast-like |
| Attaching surface | Hydrophobic surface | Hydrophilic surface |
| Cell detachment | Tightly adhered, resistant to EDTA/trypsin detachment | Sensitive to EDTA/trypsin detachment |
| **Cell surface markers** | | |
| Hematopoietic stem cell marker CD34 | Negative | Negative |
| Leukocyte common antigen CD45 | Strongly positive | Clearly negative |
| Thy-1 antigen CD90 | Negative | Positive |
| Endoglin CD105 | Negative | Positive |
| **Immunogenicity** | | |
| Class I: HLA-ABC | Very low | High |
| Class II: HLA-DR, DQ | Negative | Negative |

## 2.3 Stem Cell Educator Therapy

Based on the preclinical evidence that CB-SC possess the immune modulations,[10,38,39,41] we have developed the Stem Cell Educator Therapy in clinical trials[40,42,43] (Figure 2). Briefly, a 16-gauge IV needle is placed in the median cubital vein to isolate lymphocytes from the patient's blood by using a Blood Cell Separator. The collected lymphocytes are transferred into the device for exposure to CB-SC, and other blood components are automatically returned to the patient.[40] The stem cell educator functions as part of a closed-loop system that circulates a patient's blood through a blood cell separator, briefly cocultures the patient's lymphocytes with CB-SC in vitro, and returns the educated lymphocytes to the patient's circulation.[4,40] CB-SC tightly attached to interior surfaces in the device, and only the CB-SC-educated autologous lymphocytes are returned to the subjects. The Stem Cell Educator Therapy requires only two venipunctures with minimal pain, and does not introduce stem cells or reagents into patients in comparison with other stem cell-based therapies (e.g., MSC and HSC).[40] Additionally, CB-SC display very low immunogenicity, eliminating the need for human leukocyte antigen (HLA) matching prior to treatment.[10,35,38,40] Thus, these advantages of Stem Cell Educator Therapy may provide CB-SC-mediated immune modulation therapy while mitigating the safety and ethical concerns associated with other stem cell-based approaches and conventional immune therapies.

## 2.4 Safety of Stem Cell Educator Therapy Compared to Other Stem Cell-based Immune Therapies

Our published data[40] and unpublished data demonstrated that Stem Cell Educator Therapy was well tolerated in all participants with minimal pain from two venipunctures. Most patients experienced mild discomfort during venipuncture and some soreness of the arm during apheresis,

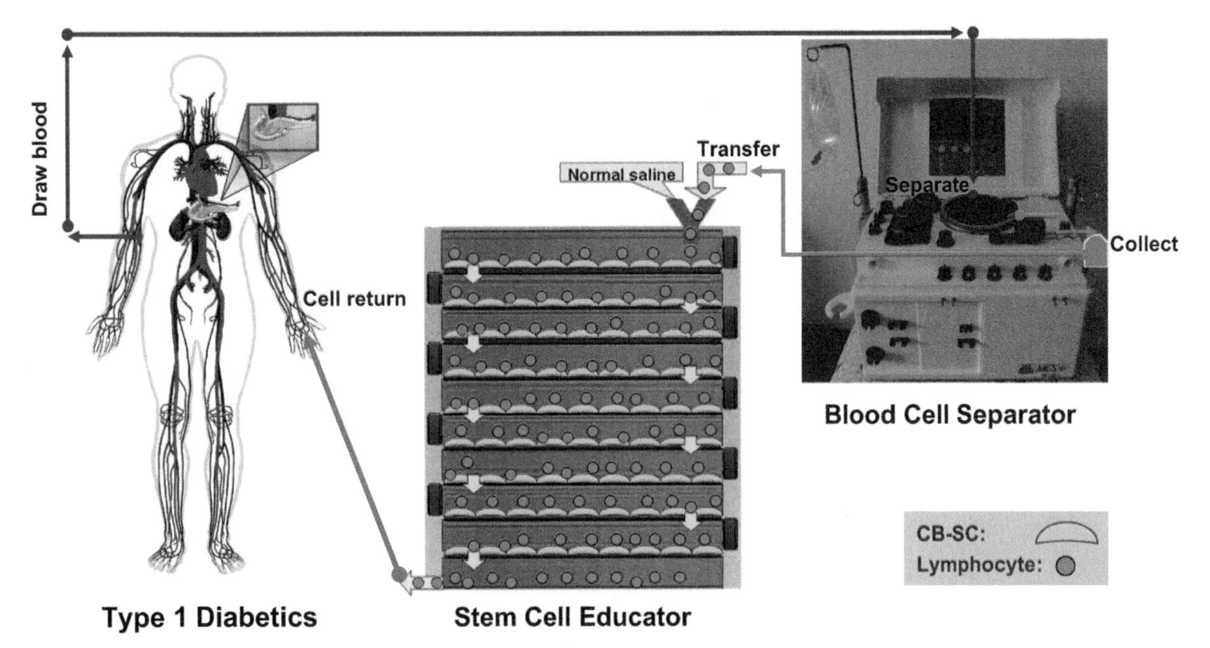


FIGURE 2 **Overview of Stem Cell Educator Therapy.** A T1D participant (left) is connected to a Blood Cell Separator (right) and the Stem Cell Educator (bottom center) to form a closed system. Lymphocytes isolated from the T1D participant by the Blood Cell Separator travel through the Stem Cell Educator where they come in contact with CB-SC attached to the interior surfaces of the device. Educated lymphocytes are returned to the patient's blood circulation. *Ref. 40.*

but discomfort and soreness resolved quickly following the conclusion of the procedure.[40] No participants experienced any significant adverse events during the course of treatment, and no adverse events occurred during one year follow-up studies. In comparison to the application of MSCs, autologous bone marrow-derived MSC has been limited for clinical applications due to the painful operations and potential infections in the procedure for harvesting bone marrow. To this end, placenta or umbilical cords are easy to access and represent valuable sources for provision of allogeneic MSCs. However, patients usually showed medium or high fever following transplant of allogeneic MSCs through interventional therapy such as intravenous delivery or direct infusion into pancreatic islets via transfemoral cannulation under angiography.

## 3. APPLICATION OF STEM CELL EDUCATOR THERAPY IN T1D

### 3.1 Clinical Efficacy in the Treatment of T1D

Findings from our clinical trials provide powerful evidence that a single treatment with the Stem Cell Educator provides lasting reversal of autoimmunity that allows regeneration of islet β cells and improvement of metabolic control in individuals with long-standing T1D.[40] In an open-label, phase I/phase II study, 15 patients with T1D received one treatment with the Stem Cell Educator. Their median age was 29 years (range, 15–41), and median diabetic history was 8 years (range, 1–21). Stem Cell Educator Therapy can markedly improve C-peptide levels, reduce the median glycated hemoglobin $A_1C$ ($HbA_1C$) values, and decrease the median daily dose of insulin in patients with some residual β cell function (n=6) and patients with no residual pancreatic islet β cell function (n=6) (Figure 3(A)). Treatment also produced an increase in basal and glucose-stimulated C-peptide levels through 40 weeks (Figure 3(B)). However, participants in the Control Group (n=3) did not exhibit significant change at any follow-up.[40] Notably, a single treatment could improve islet β function that lasts a year.[42] Individuals who received Stem Cell Educator Therapy exhibited increased expression of costimulating molecules (specifically, CD28 and inducible costimulatory molecule (ICOS)), increases in the number of $CD4^+CD25^+Foxp3^+$ Tregs, and restoration of Th1/Th2/Th3 cytokine balance.[40] Thus, findings from these trials indicate that CB-SC-mediated reversal of autoimmunity results from modulation of the immune response in multiple immune cell types, thereby meeting the expectation that successful therapies will likely address different arms of the autoimmune response and balance the immune system through the systemic and local modulations.[39,40]

### 3.2 Molecular Mechanisms Underlying the Immune Modulation of Stem Cell Educator Therapy in Type 1 Diabetes

Preclinical studies and clinical data demonstrated the immune modulation of CB-SC and the therapeutic efficacy of Stem Cell Educator Therapy in T1D. The immune modulation of CB-SC can be achieved by a variety of molecular and cellular mechanisms, which include: (1) expression of autoimmune regulator (Aire) in CB-SC plays an essential

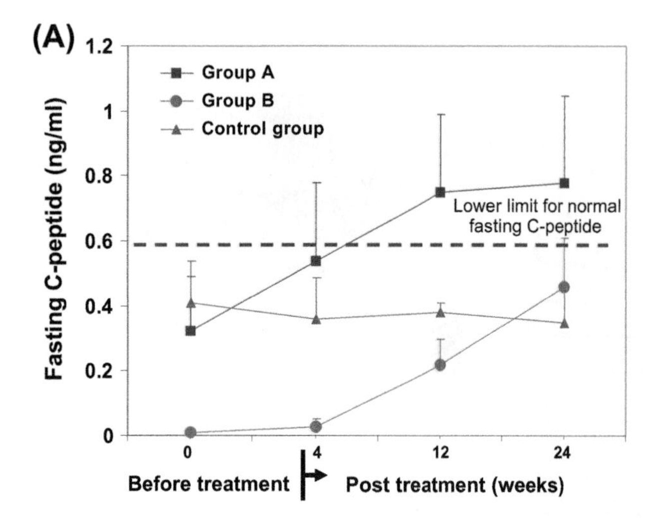


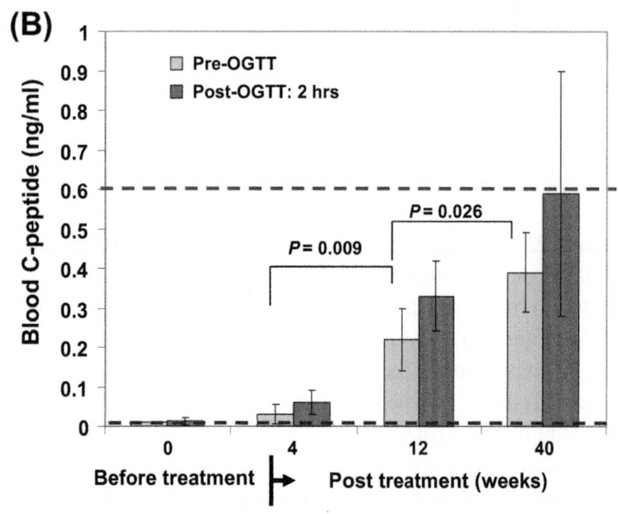


**FIGURE 3 Improvement of β-cell Function by Stem Cell Educator Therapy.** (A) Fasting C-peptide levels of T1D participants over 24 weeks. Group A and Group B participants (n=6 per group) received one Stem Cell Educator treatment. Control group participants (n=3) received sham therapy (no CB-SC in the Stem Cell Educator). (B) Comparison of C-peptide levels at glucose challenge after 40-week follow up in Group B T1D subjects. The dashed red line indicates the lower limit for normal C-peptide levels in Chinese populations. The dashed purple line indicates the minimum detectable level (sensitivity) of C-peptide by radioimmunoassay (RIA). *Ref. 40.*

role. Using human Aire-specific small interfering RNAs (siRNA) to knock down Aire expression in CB-SC, the data indicate that Aire is involved in immune modulation and induction of immune tolerance following Stem Cell Educator Therapy.[40] (2) Increase the percentage of Tregs following Stem Cell Educator Therapy.[40] (3) Correct the functional defects of Tregs.[39] (4) Directly suppress the islet β cell-specific T cell clones.[10] (5) Act through the cell–cell contacting mechanism via the surface molecule programmed death ligand 1 (PD-L1) on CB-SC.[38] (6) Act through the soluble factors released by CB-SC (e.g., nitric oxide, TGF-β1).[38] During the ex vivo coculture, T1D-derived effector T cells and/or Tregs can be educated by the favorable microenvironment created by CB-SC through cell-to-cell contact and soluble factors.[38,39,44] Quantitative real time PCR array indicated that in vitro coculture with CB-SC causes substantial modifications of gene expressions in Tregs, specifically for function-related cytokine and chemokine genes along with signaling pathway molecules and transcription factors.[39]

Specifically, our clinical trial provides evidence that CB-SC in the device educate effector T cells and/or Tregs, resulting in lasting changes in the expression of costimulating molecules, increasing the population of Tregs, and restoring Th1/Th2/Th3 cytokine balance, each of which is expected to improve control of autoimmunity of T1D.[39,45] Therapy also increases production of TGF-β1 in plasma of T1D subjects, one of the best-characterized cytokines contributing to the induction of peripheral immune tolerance.[46] Results from a nonobese diabetic (NOD) mouse study[39] demonstrated that increased plasma TGF-β1 may contribute to the formation of a "TGF-β1 ring" around pancreatic islets that protects β cells against infiltrating lymphocytes, providing a safe environment for promotion of β cell regeneration.[39,41]

## 3.3 Cellular Mechanisms Underlying the Regeneration of Islet β Cells of Stem Cell Educator Therapy

Pancreatic islets in long-standing T1D are completely destroyed by autoimmune cells.[39] Abrogation of autoimmunity without an adequate residual β-cell mass will not restore normoglycemia and improve metabolic control. Promotion of β-cell neogenesis must be part of any therapy aimed at T1D treatment. Notably, our clinical data provide powerful evidence that reversal of autoimmunity by Stem Cell Educator Therapy leads to regeneration of islet β cells and improvement of metabolic control in long-standing T1D subjects.[40] CB-SC from the device are not likely to be the source of this regeneration because they are not transferred to the patient during therapy.[40] As demonstrated in other studies, the regenerated cells may be derived from multiple endogenous resources such as transdifferentiations of duct cells or α cells,[47,48] and peripheral blood-derived insulin-producing cells.[49]

Using the ultrasensitive assay, Wang and colleagues revealed that C-peptide production persists for decades after onset of T1D.[50] Probably, the function of these residual β cells in long-standing T1D subjects may be recovered to replicate after treatment with the Stem Cell Educator Therapy. Additionally, the formation of a "TGF-β1 ring" around pancreatic islets[39] may provide a safe environment to promote the regeneration of these residual β cells and protect them against autoimmune re-attacking.[39,41]

Our previous work identified a cell population from adult human blood displaying high potential for producing insulin (designated peripheral blood-derived insulin-producing cells, PB-IPC) by using a similar approach by attaching to a plastic surface, without any genetic manipulation and any induction of differentiation.[49] In vitro characterization demonstrates that PB-IPC display the characteristics of islet β-cell progenitors, including the expression of β cell-specific insulin gene transcription factors (e.g., MafA, Nkx6.1, and PDX-1), prohormone convertases (PC1 and PC2), and the production of insulin and its by-product C-peptide. In vivo transplantation demonstrated that PB-IPC can give rise to functional insulin-producing cells after administering into the chemical streptozotocin-induced diabetic NOD-scid mice, as indicated by the production of human C-peptide in mouse plasma and reduction of hyperglycemia.[49] The data imply that PB-IPC may be a potential resource contributing the neogenesis of β cells after Stem Cell Educator Therapy.

Recently, insulin-producing cells have been generated from bone marrow-derived mesenchymal cells under in vitro conditions. Yet they remain controversial. Most bone marrow-derived stem cells originate from mesenchymal cells, such as very small embryonic-like cells characterized by Ratajczak and colleagues,[51] and the marrow-isolated adult multilineage inducible cells characterized by Schiller and colleagues.[52] Additionally, mesenchymal stem cells from adult human islet-derived precursor cells, and from Wharton's Jelly of the human umbilical cord can also give rise to insulin-expressing cells.[53] Due to lack of the hallmark leukocyte common antigen CD45, they are different from our reported PB-IPC.

Islet transplantation, drug-mediated promotion of β-cell regeneration, and stem cell transplantation have been proposed and tested as likely approaches for treating T1D. However, the continued presence of autoreactive effector T cells and B cells in the circulation may destroy insulin-producing cells generated through these approaches, thereby minimizing their therapeutic potential. An alternative approach using ex vivo coculture of immune cells through Stem Cell Educator Therapy holds promise for addressing both persistent autoimmunity and the regeneration of insulin-producing β cells.

# 4. APPLICATION OF STEM CELL EDUCATOR THERAPY IN TYPE 2 DIABETES

## 4.1 Clinical Efficacy in the Treatment of T2D

In an open-label, phase I/phase II study, patients (N=36) with long-standing T2D were divided into three groups (Group A, n=18; Group B, n=11; and Group C with impaired β-cell function, oral medications + insulin, n=7). All subjects received one treatment with the Stem Cell Educator. Clinical findings indicate that T2D patients achieve improved metabolic control and reduced inflammation markers after receiving Stem Cell Educator Therapy. Median glycated hemoglobin ($HbA_1C$) in Group A and B was significantly reduced from 8.61%±1.12 at baseline to 7.9%±1.22 at 4 weeks ($p=0.026$), 7.25%±0.58 at 12 weeks ($p=2.62E\text{-}06$), and 7.33%±1.02 at one year post treatment ($p=0.0002$).[43] Homeostasis model assessment of insulin resistance demonstrated that insulin sensitivity was improved post treatment (Figure 4(A)). Notably, the β-cell function in Group C subjects was markedly recovered, as indicated by the restoration of C-peptide levels (0.36±0.19 ng/ml at baseline vs 1.12±0.33 ng/ml at one year post treatment, $p=0.00045$) (Figure 4(B)). Mechanistic studies revealed that Stem Cell Educator Therapy reverses immune dysfunctions through immune modulation on monocytes and balancing Th1/Th2/Th3 cytokine production. Stem Cell Educator Therapy is a safe and innovative approach that produces lasting improvement in metabolic control for individuals with moderate or severe T2D.

## 4.2 Molecular and Cellular Mechanisms Underlying Stem Cell Educator Therapy in Type 2 Diabetes

### *4.2.1 Correct the Functional Defects of Monocytes/Macrophages*

Chronic inflammation of visceral adipose tissue (VAT) is a major contributor to insulin resistance mediated by adipose tissue-released adipokines (e.g., IL-6, TNFα, MCP-1, and resistin).[54,55] Growing evidence strongly demonstrated that an accumulation of macrophages by metabolic stress in the sites of affected tissues (such as vasculature, adipose tissue, muscle, and liver) has emerged as a key process in the chronic metabolic-stress-induced inflammation.[56] Monocytes/macrophages, as one of the professional antigen-presenting cells, play an essential role in control of Th1/Th2 immune responses and maintaining homeostasis through the costimulating molecules CD80/CD86 and released cytokines. Persistent destructive effects of lipid influx (e.g., fatty acids and cholesterol) cause macrophage dysfunctions (including defective efferocytosis and unresolved inflammation), resulting in recruitment and activation of more monocytes/macrophages via MCP-1 and its receptor CCR2.[56] Consequently, inflammatory cytokines (e.g., IL-6 and TNFα) produced by activated macrophages induce insulin resistance in major metabolic tissues.[56–58] To prove the action of macrophage in chronic inflammation and insulin resistance in T2D, conditional depletion of $CD11c^+$ macrophages or inhibition of macrophage recruitment via MCP-1 knockout in obese mice resulted in a significant reduction in systemic inflammation and an increase in insulin sensitivity.[59–61]

To clarify the modulation of Stem Cell Educator Therapy on blood monocytes, we found that expression of CD86 and $CD86^+CD14^+/CD80^+CD14^+$ monocyte ratios have

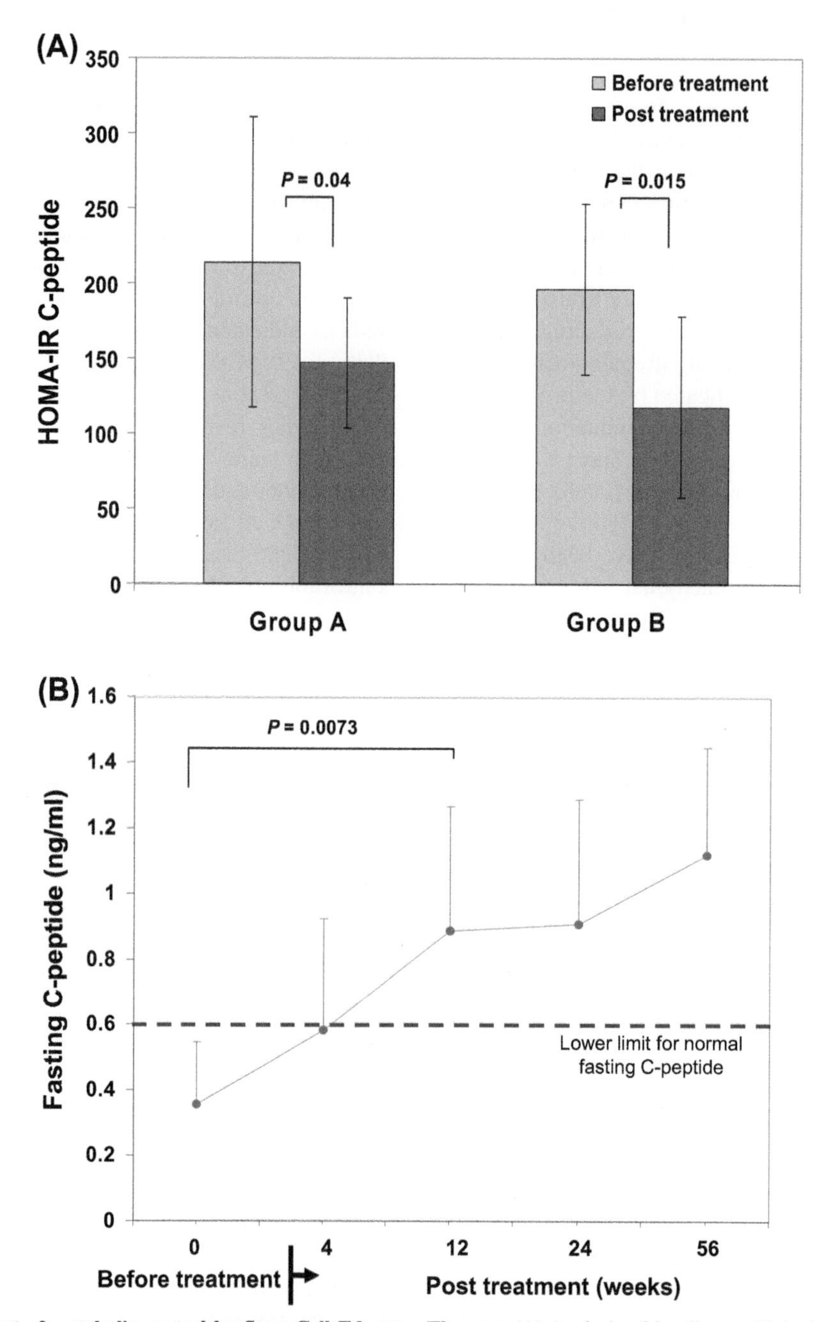


**FIGURE 4 Improvement of metabolic control by Stem Cell Educator Therapy.** (A) Analysis of insulin sensitivity by HOMA-IR C-peptide at 4 weeks post treatment with Stem Cell Educator Therapy. (B) 56-week follow-up C-peptide levels in Group C T2D subjects with impaired islet β cell function. Patients (N = 36) with long-standing T2D were divided into three groups (Group A, oral medications, n = 18; Group B, oral medications + insulin injections, n = 11; Group C having impaired β-cell function with oral medications + insulin injections, n = 7). All patients received one treatment with the Stem Cell Educator Therapy in which a patient's blood is circulated through a closed-loop system that separates mononuclear cells from the whole blood, briefly cocultures them with adherent CB-SC, and returns the educated autologous cells to the patient's circulation. *Ref. 43.*

been markedly changed after receiving Stem Cell Educator Therapy in T2D subjects. CD80 and CD86 are two principal costimulating molecules expressed on monocytes to skew the immune response toward Th1 or Th2 differentiation through their ligands CD28/CTLA4.[62,63] Due to the differences in expression levels and binding affinity between CD80 and CD86 with their ligands CD28/CTLA4, it is widely accepted that the interaction of CD86 with CD28 dominates in costimulating signals; conversely, the combination of CD80 and CTLA4 governs negative signaling.[62–65] The normalization of the $CD86^+CD14^+/CD80^+CD14^+$ monocyte ratio post treatment may favor the immune balance of Th1/Th2 responses in diabetic subjects. Taken together with our in vitro study on the direct interaction between CB-SC and purified CD14+ monocytes, these data indicate that restoration of monocyte functions (such as the expression of CD86, cytokine production, and chemokine production) mainly contributes to anti-inflammation and reversal of insulin resistance following Stem Cell Educator Therapy in T2D subjects.

### 4.2.2 Correct the Functional Defects of Other Immune Cells

Increasing animal and clinical evidence demonstrate multiple immune cells contributing to the inflammation-induced insulin resistance in T2D, such as abnormalities of lymphocytes (including T cells, B cells, and Tregs[27,66–69]), neutrophils,[70] eosinophils,[71] mast cells,[72] and dendritic cells (DCs).[73,74] Specifically, B and T lymphocytes have emerged as unexpected promoters and controllers of insulin resistance.[69] These adaptive immune cells infiltrate into the VAT, releasing cytokines (IL-6 and TNFα) and recruiting more monocytes/macrophages via MCP-1/CCR2.[56] Finally, this obese-related inflammation leads to insulin resistance.[69,75] Thus, a major challenge for treatment of T2D is to identify therapeutic approaches that fundamentally correct insulin resistance through targeting the dysfunctions of multiple immune cells. The valuable lessons from intensive research pressure over the past 25 years in T1D[6] highlight the difficulties in overcoming these multiple immune dysfunctions by utilizing the conventional immune therapy. Stem Cell Educator Therapy functions as "an artificial thymus" that circulates a patient's blood through a blood cell separator,[42] briefly cocultures the patient's blood mononuclear cells (such as T cells, B cells, Tregs, monocytes, and neutrophils) with CB-SC in vitro. During the ex vivo coculture in the device, these mononuclear cells can be educated by the favorable microenvironment created by CB-SC through: (1) the action of autoimmune regulator (Aire) expressed in CB-SC;[40] (2) the cell–cell contacting mechanism via the surface molecule programmed death ligand 1 (PD-L1) on CB-SC;[38] (3) the soluble factors released by CB-SC. Previous work[38] and current data indicate that CB-SC-derived NO mainly contributes to the immune modulation on T cells and monocytes. While monocytes and other immune cells pass through the device, NO as a free radical released by CB-SC can quickly transmit into their cellular membrane, without the aid of dedicated transporters; (4) correcting the functional defects of Tregs;[39] and (5) directly suppressing the pathogenic T cell clones.[10] During this procedure, both peripheral and infiltrated immune cells in VAT can be isolated by a blood cell separator and treated by CB-SC, leading to the correction of chronic inflammation, the restoration of the immune balance, and clinical improvements in metabolic control via increasing insulin sensitivity. Additionally, TGF-β1 is a well-recognized cytokine with a pleiotropic role in immune modulation on multiple immune cells such as the differentiation and function of Th1/Th2 cells and Tregs, as well as B cells, monocytes/macrophages, dendritic cells, granulocytes, and mast cells.[46,76] These immune cells are involved in the inflammation-induced insulin resistance in T2D.[27,66–74] Therefore, the upregulation of TGF-β1 levels in the peripheral blood of T2D subjects is another major mechanism underlying the immune modulation after receiving Stem Cell Educator Therapy.

During the procedure of Stem Cell Educator Therapy, the mononuclear cells circulating in a patient's blood are collected by a Blood Cell Separator. Additionally, patients are required to move their hips, legs, and turn to one side every 15–30 min during the treatment, in order to mobilize their immune cells from peripheral tissues (include adipose tissues) and organs entering into the blood circulation to be processed by a Blood Cell Separator. Thus, the immune cells both in peripheral blood and in tissues can be isolated by a Blood Cell Separator and treated by CB-SC. The full blood volume is processed approximately twice during Stem Cell Educator Therapy (approximately 10,000 ml whole blood),[40] which ensures a comprehensive approach to modulating essentially all circulating immune cells to address multiple immune dysfunctions and overcome global insulin resistance resulting from a variety of causes. No other current medications and/or approaches have yet been shown to achieve this unique therapy success. There are some pathogenic immune cells remaining in tissues and lymph nodes which fail to enter into the blood circulation during the procedure and may escape from treatment by CB-SC. These immune cells may migrate into the blood circulation and decrease the therapeutic effectiveness. Therefore, T2D subjects may need additional treatment 6–9 months later after receiving the first treatment; however, this is yet to be explored in the phase III clinical trial.

We observed that the improvement of islet β cell function (C-peptide levels) progresses slowly over weeks after receiving Stem Cell Educator Therapy, not vanishing again with progression of time. We reported similar data in a previous T1D trial.[40,42] If Stem Cell Educator Therapy only temporarily corrects the immune dysfunctions, the clinical efficacy in metabolic control should vanish soon after receiving Stem Cell Educator Therapy, because of the short life spans of most immune cells (e.g., 5.4 days for neutrophils,[77] 3 months for lymphocytes, 1–3 days for bone marrow-derived monocytes existing in blood and then migrating into tissues). Previous work demonstrated that CB-SC showed the marked modulation of Th1-Th2-Th3 cell-related genes including multiple cytokines and their receptors, chemokines and their receptors, cell surface molecules, along with signaling pathway molecules and transcription factors, as indicated by quantitative real time PCR array.[39] Due to these fundamental immune modulations and induction of immune balance,[42] this trial indicates that a single treatment with Stem Cell Educator Therapy can give rise to long-lasting reversal of immune dysfunction and improvement of insulin sensitivity in long-standing T2D subjects.

## 5. CONCLUSIONS

The epidemic of diabetes is creating an enormous impact on the global economy as well as the health of humans. Overcoming the immune dysfunctions is the major common target for the treatment of T1D and T2D. To date, phase I/II clinical trials in T1D reveal that a single treatment with the Stem Cell Educator provides lasting reversal of autoimmunity that allows regeneration of islet β cells and improvement of metabolic control in subjects with long-standing T1D; the phase I/II clinical study demonstrates that Stem Cell Educator Therapy can control the immune dysfunctions and restore the immune balance through the modulation of monocytes/macrophages and other immune cells both in peripheral blood and in tissues, leading to a long-lasting reversal of insulin resistance and a significant improvement in insulin sensitivity and metabolic control in long-standing T2D subjects. Stem Cell Educator Therapy functions as "an artificial thymus" that circulates a patient's blood through a blood cell separator, briefly cocultures the patient's immune cells with CB-SC in vitro, induces immune tolerance through the action of key transcription factor Aire, returns the educated autologous cells to the patient's circulation, and restores the immune balance and homeostasis. Successful immune modulation by CB-SC and the resulting clinical improvement in patient status may have important implications for other autoimmune and inflammation-related diseases without the safety and ethical concerns associated with conventional stem cell-based approaches.

## DR YONG ZHAO'S BIBLIOGRAPHY

Dr Yong Zhao has completed his M.D. and Ph.D. from Shanghai Second Military Medical University and postdoctorate at the University of Chicago. He worked as Assistant Professor at the University of Illinois at Chicago for 8 years. Currently, he is an Associate Scientist at Hackensack University Medical Center. He identified a novel type of stem cell from human cord blood and developed the Stem Cell Educator Therapy in clinical trials (phase I/II). He owns eight patents. He has published more than 30 papers. He received several national and international awards. His works were released in major media such as CNN, USA Today, Reuter, and EurekAlert.

## REFERENCES


1. Diamond J. Medicine: diabetes in India. *Nature* 2011;**469**:478–9.
2. Yang W, Lu J, Weng J, Jia W, Ji L, Xiao J, et al. Prevalence of diabetes among men and women in China. *N Engl J Med* 2010;**362**:1090–101.
3. Xu Y, Wang L, He J, Bi Y, Li M, Wang T, et al. Prevalence and control of diabetes in Chinese adults. *JAMA* 2013;**310**:948–59.
4. Zhao Y, Jiang Z, Guo C. New hope for type 2 diabetics: targeting insulin resistance through the immune modulation of stem cells. *Autoimmun Rev* 2011;**11**:137–42.
5. Nolan CJ, Damm P, Prentki M. Type 2 diabetes across generations: from pathophysiology to prevention and management. *Lancet* 2011;**378**:169–81.
6. Couzin-Frankel J. Trying to reset the clock on type 1 diabetes. *Science* 2011;**333**:819–21.
7. Bach JF. Anti-CD3 antibodies for type 1 diabetes: beyond expectations. *Lancet* 2011;**378**:459–60.
8. Mathieu C, Gillard P. Arresting type 1 diabetes after diagnosis: GAD is not enough. *Lancet* 2011;**378**:291–2.
9. Wherrett DK, Bundy B, Becker DJ, DiMeglio LA, Gitelman SE, Goland R, et al. Antigen-based therapy with glutamic acid decarboxylase (GAD) vaccine in patients with recent-onset type 1 diabetes: a randomised double-blind trial. *Lancet* 2011;**378**:319–27.
10. Zhao Y, Mazzone T. Human cord blood stem cells and the journey to a cure for type 1 diabetes. *Autoimmun Rev* 2010;**10**:103–7.
11. Bluestone JA, Herold K, Eisenbarth G. Genetics, pathogenesis and clinical interventions in type 1 diabetes. *Nature* 2010;**464**:1293–300.
12. Lehuen A, Diana J, Zaccone P, Cooke A. Immune cell crosstalk in type 1 diabetes. *Nat Rev Immunol* 2010;**10**:501–13.
13. Brooks-Worrell B, Palmer JP. Is diabetes mellitus a continuous spectrum? *Clin Chem* 2011;**57**:158–61.
14. Brooks-Worrell BM, Reichow JL, Goel A, Ismail H, Palmer JP. Identification of autoantibody-negative autoimmune type 2 diabetic patients. *Diabetes Care* 2011;**34**:168–73.
15. Goldfine AB, Fonseca V, Shoelson SE. Therapeutic approaches to target inflammation in type 2 diabetes. *Clin Chem* 2011;**57**:162–7.
16. Mathis D, Shoelson SE. Immunometabolism: an emerging frontier. *Nat Rev Immunol* 2011;**11**:81.
17. Naik RG, Palmer JP. Latent autoimmune diabetes in adults (LADA). *Rev Endocr Metab Disord* 2003;**4**:233–41.
18. Shoelson SE, Lee J, Goldfine AB. Inflammation and insulin resistance. *J Clin Invest* 2006;**116**:1793–801.
19. English K, French A, Wood KJ. Mesenchymal stromal cells: facilitators of successful transplantation? *Cell Stem Cell* 2010;**7**:431–42.
20. Uccelli A, Moretta L, Pistoia V. Mesenchymal stem cells in health and disease. *Nat Rev Immunol* 2008;**8**:726–36.
21. Nauta AJ, Fibbe WE. Immunomodulatory properties of mesenchymal stromal cells. *Blood* 2007;**110**:3499–506.
22. Olefsky JM, Glass CK. Macrophages, inflammation, and insulin resistance. *Annu Rev Physiol* 2010;**72**:219–46.
23. Schenk S, Saberi M, Olefsky JM. Insulin sensitivity: modulation by nutrients and inflammation. *J Clin Invest* 2008;**118**:2992–3002.
24. Winer S, Paltser G, Chan Y, Tsui H, Engleman E, Winer D, et al. Obesity predisposes to Th17 bias. *Eur J Immunol* 2009;**39**:2629–35.
25. Goldfine AB, Silver R, Aldhahi W, Cai D, Tatro E, Lee J, et al. Use of salsalate to target inflammation in the treatment of insulin resistance and type 2 diabetes. *Clin Transl Sci* 2008;**1**:36–43.
26. Goldfine AB, Fonseca V, Jablonski KA, Pyle L, Staten MA, Shoelson SE. The effects of salsalate on glycemic control in patients with type 2 diabetes: a randomized trial. *Ann Intern Med* 2010;**152**:346–57.
27. Winer S, Chan Y, Paltser G, Truong D, Tsui H, Bahrami J, et al. Normalization of obesity-associated insulin resistance through immunotherapy. *Nat Med* 2009;**15**:921–9.
28. Donath MY, Shoelson SE. Type 2 diabetes as an inflammatory disease. *Nat Rev Immunol* 2011;**11**:98–107.
29. Finlay D, Cantrell DA. Metabolism, migration and memory in cytotoxic T cells. *Nat Rev Immunol* 2011;**11**:109–17.

30. Haase B, Faust K, Heidemann M, Scholz T, Demmert M, Troger B, et al. The modulatory effect of lipids and glucose on the neonatal immune response induced by *Staphylococcus epidermidis*. *Inflamm Res* 2011;**60**:227–32.
31. Kragl M, Lammert E. Basement membrane in pancreatic islet function. *Adv Exp Med Biol* 2010;**654**:217–34.
32. Virtanen I, Banerjee M, Palgi J, Korsgren O, Lukinius A, Thornell LE, et al. Blood vessels of human islets of Langerhans are surrounded by a double basement membrane. *Diabetologia* 2008;**51**:1181–91.
33. Zhao Y, Guo C, Hwang D, Lin B, Dingeldein M, Mihailescu D, et al. Selective destruction of mouse islet beta cells by human T lymphocytes in a newly-established humanized type 1 diabetic model. *Biochem Biophys Res Commun* 2010;**399**:629–36.
34. Bonifacio E, Ziegler AG. Advances in the prediction and natural history of type 1 diabetes. *Endocrinol Metab Clin North Am* 2010;**39**:513–25.
35. Zhao Y, Wang H, Mazzone T. Identification of stem cells from human umbilical cord blood with embryonic and hematopoietic characteristics. *Exp Cell Res* 2006;**312**:2454–64.
36. Zhao Y, Mazzone T. Human umbilical cord blood-derived f-macrophages retain pluripotentiality after thrombopoietin expansion. *Exp Cell Res* 2005;**310**:311–8.
37. Francese R, Fiorina P. Immunological and regenerative properties of cord blood stem cells. *Clin Immunol* 2010;**136**:309–22.
38. Zhao Y, Huang Z, Qi M, Lazzarini P, Mazzone T. Immune regulation of T lymphocyte by a newly characterized human umbilical cord blood stem cell. *Immunol Lett* 2007;**108**:78–87.
39. Zhao Y, Lin B, Darflinger R, Zhang Y, Holterman MJ, Skidgel RA. Human cord blood stem cell-modulated regulatory T lymphocytes reverse the autoimmune-caused type 1 diabetes in nonobese diabetic (NOD) mice. *PLoS ONE* 2009;**4**:e4226.
40. Zhao Y, Jiang Z, Zhao T, Ye M, Hu C, Yin Z, et al. Reversal of type 1 diabetes via islet beta cell regeneration following immune modulation by cord blood-derived multipotent stem cells. *BMC Med* 2012;**10**:3.
41. Zhao Y, Lin B, Dingeldein M, Guo C, Hwang D, Holterman MJ. New type of human blood stem cell: a double-edged sword for the treatment of type 1 diabetes. *Transl Res* 2010;**155**:211–6.
42. Zhao Y. Stem cell educator therapy and induction of immune balance. *Curr Diab Rep* 2012;**12**:517–23.
43. Zhao Y, Jiang Z, Zhao T, Ye M, Hu C, Zhou H, et al. Targeting insulin resistance in type 2 diabetes via immune modulation of cord blood-derived multipotent stem cells (CB-SC) in stem cell educator therapy: phase I/II clinical trial. *BMC Med* 2013;**11**:160.
44. Abdi R, Fiorina P, Adra CN, Atkinson M, Sayegh MH. Immunomodulation by mesenchymal stem cells: a potential therapeutic strategy for type 1 diabetes. *Diabetes* 2008;**57**:1759–67.
45. Luo X, Yang H, Kim IS, Saint-Hilaire F, Thomas DA, De BP, et al. Systemic transforming growth factor-beta1 gene therapy induces Foxp3+ regulatory cells, restores self-tolerance, and facilitates regeneration of beta cell function in overtly diabetic nonobese diabetic mice. *Transplantation* 2005;**79**:1091–6.
46. Li MO, Flavell RA. TGF-beta: a master of all T cell trades. *Cell* 2008;**134**:392–404.
47. Aguayo-Mazzucato C, Bonner-Weir S. Stem cell therapy for type 1 diabetes mellitus. *Nat Rev Endocrinol* 2010;**6**:139–48.
48. Chung CH, Hao E, Piran R, Keinan E, Levine F. Pancreatic beta-cell neogenesis by direct conversion from mature alpha-cells. *Stem Cells* 2010;**28**:1630–8.
49. Zhao Y, Huang Z, Lazzarini P, Wang Y, Di A, Chen M. A unique human blood-derived cell population displays high potential for producing insulin. *Biochem Biophys Res Commun* 2007;**360**:205–11.
50. Wang L, Lovejoy NF, Faustman DL. Persistence of prolonged C-peptide production in type 1 diabetes as measured with an ultrasensitive C-peptide assay. *Diabetes Care* 2012;**35**:465–70.
51. Kucia M, Zuba-Surma E, Wysoczynski M, Dobrowolska H, Reca R, Ratajczak J, et al. Physiological and pathological consequences of identification of very small embryonic like (VSEL) stem cells in adult bone marrow. *J Physiol Pharmacol* 2006;**57**(Suppl. 5):5–18.
52. D'Ippolito G, Diabira S, Howard GA, Menei P, Roos BA, Schiller PC. Marrow-isolated adult multilineage inducible (MIAMI) cells, a unique population of postnatal young and old human cells with extensive expansion and differentiation potential. *J Cell Sci* 2004;**117**:2971–81.
53. Chao KC, Chao KF, Fu YS, Liu SH. Islet-like clusters derived from mesenchymal stem cells in Wharton's jelly of the human umbilical cord for transplantation to control type 1 diabetes. *PLoS ONE* 2008; **3**:e1451.
54. Antuna-Puente B, Feve B, Fellahi S, Bastard JP. Adipokines: the missing link between insulin resistance and obesity. *Diabetes Metab* 2008;**34**:2–11.
55. Sell H, Habich C, Eckel J. Adaptive immunity in obesity and insulin resistance. *Nat Rev Endocrinol* 2012;**8**(12):709–16.
56. Bhargava P, Lee CH. Role and function of macrophages in the metabolic syndrome. *Biochem J* 2012;**442**:253–62.
57. Rajwani A, Cubbon RM, Wheatcroft SB. Cell-specific insulin resistance: implications for atherosclerosis. *Diabetes Metab Res Rev* 2012;**28**:627–34.
58. Devaraj S, Dasu MR, Jialal I. Diabetes is a proinflammatory state: a translational perspective. *Expert Rev Endocrinol Metab* 2010;**5**:19–28.
59. Kanda H, Tateya S, Tamori Y, Kotani K, Hiasa K, Kitazawa R, et al. MCP-1 contributes to macrophage infiltration into adipose tissue, insulin resistance, and hepatic steatosis in obesity. *J Clin Invest* 2006;**116**:1494–505.
60. Kamei N, Tobe K, Suzuki R, Ohsugi M, Watanabe T, Kubota N, et al. Overexpression of monocyte chemoattractant protein-1 in adipose tissues causes macrophage recruitment and insulin resistance. *J Biol Chem* 2006;**281**:26602–14.
61. Patsouris D, Li PP, Thapar D, Chapman J, Olefsky JM, Neels JG. Ablation of CD11c-positive cells normalizes insulin sensitivity in obese insulin resistant animals. *Cell Metab* 2008;**8**:301–9.
62. Greenwald RJ, Freeman GJ, Sharpe AH. The B7 family revisited. *Annu Rev Immunol* 2005;**23**:515–48.
63. Chen L. Co-inhibitory molecules of the B7-CD28 family in the control of T-cell immunity. *Nat Rev Immunol* 2004;**4**:336–47.
64. Sethna MP, van Parijs L, Sharpe AH, Abbas AK, Freeman GJ. A negative regulatory function of B7 revealed in B7-1 transgenic mice. *Immunity* 1994;**1**:415–21.
65. Bugeon L, Dallman MJ. Costimulation of T cells. *Am J Respir Crit Care Med* 2000;**162**:S164–8.
66. Defuria J, Belkina AC, Jagannathan-Bogdan M, Snyder-Cappione J, Carr JD, Nersesova YR, et al. B cells promote inflammation in obesity and type 2 diabetes through regulation of T-cell function and an inflammatory cytokine profile. *Proc Natl Acad Sci USA* 2013;**110**:5133–8.
67. Haskell BD, Flurkey K, Duffy TM, Sargent EE, Leiter EH. The diabetes-prone NZO/HlLt strain. I. Immunophenotypic comparison to the related NZB/BlNJ and NZW/LacJ strains. *Lab Invest* 2002;**82**:833–42.
68. Winer DA, Winer S, Shen L, Wadia PP, Yantha J, Paltser G, et al. B cells promote insulin resistance through modulation of T cells and production of pathogenic IgG antibodies. *Nat Med* 2011;**17**:610–7.
69. Winer S, Winer DA. The adaptive immune system as a fundamental regulator of adipose tissue inflammation and insulin resistance. *Immunol Cell Biol* 2012;**90**:755–62.

70. Talukdar S, Oh dY, Bandyopadhyay G, Li D, Xu J, McNelis J, et al. Neutrophils mediate insulin resistance in mice fed a high-fat diet through secreted elastase. *Nat Med* 2012;**18**:1407–12.
71. Wu D, Molofsky AB, Liang HE, Ricardo-Gonzalez RR, Jouihan HA, Bando JK, et al. Eosinophils sustain adipose alternatively activated macrophages associated with glucose homeostasis. *Science* 2011;**332**:243–7.
72. Liu J, Divoux A, Sun J, Zhang J, Clement K, Glickman JN, et al. Genetic deficiency and pharmacological stabilization of mast cells reduce diet-induced obesity and diabetes in mice. *Nat Med* 2009;**15**:940–5.
73. Musilli C, Paccosi S, Pala L, Gerlini G, Ledda F, Mugelli A, et al. Characterization of circulating and monocyte-derived dendritic cells in obese and diabetic patients. *Mol Immunol* 2011;**49**:234–8.
74. Zhong J, Rao X, Deiuliis J, Braunstein Z, Narula V, Hazey J, et al. A potential role for dendritic cell/macrophage-expressing DPP4 in obesity-induced visceral inflammation. *Diabetes* 2013;**62**:149–57.
75. Kohn LD, Wallace B, Schwartz F, McCall K. Is type 2 diabetes an autoimmune-inflammatory disorder of the innate immune system? *Endocrinology* 2005;**146**:4189–91.
76. Li MO, Wan YY, Sanjabi S, Robertson AK, Flavell RA. Transforming growth factor-beta regulation of immune responses. *Annu Rev Immunol* 2006;**24**:99–146.
77. Pillay J, den B,I, Vrisekoop N, Kwast LM, de Boer RJ, Borghans JA, et al. In vivo labeling with 2H2O reveals a human neutrophil lifespan of 5.4 days. *Blood* 2010;**116**:625–7.

Section IV

# Regenerative Medicine Applications

Chapter 13

# Emerging Uses of Cord Blood in Regenerative Medicine—Neurological Applications

**Jessica M. Sun and Joanne Kurtzberg**
*The Robertson Clinical and Translational Cell Therapy Program and Carolinas Cord Blood Bank, Duke University, Durham, NC USA*

**Chapter Outline**



## 1. INTRODUCTION

The field of regenerative medicine is dedicated to the study of repairing, replacing, or regenerating damaged human cells, tissues, or organs to restore or establish normal function.[1] This could be approached through numerous strategies, from stimulating endogenous processes to repair damaged tissue to deriving or transplanting entire organs to replace those that are beyond endogenous repair. Though the field is currently in its infancy, regenerative medicine is predicted to be one of the most important disciplines in medicine to develop in the next decade, with therapeutic applications in a wide variety of medical conditions. Potential cells that could serve as source materials for regenerative medicine and cellular therapies include hematopoietic stem and progenitor cells derived from bone marrow (BM) or umbilical cord blood (CB), placental and amniotic fluid and tissues, mesenchymal stromal cells (MSCs), skin cells, and other organ-specific cells that could be engineered to perform reparative functions. This chapter will explore some of the potential regenerative applications for which CB could serve as a valuable source of cells for primary or derivative cellular therapies.

## 2. UMBILICAL CB AS A SOURCE OF STEM CELLS FOR NEUROLOGICAL APPLICATIONS

Human CB is rich in highly proliferative stem and progenitor cells of the hematopoietic and other lineages mobilized by placental signals promoting homing to developing organs.[2,3] CB is readily available, can be collected noninvasively without risk to the mother or infant donor, and can be tested, processed, and cryopreserved for several decades for future use. CB is often discarded as medical waste with the placenta after birth, but over the past 20 years, approximately 700,000 unrelated donor CB units have been collected, characterized, and banked for public use. An additional two to three million CB units have been stored privately for family use. Compared to stem cells obtained from adult BM, CB stem cells are less mature and therefore have longer telomeres and greater proliferating potential.[4] They are also less immunogenic and less likely to transmit infections via latent viruses. In over 25 years of use in allogeneic, unrelated hematopoietic stem cell transplant (HSCT), CB has not been shown to cause any teratomas or solid tumors. Recently, iPS cells have been isolated from CB with simpler methods and greater efficiency as compared to adult cell sources.[5–7] All

Cord Blood Stem Cells Medicine. http://dx.doi.org/10.1016/B978-0-12-407785-0.00013-X

of these factors favorably position CB for use as a source of cells for cellular therapies and regenerative medicine.

CB is a well-established source of stem cells for hematopoietic rescue after myeloablative HSCT. In addition, CB also contains nonhematopoietic stem cell populations with the ability to differentiate into numerous cell types throughout the body. In particular, the CB-derived unrestricted somatic stem cell (USSC) first described by Kogler et al. is a nonhematopoietic multipotent cell with the ability to differentiate into several lineages in vitro and in vivo.[8] USSCs can give rise to cell types from all three germinal layers, including osteoclasts,[8] hepatocytes,[9] and neurons,[10] among others. CB-derived cells can also differentiate into MSCs,[11] chondrocytes,[12–14] osteocytes,[13–18] adipocytes,[14–17] neural cells,[14,15,18–20] cardiac and skeletal muscle myocytes,[21] hepatocytes,[14,15] pancreatic cells,[22] skin cells, and endothelial colony forming cells.

CB donor-derived tissue-specific cells have been identified in multiple organs in both animals and humans after HSCT, including the liver,[19] lung, pancreas,[19,23] skeletal muscle,[24] and brain.[13] While CB cells have the ability to differentiate into tissue-specific cells and integrate into host organs, there is growing evidence that their therapeutic effects may be mediated by their ability to initiate tissue repair by signaling and activation of host cells via trophic and/or paracrine effects. Nonetheless, these observations indicate that transplanted CB cells are capable of repopulating more than just the hematopoietic system.[25,26] This may be due to the presence of a true embryonic-like stem cell in CB and/or small numbers of committed but tissue-specific, nonhematopoietic progenitors. It is important to note that observations of in vivo engraftment and differentiation have occurred in patients receiving myeloablative and immunoablative preparative therapies. It is not clear if, in humans, infusions of CB into an immunocompetent host will produce similar results. Observations in immunocompetent, xenogeneic models are encouraging in this regard.

The pluripotent nature of CB, as well as the relative ease of collection, processing, testing, and storage, makes it an attractive source of cells for regenerative medicine applications across many disciplines, including neurology, cardiology, orthopedics, endocrinology, and others. In this chapter, numerous preclinical, animal, and human studies evaluating the use of CB and CB-derived products for neurological conditions will be reviewed.

## 3. POTENTIAL MECHANISMS OF CB AS THERAPY FOR PATIENTS WITH NEUROLOGICAL DISEASES

Neurologic impairment can result from acquired injuries, genetic conditions, or neurodegenerative diseases of unclear etiology. Recovery from neurological injuries is typically incomplete and often results in significant and permanent disabilities. Currently, most available therapies are limited to supportive or palliative measures, aimed at managing the symptoms of the condition. Since restorative therapies targeting the underlying cause of most neurological diseases do not exist, cell therapies targeting anti-inflammatory, neuroprotective, and regenerative potential hold great promise. CB cells can induce repair through mechanisms that involve trophic or cell-based paracrine effects or cellular integration and differentiation. One or more of these effects may be operative in CB therapies for neurologic conditions. There are numerous potential applications of CB-based regenerative therapies in neurological diseases, including genetic diseases of childhood, ischemic events such as stroke, and neurodegenerative diseases of adulthood.

Multiple in vitro studies have demonstrated that neurons, astrocytes, oligodendrocytes, and microglia can be derived from CB cells via gene transfection, ex vivo culture with growth factor supplementation, through generation of iPS, and/or the use of chemical agents.[10,20,27–34] Neural differentiation has been documented in phenotypic and functional assays. The phenotype of the derived cells has been characterized by gene arrays[35] and the expression of standard neural-specific markers and proteins. Additionally, functional characteristics have been demonstrated through the presence of voltage- and ligand-gated ion channels with the ability to conduct electrical activity, indicating the development of functional characteristics of neurons.[30]

The mechanisms of CB-induced cell and tissue repair are expected to vary between indications, and several possibilities have been hypothesized.[36] Infused and/or transplanted cells may deliver trophic factors that provide anti-inflammatory and neuroprotective effects and enhance the survival potential of host cells.[37–40] They may increase the plasticity of the injured brain by enhancing synaptogenesis, angiogenesis resulting in neovascularization, and endogenous repair mechanisms, and by inducing migration and proliferation of endogenous neural stem cells.[41–43] CB stem cells may also migrate, integrate, proliferate, and differentiate into "replacement" neuronal and glial cells and play a role in remyelination.[44] Additionally, many neurologic diseases involve activation of proapoptotic signal transduction, which could be harnessed to attract cells to brain lesions in those diseases. Thus, CB-derived cells could also potentially act as a vehicle to deliver neuroprotective and restorative factors or signal endogenous cells to act in a targeted way toward damaged brain tissue.

## 4. UNRELATED DONOR CB TRANSPLANTATION FOR GENETIC BRAIN DISEASES IN CHILDREN

The first observations of the potential of CB cells to differentiate into nonhematopoietic lineages in vivo were documented in children with certain inherited metabolic diseases (IMD) undergoing unrelated donor umbilical cord blood transplant for disease correction. IMD are a heterogeneous group of genetic diseases, most of which involve a single-gene mutation resulting in deficiency of a critical enzyme necessary for production and maintenance of myelin or other cellular-based structural parts of the nervous system. In the majority of cases, the enzyme defect leads to the accumulation of substrates that are toxic and/or interfere with normal cellular function. Oftentimes, patients appear normal at birth but during infancy begin to exhibit disease manifestations, frequently including progressive neurological deterioration due to absent or abnormal brain myelination. The ultimate result is death in childhood.

Allogeneic transplantation of human CB in patients with certain genetic lysosomal and peroxisomal storage diseases is effective in preventing or ameliorating the associated neurological damage.[45–48] Engraftment of donor cells in a patient with an IMD provides a constant source of enzyme replacement, thereby slowing or halting the progression of disease. Patients with these diseases, ranging in age from newborns to young adults, transplanted early in the course of their disease derive extensive benefits from the transplant procedure, which both extends life for decades and greatly improves neurologic functioning.[49–51] High pretransplant performance status is associated with a much higher rate of survival than transplants performed in children with lower performance scores (see Figure 1). Clinical and pathological observations from these patients provide additional support for the concept that CB cells can repair nonhematopoietic tissues.

Krabbe disease (globoid-cell leukodystrophy), for example, is an autosomal recessive disorder caused by a deficiency of the lysosomal enzyme galactocerebrosidase. The resultant accumulation of galactolipids prevents myelination of the central and peripheral nervous systems, causing progressive neurological deterioration. In the early infantile form of the disease, symptoms develop within the first 6 months of life. Untreated, patients develop blindness, deafness, spasticity,

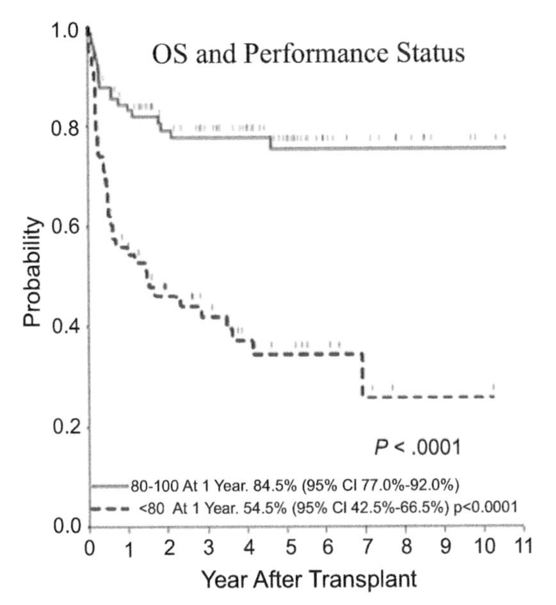


**FIGURE 1** Impact of performance status (Lansky score 80–100 vs <80) on overall survival of 159 patients transplanted with allogenic umbilical cord blood for inherited metabolic disorders.[46]

seizures, mental deterioration, and ultimately death by 2 years of age. Once symptoms have developed, CB transplantation provides little neurological benefit. In infants who are diagnosed and undergo CB transplantation before the onset of symptoms, generally in the first month of life, life is prolonged and neurologic functioning is improved. Transplanted children demonstrate progressive brain myelination and normal vision, hearing, and cognition.[51] The success of this approach has led to the implementation of newborn screening for Krabbe disease in New York State and now additional states in the US.[52] Despite these successes, a range of motor deficits persist in up to 80% of children undergoing neonatal CB transplantation, suggesting that damage to the long tracts may begin prenatally, precede clinical symptoms, and be irreversible. In a series of newborns with known early infantile Krabbe disease, evaluation of myelination of the corticospinal tracks at birth revealed defects in a significant portion of babies present in the first few days of life, strongly suggesting that damage occurred prenatally.[53] It is possible that fetal transplantation or targeted cellular therapy after standard transplantation may enhance therapeutic approaches and further improve motor function in these patients.

Autopsy studies in humans who died after intravenously administered, sex-mismatched BM and CB transplant have confirmed the engraftment of donor cells throughout the brain months after transplantation.[54–56] Most engrafting cells were nonneuronal microglial cells, but donor-derived neurons, astrocytes, and oligodendrocytes have been identified. Globoid bodies, the pathological perivascular signature of Krabbe disease, were not detected in the brain of a patient transplanted for early infantile Krabbe disease at 3 weeks of age, who died of unrelated causes at 5 years of age.[56] Based on these observations, our group hypothesized that CB contained cells capable of differentiating into oligodendrocyte and microglial-like cells. We subsequently cultured and expanded oligodendrocyte-like cells from fresh and cryopreserved CB after 3–4 weeks in tissue culture supplemented with neurotrophic growth factors.[20,28] These cells (DUOC-01 or "O-cells") grow as an adherent population that, after 21 days in culture, express surface antigens found on oligodendrocytes (O1, O4, PLP, MBP) and microglia (CD45, CD11b), make corresponding RNAs, and myelinate shiverer neuron axons in an in vitro potency assay (Figure 2). They also constitutively produce IL-6 and IL-10, and retain the ability to produce lysosomal enzymes in culture after manufacturing. Intrathecal dosing in immunodeficient newborn mice showed the best distribution of O-cells in the central nervous system as compared to IV or intracranial delivery routes. A phase I trial administering these cells intrathecally 1 month after a standard HSCT from the same CB donor is planned. This trial is one example that the availability of well characterized, screened, and HLA-typed CB coupled with its vast differentiation potential makes it an attractive source of stem cells for applications in tissue repair and regeneration, particularly in the central nervous system.

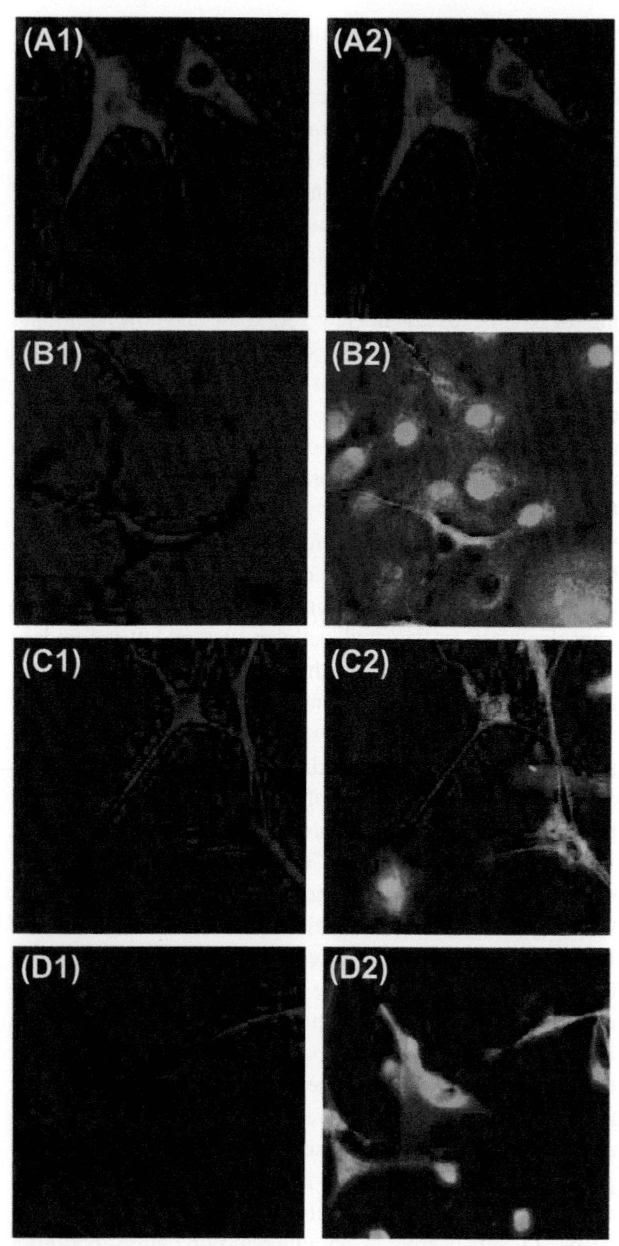


**FIGURE 2** In vitro functional assay of myelination of shiverer mouse neurons by cryopreserved CB-derived oligodendrocyte-like cells (O-cells). Shiverer neurons cocultured with O-cells were costained for BT3 (Texas Red) and MBP (fluorescein isothiocyanate). Controls stained positive for BT3 (panel A1) but not MBP (A2). When cocultured with O-cells for 1 week, BT3 (B1) and MBP (B2) were expressed. Z-stacked projection after 3 weeks in culture demonstrated BT3 expression (C1, D1) and close association between BT3-expressing neuronal cells and MBP-expressing cells (C2), with MBP expression along axonal processes (D2).[20]

## 5. ISCHEMIC INJURIES

Observations using CB to treat children with genetic conditions led to the hypothesis that CB might also be beneficial in patients with brain injury. Accordingly, CB cells have been investigated in preclinical models of stroke, neonatal hypoxic-ischemic encephalopathy (HIE), traumatic brain injury, and spinal cord injury. These injuries are typically characterized by an acute inflammatory response and immediate damage to all neural cell types within the affected region. Therefore, therapeutic strategies might involve methods to promote cell survival, and repair or regenerate the affected areas, potentially via anti-inflammatory effects, neurogenesis, synaptogenesis, and/or angiogenesis after the injury has been sustained.

### 5.1 Animal Models

Numerous animal models have demonstrated both neurological and survival benefits of CB cells in the setting of stroke, ischemia, intracranial hemorrhage, and spinal cord injury.[57–63] Neuroprotection,[57] neovascularization,[43] and neuronal regeneration[43] have all been demonstrated in various models.

Cells derived from human CB and administered via the intravenous, intraperitoneal, and intracerebral routes have demonstrated the ability to enhance functional recovery in numerous animal models of stroke.[36] Several investigators have demonstrated that CB cells administered to immunocompetent rats after stroke induced by middle cerebral artery occlusion migrate to the brain, localize to the site of injury, decrease the volume of the infarct, and improve performance on functional neurologic tests (Figure 3).

A neonatal rat model has been developed by unilateral carotid artery ligation on day 7 of life. Without intervention, these animals universally develop severe cerebral damage and contralateral spastic paresis. Meier and Jensen administered human CB mononuclear cells to these animals intraperitoneally 1 day after the hypoxic event, showing that the cells migrate to the area of brain damage and persist for at least 2 weeks. Although the extent of morphologic injury on gross pathology was not altered, animals who received CB mononuclear cells did not develop spastic paresis, indicating functional recovery.[59] In a baby rabbit model of HIE,[64] Tan demonstrated that labeled human CB cells reached the brain within 24 h, persisted for at least 1 week, and decreased the degree of brain damage on MRI. In severely affected animals, CB administration improved gross motor function in a short-term functional assay.[65] Additionally, Ballabh and colleagues developed a rabbit model of intraventricular hemorrhage (IVH) by administering glycerol intraperitoneally to premature rabbit pups.[66] In this model, IVH is followed by the development of hydrocephalus and subsequent white matter demyelination. Intraventricular administration of human CB cells 24 and 72 h after glycerol failed to prevent the hydrocephalus but did reduce subsequent demyelination (Ballabh, personal communication, 2014).

### 5.2 Human Studies

The therapeutic potential of intravenous infusions of autologous CB is currently being investigated in young children

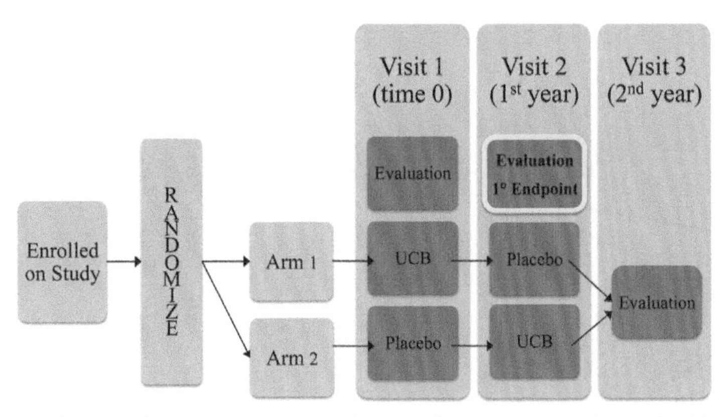


**FIGURE 3** Study design of a randomized, double-blind, placebo-controlled trial of an intravenous infusion of autologous cord blood (CB) in children with cerebral palsy. End points include changes in functional status (gross motor, fine motor, speech, cognition), quality of life, and neuroimaging (brain MRI with diffusion tensor imaging, fibertracking, and connectivity analyses[68]).

with cerebral palsy, HIE, and traumatic brain injury. In a safety study, our group treated 184 infants and children with cerebral palsy (76%), congenital hydrocephalus (12%), and other brain injuries (12%) with 198 intravenous autologous CB infusions.[67] Patients were treated in the outpatient clinic through a peripheral IV after a single dose of oral Tylenol, IV Benadryl, and Solumedrol. Approximately 1.5% of patients experienced hypersensitivity reactions (i.e., hives and/or wheezing) during the CB infusion that resolved after discontinuation of the infusion and outpatient medical management. With more than 3 years of follow-up, no additional adverse events have been reported, indicating that the procedure is safe. Parental reports of improved function were common, but it was difficult to know whether these improvements were directly related to the infusion of CB cells. Thus, a randomized, double-blind, placebo-controlled study is in progress to determine the efficacy of this approach. In this study, children aged 1–6 years are randomly assigned to the order in which they receive CB and placebo infusions, each given 1 year apart (Figure 3). Motor, cognitive, and imaging studies are performed at baseline, 1, and 2 years to evaluate any differences between CB and placebo groups. The primary end point is improvement in motor function on standardized scales, including the Gross Motor Function Measure and Peabody measures. Accrual to this trial is complete, and initial results are expected in approximately 1 year. Preliminarily, studies with the MRI biomarker of white matter connectivity have shown that clinical functional phenotype correlated with magnetic resonance imaging findings using whole-brain connectivity analysis (Figure 4).[68] A study of erythropoietin with or without allogeneic CB infusions was conducted in Korean children with cerebral palsy.[69] They reported greater improvements in cognitive and selected motor functions in children who received CB and erythropoietin versus controls. In this study, there was no CB-only group for comparison and the overall sample size was small.

In the United States, CB is being investigated in clinical trials for children with cerebral palsy, neonatal HIE, stroke, traumatic brain injury, and autism (NCT01072370, NCT01700166, NCT01988584, NCT01251003, and NCT01638819, respectively). Studies administering CB cells intravenously or intrathecally are being conducted in children with brain injury in other countries as well. Of note, intrathecal administration of allogeneic CB-derived cells, mostly MSCs, has been performed for a variety of neurologic conditions in a few small studies, primarily in China. In general, side effects are reported to be minor and transient, most commonly including fever, headache, and dizziness.[70–73] Efficacy cannot be determined at this time and further safety studies are needed.

In a phase I trial of newborns with hypoxic ischemic brain injury at birth conducted at Duke, fresh, noncryopreserved autologous CB processed for volume and RBC reduction on the Sepax 1 device (Biosafe, Geneva) was infused in one, two, or four doses of 1–5×10e7 nucleated cells/kg within the first 72h of life in babies with moderate-to-severe encephalopathy qualifying for systemic hypothermia.[74] These babies were compared with a concomitant group of babies treated at Duke who were cooled but did not receive CB cells. Infusions were found to be safe in these critically ill babies, and babies receiving cells had increased survival rates to discharge (100% vs 85%, $p=0.20$) and improved function at 1 year of age (74% vs 41% with development in the normal range, $p=0.05$). A phase II randomized trial is currently in development. If this therapy improves the outcome of babies with significant birth trauma, there could be potential implications for the current model of CB collection and banking. In order to make this therapy available to all eligible babies, at a minimum, all obstetric providers would need to be trained in CB collection, CB would need to be routinely collected at at-risk deliveries, and centers would need to either process and infuse the CB or transfer the babies and their CB to centers that have the ability to do so. Alternatively, CB collection could become a routine practice at every delivery. Units could then be stored for a limited time until it is clear whether the baby will need it. If not, they could then be discarded, used for research or

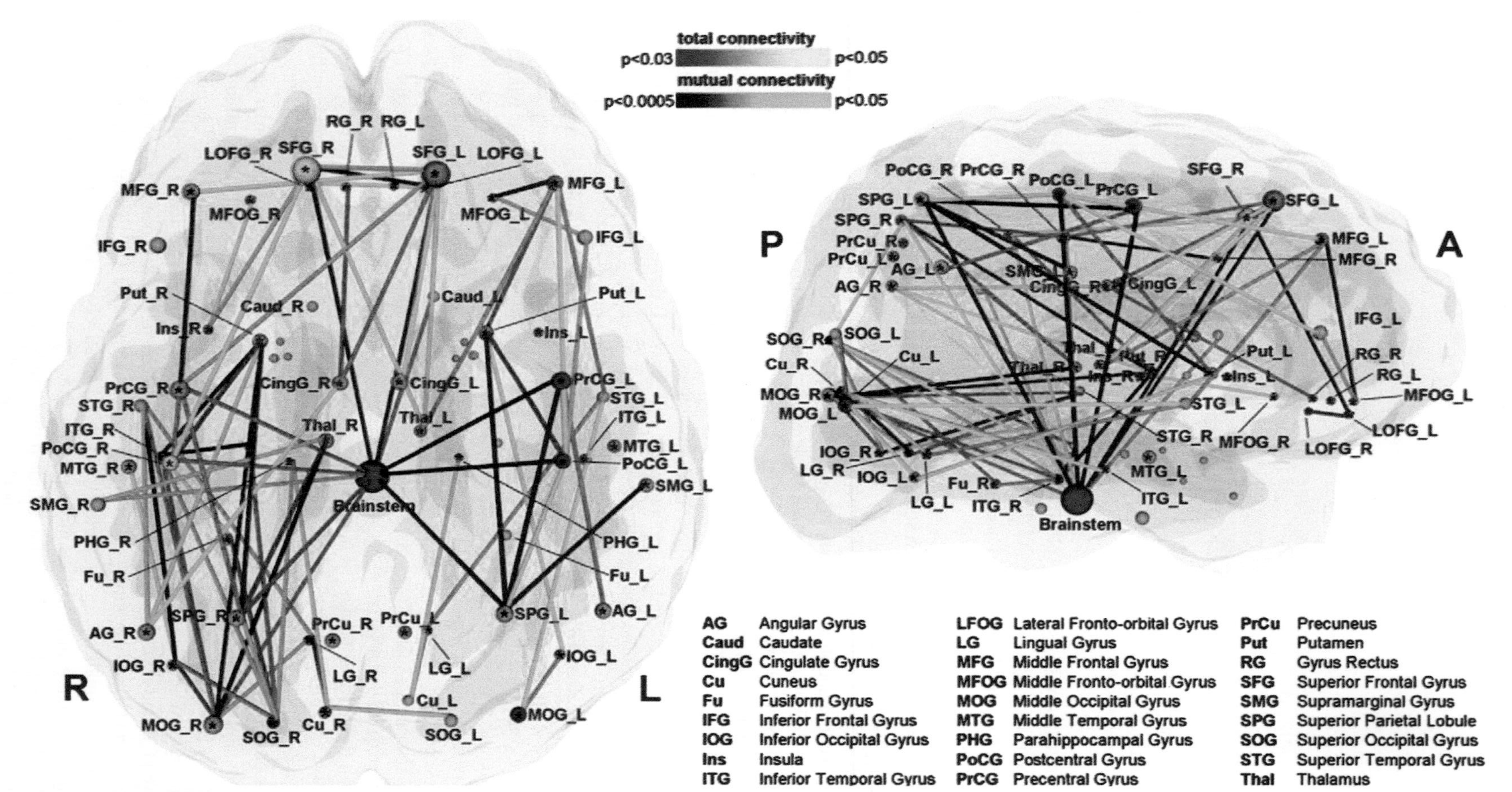


**FIGURE 4** Interregional whole-brain connectivity analysis of 17 children with cerebral palsy. Red-yellow nodes indicate significantly reduced total connectivity to all other brain regions in severe versus moderate cerebral palsy. Cool-colored connections between nodes indicate significantly reduced mutual connectivities in the severely affected group compared to the moderately affected group.[68].

other regenerative applications, or, if appropriate consent and testing was obtained, transferred to the public registry. It may also be necessary to develop safe and effective therapies using allogeneic CB donors, as many of these babies will be too sick at birth to prioritize autologous CB collection.

Repeated autologous CB infusions are also being studied in young babies born with congenital hydrocephalus. In this condition, excessive accumulation of cerebral spinal fluid (CSF) within the ventricular system of the brain results in a progressive increase in ventricular volume and intracranial pressure, leaving the brain limited space to develop and expand. The increased pressure damages the developing brain via a crush injury that causes mechanical distortion and impaired blood flow, as well as extravasation of CSF into the brain parenchyma that causes demyelination and loss of axons. While some white matter changes may be reversible after ventriculoperitoneal shunt placement to divert the flow of CSF after birth, most affected children are left with a myriad of motor, sensory, and cognitive deficits due to their early brain injury. At Duke University, more than 80 patients aged 6days–4years with congenital hydrocephalus have received autologous CB infusions. Since these babies are identified prenatally, CB collection can be planned in advance of delivery. Due to their small size, and perhaps their delivery by Cesarean-section, most autologous CB units were large enough to supply multiple doses. Thus, patients received up to four doses of autologous CB at approximately 2- to 6-month intervals (unpublished data). While the efficacy of this approach is still under investigation, there have been no safety concerns. This indicates that repeated dosing of autologous CB is safe even in very young babies, and such a dosing scheme could be potentially advantageous for other neurologic conditions as well.

Most human studies of stem cells in adults who have suffered a stroke have utilized autologous BM cells.[75,76] Though no safety concerns have been identified, the studies are too small to reliably assess efficacy, and investigations of clinical end points are currently underway. However, the majority of adult stroke victims are elderly and critically ill following their injury, compromising the likelihood and safety of autologous BM harvesting. Thus, an allogeneic CB-derived off-the-shelf product would be an attractive alternative to autologous BM, as it would avoid the need for a potentially risky BM harvest in these critically ill patients. In addition, these frequently elderly patients may also have decreased progenitor cells and other conditions that may limit the functionality of their own BM cells.[77,78] A few clinical trials of CB cells given intravenously or intraparenchymally into the brain (NCT01700166, USA; NCT01884155, Korea; NCT01673932, China) are accruing patients at the present time, but no results have been published to date.

## 6. NEURODEGENERATIVE DISEASES

Cellular approaches to neurodegenerative diseases have been studied most extensively in Parkinson's disease in which degeneration of primarily nigrostriatal dopaminergic neurons results in motor symptoms such as resting tremors, rigidity, and hypokinesia. Cell replacement strategies have been attempted in over 300 patients with Parkinson's disease using intrastriatal implantation of fetal mesencephalic tissue. While several open-label trials suggested clinical benefit, two double-blind studies did not find a significant effect.[79,80] However, some patients have demonstrated durable improvements including the ability to withdraw dopaminergic medication, and clinical benefit has been seen preferentially in less disabled patients. Given the limitations of using human fetal tissue, interest has grown in generating dopaminergic neurons from other cell sources. Neurons that express dopamine-related genes and demonstrate the ability to synthesize and release dopamine have been successfully derived from CB stem cells in vitro.[10,29] In hemiparkinsonian rats, unmodified human umbilical cord MSCs injected into the striatum can improve behavioral symptoms, and this effect is enhanced by adenovirus-mediated VEGF modification of the cells.[81] These studies indicate that CB has potential as a source of stem cells for cellular replacement strategies in Parkinson's disease.

CB cells have been evaluated in in vitro and in vivo models of Alzheimer's disease. Transgenic mice treated with CB-MSCs show a reduction in both microglial activation and beta-amyloid deposits, the pathologic signature of the disease.[82] CB-treated mice also demonstrate decreased cognitive impairment in functional assays[83] and an extended life span.[84] Although the mechanism is not entirely clear, it is possible that the CB cells mediate the microglial response to beta-amyloid deposits, promote beta-amyloid phagocytosis, and/or prevent apoptosis of host cells. Neurostem®, a CB-MSC product, has been investigated in a phase I trial in Korea, though results of that study have not yet been published.

## 7. AUTISM

Emerging data suggest that autism is caused by a complex interaction of genetic and environmental conditions, resulting in abnormal brain functioning early in life. Recently, stem cell therapy has become appealing as a potential therapy to many within the autism community. In a mouse model of autism, intraventricular administration of human adipose-derived stem cells resulted in decreased repetitive movements and improved social activity.[85] As a result of this work and early results of increased connectivity in recipients of autologous CB for cerebral palsy,[68] it has been hypothesized that CB infusion might produce clinical benefit by restoring neural connections in patients with autism. Clinical trials of autologous BM and CB in children with autism are being conducted in the United States (NCT01638819, NCT02176317), Mexico (NCT01740869), China (NCT01343511),[86] and India (NCT01974973,

NCT01836562).[87] Due to the heterogeneity in both etiology and symptomology of autism, identifying appropriate subjects and outcome measures remains particularly challenging. The initial goal of the first Duke study is to define reliable outcome measures that can objectively assess response of therapeutic interventions in these patients.

## 8. CHALLENGES

While cellular approaches, utilizing CB or other stem cell sources for the treatment of neurological diseases are promising, much remains to be done before such products and procedures may be available for clinical use in humans. Cellular therapies are much more challenging to fully characterize, develop, and refine than traditional pharmaceuticals. Current U.S. FDA regulations for drug development and manufacturing are not directly applicable to or relevant for cellular products. Many questions remain unanswered, including ideal cell source, route of administration, dose and dosing regimen, timing, and role of immunosuppression in these therapies. Risks of adverse events differ for the use of unmanipulated or manipulated products as well as autologous versus allogeneic cells. A better understanding of mechanisms of action may facilitate clinical researchers to better harness the power of CB cells as a therapeutic modality in various disease processes. The development of noninvasive methods of cell tracking, which can be used to determine whether cell integration or engraftment is required for a beneficial effect, will be essential to inform that effort.

## 9. SUMMARY

Regenerative medicine is a field with enormous potential to impact the treatment of diseases that affect millions of people and conditions for which there are currently no effective therapies. CB is an attractive source of multipotent stem and progenitor cells, which can serve as cellular therapeutic factories for many of these diseases. In addition to its ability to differentiate into numerous cell types, one of the main advantages of CB compared to other cell sources is its practicality. CB can be easily collected after most births without risk to the donor, can be cryopreserved for later use, and is not a socially or politically controversial cellular resource. In addition, generation of iPS from small numbers of cryopreserved CB cells is feasible, can serve as a tool to study disease mechanisms, and may also lead to the development of iPS-derived therapies in the future. In order to capitalize on these features and to make regenerative therapies available to the population at large, the development of allogeneic, off-the-shelf products that can be utilized without preparation with high-dose chemotherapy or long-term engraftment will be necessary.

If CB develops a significant role in regenerative medicine as expected, it could not only affect the way numerous diseases are treated, but could also promote a shift in the current paradigm of public and private CB banking. CB collection might become the standard of care for all deliveries, provided by hospitals and obstetric providers. Parents may become more inclined to bank their baby's CB privately. Public banks might therefore need to develop mechanisms to sequester banked units for the donor family for a few years and then list them on the public registry if the family did not require their use. Smaller units, not typically stored by public banks and less likely to be sufficient for standard allogeneic HSCT, might also have to be banked for a number of years and may be useful for future cell therapy and/or regenerative medicine applications.

In summary, CB has enormous potential for development as a therapeutic agent in the field of cellular therapies and regenerative medicine. Use of CB in this manner will greatly expand indications for banking in both the private and public settings. However, additional preclinical and early clinical trials are needed to fully define the safety profiles of CB therapies and to determine the clinical applications where evidence-based data justify its use.

## REFERENCES


1. Regenerative medicine. http://www.aabb.org/resources/bct/therapyfacts/Pages/regenerative.aspx. [accessed 11.12.13].
2. Broxmeyer HE, Douglas GW, Hangoc G, Cooper S, Bard J, English D, et al. Human umbilical cord blood as a potential source of transplantable hematopoietic stem/progenitor cells. *Proc Natl Acad Sci U S A* 1989;**86**(10):3828–32.
3. Broxmeyer HE, Srour EF, Hangoc G, Cooper S, Anderson SA, Bodine DM. High-efficiency recovery of functional hematopoietic progenitor and stem cells from human cord blood cryopreserved for 15 years. *Proc Natl Acad Sci U S A* 2003;**100**(2):645–50.
4. van de Ven C, Collins D, Bradley MB, Morris E, Cairo MS. The potential of umbilical cord blood multipotent stem cells for nonhematopoietic tissue and cell regeneration. *Exp Hematol* 2007;**35**(12):1753–65.
5. Broxmeyer HE, Lee MR, Hangoc G, Cooper S, Prasain N, Kim YJ, et al. Hematopoietic stem/progenitor cells, generation of induced pluripotent stem cells, and isolation of endothelial progenitors from 21- to 23.5-year cryopreserved cord blood. *Blood* 2011;**117**(18):4773–7.
6. Takenaka C, Nishishita N, Takada N, Jakt LM, Kawamata S. Effective generation of iPS cells from CD34+ cord blood cells by inhibition of p53. *Exp Hematol* 2010;**38**(2):154–62.
7. Zaehres H, Kögler G, Arauzo-Bravo MJ, Bleidissel M, Santourlidis S, Weinhold S, et al. Induction of pluripotency in human cord blood unrestricted somatic stem cells. *Exp Hematol* 2010;**38**(9):809–18. 818 e801–802.
8. Kögler G, Sensken S, Airey JA, Trapp T, Müschen M, Feldhahn N, et al. A new human somatic stem cell from placental cord blood with intrinsic pluripotent differentiation potential. *J Exp Med* 2004;**200**(2):123–35.
9. Waclawczyk S, Buchheiser A, Flogel U, Radke TF, Kogler G. In vitro differentiation of unrestricted somatic stem cells into functional hepatic-like cells displaying a hepatocyte-like glucose metabolism. *J Cell Physiol* 2010;**225**(2):545–54.

10. Greschat S, Schira J, Küry P, Rosenbaum C, de Souza Silva MA, Kögler G, et al. Unrestricted somatic stem cells from human umbilical cord blood can be differentiated into neurons with a dopaminergic phenotype. *Stem Cells Dev* 2008;**17**(2):221–32.
11. Bosch J, Houben AP, Radke TF, Stapelkamp D, Bünemann E, Balan P , et al. Distinct differentiation potential of "MSC" derived from cord blood and umbilical cord: are cord-derived cells true mesenchymal stromal cells? *Stem Cells Dev* 2012;**21**(11):1977–88.
12. de Mara CS, Duarte AS, Sartori-Cintra AR, Luzo AC, Saad ST, Coimbra IB. Chondrogenesis from umbilical cord blood cells stimulated with BMP-2 and BMP-6. *Rheumatol Int* 2013;**33**(1):121–8.
13. Bieback K, Kern S, Kluter H, Eichler H. Critical parameters for the isolation of mesenchymal stem cells from umbilical cord blood. *Stem Cells* 2004;**22**(4):625–34.
14. Lee OK, Kuo TK, Chen WM, Lee KD, Hsieh SL, Chen TH. Isolation of multipotent mesenchymal stem cells from umbilical cord blood. *Blood* 2004;**103**(5):1669–75.
15. Kang XQ, Zang WJ, Bao LJ, Li DL, Xu XL, Yu XJ. Differentiating characterization of human umbilical cord blood-derived mesenchymal stem cells in vitro. *Cell Biol Int* 2006;**30**(7):569–75.
16. Lu LL, Liu YJ, Yang SG, Zhao QJ, Wang X, Gong W, et al. Isolation and characterization of human umbilical cord mesenchymal stem cells with hematopoiesis-supportive function and other potentials. *Haematologica* 2006;**91**(8):1017–26.
17. Lu FZ, Fujino M, Kitazawa Y, Uyama T, Hara Y, Funeshima N, et al. Characterization and gene transfer in mesenchymal stem cells derived from human umbilical-cord blood. *J Lab Clin Med* 2005;**146**(5):271–8.
18. Park KS, Lee YS, Kang KS. In vitro neuronal and osteogenic differentiation of mesenchymal stem cells from human umbilical cord blood. *J Vet Sci* 2006;**7**(4):343–8.
19. Hess DA, Craft TP, Wirthlin L, Hohm S, Zhou P, Eades WC, et al. Widespread nonhematopoietic tissue distribution by transplanted human progenitor cells with high aldehyde dehydrogenase activity. *Stem Cells* 2008;**26**(3):611–20.
20. Tracy E, Aldrink J, Panosian J, Beam D, Thacker J, Reese M, et al. Isolation of oligodendrocyte-like cells from human umbilical cord blood. *Cytotherapy* 2008;**10**(5):518–25.
21. Gang EJ, Jeong JA, Hong SH, Hwang SH, Kim SW, Yang IH, et al. Skeletal myogenic differentiation of mesenchymal stem cells isolated from human umbilical cord blood. *Stem Cells* 2004;**22**(4):617–24.
22. Gao F, Wu DQ, Hu YH, Jin GX, Li GD, Sun TW, et al. In vitro cultivation of islet-like cell clusters from human umbilical cord blood-derived mesenchymal stem cells. *Transl Res* 2008;**151**(6):293–302.
23. Huang CJ, Butler AE, Moran A, Rao PN, Wagner JE, Blazar BR, et al. A low frequency of pancreatic islet insulin-expressing cells derived from cord blood stem cell allografts in humans. *Diabetologia* 2011;**54**(5):1066–74.
24. Gussoni E, Bennett RR, Muskiewicz KR, Meyerrose T, Nolta JA, Gilgoff I, et al. Long-term persistence of donor nuclei in a Duchenne muscular dystrophy patient receiving bone marrow transplantation. *J Clin Invest* 2002;**110**(6):807–14.
25. Kurtzberg J, Kosaras B, Stephens C, Snyder EY. Umbilical cord blood cells engraft and differentiate in neural tissues after human transplantation. *Biol Blood Marrow Transplant* 2003;**9**(2):128–9.
26. Hoogerbrugge P, Suzuki K, Poorthuis B, Kobayashi T, Wagemaker G, van Bekkum DW. Donor-derived cells in the central nervous system of twitcher mice after bone marrow transplantation. *Science* 1988;**239**(4843):1035–8.
27. Jeong JA, Gang EJ, Hong SH, Hwang SH, Kim SW, Yang IH, et al. Rapid neural differentiation of human cord blood-derived mesenchymal stem cells. *Neuroreport* 2004;**15**(11):1731–4.
28. Tracy ET, Zhang CY, Gentry T, Shoulars KW, Kurtzberg J. Isolation and expansion of oligodendrocyte progenitor cells from cryopreserved human umbilical cord blood. *Cytotherapy* 2011;**13**(6):722–9.
29. Fallahi-Sichani M, Soleimani M, Najafi SM, Kiani J, Arefian E, Atashi A. In vitro differentiation of cord blood unrestricted somatic stem cells expressing dopamine-associated genes into neuron-like cells. *Cell Biol Int* 2007;**31**(3):299–303.
30. Sun W, Buzanska L, Domanska-Janik K, Salvi RJ, Stachowiak MK. Voltage-sensitive and ligand-gated channels in differentiating neural stem-like cells derived from the nonhematopoietic fraction of human umbilical cord blood. *Stem Cells* 2005;**23**(7):931–45.
31. Lee MW, Moon YJ, Yang MS, Kim SK, Jang IK, Eom YW, et al. Neural differentiation of novel multipotent progenitor cells from cryopreserved human umbilical cord blood. *Biochem Biophys Res Commun* 2007;**358**(2):637–43.
32. Lee OK, Ko YC, Kuo TK, Chou SH, Li HJ, Chen WM, et al. Fluvastatin and lovastatin but not pravastatin induce neuroglial differentiation in human mesenchymal stem cells. *J Cell Biochem* 2004;**93**(5):917–28.
33. Jin W, Xing YQ, Yang AH. Epidermal growth factor promotes the differentiation of stem cells derived from human umbilical cord blood into neuron-like cells via taurine induction in vitro. *Vitro Cell Dev Biol Anim* 2009;**45**(7):321–7.
34. Naujock MSN, Reinhardt P, Stemeckert J, Haase A, Martin U, Kim KS, et al. Molecular and functional analyses of motor neurons generated from human cord blood derived induced pluripotent stem cells. *Stem Cells Dev* 2014. [E-pub ahead of print].
35. Iwaniuk KM, Schira J, Weinhold S, Jung M, Adjaye J, Müller HW, et al. Network-like impact of MicroRNAs on neuronal lineage differentiation of unrestricted somatic stem cells from human cord blood. *Stem Cells Dev* 2011;**20**(8):1383–94.
36. Bliss T, Guzman R, Daadi M, Steinberg GK. Cell transplantation therapy for stroke. *Stroke* 2007;**38**(Suppl. 2):817–26.
37. Llado J, Haenggeli C, Maragakis NJ, Snyder EY, Rothstein JD. Neural stem cells protect against glutamate-induced excitotoxicity and promote survival of injured motor neurons through the secretion of neurotrophic factors. *Mol Cell Neurosci* 2004;**27**(3):322–31.
38. Vendrame M, Gemma C, de Mesquita D, Collier L, Bickford PC, Sanberg CD, et al. Anti-inflammatory effects of human cord blood cells in a rat model of stroke. *Stem Cells Dev* 2005;**14**(5):595–604.
39. Borlongan CV, Hadman M, Sanberg CD, Sanberg PR. Central nervous system entry of peripherally injected umbilical cord blood cells is not required for neuroprotection in stroke. *Stroke* 2004;**35**(10):2385–9.
40. Arien-Zakay H, Lecht S, Bercu MM, Tabakman R, Kohen R, Galski H, et al. Neuroprotection by cord blood neural progenitors involves antioxidants, neurotrophic and angiogenic factors. *Exp Neurol* 2009;**216**(1):83–94.
41. Carmichael ST. Plasticity of cortical projections after stroke. *Neuroscientist* 2003;**9**(1):64–75.
42. Chen J, Zhang ZG, Li Y, Wang L, Xu YX, Gautam SC, et al. Intravenous administration of human bone marrow stromal cells induces angiogenesis in the ischemic boundary zone after stroke in rats. *Circ Res* 2003;**92**(6):692–9.
43. Taguchi A, Soma T, Tanaka H, Kanda T, Nishimura H, Yoshikawa H, et al. Administration of CD34+ cells after stroke enhances neurogenesis via angiogenesis in a mouse model. *J Clin Invest* 2004;**114**(3):330–8.

44. Shen LH, Li Y, Chen J, Zhang J, Vanguri P, Borneman J, et al. Intracarotid transplantation of bone marrow stromal cells increases axon-myelin remodeling after stroke. *Neuroscience* 2006;**137**(2):393–9.
45. Beam D, Poe MD, Provenzale JM, Szabolcs P, Martin PL, Prasad V, et al. Outcomes of unrelated umbilical cord blood transplantation for X-linked adrenoleukodystrophy. *Biol Blood Marrow Transpl* 2007;**13**(6):665–74.
46. Prasad VK, Mendizabal A, Parikh SH, Szabolcs P, Driscoll TA, Page K, et al. Unrelated donor umbilical cord blood transplantation for inherited metabolic disorders in 159 pediatric patients from a single center: influence of cellular composition of the graft on transplantation outcomes. *Blood* 2008;**112**(7):2979–89.
47. Boelens JJ. Trends in haematopoietic cell transplantation for inborn errors of metabolism. *J Inherit Metab Dis* 2006;**29**(2–3):413–20.
48. Staba SL, Escolar ML, Poe M, Kim Y, Martin PL, Szabolcs P, et al. Cord-blood transplants from unrelated donors in patients with Hurler's syndrome. *N Engl J Med* 2004;**350**(19):1960–9.
49. Provenzale JM, Escolar M, Kurtzberg J. Quantitative analysis of diffusion tensor imaging data in serial assessment of Krabbe disease. *Ann N Y Acad Sci* 2005;**1064**:220–9.
50. Martin PL, Carter SL, Kernan NA, Sahdev I, Wall D, Pietryga D, et al. Results of the cord blood transplantation study (COBLT): outcomes of unrelated donor umbilical cord blood transplantation in pediatric patients with lysosomal and peroxisomal storage diseases. *Biol Blood Marrow Transpl* 2006;**12**(2):184–94.
51. Escolar ML, Poe MD, Provenzale JM, Richards KC, Allison J, Wood S, et al. Transplantation of umbilical-cord blood in babies with infantile Krabbe's disease. *N Engl J Med* 2005;**352**(20):2069–81.
52. Duffner PK, Caggana M, Orsini JJ, Wenger DA, Patterson MC, Crosley CJ, et al. Newborn screening for Krabbe disease: the New York State model. *Pediatr Neurol* 2009;**40**(4):245–52; discussion 253–245.
53. Escolar ML, Poe MD, Smith JK, Gilmore JH, Kurtzberg J, Lin W, et al. Diffusion tensor imaging detects abnormalities in the corticospinal tracts of neonates with infantile Krabbe disease. *AJNR Am J Neuroradiol* 2009;**30**(5):1017–21.
54. Mezey E, Key S, Vogelsang G, Szalayova I, Lange GD, Crain B. Transplanted bone marrow generates new neurons in human brains. *Proc Natl Acad Sci U S A* 2003;**100**(3):1364–9.
55. Cogle CR, Yachnis AT, Laywell ED, Zander DS, Wingard JR, Steindler DA, et al. Bone marrow transdifferentiation in brain after transplantation: a retrospective study. *Lancet* 2004;**363**(9419):1432–7.
56. Kurtzberg J, Kosaras B, Stephens C, Snyder EY. Umbilical cord blood cells engraft and differentiate in neural tissues after human transplantation. *Biol Blood Marrow Transpl* 2003;**9**:128a.
57. Vendrame M, Cassady J, Newcomb J, Butler T, Pennypacker KR, Zigova T, et al. Infusion of human umbilical cord blood cells in a rat model of stroke dose-dependently rescues behavioral deficits and reduces infarct volume. *Stroke* 2004;**35**(10):2390–5.
58. Chen J, Sanberg PR, Li Y, Wang L, Lu M, Willing AE, et al. Intravenous administration of human umbilical cord blood reduces behavioral deficits after stroke in rats. *Stroke* 2001;**32**(11):2682–8.
59. Meier C, Middelanis J, Wasielewski B, Neuhoff S, Roth-Haerer A, Gantert M, et al. Spastic paresis after perinatal brain damage in rats is reduced by human cord blood mononuclear cells. *Pediatr Res* 2006;**59**(2):244–9.
60. Nan Z, Grande A, Sanberg CD, Sanberg PR, Low WC. Infusion of human umbilical cord blood ameliorates neurologic deficits in rats with hemorrhagic brain injury. *Ann N Y Acad Sci* 2005;**1049**:84–96.
61. Lu D, Sanberg PR, Mahmood A, Li Y, Wang L, Sanchez-Ramos J, et al. Intravenous administration of human umbilical cord blood reduces neurological deficit in the rat after traumatic brain injury. *Cell Transplant* 2002;**11**(3):275–81.
62. Zhao ZM, Li HJ, Liu HY, Lu SH, Yang RC, Zhang QJ, et al. Intraspinal transplantation of CD34+ human umbilical cord blood cells after spinal cord hemisection injury improves functional recovery in adult rats. *Cell Transpl* 2004;**13**(2):113–22.
63. Nishio Y, Koda M, Kamada T, Someya Y, Yoshinaga K, Okada S, et al. The use of hemopoietic stem cells derived from human umbilical cord blood to promote restoration of spinal cord tissue and recovery of hindlimb function in adult rats. *J Neurosurg Spine* 2006;**5**(5):424–33.
64. Derrick M, Drobyshevsky A, Ji X, Tan S. A model of cerebral palsy from fetal hypoxia-ischemia. *Stroke* 2007;**38**(Suppl. 2):731–5.
65. Ji XTE, Drobyshevsky A, Derrick M, Yu L, Liu A, Cotten M, et al. *Do human umbilical cord blood cells improve outcome in a fetal rabbit model of cerebral palsy?* Pediatric Academic Society: 2009. EPAS2008:3455.11.
66. Chua CO, Chahboune H, Braun A, Dummula K, Chua CE, Yu J, et al. Consequences of intraventricular hemorrhage in a rabbit pup model. *Stroke* 2009;**40**(10):3369–77.
67. Sun J, Allison J, McLaughlin C, Sledge L, Waters-Pick B, Wease S, et al. Differences in quality between privately and publicly banked umbilical cord blood units: a pilot study of autologous cord blood infusion in children with acquired neurologic disorders. *Transfusion* 2010;**50**(9):1980–7.
68. Englander ZA, Pizoli CE, Batrachenko A, Sun J, Worley G, Mikati MA, et al. Diffuse reduction of white matter connectivity in cerebral palsy with specific vulnerability of long range fiber tracts. *Neuroimage Clin* 2013;**2**:440–7.
69. Min K, Song J, Kang JY, Ko J, Ryu JS, Kang MS, et al. Umbilical cord blood therapy potentiated with erythropoietin for children with cerebral palsy: a double-blind, randomized, placebo-controlled trial. *Stem Cells* 2013;**31**(3):581–91.
70. Yang WZ, Zhang Y, Wu F, Min WP, Minev B, Zhang M, et al. Safety evaluation of allogeneic umbilical cord blood mononuclear cell therapy for degenerative conditions. *J Transl Med* 2010;**8**:75.
71. Dongmei H, Jing L, Mei X, Ling Z, Hongmin Y, Zhidong W, et al. Clinical analysis of the treatment of spinocerebellar ataxia and multiple system atrophy-cerebellar type with umbilical cord mesenchymal stromal cells. *Cytotherapy* 2011;**13**(8):913–7.
72. Jin JL, Liu Z, Lu ZJ, Guan DN, Wang C, Chen ZB, et al. Safety and efficacy of umbilical cord mesenchymal stem cell therapy in hereditary spinocerebellar ataxia. *Curr Neurovasc Res* 2013;**10**(1):11–20.
73. Lv YT, Zhang Y, Liu M, Qiuwaxi JN, Ashwood P, Cho SC, et al. Transplantation of human cord blood mononuclear cells and umbilical cord-derived mesenchymal stem cells in autism. *J Transl Med* 2013;**11**:196.
74. Cotten CM, Murtha AP, Goldberg RN, Grotegut CA, Smith PB, Goldstein RF, et al. Feasibility of autologous cord blood cells for infants with hypoxic-ischemic encephalopathy. *J Pediatr* 2014;**164**(5):973–79.e1.
75. Bang OY, Lee JS, Lee PH, Lee G. Autologous mesenchymal stem cell transplantation in stroke patients. *Ann Neurol* 2005;**57**(6):874–82.
76. Mendonça ML, Freitas GR, Silva SA, Manfrim A, Falcão CH, Gonzáles C, et al. [Safety of intra-arterial autologous bone marrow mononuclear cell transplantation for acute ischemic stroke]. *Arq Bras Cardiol* 2006;**86**(1):52–5.
77. Eizawa T, Ikeda U, Murakami Y, Matsui K, Yoshioka T, Takahashi M, et al. Decrease in circulating endothelial progenitor cells in patients with stable coronary artery disease. *Heart* 2004;**90**(6):685–6.

78. Siegel G, Kluba T, Hermanutz-Klein U, Bieback K, Northoff H, Schafer R. Phenotype, donor age and gender affect function of human bone marrow-derived mesenchymal stromal cells. *BMC Med* 2013;**11**:146.
79. Freed CR, Greene PE, Breeze RE, Tsai WY, DuMouchel W, Kao R, et al. Transplantation of embryonic dopamine neurons for severe Parkinson's disease. *N Engl J Med* 2001;**344**(10):710–9.
80. Olanow CW, Goetz CG, Kordower JH, Stoessl AJ, Sossi V, Brin MF, et al. A double-blind controlled trial of bilateral fetal nigral transplantation in Parkinson's disease. *Ann Neurol* 2003;**54**(3):403–14.
81. Xiong N, Zhang Z, Huang J, Chen C, Zhang Z, Jia M, et al. VEGF-expressing human umbilical cord mesenchymal stem cells, an improved therapy strategy for Parkinson's disease. *Gene Ther* 2011;**18**(4):394–402.
82. Nikolic WV, Hou H, Town T, Zhu Y, Giunta B, Sanberg CD, et al. Peripherally administered human umbilical cord blood cells reduce parenchymal and vascular beta-amyloid deposits in Alzheimer mice. *Stem Cells Dev* 2008;**17**(3):423–39.
83. Darlington D, Deng J, Giunta B, Hou H, Sanberg CD, Kuzmin-Nichols N, et al. Multiple low-dose infusions of human umbilical cord blood cells improve cognitive impairments and reduce amyloid-beta-associated neuropathology in Alzheimer mice. *Stem Cells Dev* 2013;**22**(3):412–21.
84. Ende N, Chen R, Ende-Harris D. Human umbilical cord blood cells ameliorate Alzheimer's disease in transgenic mice. *J Med* 2001;**32**(3–4):241–7.
85. Ha S-J, Lee S, Suh Y-H, Chang K-A. Therapeutic effects of human adipose-derived stem cells in VPA-induced autism mouse model. Program No. 50.01. 2013 Neuroscience Meeting Planner. San Diego, CA: Society for Neuroscience, 2013. Online.
86. Lv YT, Zhang Y, Liu M, Qiuwaxi JN, Ashwood P, Cho SC, et al. Transplantation of human cord blood mononuclear cells and umbilical cord-derived mesenchymal stem cells in autism. *J Transl Med* 2013;**11**(1):196.
87. Sharma A, Gokulchandran N, Sane H, Nagrajan A, Paranjape A, Kulkarni P, et al. Autologous bone marrow mononuclear cell therapy for autism: an open label proof of concept study. *Stem Cells Int* 2013;**2013**:623875.

Chapter 14

# Biobanks for Induced Pluripotent Stem Cells and Reprogrammed Tissues

Lee Carpenter
*NHS Blood and Transplant and Radcliffe Department of Medicine, University of Oxford, Oxford, UK*

## Chapter Outline



## 1. INTRODUCTION: A BRIEF HISTORY OF INDUCED PLURIPOTENCY

Currently there is a huge amount of effort around the world to bank human tissue for various applications in research and medicine. Traditionally, stored frozen tissues have been in the form of donated bone marrow and mobilized peripheral blood, and more recently umbilical cord blood, and used for transplantation in patients with diseases such as leukemia or aplastic anemia. Within the United Kingdom, organizations such as the National Health Service (NHS) Cord Blood Bank and British Bone Marrow Registry, including other registries and charities, are responsible for provision of stem cells from the tissues stored in these banks. More recently, with next-generation sequencing technologies and the advance of bioinformatics, tissue or "biobanks" are also being initiated as a valuable resource for medical research into human disease, such as the genome-wide association studies conducted by the Wellcome Trust Sanger Institute in Cambridge. This study linked genetic (copy number) variations across a large population (19,000 individuals) and correlated these with disease susceptibility, thus identifying the genetic traits or defects that contribute to complex diseases such as Crohn's disease and type I diabetes.[1] With the advent of reprogramming technologies, where cells of one type can be converted to another, this paves the way for a third major role of the biobanks, which will be the provision of novel cellular therapies and diagnostics that previously were not considered possible.

Reprogramming technologies have been explored in research for over 50 years, and first described by Sir Professor John Gurdon in 1962, where a nucleus from a single somatic cell was used to create adult frogs, after transplantation into an enucleated oocyte[2] (Figure 1). This work was ground-breaking since it demonstrated that adult, somatic nuclei could be reprogrammed back to an embryonic state, and thus become pluripotent and able to contribute to the whole organism. After many years, and a huge amount of effort, this phenomenon was repeated in 1996, when the announcement of Dolly the sheep was made,[3] by a process termed somatic cell nuclear transfer (SCNT). Creating cloned animals was conducted primarily for

Cord Blood Stem Cells Medicine. http://dx.doi.org/10.1016/B978-0-12-407785-0.00014-1

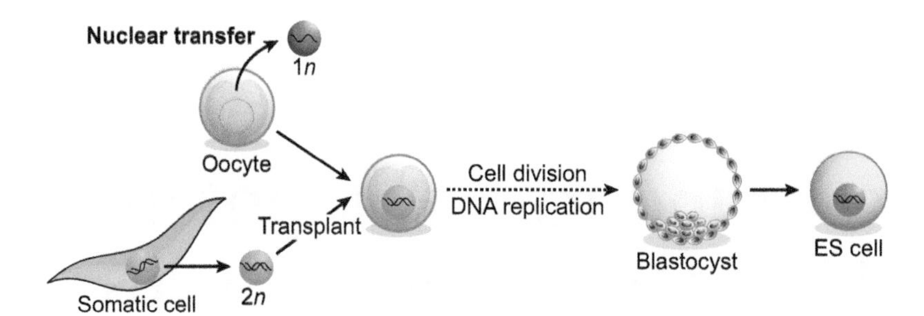


**FIGURE 1** **Somatic cell nuclear transfer.** The "traditional" approach for reprogramming the somatic nucleus into a pluripotent state, first reported by JB Gurdon in 1962, and famously repeated with "Dolly the Sheep." This allows cloning of animals when the nucleus of a somatic or differentiated cell is transferred to an oocyte that has had its own nucleus removed. This "transplanted" oocyte can then divide and form a blastocyst, whereupon implantation results in a "cloned" offspring. *From Ref. 17.*

novel therapeutics (i.e., antibody production) and better animal husbandry; however, it is an expensive and time-consuming process. The processes that occurred during SCNT represented largely a black hole in our understanding, since it was unknown which specific factors in the oocyte cytoplasm were inducing reprogramming of the transferred somatic nuclei.

Fortunately, with the isolation of mouse[4] and eventually human embryonic stem cells (hESCs)[5] scientists were able to better understand what constituted the pluripotent state, and specifically which transcription factors controlled and were most critical for maintenance of pluripotency. The pluripotent state was shown to be maintained by transcription factors such as Oct4[6] and Nanog,[7,8] so that upon their removal, pluripotency is lost, while induction of their expression is sufficient to maintain pluripotency in the absence of extrinsic factors. This work demonstrated that transcription factors alone are both necessary and sufficient to maintain the embryonic-like pluripotent state. In ground-breaking studies in 2006 and 2007, Professor Shinya Yamanaka and colleagues[9,10] were able to mimic the findings of Sir Professor John Gurdon over 50 years before, but instead to induce pluripotency using a defined set of factors and not by SCNT. This work has lead to both Sir Professor John Gurdon and Professor Shinya Yamanaka being awarded the Nobel Prize for Physiology or Medicine in 2012, and has now paved the way for huge advances in medical and basic research, in a way that was not previously possible. Diseases are being recreated in the "dish" using human induced pluripotent stem cells (hiPSCs) from patient-specific tissue, which will help to provide not only for a better understanding of genes involved in disease pathology, but also for the effective screening of drugs to improve clinical outcomes. HiPSCs from diseased and healthy individuals are also being used in basic science to understand better the role of genes in the early development of tissues and organs, and to understand the specific function of proteins, by understanding what happens when these processes go wrong. In the immediate future, novel diagnostic applications will also arise from reprogramming technologies and hiPSCs. Already stored in biobanks around the world are patient and donor tissues, with information regarding their phenotype and genotype. Lastly is the hope of reprogramming technologies to provide a novel source of cells and tissues for regenerative therapies, and it is these various applications for research and medicine that are discussed further.

## 2. WHAT DEFINES PLURIPOTENCY AND THE PLURIPOTENT STEM CELL

### 2.1 Isolation of Mouse and Human Embryonic Stem Cells

Pluripotent stem cells such as mouse embryonic stem cells (mESCs) are normally a transient cell population within the developing embryo and constitute the inner cell mass of the E3.5 blastocyst in mice, which eventually contributes to all the lineages and tissues of the adult organism. This is also true of humans, although developmental timings are delayed. Although transient in the developing embryo, when isolated in vitro under the correct conditions,[4,5] they are essentially an immortal cell line. Therefore, they can proliferate indefinitely, while retaining the potential to contribute to all the tissues of the developing embryo, and are thus referred to as embryonic stem cells. In mouse, pluripotency is demonstrated by injection of embryonic stem cells back into the developing blastocyst, whereupon they contribute to the whole animal as a chimera. The most stringent measure of pluripotency is germ-line transmission, whereupon homozygous mice can be derived.[11] Since the starting point for these "engineered" mice are embryonic stem cells that can be manipulated genetically in vitro, with genes added[12] or "knocked out,"[13] mouse lines generated this way are referred to as transgenic or knockout mice, respectively. This approach was pioneered in the 1980s by Sir Professor Martin Evans and colleagues and has been widely used in basic research to understand the role of genes in early development and disease.

Human embryonic stem cells (hESCs) were isolated in 1998, and although much was gleaned from working with mESCs, there were to be significant differences between them, which hampered progress toward their initial derivation. This most notably is due to differences in the developmental stage at which mESCs and hESCs are derived (ICM and epiblast, respectively), and the consequent variation in cytokine and growth factor requirements.[14] Since hESC lines are generally derived from unwanted fertilized embryos, and while many hESC lines have now been generated,[15] few are from disease or genotype-specific sources that make them more useful to research and regenerative medicine. Until recently, efforts to create pluripotent cell lines by SCNT had largely failed, and only by very careful manipulation of the culture conditions has this now been shown to be possible,[16] although this is unlikely to replace reprogramming approaches described herein (see Figure 1), since SCNT is still ethically contentious and technically demanding.

## 2.2 Extrinsic and Intrinsic Networks that Contribute to Pluripotency

From studies to isolate, characterize, and maintain mESC and hESCs, it has been possible to identify critical extrinsic factors such as Leukemia Inhibitory Factor (LIF)[18] and basic fibroblast growth factor (bFGF)[19] that inhibit differentiation and maintain self-renewal and pluripotency. Since constitutive expression of Signal Transducer and Activator of Transcription STAT3 downstream of pg130 and LIF signaling could maintain pluripotency in the absence of LIF[20] but only in the presence of serum, other factors were implicated in maintenance of self-renewal. Bone morphogenic protein (BMP) 4, which is present in serum,[21] has since been shown to act with LIF/STAT3 signaling to prevent differentiation and thus maintain self-renewal in mESCs.

Arguably, it has been identification of the intrinsic factors that maintain pluripotency which has lead to the ground-breaking discoveries in tissue reprogramming and induced pluripotency. Central to the pluripotent transcription factor network is the POU5F gene or Oct4 transcription factor.[22] The role for this homeodomain-containing protein was first identified in mouse embryos as being necessary for development of the inner cell mass, from which mESCs are derived.[6] Nanog is another homeodomain-containing transcription factor that was discovered independently to regulate pluripotency[7,8] that is thought to act independently of STAT3, but which may interact directly with Oct4 to provide maximal self-renewal capacity (see Figure 2). An additional transcription factor of note here is the SRY (sex determining region Y)-box 2, also known as SOX2 which is a HMG DNA-binding domain transcription factor, discovered to act synergistically with Oct4 to mediate expression of Oct4 target genes.[23]

Therefore, within the pluripotent state, there is a requirement for an extrinsic component that acts to prevent differentiation such as LIF (or bFGF in humans) and BMP4, and an intrinsic network of specific transcription factors, i.e., Oct4 and Nanog, which are critical for maintaining pluripotency (see Figure 2).

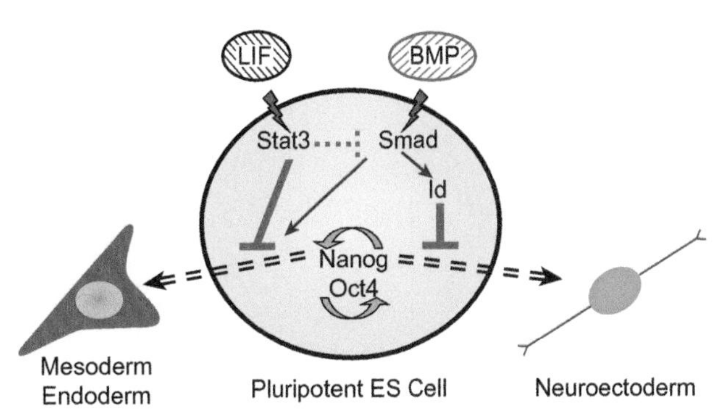


**FIGURE 2 Pluripotent networks at a glance.** Critical research using embryonic stem cells in culture was able to identify the intrinsic factors that maintain pluripotency and extrinsic factors in the surrounding environment that prevent differentiation. Central components of the pluripotent transcription factor network includes Nanog and Oct4, which maintain the stem cell in a cycle of self-renewal. Extrinsic factors include BMP4 in serum, and for mESCs, LIF, which are supplemented to promote activity of downstream signaling pathways Id and STAT3, respectively, which prevent differentiation into the first major germ layers, mesoendoderm and neuroectoderm. *From Ref. 21.*

# 3. REPROGRAMMING HUMAN SOMATIC CELLS TOWARD INDUCED PLURIPOTENT STEM CELLS USING DEFINED FACTORS

## 3.1 Initial Reprogramming Studies

In 2006, Shinya Yamanaka and colleagues made full use of advances in our understanding of pluripotency to reprogram and induce a pluripotent stem cell from mouse tail tip fibroblasts. This ground-breaking study in mouse was quickly followed up with human tissue[9,10] and by several other investigators.[24–26] Yamanaka's approach relied on the hypothesis that intrinsic factors such as Oct4 could be sufficient to reprogram a somatic nuclei toward an embryonic-like sate, and sufficient for it to regain a pluripotent "potential." This hypothesis was proved correct when after assessing 21 genes in combination, Takahashi and colleagues eventually identified just four that were critical for induction of a pluripotent state.[9,10] Using retroviral transgene expression of just Oct4, Sox2, Myc, and Klf4 but eventually omitting Myc due to its transforming potential,[27] skin fibroblasts over time acquire an embryonic-like morphology, with a large nucleus and scant cytoplasm. They also acquire an ESC phenotype in that they express ESC-specific surface markers SSEA4, TRA-1-81, and TRA-1-160 (see Figure 3), but most importantly they are functionally pluripotent, being able to contribute to chimeras in mice, and can undergo germ-line transmission.[28] Although with hiPSCs these later studies are not possible, functional pluripotency can be demonstrated in vitro by differentiation as embryoid bodies to yield cell types from the three major lineages of the body. As such Yamanaka and colleagues were able to describe specialized neurons and beating cardiac myocytes, using hiPSCs originating from human dermal fibroblasts.

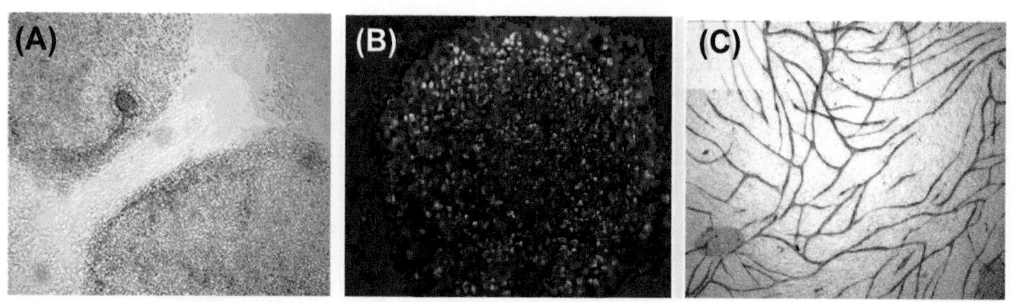


FIGURE 3 **Reprogramming of human fibroblasts to induced pluripotent stem cells with defined factors.** As first described by Yamanaka and colleagues in 2007, we used retroviral vectors containing specific transcription factors to transduce into human fibroblasts, which were subsequently cultured in embryonic stem cell media on a supportive layer of fibroblasts. Over a period of 28 days, reprogrammed induced pluripotent stem cells were derived. This is reflected by a change in appearance of the adult dermal fibroblast from a flat spindle shape to a large colony of columnar cells tightly packed and with a large nucleus with scant cytoplasm (A), that expresses a range of embryonic stem cell-specific genes such as Oct4 (green) and TRA-1-81 (red) with nucleus staining with DAPI (4′,6-diamidino-2-phenylindole) in blue (B), and which can then differentiate into a wide variety of cell types such as endothelium, shown as CD31-positive networks formed in a tubule assay (C), as well as red blood cells and B lymphocytes.[33] (For interpretation of the color in this figure legend, the reader is referred to the online version of this book.) *From: L. Carpenter et al., 2011 (Ref. 33).*

Subsequent studies by other investigators have since shown that Oct4 is the most critical component for reprogramming and the only transcription factor required for reprogramming of neuronal stem cells,[29] which can express Sox2 endogenously. Other transcription factors have been used to substitute for the original cocktail, which include Nanog and Lin28, as well as Oct4 and Sox2,[25] while the requirement for Klf4 has been shown to be replaced by other members of the Kruppel-like family of transcription factors.[30] cMyc is an oncogenic transcription factor that can be left out all together, since, although it has been shown to increase iPSC-like colonies forming in the initial reprogramming dish, the background for incompletely reprogrammed colonies is high and results in iPSC lines that are not fully pluripotent.[31] Transgene expression is generally required for the first transition toward a more epithelial state, and for initiating the pluripotent network, after which time the endogenous program of transcription factors become dominant[32] and the viral transgenes are generally silenced by epigenetic mechanisms such as DNA methylation. During this time, what were once single somatic cells become colonies that progress to become a large dense mass, although the appearance is quite different between mouse and human iPSCs (as with ESCs), with the latter being generally flatter and where single iPSCs are more easily observed.

## 3.2 Characterization of hiPSC Lines

Once colonies have formed, usually at around day 14–28 depending on starting material and reprogramming strategy, live cell staining can be performed to confirm identity. The most reliable marker for complete reprogramming at this early stage is TRA-1-160[34] and if done with live cells, then these cells can be selected directly for long-term culture and expanded into hiPSC lines for further testing. More complete characterization for hiPSC lines includes demonstrating silencing of transgene expression, determining methylation status of promoters for endogenous factors (i.e., Oct4 and Nanog), and quantification of mRNA expression of ESC-specific genes. A functional in vivo readout for pluripotency includes teratoma formation, which was all described in the original reports for induced pluripotency.[9,10] Additionally, it is now routine to describe a karyotype by G banding and/or use of single-nucleotide polymorphism arrays that identify copy-number variations, which result from genome instability, which has been observed for both hESC[35,36] and hiPSC lines.[37,38]

If hiPSCs that have been generated are to be useful for medical research and therapeutics, they need to be pluripotent and thus able to differentiate into the cell type of interest in vitro. Differentiation toward the neuroectoderm, mesoderm, and endoderm can be tested by demonstrating the presence of neurons, cardiomyocytes, and hepatocytes, respectively. Typically, the neuroectoderm lineage is most easily formed, since it represents the first lineage to form, possibly by default,[39] while the endoderm such as hepatocytes are typically more difficult to derive.[40] In the NHS Blood and Transplant, we also use these "outputs" as a measure of pluripotency for the hiPSC lines we have generated.[33] In addition, we assess the formation of a hemato-endothelial cell type, which arises from a multipotent progenitor and which can help to inform us on how differentiating hiPSCs can contribute to hematopoiesis in development. Also in work assessing the potential therapeutic use of hiPSCs for cardiovascular (CV) repair, we have been able to demonstrate that hiPSC lines generated in our laboratory are able to contribute efficiently to the CV lineage with beating tissue in culture and all relevant lineages: cardiomyocytes, endothelium, and smooth muscle being represented. Furthermore, when these cells are transplanted into animal models of CV disease, they do provide a small functional benefit[41] which not only demonstrates the overall therapeutic potential for hiPSCs, but also provides a functional readout for their pluripotency.

## 3.3 Reprogramming Technologies Now Available

Many different reprogramming technologies have been employed since Professor Shinya Yamanaka's first demonstration using retroviral vectors for transgene expression. Initially this work was reproduced by other laboratories, using similar integrating viruses such as lentivirus, although with a different set of reprogramming factors, i.e., Lin28, Nanog, Oct4, and Sox2.[25] Generally the field has moved toward safer approaches using viruses that do not integrate, such as adenovirus,[42] which are only transiently retained within the cell and therefore are considered a safer approach for downstream applications. However, this can result in reduced efficiencies of reprogramming, presumably because stable transgene expression is not maintained. Recently, Sendai virus has been reported to have greatly improved efficiencies,[43,44] and as such this method is currently popular for generating hiPSC lines. Plasmid-based approaches are varied, as too are their efficiencies for reprogramming, but here they generally adopt strategies to avoid integration or the effects of, and include excision after reprogramming[45,46] or maintenance of vectors outside of the nucleus in the episome.[47] Strategies that include the introduction of small molecules, such as HDAC inhibitors sodium butyrate and valproic acid, can greatly improve efficiencies of reprogramming.[48,49] Peptide-based delivery systems have also been explored with some success,[50,51] where the transcription factor protein is synthesized with a membrane permeant domain to promote uptake from the extracellular medium across the plasma membrane; mRNA-based delivery systems, which can introduce modified mRNA that mediates transgene expression,[52] have also been tested. However, some of these vector-free methods seem to be technically very demanding, and although in some cases commercial kits are available, they can require repeated transfections and are maybe not currently applicable for hard-to-transfect cell types or high-throughput projects. While we originally started reprogramming in Oxford using the Yamanaka retrovirus plasmids with dermal fibroblasts on mouse embryonic feeders (according to Nishikawa et al. 2007[9]), we have now adopted the OriP plasmid vectors that are "footprint free" and use the OriP/EBNA-1 vectors (pEP4EO2SET2K and pEP4EO2SCK2MEN2L) to deliver an optimized reprogramming gene cocktail. This was first reported by Thomson and colleagues,[47] and a small-molecule strategy subsequently developed by this group[53] enables iPSC reprogramming without feeder layers. Since these are mammalian expression vectors that are maintained episomally (and thus not retained within the nucleus), they are lost over time if cells are cultured without positive selection, and so are considered a safer option than other expression vectors that are integrated but subsequently excised.[45,46] Over time, as the OriP vectors are degraded, the endogenous pluripotent transcription factor network is established to maintain the hiPSC state. Initial reprogramming efficiencies of just 0.01% are not high, but are enough to yield several hiPSC colonies from $1 \times 10^6$ dermal fibroblasts when using these vectors, but this was improved by approximately 100-fold with the use of small molecules.[53] We have also successfully adopted small-molecule approaches that allow hiPSC derivation without fibroblast feeders started with a variety of blood cells (umbilical cord blood mononuclear cells, umbilical cord blood $CD34^+$ cells, umbilical cord blood erythroblasts, and adult mononuclear cells) which identify small reprogrammed colonies after just 9 days of culture. Here we used the OriP vectors described,[53] DMEM/F12 with N2 and B27 supplement, supplemented with 100 ng bFGF, 1 μM PD032590, 3 μM CHIR99021, 0.4 μM A83-01, 100 ng/ml human LIF, and 10 μM Y27632, switching to mTeSR after 10 days to stabilize the colony and establish the line.

## 3.4 Direct Reprogramming

Another intriguing aspect of the work of reprogramming towards pluripotency is the concept of direct reprogramming to the cell type of choice. This was originally termed trans-differentiation that had been explored for many years by cancer biologists to explain transformed states, and even by cell engineers trying to create rare cells for regeneration, i.e., insulin-producing beta cells from liver hepatocytes for the treatment of diabetes.[54] After this came the phenomenon of reprogramming as reported by Thomas Graf and colleagues,[55] whereby $CD19^+$ B cells were reprogrammed into macrophages expressing Mac1, by the overexpression of C/EBP, a master gene regulator for the myeloid lineage. Since the description of full reprogramming toward induced pluripotent stem cells, anything now seems possible with reprogramming technologies. Indeed, scientists have induced pancreatic islets, within exocrine tissue in experimental models of diabetes, which restored resting blood glucose levels.[56] This in itself was ground-breaking work, and since then, several other important cell types for regenerative therapies have been derived by direct conversion from fibroblasts. This includes neurons generated from tail tip fibroblasts that are shown importantly to have mature adult action potentials,[57] and cardiomyocytes that can be derived from cardiac fibroblasts in vivo.[58] Interestingly, this study demonstrated that cells do not transit through a progenitor or embryonic state, but attained an adult sate directly, and so acquire adult action potentials that are critical for normal function. It therefore seems possible to derive virtually any cell type from fibroblasts by "direct reprogramming" or "direct conversion," without the need to derive hiPSC lines initially. This generally relies on having a very good understanding of the cell type of interest, and usually requires transcript array data, while employing transcription factors that are either highly expressed or have known master gene function, in the relevant. What is of significance here, is that the phenomenon of reprogramming or direct conversion seems not to be lineage-restricted, so that fibroblasts that are stored in tissue banks all over the world, appear to be an ideal starting point for the derivation of virtually any cell type. Also, cells derived by direct conversion thus far appear to form without the need to transit through a more primitive or progenitor stage, and acquire the features of the adult cell type. This will be necessary for any prospective regenerative therapies, and is currently a severe limitation of cells differentiated from pluripotent stem cells, which instead represent a more fetal stage. The most fascinating aspect of these direct reprogramming studies is that the direct conversion can be achieved in vivo to correct disease states.[56] However, since most direct reprogramming approaches require the use of transcription factors with known oncogenic potential, these direct in-situ approaches will probably not be implemented and instead cells from biobanks will remain the most likely source of new tissues.

### 3.5 Reprogramming for Biobanked Tissues

Tissues that are routinely stored in tissue banks around the world have traditionally comprised of blood, either adult mobilized peripheral blood or more recently umbilical cord blood, stored for allogeneic transplant. From blood alone, a range of cell types can be isolated and expanded for reprogramming, which includes hematopoietic stem cells, mononuclear cells, endothelium, and mesenchymal stromal cells, while endothelium and fibroblasts are now being banked separately as well. The intention of tissue banks to store stromal cells or fibroblasts is extremely useful, since this was the source of tissue for the first "full" reprogramming studies,[9,10] and subsequent work that describes direct conversion.[57,58] Fibroblasts are convenient also, since they are easily accessible from a skin biopsy. Unfortunately, however, cell populations derived can be heterogenous, which initially led scientists to question the original phenomenon of induced pluripotency where hiPSCs could have arisen instead from rare stem cells within this heterogeneous population. However, with improved reprogramming technologies such as Sendai virus[44] and small molecules,[48,49] it is now known that a significant proportion of fibroblasts (approximately 20%) are able to start the reprogramming process and become TRA-1-60-positive. However, most do not complete this process, which suggests additional factors may influence the reprogramming process.[59,60] Endothelial cells have also represented a popular tissue for banking since it is routinely harvested from umbilical cord as human umbilical vein endothelial cells. These cells have a high proliferative capacity in culture,[61,62] which appears to contribute towards efficient and rapid reprogramming.[63]

With regards to the reprogramming of blood, that is stored in huge quantities in tissue banks around the world, the first demonstration of hiPSCs from this compartment was with cord blood stem cells (CD133-positive), and interestingly were reprogrammed with just two factors, Oct4 and Sox2.[64] Other work has since shown adult mononuclear cells can also be considered for reprogramming[53] and thus hiPSCs can also be derived from routinely donated blood.

It is now known that most cell types from the blood lineage can be reprogrammed, including B lymphocytes,[65] while hematopoietic stem cells and progenitors are more efficiently reprogrammed than differentiated cells.[66] To demonstrate this, Hochedlinger and colleagues generated a transgenic mouse that contained inducible transgenes for reprogramming in every cell. When blood from this transgenic mouse was sorted for CD34-positive, myeloid, and lymphoid progenitors, and induced under reprogramming conditions, they found that most compartments of blood were amenable to reprogramming, albeit with varying efficiencies. This remarkable use of both transgenic technologies and reprogramming helped to demonstrate the robustness of the reprogramming process from various cells within different tissue compartments,[26,59,60] so that now most cells stored in both tissue and blood banks around the world may be considered as a source of tissue reprogramming.[67]

## 4. TISSUE DIFFERENTIATION FROM hiPSCs

### 4.1 Differentiation of hiPSCs Toward Cells of Interest

Historically, differentiation of pluripotent stem cells has relied upon spontaneous differentiation upon the withdrawal of leukemia inhibitor factor (in mouse) or bFGF (in human), in the presence of fetal calf serum, where usually differentiation occurs as embryoid bodies in suspension culture. Often small molecules have been used, such as retinoic acid, that favors neuroectoderm over mesoderm and promotes formation of neurons from pluripotent stem cells.[68] However, this is still a very heterogeneous process, is difficult to control, and is unreliable. Use of animal sera is an added complication for good manufacturing practice (GMP) compliancy, while the use of embryoid bodies is not easily scalable for production and manufacturing. Another strategy for efficient differentiation of pluripotent stem cells is coculture with stroma/fibroblasts. Use of embryonic fibroblasts is useful for recreating a niche and mimicking signals received during early development, and was critical for the isolation of mouse[4] and hESCs,[5] however these have been largely replaced by the use of defined media.[69] Feeders are also popular for hematopoietic differentiation, and many embryonic and fetal stroma have been isolated for this purpose, initially useful for maintenance of adult hematopoietic stem cells[70] but subsequently for haematopoietic differentiation of pluripotent stem cells.[71] Stromal feeder layers may also be used to mimic the specific haematopoietic niche/environment so well as to also enable the formation of very rare cell types such as hemogenic endothelial cells[72] and progeny that may arise, such as B lymphocytes,[33,73] from either hESC or hiPSCs. Unfortunately, this approach of using supporting stroma for haematopoietic differentiation is still not well defined, and can be unreliable if great care is not taken to ensure culture conditions are consistent.

As with maintenance of hESCs and hiPSCs, defined or directed approaches are also being employed more routinely for differentiation of pluripotent stem cells. Usually this is possible with information gleaned from transgenic studies and mapping the involvement of genes in early embryonic development. Here, directed differentiation relies on the use of growth factors and cytokines to recapitulate signals and the correct environment, which help to determine the fate of the pluripotent stem cell. It was unclear for many years how supplementation of growth factors to differentiating cultures really acted to promote specific fates, whether they "pushed" or "pulled" cells toward the lineage of choice. However, stochastic models at a single cell level show that differentiating pluripotent and multipotent stem cells generally respond to the environmental signals in a self-perpetuating transcriptional response that forms the basis of directed differentiation.[74]

The pluripotent stem cell has a choice of three germ layers to differentiate toward: neuroectoderm, mesoderm, and endoderm. Formation of the ectoderm is considered a default pathway,[39,75] which is formed in the absence of BMP4[21,76] or where BMP4 signaling is inhibited.[77] However, if members of the transforming growth factor-β (TGFβ) superfamily (including activin/nodal and BMPs) are present, this promotes mesoderm formation[78] and subsequent CV phenotypes. An example of the work toward defined conditions for the differentiation of specific cell types would be the multistep process described by Keller and colleagues for hESCs,[79] and subsequently with hiPSCs.[41,80] Here, multipotent platelet derived growth factor receptor (PDGFR)-positive CV cells are formed upon supplementation with specific amounts of bFGF, BMP4, activin A, and vascular endothelial growth factor (VEGF), and eventually form smooth muscle, endothelial cells, and cardiomyocytes (see Figure 4).

These were shown to have a functional benefit in animal models of myocardial infarction and to be retained for up to 12 weeks posttransplant.[41] The important aspects for any potential cellular therapy to reach clinical trials are the efficiency, reliability, and ease of such protocols, which are required for scale-up and manufacture. These cells were generated in "one dish," without extensive manipulation or selection, where resulting cells constituted nearly 50% of the population. Of note was that the CV population formed a sheet of contractile tissue, which was interspersed with endothelial networks and smooth muscle, which may also be ideal for toxicology screening of new pharmaceutical drugs prior to costly clinical trials, and may possibly help to reduce the number of animals currently used for this purpose. Other examples of defined protocols for deriving cells from pluripotent sources include neurons[81] and endoderm such as insulin-producing pancreatic beta cells[82] as well as hepatocytes.[83]

An alternative approach to cytokine-mediated "directed differentiation" is to block alternative pathways, so that fate choice is restricted to the desired cell type. Examples here include use of small inhibitory molecules such as SB431542 that block TGFβ-1 signaling[84] and/or the use of RNA interference,[77] both of which inhibit BMP signaling to efficiently induce neuronal differentiation. It is clear that a directed differentiation approach will be critical for improved efficiency and reproducibility of differentiation from hESCs and hiPSCs. However, it may well be that the method of choice for scale-up and manufacture is instead determined by issues such as GMP compliancy, availability in bulk, and cost of the reagents used.

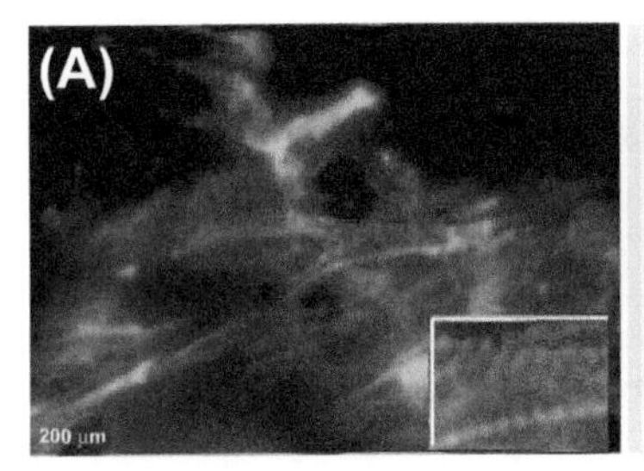


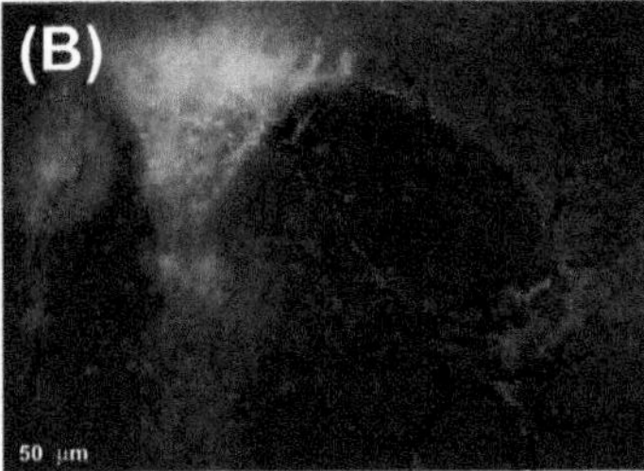


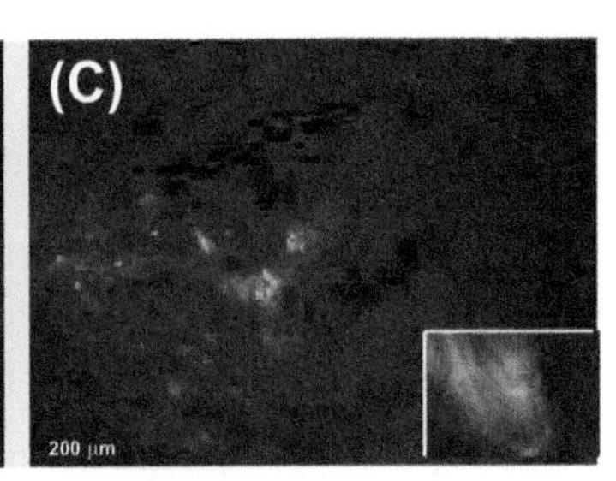


FIGURE 4 **Directed approaches for tissue-specific differentiation from pluripotent stem cells.** Here we have shown that with a specific cocktail of cytokines, hiPSC can be differentiated as monolayers, which over 14 days become beating sheets of cardiac myocytes that are positive for troponin (green) and which appear as triations (A), also, networks of CD31-positive endothelium (in red) can form (B), as well as smooth muscle that is positive for smooth muscle actin (green), with inset showing typical cytoskeletal staining (C). Nuclear staining with DAPI is in blue. (For interpretation of the color in this figure legend, the reader is referred to the online version of this book.) *From L. Carpenter et al., 2012*[41].

## 4.2 Phenotyping and Functional Tests

For applications in toxicology testing, disease modeling, and any potential therapies, it essential to not only undertake extensive phenotyping to ensure the correct identity, but also show that cells derived from pluripotent sources can function as expected. Without extensive functional tests, cells may behave unreliably during in vitro toxicology testing, or not provide a therapeutic benefit in disease models. For instance, pancreatic islets secrete glucagon, somatostatin, and most importantly insulin in response to elevated blood glucose, which then induces uptake and storage of glucose in sensitive tissues such as liver and muscle. Islets from pluripotent sources will therefore need to be glucose responsive, within a physiological range of glucose,[82] for them to be useful in the clinical setting for treatment of diabetes. Electrophysiological measurements can be used to query the functionality of cells such as neurons and cardiomyocytes, and their relationship to adult somatic cells of the same type, which has been described recently for cells obtained by direct conversion.[57,58] Cells can also be tested for their ability to integrate and be retained within the appropriate anatomical setting, i.e., when CV progenitors, derived from hiPSCs sources, are injected into the damaged myocardium, that they integrate into the CV tissue and are retained.[41] Most importantly for regenerative therapies, transplanted cells from pluripotent sources should also provide a functional benefit in the disease model and not be tumorigenic.

Where testing in animal models is not possible, or unnecessary, in vitro 3-D reconstruction of whole cellular compartments may be an alternative approach, particularly if cells are to be used for research, i.e., modeling disease and development in a dish. We have shown previously that when hiPSCs are differentiated down the CV and hematopoietic lineages, they can give rise to most of the cell types expected for these compartments. Directed differentiation of hiPSCs can be used to derive smooth muscle, endothelium, and CV cells, which form sheets of beating tissue that mimic the in vivo setting,[41] and thus offer a good starting point for toxicology testing or for basic research. On OP9 stroma we have shown that hiPSCs yield hematopoietic progenitors that have a rare potential to form B lymphocytes, as well as myeloid, erythroid, and endothelial cells, and as such this acts as a functional output for the existence of a multipotent hematopoietic progenitor.[33] This acts as a valuable system to better understand early events during hematopoietic development and to describe formation of the hematopoietic stem cell in vitro.

# 5. POTENTIAL APPLICATIONS FOR hiPSCs IN FUTURE REGENERATIVE MEDICINES

## 5.1 Novel Diagnostics

Diagnostics will be a very exciting area for novel applications of reprogramming technologies. Many examples can be suggested, but of particular interest in transfusion medicine for example generation of red cell panels for testing antibody reactivity in sera of patients receiving blood transfusions. Currently these are provided to hospitals in the United Kingdom by the NHS Blood and Transplant as "kits" with blood from donors with specific antigen profiles. Typically, panels can consist of 3 or 10 donor samples (see Figure 5), which are constructed to cover the antigen

FORM FRM833/1.1

Effective: 25/08/09

**3 Cell Screen Profile Product PR121 & PR122**

CE 0843

IVD

NHS Blood and Transplant

NBS REAGENTS

| Product | Lot No | Product | Lot No. | Expiry Date |
|---|---|---|---|---|
| Alsevers | R121 3287 | CellStab | R122 3287 | 2009.10.08 |

**Unless otherwise indicated, all cells are positive for $Kp^b$ and $Lu^b$ and negative for $Wr^a$, $Lu^a$ and $Co^b$**
Instructions for use can be found at http://www.blood.co.uk/hospitals/diagnostic_services/reagents/index.asp#Pro

| | Rh | C | D | E | c | e | $C^w$ | M | N | S | s | P1 | K | k | $Kp^a$ | $Le^a$ | $Le^b$ | $Fy^a$ | $Fy^b$ | $Jk^a$ | $Jk^b$ | Other |
|---|---|---|---|---|---|---|---|---|---|---|---|---|---|---|---|---|---|---|---|---|---|---|
| 1 | $R_1{}^WR_1$ | + | + | 0 | 0 | + | + | + | 0 | 0 | + | + | 0 | + | 0 | + | 0 | + | 0 | + | 0 | |
| 2 | $R_2R_2$ | 0 | + | + | + | 0 | 0 | + | 0 | + | 0 | 0 | + | + | 0 | 0 | + | 0 | + | + | + | |
| 3 | rr | 0 | 0 | 0 | + | + | 0 | 0 | + | 0 | + | + | 0 | + | + | 0 | + | + | 0 | 0 | + | |

**FIGURE 5 An example of a reagents testing panel provided by the NHS Blood and Transplant.** Here is an example of a 3 Cell Screen Profile Product, which is normally generated from a specific set of donors who fit the requirements of the "Red Book." These donors are difficult to identify, and are not always available for continuous supply. As such, hiPSCs from these individuals could be generated and used for an indefinite supply of red blood cells for this type of kit.

profile defined by the Guidelines for the Blood Transfusion Services in the United Kingdom (8th Edition 2013), otherwise known as the Red Book. While donor blood is optimal, it is an extremely demanding challenge to identify and recruit the specific donors that are required for the creation of antibody testing panels, and it can also be difficult to maintain a suitable supply of this material to hospitals.

Clearly an indefinite supply of these red blood cells would be beneficial, and this can be provided if hiPSCs were generated from donors that are used for the testing kits. It is known that red blood cells arising from hiPSCs can acquire the surface expression profile expected of red blood cells that are CD235 positive but CD45a negative, and can expand and enucleate with similar kinetics of that observed for hESCs.[85] Other applications may include commercial production of rare antibodies against human pathogens, where rare donors with disease resistance can be identified. From these donors, specific B-cell clones may be isolated for reprogramming[65] for an unlimited supply of cells, which upon differentiation back to B cells[33] can produce a continuous supply of rare antibodies for therapy.

## 5.2 Modeling Development and Disease

For nearly 30 years, mESCs have been used to model early embryonic development and disease. mESCs retain pluripotency and recapitulate early development so well that they can be used to generate living embryos[11,12] and after manipulating genes to generate either transgenic or knockout mice,[12,86,87] which also include inducible genes and/or lineage specific expression.[88,89] By studying the impact of these changes in the whole organism or in a dish, these genetic approaches help to decipher the role of genes in development.

Unfortunately, such approaches are not always able to recapitulate complex human diseases faithfully, particularly when the causative factors are unknown or are multifactorial. This was the primary driving force for achieving SCNT, which has only recently been reported in humans.[16] With the advent of hiPSC reprogramming in 2007, suddenly disease phenotypes can with relative ease, be recreated in a dish.[90] Since hiPSCs are pluripotent, virtually any disease cell type can be derived for study.[33,81–83]

A hindrance to the aims of disease modeling, is, however, faithfully recapitulating development in a dish. Cells from pluripotent sources are generally considered to be fetal in characteristics[91] and possibly a consequence of short-term culture. Also, as exemplified by formation of the hematopoietic lineage, a complicated sequence of events must occur before the formation of the adult blood system. Developmental hematopoiesis is thought to proceed in successive waves, each yielding hematopoietic progeny with increased multilineage potential. Ultimately, this results in a hematopoietic stem cell that will supply the blood lineage throughout adult life.[92–94] Adult hematopoietic stem cells are thought to arise from the aorta-gonad-mesonephros and other major arteries at around E10.5[95] by a process of endothelial to hematopoietic transition.[96–99] Prospective hematopoietic stem cells then bud off and colonize the spleen and later the bone marrow, and mature into "adult" hematopoietic stem cells that can then support adult hematopoiesis. Pluripotent stem cells can generally recapitulate the earliest events and undergo primitive hematopoiesis;[100] however, a recent breakthrough shows us that it is possible to recapitulate later events also with pluripotent stem cells. Here the hemogenic endothelium and the hemogenic to endothelium transition have been shown in both hESCs and hiPSCs[72] to yield hematopoietic progeny. However, to date no reports demonstrate the derivation of the adult human hematopoietic stem cell from pluripotent sources, which remains a long-standing goal in the field.

## 5.3 Pharmaceutical Testing

HiPSC lines, either from normal or diseased individuals, represent a unique and unparalleled tool for toxicology testing. It is known that CV toxicity for new drugs is extremely high, and can cause arrhythmias, coronary and cardiac disorders, and ultimately heart failure. Use of hiPSC-derived cardiomyocytes for toxicology screens is a highly attractive tool for reducing drug withdrawals from the market due to CV toxicity. Now that beating contractile sheets can be formed efficiently and reliably from hiPSCs,[41] pharmaceutical companies are able to test promising new drugs against many different hiPSC lines from different genetic backgrounds (predicative toxicology). HiPSCs from diseased lines may also be useful to identify novel compounds to prevent disease, such as arrhythmias from congenital defects, or other inherited disorders. Generating hepatocytes for predictive toxicology will also be of great value, since as the major cell population in the liver, the hepatocyte is responsible for, among other things, the metabolism and breakdown of drugs. Not only can this affect the pharmacokinetics of drugs in the body, but also it is the first cell to respond to toxic side effects of such drugs, resulting in drug-induced liver toxicity. In vitro cytotoxicity screens and preclinical in vivo testing have been used traditionally for testing pharmacokinetics and toxicity; however, since they are derived from a limited number of cell-lines, they are generally not reliable and may be responsible for the high failure rates. Unsurprisingly, the potential for high-throughput screening programs of novel drugs, many hiPSC-derived hepatocyte lines provides a much greater power for predicting potential toxicity of new drugs and is clearly a very attractive option to the pharmaceutical industry.

## 5.4 HiPSCs for Transplantation or Transfusion

HiPSCs and hESCs hold great promise of unlimited cell supply for regenerative therapies. Virtually any cell type can now be derived from pluripotent stem cells or by direct conversion from somatic tissue with reprogramming technologies.

Issues around safety and manufacture are discussed later, but another major stumbling block toward translation of this new technology into real clinical options is that of graft rejection and human leukocyte antigen (HLA) matching. Graft rejection is generally observed in response to poor matching for class I molecules HLA-A and HLA-B, and the class II molecule HLA-DR, whereas matching to these loci can help prevent acute rejection. Typically, cord blood banks and bone marrow registries have these "tissue types" recorded and stored electronically, so that when combined with hiPSC technologies, the task of providing tissues with a beneficial HLA match to patients is now much easier. Studies have shown that in the United Kingdom, 150 hiPSC lines would be sufficient to provide a complete or beneficial match for over 80% of the UK population,[101] and a similar study in Japan predicted improved matching again, due to lower ethnic diversity.[102] When the UK registry of 10,000 donors was considered, it was estimated that only 10 very highly selected homozygous donors should be required to provide a beneficial match to nearly 70% of the population.[103] Before the availability of reprogramming technologies, the reliance on randomly tested ESC lines for HLA matching meant that many hundreds of lines would need to be created, which was a severe limitation in the field of regenerative therapies.

To date, Geron has used hESC-derived oligodendrocytes for spinal injury in phase I trials that have been completed. While Geron has since withdrawn from hESC-based cellular therapies, and although no official results were published, no serious adverse effects were reported from the use of these cells, which in itself is a worthy outcome. Other trials have been approved by the US Food and Drug Administration (FDA) for the use of hESC-derived retinal pigment epithelial for the treatment of Stargardt macular dystrophy and severe myopia, both to be conducted by Advanced Cell Technologies, while Viacyte (formerly NovoCell) is progressing towards the treatment of both type 1 and type 2 diabetes, using encapsulated pancreatic precursor cells, beta cells, from a well-characterized hESC line.

Feasibility studies are also underway with a view to providing blood cell products, and Advanced Cell Technologies is currently applying (at the time of writing) to the FDA for a license to conduct the first human phase I clinical trials with hiPSC-derived platelets. Trials for red blood cell transfusion may be some way off using hiPSCs, since several hurdles will need to be overcome, not least enucleation rates that are generally poor, and as such represent a significant safety barrier. Logistics of scale-up of red blood cell production for transfusion currently also represent a huge engineering challenge.[104] In one unit of blood, typically there are $1 \times 10^{12}$ red blood cells. This has not yet been achieved in vitro, although advances are being made toward this goal.[105] Additional experimental strategies using knowledge gleaned from basic research and developmental biology will help to achieve this aim.

Another intriguing facet of hiPSC technologies for therapy is the correction of diseases such as sickle cell anemia[106] or Falconi anemia, which may offer an alternative to gene therapy. Also combined with organ replacement as described for liver bud formation with hiPSCs,[107] one day, there may be an unlimited supply of genetically engineered organs and tissues to treat a huge variety of diseases.

## 6. CONSIDERATIONS TOWARD DEVELOPMENT OF hiPSC-DERIVED THERAPIES

### 6.1 Genomic Stability and Characterization

Genetic stability and changes that may lead to tumor formation are of great concern and detrimental to any future therapeutic application of reprogramming strategies and hiPSCs. Particularly for hiPSCs that can now be generated in a variety of ways are various steps that may encourage genome instability and a transformed phenotype. These include: (1) the use of a combination of transgenes that have transforming potential such as Oct4; (2) the use small molecules; (3) extensive manipulation in culture; and (4) isolation from somatic cells that already harbor genetic mutations. On this first point, the reprogramming field has moved away from the use of integrating retroviruses as first described[9,10] that induces insertional mutagenesis that have been shown to cause leukemias.[108] Fortunately, there are now many alternative approaches (see Section 3.3) that do not require stable integration to induce reprogramming more safely.

The point worthy of further discussion would be mutagenesis that occurs in routine culture or those mutations that are present in the original starting tissue. There have now been several reports to show that hiPSC lines harbor copy number variations (CNVs), some which are designated "class I" or somatic CNVs apparent across hiPSC lines, and which originate from a single donor.[109,110] Indeed this is something that we also observe. The possibility of such somatic mutations that may be present in banked tissue, and hence persist in hiPSC lines arising, will need to be considered carefully before any future applications in cellular therapies are progressed toward the clinic. A further consideration is the inherent genomic instability of cells that are cultured in vitro for extended passage. G-banded karyotyping can reveal large chromosomal aberrations that include aneuploidy, translocations, and trisomy. These are well described for hESCs and commonly affect chromosomes 1, 12, 17, and 20, while acquired mutations (class II) have also been reported in both hESCs[36,111–113] and more recently hiPSCs.[114,115] Of note, class II CNVs are also observed in adult stem cells,[116] while on the whole hiPSC lines have been shown to be genetically stable, maintaining "normal" karyotype in 87% lines investigated.[110] Therefore, problems of culture-induced karyotypic changes are of concern for all cells derived from extensive in vitro manipulation. Aberrant epigenetic status for hiPSCs has also been reported.[117–120] Of less concern is the retention of epigenetic memory or methylation status inherited from the somatic cell of origin,[121,26] since this may be overcome with simple culturing strategies such as the inclusion of L-ascorbic acid, which not only improves reprogramming efficiencies, but also restores imprinted epigenetic states.[122]

Regardless of genomic instability and epigenetic status, the safety of resulting lines can be assessed independently, and the bar raised or lowered accordingly. Hard-to-treat or life-threatening diseases may warrant early intervention with hiPSC-derived tissues, and in some cases such as diabetes, they may be encapsulated to prevent possible spread of any rare malignant cells. Vital for assessing the transforming potential of hiPSC lines generated will be rigorous testing such as transplantation of ex vivo differentiated hiPSCs into animal models,[123–125] which forms the basis of any FDA and European approvals prior to clinical trials.

## 6.2 GMP and Scale-up

GMP is a quality assurance system that covers both manufacturing and testing of cellular products for the clinic. It requires traceability of raw materials and ensures that production follows validated standard operating procedures. In Europe, this is currently covered by several directives and guidelines (Directive 2004/23/EC, Commission Directives 2006/17/EC and 2006/86/EC), while in the United States, approval is by the FDA and is covered by the US Draft Guidance for Reviewers, which requires an Investigational New Drug Application. These set standards of quality and safety for donation, procurement, testing, processing, preservation, storage, and distribution of human tissues and cells. Nevertheless, the legislation is still under development and new directives and guidelines are being issued regularly. However, of primary importance for any hiPSC-related applications will be the quality (manufacturing details) and safety.

Regarding general GMP compliancy, animal products are usually omitted since they carry the risk of xenoviruses from the animal, and may induce an immune reaction in the recipient. However, for some pluripotent stem cell procedures, a GMP-"certified" fibroblast may be considered and has been used for derivation of several GMP-compliant hESC lines.[126] Other approaches to isolate hESCs and hiPSCs may include animal products such as Matrigel that can be heat treated first, or entirely animal free products such as Cellstart or human Laminin, which are recombinant proteins. GMP-compliant media have included KnockOut Serum Replacement (Invitrogen Inc) instead of animal sera; mTeSR, which contains bovine serum albumin;[127] and more recently E8 media that does not contain any albumin.[128] Passaging of hESCs and hiPSCs for GMP compliancy is currently restricted to use of collagenase IV, although enzyme-free methods have been recently reported.[129]

The fully defined media (E8)[128] for the derivation and culture of hiPSCs that is widely available commercially was the first description of its kind to remove the need for BSA, a major stumbling block toward GMP compliancy. By simplifying the components to just eight factors (DMEM/F12, insulin, selenium, transferrin, L-ascorbic acid, FGF2, TGFβ, and $NaHCO_3$), this greatly improves development towards GMP compliancy for hiPSCs prior to potential clinical trials. HiPSC cells in this study were also cultured on recombinant vetronectin, a substitute for extracellular matrix provided by fibroblast feeders, and a subsequent report by the same group alleviated the need for collagenase IV.[129]

Currently, the approaches developed by the Thomson group most adequately fulfill the requirements for the provision of clinical grade hiPSCs, while also permitting scale-up for manufacture. This process typically involves generation of a master cell bank, it needs to be from a clonal population, scalable, and undergo cryopreservation (both of hiPSCs and their progeny). During scale-up and manufacture, the phenotype also needs to be stable over extended passage, and this is currently fulfilled with the E8 approach.[128,129] Another aspect to scale-up and manufacture are the costs involved. Generally, GMP-compliant reagents are considerably more expensive than research grade materials, so approaches that minimize components, such as the E8 approach, will not only be easier to have approved by the FDA or appropriate authority, but will also enable a cost-effective manufacture.

## 6.3 Ethics and Consent

Although consenting can be fairly straightforward, it clearly requires a comprehensive, multistage donor consenting process, and is needed for the creation of all hiPSC lines, both for research and commercial activities. This generally requires a donor information pack, informing the donor of the need to create such lines, information relating to its storage and potential uses, and whether it is purely for research or commercial activities. Retrospective consent for tissues already banked is currently not an option in the United Kingdom, generally because informed consent was not given at the time of tissue collection, however, retrospective consent can be given in other European countries. Informed consent can only be given when appropriate ethical permission has been granted, and the research/activity has gained ethical approval, via an appropriate body that is usually locally administered by the University or Institution. Ethics for the creation of such lines from human donors is considered on a case-by-case basis, and a detailed outline of the intended use and procedures to maintain anonymity must be considered. Generally, however, creation of hiPSC lines from donors is less ethically contentious than for the creation of hESC lines, which require donation of eggs from female donors, and the destruction of a human embryo, all without informed consent from the donor. This dilemma in fact provided the impetus Shinya Yamanaka's ground-breaking work to reprogram somatic cells and create hiPSCs.[9] One critical ethical distinction for hiPSCs and their mouse counterparts is that hiPSCs are not being created for cloning purposes; but for basic research and therapeutics only, and so, there is generally much less ethical concern for hiPSCs in this area.

## 6.4 Banking hiPSCs and Multicenter Tissue Banks

Multicentered and internationally coordinated tissue banks will be needed to deal with the large number of hiPSC lines created for applications in research, but probably more so for any future therapeutic applications. For HLA cross-matching to prevent graft rejection, it has been estimated that just 10 highly selected lines would suffice for a beneficial match of nearly 70% of the population.[103] Unfortunately, however, it may be necessary to bank up to 150 lines for a 93% match across the United Kingdom. Although this is a significant number of lines to be generating, it is a vast improvement upon numbers predicted for similar coverage from hESC lines. With over 10,000 donors already HLA-typed on the British Bone Marrow Registry, these donors could be selected far more easily for the creation of a selected hiPSC/HLA tissue bank, and this could be replicated for several countries around the world where these HLA databases already exist.

Individual banks would need to comply with their domestic laws and regulations, while also agreeing to comply with an international framework for quality and assurance, as outlined by the International Society for Stem Cell Research (ISSCR) and International Stem Cell Banking Initiative (ISCBI), for a gold standard in best practices. Although a centralized initiative for supply of hiPSCs to academia, industry, and biotech is favored, an agreed operating framework among participating tissue banks may be a more realistic option. Cell donor identity will need to be anonymized, but both phenotype (healthy, disease, drug treatment) and genotype (source tissue, pluripotency status, gene expression profiles) data will be recorded electronically, and provided on a searchable database. These hiPSC lines could then be made freely available for wider dissemination. HiPSC lines submitted to the bank would need to be tested against a rigorous set of criteria that would include microbiological testing and characterization for pluripotency (as outlined in Section 3.2.) and then cells stored at all stages of the process (deposited cells, master stocks, and distributed cells), and passage kept to a minimum, in accordance with accepted practice[130] and that of the iso.org./(i.e., ISO9001:2000 for general quality management and ISO17025 for laboratory testing and monitoring). Quality assurance and information provided should include microbiological testing, characterization, and information relating to growth and storage criteria.

Working within these current frameworks, many consortia are being developed for banking of hiPSCs for academic research, with full public access (i.e., the Cambridge BioResource in the United Kingdom). However, it is health organizations such as the NHS Cord Blood Bank and British Bone Marrow Registries in the United Kingdom, and other public health service lead initiatives across Europe, that are uniquely positioned to bank hiPSCs for therapeutics, and with its international partners are able to coordinate efforts to provide hiPSCs of clinical grade for medical applications.

# 7. FINAL CONSIDERATIONS

As medical research continues to make breath-taking advances in regenerative medicine, it seems that there is almost unlimited potential for hiPSCs stored in tissue banks around the world. In the short years since the phenomenal discovery of induced pluripotency,[9,10] an endless number of opportunities are available in basic research, diagnostics, and regenerative therapies. Difficulties for translation toward the clinic are highlighted here in regards to practicalities for tissue banking, differentiation, and safety of cells and tissues created; however, the most challenging aspect to all of this work may yet be the coordination of a multicentered, internationally accredited network of tissue banks for storage and distribution of hiPSCs and reprogrammed tissues.


# ACKNOWLEDGMENT

My sincere apologies go to authors whose studies could not be cited here owing to space limitations. Since this is a broad-ranging review, I have needed to limit references to generally the first or most informative publication.


# REFERENCES


1. Wellcome Trust Case Control Consortium, Craddock N, Hurles ME, Cardin N, Pearson RD, Plagnol V, et al. Genome-wide association study of CNVs in 16,000 cases of eight common diseases and 3,000 shared controls. *Nature* 2010;**464**(7289):713–20.
2. Gurdon JB. Adult frogs derived from the nuclei of single somatic cells. *Dev Biol* 1962;**4**:256–73.
3. Campbell KH, McWhir J, Ritchie WA, Wilmut I. Sheep cloned by nuclear transfer from a cultured cell line. *Nature* 1996;**380**(6569):64–6.
4. Evans MJ, Kaufman MH. Establishment in culture of pluripotential cells from mouse embryos. *Nature* 1981;**292**(5819):154–6.
5. Thomson JA, Itskovitz-Eldor J, Shapiro SS, Waknitz MA, Swiergiel JJ, Marshall VS. et al. Embryonic stem cell lines derived from human blastocysts. *Science* 1998;**282**(5391):1145–7.
6. Nichols J, Zevnik B, Anastassiadis K, Niwa H, Klewe-Nebenius D, Chambers I, et al. Formation of pluripotent stem cells in the mammalian embryo depends on the POU transcription factor Oct4. *Cell* 1998;**95**(3):379–91.

7. Chambers I, Colby D, Robertson M, Nichols J, Lee S, Tweedie S, et al. Functional expression cloning of Nanog, a pluripotency sustaining factor in embryonic stem cells. *Cell* 2003;**113**(5):643–55.
8. Mitsui K, Tokuzawa Y, Itoh H, Segawa K, Murakami M, Takahashi K, et al. The homeoprotein Nanog is required for maintenance of pluripotency in mouse epiblast and ES cells. *Cell* 2003;**113**(5):631–42.
9. Takahashi K, Tanabe K, Ohnuki M, Narita M, Ichisaka T, Tomoda K, et al. Induction of pluripotent stem cells from adult human fibroblasts by defined factors. *Cell* 2007;**131**(5):861–72.
10. Takahashi K, Yamanaka S. Induction of pluripotent stem cells from mouse embryonic and adult fibroblast cultures by defined factors. *Cell* 2006;**126**(4):663–76.
11. Bradley A, Evans M, Kaufman MH, Robertson E. Formation of germ-line chimaeras from embryo-derived teratocarcinoma cell lines. *Nature* 1984;**309**(5965):255–6.
12. Robertson E, Bradley A, Kuehn M, Evans M. Germ-line transmission of genes introduced into cultured pluripotential cells by retroviral vector. *Nature* 1986;**323**(6087):445–8.
13. White JK, Gerdin AK, Karp NA, Ryder E, Buljan M, Bussell JN, et al. Genome-wide generation and systematic phenotyping of knockout mice reveals new roles for many genes. *Cell* 2013;**154**(2):452–64.
14. Brons IG, Smithers LE, Trotter MW, Rugg-Gunn P, Sun B, Chuva de Sousa Lopes SM, et al. Derivation of pluripotent epiblast stem cells from mammalian embryos. *Nature* 2007;**448**(7150):191–5.
15. Andrews PW, Benvenisty N, McKay R, Pera MF, Rossant J, Semb H, et al. The International Stem Cell Initiative: toward benchmarks for human embryonic stem cell research. *Nat Biotechnol* 2005;**23**(7):795–7.
16. Tachibana M, Amato P, Sparman M, Gutierrez NM, Tippner-Hedges R, Ma H, et al. Human embryonic stem cells derived by somatic cell nuclear transfer. *Cell* 2013;**153**(6):1228–38.
17. Yamanaka S, Blau HM. Nuclear reprogramming to a pluripotent state by three approaches. *Nature* 2010;**465**(7299):704–12.
18. Smith AG, Nichols J, Robertson M, Rathjen PD. Differentiation inhibiting activity (DIA/LIF) and mouse development. *Dev Biol* 1992;**151**(2):339–51.
19. Xu RH, Peck RM, Li DS, Feng X, Ludwig T, Thomson JA. Basic FGF and suppression of BMP signaling sustain undifferentiated proliferation of human ES cells. *Nat Methods* 2005;**2**(3):185–90.
20. Niwa H, Burdon T, Chambers I, Smith A. Self-renewal of pluripotent embryonic stem cells is mediated via activation of STAT3. *Genes Dev* 1998;**12**(13):2048–60.
21. Ying QL, Nichols J, Chambers I, Smith A. BMP induction of Id proteins suppresses differentiation and sustains embryonic stem cell self-renewal in collaboration with STAT3. *Cell* 2003;**115**(3):281–92.
22. Niwa H, Miyazaki J, Smith AG. Quantitative expression of Oct-3/4 defines differentiation, dedifferentiation or self-renewal of ES cells. *Nat Genet* 2000;**24**(4):372–6.
23. Masui S, Nakatake Y, Toyooka Y, Shimosato D, Yagi R, Takahashi K, et al. Pluripotency governed by Sox2 via regulation of Oct3/4 expression in mouse embryonic stem cells. *Nat Cell Biol* 2007;**9**(6):625–35.
24. Park IH, Zhao R, West JA, Yabuuchi A, Huo H, Ince TA, et al. Reprogramming of human somatic cells to pluripotency with defined factors. *Nature* 2007;**451**(7175):141–6.
25. Yu J, Vodyanik MA, Smuga-Otto K, Antosiewicz-Bourget J, Frane JL, Tian S, et al. Induced pluripotent stem cell lines derived from human somatic cells. *Science* 2007;**318**(5858):1917–20.
26. Maherali N, Sridharan R, Xie W, Utikal J, Eminli S, Arnold K, et al. Directly reprogrammed fibroblasts show global epigenetic remodeling and widespread tissue contribution. *Cell Stem Cell* 2007;**1**(1):55–70.
27. Nakagawa M, Koyanagi M, Tanabe K, Takahashi K, Ichisaka T, Aoi T, et al. Generation of induced pluripotent stem cells without Myc from mouse and human fibroblasts. *Nat Biotechnol* 2008;**26**(1):101–6.
28. Okita K, Ichisaka T, Yamanaka S. Generation of germline-competent induced pluripotent stem cells. *Nature* 2007;**448**(7151):313–7.
29. Kim JB, Sebastiano V, Wu G, Araúzo-Bravo MJ, Sasse P, Gentile L, et al. Oct4-induced pluripotency in adult neural stem cells. *Cell* 2009;**136**(3):411–9.
30. Nandan MO, Yang VW. The role of Kruppel-like factors in the reprogramming of somatic cells to induced pluripotent stem cells. *Histol Histopathol* 2009;**24**(10):1343–55.
31. Nakagawa M, Koyanagi M, Tanabe K, Takahashi K, Ichisaka T, Aoi T, et al. Generation of induced pluripotent stem cells without Myc from mouse and human fibroblasts. *Nat Biotechnol* 2008;**26**(1):101–6.
32. Stadtfeld M, Maherali N, Breault DT, Hochedlinger K. Defining molecular cornerstones during fibroblast to iPS cell reprogramming in mouse. *Cell Stem Cell* 2008;**2**(3):230–40.
33. Carpenter L, Malladi R, Yang CT, French A, Pilkington KJ, Forsey RW, et al. Human induced pluripotent stem cells are capable of B-cell lymphopoiesis. *Blood* 2011;**117**(15):4008–11.
34. Chan EM, Ratanasirintrawoot S, Park IH, Manos PD, Loh YH, Huo H, et al. Live cell imaging distinguishes bona fide human iPS cells from partially reprogrammed cells. *Nat Biotechnol* 2009;**27**(11):1033–7.
35. International Stem Cell Initiative, Adewumi O, Aflatoonian B, Ahrlund-Richter L, Amit M, Andrews PW, Beighton G, Bello PA, Benvenisty N, Berry LS, Bevan S, Blum B, Brooking J, Chen KG, Choo AB, Churchill GA, et al. Characterization of human embryonic stem cell lines by the International Stem Cell Initiative. *Nat Biotechnol* 2007;**25**(7):803–16.
36. Närvä E, Autio R, Rahkonen N, Kong L, Harrison N, Kitsberg D, et al. High-resolution DNA analysis of human embryonic stem cell lines reveals culture-induced copy number changes and loss of heterozygosity. *Nat Biotechnol* 2010;**28**(4):371–7.
37. Gore A, Li Z, Fung HL, Young JE, Agarwal S, Antosiewicz-Bourget J, et al. Somatic coding mutations in human induced pluripotent stem cells. *Nature* 2011;**471**(7336):63–7.
38. Hussein SM, Batada NN, Vuoristo S, Ching RW, Autio R, Närvä E, et al. Copy number variation and selection during reprogramming to pluripotency. *Nature* 2011;**471**(7336):58–62.
39. Munoz-Sanjuan I, Brivanlou AH. Neural induction, the default model and embryonic stem cells. *Nat Rev Neurosci* 2002;**3**(4):271–80.
40. Roelandt P, Vanhove J, Verfaillie C. Directed differentiation of pluripotent stem cells to functional hepatocytes. *Methods Mol Biol* 2013;**997**:141–7.
41. Carpenter L, Carr C, Yang CT, Stuckey DJ, Clarke K, Watt SM. Efficient differentiation of human induced pluripotent stem cells generates cardiac cells that provide protection following myocardial infarction in the rat. *Stem Cells Dev* 2012;**21**(6):977–86.
42. Stadtfeld M, Nagaya M, Utikal J, Weir G, Hochedlinger K. Induced pluripotent stem cells generated without viral integration. *Science* 2008;**322**(5903):945–9.
43. Ban H, Nishishita N, Fusaki N, Tabata T, Saeki K, Shikamura M, et al. Efficient generation of transgene-free human induced pluripotent stem cells (iPSCs) by temperature-sensitive Sendai virus vectors. *Proc Natl Acad Sci USA* 2011;**108**(34):14234–9.
44. Fusaki N, Ban H, Nishiyama A, Saeki K, Hasegawa M. Efficient induction of transgene-free human pluripotent stem cells using a vector based on Sendai virus, an RNA virus that does not integrate into the host genome. *Proc Jpn Acad Ser B Phys Biol Sci* 2009;**85**(8):348–62.

45. Kaji K, Norrby A, Paca A, Mileikovsky M, Mohseni P, Woltjen K. Virus-free induction of pluripotency and subsequent excision of reprogramming factors. *Nature* 2009;**458**(7239):771–5.
46. Woltjen K, Michael IP, Mohseni P, Desai R, Mileikovsky M, Hämäläinen R, et al. piggyBac transposition reprograms fibroblasts to induced pluripotent stem cells. *Nature* 2009;**458**(7239): 766–70.
47. Yu J, Hu K, Smuga-Otto K, Tian S, Stewart R, Slukvin II, et al. Human induced pluripotent stem cells free of vector and transgene sequences. *Science* 2009;**324**(5928):797–801.
48. Huangfu D, Osafune K, Maehr R, Guo W, Eijkelenboom A, Chen S, et al. Induction of pluripotent stem cells from primary human fibroblasts with only Oct4 and Sox2. *Nat Biotechnol* 2008;**26**(11): 1269–75.
49. Silva J, Barrandon O, Nichols J, Kawaguchi J, Theunissen TW, Smith A. Promotion of reprogramming to ground state pluripotency by signal inhibition. *PLoS Biol* 2008;**6**(10):e253.
50. Chang MY, Kim D, Kim CH, Kang HC, Yang E, Moon JI, et al. Direct reprogramming of rat neural precursor cells and fibroblasts into pluripotent stem cells. *PLoS One* 2010;**5**(3):e9838.
51. Kim D, Kim CH, Moon JI, Chung YG, Chang MY, Han BS, et al. Generation of human induced pluripotent stem cells by direct delivery of reprogramming proteins. *Cell Stem Cell* 2009; **4**(6):472–6.
52. Warren L, Manos PD, Ahfeldt T, Loh YH, Li H, Lau F, et al. Highly efficient reprogramming to pluripotency and directed differentiation of human cells with synthetic modified mRNA. *Cell Stem Cell* 2010;**7**(5):618–30.
53. Yu J, Chau KF, Vodyanik MA, Jiang J, Jiang Y. Efficient feeder-free episomal reprogramming with small molecules. *PLoS One* 2011;**6**(3):e17557.
54. Horb ME, Shen CN, Tosh D, Slack JM. Experimental conversion of liver to pancreas. *Curr Biol* 2003;**13**(2):105–15.
55. Xie H, Ye M, Feng R, Graf T. Stepwise reprogramming of B cells into macrophages. *Cell* 2004;**117**(5):663–76.
56. Zhou Q, Brown J, Kanarek A, Rajagopal J, Melton DA. In vivo reprogramming of adult pancreatic exocrine cells to beta-cells. *Nature* 2008;**455**(7213):627–32.
57. Vierbuchen T, Ostermeier A, Pang ZP, Kokubu Y, Südhof TC, Wernig M. Direct conversion of fibroblasts to functional neurons by defined factors. *Nature* 2010;**463**(7284):1035–41
58. Ieda M, Fu JD, Delgado-Olguin P, Vedantham V, Hayashi Y, Bruneau BG, et al. Direct reprogramming of fibroblasts into functional cardiomyocytes by defined factors. *Cell* 2010;**142**(3):375–86.
59. Hanna J, Saha K, Pando B, van Zon J, Lengner CJ, Creyghton MP, et al. Direct cell reprogramming is a stochastic process amenable to acceleration. *Nature* 2009;**462**(7273):595–601.
60. Tanabe K, Nakamura M, Narita M, Takahashi K, Yamanaka S. Maturation, not initiation, is the major roadblock during reprogramming toward pluripotency from human fibroblasts. *Proc Natl Acad Sci USA* 2013;**110**(30):12172–9.
61. Ingram DA, Mead LE, Moore DB, Woodard W, Fenoglio A, Yoder MC. Vessel wall-derived endothelial cells rapidly proliferate because they contain a complete hierarchy of endothelial progenitor cells. *Blood* 2005;**105**(7):2783–6.
62. Zhang Y, Fisher N, Newey SE, Smythe J, Tatton L, Tsaknakis G, et al. The impact of proliferative potential of umbilical cord-derived endothelial progenitor cells and hypoxia on vascular tubule formation in vitro. *Stem Cells Dev* 2009;**18**(2):359–75.
63. Panopoulos AD, Ruiz S, Yi F, Herrerías A, Batchelder EM, Izpisua Belmont JC. Rapid and highly efficient generation of induced pluripotent stem cells from human umbilical vein endothelial cells. *PLoS One* 2011;**6**(5):e19743.
64. Giorgetti A, Montserrat N, Aasen T, Gonzalez F, Rodríguez-Pizà I, Vassena R, et al. Generation of induced pluripotent stem cells from human cord blood using OCT4 and SOX2. *Cell Stem Cell* 2009;**5**(4):353–7.
65. Hanna J, Markoulaki S, Schorderet P, Carey BW, Beard C, Wernig M, et al. Direct reprogramming of terminally differentiated mature B lymphocytes to pluripotency. *Cell* 2008;**133**(2):250–64.
66. Eminli S, Foudi A, Stadtfeld M, Maherali N, Ahfeldt T, Mostoslavsky G, et al. Differentiation stage determines potential of hematopoietic cells for reprogramming into induced pluripotent stem cells. *Nat Genet* 2009;**41**(9):968–76.
67. Ohnishi H, Oda Y, Aoki T, Tadokoro M, Katsube Y, Ohgushi H, et al. A comparative study of induced pluripotent stem cells generated from frozen, stocked bone marrow- and adipose tissue-derived mesenchymal stem cells. *J Tissue Eng Regen Med* 2012;**6**(4): 261–71.
68. Andrews PW. Retinoic acid induces neuronal differentiation of a cloned human embryonal carcinoma cell line in vitro. *Dev Biol* 1984;**103**(2):285–93.
69. Ludwig TE, Bergendahl V, Levenstein ME, Yu J, Probasco MD, Thomson JA. Feeder-independent culture of human embryonic stem cells. *Nat Methods* 2006;**3**(8):637–46.
70. Dorshkind K, Johnson A, Collins L, Keller GM, Phillips RA. Generation of purified stromal cell cultures that support lymphoid and myeloid precursors. *J Immunol Methods* 1986;**89**(1):37–47.
71. Nakano T, Kodama H, Honjo T. Generation of lymphohematopoietic cells from embryonic stem cells in culture. *Science* 1994;**265**(5175): 1098–101.
72. Choi KD, Vodyanik MA, Togarrati PP, Suknuntha K, Kumar A, Samarjeet F, et al. Identification of the hemogenic endothelial progenitor and its direct precursor in human pluripotent stem cell differentiation cultures. *Cell Rep* 2012;**2**(3):553–67.
73. Vodyanik MA, Bork JA, Thomson JA, Slukvin II. Human embryonic stem cell-derived $CD34^+$ cells: efficient production in the coculture with OP9 stromal cells and analysis of lymphohematopoietic potential. *Blood* 2005;**105**(2):617–26.
74. Rieger MA, Hoppe PS, Smejkal BM, Eitelhuber AC, Schroeder T. Hematopoietic cytokines can instruct lineage choice. *Science* 2009;**325**(5937):217–8.
75. Wiles MV, Johansson BM. Analysis of factors controlling primary germ layer formation and early hematopoiesis using embryonic stem cell in vitro differentiation. *Leukemia* 1997;**11**(Suppl. 3): 454–6.
76. Wilson PA, Hemmati-Brivanlou A. Induction of epidermis and inhibition of neural fate by Bmp-4. *Nature* 1995;**376**(6538):331–3.
77. Carpenter L, Zernicka-Goetz M. Directing pluripotent cell differentiation using "diced RNA" in transient transfection. *Genesis* 2004;**40**(3):157–63.
78. Johansson BM, Wiles MV. Evidence for involvement of activin A and bone morphogenetic protein 4 in mammalian mesoderm and hematopoietic development. *Mol Cell Biol* 1995;**15**(1):141–51.
79. Yang L, Soonpaa MH, Adler ED, Roepke TK, Kattman SJ, Kennedy M, et al. Human cardiovascular progenitor cells develop from a KDR+ embryonic-stem-cell-derived population. *Nature* 2008;**453**(7194):524–8.

80. Kattman SJ, Witty AD, Gagliardi M, Dubois NC, Niapour M, Hotta A, et al. Stage-specific optimization of activin/nodal and BMP signaling promotes cardiac differentiation of mouse and human pluripotent stem cell lines. *Cell Stem Cell* 2011;**8**(2):228–40.
81. Carpenter MK, Inokuma MS, Denham J, Mujtaba T, Chiu CP, Rao MS. Enrichment of neurons and neural precursors from human embryonic stem cells. *Exp Neurol* 2001;**172**(2):383–97.
82. Kroon E, Martinson LA, Kadoya K, Bang AG, Kelly OG, Eliazer S. Pancreatic endoderm derived from human embryonic stem cells generates glucose-responsive insulin-secreting cells in vivo. *Nat Biotechnol* 2008;**26**(4):443–52.
83. Rambhatla L, Chiu CP, Kundu P, Peng Y, Carpenter MK. Generation of hepatocyte-like cells from human embryonic stem cells. *Cell Transpl* 2003;**12**(1):1–11.
84. Chambers SM, Fasano CA, Papapetrou EP, Tomishima M, Sadelain M, Studer L. Highly efficient neural conversion of human ES and iPS cells by dual inhibition of SMAD signaling. *Nat Biotechnol* 2009;**27**(3):275–80.
85. Dias J, Gumenyuk M, Kang H, Vodyanik M, Yu J, Thomson JA, et al. Generation of red blood cells from human induced pluripotent stem cells. *Stem Cells Dev* 2011;**20**(9):1639–47.
86. McMahon AP, Bradley A. The Wnt-1 (int-1) proto-oncogene is required for development of a large region of the mouse brain. *Cell* 1990;**62**(6):1073–85.
87. Dudley AT, Lyons KM, Robertson EJ. A requirement for bone morphogenetic protein-7 during development of the mammalian kidney and eye. *Genes Dev* 1995;**9**(22):2795–807.
88. Gu G, Brown JR, Melton DA. Direct lineage tracing reveals the ontogeny of pancreatic cell fates during mouse embryogenesis. *Mech Dev* 2003;**120**(1):35–43.
89. Gu G, Dubauskaite J, Melton DA. Direct evidence for the pancreatic lineage: NGN3+ cells are islet progenitors and are distinct from duct progenitors. *Development* 2002;**129**(10):2447–57.
90. Maehr R, Chen S, Snitow M, Ludwig T, Yagasaki L, Goland R, et al. Generation of pluripotent stem cells from patients with type 1 diabetes. *Proc Natl Acad Sci USA* 2009;**106**(37): 15768–73.
91. Doevendans PA, Kubalak SW, An RH, Becker DK, Chien KR, Kass RS. Differentiation of cardiomyocytes in floating embryoid bodies is comparable to fetal cardiomyocytes. *J Mol Cell Cardiol* 2000;**32**(5):839–51.
92. Dieterlen-Lievre F, Beaupain D, Martin C. Origin of erythropoietic stem cells in avian development: shift from the yolk sac to an intraembryonic site. *Ann Immunol Paris* 1976;**127**(6):857–63.
93. Cumano A, Ferraz JC, Klaine M, Di Santo JP, Godin I. Intraembryonic, but not yolk sac hematopoietic precursors, isolated before circulation, provide long-term multilineage reconstitution. *Immunity* 2001;**15**(3):477–85.
94. Cumano A, Godin I. Ontogeny of the hematopoietic system. *Annu Rev Immunol* 2007;**25**:745–85.
95. de Bruijn MF, Speck NA, Peeters MC, Dzierzak E. Definitive hematopoietic stem cells first develop within the major arterial regions of the mouse embryo. *Embo J* 2000;**19**(11):2465–74.
96. Bertrand JY, Chi NC, Santoso B, Teng S, Stainier DY, Traver D. Haematopoietic stem cells derive directly from aortic endothelium during development. *Nature* 2010;**464**(7285):108–11.
97. Boisset JC, van Cappellen W, Andrieu-Soler C, Galjart N, Dzierzak E, Robin C. In vivo imaging of haematopoietic cells emerging from the mouse aortic endothelium. *Nature* 2010;**464**(7285):116–20.
98. Kissa K, Herbomel P. Blood stem cells emerge from aortic endothelium by a novel type of cell transition. *Nature* 2010;**464**(7285): 112–5.
99. Eilken HM, Nishikawa S, Schroeder T. Continuous single-cell imaging of blood generation from haemogenic endothelium. *Nature* 2009;**457**(7231):896–900.
100. Kennedy M, Firpo M, Choi K, Wall C, Robertson S, Kabrun N, et al. A common precursor for primitive erythropoiesis and definitive haematopoiesis. *Nature* 1997;**386**(6624):488–93.
101. Taylor CJ, Bolton EM, Pocock S, Sharples LD, Pedersen RA, Bradley JA. Banking on human embryonic stem cells: estimating the number of donor cell lines needed for HLA matching. *Lancet* 2005;**366**(9502):2019–25.
102. Nakajima F, Tokunaga K, Nakatsuji N. Human leukocyte antigen matching estimations in a hypothetical bank of human embryonic stem cell lines in the Japanese population for use in cell transplantation therapy. *Stem Cells* 2007;**25**(4):983–5.
103. Taylor CJ, Peacock S, Chaudhry AN, Bradley JA, Bolton EM. Generating an iPSC bank for HLA-matched tissue transplantation based on known donor and recipient HLA types. *Cell Stem Cell* 2012;**11**(2):147–52.
104. Anstee DJ, Gampel A, Toye AM. Ex-vivo generation of human red cells for transfusion. *Curr Opin Hematol* 2012;**19**(3):163–9.
105. Lu SJ, Feng Q, Park JS, Vida L, Lee BS, Strausbauch M, et al. Biologic properties and enucleation of red blood cells from human embryonic stem cells. *Blood* 2008;**112**(12):4475–84.
106. Hanna J, Wernig M, Markoulaki S, Sun CW, Meissner A, Cassady JP, et al. Treatment of sickle cell anemia mouse model with iPS cells generated from autologous skin. *Science* 2007;**318**(5858):1920–3.
107. Takebe T, Sekine K, Enomura M, Koike H, Kimura M, Ogaeri T, et al. Vascularized and functional human liver from an iPSC-derived organ bud transplant. *Nature* 2013;**499**(7459):481–4.
108. Kang EM, Tisdale JF. The leukemogenic risk of integrating retroviral vectors in hematopoietic stem cell gene therapy applications. *Curr Hematol Rep* 2004;**3**(4):274–81.
109. Martins-Taylor K, Nisler BS, Taapken SM, Compton T, Crandall L, Montgomery KD, et al. Recurrent copy number variations in human induced pluripotent stem cells. *Nat Biotechnol* 2011;**29**(6):488–91.
110. Martins-Taylor K, Xu RH. Concise review: genomic stability of human induced pluripotent stem cells. *Stem Cells* 2012;**30**(1):22–7.
111. Baker DE, Harrison NJ, Maltby E, Smith K, Moore HD, Shaw PJ, et al. Adaptation to culture of human embryonic stem cells and oncogenesis in vivo. *Nat Biotechnol* 2007;**25**(2):207–15.
112. Harrison NJ, Baker D, Andrews PW. Culture adaptation of embryonic stem cells echoes germ cell malignancy. *Int J Androl* 2007;**30**(4):275–81. discussion 281.
113. International Stem Cell Initiative, Amps K, Andrews PW, Anyfantis G, Armstrong L, Avery S, et al. Screening ethnically diverse human embryonic stem cells identifies a chromosome 20 minimal amplicon conferring growth advantage. *Nat Biotechnol* 2011;**29**(12):1132–44.
114. Ben-David U, Benvenisty N, Mayshar Y. Genetic instability in human induced pluripotent stem cells: classification of causes and possible safeguards. *Cell Cycle* 2010;**9**(23):4603–4.
115. Mayshar Y, Ben-David U, Lavon N, Biancotti JC, Yakir B, Clark AT, et al. Identification and classification of chromosomal aberrations in human induced pluripotent stem cells. *Cell Stem Cell* 2010;**7**(4):521–31.
116. Ben-David U, Mayshar Y, Benvenisty N. Large-scale analysis reveals acquisition of lineage-specific chromosomal aberrations in human adult stem cells. *Cell Stem Cell* 2011;**9**(2):97–102.

117. Lister R, Pelizzola M, Dowen RH, Hawkins RD, Hon G, Tonti-Filippini J, et al. Human DNA methylomes at base resolution show widespread epigenomic differences. *Nature* 2009;**462**(7271):315–22.
118. Lister R, Pelizzola M, Kida YS, Hawkins RD, Nery JR, Hon G, et al. Hotspots of aberrant epigenomic reprogramming in human induced pluripotent stem cells. *Nature* 2011;**471**(7336):68–73.
119. Doi A, Park IH, Wen B, Murakami P, Aryee MJ, Irizarry R, et al. Differential methylation of tissue- and cancer-specific CpG island shores distinguishes human induced pluripotent stem cells, embryonic stem cells and fibroblasts. *Nat Genet* 2009;**41**(12):1350–3.
120. Kim K, Doi A, Wen B, Ng K, Zhao R, Cahan P, et al. Epigenetic memory in induced pluripotent stem cells. *Nature* 2010;**467**(7313): 285–90.
121. Polo JM, Liu S, Figueroa ME, Kulalert W, Eminli S, Tan KY, et al. Cell type of origin influences the molecular and functional properties of mouse induced pluripotent stem cells. *Nat Biotechnol* 2010;**28**(8):848–55.
122. Stadtfeld M, Apostolou E, Ferrari F, Choi J, Walsh RM, Chen T, et al. Ascorbic acid prevents loss of Dlk1-Dio3 imprinting and facilitates generation of all-iPS cell mice from terminally differentiated B cells. *Nat Genet* 2012;**44**(4):398–405. S1-2.
123. Miura K, Okada Y, Aoi T, Okada A, Takahashi K, Okita K, et al. Variation in the safety of induced pluripotent stem cell lines. *Nat Biotechnol* 2009;**27**(8):743–5.
124. Okano H, Nakamura M, Yoshida K, Okada Y, Tsuji O, Nori S, et al. Steps toward safe cell therapy using induced pluripotent stem cells. *Circ Res* 2013;**112**(3):523–33.
125. Tsuji O, Miura K, Okada Y, Fujiyoshi K, Mukaino M, Nagoshi N, et al. Therapeutic potential of appropriately evaluated safe-induced pluripotent stem cells for spinal cord injury. *Proc Natl Acad Sci USA* 2010;**107**(28):12704–9.
126. Crook JM, Peura TT, Kravets L, Bosman AG, Buzzard JJ, Horne R, et al. The generation of six clinical-grade human embryonic stem cell lines. *Cell Stem Cell* 2007;**1**(5):490–4.
127. Ludwig TE, Levenstein ME, Jones JM, Berggren WT, Mitchen ER, Frane JL, et al. Derivation of human embryonic stem cells in defined conditions. *Nat Biotechnol* 2006;**24**(2):185–7.
128. Chen G, Gulbranson DR, Hou Z, Bolin JM, Ruotti V, Probasco MD, et al. Chemically defined conditions for human iPSC derivation and culture. *Nat Methods* 2011;**8**(5):424–9.
129. Beers J, Gulbranson DR, George N, Siniscalchi LI, Jones J, Thomson JA, et al. Passaging and colony expansion of human pluripotent stem cells by enzyme-free dissociation in chemically defined culture conditions. *Nat Protoc* 2012;**7**(11):2029–40.
130. Coecke S, Balls M, Bowe G, Davis J, Gstraunthaler G, Hartung T, et al. Guidance on good cell culture practice. A report of the second ECVAM task force on good cell culture practice. *Altern Lab Anim* 2005;**33**(3):261–87.

Section V

# Cord Blood Banking: A Current State of Affairs

Chapter 15

# Cord Blood Banking: Operational and Regulatory Aspects

Cristina Navarrete

*Histocompatibility and Immunogenetic Services and NHS-Cord Blood Bank, National Blood and Transplant (NHSBT), England, UK; Division of Infection and Immunity, University College London, London, UK*

## Chapter Outline



## 1. INTRODUCTION

The demonstration that umbilical cord blood (UCB) contained cells capable of reproducing hematopoiesis in vitro and that they could be cryopreserved for long periods without significant loss in their function paved the way for the use of this new source of stem cells in the field of clinical hematopoietic stem cell (HSC) transplantation.[1–5] Although the first attempt to transplant UCB was reported in 1972,[6] the first successful related UCB transplant was performed in 1988 in a patient with Fanconi anemia.[7] The first unrelated UCB transplants were then performed in 1993 and the report of a large series of patients was published in 1996.[8]

These results prompted the establishment of large repositories of frozen unrelated UCB units, i.e., cord blood banks (CBBs) to support this clinical development. The first UCB bank was set up in 1991 by P. Rubinstein at the New York Center in the United States,[9] and this was followed by the establishment of CBBs in Dusseldorf, Milan, Paris, and London.[10–13] At present, there are over 130 CBBs all over the world storing over 650,000 unrelated UCB units readily available for transplantation.[14,15] These CBBs have enabled the performance of over 30,000 unrelated UCB transplants in children and adults with both malignant and nonmalignant diseases.[16–21]

With the increased clinical use and exchange of these UCB units it became clear that it was necessary to standardize practices across the various CBBs. In order to contribute to this, a group of experts involved in UCB banking established the NetCord organization in 1998.[22] The initial remit of NetCord was to set up an international registry of UCB and to develop procedures and quality standards for the safe collection, exchange, and clinical use of these banked units. These efforts culminated with the development of the NetCord-FACT (Foundation for the Accreditation of Cellular Therapy) International Standards for Cord Blood Collection, Processing, Testing, Banking, Selection, and Release in 2000 with the last version (5th) published in 2013.[23] These standards form the basis for the NetCord-FACT accreditation program. The American Association of Blood Banks (AABB) has also developed standards and an accreditation

Cord Blood Stem Cells Medicine. http://dx.doi.org/10.1016/B978-0-12-407785-0.00015-3

scheme, and since 2004 have been incorporated into the Standards for Cellular Therapy Services.[24]

Different types of UCB banking programs have been established depending on the genetic relationship of the donated UCB unit with the potential recipient, i.e., allogeneic or autologous, and on the funding sources, i.e., public or private.[25–27]

In the allogeneic setting, an additional distinction needs to be drawn between unrelated (also called altruistic) and related donations. Unrelated allogeneic UCB banking includes the collection, processing, and storage of altruistically donated UCB, in order to create an inventory of HSCs that can be searched for any patient in any part of the world and in need of an unrelated allogeneic HSC donor. These programs are also referred to as public UCB banking.

In the allogeneic-related setting, the UCB is collected from a healthy sibling of a patient with a disease that can potentially be treated with a UCB transplant. These collections are performed at the request of the physician treating the patient, with the agreement of the obstetrician looking after the mother.[28–31]

Although the standards for the collection of unrelated and related UCB are similar, in the banking of related UCB units, no threshold values for minimum volume collected are required and there is no operational need for volume reduction (VR) of these units, although clinically VR may be beneficial (see below). Also, the exclusion from banking due to microbial contamination does not always apply since antibiotic sensitivity tests can be performed if and when the collection is required for transplantation.

An important consideration in related UCB banking is that in approximately 70% of cases, the collected UCB unit is not fully human leukocyte antigen (HLA)-matched with the patient and it is unlikely that it will ever be used for the intended patient. These units will have to be kept frozen indefinitely unless clear policies regarding their disposal are put in place. With the availability of prenatal genetic diagnosis, including HLA typing, it will be possible to collect units selected only from HLA-identical siblings, as has already proved possible.[32]

So far, the majority of the transplanted related UCB units have been fully HLA-matched and in patients with hemoglobinopathies. In fact, in some centers, related UCB transplantation is the first line of treatment for patients with thalassemia major.[33]

In most European countries, these two types of allogeneic UCB banking (unrelated and related) are carried out by government-administered institutions and funded by the national health systems available in each country.

In the autologous or family UCB banking setting, the collections are normally performed by privately funded institutions at the request of the family of the potential donor, and as its name indicates these collections are mainly for autologous use or for a named recipient normally within the family. Worldwide, many private CBBs are collecting and storing UCB for eventual autologous or family use. Although there are more than 1 million of these units stored in private CBBs, a very small number have been transplanted, mostly with unknown outcomes. There are anecdotal cases of autologous UCB transplantation, but in general the scientific and clinical arguments for the banking of these units are not universally accepted.[34] There are also a number of ethical issues associated with this practice, which have been extensively reviewed.[35–39] Furthermore, it appears that quality parameters for privately stored units seem to be inferior to those measured in those stored in public CBB, highlighting the fact that if the clinical value of autologous UCB transplantation is established, the issue of the quality of these units will become even more relevant.[40]

Alternative models of CBB, called hybrid or mixed, have been developed and in these programs the UCB collection is carried out on behalf of (and paid for) the family requesting the collection for either autologous use or for a named person within the family. In one of these models, the collection is split into three and 1/3 is donated to the unrelated public bank and the other 2/3 is stored as a private collection. However, with the increasing evidence on the impact of high total nucleated cell (TNC) and CD34+ cell content on the clinical outcome of UCB transplantation, which is directly correlated with the volume collected, the value of these small volumes of UCB is limited. The other "mixed" model involves the collection and storage of the UCB, for a fee, for autologous or for a named family member, but in this case, the relevant information on the UCB unit is made accessible to unrelated donor registries. Once or if the unit is selected for an unrelated patient, the family is then asked to release the stored unit with the consequent refund. To date, no evidence on the application of this model has been presented.

The majority, if not all of, the UCB banking procedures and standards were initially developed primarily for the collection and banking of allogeneic unrelated cord blood units (CBUs) from altruistic donors. However, in order to ensure the quality and efficacy of all collected units and to safeguard the eventual recipients of these products, NetCord-FACT and AABB have also developed standards that are applicable to the collection and storage and release of allogeneic related and autologous UCB.

The main aspects of an unrelated UCB banking program include the following:

1. Promotion, recruitment, donor selection, informed consent, collection, and transportation to processing facilities and donor follow-up
2. Processing, testing, cryopreservation, and storage
3. Listing, searches, selection, testing, and distribution to a transplant program and posttransplant clinical follow-up

All these procedures and activities of a CBB program are supported by comprehensive inventory and quality management systems covering all the various components of UCB banking as described below (see Figure 1).

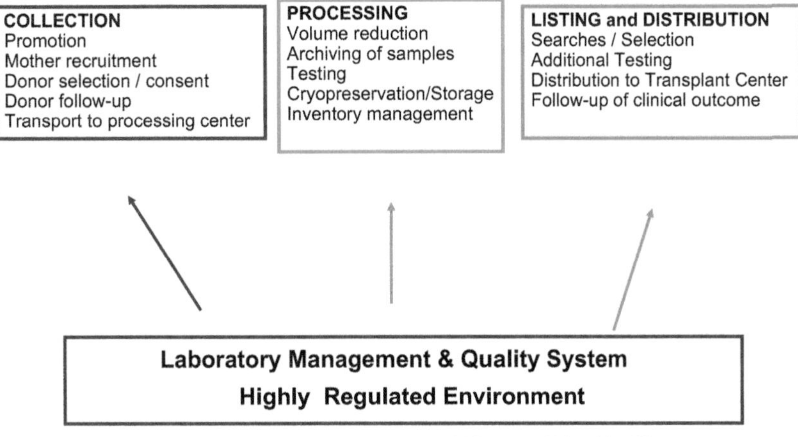


FIGURE 1 Operational aspects of umbilical cord blood banking.

## 1.1 Promotion, Recruitment, Donor Selection, Consent, Collection, Transportation to Processing Facilities, and Donor Follow-up

### 1.1.1 Promotion

The promotion and model of collection of UCB vary from country to country and depend largely on the nature of the funding and health-care system available in each country. In CBBs funded with public or government monies, the promotion is generally carried out and restricted to antenatal clinics selected as collection sites. The material used in the promotion includes leaflets, videos, seminars, etc. These should provide as much information as possible about the program including the need for consent, the right to withdraw at any step of the process, the clinical use of the collection, and about its potential use in research if the collected unit is not suitable for clinical banking. The promotional material should be clearly understood by the potential mother donors, and it should be translated into various languages if required. The latter is particularly important when recruiting donors are from an ethnic minority background and are not fluent in the language of the country where the recruitment is taking place.[41]

### 1.1.2 Recruitment

The majority of the banked units so far have been collected in countries with a population of predominantly European Caucasoid ethnic background and expressing the HLA profile of this ethnic group. There is therefore a need to increase the HLA diversity of the banked UCB and one way of achieving this is to recruit donors in maternity units with high numbers of deliveries from an ethnically diverse population.

A number of CBBs have now been established in countries with an ethnic and HLA profile not historically represented in the international registries of unrelated HSC and these units may contribute to expand the HLA profile of the internationally available pool of unrelated UCB units.

A number of publications have indicated that the volume and TNC content of the units collected are smaller in mothers from Afro-Caribbean and Asian ethnic background compared to those collected from European Caucasoid donors.[42,43] It is therefore crucial to carefully select these donors in order to minimize the waste of these units.

The National Health Service Cord Blood Bank (NHS-CBB), formerly known as the London Cord Blood Bank, was set up in 1996 with the aim of enriching the national and international HSC donor pool with units from ethnic minorities (EM).[44,45] At present, nearly 38% of the banked CBUs are from EM and ethnically mixed genetic backgrounds, expressing unique HLA haplotypes. This is of great benefit to EM patients as reflected by the fact that nearly 36% of the units issued for transplantation by the NHS-CBB are from EM donors.

Looking into the future it will be important to try to enrich the CBBs with units from a mixed genetic background. In fact, the majority of patients who have difficulties in finding an HLA-matched stem cell donor express a combination of a common and a rare HLA haplotype.

### 1.1.3 Donor Selection

This is one of the most important aspects of CBB and detailed and comprehensive procedures and policies for the selection and acceptance of mothers donating their UCB are crucial to ensure the quality and safety of the collected units. The responsibility for the selection of an UCB donor lies with the Medical Director of the CBB who should ensure that an appropriate medical and social history is obtained from the mother in order to prevent the transmission of microbiological infection and/or genetic, malignant, or degenerative disease. Donors with a family or personal history of genetic disease, particularly relating to the hematopoietic or immune system, should be asked for details of those suffering from such diseases and their family relationship to the infant donor. Further details may be required from the general physician or other professionals. The communicable disease risk history of a surrogate mother who carries an infant not genetically related to her, and of a sperm, egg, or embryo donor shall also be obtained and documented if applicable. Travel history of potential donors is also important to assess risks. More recently, a number of publications have indicated other factors such as the ethnicity, gender, and age of gestation which may affect the size/volume of the collections.

### 1.1.4 Consent

Initially and in order to make the best use of resources, consent was only obtained from mothers from whom a successful collection had been obtained. However, in Europe following the implementation of the European Union Tissues and Cells Directive (EUTCD) 2004/23/EC in April 2006, stating that "procurement of human tissues or cells shall be authorised only after all mandatory consent or authorisation requirements have been met," all collected UCB units need to have a signed consent obtained prior to delivery.[46] At present, in most CBB programs, consent is obtained when the mothers attend their hospital at around 30 weeks of pregnancy or via a "mini consent" form completed before the mother is in established labor. The introduction of the "mini consent" form, which allows for the collection of the CBU, has facilitated the implementation of the EUTCD, and it has also increased the efficiency of collections by decreasing the wastage of CBUs discarded due to the lack of consent.[47] Following the initial mini consent, a full more detailed consent is required and this provides the basis for proceeding to the processing and testing of the collected unit.

An important aspect of the consent process is to provide detailed and clear information about the tests required, the intended use of the unit, particularly in relation to the altruistic nature of the donation, and about the potential use of the clinically unsuitable units for research and development. Also, it needs to include the consent to contact relevant health professionals in the event of a positive result relevant to their health, to obtain and store samples for future testing, and to store personal information.

This also means that for mothers who do not speak the language of the country where they are giving birth, all relevant information and the process of obtaining consent should be performed not only in their own language, but also by somebody able to communicate and answer the relevant questions of the mother. Consent should be taken when the mother is able to concentrate on the process and certainly not when the mother is in labor. Importantly, the NetCord-FACT Standards also mention that regardless of whether the unit is collected for unrelated or related use, if this unit may potentially be used for reasons other than the initial clinical intent, not only this should be mentioned in the informed consent but also the donor should have given consent with documents and information related to the potential related or unrelated use of the unit.

### 1.1.5 Collection

The collection of the UCB is carried out by suspending the placenta, cannulating the vein and allowing the blood to drain by gravity into a specially designed UCB collection bag placed on a shaker in order to avoid the formation of clots.

Collections can be carried out at fixed and nonfixed sites and in either case an agreement between the CBB and the collection site is required.[23]

*Fixed sites*. In this model, the UCB units are collected by trained staff employed by the CBB or by the maternity units in each hospital. In Europe, the collection of UCB can only be performed in sites that comply with the regulatory requirement of the EUTCD, i.e., in licensed or fixed sites and by trained staff.

*Nonfixed sites*. In this case, the collection is performed at any maternity unit by either their own staff or by agency staff. The CBB provides the appropriate kit and instructions for the UCB collection. Although this practice facilitates and allows for the collection of altruistically donated units anywhere in the country, it is also associated with a reduced number of units suitable for banking due to an increase in the number of bacterial infections, and low volumes and TNC of the collected units. This may be due to the lack of training or experience of the personnel performing these collections. This practice is not very common in Europe due to the stringent training requirement of the EUTCD. The current revised version of the NetCord-FACT Standards covers the collection at both fixed and nonfixed sites.

When selecting the collection sites, it is important to consider the number of deliveries per year in order to maximize the resources and to maintain the training of the collection staff.

#### 1.1.5.1 In Utero versus Ex Utero Collections

The UCB collection can be performed in utero or ex utero, following full-term normal delivery or cesarean section. The minimum gestation period for collection is 34 weeks.

In utero collections are performed by a trained member of the delivery team during the third stage of labor before the placenta is delivered. Ex utero collections are carried out by CBB trained staff, normally outside the delivery room to avoid interference with the delivery process. In these collections, the risk to the mother or infant is minimal, but the risk of microbial contamination may be higher.

Initial studies had indicated that in utero collections yielded larger volumes (and TNC doses) than ex utero collections, but more recent studies have shown that, if appropriately trained staff are involved in the collection, there is no significant difference in the volume, or indeed in the contamination rate, with either of these two methods.[47,48]

Since the safety of mother and child are paramount and because of the possible diversion of the attention from the mother and newborn to the UCB collection, both the UK Royal College of Obstetricians and Gynaecologists (RCOG) (Scientific Advisory Committee for the Royal Colleges of Obstetricians and Gynaecologists, 2001) and the Royal College of Midwives in the United Kingdom have recommended that all UCB collections be carried out ex utero.[49] Also, following reports on the effect of delayed clamping on infant development and iron status, the UK RCOG has also issued guidelines regarding the timing of clamping.[50–52]

Once a successful UCB collection has been obtained, blood samples from the mother are taken for communicable disease testing within 7 days before or after collection. A history of the current pregnancy and delivery, and the infant's donor birth data should be obtained and documented including gender, gestational age, and results of any other relevant test. Information about the clinical examination or any finding suggestive of disease potentially transmitted through transplantation should also be recorded.

The collected units and the associated maternal samples are then transported to the CBB processing centers in a temperature-controlled environment as soon as possible following the collection. An agreement between the collection site and the staff responsible for the transport of these units is required, as well as a documented evidence of training the staff involved in this task.

### *1.1.6 Donor Follow-up*

In some CBBs, a follow-up telephone interview is carried out 8–12 weeks postcollection in order to check the health status of the mother and the newborn from whom the UCB was collected. Other CBB programs have a policy of contacting the mother when a unit has been reserved or prior to its release to the transplant program. All CBUs are quarantine frozen in temporary containers until all relevant test results are reviewed and the units are medically released for long-term storage. Counseling resources should be in place to support the donor and family in case of a positive infectious disease marker other than cytomegalovirus.

## 1.2 Processing, Cryopreservation, Storage, and Testing

An appropriately signed consent authorizing the processing, testing, and storage of the units and associated samples is required before commencing these procedures.

Although the viability and functionality of the collected stem cells seem to be preserved for up to 96 h, the majority of publications have shown the benefits of processing the UCB units as soon as possible and ideally within 48 h of collection.[53–56] The current NetCord-FACT Standards indicate that the cryopreservation of unrelated UCB units should be performed within 48 h of collection in either a closed system or in an environmentally controlled clean room. For related UCB, the cryopreservation should commence within 72 h of collection.[23]

### *1.2.1 Processing*

The TNC and CD34+ cell content per kilogram of patient's weight (as well as HLA matching) are important factors influencing the outcome of UCB transplantation, particularly in improving engraftment rates.[19,25,57,58] As a result, a minimum dose of TNC and CD34+ cell content of the units has been proposed for transplantation into patients with malignant and nonmalignant diseases.[59–61]

Furthermore, World Marrow Donor Association (WMDA) data indicate that the minimum cut-off TNC and CD34+ count of the transplanted units have increased throughout the years.[14]

Thus, in order to ensure that the stored units can meet the requirements of transplant center acceptable values for volume, TNC and CD34+ cell content of the UCB units collected (i.e., preprocessing) are established by each CBB program in order to compensate for the cell loss occurring during the processing of the units at various stages of the procedure (approximately 10%). Therefore, in order to improve the quality of the units banked and to minimize the costs of CBB, most UCB banks evaluate the collected units for a number of parameters before processing a unit, and these include volume, TNC, CD34, and colony forming unit (CFU) content.[62–68]

The volume of the collected unit was the original parameter used to select those collections that should proceed to processing and subsequent banking and it remains a simple, fast, and cost-effective surrogate marker to be used in the assessment of the collected units. Although the collected volume has a strong correlation with the TNC content, most CBBs use the TNC parameter to inform decisions throughout the CBB process and for the selection of the UCB unit for transplantation.

The benefits of measuring the CD34+ cell content of the units has been highlighted by studies showing that this marker was a better correlate for engraftment than the TNC dose.[19,69] However, in the past, not all banked units had CD34 counts performed at banking, and there was no standardized test for the identification and measurement of CD34+ cells. Today, this test has become more standardized and it is now also possible to assess simultaneously both the percentage of CD34+ cells and viability.[70]

It has been suggested that the potency of the UCB measured by the number of CFUs is strongly positively associated with engraftment rates in children.[71] Page et al. have shown that the potency of the units (measured by the CFU content) was a strong predictor of engraftment.[72] Current methods for assessing the number of CFUs are complex and time-consuming, and unless these assays are performed on every single unit banked at a considerable cost bearing in mind that only about 1–2% of banked units are issued for transplantation. Also, the results may not be available when the unit is requested (they can take up to 14 days). Alternative methods to assess the function or potency of the UCB units have been recently described and are currently undergoing a more extensive clinical evaluation. Our own studies have shown a very good correlation between the amount of volume collected with the TNC, CD34, and CFU content of the units (see Figure 2).

Because the processing procedures can impact on the recovery of the TNC and CD34 (and CFUs), all these parameters have to be measured pre- and postprocessing (or before freezing), and also on the finally selected units released for transplantation.

Other obstetric factors including birth weight also seem to affect the characteristics of the collected units.[73–75]

More recently, Kurtzberg and colleagues have developed a scoring system called the Cord Blood Apgar to optimize UCB unit selection for transplantation.[76] This system considers a number of characteristics of the UCB unit such as volume, TNC, CD34, and CFUs before and after processing, and it also includes thawing techniques and other donor or patient variables.

#### 1.2.1.1 Volume Reduction

Initially, all UCB units were frozen without any manipulation but it soon became clear that the long-term storage of large numbers of frozen units would create a space issue. These considerations led to the development of protocols to reduce the volume of the collected UCB units prior to storage. A number of VR methods are currently being used, the majority of which deplete the unit of red cells and plasma, leaving the buffy coats in a standard volume.[77,78] An important consideration with any VR method is the preservation of a maximum number of viable TNCs and CD34+ cells in the stored buffy coat layer.

The first semiautomated system for VR, the OptiPress, was introduced in 1999. At present, most CBUs are reduced to a standard volume of 21 ml prior to freezing, using automated systems such as SEPAX 540 or the AutoXpress. Through the introduction of two new filters, one for the hydroxyethyl starch/anticoagulant (complete with a small sample bulb to allow for resampling) and a second to add the dimethyl sulfoxide (DMSO), the SEPAX 540 system remains sterile and is referred to as "closed."[79–81] This means that processing of these units can be undertaken in a grade C room, under a laminar flow cabinet.

An additional clinical benefit of VR is that it reduces the amount of DMSO contained in the unit, which is particularly beneficial for units that will be infused to small children. Initially, due to the large volume of DMSO, CB cells had to be washed prior to infusion, especially in the case of small children. Nowadays, washing is not essential for volume-reduced units.

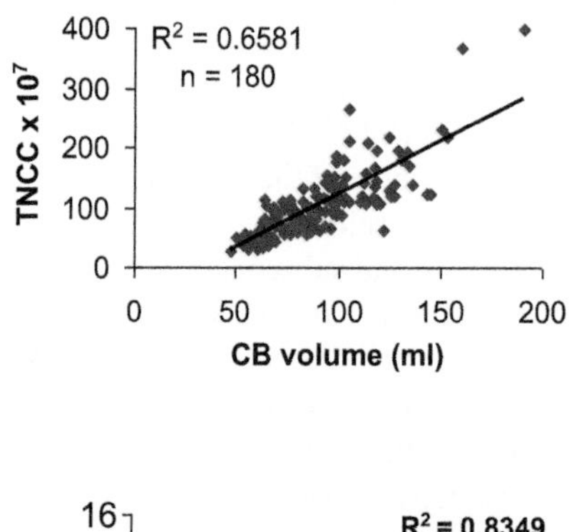


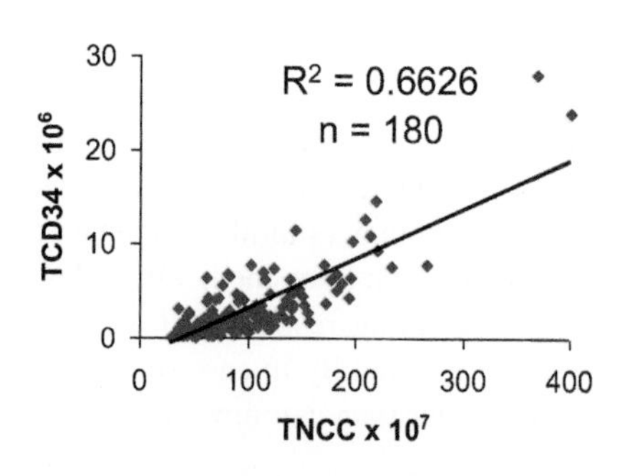


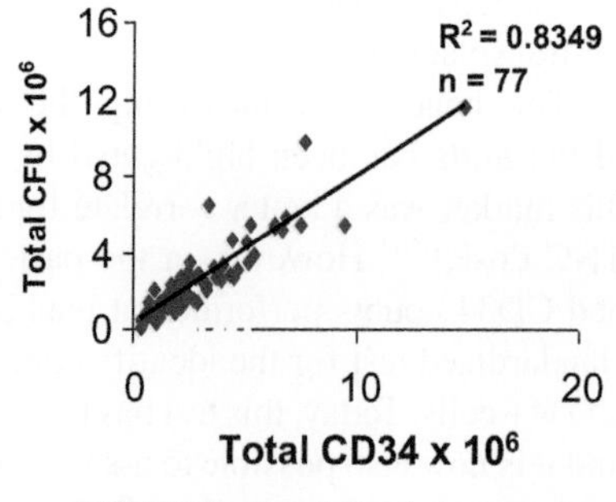


FIGURE 2 Correlation between volume, TNC, CD34, and CFU content of UCB units.

With the introduction of VR, it is now necessary to perform full blood count, TNC, nucleated red blood cell, and CD34 (and CFUs) counts before and after processing, in order to assess the effect of the manipulation on the viability and quality of the unit prior to its long-term storage.

### *1.2.2 Cryopreservation and Storage*

The cryopreservation of buffy coats containing the HSC can be performed using manually or automated (BioarchiveSystem) controlled rate freezer equipment. The automated system provides a platform to freeze and store cells in the same place minimizing exposure to temperature changes and also allowing the electronic identification of the archived units. Thus, when a unit is required for issue, it can be automatically retrieved, through a periscope, without exposing the other units to temperature changes. The VR UCB units are resuspended in 10% DMSO cryoprotectant (50% DMSO diluted in dextran 40) in a freezing bag with two compartments, which is placed in a metal container for cryopreservation and long-term storage However, both the automated and the manual systems are perfectly adequate, provided the temperatures are regularly monitored and the process is fully evaluated and quality controlled. Long-term viability of the frozen cells was also of concern but it is now known that the standard cryopreservation protocols of freezing the cells in 10% DMSO in controlled rate freezers and storage below −135 °C give an average of 80% recovery of nucleated cells and >90% recovery of progenitor cells, as measured by stem cell surrogate markers, CD34+ cells, and CFU assays.[82–85]

#### 1.2.2.1 Archiving of Samples

In order to maximize the amount of cells stored, all the "waste" components produced during the processing of the units are utilized for testing and archiving. Archiving of samples is crucial, in order to be able to perform additional tests in future when a unit is selected for transplantation and, if required, to test for any new marker that may affect the use of the units. At the NHS-CBB, a blood film is prepared from the fresh CB to perform an initial hematological screening of the unit. In addition, a small piece of cord tissue is collected and frozen as a source of DNA for future testing, if required.

### *1.2.3 Testing*

The algorithm for testing the UCB collection is complex. Some tests need to be performed upfront before banking (pre- and postprocessing) and others are carried out when the units are listed for searches. Some tests are performed on the mothers, others in the UCB unit, and others on both. Other tests can (and some must) be performed once the unit has either been reserved or selected for transplantation (see Figure 3).

Among the tests required at banking are those performed on the mother's blood, which in the United Kingdom at least, is the same as those required for blood donors. With the shortening of the window period of infectivity by the introduction of nucleic acid testing for human immunodeficiency virus/hepatitis B virus/hepatitis C virus, it should be possible to eliminate the need for a second 6-month follow-up sample from the mother to retest for infectious disease markers. This requirement was one of the important reasons why significant numbers of units had to be discarded, in spite of their compliance with the banking requirements. Also, mostly depending on the country of origin of the mother-donors, additional screening, such as tests for malaria, Chagas disease, and, more recently, West Nile virus and severe

***Prior to banking***

**Maternal sample**

- **HIV (Ab + PCR), HCV (Ab + PCR), HBV (Ab + PCR), (HBsAg + anti HBcore), HTLV 1 + 2 Ab, TPHA, CMV IgG, ±Malaria Ab, ± Chagas Ab**

**Cord blood Sample**

- **Bacteriology; post processing**
- **HLA-A, -B, -DRB1 (DNA typing)**
- **ABO/Rh**
- **FBC pre & post process**
- **CD34/viability; post processing**
- **TNC/MNC/nRCC – post processing**
- **Medical Review & Quality Checked**

***Reservation/release***

**Maternal sample**

- **HLA type**
- **Discretionary tests as necessary**

**Cord blood Sample**

- **HIV (Ab + PCR), HCV (Ab + PCR), HBV (Ab + PCR), (HBsAg + anti HBcore), HTLV 1 + 2 Ab, TPHA, CMV (IgG + PCR), ±Malaria Ab, ± Chagas Ab**
  - **Others as necessary**
- **Blood film examination**
- **Cryovial (reservation)**
  - **CFU assay**
  - **CD34 count + viability**
- **Bleedline**
  - **TNC/MNC**
  - **HLA typing (HR)**
  - **STR analysis**
  - **CFU assay**
  - **CD34 count + viability**
- **Donor follow up**
- **Medical Review & Results Checked**

FIGURE 3 Cord blood unit and maternal sample testing.

acute respiratory syndrome, are required to comply with regulations in each country. The UCB unit is also tested for these markers, once selected for transplantation.

The finally processed unit is also tested for both aerobic and anaerobic cultures prior to freezing to assess the presence of bacterial and/or fungal cross-contamination from the birth canal or systemic sepsis in the donor-mother or infant. Initially, lightly contaminated units were kept in the bank provided an antibiotic sensitivity test was performed and the results communicated to the transplant center if required. However, the current NetCord-FACT Standards mandate that bacterially contaminated unrelated units should be discarded. UCB units collected for directed use, either related or autologous, can still be banked provided the above-mentioned tests are performed.

All UCB units are tested for ABO/Rh, TNC, and CD34, and some CBBs also perform CFU on all units post processing. The HLA typing of the units is also carried out at the time of banking. Current standards indicate that all HLA typing should be carried out using DNA-based molecular techniques. Low-resolution HLA–ABC and high-resolution HLA-DRB1 typing should be performed prior to the listing of the units with the relevant UCB or adult unrelated HSC donor registries. A number of recent publications have indicated that if the transplanted UCB unit is a mismatch at one of the noninherited maternal HLA alleles, there is an improvement in the outcome of this transplant.[86,87] As a result, a number of UCB banks are now typing and reporting the maternal HLA type of the listed units. The role of the KIR receptor/ligand mismatching is still controversial;[88,89] therefore, the majority of CBBs do not routinely test their units for these markers before listing them in the national or international HSC registries.

## 1.3 Listing, Searches, Selection, Testing, Distribution, and Follow-up

### *1.3.1 Listing, Searches, Additional Testing, and Selection of UCB Units*

On completion of the processing and testing, all the information regarding the mother and the CBU must be reviewed by the Medical Director of the CBB (or a designee) to assess the suitability of the unit for inclusion into the bank. Once the units are medically released, they can be listed for searches with both the national and international registries. All units are listed under a unique identifier with the following information: HLA type, volume of collected UCB, and TNC (and CD34+, CFUs, and maternal HLA typing if available) of the final product. The issue as to whether the CD34 count should be included at registration is currently under discussion. Some CBBs are also registering the HLA typing of the mother of the CBU in order to provide the option of selecting mismatched CBU based on the noninherited maternal antigens (NIMAs.)

At present, there are several international registries, NetCord listing only UCB units and AsiaCord and Bone Marrow Donors Worldwide (BMDW), which contain both adult HSC donors and CBUs (http://www.bmdw.org). The National Marrow Donor Program (NMDP) also lists units of its associated partners for searches.[90] There are approximately 200,000 CBUs in NetCord, and over 600,000 registered with BMDW.[15] Most units registered with NetCord are also in BMDW (see Figure 4). In future, with the implementation of the European Marrow Donor Information System (EMDIS) Cord, an electronic system designed for the rapid exchange of information and requests between transplant centers and donor registries, this system should speed up the whole process of donor searches and selection, since all relevant information about a UCB unit will be readily available.

Some of the first CBBs that were established operate as independent registries. However, today, the vast majority of CBBs work through their national registries due to the fact that most transplant centers prefer a combined search report, listing all available, suitably matched adult donors and CBUs at the same time.

The current NetCord-FACT Standards indicate that all registry aspects of the CBB programs need to operate under the guidelines of the WMDA,[91,92] and that these registries should be WMDA accredited or in the process of obtaining accreditation. Transplant centers initiate a search for CBUs in the same way as for adult stem cell donors and once a transplant center receives a match report, it contacts the relevant UCB bank directly to request further information and/or additional tests such as high-resolution HLA data, CFU content, or microbiological markers.

The CBB program should have a fully validated electronic record system to enable the listing, searches, and distribution of UCB to the transplant program.

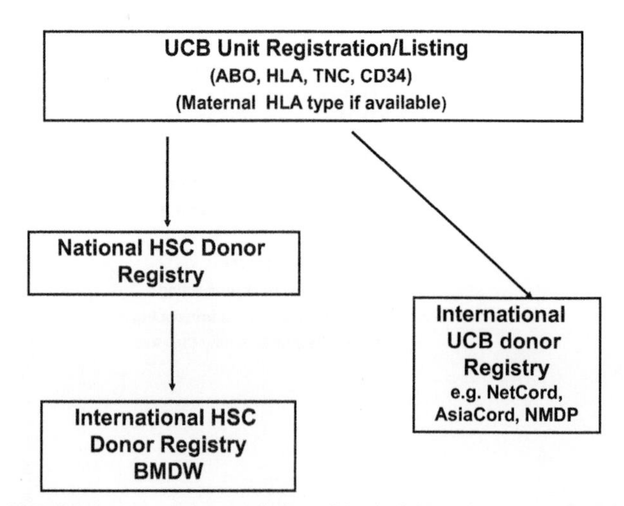


FIGURE 4 Description on listing of banked CB units nationally and internationally.

### *1.3.2 Additional Testing*

When a UCB unit is reserved or released for transplantation, a number of additional tests are performed at the request of the transplant centers. The type and resolution of tests required at this point have changed with the years as a result of the clinical outcome analyses. For instance, the range of required tests for infectious disease markers is expanding and now includes Epstein Barr virus, human herpesvirus 6, 7, and 8, and toxoplasmosis.

The request for CFU assays to assess the functionality or potency of the UCB cells is not consistent and many transplant centers are prepared to go ahead with the procedure in the absence of these results. Due to the high cost of CFU assays and since these results can take up to 14 days, most CBBs perform this test at the stage of reservation of the unit. High-resolution HLA-A, B, C, and DRB1 typing is also performed prior to the release of the unit and on a contiguous segment, if this is not available then another method needs to be used to confirm the identity of the UCB unit. Screening of the selected UCB unit for abnormal hemoglobins has also become an additional requirement prior to their release.

### *1.3.3 Distribution to a Clinical Program and Follow-up of the Transplanted UCB Units*

Clear documentation and procedure for the transport of the frozen selected UCB units to the transplant center should be in place. The units are transported in shipment containers by accredited and trained carriers.

The clinical follow-up of the released units is an important quality aspect of the operation of CBB. This is normally carried out by a national registry and/or by Eurocord[93] and the Center for International Blood and Marrow Transplant Research (CIBMTR).[94] Eurocord was established in 1999 and is responsible for collecting and analyzing all clinical outcome data on CB transplantation on behalf of the European Blood and Marrow Transplant Group.[95] CIBMTR fulfills a similar role for the transplant activity in the United States and other North and South American countries. Eurocord and CIBMTR have recently agreed to share information and analyses in order to avoid duplication of the reported data.

As mentioned above, all the activities of a CBB program need to be supported by robust electronic inventory and quality management systems and programs. All policies and procedures should be documented and updated regularly to incorporate changes in the relevant standards and/or to the outcome of internal or external audits.

The label of the UCB and all associated samples including maternal samples produced throughout the various stages of the process have to be International Society of Blood Transfusion 128 compatible for traceability.

#### 1.3.3.1 Accreditation/Licensing and Regulations

Since CBB and UCB transplantation activities involve the import and export of a cellular product across different countries, they need to operate within a highly regulated environment in order to ensure that the donations released are safe and of high quality. NetCord-FACT and the AABB have developed standards and accreditation schemes to support this activity. These standards also state that all laboratories supporting CBB activities need to have the relevant additional accreditations in place, e.g., European Federation for Immunogenetics or American Society of Histocompatibility and Immunogenetics for the HLA aspects and WMDA for the registry aspects.[23,92,96] Internationally, all aspects related to the clinical transplantation of UCB cells are covered by the FACT-JACIE (Joint Accreditation Committee ISCT & EBMT) Standards and not by the NetCord-FACT or AABB Standards.

The regulatory aspects covering the activity of CBB have also increased significantly in the past years. In the European Union, the EU Directive 2006/17/Ec and 2006/86/EC regulates on the quality and safety issues covering the donation, procurement, testing, processing, preservation, storage, and distribution of human tissues and cells.[97] These directives require all member states to have inspection and accreditation systems in place ensuring that all banks providing these services comply with an agreed set of standards. In the United Kingdom this is implemented by the Human Tissue Act, set up in 2004 and implemented in April 2006.[98] Also, locally in the United Kingdom, the Code of Practice for Tissue Banks published in 2001 covers all establishments providing tissues and cells of human origin for therapeutic use. This forms the basis for the Department of Health accreditation scheme to which all CBBs within the United Kingdom are required to be licensed, with inspections carried out by the Medicines and Healthcare products Regulatory Agency.[99]

In the United States, the Food and Drug Administration (FDA) in 2005 introduced the regulation of the manufacture of unrelated UCB to support the compliance of the Current Good Tissue Practices 21 CFR 1271.210. Later on in 2007, the FDA issued a draft guidance recommending the licensure of CBB for the manufacture of UCB units. This was finally implemented in 2011 and requires that all manufacturers of UCB units need to have an approved Biologics License Application or Investigational New Drug Application in order to be able to import a UCB into the United States. This regulation treats a UCB unit as a biological drug.[100,101]

#### 1.3.3.2 Optimal Size of a Cord Blood Bank

Discussions around cost efficiency and optimal size of UCB units required to provide donors for the majority (80%) of patients in need of an unrelated donor have been ongoing since adult HSC donor registries were first established. The probability of finding an HLA-matched unrelated donor depends not only on the number of loci (i.e., 6/6 or 10/10 loci) and resolution (medium vs. high) of the HLA matching required, but also on the ethnic background of the patient and the pool of donors to be searched. Since the vast majority of donors currently available in the international registries are of European Caucasoid ethnic background, the probability of finding a 6/6 (or a 10/10) HLA-matched donor for patients from EM backgrounds is significantly reduced.[102]

In UCB transplantation, a higher degree of HLA NetCord mismatches could be tolerated and these transplants could be performed with as little as 3/6 HLA loci matching between the recipient and the CBU. Also, several studies have confirmed that the outcomes of CB transplantation between 4/6 and 5/6 matched donors and recipients seem to be comparable to those seen between fully NetCord matched adult donors.

These results led to the proposition that the required size of UCB inventory could be smaller than that of adult unrelated HSC donors. Most patients could find at least one 4/6 matched donor from the current global CBU inventory.

A study published in 2009 demonstrated that, at least for the United Kingdom, a minimum number of 50,000 banked UCB should be sufficient to provide a UCB NetCord matched unit to approximately 80% of patients.[103–107] The role of high resolution (HR) HLA matching was found not to be significant.[108]

However, it has now been shown that high-resolution HLA-A, B, C, and DRB1 matching may also influence the clinical outcome of UCB transplantation and if these results are confirmed, a larger pool of donors to select the best compatible UCB will be required.[109,110] Therefore, the question as to whether the exact number of banked UCB units would suffice requires further evaluation.

Recent data have shown improved transplant outcomes after HLA-mismatched UCB transplantation where the mismatched antigen in the patient is matched to the donor NIMA, suggesting that when a fully HLA-matched CBU donor is unavailable, a NIMA-matched donor could be chosen.[86,87] If NIMAs are taken into account, additional "virtual" HLA phenotypes of the CBU are available for matching consideration. One locus NIMA substitution for the current matching guidelines for HLA-A, B, or DRB1 loci would add six "virtual" phenotypes, two substitutions would provide 12 "virtual" phenotypes, and three substitutions would provide eight "virtual" phenotypes. One CBU could then provide a maximum of 26 "virtual" phenotypes if all HLA-A, B, and DRB1 loci are included. The NHS-CBB currently has 3000 UCB units listed with maternal HLA phenotypes in the British Bone Marrow Registry (BBMR) and BMDW. These UCB units have the potential of adding 18,000 new phenotypes with one NIMA substitution, 36,000 new phenotypes with two, and 24,000 with three substitutions, giving a total of 78,000 new phenotypes for the CBU registered. If maternal HLA typing was performed on the 15,000 NHS-CBB banked and registered CBUs, a potential 390,000 additional "virtual" CBU phenotypes would be added to the BBMR. By using information on the HLA types of NIMAs it is possible to increase the number of CBU phenotypes available for searching and consequently increase the probability of finding an appropriately matched donor for a patient.

#### 1.3.3.3 Future Challenges

One of the important remaining challenges in UCB transplantation is how to improve engraftment, associated with a slow and often incomplete immune recovery, particularly in adult patients. This limitation, which seems to be primarily due to the insufficient number of HSC and to immunological naïvety of the immunological effectors, such as T cells in the UCB collections, has led to the development of a number of approaches to improve these outcomes. Of these the most successful so far has been the use of two UCB units for an individual patient in order to increase the number of transplanted HSCs.[111–113] This approach has yielded comparable results to those using a single CBU, and has opened the way for using UCB transplantation in older and heavier patients. The initial protocol developed by the Minneapolis group has now been adopted by many centers with or without modifications and has allowed the performance of UCB transplantation in patients not previously considered for the procedure due to their age or weight.

Other developments include the in vitro expansion of UCB HSC but early attempts were not very successful, since it appears that the majority of the protocols used led to the expansion of mainly mature progenitors. More recently, in vitro and in vivo expansion using SDF-1/CXCL12 associated to diprotin A and/or other cytokines, or using Notch-ligand Delta 1, or mesenchymal stem cells (MSCs) has been described.[114,115]

Enhancing the homing capacity of UCB cells by inhibiting the enzymatic activity of CD26/dipeptidylpeptidase IV or by in vivo direct injection of CB cells into the iliac crest has been published and has also gone into phase II clinical trials.[116]

Some investigators have now attempted the infusion of UCB intrabone or in conjunction with CD34+ or third party bone marrow-derived MSCs, with or without CD34+ cells, with limited improvement in engraftment rates.[117,118]

Regardless of these potential new developments, the majority of CBBs are now banking UCB units with greater TNC and a high number of CD34+ cells.

The identification of modifiable prognostic factors for engraftment such as choosing the "best" CBU based on cell dose such as HLA, diagnosis, and presence of HLA antibodies may also contribute to the improvement of this procedure.

Another challenge is to try to improve the immune reconstitution of cord blood transplant patients in order to reduce infections and/or viral reactivation.[119]

## 2. CONCLUSIONS

Unrelated UCB banking is a complex and highly regulated procedure and involves the collection, processing, testing, banking, listing, selection, and release of frozen UCB units to patients in need of an HSC transplant. Since its inception, it has undergone a significant evolution driven primarily by the clinical results obtained with the use of the banked units.

On the other hand, despite the initial skepticism of many transplant physicians, the success of UCB transplantation that we see today has been aided by development and implementation of stringent standards and international accreditation programs to ensure the safety, quality, and efficacy of these UCB units. In future and if the new experimental protocols for the expansion of HSCs and/or immune effectors prove to be successful, further development of the procedures and standards currently used in the banking of UCB will be required.

As UCB transplantation continues to increase and with the introduction of new clinical protocols, other genetic and epigenetic factors related to the quality and the efficacy of the UCB units and/or related to the recipient of these units may begin to emerge. Also, some of the immunotherapy protocols derived from adult unrelated HSC donors such as expansion of viral-specific T cell, regulatory T cells, NK cells, or MSCs could be applied to UCB transplantation. If this is the case, CBBs will have to consider the operational changes that will be required to collect, process, store, and release these new associated products.

## REFERENCES


1. Knundtzon S. In vitro growth of granulocytic colonies from circulating cells in human cord blood. *Blood* 1974;**43**:357–61.
2. Fauser AA, Messner HA. Granuloerythropoietic colonies in human bone marrow, peripheral blood and cord blood. *Blood* 1978;**52**: 1243.
3. Prindull G, Prindull B, Meulen N. Haematopoietic stem cells (CFUc) in human cord blood. *Acta Paediatr Scand* 1978;**67**:413–6.
4. Broxmeyer HE, Douglas GW, Hangoc G, Cooper S, Bard J, English D, et al. Human umbilical cord blood as a potential source of transplantable hematopoietic stem/progenitor cells. *Proc Natl Acad Sci U S A* 1989;**86**:3828–32.
5. Broxmeyer HE, Hangoc G, Cooper S, Ribeiro RC, Graves V, Yoder M, et al. Growth characteristics and expansion of human umbilical cord blood and estimation of its potential for transplantation in adults. *Proc NatlAcad Sci U S A* 1992;**89**:4109–13.
6. Ende M, Ende N. Hematopoietic transplantation by means of fetal (cord) blood. A new method. *Va Med Mon* 1972;**99**:276–80.
7. Gluckman E, Broxmeyer HE, Auerbach AD, Friedman HS, Douglas GW, Devergie A, et al. Hematopoietic reconstitution in a patient with Fanconi's anemia by means of umbilical-cord blood from an HLA-identical sibling. *N Engl J Med* 1989;**321**:1174–8.
8. Kurtzberg J, Laughlin M, Graham ML, Smith C, Olson JF, Halperin EC, et al. Placental blood as a source of hematopoietic stem cells for transplantation into unrelated recipients. *N Engl J Med* 1996;**335**:157–66.
9. Rubinstein P, Taylor PE, Scaradavou A, Adamson JW, Migliaccio G, Emanuel D, et al. Unrelated placental blood for bone marrow reconstitution: organization of the placental blood program. *Blood Cells* 1994;**20**:587–600.
10. Gluckman E, Rocha V, Boyer-Chammard A, Locatelli F, Arcese W, Pasquini R, et al. Outcome of cord-blood transplantation from related and unrelated donors. *N Engl J Med* 1997;**337**:373–81.
11. Navarrete C, Warwick R, Armitage S, Fehily D, Contreras M. The London Cord Blood Bank. *Bone Marrow Transpl* 1998;**22**:6–7.
12. Kogler G, Callejas J, Hakenberg P, Enczmann J, Adams O, Daubener W, et al. Hematopoietic transplant potential of unrelated cord blood: critical issues. *J Hematother* 1996;**5**:105–16.
13. Lazzari L, Corsini C, Curioni C, Lecchi L, Scalamogna M, Rebulla P, et al. The Milan Cord Blood Bank and the Italian Blood Network. *J Hematother* 1996;**5**:117–22.
14. *WMDA annual report cord blood Banks/Registries*. 2012. Available on request at mail@wmda.info.
15. Bone Marrow Donors Worldwide. *Number of donors/CB per registry in BMDW*. Cited June 2014. Available from: http://www.bmdw.org.
16. Barker JN, Scaradavou A, Stevens CE. Combined effect of total nucleated cell dose and HLA match on transplantation outcome in 1061 cord blood recipients with hematologic malignancies. *Blood* 2010;**115**:1843–9.
17. Rubinstein P, Carrier C, Scaradavou A, Kurtzberg J, Adamson J, Migliaccio AR, et al. Outcomes among 562 recipients of placental-blood transplants from unrelated donors. *N Engl J Med* 1998;**339**:1565–77. Published by FACT (Foundation for the Accreditation of Cellular Therapy), University of Nebraska Medical Center.
18. Allan D, Petraszko T, Elmoazzen H, Susan S. A Review of Factors Influencing the Banking of Collected Umbilical Cord Blood Units. *Stem Cells International* 2013;**vol. 2013**,Article ID 463031, 7 pages, doi:10.1155/2013/463031.
19. Akel S, Regan D, Wall D, Petz L, McCullough J. Current thawing and infusion practice of cryopreserved cord blood: the impact on graft quality, recipient safety, and transplantation outcomes. *Transfusion* 2014. doi: 10.1111/trf.12719.
20. Prasad VK, Mendizabal A, Parikh S, Szabolcs P, Driscoll TA, Page K, et al. Unrelated donor umbilical cord blood transplantation for inherited metabolic disorders in 159 pediatric patients from a single center: influence of cellular composition of the graft on transplantation outcomes. *Blood* 2008;**112**:2979–89.
21. Takahashi S, Ooi J, Tomonari A, Konuma T, Tsukada N, Oiwa-Monna M, et al. Comparative single-institute analysis of cord blood transplantation from unrelated donors with bone marrow or peripheral blood stem-cell transplants from related donors in adult patients with hematologic malignancies after myeloablative conditioning regimen. *Blood* 2009;**109**:1322–30.
22. Wernet PW. The international NETCORD Foundation. In: Broxmeyer H, editor. *Cord blood: biology, immunology, banking and clinical transplantation*. Bethesda, MD: AABB Press; 2004. p. 429–35.
23. NetCord-FACT International Standards for Cord Blood Collection, Banking and release for administration accreditation manual. 5th ed. July 2013.
24. AABB standards for cellular therapy product services. Available from: www.aabb.org.
25. Brunstein CG, Wagner JE. Umbilical cord blood transplantation and banking. *Ann Rev Med* 2006;**57**:403–17.
26. Stravropoulos-Giokas C, Papassavas AC. Cord blood banking and transplantation: a promising reality. *Haema* 2006;**9**:1–21.
27. Ballen KK, Barker JN, Stewart SK, Greene MF, Lane TA. Collection and preservation of cord blood for personal use. *Biol Blood Marrow Transpl* 2008;**14**:356–63.
28. Reed W, Smith R, Dekovic F, Lee JY, Saba JD, Trachtenberg E, et al. Comprehensive banking of sibling donor cord blood for children with malignant and non-malignant disease. *Blood* 2003;**101**:351–7.
29. Smythe J, Armitage S, McDonald D, Pamphilon D, Guttridge M, Brown J, et al. Directed sibling cord blood banking for transplantation: the ten year experience in the national blood service in England. *Stem Cells* 2007;**25**:2087–93.

30. Gluckman E, Ruggeri A, Vanderson R, Baudoux E, Boo M, Kurtzberg J, et al. Family-directed umbilical cord blood banking. *Haematologica* 2011;**96**:1701–7.
31. Herr AL, Kabbara N, Bonfim CM, Teira P, Locatelli F, Tiedemann K, et al. Long-term follow-up and factors influencing outcomes after related HLA-identical cord blood transplantation for patients with malignancies: an analysis on behalf of Eurocord-EBMT. *Blood* 2010;**116**:1849–56.
32. Grewal SS, Kahn JP, MacMillan ML, Ramsay NK, Wagner JE. Successful hematopoietic stem cell transplantation for Fanconi's anemia from an unaffected HLA-genotype-identical sibling selected using preimplantation genetic diagnosis. *Blood* 2004;**103**:1147–51.
33. Locatelli F, Rocha V, Reed W, Bernaudin F, Ertem M, Grafakos S, et al. Related umbilical cord blood transplantation in patients with thalassemia and sickle cell disease. *Blood* 2003;**101**:2137–43.
34. Frutchman SM, Hurlet A, Dracker R, Isola L, Goldman B, Schneider BL, et al. The successful treatment of severe aplastic anemia with autologous cord blood transplantation. *Biol Blood Marrow Transpl* 2004;**10**:741–2.
35. Lind SE. Ethical considerations related to the collection and distribution of cord blood stem cells for transplantation to reconstitute hematopoietic function. *Transfusion* 1994;**34**:828–34.
36. Katz-Benichou G. Umbilical cord blood banking: economic and therapeutic challenges. *Int J Healthc Technol Manag* 2007;**8**:464–77.
37. Thornley I, Eapen M, Sung L, Lee SJ, Davies SM, Joffe S. Private cord blood banking: experiences and views of pediatric hematopoietic cell transplantation physicians. *Pediatrics* 2009;**123**:1011–7.
38. Wagner AM, Krenger W, Suter E, Hassem DB, Surbeck DV. High acceptance rate of hybrid allogeneic-autologous umbilical cord blood banking among actual and potential Swiss donors. *Transplant Cell Eng Transfus* 2013;**53**:1510–9.
39. Petrini C. Ethical issues in umbilical cord blood banking: a comparative analysis of documents from national and international institutions. *Transfusion* 2013;**53**:902–10.
40. Sun J, Allison J, McLaughlin C, Sledge L, Waters-Pick B, Wease S, et al. Differences in quality between privately and publicly banked umbilical cord blood units: a pilot study of autologous cord blood infusion in children with acquired neurologic disorders. *Transfusion* 2010;**50**:1980–7.
41. Armitage S, Warwick R, Fehily D, Navarrete C, Contreras M. Cord blood banking in London: the first 1000 collections. *Bone Marrow Transpl* 1999;**24**:139–45.
42. Ellis J, Regan F, Cockburn H, Navarrete C. Does ethnicity affect cell dose in cord blood donation? *Transfus Med* 2007;**17**(Suppl. 1):50.
43. Ballen KK, Kurtzberg J, Lane TA, Lindgren BR, Miller JP, Nagan D, et al. Racial diversity with high nucleated cell counts and CD34 achieved in a national network of cord blood banks. *Biol Blood Marrow Transpl* 2004;**10**:269–75.
44. Davey S, Armitage S, Rocha V, Garnier F, Brown J, Brown CJ, et al. The London Cord Blood Bank: analysis of banking and transplantation outcome. *Br J Haematol* 2004;**125**:358–65.
45. Dew A, Collins D, Artz A, Rich E, Stock W, Swanson K, et al. Paucity of HLA-identical unrelated donors for African-Americans with hematologic malignancies. The need for new donor options. *Biol Blood Marrow Transpl* 2008;**14**:938–41.
46. Kidane L, Kawa S, Pushpanathan P, Cockburn H, Teesdale P, Navarrete C, et al. Introduction of mini consent process for the collection of cord blood following implementation of EU Directive 2004/23/EC. *Transfus Med* 2007;**17**(Suppl. 1):49.
47. Solves P, Moraga R, Saucedo E, Perales A, Soler MA, Larrea L, et al. Comparison between two strategies for umbilical cord blood collection. *Bone Marrow Transplant* 2003;**31**:269–73.
48. Lasky LC, Lane TA, Miller JP, Lindgren B, Patterson HA, Haley NR, et al. In utero or ex utero cord blood collection: which is better? *Transfusion* 2002;**42**:1261–7.
49. Royal College of Obstetricians and Gynaecologists Umbilical Cord Blood Banking. *Scientific advisory committee opinion paper 2*. 2006. revised June 2006. http://www.rcog.org.uk/files/rcog-corp/uploaded-files/SAC2UmbilicalCordBanking2006.pdf.
50. Andersson O, Hellstrom-Westa L, Anderson A, Domellof M. Effect of delayed versus early umbilical cord clamping on neonatal outcomes and iron status at 4 months: a randomised controlled trial. *BMJ* 2011;**343**:d7157.
51. McDonald SJ, Middleton P. The Cochrane Collaboration. *Effect of timing of umbilical cord clamping of term infants on maternal and neonatal outcomes*. John Wiley and Sons Ltd. Publishers; 2009 (Review).
52. *Scientific advisory committee's opinion paper no. 14*. 2009. http://www.rcog.org.uk/clamping-umbilical-cord-andplacental-transfusion.
53. Shlebak AA, Marley SB, Roberts IAG, Davidson RJ, Goldman JM, Gordon MY. Optimal timing for processing and cryopreservation of umbilical cord haematopoietic stem cells for clinical transplantation. *Bone Marrow Transpl* 1999;**23**:131–6.
54. Fernanda G, Pereira-Cunha, Duarte ASS, Costa FF, Saad STO, Lorand-Metze I, Luzo ACM. Viability of umbilical cord blood mononuclear cell subsets until 96 hours after collection. *Transfusion* 2013;**53**:2034–42.
55. Guttridge MG, Soh TG, Belfield H, Sidders C, Watt SM. Storage time affects umbilical cord blood viability. *Transfusion* 2014;**54**:1278–85.
56. Pereira-Cunha FG, Duarte AS, Costa FF, Saad ST, Lorand-Metze I, Luzo AC. Viability of umbilical cord blood mononuclear cell subsets until 96 hours after collection. Transplantation and cellular engineering. *Transfusion* 2013;**53**:2034.
57. Gluckman E, Rocha V, Arcese W, Michel G, Sanz G, Chan KW, et al. Factors associated with outcomes of unrelated cord blood transplant: guidelines for donor choice. *Exp Hematol* 2004;**32**:397–407.
58. Eapen M, Rubinstein P, Zhang M-J, Stevens C, Kurtzberg J, Scaradavou A, et al. Outcomes of transplantation of unrelated donor umbilical cord blood and bone marrow in children with acute leukaemia: a comparison study. *Lancet* 2007;**369**:1947–54.
59. Gluckman E, Rocha V. History of the clinical use of umbilical cord blood hematopoietic cells. *Cytotherapy* 2005;**7**:219–27.
60. Shaw BE, Veys P, Pagliuca A, Addada J, Cook G, Craddock CF, et al. Recommendations for a standard UK approach to incorporating umbilical cord blood into clinical transplantation practice: conditioning protocols and donor selection algorithms. *Bone Marrow Transplant* July 2009;**44**:7–12.
61. Barker JN, Byam C, Scaradavou A. How I treat: the selection and acquisition of unrelated cord blood grafts. *Blood* 2011;**117**:2332–9.
62. Rogers I, Sutherland DR, Holt D, Macpate F, Lains A, Hollowell S, et al. Human UC-blood banking: impact of blood volume, cell separation and cryopreservation on leukocyte and CD34+ cell recovery. *Cytotherapy* 2001;**3**:269–76.
63. George TJ, Sugrue MW, George SN, Wingard JR. Factors associated with parameters of engraftment potential of umbilical cord blood. *Transfusion* 2006;**46**:1803–12.

64. Meyer-Monard S, Tichelli A, Troeger C, Arber C, de Faveri GN, Gratwohl A, et al. Initial cord blood unit volume affects mononuclear cell and CD34+ cell-processing efficiency in a non-linear fashion. *Cytotherapy* 2012;**14**:215–22.
65. Jaime-Perez JC, Monteal-Robles R, Rodriguez-Romo LN, Herrera-Garza JL. Evaluation of volume and total nucleated cell count as cord blood selection parameters. *Am J Clin Pathology* 2011;**136**:721–6.
66. Keersmaekers CL, Mason BA, Keersmaekers J, Ponzini M, Mlynarek RA. Factors affecting umbilical cord blood stem cell suitability for transplantation in an in utero collection program. *Transfusion* 2014;**54**:545–9.
67. Page KM, Mendiazabal A, Betz-Stablein WS, Shoulars K, Tracy G, Prasad VK, et al. Optimizing donor selection for public cord blood banking: influence of maternal, infant and collection characteristics on cord blood unit quality. *Transfusion* 2014;**54**:340–52.
68. Allan D, Petraszzko T, Elmoazzen H, Smith SA. Review of factors influencing the banking of collected umbilical cord blood units. *Stem Cells Int* 2013:7. Article ID 463031. http://dx.doi.org/10.1155/2013/463031.
69. Van haute I, Lootens N, De Smet S, De Buck, Verdegem L, Vanheusden K, et al. Viable CD34+ stem cell content of a cord blood graft: which measurement performed before transplantation is most representative? *Transfusion* 2004;**44**:547–54.
70. Keeney M, Chin-Yee I, Weir K, Popma J, Nayar R, Sutherland DR. Single platform flow cytometric absolute CD34+ cell counts based on the ISHAGE guidelines. International society of hematotherapy and graft engineering. *Cytometry* 1998;**34**:61–70.
71. Yoo KH, Lee SH, Kim HJ, Sung KW, Jung HL, Cho EJ, et al. The impact of post-thaw colony-forming units-granulocyte/macrophage on engraftment following unrelated cord blood transplantation in pediatric recipients. *Bone Marrow Transplant* 2007;**39**:515–21.
72. Page KM, Zhang L, Mendizabal A, Wease S, Carter S, Gentry T, et al. Total colony-forming units are a strong, independent predictor of neutrophil and platelet engraftment after unrelated umbilical cord blood transplantation: a single-centre analysis of 435 cord blood transplants. *Biol Blood Marrow Transplant* 2011;**17**:1362–74.
73. Jones J, Stevens CE, Rubinstein P, Robertazzi RR, Kerr A, Cabbad MF. Obstetric predictors of placental/umbilical cord blood volume for transplantation. *Am J Obstet Gynecol* 2003;**188**:503–9.
74. Ballen KK, Wilson M, Wuu J, Ceredona AM, Hsieh C, Stewart FM, et al. Bigger is better: maternal and neonatal predictors of hematopoietic potential of umbilical cord blood units. *Bone Marrow Transplant* 2001;**27**:7–14.
75. Wen SH, Zhao WI, Lin PY, Yang KL. Association among birth weight, placental weight, gestational period and product quality indicators of umbilical cord blood units. *Transfus Apher Sci* 2012;**46**:39–45.
76. Page KM, Zhang L, Mendizabal A, Wease S, Carter S, Shoulars K, et al. The Cord Blood Apgar: a novel scoring system to optimize selection of banked cord blood grafts for transplantation. *Transfusion* 2012;**52**:272–83.
77. Armitage S, Fehily D, Dickinson A, Chapman C, Navarrete C, Contreras M. Cord blood banking: volume reduction of cord blood units using a semi-automated closed system. *Bone Marrow Transpl* 1999;**23**:505–9.
78. Armitage S. Cord blood processing: volume reduction. *Cell Preserv Technol* 2006;**4**:9–16.
79. Rubinstein P. Cord blood banking for clinical transplantation. *Bone Marrow Transplant* 2009;**44**:635–42.
80. Zinno F, Landi F, Aureli V, Caniglia M, Pinto RM, Rana I, et al. Pre-transplant manipulation processing of umbilical cord blood units: efficacy of Rubinstien's thawing technique used in 40 transplantation procedures. *Transfus Apher Sci* 2010;**43**:173–8.
81. Solves P, Planelles D, Mirabet V, Blanquer A, Carbonell-Uberos F. Qualitative and quantitative cell recovery in umbilical cord blood processed by two automated devices in routine cord blood banking: a comparative study. *Blood Transfus* 2013;**11**:405–11.
82. Regan DM, Wofford JD, Wall DA. Comparison of cord blood thawing methods on cell recovery, potency and infusion. *Transfusion* 2010;**50**:2670.
83. Antoniewicz-Papis J, Lachert E, Wozniak J, Janik K, Letowska M. Methods of freezing cord blood hematopoietic stem cells. *Transfusion* 2014;**54**:194–202.
84. Anagnostakis I, Papssavas AC, Michalopoulos E, Chatzistamatiou T, Andriopoulou S, Athanassios T, et al. Successful short-term cryopreservation of volume-reduced cord blood units in a cryogenic mechanical freezer: effects on cell recovery, viability, and clonogenic potential. *Transfusion* 2014;**54**:211–23.
85. Akel S, Regan D, Wall D, Petz L, McCullough J. Current thawing and infusion practice of cryopreserved cord blood: the impact on graft quality, recipient safety and transplantation outcomes. *Transfusion* 2014. http://dx.doi.org/10.1111/trf.12719.
86. van Rood JJ, Steavens CE, Smits J, Carrier C, Carpenter C, Scaradavou A. Reexposure of cord blood to noninherited maternal HLA antigens improves transplant outcome in hematological malignancies. *PNAS* 2009;**106**:19952–7.
87. Rocha V, Spellman S, Zhang M-J, Ruggeri A, Purtill D, Brady C, et al. Effect of HLA-matching recipients to donor non-inherited maternal antigens on outcomes after mismatched umbilical cord blood transplantation for hematologic malignancy. *Biol Blood Marrow Transpl* 2012;**18**:1890–6.
88. Willemze R, Rodrigues CA, Labopin M, Sanz G, Michel G, Socie G, et al. KIR-ligand incompatibility in the graft-versus-host direction improves outcomes after umbilical cord blood transplantation for acute leukemia. *Leukemia* 2009;**23**:492–500.
89. Brunstein CG, Wagner JE, Welsdorf DJ, Cooley S, Noreen H, Barker JN, et al. Negative effect of Kir alloreactivity in recipients of umbilical cord blood transplant depends on transplantation conditioning intensity. *Blood* 2009;**113**(22):5628–34.
90. NMDP. http://www.bethematch.org.
91. http://www.worldmarrow.org.
92. Welte K, Foeken L, Gluckman E, Navarrete C. International exchange of cord blood units: the registry aspects. *Bone Marrow Transplant* 2010;**45**:825–31.
93. Eurocord. www.eurocord.org.
94. Center for International Blood and Marrow Transplant Research. www.cibmtr.org/.
95. EBMT. www.ebmt.org/.
96. A gift for life. WMDA handbook for blood stem cell donations. ISBN:978-90-821221-0-7.
97. EU Directive Directive 2004/23/EC/2006 for setting standards of quality and safety for the donation, procurement, testing, processing, preservation, storage and distribution of human tissues and cells.
98. Human Tissue Authority. www.hta.gov.uk/.
99. MHRA. www.mhra.gov.uk/.
100. Reems JA, Fujita D, Tyler T, Moldwin R, Smith SD. Obtaining an accepted investigational new drug application to operate an umbilical cord blood bank. *Transfusion* 1999;**39**:357–63.
101. *FDA guidance for industry and FDA staff investigational new drug applications (INDs) for minimally manipulated, unrelated allogeneic placental/umbilical cord blood intended for hematopoietic reconstitution for specified indications.* Rockville, MD: CBER Office of Communication, Outreach, and Development; October 2009.

102. Beatty PG, Boucher KM, Mori M, Milford EL. Probability of finding HLA mismatched related or unrelated marrow or cord blood donors. *Hum Immunol* 2000;**61**:834–40.
103. Querol S, Mufti GJ, Marsh SGE, Pagliuca A, Little AM, Shaw BE, et al. Cord blood stem cells for hematopoietic stem cell transplantation in the UK: how big should the bank be? *Haematologica* 2009;**94**:1–6.
104. Querol S, Gomez SG, Pagliuca A, Torrabadella M, Madrigal JA. Quality rather than quantity: the cord blood bank dilemma. *Bone Marrow Transplant* 2010;**45**:970–8.
105. Howard DH, Meltzer D, Kollman C, Maiers M, Logan B, Gragert L, et al. Use of cost-effectiveness analysis to determine inventory size for a national cord blood bank. *Med Decis Mak* Mar–Apr 2008:243–53.
106. Kato S, Nishihira H, Hara H, Kato K, Takashi T, Sato N, et al. Cord blood transplantation and cord blood bank in Japan. *Bone Marrow Transplant* 2000;**25**(Suppl. 2):S68–70.
107. Yoon JH, Oh S, Shin S, Park JS, Roh EY, Song EY, et al. The minimum number of cord blood units needed for Koreans is 51,000. *Transfusion* 2014;**54**:504–8.
108. Kogler G, Enczmann J, Rocha V, Gluckman E, Wernet P. High-resolution HLA typing by sequencing for HLA-A, -B, -C, -DR, -DQ in 122 unrelated cord blood/patient pair transplants hardly improves long-term clinical outcome. *Bone Marrow Transplant* 2005;**36**:1033–41.
109. Eapen M, Klein JP, Sanz GF, Spellman S, Ruggeri A, Anasetti C, et al. Effect of donor-recipient HLA matching at HLA A, B, C and DRB1 on outcomes after umbilical-cord blood transplantation for leukaemia and myelodysplastic syndrome: a retrospective analysis. *Lancet Oncol* 2011;**12**:1214–21.
110. Song EY, Huh JY, Kim SY, Kim TG, Oh S, Yoon JH, et al. Estimation of size of cord blood inventory based on high-resolution typing of HLAs. *Bone Marrow Transplant* 2014;**49**:977–9.
111. Barker JN, Weisdorf DJ, Wagner JE. Creation of a double chimera after the transplantation of umbilical-cord blood from two partially matched unrelated donors. *N Engl J Med* 2001;**344**:1870–1.
112. Barker JN, Weisdorf DJ, DeFor TE, Blazer BR, McGlave PB, Miller JS. Transplantation of 2 partially HLA-matched umbilical cord blood units to enhance engraftment in adults with hematologic malignancy. *Blood* 2005;**105**:1343–7.
113. Bradstock K, Hertzberg M, Kerridge I, Svennilson J, George B, McGurgan M, et al. Single versus double unrelated umbilical cord blood units for allogeneic transplantation in adults with advanced hematological malignancies: a retrospective comparison of outcomes. *Intern Med J* 2008;**39**:744–51.
114. Robinson S, Niu T, de Lima M, Yang H, McMannis J, Karandish S, et al. Ex vivo expansion of umbilical cord blood. *Cytotherapy* 2005;**7**:243–50.
115. Delaney C, Shelly H, Brashem-Stein C, Voorhies H, Manger RL, Bernstein ID. Notch-mediated expansion of human cord blood progenitor cells capable of rapid myeloid reconstitution. *Nat Med* 2010;**16**:232–7.
116. Intrabone Frassoni F, Gualandi F, Podestà M, Raiola AM, Ibatici A, Piaggio G, et al. Direct intrabone transplant of unrelated cord-blood cells in acute leukaemia: a phase I/II study. *Lancet* 2008;**9**:831–9.
117. Gonzalo-Daganzo R, Regidor C, Martín-Donaire T, Rico MA, Bautista G, Krsnik I, et al. Results of a pilot study on the use of third-party donor mesenchymal stromal cells in cord blood transplantation in adults. *Cytotherapy* 2009;**11**:278–88.
118. Fernández MN, Regidor C, Cabrera R, García-Marco JA, Forés R, Sanjuán I, et al. Unrelated umbilical cord blood transplants in adults: early recovery of neutrophils by supportive co-transplantation of a low number of highly purified peripheral blood CD34+ cells from an HLA-haploidentical donor. *Exp Hematol* 2003;**31**:535–44.
119. van Burik J-A, Brunstein CG. Infectious complications following unrelated cord blood transplantation. *Vox Sang* 2007;**92**:289–96.

Chapter 16

# Cord Blood Unit Selection for Unrelated Transplantation

Andromachi Scaradavou

*National Cord Blood Program, New York Blood Center, New York, NY, USA; Department of Pediatrics, Memorial Sloan-Kettering Cancer Center, New York, NY, USA*

## Chapter Outline



## 1. OVERVIEW: SEARCH AND CORD BLOOD UNIT SELECTION

Cord blood (CB) grafts extend the availability of allogeneic hematopoietic stem cell transplantation for patients who do not have suitable adult donors, particularly those from ethnic minorities.[1,2] In some selected patient populations, results of CB grafts may be even superior to those of unrelated Bone Marrow (BM) or Peripheral Blood Stem Cell (PBSC) donors.[3] In most cases, careful consideration of the patient's graft options, as well as the urgency of the transplant, leads to the final decision.[4,5] Critical in the success of the overall transplantation is the selection of the cord blood unit (CBU) for the graft.

One practical consideration is the logistics of performing a CB search: with the growth of the global CB inventory (approximately 650,000 CBU currently, according to the WMDA data; Figure 1),[6] it is important to evaluate domestic as well as international CBU choices for each patient. This approach requires search engines and registries that cover the worldwide Inventory.[5] As a result, transplant centers and physicians are faced with a multitude of public CB banks in different countries, with variable banking procedures and testing assays. In this context, comparison among different CBU can be challenging, particularly since banking practices clearly have an effect on CBU quality and engraftment potential.

Cord Blood Stem Cells Medicine. http://dx.doi.org/10.1016/B978-0-12-407785-0.00016-5

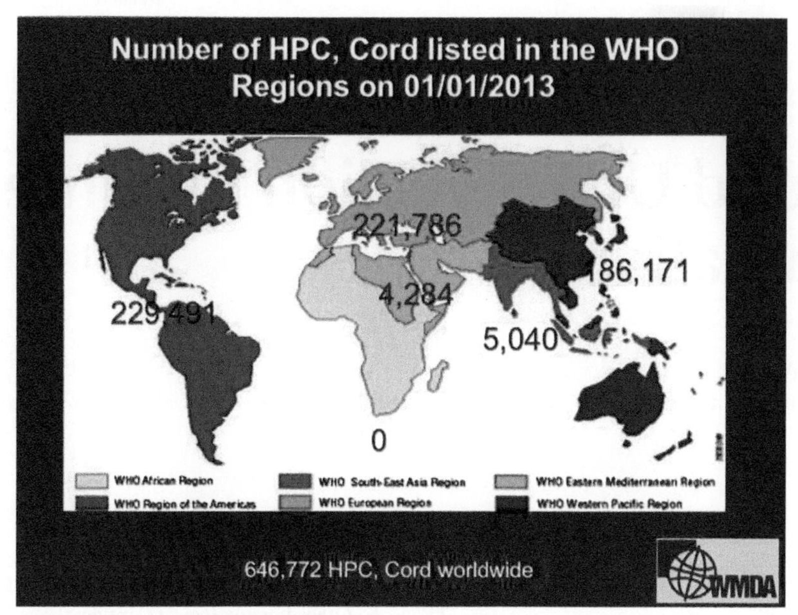


**FIGURE 1** Numbers of cord blood (CB) units (HPC, Cord Blood) in the world, WMDA Report 2012.

For the majority of transplant candidates a search will display many potential CBU matches. The first step is to "screen" these CBU, mostly based on total nucleated cells (TNC) and HLA match level (described in detail in Sections 1–3), as well as the CB bank of origin, so that a smaller list of potential units can be generated. Detailed evaluation of those will lead to the final selection of CBU for the graft.

The present review, based on knowledge of CB banking and CB testing standards, experience with CB searches and review of relevant publications, aims to address several aspects of CBU selection, including assays to evaluate the quality/potency of the CB graft, interactions between CBU TNC dose and HLA mismatch, selection of CBU with "permissible" mismatches, and other graft characteristics that need to be considered. Most of the information discussed comes from retrospective studies of single unit CB transplants but important data on double unit CB grafts are also presented. We also refer to different banking practices and testing procedures, as well as presenting some new data from the New York Blood Center's (NYBC) National Cord Blood Program. The selection information pertains to CBU to be used as single or double unit grafts for hematopoietic reconstitution of unrelated recipients with hematologic and nonhematologic diseases, but does not apply fully to other uses of CB, as in regenerative medicine.

## 2. QUALITY/POTENCY OF THE CBU

Quality, a broad term, indicating the engraftment potential or potency of the product, can be affected by a variety of events at any point during the "life" of a CB unit from collection to the infusion to the patient. Since there can be considerable variability in the quality of the different CBU, the need for a reliable "potency assay" is clear in the field. In the absence of this, surrogate markers for CBU quality/potency are being used. $CD34^+$ cell viability has emerged as a reliable surrogate of CBU quality.

Ideally, CBU potency should be measured with an assay for the true hematopoietic stem cell. In the absence of such an assay, other cells have been used as surrogates for estimating engraftment potential and comparing CB graft characteristics.

Counting the number of total nucleated cells (TNC, which includes both white blood cells and nucleated red blood cells) is technically well standardized, reproducible, and accurate. TNC correlates significantly with hematopoietic progenitors such as colony-forming units (CFU) or $CD34^+$ cells, and with transplant outcome endpoints such as myeloid engraftment and transplant-related mortality (TRM). TNC dose, therefore, has been the most universally accepted measurement of potency of a CBU. However, significant limitations exist: the pre-cryopreservation TNC of a CB unit varies depending on the method of processing: for example, CB units enriched for mononuclear cells, processed by the AXP semiautomated processing system have over 95% recovery of mononuclear cells, $CD34^+$, and CFU cells, but tend to have a lower TNC recovery (80%).[7] As a result, comparison of TNC among CBU that have undergone processing with different methods (e.g., no red blood cell depletion) may not be accurate. Additionally, there is considerable variability within the numbers of hematopoietic progenitor cells among CBU with similar TNC.

Hematopoietic progenitor cells, while not "true" stem cells, are better predictors of engraftment and survival. Early single-institution studies showed that time to engraftment correlated with the post-thaw $CD34^+$ cell dose of the CBU.[8,9]

Performing a more detailed evaluation in patients who received double unit grafts after myeloablative chemotherapy, a recent analysis showed very good correlation of time to neutrophil recovery with the infused viable CD34+ cell dose of the engrafting CBU.[10] The findings indicate that not only the number of CD34+ cells after thawing, but their quality also become important for engraftment.[10] CD34+ cell viability, measured by flow cytometry with 7-AAD exclusion, was shown to be the critical determinant of engraftment in double unit grafts, since units with CD34+ cell viability below 75% had a very low probability of engraftment.[11] Furthermore, low CD34+ cell viability correlated with low numbers of colony-forming cells.[11] These findings indicate that CD34+ cell viability can be a surrogate of overall CBU quality: CBU with low percentages of viable CD34+ cells have a significant proportion of the CD34+ cells destroyed; the remaining cells, although "viable" (i.e., not dead as determined by 7-AAD staining), are likely damaged also. As a result, the engraftment potential of the entire CBU is compromised.

Studies evaluating CFU pre-cryopreservation[12] or post-thaw[13,14] have identified this measurement as the primary correlate with engraftment.

However, assays for CFU and CD34+ cells, unfortunately, have sufficient interlaboratory variability to make it difficult to use these indices to compare units from different CB banks. For CFU in particular, the traditional assay is operator dependent and the results cannot be reproduced. To this end, significant progress has been made with the NYBC CFU strategy, an approach that combines the traditional assay with high-resolution digital imaging and storage of the electronic images, so that the colonies can be classified, enumerated, and reviewed at any later point.[15,16] This approach has been tried successfully for thousands of CBU prior to cryopreservation, as well as for segment evaluation (see Section 5). Furthermore, standardization was achieved and results between two testing laboratories on the same samples correlated closely.[17] However, this method is not widely used at this time.

Additional approaches to overcome the technical challenges of CFU assays have only been tried in very limited numbers of samples. For example, the 7-day CFU assay[18] has similar technical limitations as the traditional 14-day test and provides less information, and the HALO functional assay, based on intracellular ATP levels related to cellular proliferation,[19] only provides an indirect assessment of hematopoietic stem cell function with very limited data so far. Other indices, such as ALDH-bright cell counts[20] have been shown to correlate with engraftment in small numbers of patients, but they are even more difficult to standardize among banks.

In summary, despite important limitations, TNC dose remains the potency measure used during the initial CBU assessment from the review of a search, so that some potential units can be identified and subsequently evaluated in detail. Additional potency indicators such as CD34+ cell count and viability and CFU assays, if present, need to be considered, particularly when evaluating CBU with a similar TNC. Recent studies highlight the CD34+ cell viability as indicator of CBU quality/potency.

## 3. SELECTION OF CB UNITS FOR TRANSPLANT: INTERACTION OF TNC AND HLA

The interaction of TNC and HLA remains a very important consideration for CBU selection. CBU-recipient HLA match has been evaluated at low/intermediate resolution for HLA-A, -B, and at allele level for -DRB1, or more recently, at the allele level for HLA-A, -B, -C, and -DRB1. In all studies, higher TNC dose can overcome, to some extent, HLA disparities.

### 3.1 TNC–HLA Match at HLA-A, -B, -DRB1

As described in early single-institution studies, the speed of myeloid engraftment correlated with the TNC dose of the CB unit, that is, the number of TNC per kilogram of the patient's body weight.[8,9,21–23] In the original NYBC analysis of transplant outcomes, TNC dose was the most significant graft characteristic correlating with engraftment: a stepwise improvement in the time and probability of engraftment was seen with increasing TNC doses.[24] In the same analysis, the role of HLA match was identified as affecting time to engraftment as well as transplant-related events and survival.[24]

Subsequent NYBC analyses have highlighted the combined effect of HLA and TNC. In the NYBC data set, for 5/6 matched CBU, pre-cryopreservation TNC doses $>2.5\times10^7$/kg would lead to overall survival (OS) above 50%. However, for a similar survival rate, higher TNC dose would be required ($>5\times10^7$/kg) when 4/6 matched grafts were used.

Eapen et al.[3] compared the outcome of 503 children with leukemia transplanted with unrelated CB grafts to that of 282 recipients of unrelated BM. In that study higher TRM was seen for patients who received 1 MM CB units with low TNC (defined as $<3\times10^7$/kg) or 2 MM CB grafts independently of the cell dose infused. In contrast, 6/6 M and 5/6 M grafts with TNC $>3\times10^7$/kg had outcomes similar to those of matched BM grafts.

Similar analyses from Eurocord described a log-linear relationship between cell dose and probability of engraftment.[25]

Based on these studies, it was recommended to use CB units with ≤2 MM and a TNC $>2.5–3.0\times10^7$ cells/kg. These recommendations reflect a clear emphasis on TNC, given the higher TRM with the low cell doses, rather than on HLA matching.

The analysis of 1061 patients (children and adults, median age: 9.3 years, range: 0.1–64 years) who received single unit grafts from the NYBC during the period 1993–2006 for the treatment of leukemia or myelodysplasia after myeloablative cytoreduction, by Barker et al. aimed to evaluate how we can "trade" HLA MM and TNC dose.[26] For this analysis the geometric TNC dose was used, and patients were grouped in four dose categories depending on the pre-cryopreservation TNC (namely TNC dose $<2.5\times10^7$/kg, dose of $2.5–4.9\times10^7$/kg; $5.0–9.9\times10^7$/kg and TNC dose $>10.0\times10^7$/kg, respectively). Mismatch for HLA-A and -B was defined at the intermediate level of resolution, MM for -DRB1 was defined at high resolution, allele level.

TNC dose was associated with neutrophil and platelet engraftment, in a dose–response relationship, with progressively faster and higher engraftment rates as the dose increased. HLA-matched grafts (0 MM, 5% of total) were associated with faster neutrophil engraftment; however, no difference in the time to engraft was seen between recipients of 1 or 2 MM grafts.

Consistent with prior reports, pre-cryopreservation TNC dose and HLA match level each affected TRM and survival. The lowest TRM was seen in recipients of 0 MM units and was, interestingly, no different in children and adults given 0 MM grafts.

To evaluate the effect of combined TNC and HLA match on TRM, the patients who received 1 MM grafts with TNC dose $2.5–4.9\times10^7$/kg were used as the reference group. TRM was not different in recipients of CB units with TNC dose $>5\times10^7$/kg and either 1 or 2 MM, compared with the reference. However, patients given units with 2 MM and TNC range $2.5–4.9\times10^7$/kg had higher TRM than those receiving the same cell dose with 1 MM. Furthermore, no difference in survival was seen among patients who received units with 1 MM and TNC $2.5–4.9\times10^7$/kg and those with 2 MM and TNC $>5\times10^7$/kg.

Notably, neither TNC dose nor HLA match were associated with an effect on relapse. As a result, lower TRM was achieved with better HLA match without an increase in relapse rates. In other words, no advantage in selecting 4/6 M units to increase the anti-leukemic effect of the graft was seen in this study.

These results suggest a selection algorithm that gives priority to 0 HLA MM units. Although a wide range of TNC doses were used for this group (range: $0.7–19.4\times10^7$/kg, geometric mean $4.4\times10^7$/kg), the number of patients was relatively small, thus, the "threshold" dose for this group has not been firmly established.

In the absence of HLA-matched grafts, the recommendation was the selection of a 1 MM CBU with TNC $>2.5\times10^7$/kg or a 2 MM unit with TNC $>5.0\times10^7$/kg. Furthermore, CB units with TNC $<2.5\times10^7$/kg, with either 1 or 2 MM need to be avoided.[26,27]

## 3.2 TNC–HLA Match at Allele Level HLA-A, -B, -C, -DRB1

Initial analyses involving relatively small numbers of patients that evaluated the effect of high-resolution class I typing on the outcome of CB transplantation did not show a clear effect on survival, although an impact on the incidence of acute GvHD was seen.[28,29]

However, analyzing a larger cohort of single CBU recipients (n = 803 patients, 49% of them below the age of 10 years) for hematologic malignancies, Eapen and CIBMTR colleagues were able to see effects on TRM by including matching for HLA-C.[30] In this analysis, class I HLA typing was at least at antigen-split level (i.e., intermediate resolution for HLA-A, -B, and -C) and allele-level resolution for -DRB1. Compared with 8/8 matched donor–recipient pairs (n = 69), those with a single MM at HLA-C (n = 23) had a higher TRM (HR 3.97, p = 0.018). Furthermore, compared with patients who had 1 MM at HLA-A, -B, -DRB1 (n = 127), those with 1 additional MM at HLA-C (n = 234) showed a higher TRM (HR 1.7, p = 0.029). The study recommended evaluating HLA-C matching, and avoiding HLA-C MM if there was already an MM at -DRB1, to reduce TRM.

Furthermore, Eapen and colleagues recently published results of a large retrospective analysis of 1568 single unit CB recipients with hematologic malignancies treated with myeloablative regimens, and they evaluated allele-level matching for HLA class I and the effects on Non-Relapse Mortality (NRM).[31] The population was primarily pediatric (only 29% of the patients were older than 16 years) still, only 7% of the patients received 8/8 allele-matched CBU. The remaining had MM at various alleles: 15% had MM at one, 26% at two, 30% at three alleles, 16% at four and 5% were mismatched at five alleles. It is of note that, 50% of the HLA typings of the study, primarily those of transplants prior to 2005, were not available at the allele level, but they were imputed using the Haplogic III (high-resolution imputation algorithm). Compared with the results of the 8/8 matched grafts, NRM was not significantly higher with MM at one or two alleles (HR 2.79 and 2.69, respectively; p = 0.002, both). However, compared with the grafts with one or two allele MM, significantly higher NRM was seen in those with three or four allele MM (HR 1.31, p = 0.01) or those with five allele MM (HR 1.69, p = 0.006). It is important to interpret the incidence of NRM with the consideration that the study evaluated registry data over a very wide period of time. On the other hand, in the same study, OS was not significantly different among the groups except for grafts with five mismatched alleles.

Additionally, MM at three or more alleles increased the risk of graft failure.

Importantly, the risk of relapse was not associated with HLA matching, so there was no advantage of using CBU with a higher number of mismatches for anti-leukemic effect.

The authors concluded that in the absence of a fully allele-level matched CB graft, units with one or two allele MM are well tolerated, but units with ≥3 allele MM should be avoided. Based on this recommendation, however, almost 50% of the patients in the study would not have been transplanted, since they did receive CBU with ≥3 allele MM. The same study also evaluated TNC dose (86% of the patients had TNC doses $>3 \times 10^7$/kg) and its impact on NRM, and the effect was independent of HLA. In fact, patients with grafts with ≥3 allele MM and high TNC doses ($>5 \times 10^7$/kg) had lower NRM compared with those with similar MM and lower TNC. As a result, higher CBU cell dose could overcome the HLA disparity, a finding similar to the results seen with low-resolution matching for HLA-A and -B in the NYBC studies.

Altogether the data suggest that CBU must have a minimal TNC cell dose for engraftment ($>3.0 \times 10^7$/kg pre-cryopreservation for single unit CB transplants according to the CIBMTR data) but above that "threshold" TNC dose, HLA allele-level matching should be prioritized.

These results modify the "landscape" of HLA matching for unrelated CB transplants. It becomes clear that high-resolution HLA typing has to be performed for patient and CBU, and that allele-level matching has to be *evaluated*. Prospective studies will need to verify the role of allele-level matching on outcomes. Furthermore, significant increases in the TNC dose can overcome, to some extent, HLA disparities. Importantly, CBU selection has to prioritize HLA matching above a "threshold" TNC dose. Moreover, use of mismatched CBU does not decrease the risk of relapse but could increase NRM.

Overall, the results highlight the need for larger CB inventories worldwide, as well as high-resolution typing of the existing CBU.

## 4. SELECTION OF CBU WITH "PERMISSIBLE" HLA MISMATCHES

If no fully-matched CB grafts are available, CBU with "permissible" mismatches, that is, those that will not adversely affect patient survival, can be selected.

Despite the size of the worldwide CB Inventory estimated at 650,000 CBU (WMDA), only small numbers of patients (less than 10% in the studies described) will have a fully matched CBU. The vast majority of the CB recipients will still receive mismatched CB grafts. So there has been a lot of interest in approaches identifying CBU with "permissible" HLA mismatches.

Two areas have been studied extensively.

### 4.1 Direction of HLA MM and Effects on Transplant Outcomes

The effect of the direction of HLA MM was evaluated in a study by the NYBC,[32] with the hypothesis that vector effects, if any, should be most apparent in unidirectional mismatches. By definition, when a mismatched HLA antigen is present in both recipient and donor, the mismatch is bidirectional. In contrast, if either the donor or the recipient is homozygous in one locus, the MM is unidirectional. Thus, if the *donor* is homozygous at the mismatched locus but the patient is not, only donor cells have an HLA target (GvHD direction). Conversely, if the *patient* is homozygous at the mismatched locus but the donor is not, only host cells have an HLA target on the graft (HvG or rejection direction). Analysis of 1207 patients transplanted with single CB units from the NYBC during the period 1993–2006 revealed 98 patient–donor pairs (8.1%) that had unidirectional MM: 58 in the GvHD direction only (GvH-Only MM) and 40 in the rejection direction only (Rejection-Only MM). Seventy patients (6% of total) had 0 MM. The remaining patients had bidirectional, or combination of MM. Mismatch for HLA-A and -B was defined at the intermediate level of resolution, MM for -DRB1 was defined at high-resolution allele level. GvH-Only MM grafts had engraftment and TRM outcomes that were as good as those that had 0 MM, and significantly better that those with 1 bidirectional MM (HR 1.6, $p=0.003$ for engraftment; HR 0.5, $p=0.016$ for treatment failure). Rejection-Only MM grafts, on the other hand, carried a higher risk of relapse (HR 2.4, $p=0.013$) and lower engraftment rate.

In agreement with these results, two other CB transplant studies showed relationships between the number of mismatches in the GvH direction and myeloid engraftment,[33,34] but those did not report on GvHD, relapse, or survival endpoints. Moreover, a recent single institution study by Sanz et al. evaluating 79 patients with acute myeloid leukemia (AML) who received single unit CB transplants also highlights the MM in the GvHD direction having a significant impact on relapse.[35]

On the other hand, the Eurocord Registry data analysis by Cunha et al.[36] did not support these findings. The investigators evaluated outcomes of 1565 recipients of single unit CB grafts; of those 10% had 0 MM grafts. Using the 5/6 matched (i.e., 1 MM) recipient group as "reference," no difference was seen with the unidirectional MM. However, this may not be surprising since no association between HLA and any outcomes were seen in this study. These results are very different from the NYBC study: different population characteristics and diseases, and analytical approaches may account for that.

The implication of these observations is that the direction of the HLA mismatch needs be evaluated in CB selection algorithms so that grafts with GvH-Only MM can be identified and given priority over other types of mismatches. The potential benefit of such an approach can be evaluated by the numbers of CBU that are homozygous at one or more loci. For example, among the NYBC CB Inventory (now approximately 60,000 units), almost 1/5 of all units were homozygous at one HLA-A, -B, or -DRB1 locus. Importantly, patients who have difficulty finding suitable grafts because they have a "rare" antigen at a given locus, along with a common one, would benefit especially from GvH-Only MM grafts with a "blank" instead of a bidirectional MM at the uncommon antigen.

## 4.2 Fetal–Maternal Interactions During Pregnancy: NIMA and IPA Effects

Another area of exploration for "permissible" mismatches is the immunologic fetal–maternal interactions during pregnancy and the subsequent effects on the transplant recipients.

The fetus inherits one HLA haplotype from the father (Inherited Paternal Antigens—IPA) and one from the mother (Inherited Maternal Antigens—IMA). During pregnancy, bidirectional transplacental trafficking of cells exposes the fetus to the maternal cells, expressing both the IMA as well as NIMA (Non-Inherited Maternal Antigens), resulting in tolerogenic as well as immunogenic effects (NIMA-specific responses).[37,38] Furthermore, fetal cells enter the maternal circulation and the mother gets sensitized to the IPA of the fetus.

The so-called "NIMA effect" has been studied extensively, and the proposed mechanism described by Mold et al.[39] is the development of $CD4^+CD25^+$ $Fox^+$ regulatory T cells that suppress fetal responses specifically to NIMA. The presence of regulatory T cells has been implicated in the role of NIMA in related kidney[40] and hematopoietic stem cell transplants.[41] An additional mechanism, proposed by Mommaas et al.[42] is that CB carries NIMA-specific cytotoxic $CD8^+$ T cells, which can be present at birth or can be generated after ex vivo priming, that are capable of lysing NIMA-specific targets in vitro.

The first study to evaluate the impact of fetal exposure to NIMA on the outcome of unrelated CB transplants was published in 2009 by van Rood et al.[43] The hypothesis was that exposure to NIMA during fetal life would have an effect on transplant outcomes in cases where there was a NIMA match between recipient and CB donor (example shown in Figure 2). This was evaluated in 1121 patients with hematologic malignancies that received single CBU from the NYBC.[43]

Patients were assigned in three groups: (1) those with 0 HLA-mismatched grafts (N=62, 6% of total); (2) those with HLA MM, NIMA-matched grafts (N=79, 7% of total); and (3) those with HLA MM, NIMA MM grafts (n=980). It was of note that NIMA matching was assigned retrospectively, so matches happened only by chance; CBU were not selected based on NIMA at the time of transplant. The analysis showed statistically significant improvements in TRM for the NIMA-matched grafts ($p=0.034$ for all patients; $p=0.012$ for those older than 10 years) compared with those of NIMA MM. Furthermore, overall mortality and treatment failure for HLA MM, NIMA-matched grafts were significantly improved ($p=0.022$ and 0.002, respectively, for patients above 10 years), and engraftment was improved, particularly for patients who received lower cell dose grafts. Overall, outcomes of 1 HLA MM, NIMA-matched grafts were similar to those of 0 MM (i.e., 6/6 HLA-matched) grafts. Importantly, posttransplant relapse tended to be lower in patients with AML who received 1 HLA MM, NIMA-matched CB units. Furthermore, there was no increased incidence of GvHD in recipients of HLA MM, NIMA-matched CB grafts.[43]

A subsequent study by Rocha at al,[44] combining National Marrow Donor Program (NMDP) and Eurocord data, aimed to confirm the superior outcomes of HLA MM, NIMA-matched CB grafts in patients with hematologic malignancies. Using a smaller patient cohort and a different type of analysis, the authors compared the results of 48 HLA MM, NIMA-matched CB grafts to those patients who received HLA MM, NIMA MM CBU. This study also assigned NIMA matches retrospectively. The frequency of NIMA-matched CB grafts was 10% among the 508 eligible patients. Importantly, in this study also, the TRM was lower after NIMA-matched grafts (RR 0.48, $p=0.05$); consequently OS was shown to be higher after NIMA-matched CB transplants: 5-year probability of OS was 55% after NIMA-matched grafts versus 38% after NIMA MM units ($p=0.04$). No effects on engraftment, incidence of GvHD or relapse were detectable in this data set.[44]

In summary, using different analytical approaches, these two large retrospective studies showed a beneficial role for NIMA-matched CB grafts, leading to significant improvement in posttransplant survival. Therefore, in the absence of a fully-matched donor, HLA-mismatched, NIMA-matched CB grafts can be the graft of choice for patients with hematologic malignancies.

Figure 3 illustrates the potential impact of including the NIMA in the CB search: by substituting a single NIMA in the CB phenotype we can generate six *potential new phenotypes*. The new phenotypes have been named "virtual" phenotypes.[45] Further on, by substituting one or two NIMA antigens we may increase the number of potential phenotypes to 18. Moreover, if we were to substitute one, two, or three NIMA, the maximum potential phenotypes that can be generated can be as many as 28. Although this is not the case in every mother–CB pair (because of homozygosity or shared HLA antigens between them), it is evident from the example that the number of potential "virtual" phenotypes that can be derived from a single CB unit by including the NIMA is fairly large. Furthermore, in search simulation studies, the number of patients that could find optimal matches using the NIMA-generated phenotypes improves significantly.[45] The numerical improvement of the probability of finding donor grafts by including the NIMA in CB searches is under evaluation,[45] but overall, such a strategy

(A) HLA-mismatched, NIMA-matched CB unit

| | | | |
|---|---|---|---|
| Mother | A1, A2 | B7, B8 | DRB1*03:01, DRB1*15:02 |
| CB unit | A1, A3 | B7, B44 | DRB1*03:01, DRB1*11:01 |
| Patient | A1, A2 | B7, B44 | DRB1*03:01, DRB1*11:01 |

NIMA: Non-Inherited Maternal Antigen

(B) Sharing of IPA targets between patient and CB unit

| | HLA-A | IPA Target | Shared IPA |
|---|---|---|---|
| Mother | A2, A68 | | |
| CB unit | A1, A68 | A1 | |
| Patient | A1, A36 | | Yes |

IPA: Inherited Paternal Antigen

| | HLA-A | IPA Target | Shared IPA |
|---|---|---|---|
| Mother | A2, A24 | | |
| CB unit | A1, A24 | A1 | |
| Patient | A3, A24 | | No |

HLA Matching at low/intermediate resolution for -A, -B, and allele level for-DRB1

FIGURE 2 **Maternal–Cord blood (CB) unit HLA phenotypes and NIMA/IPA assignments.** (A) Example of HLA mismatched, NIMA-matched CB unit: The HLA assignments of CB unit and mother are shown; NIMA are shown in blue. The patient has one HLA-A locus MM with the CB unit (CB has A3 while patient has A2); however the NIMA at that locus "matches" the patient's mismatched antigen (A2). (B) Example of Shared or Not Shared IPA targets: The HLA assignments of CB unit and mother are shown for the HLA-A locus; the IPA can be inferred, that is, the CB HLA-A antigen not present in the mother, shown in green (A1). Maternal cells have been sensitized to the IPA target. Patient and CB unit match at A1, so the patient "shares" the IPA target of the maternal cells. In contrast, in the example on the R, the patient does not have A1, so the sensitized maternal cells have no target in the recipient.

| | | | | | | |
|---|---|---|---|---|---|---|
| CBU: | A2 | A24 | B7 | B65 | DRB1*01:02 | DRB1*15:01 |
| Mother: | A1 | A24 | B57 | B65 | DRB1*01:02 | DRB1*13:05 |

Substitution of one HLA antigen with a NIMA increases the potential phenotypes of a CB unit by 6-fold

| | | | | | | |
|---|---|---|---|---|---|---|
| VP1: | A1 | A2 | B7 | B65 | DRB1*01:02 | DRB1*15:01 |
| VP2: | A1 | A24 | B7 | B65 | DRB1*01:02 | DRB1*15:01 |
| VP3: | A2 | A24 | B7 | B57 | DRB1*01:02 | DRB1*15:01 |
| VP4: | A2 | A24 | B57 | B65 | DRB1*01:02 | DRB1*15:01 |
| VP5: | A2 | A24 | B7 | B65 | DRB1*01:02 | DRB1*13:05 |
| VP6: | A2 | A24 | B7 | B65 | DRB1*13:05 | DRB1*15:01 |

Substitution of 1 or 2 HLA antigens with NIMA could increase the potential HLA phenotypes up to 18-fold

FIGURE 3 Impact of including the NIMA in searches: generation of alternative ("virtual" phenotypes) from a single cord blood unit (CBU) by substituting one NIMA.

increases substantially the probability of finding optimal, "virtual" matches for the patients.

To select CB grafts with NIMA matches or shared IPA with the recipients, the maternal (CB donor mother) HLA typings need to be included in search algorithms.[46] BM Donors Worldwide (BMDW) has now implemented a search strategy for HLA MM but NIMA MM CB units based on the maternal HLA typings provided by the NYBC and other European CB Banks. In addition to the regular search, an option exists to identify NIMA-matched CB units (at this point, the search identifies CB units with all mismatches being NIMA matches).[47] An alternative strategy is to evaluate individual patient searches, and, if no fully-matched CB graft can be identified, search for HLA-mismatched but NIMA-matched CBU by typing selected mothers' samples and evaluating maternal/CBU HLA phenotypes.[48]

## 4.3 Improved Outcomes with CB Units that Share IPA Targets with the Recipients

Another important biological aspect of the fetal–maternal interactions is the presence of maternal microchimerism in the fetus, and in the CB.[49] van Rood et al. hypothesized that the maternal cells sensitized to the fetal IPA, when transplanted with the CB, may have an effect on outcome when the patient has the same antigen as the IPA (example shown in Figure 2). In those cases, patient and CB donor "share" an IPA target for the maternal cells.[50]

A total of 845 recipients with AML or acute lymphoblastic leukemia (ALL) and the NYBC CB units they received were retrospectively assigned in two groups, those with shared IPA targets at one, two, or all HLA three loci (n=751), or those with no shared IPA targets (n=64), representing 6% of the total patient–unit pairs.[50] All recipients received single unit grafts. The two groups were similar with regard to patient and disease characteristics and TNC doses. The incidence of acute GvHD grade III and IV was not different among the groups. On the contrary, there were significantly lower relapse rates in the group of HLA MM but shared IPA grafts. In particular, relapse reduction was most significant in patients receiving one HLA MM CB graft with shared IPA target ($HR=0.15$, $p<0.001$). The strong Graft-versus-Leukemia effect was mediated by the maternal microchimeric cells and it was independent of other HLA associations.[50–52]

The above findings support avoiding CB units with no IPA targets, if possible, for patients with hematologic malignancies. They also highlight the need for having the maternal HLA typing (so that IPA can be inferred) at the time of CBU selection.[46]

These analyses used HLA donor–recipient matching at the antigen level (including the "antigen-level splits") for HLA-A and -B, and allele-level typing for HLA-DRB1. The effects on OS and TRM were significant even without considering HLA-C or class I allele-level matching.[30,31] Future studies will need to address NIMA matches at HLA-C also.

## 5. APPROACHES TO OVERCOME THE TNC LIMITATIONS OF SINGLE CB GRAFTS

Several novel strategies are being evaluated for their effects on CB transplant outcomes. In these treatments CBU selection depends, to some extent, on the combination of grafts and stem cell sources used.

In young pediatric patients, the CBU TNC dose is not a major obstacle: most centers aim for a pre-cryopreservation cell dose above $2.5\times10^7$/kg with no well-defined upper limit. It needs to be noted though that merely increasing the TNC dose of unmanipulated CBU does not shorten the time to neutrophil recovery significantly. For larger pediatric patients dose limitations exist, as is the case with most adults. In addition, even with moderate cell doses the post-transplant neutropenia is long. To improve engraftment as well as decrease the period of posttransplant pancytopenia and related medical complications including prolonged hospital stay, several approaches have been investigated such as use of double unit grafts,[53–56] intrabone marrow injections, ex vivo expansion of the hematopoietic stem cells, systemic addition of mesenchymal stem cells, use of haplo-identical T cell-depleted grafts in combination with CB unit(s), and use of agents to enhance homing of CB cells to the BM. These novel strategies, reviewed in[57,58], are aiming to improve the overall outcomes of unrelated CB transplantation. Their advantages and indications are outside the scope of this review; the points related to CBU selection for these different approaches are discussed briefly below.

### 5.1 Double Unit CB Grafts

In double unit grafts only one unit gives rise to long-term hematopoiesis in the majority of cases. The time to neutrophil engraftment has been shown to correlate with the TNC of the engrafting CBU, and most recently, with the viable infused $CD34^+$ cell dose of the engrafting CBU. Since we cannot accurately predict which unit will engraft, selection of *each* of the CB units of the graft empirically follows the criteria of single unit grafts with a minimum TNC $>1.5–2.0\times10^7$/kg for 5/6 or 4/6 matched CBU, and consideration of $CD34^+$ cell dose and viability (see Section 1). The role of allele-level HLA matching has not been evaluated in double CBU grafts specifically, but most transplant physicians would apply the criteria of single unit CBU selection, including evaluation at the allele level for HLA class I. Importantly, there are no data indicating that the level of mismatch between the two units of the graft has any effect on engraftment. This question was evaluated in 84 double CB recipients transplanted for hematologic malignancies following myeloablative cytoreduction by Avery and colleagues.[59] There was no association between unit-to-unit match and incidence of graft failure or speed of engraftment.

### 5.2 Combination Grafts

The use of another cell source providing hematopoietic progenitor cells, in most cases T cell-depleted, until the CB-derived cells are seen in the peripheral blood is being evaluated in several studies. Initial studies combining haplo-identical $CD34^+$ selected BM with the CBU had promising results: the Spanish experience is reviewed in[60]; studies in the United States are described in [61–63]. More recently, CB-expanded grafts are used as a "bridge" for neutrophils.[64,65] The advantage of this approach with regard to CBU selection is that CBU TNC may not be a limiting factor since the other graft is supposed to provide neutrophils during the early posttransplant period. As a result the best HLA-matched CBU can be selected.[60,66] In fact, good engraftment and outcomes have been reported with a combination of haplo-identical stem cell grafts and CBU with low TNC doses: TNC doses of $1.1–4.3\times10^7$/kg were used with myeloablative cytoreduction in the studies by Fernandez et al.;[60]

similarly, TNC doses of $1.24–2.1 \times 10^7$/kg were used with reduced intensity regimen.[66] Ongoing analyses will have to confirm if this is the case in all clinical scenarios.

# 6. QUALITY OF CBU—BANKING PRACTICES

## 6.1 Standardization of Banking Practices and Oversight

Standardization of banking practices is crucial for consistent high quality of the products and reliability of testing results.

With the growth of CB banking worldwide, it is important to evaluate domestic as well as international CBU for each patient.[5] International search systems and networks for adult volunteer donors have included CB searches to provide transplant physicians with access to the international CB Inventory. The largest are NMDP, BMDW, and EMDIS (European Marrow Donor Information System) which report information from a very large number of international CB banks.[67,68]

Standardization of banking practices is crucial for consistent high quality of the products and reliability of testing results. To that end, accreditation agencies (NetCord/FACT—Foundation for the Accreditation of Cellular Therapy and AABB—American Association of Blood Banks) inspect public CB banks to evaluate compliance with the Standards and ensure their optimal function. It is of note that a recent study evaluated CD34+ cell viability for a number of CBU obtained from domestic and international CB banks that were thawed at a single transplant center. Post-thaw CD34+ cell viability varied significantly among the CBU from different CB banks[11] (Figure 4). Updated, more

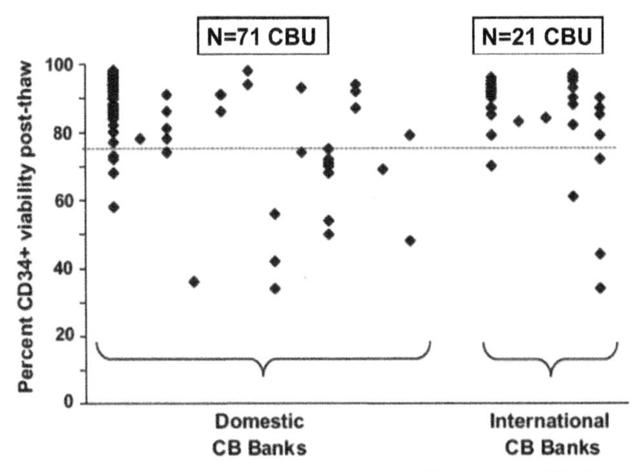


FIGURE 4 **Post-thaw CD34+ cell viability and Cord Blood (CB) Bank of origin.** CB units from domestic (n=71) and international (n=21) CB banks were evaluated at thaw at a single institution (details provided in Ref. 11).

extensive analysis included some of the banking practices that may affect the result.[10] CB bank accreditation by FACT was shown to be an independent prognostic variable. So it is desirable to obtain CBU from accredited banks to optimize transplant outcomes.

The US FDA (Food and Drug Administration) license procedures for the CB banks need to be viewed in the same context.[69] The FDA regulations focus on Current Good Manufacturing Practices (cGMP) that ensure safety, quality, identity, potency, and purity of the product. They provide assurance that all steps from CBU collection to CBU release for transplant have undergone close monitoring and review and the results meet predetermined standards. Different agencies in European counties have similar high standards and perform rigorous evaluations.

## 6.2 Evaluation of the Attached Segment: "Prediction" of CBU Quality Post-thaw

Studies are evaluating the use of cells from the segment for CD34+ cell content and viability, and CFU assays, as indicators of the quality of the cryopreserved product.

CBU selection is based on information provided by the banks, and the potency assays reflect the values at the time of CBU cryopreservation. The post-thaw results at the transplant centers do not always correlate with the bank data:[70,71] to some degree this is expected based on different laboratory practices and assays. However, the ability to accurately predict the potency of the CB graft from the pre-cryopreservation information is vital to optimal CBU selection.

Ongoing studies are evaluating the use of cells from the segment, the tubing integrally attached to the cryopreservation bag (Figure 5) for evaluation of CD34+ cell content and viability, and CFU assays, as indicators of CBU quality post-cryopreservation.[16,72] There are some considerations to the use of the frozen–thawed segment: it is a delicate sample, it has to be handled carefully, and the laboratory performing the assays needs to have experience with such samples and with interpretation of the results. Furthermore, the freezing conditions of the cryopreservation bag and segment must be similar for the quality of the segment to be representative of that of the CBU bag. It is of note that the evaluation of the segment is not standardized and is not a requirement for CBU release for many banks. Additionally, intralaboratory variability applies to those results as well.

NYBC recently presented results for a total of 1214 segments of 956 CBU processed with the AXP system; some CBU are cryopreserved in two bags and in those cases, both segments were evaluated. The CBU were processed over a period of 6 years (2006–2013) and all have been released for transplantation.[73] Evaluation by flow cytometry with 7-AAD exclusion showed that the mean segment CD34+ viability was 96%; median was 96.5%, and the standard deviation (SD) was 3%. These analyses

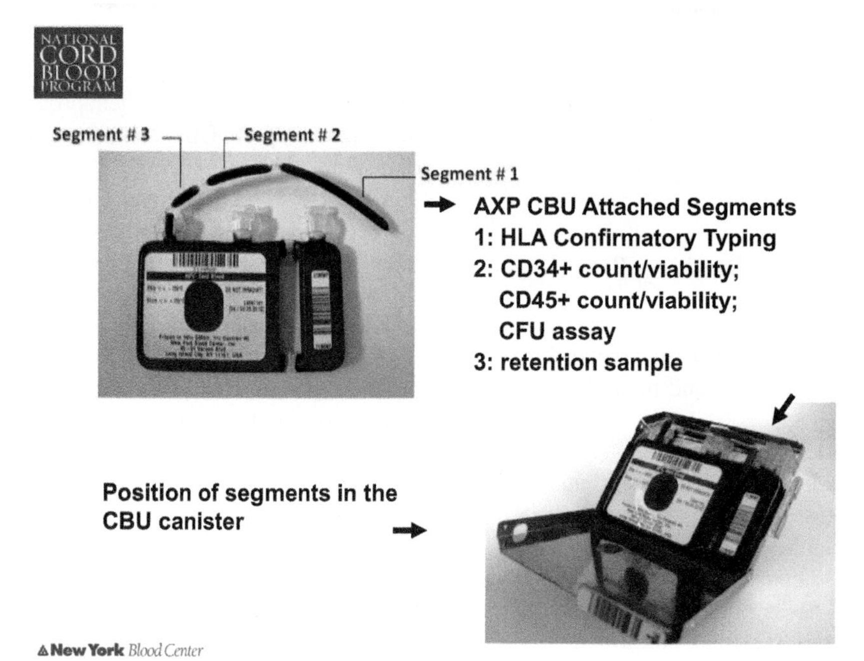


FIGURE 5 Cord blood units (CBU) in cryopreservation bag with the attached segment.

also showed that segment CD34+ cell viability correlated with the CFU output of the hematopoietic progenitor cells from the segment and, therefore, indicated that the CD34+ cell viability is a reliable assay for CBU potency.[73] Additionally, using the NYBC CFU strategy,[16] functional evaluation of the hematopoietic progenitor cells could be performed in all samples. Segment evaluation, therefore, is feasible on a large-scale basis, but the results have to be interpreted with caution. It is important to note that CD34+ cell viability is the indicator of quality, while CD45+ cell viability is usually lower since it reflects the granulocytes that are dying during the freezing–thawing procedures. Importantly, CD45+ cell viability has not been shown to have any effect on CBU engraftment.[11]

Good correlation of the segment results with those obtained from the CBU after thawing has been shown for a small number of samples in bank studies.[74]

However, the important question for the clinical transplants is whether the segment evaluation performed at the bank could predict the results at the transplant center after thawing the CBU.[73]

To address this, the post-thaw results of 48 CBU from the NYBC, that were selected for transplantation at MSKCC and were thawed there, were compared with the segment evaluation of those units that had been performed at the NYBC prior to CBU release.[73] The comparison showed very good correlation between the viable CD34+ counts ($R^2$: 0.73, $p<0.0001$), no clinically significant difference in CD34+ cell viability, but poor correlation with CFU. The findings illustrated that the CD34+ cell viability of the segment, as well as that of the CBU, is a reliable assay for the evaluation of CBU quality post-thaw. In the same center's experience, CD34+ cell viability also correlated with CBU engraftment (described in Section 1).

CB banks have modified their practices so that more segments are available for the recently processed CBU, to allow for prerelease testing. Occasional CBU, however, may not have an extra segment available for potency evaluation, especially if they were cryopreserved years ago and their (only) segment was used for HLA confirmatory typing. These CBU can be used for transplantation provided they meet the other selection criteria. Evaluation of the overall performance of the CB bank may help the decision, as well as back-up plans in case of unexpected results on the day of thawing (see Section 9).

## 7. PATIENT DIAGNOSIS, RELAPSE RISK, AND CBU SELECTION

No advantage in selecting CBU with higher HLA mismatch for a more potent anti-leukemic effect has been shown in large retrospective single unit CB studies.

Overall, there is no evidence to suggest a higher risk of relapse after CB transplantation. Relapse risks were not different in the comparison studies of unrelated CB or BM grafts in both pediatric[3] and adult patients,[75] performed by CIBMTR. Furthermore, there was no indication that HLA-mismatched grafts lead to lower posttransplant relapse rates for patients with hematologic malignancies in the NYBC analyses[24,26] and the recent CIBMTR evaluations[31] of single unit CB grafts. The studies however, clearly showed an increase in TRM with higher MM. As a result, there is no clinical advantage in selecting preferentially CBU with higher number of HLA mismatches for patients with hematologic malignancies.

In addition, when only mismatched grafts are available, CBU with certain "permissible" MM may have better outcomes than others: TRM was improved in the two large retrospective studies that evaluated HLA MM but NIMA-matched CBU.[43] Furthermore, relapse rates were lower in recipients of 1 HLA MM grafts with shared IPA targets with the recipient.[50] Finally, in the unidirectionally mismatched CBU-recipient pairs, CBU with Rejection-Only MM had inferior outcomes in patients with hematologic malignancies,[32] (as described in Section 3). The effect of double unit CB grafts on relapse is currently under evaluation in retrospective as well as prospective studies.

Patients with non-malignant diseases have a higher overall probability of graft failure for a variety of disease- or prior treatment-related reasons. In a recent Eurocord analysis HLA disparity was found to have a major impact on engraftment, GvHD, TRM, and survival, and the study suggested higher TNC dose for these patients ($>3.5\times10^7$/kg).[25]

# 8. OTHER IMMUNOLOGICAL CONSIDERATIONS FOR CBU SELECTION

## 8.1 KIR-L Compatibility

Conflicting data exist on the effect of Killer-immunoglobulin receptor-ligand (KIR-L) matching in CB transplantation.

The European group evaluated 218 recipients with acute leukemia in complete remission, transplanted with single CB units in a total of 57 transplant centers.[76] Of those, 69 donor–recipient pairs were KIR-L incompatible, based on high-resolution typing of donor and recipient for HLA-A, -B, and -C alleles. The number of HLA mismatches, stage of disease, CMV seropositivity, and frequency of myeloablative or reduced intensity cytoreduction regimens were not different among the two groups. With a median follow-up of 15 months, improved leukemia-free survival and OS was seen in the recipients of KIR-L-incompatible grafts in the GvHD direction, as well as decreased incidence of relapse, particularly in patients with AML.[76]

In contrast, the Minnesota group evaluated 257 recipients of single (n=91) or double unit (n=166) HLA-mismatched CB grafts after myeloablative (n=155) or reduced intensity (n=102) cytoreduction and found no advantage in using KIR-L-mismatched CB grafts.[77] In that single-institution study, KIR-L-mismatch was assigned in a similar way as the European study, based on high-resolution HLA. There was no reduction in relapse in any of the patient groups after transplantation with KIR-L-mismatched units. For recipients of myeloablative regimen specifically, KIR-L MM grafts had no impact in any of the endpoints evaluated. On the other hand, patients who received reduced intensity cytoreduction and were engrafted with a KIR-L-mismatched CB unit had significantly higher incidence of acute GvHD and TRM, and poorer survival.[77] Recently, a Japanese study evaluated 643 single unit transplants in patients with acute leukemia[78] and assigned KIR ligand incompatibility as in the previous studies. The effects on outcomes were analyzed taking into consideration the vector of incompatibility also. There were no effects when KIR ligand incompatibility was in the GvH direction, but incompatibility in the HvG (or rejection) direction was found to have an effect on engraftment in patients with ALL.

## 8.2 Donor-specific HLA Antibodies

Controversial data exist for the role of donor-directed specific anti-HLA antibodies (DSA). Their impact on CB engraftment has been evaluated in several studies with somewhat conflicting results.

In single CB transplants after myeloablative chemotherapy Takahashi et al.[79] showed significantly lower incidence of engraftment in 20 CB recipients with CBU-specific antibodies compared with those with no donor-specific antibodies, or no antibodies at all.

Brunstein et al.[80] evaluated a total of 126 recipients of double CB grafts and reported a comparable cumulative incidence of engraftment in DSA and non-DSA patients: 83% in the 18 patients with DSA versus 78% in those with no DSA, and no association with CB unit dominance. Moreover, Dahi et al.[81] recently found no effect on engraftment for patients with DSA antibodies that received double unit CB transplants after myeloablative conditioning: in a cohort of 82 recipients, sustained donor engraftment was observed in 95% without HLA-Abs, 100% with non-specific HLA-Abs (n=16), and 92% with DSA (n=12).

On the contrary, Cutler et al.[82] found a negative effect of DSA on engraftment of double CBU transplants in 73 patients most of whom received ATG. The incidence of graft failure was 5.5% in patients with no detectable DSA, 18.2% in the 11 patients with DSA against one of the two units, and 57% in the nine recipients with DSA against both CBU ($p=0.0001$). Furthermore, patients with DSA had delayed neutrophil recovery and higher 100-day mortality.

Additionally, Ruggeri et al.[83] reviewed the European experience of 294 patients after reduced intensity conditioning and the impact of DSA: the incidence of engraftment in the 14 patients with DSA was 44% compared with 81% for those without DSA ($p=0.006$). In addition to differences in patient populations, CB unit selection, conditioning and immunosuppression regimens, and use of ATG or not, the conflicting results may also be explained by variations in HLA antibody assays and mean fluorescence intensity of those with positive DSA.

## 9. OTHER GRAFT CHARACTERISTICS AFFECTING CBU QUALITY AND SAFETY

### 9.1 Attached Segment

NetCord/FACT (Foundation of the Accreditation of Cellular Therapy) and FDA (Food and Drug Administration) require confirmation of the identity of the CBU by HLA typing of an attached segment prior to release for transplantation. Since mislabeling errors still can occur,[84] the presence of an attached segment remains an important consideration for the selection of CB units. CB banks use DNA from the segment for confirmatory HLA typing and verification of unit identity. Some banks perform STR identity testing of segment and initial CBU sample prior to CBU release.

Use of the segment for the evaluation of CBU quality is discussed in Section 5.

### 9.2 Time in Storage—"Expiration Date"

The significance of unit "age" (i.e., length of time in the freezer) is an area of active investigation. While there is no apparent decrease in the hematopoietic potential of CB cells that have been cryopreserved for over a decade, as evaluated by in vitro assays and mouse models,[85] the importance of the collection/processing year may relate more to changing banking practices, equipment, standards, and testing assays (cryopreservation and testing for hematopoietic progenitors, infectious disease markers (IDM), and other) over time.

Several CB banks are evaluating the long-term effects of cryopreservation on the stored products.

NYBC is performing annual stability studies to compare the post-thaw results of CBU cryopreserved for several years to those of newly cryopreserved CBU. For these studies clinical CBU from the different manufacturing periods are thawed and evaluation of segment and bag address aspects of potency, bacterial contamination, and identity of the products, as well as container and label stability. So far, CBU cryopreserved as long as 20 years do not show a decrease in potency by in vitro assays. Based on the ongoing evaluations, the expiration time of the products (an FDA requirement) is being extended annually.

Furthermore, post-thaw flow cytometric evaluation of 133 segments from "old" CBU, cryopreserved for a median of 9 years (range 6–16 years) showed a median $CD34^+$ cell viability of 96.2% (mean: 95.2%, SD: 3.2%).[73] These results were not different from the $CD34^+$ cell viabilities of 1214 segments from recently cryopreserved CBU with median storage time of only 1.9 years (mean: 2.1 years, range: 0.1–6.4 years); their median $CD34^+$ cell viability post-thaw was 96.5% (mean: 95.8%, SD: 3.0%, as described in Section 5). Moreover, excellent correlation between the viable $CD34^+$ cell counts and the CFU from the segment was seen ($R^2$: 0.79, $p<0.001$).

Moreover, outcome data analysis from the NYBC showed no difference in time to engraftment, graft failure rate, and OS for CBU infused after cryopreservation and storage for >8 years (n=43, median time of storage: 9.2 years), compared with those transplanted within 2 years from collection (n=300, median storage time: 1.1 years).[86]

These results indicate that long-term cryopreservation is feasible without compromising the quality and engraftment ability of the CBU. It should be noted, however, that freezing and storage procedures, equipment and devices may vary significantly over time and from bank to bank, and NYBC results may not necessarily apply to other banks.

### 9.3 Red Cell Content

The red cell content of the CB unit (partially RBC depleted or not, i.e., RBC-replete) may influence unit choice. The red cell content can be evaluated by the postprocessing hematocrit of the CBU or the total number of RBCs remaining in the product. For example, CBU processed with the AXP semiautomated system typically have hematocrit below 50%, in most cases below 30%. In contrast, manually processed CBU have a wide range of hematocrit values, as high as 60%, in some cases. Plasma-depleted but RBC-replete CBU can have hematocrits as high as 70% and usually have higher cryopreservation volumes also, so the total RBC load is substantial. Although data on direct comparison of RBC-depleted and RBC-replete units may be difficult to obtain, analysis of the outcomes of RBC-replete but plasma-depleted units has shown similar results to the partially RBC-depleted grafts.[87] However, important concerns remain about the significant load of red cell debris and free hemoglobin of these units upon thawing,[88] and the serious, sometimes fatal, infusion adverse events that have been reported, if the products are not washed prior to infusion. Current recommendations require washing of these CBU prior to infusion, and hydration and careful monitoring of the patients. On the other hand, washing post-thaw may result in high WBC losses due to the difficulty of separating the supernatant from the mononuclear cells after centrifugation. So, this graft characteristic becomes an important consideration in heavily pretreated patients with renal compromise.

### 9.4 Nucleated Red Blood Cells

Nucleated Red Blood cells (NRBC) can be present in substantial proportions in CB.[89] Most automated hematology analyzers enumerate NRBC and WBC when performing counts, and the TNC count of the CBU includes both populations. The presence of NRBCs in CB TNC evaluation has two practical implications: firstly, NRBCs may lyse more easily that WBC and may account for an overall lower cell recovery post-thaw. Secondly, there is a misconception that, because they are not WBC, the engraftment ability of the CBU based on TNC may be "overestimated." The influence of the NRBC content on engraftment was evaluated by a retrospective study on 1112 recipients that received single CBU grafts provided by the NYBC. The evaluation showed that NRBC numbers correlated with CFU indicating an overall BM response and release of immature cells in the peripheral blood, and most importantly, they did not reduce the engraftment potential of the CBU.[89] So their presence, even in high numbers, does not imply a disadvantage for the CBU.

## 9.5 Hemoglobinopathy Screening

CBU hemoglobinopathy screening must utilize methodology that distinguishes hemoglobin A, A2, S, C, F, and H. If there is a family history of hemoglobinopathy, or, if any hemoglobin types except HbA and HbF are detected, further testing is required. Units reported as normal or "AF" are acceptable. Presence of S hemoglobin in addition to HbA and HBF indicates sickle cell trait. CBU homozygous for either sickle cell disease or thalassemia, or heterozygous for both sickle cell and thalassemia, cannot be used for transplantation. Units heterozygous for either sickle cell trait or thalassemia can be used if other donor options are limited.

It is of note that with current molecular testing assays for thalassemia, heterozygotes for alpha thalassemia, evaluated because of elevated hemoglobin H on the screening assay or low Mean Corpuscular Volume in the complete blood count, are relatively frequent. Most are heterozygotes for a single alpha globin gene deletion and therefore are "silent" carriers for a-thalassemia, so this molecular finding does not have clinical implications, and CBU have no reason not to be used for transplantation.

## 9.6 Blood Group

ABO blood group and Rh typing of the CBU is considered part of the identity testing.

Considering infusion reactions, ABO-incompatible CB grafts have not been shown to have a higher incidence of events in RBC-depleted CB units,[90] probably because a large proportion of the RBCs lyse with the freezing and thawing procedure, and the patients are well hydrated.

Although small studies have shown some effect of ABO incompatibility on acute GvHD after CB transplants,[91] larger analyses have not confirmed these results. There was no impact of ABO incompatibility on the incidence of acute and chronic GvHD in 503 recipients of single or double unit grafts analyzed at the University of Minnesota.[92] Furthermore, a Japanese study reviewed the outcomes of 191 adult recipients of single CB unit grafts after myeloablative cytoreduction for malignant diseases and found no effect of ABO mismatch on GvHD or TRM.[93]

Based on these data, there appears to be no reason to include ABO/Rh type in CBU selection criteria.

## 9.7 CBU Cryopreservation Volume

Most automated CBU processing systems have a predefined final volume (e.g., 20ml). In contrast, manually processed CBU may have variable final volumes. Cryoprotectant concentration and cooling rates during freezing procedures are important for the quality of the product and these may vary, if the final volume is not standardized.

Registries report on the final cryopreservation volume of the CBU and whether the product is in one or two bags (or more). The method of CBU preparation for infusion has to be considered when evaluating the CBU product volume. If albumin–dextran dilution is used, and most banks recommend dilution seven- to eight-fold, then the infusion volume may be large for small pediatric patients.

## 9.8 Bacteriology Screening

NetCord/FACT and FDA requirements mandate that CB grafts need to be sterile (i.e., bacterial and fungal cultures should be negative).

## 9.9 Infectious Disease Markers

In the United States, infectious disease screening tests are performed on the maternal sample (collected within 7days from CBU collection), as outlined by FDA requirements.[94] Complete maternal IDM testing in a Clinical Laboratory Improvement Amendments (CLIA) certified laboratory is the standard. Additional IDM requirements exist in other countries and tests (on the maternal and/or the CB sample) need to be performed prior to CBU acceptance in appropriately accredited testing laboratories. Information about available stored samples for testing is important. It is of note that the testing requirements and approved screening assays change over time, so stored CBU may not have all the currently required tests performed at the time of collection. Regulatory requirements and practice of medicine should guide the decisions about additional testing and acceptance of older CBU.

## 9.10 Donor Eligibility

Donor eligibility of the CBU is based on the history and risk factors of the mother according to FDA guidelines, and the results of the screening assays of the maternal sample.[94] CB units from ineligible donors can be used for transplantation, based on US FDA requirements of "Urgent Medical Need" after evaluating the potential risk associated with the reason for ineligibility versus the potential benefit of the TNC and HLA match of the respective CBU relative to other graft options for the patient. The requirements many vary in other countries.

## 10. "BACK-UP" CB GRAFTS

Many transplant centers have implemented a policy to have at least one back-up CBU identified pretransplant in the event that there are problems with unit shipping, thaw, infusion, or graft failure.[95] These back-up units should have confirmatory HLA typing completed, and need to be ready for shipment on the transplant day in case there are unexpected problems with the thawing of the graft. The CB units remain reserved until the patient has engrafted. It is of note that several non-US banks may require fees if reserved CBU are not used.

It is the transplant center's decision what procedures to have in place in case of emergency for the acquisition of additional CB units.

## 11. CONCLUSIONS—SELECTION GUIDELINES

The CB transplantation field has evolved tremendously over the past 20 years, and the efficacy and safety of the unrelated CB grafts have improved significantly. Still, significant work lies ahead. With the expanding global inventory, more precise evaluation of the cellular contents and the quality of the CB units become necessary.

Selection of CBU for transplantation remains a complex decision and no algorithm "fits" all, since there are many considerations related to graft characteristics, the patient's disease and clinical condition, as well as the treatment and type of transplant planned. While recommendations can be provided for selection of CBU for single or double unrelated grafts (Figure 6), transplant centers need to develop their own algorithms based on their specific studies as well as their results of post-thaw CBU evaluation and outcomes. Furthermore, selection of CBU to be used with novel expansion or homing strategies or with haplo-identical grafts may be somewhat different, particularly regarding the TNC dose.

For most patients, a search through domestic and international registries will display many potentially matched CBU.

1. "Screen" CBU of the search by TNC dose: selection algorithms need to establish a TNC dose "threshold," *below which CBU will not be evaluated.* The TNC dose depends on the use of single or double CBU graft or additional stem cell sources, and on specific center studies. Most would recommend a minimum TNC dose of $2–3\times10^7$/kg for single CBU and $1.5–2.0\times10^7$/kg for each of the CBU in a double graft.
2. For CBU above the "threshold" TNC dose:
   a. Evaluate HLA match level (at six and eight alleles, preferably)
   Since large retrospective studies have shown best outcomes with fully-matched CBU, a 6/6 (or 8/8 allele)-matched CBU would be the first choice. It needs to be considered, however, that the lower TNC threshold for a fully-matched graft is not established. Furthermore, CBU quality has to be considered as in all other CBU.
   b. Avoid CBU with <3/8 match, if possible
   If patient has only HLA MM CBU (as is the case in the vast majority of the searches) consider "permissible" mismatches particularly in patients with hematologic malignancies: unidirectional

CBU selection algorithm for most patients, a search through domestic and international registries will display many potentially matched CBU

1. **"Screen" CBU by TNC dose: Establish a TNC dose "threshold"**
   depending on graft: single or double CBU and/or other stem cell sources
   minimum TNC dose: 2-3 $\times10^7$/kg for single CBU
   1.5-2 $\times10^7$/kg for <u>each</u> of the CBU in a double graft
2. **For CBU above the "threshold" TNC dose:**
   - evaluate HLA match level (at 6 and 8 alleles preferably):
     - If fully matched CBU: best choice (CBU quality needs to be considered)
     - avoid CBU with < 3/8 match, if possible
     - consider "permissible" mismatches in patients with hematologic malignancies (unidirectional HLA MM, maternal HLA phenotype for NIMA/IPA assignments)
     - do not limit selection based on unit-unit match
   - evaluate potency assays, if available; presence of CBU segment
   - evaluate CB Bank of origin; overall quality of products
   - consider patient-related variables (DSA, RBC content, CBU volume)
3. **Identify CBU for the graft and back-up**

FIGURE 6 Cord blood unit (CBU) selection for transplantation.

MM, maternal HLA phenotype for NIMA/IPA assignments.

c. Do not limit selection based on unit–unit match, for double CBU grafts
d. Evaluate potency assays if present; presence of CBU segment
e. Evaluate CB bank of origin; overall quality of products
f. Evaluate potential patient-related variables (DSA, RBC content, CBU volume)

3. Identify CBU for the graft; also identify back-up graft.


## ACKNOWLEDGMENT

We thank the NYBC National Cord Blood Program staff who perform all of the tasks needed to ensure the quality of the CB units and Dr Pablo Rubinstein, Director of the National Cord Blood Program, for his support and constructive criticism. We are grateful to the obstetricians in the collaborating hospitals who support the National Cord Blood Program, and to the mothers who generously donate their infant's cord blood to be used by any patient that might need it. We also thank Professor Jon J. van Rood for his insight and support.

We thank the Memorial Sloan-Kettering Cancer Center hospital staff, clinical teams, unrelated search coordinators and laboratories for excellent patient care, detailed evaluation of CBU searches and post-thaw CBU studies.


## REFERENCES


1. Barker JN, Byam CE, Kernan NA, Lee SS, Hawke RM, Doshi KA, et al. Availability of cord blood extends allogeneic hematopoietic stem cell transplant access to racial and ethnic minorities. *Biol Blood Marrow Transpl* 2010;**16**(11):1541–8.
2. Scaradavou A, Brunstein CG, Eapen M, Le-Rademacher J, Barker JN, Chao N, et al. Double unit grafts successfully extend the application of umbilical cord blood transplantation on adults with acute leukemia. *Blood* 2013;**121**:752–8.
3. Eapen M, Rubinstein P, Zhang M-J, Stevens C, Kurtzberg J, Scaradavou A, et al. Outcomes of transplantation of unrelated donor umbilical cord blood and bone marrow in children with acute leukemia: a comparison study. *Lancet* 2007;**369**:1947–54.
4. Ponce DM, Zheng J, Gonzales AM, Lubin M, Heller G, Castro-Malaspina H, et al. Reduced late mortality risk contributes to similar survival after double-unit cord blood transplantation compared with related and unrelated donor hematopoietic stem cell transplantation. *Biol Blood Marrow Transpl* 2011;**17**:1316–26.
5. Barker JN, Byam C, Scaradavou A. How I treat: the selection and acquisition of unrelated cord blood grafts. *Blood* 2011;**117**:2332–9.
6. World Marrow Donors Association. *Annual report* 2012.
7. Scaradavou A, Stevens C, Dobrila L, Zhu T, Jiang S, Daniels D, et al. Cord blood (CB) unit mononuclear cell (MNC) dose: effect on transplantation outcome and relevance to processing method and CBU selection. *Blood* 2008;**112**:1969 [abstract].
8. Wagner JE, Barker JN, DeFor TE, Baker KS, Blazar BR, Eide C, et al. Transplantation of unrelated donor umbilical cord blood in 102 patients with malignant and nonmalignant diseases: influence of CD34 cell dose and HLA disparity on treatment-related mortality and survival. *Blood* 2002;**100**:1611–8.
9. Laughlin MJ, Barker J, Bambach B, Koc ON, Rizzieri DA, Wagner JE, et al. Hematopoietic engraftment and survival in adult recipients of umbilical cord blood from unrelated donors. *N Eng J Med* 2001;**344**:1815–22.
10. Purtill D, Smith KM, Tonon J, Evans KL, Lubin MN, Byam C, et al. Analysis of 402 cord blood units to assess factors influencing infused viable CD34⁺ cell dose: the critical determinant of engraftment. *Blood* 2013;**122**:296 [abstract].
11. Scaradavou A, Smith KM, Hawke R, Schaible A, Abboud M, Kernan NA, et al. Cord blood units with low CD34⁺ cell viability have a low probability of engraftment after double unit transplantation. *Biol Blood Marrow Transpl* 2010;**16**:500–8.
12. Migliaccio AR, Adamson JW, Stevens CE, Dobrila NL, Carrier CM, Rubinstein P. Cell dose and speed of engraftment in placental/umbilical cord blood transplantation: graft progenitor cell content is a better predictor than nucleated cell quantity. *Blood* 2000;**96**:2717–22.
13. Prasad VK, Mendizabal A, Parikh SH, Szabolcs P, Driscoll TA, Page K, et al. Unrelated donor umbilical cord blood transplantation for inherited metabolic disorders in 159 pediatric patients from a single center: influence of cellular composition of the graft on transplantation outcomes. *Blood* 2008;**112**:2979–89.
14. Page KM, Zhang L, Mendizabal A, Wease S, Carter S, Gentry T, et al. Total colony-forming units are a strong, independent predictor of neutrophil and platelet engraftment after unrelated umbilical cord blood transplantation: a single-center analysis of 435 cord blood transplants. *Biol Blood Marrow Transpl* 2011;**17**:1362–74.
15. Albano MS, Rothman W, Watanabe C, Gora A, Scaradavou A, Rubinstein P. Hematopoietic colony forming unit: development of a high-throughput CFU assay strategy by the use of high-resolution digital images stored in a laboratory information system. *Blood* 2008;**112**:2306 [abstract].
16. Albano MS, Stevens CE, Dobrila LN, Scaradavou A, Rubinstein P. Colony-Forming-Unit (CFU) assay with high-resolution digital imaging: a reliable system for cord blood (CB) CFU evaluation. *Biol Blood Marrow Transpl* 2009;**15**:45 [abstract].
17. Albano MS, Rothman W, Scaradavou A, Blass DP, McMannis JD, Rubinstein P. Automated counting of Colony Forming Units (CFU): towards standardization of the measurement of potency in cord blood cell therapy products. *Blood* 2011;**118**:485 [abstract].
18. Nawrot M, McKenna DH, Sumstad D, McMannis JD, Szczepiorkowski ZM, Belfield H, et al. Interlaboratory assessment of a novel colony-forming unit assay: a multicenter study by the cellular team of Biomedical Excellence for Safer Transfusion (BEST) collaborative. *Transfusion* 2011;**51**:2001–5.
19. Reems JA, Hall KM, Gebru LH, Taber G, Rich IN. Development of a novel assay to evaluate the functional potential of umbilical cord blood progenitors. *Transfusion* 2008;**48**:620–8.
20. Kurtzberg J, Balber A, Mendizabal A, Reese M, Kaestner A, Gentry T. Preliminary results of a pilot trial of unrelated umbilical cord blood transplantation (UCBT) augmented with cytokine-primed aldehyde dehydrogenase-bright (ALDHbr) cells. *Blood* 2006;**108**:3641 [abstract].
21. Rocha V, Wagner JE, Sobocinski KA, Klein JP, Zhang M-J, Horowitz MM, et al. Graft-versus-host disease in children who have received a cord blood or bone marrow transplant from an HLA-identical sibling. *N Eng J Med* 2000;**342**:1846–54.

22. Laughlin MJ, Eapen M, Rubinstein P, Wagner JE, Zhang MJ, Champlin RE, et al. Outcomes after transplantation of cord blood or bone marrow from unrelated donors in adults with leukemia. *N Eng J Med* 2004;**351**:2265–75.
23. Rocha V, Labopin M, Sanz G, Arcese W, Schwerdtfeger R, Bosi A, For the Acute Leukemia Working Party of European Blood and Marrow Transplant Group and the Eurocord-Netcord Registry, et al. Transplants of umbilical-cord blood or bone marrow from unrelated donors in adults with acute leukemia. *N Eng J Med* 2004;**351**:2276–85.
24. Rubinstein P, Carrier C, Scaradavou A, Kurzberg J, Adamson J, Migliaccio AR, et al. Initial results of the placental/umbilical cord blood program for unrelated bone marrow reconstitution. *N Eng J Med* 1998;**339**:1565–77.
25. Gluckman E, Rocha V. Cord blood transplantation: state of the art. *Hematologica* 2009;**94**(4):451–4.
26. Barker JN, Scaradavou A, Stevens CE. Combined effect of total nucleated cell dose and HLA-match on transplant outcome in 1061 cord blood recipients with hematological malignancies. *Blood* 2010; **115**:1843–9.
27. Scaradavou A. Unrelated umbilical cord blood unit selection. *Seminars Hematol* 2010;**47**:13–21.
28. Kogler G, Enszmann J, Rocha V, Gluckman E, Wernet P. High-resolution HLA typing by sequencing for HLA-A, -B, -C, -DR, -DQ in 122 unrelated cord blood/patient pair transplants hardly improves long-term clinical outcome. *Bone Marrow Transpl* 2005;**36**:1033–41.
29. Kurtzberg J, Prasad VK, Carter SL, Wagner JE, Baxter-Lowe LA, Wall D, et al. COBLT steering committee results of the cord blood transplantation study (COBLT): clinical outcomes of unrelated donor umbilical cord blood transplantation in pediatric patients with hematologic malignancies. *Blood* 2008;**112**:4318–27.
30. Eapen M, Klein PK, Sanz GF, Spellman S, Ruggeri A, Anasetti C, et al. Effect of donor-recipient HLA matching at HLA A, B, C, and DRB1 on outcomes after umbilical-cord blood transplantation for leukaemia and myelodysplastic syndrome: a retrospective analysis. *Lancet* 2011;**12**:1214–21.
31. Eapen M, Klein PK, Ruggieri A, Spellman S, Lee SJ, Anasetti C, et al. Impact of allele-level HLA matching on outcomes after myeloablative single unit umbilical cord blood transplantation for hematologic malignancy. *Blood* 2014;**123**:133–40.
32. Stevens CE, Carrier C, Carpenter C, Sung D, Scaradavou A. HLA mismatch direction in cord blood transplantation: impact on outcome and implications for cord blood unit selection. *Blood* 2011;**118**:3969–78.
33. Ottinger HD, Ferencik S, Beelen DW, Lindemann M, Peceny R, Elmaagacli AH, et al. Hematopoietic stem cell transplantation: contrasting the outcome of transplantations from HLA-identical siblings, partially HLA-mismatched related donors, and HLA matched unrelated donors. *Blood* 2003;**102**:1131–7.
34. Matsuno N, Wake A, Uchida N, Ishiwata K, Aroaka H, Takagi S, et al. Impact of HLA disparity in the graft-versus-host direction on engraftment in adult patients receiving reduced-intensity cord blood transplantation. *Blood* 2009;**114**:1689–95.
35. Sanz J, Jaramillo FJ, Planelles D, Montesinos P, Lorenzo I, Moscardó F, et al. Impact on outcomes of human leukocyte antigen matching by allele-level typing in adults with acute myeloid leukemia undergoing umbilical cord blood transplantation. *Biol Blood Marrow Transpl* 2014;**1**:106–10.
36. Cunha R, Loiseau P, Ruggeri A, Sanz G, Michel G, Paolaiori A, et al. Impact of HLA mismatch direction on outcomes after umbilical cord blood transplantation for hematological malignant disorders: a retrospective Eurocord-EBMT analysis. *Bone Marrow Transpl* 2014;**49**:24–9.
37. van Rood JJ, Oudshoorn M. When selecting a HLA mismatched stem cell donor consider donor immune status. *Curr Opin Immunol* 2009;**21**:1–6.
38. Van Rood JJ, Eemisse JG, van Leeuwen A. Leukocyte antibodies in sera from pregnant women. *Nature* 1958;**181**:1735–6.
39. Mold JE, Michaëlsson J, Burt TD, Muench MO, Beckerman KP, Busch MP, et al. Maternal alloantigens promote the development of tolerogenic fetal regulatory T cells in utero. *Science* 2008;**322**:1562–5.
40. Burlingham WJ, Grailer AP, Heisey DM, Claas FH, Norman D, Mohanakumar T, et al. The effect of tolerance to non-inherited maternal HLA antigens on the survival of renal transplants from sibling donors. *N Engl J Med* 1998;**339**:1657–64.
41. van Rood JJ, Loberiza Jr FR, Zhang MJ, Oudshoorn M, Claas F, Cairo MS, et al. Effect of tolerance to noninherited maternal antigens on the occurrence of graft-versus-host disease after bone marrow transplantation from a parent or an HLA-haploidentical sibling. *Blood* 2002;**99**:1572–7.
42. Mommaas B, Stegehuis-Kamp JA, van Halteren AG, Kester M, Enczmann J, Wernet P, et al. Cord blood comprises antigen-experienced T cells specific for maternal minor histocompatibility antigen HA-1. *Blood* 2005;**105**:1823–7.
43. van Rood JJ, Stevens CE, Schmits J, Carrier C, Carpenter C, Scaradavou A. Re-exposure of cord blood to non-inherited maternal HLA antigens improves transplant outcome in hematological malignancies and might enhance its anti-leukemic effect. *Proc Natl Acad Sci USA* 2009;**106**:19952–7.
44. Rocha V, Spellman S, Zhang MJ, Ruggeri A, Purtill D, Brady C, Eurocord-European Blood and Marrow Transplant Group and the Center for International Blood and Marrow Transplant Research, et al. Effect of HLA matching recipients to donor noninherited maternal antigens on outcomes after mismatched umbilical cord blood transplantation for hematologic malignancy. *Biol Blood Marrow Transpl* 2012;**18**:1890–6.
45. Van der Zanden HG, Van Rood JJ, Oudshoorn M, Bakker JN, Melis A, Brand A, et al. Noninherited maternal antigens identify acceptable HLA mismatches: benefit to patients and cost-effectiveness for cord blood banks. *Biol Blood Marrow Transplant* 2014;**20**:1791–5.
46. Scaradavou A. HLA-mismatched, noninherited maternal antigen-matched unrelated cord blood transplantations have superior survival: how HLA typing the cord blood donor's mother can move the field forward. *Biol Blood Marrow Transpl* 2012;**18**:1773–5.
47. Bone Marrow Donors Worldwide. http://www.bmdw.org.
48. Scaradavou A, Rubinstein P. Personal communication.
49. Scaradavou A, Carrier C, Mollen N, Stevens C, Rubinstein P. Detection of maternal DNA in placental/umbilical cord blood by locus-specific amplification of the non-inherited maternal HLA gene. *Blood* 1996;**88**:1494–500.
50. van Rood JJ, Scaradavou A, Stevens CE. Indirect evidence that maternal microchimerism in cord blood mediates a graft-versus-leukemia effect in cord blood transplantation. *Proc Natl Acad Sci USA* 2012;**109**:2509–14.
51. Burlingham WJ, Nelson LJ. Microchimerism in cord blood: mother as anticancer drug. *Proc Natl Acad Sci USA* 2012;**109**:2190–1.
52. Milano F, Nelson LJ, Delaney C. Fetal maternal immunity and anti-leukemia activity in cord-blood transplant recipients. *Bone Marrow Transpl* 2013;**48**:321–2.
53. Barker JN, Weisdorf DJ, DeFor TE, Blazar BR, Miller JS, Wagner JE. Rapid and complete donor chimerism in adult recipients of unrelated donor umbilical cord blood transplantation after reduced-intensity conditioning. *Blood* 2003;**102**:1915–9.

54. Barker JN, Weisdorf DJ, DeFor TE, Blazar BR, McGrave PB, Miller JS, et al. Transplantation of two partially HLA-matched umbilical cord blood units to enhance engraftment in adults with hematologic malignancy. *Blood* 2005;**105**:1343–7.
55. Brunstein CG, Barker JN, Weisdorf DJ, DeFor TE, Miller JS, Blazar BR, et al. Umbilical cord blood transplantation after non-myeloablative conditioning: impact on transplant outcomes in 110 adults with hematological disease. *Blood* 2007;**110**:3064–70.
56. Brunstein CG, Gutman JA, Weisdorf DJ, et al. Allogeneic hematopoietic cell transplantation for hematological malignancy: relative risks and benefits of double umbilical cord blood. *Blood* 2010;**116**:4693–9.
57. Oran B, Shpall E. Umbilical cord blood transplantation: a maturing technology. *ASH Educ Program* 2012;**2012**:215–22.
58. Ballen KK, Gluckman E, Broxmeyer HE. Umbilical cord blood transplantation: the first 25 years and beyond. *Blood* 2013;**122**:491–8.
59. Avery S, Shi W, Lubin M, Gonzales AM, Heller G, Castro-Malaspina H, et al. Influence of infused cell dose and HLA match on engraftment after double-unit cord blood allografts. *Blood* 2011;**117**:3277–85.
60. Sebrango A, Vicuna I, de Laiglesia A, Millan I, Bautista G, Martin-Donaire T, et al. Haematopoietic transplants combining a single unrelated cord blood unit and mobilized hematopoietic stem cells from an adult HLA-mismatched third party donor. Comparable results to transplants from HLA-identical related donors in adults with acute leukemia and myelodysplastic syndromes. *Best Pract Res Clin Haematol* 2010;**23**:259–74.
61. van Besien K, Liu H, Jain N, Stock W, Artz A. Umbilical cord blood transplantation supported by third-party donor cells: rationale, results, and applications. *Biol Blood Marrow Transpl* 2013;**19**:682–91.
62. Barker JN, Ponce DM, Devlin S, Evans KL, Lubin MN, Meagher R, et al. Double-unit cord blood (CB) transplantation combined with haplo-identical CD34+ cell-selected PBSC results in 100% CB engraftment with Enhanced myeloid recovery. *Blood* 2013;**122**:298 [abstract].
63. Kotecha R., Tian X., Wilder J., Gormley N., Khuu N., Stroncek D., et al. NK cell KIR ligand mismatches influence engraftment following combined haploidentical and umbilical cord blood (UCB) transplantation in patients with severe aplastic anemia (SAA). *Blood* 2013;**122**:2038 [abstract].
64. Delaney C, Heimfeld S, Brashem-Stein C, Voorhies H, Manger RL, Bernstein ID. Notch-mediated expansion of human cord blood progenitor cells capable of rapid myeloid reconstitution. *Nat Med* 2010;**16**:232–6.
65. Norkin M, Lazarus HM, Wingard JR. Umbilical cord blood graft enhancement strategies: has the time come to move these into the clinic? *Bone Marrow Transpl* 2013;**48**:884–9.
66. Liu H, Rich ES, Godley L, Odenike O, Joseph L, Marino S, et al. Reduced-intensity conditioning with combined haploidentical and cord blood transplantation results in rapid engraftment, low GVHD, and durable remissions. *Blood* 2011;**118**:6438–45.
67. Stavropoulos-Giokas C, Dinou A, Papassavas A. The role of HLA in cord blood transplantation. *Bone Marrow Res* 2012;**2012**:485160.
68. Gluckman E, Ruggeri A, Volt F, Cunha R, Boudjedir K, Rocha V. Milestones in umbilical cord blood transplantation. *Br J Haematol* 2011;**154**:441–7.
69. FDA Guidance for Industry. *Minimally manipulated, unrelated allogeneic placental/umbilical cord blood intended for hematopoietic reconstitution for specified indications*. October 2009.
70. Wagner E, Duval M, Dalle JH, Morin H, Bizier S, Champagne J, et al. Assessment of cord blood unit characteristics on the day of transplant: comparison with data issued by cord blood banks. *Transfusion* 2006;**46**:1190–8.
71. NYBC analyses. Scaradavou A, Rubinstein P. Personal communication.
72. Rodriguez L, Garcia J, Querol S. Predictive utility of the attached segment in the quality control of a cord blood graft. *Biol Blood Marrow Transpl* 2005;**11**:247–51.
73. Scaradavou A, Albano MS, Dobrila NL, Smith K, Lubin MN, Tonon J, et al. Analysis of cord blood unit (CBU) segment for CD34+ cell viability and hematopoietic progenitor cell content correlates with the Post-Thaw CBU after albumin-dextran reconstitution. *Blood* 2012;**120**:3023 [abstract].
74. Querol S, Gomez SG, Pagliuca A, Torrabadella M, Madrigal JA. Quality rather than quantity: the cord blood bank dilemma. *Bone Marrow Transpl* 2010;**45**:970–8.
75. Eapen M, Rocha V, Sanz G, Scaradavou A, Zhang MJ, Arcese W, et al. Effect of graft source on unrelated donor haemopoietic stem-cell transplantation in adults with acute leukaemia: a retrospective analysis. *Lancet Oncol* 2010;**11**(7):653–60.
76. Willemze R, Rodrigues CA, Labopin M, Sanz G, Michel G, Socie G, et al. KIR-ligand incompatibility in the graft-versus-host direction improves outcomes after umbilical cord blood transplantation for acute leukemia. *Leukemia* 2009;**23**:492–500.
77. Brunstein CG, Wagner JE, Weisdorf DJ, Cooley S, Noreen H, Barker JN, et al. Negative effect of KIR alloreactivity in recipients of umbilical cord blood transplant depends on transplantation conditioning intensity. *Blood* 2009;**113**:5628–34.
78. Tanaka J, Morishima Y, Takahashi Y, Yabe T, Oba K, Takahashi S, et al. Effects of KIR ligand incompatibility on clinical outcomes of umbilical cord blood transplantation without ATG for acute leukemia in complete remission. *Blood Cancer J* 2013;**3**:e164.
79. Takanashi M, Atsuta Y, Fujiwara K, Kodo H, Kai S, Sato H, et al. The impact of anti-HLA antibodies on unrelated cord blood transplantations. *Blood* 2010;**116**:2839–46.
80. Brunstein CG, Noreen H, DeFor TE, Maurer D, Miller JS, Wagner JE. Anti-HLA antibodies in double umbilical cord blood transplantation. *Biol Blood Marrow Transpl* 2011;**17**(11):1704–8.
81. Dahi PB, Barone J, Devlin SM, Byam C, Lubin M, Ponce DM, et al. Sustained donor engraftment in recipients of double-unit cord blood transplantation is possible despite donor-specific HLA-antibodies. *Biol Blood Marrow Transpl* 2014;**20**(5):735–9.
82. Cutler C, Kim HT, Sun L, Sese D, Glotzbecker B, Armand P, et al. Donor-specific anti-HLA antibodies predict outcome in double umbilical cord blood transplantation. *Blood* 2011;**118**:6691–7.
83. Ruggeri A, Rocha V, Masson E, Labopin M, Cunha R, Absi L, et al. Impact of donor-specific anti-HLA antibodies on graft failure and survival after reduced intensity conditioning-unrelated cord blood transplantation: a Eurocord, Société Francophone d'Histocompatibilité et d'Immunogénétique (SFHI) and Société Française de Greffe de Moelle et de Thérapie Cellulaire (SFGM-TC) analysis. *Haematologica* 2013;**98**:1154–60.
84. McCullough J, McKenna D, Kadidlo D, Maurer D, Noreen HJ, French K, et al. Mislabeled units of umbilical cord blood detected by a quality assurance program at the transplantation center. *Blood* 2009;**114**:1684–8.
85. Broxmeyer HE, Srour EF, Hangoc G, Cooper S, Anderson SA, Bodine DM. High-efficiency recovery of functional hematopoietic progenitor and stem cells from human cord blood cryopreserved for 15 years. *Proc Natl Acad Sci USA* 2003;**100**:645–50.
86. Scaradavou A, Stevens CE, Dobrila L, Sung D, Rubinstein P. "Age" of the cord blood (CB) unit: impact of long-term cryopreservation and storage on transplant outcome. *Blood* 2007;**110**:2033 [abstract].

87. Chow R, Nademanee A, Rosenthal J, Karanes C, Jaing T-H, Graham M, et al. Analysis of hematopoietic cell transplants using plasma-depleted cord blood products that are not red cell reduced. *Biol Blood Marrow Transpl* 2007;**13**:1346–57.
88. Barker JN, Scaradavou A. The controversy of red blood cell-replete cord blood units. *Blood* 2011;**118**:480 [Letter to Blood; Response].
89. Stevens CE, Gladstone J, Taylor PE, Scaradavou A, Migliaccio AR, Visser J Dobrila NL, et al. Placental/umbilical cord blood for unrelated-donor bone marrow reconstitution: relevance of nucleated red blood cells. *Blood* 2002;**100**:2662–4.
90. Dahi PB, Lubin MN, Evans KL, Gonzales AMR, Schaible A, Tonon J, et al. "No wash" albumin-dextran dilution for double-unit cord blood transplantation (DCBT) is safe with appropriate management and results in high rates of sustained donor engraftment. *Biol Blood Marrow Transpl* 2014;**20**(4):490–4.
91. Berglund S, Le Blanc K, Remberger M, Gertow J, Uzunel M, Svenberg P, et al. Factors with an impact on chimerism development and long-term survival after umbilical cord blood transplantation. *Transplantation* 2012;**94**:1066–74.
92. Romee R, Weisdorf DJ, Brunstein C, Wagner JE, Cao Q, Blazar BR, et al. Impact of ABO-mismatch on risk of GVHD after umbilical cord blood transplantation. *Bone Marrow Transpl* 2013;**48**:1046–9.
93. Konuma T, Kato S, Ooi J, Oiwa-Monna M, Ebihara Y, Mochizuki S, et al. Effect of ABO blood group incompatibility on the outcome of single-unit cord blood transplantation after myeloablative conditioning. *Biol Blood Marrow Transpl* 2013:S1083–8791 [abstract].
94. FDA Guidance for Industry: eligibility determination of donors of human cells, tissues, and cellular and tissue-based products (HCT/Ps). U.S. Department of Health and Human Services, Food and Drug Administration, Center for Biologics Evaluation and Research, August 2007.
95. Ponce DM, Lubin M, Gonzales AM, Byam C, Wells D, Ferrante R, et al. The use of back-up units to enhance the safety of unrelated donor cord blood transplantation. *Biol Blood Marrow Transpl* 2012;**18**:648–51.

Chapter 17

# Quality Management Systems Including Accreditation Standards

**Andreas Papassavas, Theofanis K. Chatzistamatiou, Efstathios Michalopoulos, Markella Serafetinidi, Vasiliki Gkioka, Elena Markogianni and Catherine Stavropoulos-Giokas**

*Hellenic Cord Blood Bank, Biomedical Research Foundation Academy of Athens (BRFAA), Greece*

## Chapter Outline



**Cord Blood Stem Cells Medicine. http://dx.doi.org/10.1016/B978-0-12-407785-0.00017-7**

## 1. INTRODUCTION

After more than 20 years of clinical experience, cord blood (CB) is now generally accepted as being equivalent to or even better than bone marrow as a source of hematopoietic cells for transplantation. CB is being used increasingly not only for pediatric patients but also for adult patients, with more than 2500 unrelated CB transplantation procedures performed each year. Recognizing the potential clinical benefit of banked cord blood units (CBUs), the first cord blood bank (CBB) was created by the New York Blood Center in 1991. Other banks soon followed, and there are now over 608,000 publicly available CBUs listed in the registries worldwide. Furthermore, there are estimated to be at least 360 private CBBs throughout the world storing CBUs for autologous and related-party use.

This chapter will try to show that the development of a quality management program (QMP) in a CBB, along with the never-ending search for improved quality, can set the framework for the establishment of effective promotional strategies, which will control and certify all the possible factors which lead to a successful CB transplantation. In addition, it is intended to be shown that the implementation to a CBB of a QMS/P certifies the security of the patients' lives and permits the CBB to continue its function properly and efficiently.

## 2. QUALITY MANAGEMENT

### 2.1 Definition and Important Components

Quality management (QM) is a recent phenomenon, has a specific meaning within many business sectors, and is focused not only on product/service quality, but also on the means to achieve it. QM therefore uses quality assurance (QA) and control of processes as well as products to achieve more consistent quality.[1] For more than two decades "quality" issues in business have been responsible for the development of new organizations and even industries, for instance, the "American Society for Quality" and "Six Sigma" (a statistically oriented approach to process improvement that uses a variety of tools, including statistical process control, total quality management (TQM), and design of experiments) consulting.

In general, QM:

- Is focused on the effectiveness of services provided by an organization and its produced outcomes.
- Aims to establish new benchmarks for "best practice."
- Is interested in continuous improvement of systems and processes.
- Improves quality.

### 2.2 What is "Quality"?

"Quality" is not something that occurs as an independent activity. It needs to be designed into all elements and functions of the organization, and then systematically controlled. The literature on QM provides a broad range of definitions of quality. In particular, the literature notes that quality is a subjective term and that individuals and organizations have their own perceptions and definitions. However, the common theme or focus of each of these definitions reflects the need for the total characteristics and features of a product or service to satisfy a specified need or use. Thus, the *International Organization for Standardization* (*ISO*) provides this in its definition of "quality:" the "degree to which a set of inherent characteristics fulfils requirements." "*Requirement*" signifies "need or expectation that is stated generally implied or obligatory;" "inherent" signifies "quality is relative to what something should be and what it is, especially as a permanent characteristic." In addition, in the literature on quality according to the *American Society for Quality*, "quality" can be defined in the following ways:

- It is based on customer's (e.g., transplants) perceptions of a product (e.g., CBU)/service's (CBB) design and how well the design matches the original specifications.
- It shows the ability of a product (e.g., CBU)/service (CBB) to satisfy stated or implied needs.
- It is achieved by conforming with established requirements within an organization (e.g., CBB).

Another influential individual in the development of "quality" was *Joseph M. Juran*, who defined "quality" as "fitness for use," meaning that the users of products or services should be able to rely on that product or service 100% of the time without any concern about defects.

Nowadays, the influence of "quality" thinking has spread to nontraditional applications outside the walls of manufacturing, extending into service sectors and into areas such as sales, marketing, customer service and, recently, in the health service concerning the biological products used for cellular therapies. Thus, quality in medical and laboratory management for cellular therapies includes:

- Any feature or characteristic of a product (e.g., CBU) that is needed to satisfy a user's (e.g., transplant center) needs.
- Measures to investigate, detect, assess, correct, and prevent errors.
- Significant reduction of errors.
- Credibility of outcomes.
- Improvement of patient safety and quality of processes.

It is important to recognize that, when an organization/institute adopts a QM approach, "quality" is not something that can be put in place and just forgotten. It is not like an obsolete policy document on the shelf. QM is a journey, not a destination.

### 2.3 Quality Characteristics/Requirements

Any feature or characteristic of a product (e.g., CBU) or service (e.g., CBB) that is needed to satisfy user needs or achieve fitness for use is a "quality characteristic/requirement." When dealing with products (e.g., CBUs), the characteristics are mostly technical ("product technical characteristic"), for example accessibility, availability, operability, and durability, whereas "service quality characteristics" have a human dimension, for example, waiting time, delivery time accuracy, and accessibility.

## 2.4 QM Principles

The quality management principles (QMPr) reflect best practice and are designed to enable continual improvement of the QM. These QMPr can be used as a framework to guide an organization/institute toward improved performance. These QMPr include:

1. "Customer" *focus*. The organization/institute should document "customer" requirements and monitor the quality of services as perceived by them. All "customer" feedback and complaints should be formally recorded and followed up without delay. Details of action taken and recommendations for improvement should be documented.
2. *Leadership*. Leaders establish unity of purpose and direction of the organization/institute. They should create and maintain the internal environment in which staff can become fully involved in achieving the organization/institute objectives. The implementation of QM will hardly be successful if there is lack of commitment from top management. As such, it is critical that top management has a sound appreciation and understanding of all facets of QM and, in particular, issues pertaining to QA. This understanding and appreciation should be obtained through appropriate training and experience. It must also be remembered that leadership can be found at all levels within an organization and identifying this quality may be of great benefit in establishing a quality culture within a specific section of an organization or throughout the organization as a whole.
3. *Involvement of people*. People at all levels are the essence of the organization/institute and their full involvement enables their abilities to be used for the benefit of the organization/institute. Staff must be suitably qualified and competent in their jobs, as the quality of their work directly affects the quality of service. This can be achieved through the provision of appropriate training and evaluation. Quality awareness training should also be provided to all relevant staff to heighten responsibility, accountability, and quality consciousness, that is, to assist in building a quality-focused culture. With the implementation of QM, staff needs to take on additional responsibilities such as the day-to-day consistency checks as part of the data for product QA and control processes.
4. *Process approach*. A desired result is achieved more efficiently when activities and related resources are managed as a process. A "process" is a set of interrelated or interacting activities that transform inputs into outputs. QM can be thought of as a single large process that uses many inputs to generate many outputs. In turn, this large process is made up of many smaller processes.
5. *System approach to management*. The effectiveness and efficiency of an organization/institute in achieving its quality objectives are contributed by identifying, understanding, and managing all interrelated processes as a system.
6. *Continual improvement*. Continual improvement of the organization's overall performance should be a permanent objective of the organization. Specifically, the effectiveness and suitability of the QM has to be evaluated, and areas for improvement identified and rectified. Management reviews have to be conducted regularly using the data collected from the monitoring and measurement process to identify areas for further improvement. Channels may need to be established to allow all staff in the organization/institute to make suggestions on ways to improve the service.
7. *Factual approach to decision-making*. Effective decisions are based on the analysis of data and information.
8. *Mutually beneficial supplier relationships*. An organization/institute and its suppliers are interdependent and a mutually beneficial relationship enhances the ability of both to create value. Suppliers should be evaluated and selected on the basis of their ability to meet purchase order requirements and on their past performance.

## 2.5 QM Components

QM can be considered to have three main components:

- Quality control (QC).
- Quality assurance (QA).
- Quality improvement (QI).

### *2.5.1 Quality Control*

The QC function of an organization first evolved when inspectors were hired to inspect products to differentiate between the good and the bad. The 100% inspection later evolved into sampling inspection. QC is a part of QM focused on fulfilling quality requirements. In other words, the operational techniques and activities, such as the sampling inspection mentioned above, are used to fulfill the requirements for quality. The nature of this approach remains more or less detection and is considered a reactive downstream approach, i.e., correction only after problems occur.

### *2.5.2 Quality Assurance*

The QA is also part of QM, but it is focused on providing confidence that quality requirements (particular standards) will be fulfilled. QA pertains to all those planned and systematic actions necessary to provide adequate confidence that a product (e.g., CBU) will satisfy the requirements for quality. This is a fundamental shift in concept from the reactive downstream approach of QC by means of detection, to a proactive upstream approach that controls and manages the upstream activities to prevent problems from arising.

Components of QA include:

- Focus on outputs.
- Use of a standard as a benchmark of quality.
- Control of systems and processes.
- Focus on efficiency.
- Driven from the top.
- "Assures" quality.

"Quality" viewed as a destination is really QA, where it is aiming to meet a required standard, with no commitment to improvement. QA is more applicable when the output is a product. However, where the output is the delivery of a service, "quality" becomes an ongoing journey.

#### *2.5.3 Quality Improvement*

Quality improvement (QI) is another part of QM that is often called *continuous improvement process* (*CIP*), in order to characterize the ongoing effort to improve *efficiency* (by identifying critical points and controlling for variation) of products, services, or processes by maximizing the use of resources, and to improve *effectiveness* (by identifying individual needs, and developing appropriate responses to achieve improved outcomes).[2] It is not connected with corrective errors but is concerned with doing things better to improve system efficiency and effectiveness. Delivery processes are constantly evaluated and improved in the light of their efficiency, effectiveness, and flexibility.

The concept of QI needs to be embedded in all levels of an organization/institute, and drive decision-making and resource allocation. It needs to become part of the QM of the organization/institute. It is a way of operating. A good starting point is to include QI as a standing agenda item on all management committee meetings and team meetings.

In health-care the CIP has taken on the acronym "FOCUS-PDCA:"

- *Find* a process to improve.
- *Organize* to improve a process.
- *Clarify* what is known.
- *Understand* variation.
- *Select* a process improvement.

Then move through the process improvement plan:

- *Plan*: create a timeline, including all resources, activities, dates, and personnel training.
- *Do*: implement the plan and collect data.
- *Check*: analyze the results of the plan.
- *Act*: act on what was learned and determine the next steps.

The "FOCUS-PDCA" acronym is an easy system for management to communicate to teams, and it helps them stay organized and on track with the end result in mind. The system has proven to be very successful for the CIP team approach.

## 3. QM SYSTEMS

### 3.1 Definition

The definition of a quality management system (QMS) is evolving into a definition of *good management*. A QMS can be seen as a complex system consisting of all the parts and components of an organization dealing with the quality of processes and products. There are many definitions of a QMS, but most definitions do not provide any more information than the words "quality management system-program." It is not an addition to an organization. It is an integral part of its management and production.

A QMS can be defined as:

- "A set of coordinated activities to direct and control an organization in order to continually improve the effectiveness and efficiency of its performance."[3]
- "The managing structure, responsibilities, procedures, processes, and management resources to implement the principles and action lines needed to achieve the quality objectives of an organization."
- "A management technique used to communicate to staff what is required to produce the desired quality of products and services, and to influence staff's actions to complete tasks according to the quality specifications."

A QMS would not in itself solve an organization's problems, but it does significantly increase the chances of identifying and removing the causes of error and waste, and thence of improving processes and information. A good QMS does not in itself make an organization more profitable and efficient but it will give an organization the ability to do anything better, from production to sales.

### 3.2 Benefits and Risks of Implementing a Quality Management System

The organization shall establish, document, implement, and maintain a QMS and continually improve its effectiveness. In "good reasons for implementing a QMS," a number of "direct benefits" of a QMS are stated:

- Improved "customer" satisfaction.
- Improved quality of products and services.
- Staff's satisfaction and more commitment to the organization.
- Better management and a more effective organization.
- Improved relations with suppliers.
- Improved promotion of corporate image.

Besides these direct benefits, there are also several "indirect benefits" to identify, which give opportunities to:

- Review business goals, and assess how well organization is meeting those goals.
- Identify processes that are unnecessary or inefficient, and then remove or improve them.
- Review the organizational structure, clarifying managerial responsibilities.
- Improve internal communication, and business and process interfaces.
- Improve staff morale by identifying the importance of their output to the business, and by involving them in the review and improvement of their work.

The main risks of implementing and maintaining a QMS are now well known. Although they cannot necessarily be eliminated they can be managed and their impact reduced. Briefly, the main identified risks include the following:

- Short-term increase in production costs during training and implementation of the QMS.
- Dissatisfaction of staff because of new procedures-methodology (e.g., resistance to change and perceived risk of "exposure").
- Another set of policies and documents without actual results (e.g., documents that reflect what management think is happening, not what is really happening).
- No improvement of the quality level in the final product (additional bureaucratic effort with no gain).

## 3.3 Elements of a Quality Management System

There are several elements to a QMS and each organization is going to have a unique system. The most important elements of a QMS are the following:

### 3.3.1 Plan for the Development of the Quality Management System—"Vision and Values"

The starting point for the management and leadership is the formation of a well-defined vision and value statement, which will be used to establish the importance of the QMS, in order to build motivation for the changes that need to take place. The exact form of the vision and values is not as important as the fact that it is articulated and known by everyone involved. This vision and value statement is going to be a driving force to help mold the culture that is needed throughout the organization in the drive for quality. It is not the words of the value statement that produce quality products and services; it is the people and processes that determine if there is going to be a change in quality. The vision and value goals will be very important statements to set agendas for all other processes used to manage the quality system.

Although the plan for the QMS is usually different for every organization, there are similar characteristics, which must include the following:

- Clear and measurable goals.
- Financial resources available for quality.
- QMS consistent with the organization's vision and values.
- A flexible Quality Management Plan (QMP) for employee improvement, because, as has been demonstrated, the most successful QMS allow employees at all levels to provide input.

The plan for the quality system might also include pilot projects that would entail setting up small quality projects within the organization. This will allow management to understand how well the quality system is accepted, learn from mistakes, and have greater confidence in launching an organization-wide quality system. The plan should provide some flexibility for employee empowerment, because, as has been demonstrated, the most successful quality systems allow employees at all levels to provide input.

### 3.3.2 Organizational Structure

An organizational structure can be considered as the viewing glass or perspective through which individuals see the activities (task allocation, coordination, and supervision) of their organization, which are directed toward the achievement of organizational aims.[4] An organization can be structured in many different ways, depending on their objectives. The structure of an organization will determine the methods by which it operates and performs.

Organizational structure allows the expressed allocation of responsibilities for different functions and processes to different entities such as the branch, department, section, workgroup, and individual. Furthermore, it determines which individuals get to participate in which decision-making processes, and thus to what extent their views shape the organization's actions.[4–7]

Organizational structure may not coincide with facts evolving in operational action, and such divergence decreases performance when growing. Thus, wrong organizational structure may hamper cooperation and thus hinder the completion of orders in due time and within limits of resources and budgets. Organizational structure should be adaptive to process requirements, aiming to optimize the ratio of effort and input to output.

### 3.3.3 Data Management

In an increasing number of scientific disciplines, large data collections are emerging as important community resources.[8] The definition provided in *Data Management (DAMA)—Data Management Body of Knowledge*

(*DAMA-DMBOK*) is: "Data management is the development, execution and supervision of plans, policies, programs and practices that control, protect, deliver and enhance the value of data and information assets."[9]

The combination of large data set size, geographic distribution of users and resources, and computationally intensive analysis results in complex and stringent performance demands that are not satisfied by any existing data management infrastructure. A large scientific collaboration may generate many queries, each involving access to—or supercomputer-class computations on—gigabytes or terabytes of data. Efficient and reliable execution of these queries may require careful management of terabyte caches, gigabits per second data transfer over wide area networks, coscheduling of data transfers and supercomputer computation, accurate performance estimations to guide the selection of data set replications, and other advanced techniques that collectively maximize use of scarce storage, networking, and computing resources.[8]

CBB information technology has been touted as a promising strategy for preventing errors.[10] For example, computerized order entry by the CBB Laboratory Manager or/and the Technical Supervisor has been shown to reduce the incidence of serious reagents' errors by 55%. Bar-code verification technology, ubiquitous in industries outside the field of health-care, is another example. Thus, at the processing laboratory, the use of bar-code technology to verify a CBU's identity is an important strategy for preventing errors.[11] Bar-code verification of supplies–reagents at the processing laboratory is usually implemented in conjunction with an electronic administration system (eSAR), allowing staff to automatically document the administration of supplies and reagents by means of bar-code scanning. Because the eSAR imports electronically orders of supplies and reagents from either the Laboratory Manager/Technical Supervisor or the CBB staff, its implementation may reduce errors. Given its potential to improve safety, bar-code eMAR technology is being considered as a criterion for achieving "meaningful use" of health information technology in CBBs.

### 3.3.4 Statistics

Statistical analysis is a very important measurement portion of quality systems which allow it to be managed. It could be considered a cornerstone of the quality improvement process and is very closely tied to auditing a quality system. A very common saying in management, which relates well to quality, is "you cannot manage what you cannot measure," and statistical analysis will provide the organization with the necessary measurements to make management decisions.

### 3.3.5 Control Charts

Control charts are the most widely used tool in quality systems. Control charts communicate a lot of information effectively.

The exact use of statistical measures is going to be different for each organization/institute. Some statistical analysis will be very easy to set up and use. For example, the length or weight of a particular part can be measured and analysis can show if the parts are within the required specifications.

### 3.3.6 Education and Training

The education of the quality supervisor (QS) on quality issues should start with a general discussion of quality systems and the roles that quality supervisors (QSs) play in quality programs. More specifically, QSs must know how quality programs have affected their specific field in the past, and they should have an idea of what role quality programs play in the future of their industry. Management must also keep abreast of new developments in quality. The discussion of the roles that the QSs must play in a quality system is the most important aspect of their education. They must understand how employees view their actions or inactions, how their individual actions and jobs impact quality, and the overall importance of dedication to quality by them. The QSs must understand that without strong leadership and reinforcing dedication to quality, a quality program will not be meaningful.

The education of employees for a quality program will include a discussion of how these programs will affect their jobs on a daily basis. It should also include a brief overview of quality as well as the tools employees will use in order to ensure outputs and how their roles add to the overall quality goals of the organization.

### 3.3.7 Auditing

Audit is "the process of systematic examination of a QMS carried out by an internal or external quality auditor or an audit team. It is an important part of organization's QMS."[12] Traditionally, audits were mainly associated with gaining information about financial systems and the financial records of a company or a business.

Auditing includes a systematic and independent examination of data, statements, records, operations, and performances of an organization/institute for a stated purpose. In any auditing the auditor perceives and recognizes the propositions before him for examination, collects evidence, evaluates the same, and on this basis formulates his judgment which is communicated through his audit report.

Audits are typically performed at predefined time intervals and ensure that the organization/institution has clearly defined internal system monitoring procedures linked to effective action. This can help to determine whether the organization/institute complies with the defined quality processes or results-based assessment criteria.

Auditing a QMS is just as important as any other aspect of the system. The audit process allows everyone involved to see if the QMS is working correctly and if the goals and objectives are being reached. Auditing also plays major roles in motivating employees and allows for rewards and acknowledgment measures to be assessed, as well as possible compensation.

To benefit an organization/institute, quality auditing should be performed. Quality audits are performed to verify conformance to standards through review of objective evidence. A system of quality audits may verify the effectiveness of a QMS. Quality audits are essential to verify the existence of objective evidence showing conformance to required process to assess how successfully processes have been implemented, and to judge the effectiveness of achieving any defined target levels. Quality audits are also necessary to provide evidence concerning reduction and elimination of problem areas, and they are a hands-on management tool for achieving continual improvement in an organization.

Auditing of QMS can take many forms, and each organization will have a unique auditing process that fits its system. Health-care organizations will have a very different auditing system than a manufacturing organization, but the end result of the systems is going to be the same.

### *3.3.8 Corrective and Preventive Actions*

*Corrective and preventive actions* are part of the overall QMS and include the systematic investigation of the root causes of nonconformities in an attempt to prevent their recurrence (for corrective action) or to prevent occurrence (for preventive action). Thus, their goal is the improvements to an organization's processes taken in order to eliminate the causes of nonconformities or other undesirable situations. To ensure that corrective and preventive actions are effective, the systematic investigation of the root causes of failure is pivotal.

### *3.3.9 Purchasing*

Purchasing is an area in an organization/institute where substantial gains in quality can be realized through the implementation of just a few policies and procedures designed around quality. Today's suppliers need to be partners in the quality effort. An organization's/institute's products or services are only as good as the combination of all the inputs.

Steps in molding the purchasing system include the following:

1. Collaboration with the entire quality system, in order to take all the standards developed for all incoming materials that can be qualified as an input to routine process or activity. If the quality system's performance standards and procedures are completed as described in the design phase these standards should already be established.
2. Education of the purchasing personnel on how the standards are important to the process flows of the organization. If standards are not upheld, the quality of the product or service will be jeopardized. The employees should be educated on how to measure and communicate the required standards. This may involve materials or statistical processes that control education, and it could even be as simple as cross-training the purchasing personnel so that they know exactly how the inputs fit into the organization. Once the purchasing area knows how the products are used and what problems can arise, they will have a better chance of procuring inputs that meet all the specifications.

Once steps one and two are complete it will be the purchasing department's responsibility to communicate the requirements to suppliers and hold them accountable for the quality. This sometimes may not be a simple task and could involve finding new suppliers or working with current suppliers to develop higher-quality standards.

### *3.3.10 Continual Improvement Process*

Leading and operating an organization successfully requires managing it in a systematic and visible manner. Success should result from implementing and maintaining a management system that is designed to continually improve the effectiveness and efficiency of the organization's performance by considering the needs of the interested parties.

The organization shall establish, document, implement, and maintain a QMS and continually improve its effectiveness in order to keep with and meet current quality levels, meet the customer's requirements for quality, retain staff through competitive compensation programs, and keep up with the latest technology.

The entire quality process (a dynamic concept, which uses a series of steps taken to complete a task), once started, will be an ongoing dynamic effort of the organization to improve products, services or processes. These efforts can seek "incremental" improvement over time or "breakthrough" improvement all at once.[13] The implementation of the successful quality system shall involve many different aspects that must be addressed on a continuous basis.

Among the most widely used tools for continuous improvement is a four-step quality model: the "Plan-Do-Check-Act (PDCA) cycle," also known as the *Deming Cycle* or *Shewhart Cycle*, which is briefly described below:[14]

- *Plan*: Identify an opportunity and plan for change.
- *Do*: Implement the change on a small scale.
- *Check*: Use data to analyze the results of the change and determine whether it made a difference.
- *Act*: If the change was successful, implement it on a wider scale and continuously assess your results. If the change did not work, begin the cycle again.

The "PDCA" cycle can be used in the following situations:

- As a model for continuous improvement.
- When starting a new improvement project.

- When developing a new or improved design of a process, product, or service.
- When defining a repetitive work process.
- When planning data collection and analysis in order to verify and prioritize problems or root causes.
- When implementing any change.

*William Edwards Deming*, a pioneer of the field (best known for the "Plan-Do-Check-Act" cycle popularly named after him), considered the "continuous improvement process" (abbreviated as "CIP" or "CI"), as part of the QMS whereby feedback from the process and customer were evaluated against organizational goals.[14]

### 3.3.11 Rewards and Acknowledgments

Rewards, compensation, and acknowledgment for achievements in quality are very effective ways to motivate employees and they may also be seen as a form of communication. They could be seen as tangible methods that a QMS of an organization/institute could use in order to let employees know that quality is important. This could come in the form of individual rewards or team rewards.

### 3.3.12 Communication

Communication:

- Is the vital link between management, employees, consumers, and stakeholders.
- Needs to allow employees to give feedback and provide possible solutions to issues the organization/institute must face. Thus, management needs to allow communication, in both formal and informal ways, such as employee feedback slips and feedback roundtable meetings.

## 3.4 Setting up a Quality Management System

The adoption of a QMS needs to be a strategic decision of an organization, and is influenced by varying needs, objectives, the products/services provided, the processes employed, the size and structure of the organization. The QMS of an organization can be related to the institutional program in several ways:

- Nested within the institutional QMS.
- Shared components with the institutional QMS.
- Completely separate from the QMS.

A QMS must ensure that the products/services conform to customer needs and expectations, and the objectives of the organization. Issues to be considered when setting up a QMS include the following:[3]

- *Design* and *build* must be included in the structure of the QMS, the process, and its implementation. Its design must suit the needs of the organization, and this is ideally done using a framework to lead the thinking. Furthermore, the design of the QMS should come from determining the organization's core processes and well-defined goals and strategies, and be linked to the needs of one or more stakeholders. The process for designing and building the QMS must also be clear, with the quality function playing a key role, but involvement and buy-in to the system must also come from all other functions.
- *Deployment* and *implementation* are best achieved using process packages, where each core process is broken down into subprocesses, and described by a combination of documentation, education, training, tools, systems, and metrics. Electronic deployment via intranets is increasingly being used.
- *Control* of the QMS will depend on the size and complexity of the organization. Local control, where possible, is effective, and good practice is found where key stakeholders are documented within the process and where the process owner is allowed to control all of the process. Ideally, process owners/operators are involved in writing procedures.
- *Measurement* is carried out to determine the effectiveness and efficiency of each process toward attaining its objectives. It should include the contribution of the QMS to the organization's goals; this could be achieved by measuring the following:
  - policy definition completeness;
  - coverage of business;
  - reflection of policies;
  - deployment;
  - usage;
  - whether staff find the QMS helpful in their work;
  - speed of change of the QMS; and
  - relevance of QMS architecture to the job in hand.

A form of scorecard deployed through the organization down to individual objective level can be employed, and the setting of targets at all levels is vital.

- *Review* of the effectiveness, efficiency, and capability of a QMS is vital, and the outcome of these reviews should be communicated to all employees. Reviewing and monitoring should be conducted whether or not improvement activities have achieved their expected outcomes.
- *Improvement* should follow as a result of the review process, with the aim of seeking internal best practice. It is part of the overall improvement activities and an integral part of managing change within the organization.

## 3.5 Total Quality Management

The "TQM" is a management approach in which quality is emphasized in every aspect of the organization/institute. Its goals are aimed at long-term development of quality products and services. TQM breaks down every process or activity and emphasizes that each contributes or detracts from the quality and productivity of the organization as a whole.

The role of the quality managers in TQM is to develop a quality strategy that is flexible enough to be adapted to every department, aligned with the organizational objectives, and based on stakeholder needs. Once the strategy is defined, it must be the motivating force to be deployed and communicated for it to be effective at all levels of the organization. Some degree of employee empowerment is also encompassed in the TQM strategy and this usually involves both departmental and cross-functional teams to develop strategies to solve quality problems and make suggestions for improvement.

# 4. CREATING A DOCUMENTED QUALITY MANAGEMENT SYSTEM (PROGRAM): QUALITY MANAGEMENT PLAN

## 4.1 Quality Management Plan: Definition, Purpose

The QMP is a written document that describes the systems in place to implement the QMS. The QMP shall include several categories of documents ("master documents") needed by the organization/institute to ensure the effective planning, operation, and control of its process. The main three categories of the master documents are mentioned below:

1. *Policies*: documents that define the scope of an organization/institute, explain how the goals of the organization will be achieved, and/or serve as a means by which authority can be delegated.
2. *Standards operating procedures (SOPs):* documents that describe in detail the process or chronological steps taken to accomplish specific tasks; a procedure is more specific than a policy.
3. *Supporting documentation:* worksheets, forms, and templates through which performance of a procedure is documented.

The purpose of the QMP is:

- To define and establish the structure of the QMS to be implemented in an organization/institute.
- To provide a description of the activities carried out from the organization/institute, according to the requirements of an appropriate accreditation body (e.g., NetCord-FACT).
- To improve standards in those areas where performance is wanting.
- To identify the organizational framework, policy, and procedures, functional responsibilities of management and staff, lines of authority, and its processes for planning, implementing, documenting, and assessing activities conducted within the organization/institute.
- To harmonize policies and practices across all departments.
- To create stability and minimize variances.
- To eliminate complexity and reduce processing time.
- To produce consistent quality products through a better understanding of the organization's/institute's structure and activities.
- To integrate essential systems that must be in place to meet specific objectives.
- To inform the personnel on the techniques and the procedures required in order to obtain, to maintain, and to improve adequate standards of quality.

## 4.2 Management Requirements of a Quality Management Plan

### *4.2.1 Structure of the Organization/Institute*

The QMP should define the following:

- The management structure of the organization/institute and its place in any larger organization, together with the relationships which exist between QM, technical operations, and support services. In addition, if the organization/institute is part of a larger organization, the responsibilities of any person in that organization who is involved in, or can influence, the work of the organization should be defined.
- Who is legally responsible for the work of the organization/institute in the event of legal action being taken (for example, by a "client"—transplant center). Legal responsibility might lie with, for example, the director of the organization, the head of the laboratory, or with individual scientists.
- The responsibility, authority, and interrelationships of all personnel who manage, perform, or verify work affecting the quality of the results.
- Which personnel have managerial and technical authority for QM, what their specific responsibilities are, nominated deputies, and what resources are available to them to carry out their duties. Thus, it should be included that a *quality supervisor* (*QS*) who, irrespective of other duties and responsibilities, has defined responsibility and authority for ensuring that the QMS is implemented and followed at all times; has direct access to the highest level of management at which decisions are made on organization policy or resources; maintains and updates the QMP; monitors laboratory practices; ensures the validation of new technical procedures; selects, trains, and evaluates internal auditors; recommends training to improve the quality of laboratory staff; and proposes improvements in the already existing quality system.

### 4.2.2 Document Control

All master documents of an organization/institute should be controlled from the Quality Management Unit (QMU) and archived documents, including SOPs, policies, reports, worksheets, protocols, validation/qualification plans/reports, agreements, informative materials, and labels should be maintained indefinitely. It is obvious that a QMS can only operate effectively if all the above-mentioned master documents are archived, kept up-to-date, and a procedure for their annual review is scheduled.

"Document control" is the mechanism by which the QMS master documents are created, amended, reviewed, approved, distributed, and archived to ensure that all staff uses the latest authorized versions. The quality supervisor should ensure that all aspects of the organization's/institute's work are documented in the QMS and that new documents are created by competent personnel and authorized by designated staff before being issued. The quality supervisor should also ensure that all master documents are subject to periodical review and, where necessary, revised to take account of changing circumstances and incorporate best practice. Once approved, the quality supervisor should arrange for documents to be made available to all relevant staff in their workplace. When changes have been made to existing documents, the quality manager should make sure that the changes are highlighted in the latest versions.

All QMS documents should be uniquely identified and should bear the name/signature of the authorizing person. Each page of a document should be individually numbered as "page x of y pages" and should include the unique QMS document identifier, date of issue, and version. This system for page identification minimizes the risk of undetected omission of current pages and undetected retention of obsolete pages. If documents are held and distributed electronically, they should be read-only versions which may only be edited by authorized staff. A master list bearing the date of issue and, where appropriate, a complete record of all versions, the dates on which they were made and the name of the authorizing person and distribution list should be maintained. This may be supplemented with a sheet attached to each document identifying the same details.

The organization/institute may allow minor changes to printed documents, such as correction of typing errors, to be handwritten on the document in permanent ink, dated and authorized. However, a revised version should be issued as soon as practicable. Invalid or obsolete documents should be promptly removed from all locations to prevent their accidental use. A copy of each obsolete document should be retained for either legal or knowledge preservation purposes and suitably marked (for example, "not valid").

### 4.2.3 Third-party Agreements

The organization/institute should have procedures to ensure that the requirements of its third party are adequately defined, documented, and understood, and that it has the capability of meeting these requirements before agreeing (making a contract) to do the work. The agreement may be written or oral, but if reached orally it should subsequently be documented. In case the organization/institute does not have the capability, it should attempt to reach agreement with the third party on what work it could carry out or would subcontract before any work commences. Any revised agreement should then be documented.

The agreement should be kept under review by the third parties and the organization/institute. If new requirements are requested by a third party, or the organization/institute is unable to meet the original agreement or the new requirements, this should be communicated between the parties, discussed, and a revised agreement reached. Records should also be kept of all communications between the third party and the organization/institute related to the work being carried out. This will ensure the common understanding of requirements, responsibilities, and work to be performed by the laboratory, its clients, and all other parties involved.

### 4.2.4 Subcontracting of Analytical Work

If the organization/institute uses another party ("subcontractor") to undertake work on its behalf, it should have documented policies and processes to ensure that the other party is competent to do the work.

### 4.2.5 Purchasing Services and Supplies

The organization/institute should have a policy and procedure(s) for the selection and purchase of services, reagents, and laboratory consumable materials which might affect the quality of its work, for the reception and storage of the reagents and consumables, and for ensuring that the services, reagents, and consumables comply with the technical specifications described in the QMS before they are purchased and used.

### 4.2.6 Complaints

The organization/institute should have a system for dealing effectively with complaints. This should include a requirement to inform staff and/or "clients" (e.g., transplant centers) of any actions taken to resolve the issue and prevent any recurrence. Records should be maintained of all complaints and corrective actions, and used as an opportunity to improve QM in the organization/institute.

Failure to handle complaints from staff and/or "clients" to their satisfaction can adversely affect the relationship between the organization/institute and the staff, and/or the "clients."

### 4.2.7 Corrective and Preventive Actions

When the work carried out by a section of the organization/institute is inconsistent with its QMS (for example, the work deviates from an operating procedure, or the requirements of the transplant center), this is termed "nonconformance." The organization/institute should have systems in place (for example, for checking work within the section, case file or QMS audit, staff/customer feedback) to recognize when nonconformance has occurred and how it should be managed. Implementation of corrective and preventive actions is the path toward improvement and effectiveness of QMS. Corrective actions are the action(s) taken based on the problem identification. The problem or a nonconformance can be identified internally through staff suggestions, management reviews, document reviews, or audits (internal or/and external). Every effort should be made to detect nonconformities before the work is released to the transplant center, as release of incorrect results can severely damage the organization–transplant center relationship and could lead to a miscarriage of justice. If the nonconformance could recur without wider action being taken (for example, amending the QMS or staff re-training), then preventive actions should be authorized at an appropriate level to prevent any recurrence.

All nonconformances, corrective actions, and preventive actions should be recorded. Corrective actions should be checked for their effectiveness by authorized staff to show that the work now conforms to the requirements of the QMS/transplant center. Preventive actions should also be monitored to ensure that the nonconformances are not recurring (for example, through audit).

The recognition of nonconformances and the implementation of corrective and preventive actions are essential elements in continuously improving the effectiveness of the organization's performance.

### *4.2.8 Control of Records/Chain of Custody*

The organization/institute should have systems in place for the creation, identification, management, storage, movement/transmission, retrieval, and disposal of all records, in both paper and electronic formats. All paper records should be easily readable, uniquely identifiable (for example, with the date, author, and page number) and made in a permanent medium such as ink. Pencil should not be used. No record should be deleted. Alterations and correction of mistakes which have been made by hand should not obscure the original record and should be signed/initialed and dated. The organization/institute should also have measures in place to safeguard original electronic records (for example, by creating backups of computer files), to identify any alterations to them (for example, through the electronic audit trails some manufacturers provide in their software) and to ensure their integrity and confidentiality. All records should be filed systematically to make retrieval easy. At the same time, they should be treated as confidential, the data protection and legislation requirements for the rights of an individual to privacy, etc., should be observed and access should be restricted to authorized personnel.

Records may be subdivided into "quality" and "technical" records.

- "Quality" records include audit reports, proficiency tests, transplant center(s) feedback, corrective and preventive actions, and management reviews. The records should be uniquely identifiable (for example, with the date, author, etc.) and held in a safe and secure location which is accessible to relevant staff. The organization/institute should have a policy for how long these records should be kept (for example, based on legal requirements). Records for disposal should be treated as confidential waste and incinerated or shredded.
- "Technical" records include all materials relating to cases, including sample submission forms, chain of custody documents, case notes (including drawings and diagrams), records of telephone conversations, spectra, calibration and other QC data, instrumental operating parameters and printouts, reports, statements, etc., instrument maintenance records, and staff training, competency and authorization records. They should be made at the time the work is done.

Each entry of every record should be traceable to the analyst/examiner and, where appropriate, to a uniquely identified case or exhibit. It should be clear from the case record who has performed all stages of the analysis/examination and when each stage of the analysis/examination was performed (for example, the relevant dates). All records should contain sufficient information to allow an audit trail to be established showing who did what work and how and when it was done.

Critical findings (for example, calculations and data transfers which do not form part of a validated electronic process) should be checked, preferably by a second authorized person. The organization/institute should have documented procedures for the overall review of case records by authorized persons. The case record should include an indication that such checks and reviews have been carried out, when and by whom. This may be indicated in a number of ways (for example, by entries against each finding, entry on a summary of findings, or a statement to this effect in the records). If the checker or case reviewer disagrees on any point in the initial record, the reason(s) for the disagreement and any action taken as a result should be recorded.

In general, the records required to support conclusions should be such that, in the absence of the analyst/examiner, another competent analyst/examiner could evaluate what had been performed, interpret the data and, where appropriate, repeat the work. When a test result or observation is rejected, the reason(s) should be recorded. This information is necessary for another analyst to understand how the case was managed.

Technical records should be kept in a safe and secure place, to prevent damage, deterioration, unauthorized access or loss, for a period depending on the needs of the organization/institute. The length of retention may also be defined in a policy of the organization/institute. This could result in donor records being retained indefinitely, being destroyed, or some intermediate arrangement. Records for disposal should be treated as confidential waste and incinerated or shredded.

### 4.2.9 Internal Audit

The organization/institute should have a schedule and procedure for periodic audit of all elements of its activities to verify that its operations comply with the requirements of its QMS. The audits should be organized by the quality supervisor at least once per year and carried out and reported by trained, qualified personnel who are, whenever resources permit, independent of the activity to be audited. When audit findings cast doubt on the effectiveness of the operations or on the correctness or validity of the laboratory's work, the findings should be discussed by the auditor, the quality manager and relevant staff, and appropriate corrective actions agreed.

The area of activity audited, the audit findings and corrective actions should be recorded. Future audits of the same activity should record how effective the corrective actions have been in order to identify lessons learned and consequently improve the organization's performance.

### 4.2.10 Management Reviews

The quality supervisor should organize an annual review of the organization's/institute's QMS to ensure that it remains suitable and effective or to identify any changes and improvements required. The review should be carried out by high level management of the organization/institute, the quality supervisor, and other relevant staff. The review and any recommended actions should be recorded and the actions carried out within an appropriate timescale, agreed between the reviewers and the quality supervisor, taking into account availability of resources.

The management review should cover the following areas:

- QC activities, resources and staff training, and other relevant factors.
- The suitability of policies and procedures.
- Reports from managers and supervisors.
- The outcome of recent internal audits.
- Assessments of external bodies.
- Feedback from transplant center(s).
- Complaints.
- The results of interlaboratory comparisons or proficiency tests (in the case of CBBs).
- Corrective and preventive actions.
- Recommendations for improvement.

## 4.3 Technical Requirements of a Quality Management Plan

Various factors, such as personnel, facilities, health and safety, test methods, method validation and procedures, equipments, reference standards, materials and reagents, handling of test items, reporting the results, QC and proficiency testing contribute toward the accuracy and reliability of results and also determine to a large extent the uncertainty of the measurement. These factors should all be taken into account when developing methods and procedures, in the training of personnel, and in the choice and use of equipment.

### 4.3.1 Personnel

The organization/institute should foster an atmosphere wherein employees are encouraged to improve their knowledge and skills, to grow as individuals, and to fully develop their potential. It should only use personnel with whom the organization/institute has an employment contract, including temporary staff.

The organization/institute should ensure that the staff has the appropriate education, training, experience, knowledge, skills and abilities (i.e., competence) to do the work assigned, is appropriately supervised, and works in accordance with organization's/institute's QMS. It should have in place policies and procedures for identifying training needs and providing training for staff to help them achieve and maintain competence (for example, through structured on-the-job training programs, participation in scientific meetings, conferences and workshops, technical training courses, instrument operation and maintenance courses taught by vendors, in-house technical meetings, courses, seminars, and further education). While staff is being trained they should be more closely supervised and the effectiveness of their training should be monitored and evaluated. Where test- or technique-specific training is given, acceptance criteria should be assigned (for example, observation of the relevant tests or analyses by an experienced officer, or their satisfactory performance in the analysis of QC/QA samples and correlation of results with those obtained by other trained staff).

Records should be maintained for each member of staff of their education, qualifications and training, together with a list of tasks they are competent to perform and authorized to carry out (for example, to perform particular types of test, to issue test reports, to give opinions and interpretations, and to operate particular types of equipment). This information should be readily available to staff and should include the date on which competence and authorization are confirmed, so all staff has a clear understanding of the scope of their assigned tasks and their responsibilities. Each member of staff should have a current job description which they have agreed with their designated line manager. This should include their responsibilities, duties, and required competencies.

### 4.3.2 Facilities: Accommodation and Environmental Conditions

The organization/institute premises and environmental conditions should permit the work to be carried out to the required quality standards. Thus, for example, in the processing facility of a CBB the following factors should be considered: space, security, health and safety of staff, temperature and humidity control, lighting, air flow and ventilation, in addition to the provision of basic laboratory facilities (for example, electricity, gas, water, telephone and computer connections, laboratory benches, safety cabinets, refrigerators and freezers). Furthermore, processing facilities should enable correct performance of the work, and where specific environmental conditions are critical they should be specified, documented, and monitored (for example, storage temperatures of samples). Environmentally sensitive equipment should be in low-access areas and microbalances should be protected from vibration and chemical corrosion.

Adequate and appropriate space should be allocated for each activity/function and employee. Suitable accommodation is required for the storage of evidential materials to prevent loss, deterioration, or contamination and so maintain the integrity and identity of the evidence, both before and after it has been examined.

The areas of the processing facility should be sufficiently clean and tidy to minimize the risk of contamination and ensure that the quality of the work carried out is not compromised. There should thus be effective spatial separation of incompatible activities (for example, the processing of CB and DNA extraction for human leukocyte antigen typing) which should not be carried out using the same facilities. Measures should also be taken to prevent cross-contamination (for example, between two different bulk seizures or between reference materials and case samples). These measures might include control of staff movement, flow of samples, and sharing of equipment (for example, opened biologically hazardous specimens and dirty glassware should not be transported through unprotected areas).

Processing facility and evidence storage areas should be kept secure at all times to prevent theft or interference and there should be limited, controlled access. There should be controls at both entry and exit points of buildings and between different secure areas (for example, by use of keys or magnetic cards distributed to authorized personnel for specific purposes). In this way, no unauthorized personnel will be able to handle samples or gain access to restricted areas where reagents, specimens, or records are stored.

The organization/institute should hold on record a list of all staff that is authorized to enter the secure areas. This list should be reviewed and updated on a regular basis. Unauthorized persons needing to enter secure areas (for example, other laboratory personnel, clients, service engineers, cleaners, administrative personnel, and visitors) should be escorted at all times by authorized personnel and a record of these entries should be maintained.

### 4.3.3 Health and Safety

The organization/institute should have a safety manual containing procedures that address issues affecting the health and safety of the staff and which are designed to safeguard employees from service-related injury and health problems. These should be based on risk assessments of all activities and documented safe systems of work (for example, the need to handle hazardous chemicals).

Details of the following should be contained in the safety manual:

- Designated staff with responsibility for aspects of safety (for example, safety officer, biological safety officer, fire safety officer, first aid staff). These responsibilities can be held by several people or by a single member of staff.
- Emergency procedures and contact information (for example, what to do in the event of fire, chemical spillages, personal injury).
- Staff training (for example, fire drills and first aid).

Accommodation (for example, hand wash facilities, emergency showers, first aid cabinets, eye wash bottles, safety cabinets/exhaust hoods, autoclaves, fire extinguishers, solvent and chemical stores, disposal facilities for waste, chemicals, sharps and radioactive material, safety signs/hazard warnings, notices for fire exits and location of safety equipment, and emergency telephone numbers).

Personal protective equipment (PPE) (for example, laboratory coats, disposable gloves, goggles/safety glasses, face protectors, ear protectors, and radiation safety badges).

- General laboratory hygiene/safety (for example, cleaning and disinfection of surfaces, autoclaving of biologically contaminated equipment, the wearing of PPE, prohibition of eating, drinking and smoking in the laboratory, prohibition of laboratory clothing in designated clean areas, and prohibition of lone working in laboratories).
- Specific biological hazards (for example, the use of microbiological safety cabinets, immunization of staff, safe disposal of clinical waste, sterilization of equipment, and PPE).
- Radioactivity hazards (for example, reference to relevant regulations dealing with the use of radioactive materials).

### 4.3.4 Test Methods, Method Validation, and Procedures

The laboratory should use appropriate methods and procedures for all of its work: sampling, handling, transport and storage of evidence, use of equipment, testing, evaluation and interpretation of results, and reporting. The methods and procedures should be up-to-date, fully documented, and readily available to relevant personnel.

The documentation of methods should record the name/reference number of the method, the scope of the method, the principle of the method; a summary of the validation parameters; the apparatus and equipment required, including technical specifications; the reference materials required; the environmental conditions required (for example, room temperature); and a step-by-step description of the procedure, including the following:

- Special precautions which should be observed (for example, health and safety issues).
- Requirements for sampling, labeling, packaging, transporting, and storing of samples.
- Preparation of samples, reference materials, controls, and calibrators for analysis.
- Requirements for equipment checks and calibration (for example, running a QC sample, tuning and calibration of a mass spectrometer).
- Analysis process/test procedure and QC (for example, use of blanks, controls, and calibrators).
- Recording and processing of results (for example, calculations, preparation of calibration curves and charts), including the criteria and/or requirements for acceptance/rejection (for example, if results are outside the calibration range or QCs give unacceptable results).
- Requirements for the reporting of results and the uncertainty of the method.

Any departure from these methods and procedures should only be allowed if they are justified, authorized, agreed with the client, where appropriate, and documented.

### 4.3.5 Equipment

Laboratories should ensure the reliability and performance of the equipment used. Equipment and software required for the work carried out should be fit for its intended purpose and should preferably be available within the laboratory. If equipment outside the laboratory is used, it must comply with the QMS standards. An inventory of equipment should be maintained along with records of their location, date of purchase, service history, and maintenance. Major items of equipment (for example, instruments such as spectrometers) should have their own logbook for recording this information kept nearby. Staff should be trained to use the equipment and only be authorized to use it when they have been found to be competent. Training and authorization should be documented in staff records.

When equipment is procured, the manufacturer's specifications should match or exceed the laboratory's requirements. It should be checked on installation to ensure that it meets the manufacturer's specifications (performance verification, usually carried out by the supplier's installation engineer, also known as the "before use check" or "equipment validation"). If an instrument is subsequently moved, the installation check should be repeated and its performance certified if necessary (for example, a balance will usually need to have its calibration checked if it is moved to a new location).

Equipment should also be checked regularly when it is in use, using documented procedures to show that the performance continues to be acceptable (for example, an instrument might be checked before each set of samples is analyzed to ensure that it is working properly). This might include checking temperatures, gas pressures, tuning, calibration, etc., depending on the instrument involved. Test samples can also be analyzed for checking purposes. Appropriate corrective action is taken when necessary. Up-to-date instructions on the use and maintenance of equipment (for example, abbreviated operating instruction sheets prepared in-house as well as instruction manuals supplied by the manufacturer) should be readily available, preferably adjacent to the instrument, for use by the appropriate laboratory personnel.

Equipment having a significant influence on the accuracy of test results should be calibrated according to a schedule, using documented procedures which are available to authorized users. Equipment should be labeled to indicate the status of calibration (for example, the date when last calibrated and the date or expiration criteria when recalibration is due) to ensure that it is not confused with uncalibrated equipment.

"Critical equipment" (e.g., "critical equipment" for CBBs is any piece of equipment used for the processing, cryopreservation, storage, and testing of CB that might adversely affect the quality of the CBU or lead into erroneous donor eligibility determination) (for example, balances, devices for CBU volume reduction) should be uniquely identified and calibration records, including certificates if available, should be kept. Staff members responsible for equipment should ensure that regular documented check samples, calibrators, and blanks are analyzed and that the performance specifications are being maintained. Records should be kept of all calibration, maintenance, and servicing, whether by in-house personnel or an outside agency. These should include the unique identifier of the item of equipment and data system (if any), the manufacturer, model, serial number, the location (if appropriate), performance checks, operating instructions prepared in-house, the manual supplied by the manufacturer or reference to its location, dates, results, copies of reports, certificates of all calibrations, acceptance criteria relating to performance, the due date of the next calibration, the maintenance schedule (where appropriate), maintenance carried out to date, any damage, faults, and modification or repair to the equipment.

Identified faults should be brought to the attention of the person responsible for the equipment and for taking corrective actions. If the fault is critical (for example, if the light source of a microscope is unstable or a vacuum pump is not holding a sufficient vacuum pressure), the equipment should be withdrawn from service until the problem has been rectified, and the manner in which the problem is resolved, and the time and date should be recorded in the logbook for the equipment. Similarly, equipment that has been operated incorrectly, such that it no longer gives reliable results or has been found to be malfunctioning, should be taken out of service. It should be isolated to prevent its use or clearly labeled or marked as being out of service until it has been repaired and shown by calibration or test to be working properly. Analytical results that were obtained during the period of equipment malfunction should be checked and the procedure for control of nonconformities should be initiated.

### 4.3.6 Reference Samples, Materials, and Reagents

Reference samples, materials, and reagents should be adequate for the procedure used and they should comply with the quality specifications of the method. Lot/batch numbers of reference samples and critical reagents should be recorded and they should be tested for their reliability and should be labeled with: name, concentration (where appropriate), preparation date and/or expiry date, identity of preparer, storage conditions, if relevant biohazard warning, where necessary.

Reference samples, reagents, and other materials should be correctly stored to ensure their stability and integrity. An expiry date furnished by a vendor/manufacturer determines the useful lifetime of the sample/reagent/material unless it can be verified beyond that date. Where appropriate, "opened" dates and "use by" dates should also be recorded. It is important to note that problems with standards and reagents can arise after they have been received in the laboratory.

"Certified reference materials" (CRMs) can be used if available.[4] A CRM is a reference sample, usually obtained commercially, for which the analyte concentration(s) has been certified by analysis, accompanied by, or traceable to, a certificate or other documentation that is issued by a certifying body. These commercial reference samples are supplied with a description of their chemical identity, purity, and concentration (for example, a "specification and certificate of analysis"). However, it is recommended that the laboratory should independently verify their identity and purity (or concentration) before putting them to use (for example, by interlaboratory comparisons or within the laboratory using a previously used reference sample).

The reference samples required for each procedure/method used by the laboratory should be documented and should be available in the laboratory. They should be appropriate for the test being performed. A record should be maintained describing their source, the date of acquisition, and the quantity held in the laboratory. When reference samples need to be imported or exported, import/export certificates should be obtained from the relevant national competent authority.[15]

The responsibility for acquiring and maintaining reference standards should be given to a designated person, who should keep a central register of these materials. The register should include all official reference substances and reference preparations.

### 4.3.7 Handling of Test Items

The organization/institute (e.g., a CBB) should have procedures for the collection of CBUs and these procedures should be available at the collection sites where the CBUs are collected. On the other hand, the sampling procedure should ensure that the portion taken for analysis is representative of the whole. The sampling plan and/or sampling procedure, the identification of the person taking the sample and the environmental conditions, if relevant, should be recorded. The sampling plan should be practical and easy to carry out by nonscientists, avoid unnecessary additional work for the laboratory, be easy to explain, and be defensible in court.

The QMS should provide guidelines to ensure that the materials to be tested are properly sampled, labeled, packaged, preserved, and stored. The labeling should be sufficient to allow unique identification of the samples and subsamples and their relationship with their original source. It is important that the packaging at the collection sites will prevent unauthorized access to and loss or contamination of the collected CBUs during transportation. At the processing laboratory, an authorized person should receive and carefully check the samples and documents. One or more identified individuals should also be authorized to reject CBU collections that do not accord with the CBB's acceptance policies. They should inform the laboratory manager of any such rejections. Any remedial action taken should always be documented.

The QMS should have an effective documented system for the secure storage of collected CBUs and accompanying samples, both before and after examination, correlating the CBUs and the samples to other information provided with them (for example, the signed informed consent), identifying any subsamples prepared from the CBUs and their samples, and showing the progress of analysis, date of issue of the report of analysis, and the date and subsequent means of disposal of any remaining sample after analysis. The system should be designed and operated to ensure that CBUs and samples cannot be confused physically or when referred to in records or other documents. The processing laboratory should also have documented procedures and appropriate facilities to minimize deterioration and avoid loss or damage to the CBUs and samples during storage, handling, and analysis.

There should be documented procedures for the disposal of CBUs which comply with the applicable laws and procedures of the jurisdiction. Records should be kept of each disposal.

### 4.3.8 Reporting the Results

The results of analyses carried out by the processing laboratory and the reference laboratories should be reported accurately, clearly, unambiguously, and objectively, and meet the quality requirements of the CBB. The report format should be designed to accommodate each type of analysis carried out and to minimize the possibility of misunderstanding or misuse. Where required and appropriate, an interpretation of the significance of the analysis results in the context of the case should also be provided.

### 4.3.9 QC, Proficiency Testing, and Interlaboratory Comparisons

There should be an appropriate level of QC for each analysis. QC check samples should be analyzed by the defined procedures, at the required frequency. Where control charts are used, a record should be kept of performance outside the acceptable criteria. The processing laboratory should participate in proficiency testing and interlaboratory comparisons.

Where deficiencies or opportunities for improvement are identified there should be a process for taking appropriate action. Improvement and corrective actions should be recorded. There should also be an effective system for linking proficiency testing performance with day-to-day QC.

## 5. STANDARDIZED SYSTEMS

### 5.1 International Organization for Standardization

The ISO, established in 1947 and based in Geneva, Switzerland, prescribes procedures controlling the basic process whereby an ISO International Standard is updated and released. The ISO is a worldwide federation of national standards bodies, which are responsible for standards in some 155 countries, many of which are government organizations. The objective of the ISO is to promote the development of standardization and related activities globally with a view to facilitating international exchange of goods and services, and to developing cooperation in the spheres of intellectual, scientific, technological, and economic activity.

The ISO is an excellent tool for allowing an organization/institute to have a set of standards (serve the purpose to assuring the quality of the products) and a process which describes how it will meet those standards. The organization/institute is audited and certified as to the effectiveness of its process.

The work of preparing international standards is normally carried out through ISO specialized technical committees (TCs). ISO/TC 176—QM and QA, the secretariat of which is held by the Standards Council of Canada, is the ISO TC responsible for developing and maintaining a universally accepted set of QM standards.

### 5.2 The ISO 9000 QM system

The ISO[16] developed a series of standards for service industries (ISO 9000) that has been used to assess quality systems in specific aspects of health services. Since these standards largely relate to administrative procedures rather than to clinical results, ISO standards have been used more frequently in mechanical departments such as laboratories (EN 45001), radiology, and transport, but have also been applied to whole hospitals and clinics. The ISO has also published a glossary of QM in ISO 8402.

The ISO 9000 QMS, as it became known, was first published in 1987 but it was not until 1994 that the first revisions were published. The reason was that management systems were new to many of the organizations engaged in establishing quality systems on the basis of the ISO 9000 standards. In this situation, ISO/TC 176 believed that making major changes in the standards could run the risk of disrupting such efforts. Consequently, the 1994 revisions were relatively minor and mostly related to the removal of internal inconsistencies.

There are many reasons why a new series of the standards was published in 2000. First, ISO International Standards have a normal review cycle of five years. Secondly, the user community requested it. The year 2000 revisions represented a thorough overhaul of the standards to take into account developments in the field of quality and the considerable body of experience that had built up as a result of implementing ISO 9000. The users demanded a process-oriented approach and a defined route for performance improvement. A revised version of the ISO 9000 series of standards, issued in 2000, is moving closer to the development model of the European Foundation for Quality Management (EFQM).[15] A further updatable ISO 9000 series of standards was introduced by 2008 in order to provide clarification and to enhance compatibility with the ISO 1400:2004 standards. This update included ISO 9000:2005 containing the fundamentals and vocabulary, and ISO 9001:2008 providing the requirements.

In order to implement ISO 9000 QMS successfully, there are 14 essential steps, which must be followed. These steps are the following: top management commitment, establishment of an implementation team, awareness programs for ISO 9000, staff training, initial status survey should be conducted, creation of a documented implementation plan, development of a QMS documentation, document control, implementation, internal quality audit, management review, preassessment audit, certification and registration, continual improvement.

In each country, a national body tests and recognizes—i.e., accredits—independent agencies as competent to certify organizations that comply with the standards. The audit process tests compliance with standards and is not intended in itself to be a tool for organizational development.

## 6. ACCREDITATION

### 6.1 Accreditation: An Overview

*Accreditation* is an evaluation process used to assess the quality of the health-care services provided by the entity seeking accreditation. Specifically, accreditation is the process designed to improve the quality, efficiency, effectiveness of a health-care organization and the products or services it provides, including its structures, processes, and outcomes. It is based on the premise that adherence to evidence-based standards will reliably produce higher-quality health products and services in a safer environment than would be the case without such standards.[17] Furthermore, accreditation is a process in which certification of competency, authority, or credibility is presented.

The term "accreditation" (applied to organizations rather than specialty clinical training) reflects the origins of systematic assessment of hospitals against explicit standards; it began in the United States in 1917 as a mechanism for recognition of training posts in surgery. This independent, voluntary program developed from a focus on training into multidisciplinary assessments of health-care functions, organizations, and networks.[18]

The Joint Commission model spread first to other English-speaking countries and Europe, Latin America, Africa, and the Western Pacific during the 1990s. Mandatory programs have been adopted in France, Italy, and Scotland. At least 28 countries now have an operational accreditation program. Several of the established programs provide development support to other countries, but only *Joint Commission International (JCI)* also offers external accreditation. An international taskforce developed the JCI standards for application worldwide.

Accreditation can serve as a risk mitigation strategy, and it can also measure performance; it provides key stakeholders with an unbiased, objective, and third-party review. It can constitute a management tool for diagnosing strengths and areas for improvement, as well as for facilitating the merger of health-care organizations by stimulating the emergence of common organizational identity, culture, and practices.[19,20]

Organizations that participate in accreditation confirm their commitment to quality improvement, risk mitigation, patient safety, improved efficiency, and accountability; it sends a powerful message to key decision-makers and the public. This performance measure contributes to the sustainability of the health-care system.[17] Finally, accreditation or certification of health-care organizations should be strongly encouraged with incentives, or indeed made mandatory, but choice of accreditation/certification/award approaches should be allowed.[17] Whatever the mechanisms adopted, accreditation of health-care organizations should require processes for continuous improvement, and achievement of quality enhancement outcomes.

### 6.2 The Benefits of Accreditation

Accreditation organizations are uniquely positioned to provide a comprehensive look at the challenges and successes of health-care organizations, and to identify themes and trends in the delivery of health-care services. The data collected through accreditation is an invaluable resource for health-care providers, governments, policy-makers, and other health-care leadership organizations, as it can contribute to effective and efficient decision-making to ensure ongoing quality improvement and reduce costs through risk mitigation.[17]

In the current era of heightened fiscal responsibility, transparency, accountability, and escalating health-care complexity and risk, accreditation contributes to ensuring that care meets the highest standards of health-care decision-making and provision.[17] Furthermore, "accreditation:"

1. *Provides a framework to help create and implement systems and processes that improve operational effectiveness and advance positive health outcomes.*[21]
2. *Improves communication and collaboration internally and with external stakeholders.*[22]
3. *Strengthens interdisciplinary team effectiveness.*[23]
4. *Demonstrates credibility and a commitment to quality and accountability.*[24]
5. *Decreases liability costs; identifies areas for additional funding for health-care organizations and provides a platform for negotiating this funding.*[20]
6. *Mitigates the risk of adverse events.*[24]
7. *Sustains improvements in quality and organizational performance.*[25]
8. *Supports the efficient and effective use of resources in health-care services.*[26]
9. *Enables ongoing self-analysis of performance in relation to standards.*[27]

10. *Ensures an acceptable level of quality among health-care providers.*[28]
11. *Enhances the organization's understanding of the continuum of care.*
12. *Improves the organization's reputation among end-users and enhances their awareness and perception of quality care, as well as their overall satisfaction level.*[22,29]
13. *Promotes capacity-building, professional development, and organizational learning.*[30]
14. *Codifies policies and procedures.*[31]
15. *Promotes the use of ethical frameworks.*
16. *Drives compliance with medication reconciliation.*[32]
17. *Decreases variances in practice among health-care providers and decision-makers.*
18. *Provides health-care organizations with a well-defined vision for sustainable quality improvement initiatives.*[33]
19. *Stimulates sustainable quality improvement efforts and continuously raises the bar with regard to quality improvement initiatives, policies, and processes.*[33]
20. *Leads to the improvement of internal practices.*[23]
21. *Increases health-care organizations' compliance with quality and safety standards.*[20]
22. *Enhances the reliability of laboratory testing.*
23. *Improves patients' health outcomes.*[34]
24. *Provides a team-building opportunity for staff and improves their understanding of their coworkers' functions.*[35]
25. *Promotes an understanding of how each person's job contributes to the health-care organization's mission and services.*[35]
26. *Contributes to increased job satisfaction among physicians, nurses, and other providers.*[29]
27. *Engenders a spill-over effect, whereby the accreditation of one service helps to improve the performance of other service areas.*[20]
28. *Highlights practices that are working well.*[33]
29. *Promotes the sharing of policies, procedures, and best practices among health-care organizations.*
30. *Promotes a quality and safety culture.*[36]

## 6.3 The Uniformity of Accreditation—Accreditation Programs

Although empirical evidence that accreditation provides a higher-quality product or service is limited,[17] its near universal acceptance as a mechanism to ensure quality supports the assumption that the development and use of accreditation is an important mechanism to ensure the quality of products available in clinical practice.

Accreditation programs are varied in approach and content, thus comparisons are at times difficult or inappropriate.[37] While there is no conclusive evidence about the direct impact of accreditation on health-care, there is some indication that if accreditation strengthens interdisciplinary team effectiveness and communication and enhances the use of indicators leading to evidence-based decision-making, then it contributes to improving health outcomes.

Few studies have attempted to draw causal inferences about the direct influence of accreditation on patients' health outcomes, so further research is warranted. For this purpose, Donabedian's "structure–process–outcomes model" or the resultant Quality Health Outcomes Model (QHOM) could be particularly useful (Thornlow & Merwin, 2009).

Many countries have adopted external accreditation of health services as a vehicle for disseminating national standards and for public accountability. Traditionally, in Australia, Canada, and the United States these programs were begun by voluntary collaboration of clinical associations (especially medical) and hospital administrators as a means of organizational development. More recently, they have also been driven by reimbursement schemes, central control, and an emphasis on primary care, health networks and community-based services.

## 6.4 Accreditation Programs in Cellular Therapy

Accreditation has been embraced in the blood and bone marrow transplant field. Accreditation programs in cellular therapy are usually performed by a multidisciplinary team of health professionals and are assessed against published standards for the environment in which clinical care is delivered. The standards adopted nationally usually derive from an amalgamation of national statutes, governmental guidance, independent reports, overseas accreditation standards, and biomedical and health service research. Their content and structure can be applied in many settings, as described in a WHO report in 1993. In general, standards are tailored to individual countries, but there is a growing trend toward consistency with other countries and with other standards such as ISO.

Currently, accreditation programs for transplant programs are provided by organizations such as the *Foundation for the Accreditation of Cellular Therapy (FACT), the Joint Accreditation Committee of the International Society for Cellular Therapy (ISCT)*, and *the European Group for Blood and Marrow Transplantation (EBMT)*, the combination being abbreviated as *JACIE*. Accreditation for cellular laboratories is offered by organizations such as the AABB, (*American Association of Blood Banks*) *FACT, JACIE*.[38] National donor registries that facilitate access to adult donor and CB registries may be accredited by the *World Marrow Donor Association (WMDA)*.

### 6.4.1 American Association of Blood Banks

In 1958, the AABB published its first set of standards for blood banking and began an independent accreditation program. These standards evolved as the *Federal Drug Administration* (now *Food and Drug Administration*, or *FDA*) introduced additional regulation in the blood banking industry through the application of good manufacturing practice recommendations and the adoption of the *Clinical Laboratory Improvement Amendments* of 1988 (*CLIA*).[39]

In 2001, the AABB published the first edition of the *Standards for Cord Blood Services* and consolidated these subsequently with the standards for hematopoietic progenitor cell services into a unified set, the *Standards for Cellular Therapy Product Services*. The fifth edition of those standards became effective on September 1, 2011. CBB accreditation certification by AABB is valid for 2 years.

AABB Accreditation is granted for collection/procurement, processing, testing, storage, distribution, and administration for transfusion medicine activities (donor center, transfusion, perioperative, immunohematology reference laboratories, and molecular testing laboratories); cellular therapy activities (clinical, hematopoietic stem cells, CB, and somatic cell); relationship testing activities; and Specialist in Blood Banking Technology (SBB) schools.

### *6.4.2 NetCord-FACT*

The NetCord-FACT standards were first published in 2000 as a collaboration between the International NetCord Foundation (NetCord) and FACT. NetCord was founded in 1998 by a group of CBBs with a mission to promote high-quality CB banking and uniform clinical use of CB for allogeneic progenitor cell transplantation. Its membership created a set of standards as a common set of banking procedures among the members.[40] FACT was founded in 1996 and is the accreditation arm of the American Society of Blood and Marrow Transplantation and the ISCT. It has created accreditation programs for transplant centers and cell processing labs. The first draft of set of standards was published by FACT in 1996 for hematopoietic progenitor cell collection, processing, and transplantation.

In 1997, FACT recognized the need for standards to address CB banking recognizing CB as an important cell source for transplantation. Through its collaboration with NetCord, it published the first *NetCord-FACT International Standards for Cord Blood Collection, Processing, Testing and Release* in 2000. New editions of the standards are published approximately at 3-year intervals to keep abreast of the latest developments and requirements of high-quality banking.[41]

CBBs are accredited for unrelated allogeneic, directed allogeneic, and/or autologous banking, and inspections began in 2001.

## 7. CONCLUSION

The development/implementation of a quality management program and the final accreditation of CBBs/transplant programs have been critical operations. Since CBUs are distributed worldwide, it is common that transplant programs will be utilizing units from programs with which they do not have established working relationships. The transplant program needs to be assured that the unit has been properly collected, processed, tested, and stored since they will be accepting full responsibility for that product once it is transferred to them. The only way to have that assurance is for the transplant program to know that the bank has the proper systems in place, something that can only be ascertained by on-site inspection by an accrediting program/body that is familiar with the field.

Voluntary accreditation of cells, tissues, and cellular and tissue-based products intended for human transplantation is an important mechanism for improving quality in cellular therapy.

## ABBREVIATIONS

**AABB** American Association of Blood Banks
**CBB** Cord blood bank
**CB** Cord blood
**CBU** Cord blood unit
**CRM** Certified reference material
**CIP** Continuous improvement process
**CLIA** Clinical Laboratory Improvement Amendments
**EFQM** European Foundation for Quality Management
**eSAR** Electronic administration system
**FACT** Foundation for the Accreditation of Cellular Therapy
**FDA** Food and Drug Administration
**ISCT** International Society for Cellular Therapy
**ISO** International Organization for Standardization
**QMPr** Quality management principles
**PPE** Personal protective equipment
**TC** Technical committee
**TQM** Total quality management
**SOP** Standard operating procedure
**QA** Quality assurance
**QC** Quality control
**QI** Quality improvement
**QM** Quality management
**QS** Quality supervisor
**QSs** Quality Supervisors
**QMP** Quality Management Plan
**QMS** Quality Management System
**QMU** Quality Management Unit
**QHOM** Quality Health Outcomes Model
**SBB** Specialist in Blood Banking Technology
**WMDA** World Marrow Donor Association

## REFERENCES

1. http://en.wikipedia.org/wiki/Quality_management_system; [accessed September 2013].
2. http://en.wikipedia.org/wiki/Continious_improvement; [accessed July 2013].
3. www.dti.gov.uk/quality/qms; [accessed July 2013].
4. Jacobides MG. The inherent limits of organizational structure and the unfulfilled role of hierarchy: lessons from a near-war. *Organ Sci* 2007;**18**:455–77.

5. Amaral LAN, Uzzi B. Complex systems - a new paradigm for the integrative study of management, physical, and technological systems. *Manag Sci* 2007;**53**:1033–5.
6. Braha D, Bar-Yam Y. The statistical mechanics of complex product development: empirical and analytical results. *Manag Sci* 2007;**53**:1127–45.
7. Kogut B, Urso P, Walker G. Emergent properties of a new financial market: American venture capital syndication, 1960–2005. *Manag Sci* 2007;**53**:1181–98.
8. Chervenak A, Foster I, Kesselman C, Salisbury C, Tuecke S. The data grid: towards an architecture for the distributed management and analysis of large scientific datasets. *J Netw Comput Appl* 2000;**23**:187–200.
9. http://www.dama.org/files/public/DI_DAMA_DMBOK_Guide_Presentation_2007; [accessed July 2013].
10. Bates DW, Gawande AA. Improving safety with information technology. *N Engl J Med* 2003;**348**:2526–34.
11. Wright AA, Katz IT. Bar coding for patient safety. *N Engl J Med* 2005;**353**:329–31.
12. http://en.wikipedia.org/wiki/Quality_audit; [accessed August 2013].
13. http://www.asq.org/learn-about-quality/continuous-improvement/overview/overview.html; [accessed August 2013].
14. Scherkenbach WW. *Demings Road to Continual Improvement*. Knoxville, Tennessee: SPC Press, Inc.; 1991. ISBN 0-945320-10-8. OCLC 24791076.
15. Cranovsky R, Schilling J, Straub R. *Quality management system for health care institutions*. Frankfurt: PMI Verlag; 2000.
16. International organization for standardization (available on the Internet at: http://www.iso.org/iso/home.html; [accessed May 2014]).
17. Nicklin W. *The value and impact of health care accreditation: a literature review*. Accreditation Canada; 2014. Available at: http://accreditation.ca/; [accessed May 2014].
18. Novaes Hde M. Implementation of quality assurance in latin american and caribbean hospitals through standards and indicators. In: *Applicability of different quality assurance methodologies in developing countries. Proceedings of a pre-ISQua meeting, Newfoundland, Canada, 29–30 May 1995*. Geneva: World Health Organization; 1996. document WHO/SHS/DHS/96.2; available on the Internet at: http://whqlibdoc.who.int/hq/1996/WHO_SHS_DHS_96.2.pdf.
19. Peer KS, Rakich JS. Accreditation and continuous quality improvement in athletic training education. *J Athl Train* 2000;**35**:188–93.
20. Peter TF, Rotz PD, Blair DH, Khine A-A, Freeman RR, Murtagh MM. Impact of laboratory accreditation on patient care and the health system. *Am J Clin Pathol* 2010;**134**:550–5.
21. Alkhenizan A, Shaw C. Impact of accreditation on the quality of healthcare services: a systematic review of the literature. *Ann Saudi Med* 2011;**3**:407–16.
22. El-Jardali F, Jamal D, Dimassi H, Ammar W, Tchaghchaghian V. The impact of hospital accreditation on quality of care: perception of Lebanese nurses. *Int J Qual Health Care* 2008;**20**:363–71.
23. Pomey M-P, Lemieux-Charles L, Champagne F, Angus D, Shabah A, Contandriopoulos A-P. Does accreditation stimulate change? A study of the impact of the accreditation process on Canadian health care organizations. *Implement Sci* 2010;**5**:31–44.
24. Kaminski V. Accreditation - a roadmap to healing in Newfoundland and Labrador. *Qmentum Q* 2012;**4**:10–3.
25. Flodgren G, Pomey MP, Taber SA, Eccles MP. Effectiveness of external inspection of compliance with standards in improving healthcare organization behaviour, healthcare professional behaviour or patient outcomes. *Cochrane Database Syst Rev* 2011. Issue 11. Art. No.: CD008992.
26. Martin LA, Neumann CW, Mountford J, Bisognano M, Nolan TW. *Efficiency and Enhancing Value in Health Care: Ways to Achieve Savings in Operating Costs per Year. IHI Innovation Series white paper*. Cambridge, Massachusetts: Institute for Healthcare Improvement; 2009; p:1–24. (www.IHI.org, accessed October 2014).
27. Newhouse RP. Selecting measures for safety and quality improvement initiatives. *J Nurs Adm* 2006;**36**:109–13.
28. René A, Bruneau C, Abdelmoumene N, Maguerez G, Mounic V, Gremion C. *Improving patient safety through external auditing: the SIMPATIE (Safety Improvement for Patients in Europe) project*. Saint-Denis La Plaine: Haute Autorité de Santé; 2006.
29. Al Tehewy M, Salem B, Habil I, El Okda S. Evaluation of accreditation program in non-governmental organizations' health units in Egypt: short-term outcomes. *Int J Qual Health Care* 2009;**21**: 183–9.
30. Touati N, Pomey M-P. Accreditation at a crossroads: are we on the right track? *Health Policy* 2009;**90**:156–65.
31. Bird SM, Cox D, Farewell VT, Goldstein H, Holt T, Smith PC. Performance indicators: good, bad, and ugly. *J R Stat Soc* 2005;**168**: 1–27.
32. Colquhoun M, Owen M. A snapshot of medication reconciliation in Canada. *Qmentum Q* 2012;**4**:10–3.
33. Baskind R, Kordowicz M, Chaplin R. How does an accreditation programme drive improvement on acute inpatient mental health wards? An exploration of members' views. *J Ment Health* 2010;**19**:405–11.
34. Thornlow DK, Merwin E. Managing to improve quality: the relationship between accreditation standards, safety practices, and patient outcomes. *Health Care Manag Rev* 2009;**34**:262–72.
35. Davis MV, Reed J, Devlin LM, Michalak CL, Stevens R, Baker E. The NC accreditation learning collaborative: Partners enhancing local health department accreditation. *J Public Health Manag Pract* 2007;**13**:422–6.
36. Greenfield D, Pawsey M, Braithwaite J. What motivates professionals to engage in the accreditation of healthcare organizations? *Int J Qual Health Care* 2011;**23**:8–14.
37. Shaw CD. Evaluating accreditation. *Int J Qual Health Care* 2003;**15**:455–6.
38. Caunday O, Bensoussan D, Decot V. Regulatory aspects of cellular therapy product in Europe: JACIE accreditation in processing facility. *Biomed Mater Eng* 2009;**19**:373–9.
39. http://www.aabb.org/; [accessed May 2014].
40. http://www.factwebsite.org/; [accessed May 2014].
41. Hess JR. Conventional blood banking and blood component storage regulation: opportunities for improvement. *Blood Transfus* 2010;**8**(Suppl. 3):s9–15.

Chapter 18

# Regulation Across the Globe

Maria Mitrossili[1], Amalia Dinou[2], Vasiliki Gkioka[2,3] and Catherine Stavropoulos-Giokas[2]
[1]*Institutional Technological Institute of Athens, Athens, Greece;* [2]*Hellenic Cord Blood Bank, Biomedical Research Foundation of Academy of Athens, Athens, Greece;* [3]*Hellenic Transplant Organization, Athens, Greece*

## Chapter Outline



## 1. INTRODUCTION

The development of biomedicine in the field of life sciences in modern times constitutes a complex reality in which the international community is called to regulate the freedom of scientific research and the protection of public health based on respect for human rights.

However, greater awareness is necessary regarding the implementation of scientific achievements during clinical practice, as medicine intervenes with suffering individuals through medical acts, using technical and pharmaceutical products, as well as human tissues and stem cells. In some cases, however, these acts have not been established through long-lasting practice by the medical community, or safe research results suitable for implementation on humans have not been reached yet. This is with regard to diagnostic and therapeutic acts that concern mainly regenerative medicine and not the established and increasing use of units of umbilical cord blood (UCB) for transplantation. Recently, cell therapies and particularly advanced or innovative ones in the field of regenerative medicine promise to treat chronic illnesses and malfunctions of the human body that result from diseases, accidents, or even unsuccessful treatments.

To the extent that scientific developments and the implementations of the use of hematopoietic stem cells (HSCs) from UCB, mesenchymal stem cells, and iPs for therapeutic purposes advance rapidly, the European Union (EU), the USA, and other countries of the international community hasten to promote regulations and standards, managing the specific issue for reasons regarding the safety of public health and the dismantling of barriers in order to ensure free circulation among the countries of stem cells which come

Cord Blood Stem Cells Medicine. http://dx.doi.org/10.1016/B978-0-12-407785-0.00018-9

from donation, testing, processing, preservation, and storage of UCB in the interest of suffering individuals.[1] Regulation and accreditation of banks for the importation and exportation of UCB and for further "healthy" development of the cord blood industry across the globe are necessary.[2]

## 2. REGULATION IN THE EU

It is known that there is a variety of policies and legislations in the interior of the national states of the EU that regulate in a different manner issues which pertain to the collection, storage, and transplantation of UCB, and also to cell therapies in the field of regenerative medicine. It is due to this legal inflation that during the last 20 years the EU has focused on the establishment of common policies and the formation of a converging European law among its member states. The law of the EU comprises mostly of the rules that are founded on conventions and binding legal acts, such as the regulations, the directives, and the decisions that prevail in the internal law of member states, while member states can enact stricter rules provided that they do not oppose the conventions.[3] Even though there is a significant difference between regulation or directive of the EU and regulation of the American legal order[a], the case remains that both of them provide the minimum set of standards to be able to provide a product of service within the regulated jurisdiction.[4]

In light of the European law[b] and in relation to UCB, the EU sets the following general and specific rules:

- The protection of Public Health with the Maastricht Treaty (article 152) and the Protection of Fundamental Rights with the Charter of Nice in 2000, which was incorporated in the Treaty of Lisbon
- Article 12, Directive 2004/23/EC according to which member states shall take all necessary measures to ensure that procurement of cells and tissues takes place on a non-profitable basis
- The establishment of quality and safety protocols
- The strengthening of the principles of informed consent, anonymity, and protection of personal data
- The regulation of cell therapies from "unprocessed" cells and tissues for allogeneic use
- The regulation of cell therapies from fundamentally "processed" cells and tissues for allogeneic use. This "processing" should not induce substantial modification and the use of cells should remain the same (e.g., HSCs should be used for blood therapies)
- The prediction of the traceability and the disclosure of serious and unwanted incidents and reactions, as well as the establishment of a mechanism of biovigilance

### 2.1 Cord Blood Banking in the EU: Directives 2004/23/EC, 2006/17/EC, and 2006/86/EC

Already since the 2000s, the three directives of the EU related to "tissues and cells" to be implemented and incorporated by the member states pose a minimum of regulations regarding UCB and banking.

To begin with, Directive 2004/23/EC of the European Parliament and of the Council of March 31, 2004 (Directive 2004/23/EC of the European Parliament and of the Council of March 31, 2004 on setting standards of quality and safety for the donation, procurement, testing, processing, preservation, storage, and distribution of human tissues and cells)[5] concerns the establishment of standards of quality and safety for the donation, procurement, testing, processing, preservation, storage, and distribution of UCB and other tissues and cells, as well as modified products that come from tissues and cells to be used on humans (Directive 2004/23/EC, article 2 and Directive 2006/86/CE, article 1). Directive 2004/23/CE does not cover research using human tissues and cells (Article 1).

For the implementation of the abovementioned activities, it is required that the institutions providing tissues have been accredited, defined, approved, or licensed by a competent authority, while specific tissues and cells may be given directly to the recipient by the institution/supplier for direct transplantation with the agreement of the competent authority.

Member states must assign competent authorities the inspection and regular control of tissue institutions in order for them to determine whether the directives and the standards required for the procurement of human tissues and cells are observed. Controls include inspections of tissue institutions and facilities of third parties, evaluation of the procedures, and the acts and examination of documentation and of other elements. More specifically, inspections and controls are organized in the case of a serious unwanted reaction or a serious unwanted incident, especially following the justified request of a competent authority of another member state.

Moreover, member states are accountable for the insurance of the traceability of all tissues and cells throughout the stages of the procedure, from the donor to the recipient and vice versa. This insurance is extended to all elements related to pertinent products and materials, and a system of donor identification with a unique code for every donation and product must be secured. Tissues and cells are labeled and classified in cell institutions, which keep the data for a period of at least 30 years after their clinical use, while digital storage of data is also allowed.

Import and export of human tissues and cells from and to Third World countries are possible provided that they

[a] "Regulations in the United States are requirements mandated by the governing body." In general, regulation and accreditation are conceptually different.

[b] WHO has given special attention to human cell and tissue transplantation, and so have the Council of Europe and other national organizations.

take place among tissue institutions that are legitimate and meet quality and safety criteria.

Tissue institutions record all activities and submit an annual report, which is published. Member states and the committee form an interconnected network of registries of national institutions. Their primary concern is the publication of serious and unwanted incidents and reactions, which may concern stages from collection to distribution but also the stage of implementation on humans, in order to ensure quality and safety, while the possibility of withdrawal from product distribution is assured.

Directive 2006/17/CE of the February 8, 2006 (Directive 2006/17/EC of February 8, 2006 implementing Directive 2004/23/EC of the European Parliament and of the Council as regards certain technical requirements for the donation, procurement, and testing of human tissues and cells)[6] refers to the implementation of Directive 2004/23/CE and is related to technical requirements that concern donation, procurement, and testing of tissues and cells of human origin.

Given the fact that the use of tissues and cells for implementation on the human body poses the risk of disease transmission and other likely side effects, this risk can be reduced through a number of measures. Examples of such measures are the selection of the donor, laboratory testing of every donation, and the implementation of procedures regarding the procurement of tissues and cells according to the appropriate regulations and procedures. In this regard, all tissues and cells, including those to be used as original material for the production of medical products to be used in the EU, must meet the safety and quality requirements of this directive.

Standardized procedures of operation that determine the donor's identity, family's consent and authorization, careful selection of the donor based on specific criteria, evaluation of laboratory testing, coding of data in a registry, and the ability of identification with the donor, especially regarding issues of traceability, are important aspects in this regard.

Directive 2006/86/EC of the October 24, 2006 (Directive 2006/86/EC of October 24, 2006 implementing Directive 2004/23/EC of the European Parliament and of the Council as regards traceability requirements, notification of serious adverse reactions and events, and certain technical requirements for the coding, processing, preservation, storage, and distribution of human tissues and cells)[7] focuses on and defines traceability requirements, notification of serious adverse reactions and events, as well as certain technical requirements regarding the coding, processing, preservation, storage, and distribution of human tissues and cells. It follows the trace of UCB from donation to distribution of tissues and cells that are to be given for implementation on humans. In general, emphasis is put on the development of a quality system, "standard operating procedures" (SOPs)[c] and the creation of a safety system that will enable the tracking and identification of cells and tissues throughout the stages of the procedure. Notification with the competent authority is necessary even in cases regarding research results, which concern the analysis of causes and effects. Processed products from cells and tissues that are not subject to other directives or regulations are also in its jurisdiction.

Within the European legal framework, our goal is to ensure high standards of protection of human health. For this reason it is necessary that member states of the EU predict procedures that accredit, define, approve, or license institutions and production methods. In this context, issues regarding organization, administration, staff, equipment, documentation, file observation, creation of a single European code for the identification of the donor and the traceability of the donated material, use of technology and solutions regarding e-government for the transmission and processing of data in standardized format for the exchange of information must be incorporated in the legislation of member states. The information is annually submitted by the competent national authorities to the commission at a European level. Following that, the commission sends a summary of the report to the other member states. At the same time, the notification of serious, unwanted actions and events among the competent authorities is ensured.

The following table presents the main regulations among the three directives regarding "tissues and cells" of the EU.

[c] "SOPs mean written instructions describing the steps in a specific process, including the materials and methods to be used and the expected end product."

| | Definition | Source |
|---|---|---|
| Audit | Documented review of procedures, records, personnel functions, equipment, materials, facilities, and/or vendors in order to evaluate compliance with written SOPs, standards, or government laws and regulations, conducted by professional peers, internal quality system auditors, or certification body auditors | Adapted from the Council of Europe Guide for Safety and Quality Assurance for Organs, Tissues and Cells for Transplantation, 3rd Edition, Council of Europe publishing, January 2007 |
| Cells | Individual human cells or a collection of human cells when not bound by any form of connective tissue | Directive 2004/23/EC |

*Continued*

—cont'd

| | Definition | Source |
|---|---|---|
| Critical | Potentially having an effect on the quality and/or safety of or coming into contact with the cells and tissues | Directive 2006/86/EC |
| Distribution | Transport and delivery of tissues or cells intended for human applications | Directive 2004/23/EC |
| Donation | Donating human tissues or cells intended for human applications | Directive 2004/23/EC |
| Donor | Every human source, whether living or deceased, of human cells or tissues | Directive 2004/23/EC |
| Expert | Individual with appropriate qualifications and experience to provide technical advice to a Competent Authority (CA) inspector | Guidelines Drafting Group |
| Human application | Use of tissues or cells on or in a human recipient/patient and extracorporal applications | Directive 2004/23/EC |
| Organization responsible for application of human tissues and cells | A health-care establishment or unit of a hospital or another body that carries out human application | Directive 2006/86/EC |
| Partner donation | Donation of reproductive cells between a man and a woman who declare that they have an intimate physical relationship | Directive 2006/86/EC |
| Quality system | The organizational structure, defined responsibilities, procedures, processes, and resources for implementing quality management, including all activities that contribute to quality, directly or indirectly | Directive 2006/86/EC |
| Reproductive cells | All tissues and cells intended to be used for the purpose of assisted reproduction | Directive 2006/86/EC |
| Serious adverse event | Any untoward occurrence associated with the procurement, testing, processing, storage, and distribution of tissues and cells that might lead to transmission of a communicable disease, to death, or to life-threatening, disabling, or incapacitating conditions for patients, or that might result in, or prolong, hospitalization or morbidity | Directive 2004/23/EC |
| Serious adverse reaction | An unintended response, including a communicable disease, in the donor or in the recipient associated with the procurement or human application of tissues and cells that is fatal, life-threatening, disabling, incapacitating, or that results in, or prolongs, hospitalization or morbidity | Directive 2004/23/EC |
| Standard operating procedures | Written instructions describing the steps in a specific process, including the materials and methods to be used and the expected properties of the tissues or cells to be distributed | Adapted from Directive 2006/86/EC |
| Storage | Maintaining the tissues or cells under appropriate controlled conditions until distribution | Directive 2004/23/EC |
| Third country | Any country that is not a member state of the EU | European Commission: ec.europa.eu |
| Third party | Any organization that provides a service to a procurement organization or a Tissue Establishment (TE) on the basis of a contract or written agreement. Includes donor- or tissue-testing laboratories, contract sterilizers, and user hospitals, which store tissues or cells pending human application | Operational Manual Drafting Group |
| Tissue | All constituent parts of the human body formed by cells | Directive 2004/23/EC |
| Tissue establishment | A tissue bank or a unit of a hospital or another body where processing, preservation, storage, or distribution of human tissues and cells are undertaken. It may also be responsible for procurement or testing of tissues and cells. | Directive 2004/23/EC |
| Traceability | The ability to locate and identify the tissues/cells during any step from procurement, through processing, testing, and storage, to distribution to the recipient or disposal. This also implies the ability to identify the donor and the tissue establishment or the manufacturing facility receiving, processing, or storing the tissues/cells and to identify the recipient(s) at the medical facility or facilities applying the tissues/cells to the recipient(s). Traceability also covers the ability to locate and identify all relevant data relating to products and materials coming into contact with those tissues/cells. | Directive 2006/86/EC |

—cont'd

| | Definition | Source |
|---|---|---|
| Validation (or "qualification" in the case of equipment or environments) | Establishing documented evidence that provides a high degree of assurance that a specific process, piece of equipment, or environment will consistently produce tissues or cells meeting its predetermined specifications and quality attributes; a process is validated to evaluate the performance of a system in terms of its effectiveness for its intended use. | Directive 2006/86/EC |

Inspection of Tissue and Cell Procurement and Tissue Establishments - Operational Manual for Competent Authorities; Version 1.0; European Commission Health and Consumers Directorate-General

## 2.2 Regulation (EC) No 1394/2007 and Regulation (EU) No 1235/2010 regarding Advanced or Innovative Cell Therapies

Relatively recently, the development of cell therapies that come from UCB as a source of HSCs and their use in regenerative medicine pose additional issues to be addressed by the EU, which cannot be solved by the abovementioned directives of "tissues and cells" regarding UCB and cord blood banking. Cell therapies derive from living cells and cure human diseases or malfunctions through their medical application on humans.

In the light of this, medicinal products that are produced from stem cells, including those from UCB, are classified at an EU level and with reference to legislation, regulatory acts, and administrative procedures as follows:

1. Advanced therapy medicinal products (ATMPs) are considered medicines, provided that they have been industrially manipulated. They are covered by Regulation (EC) no 1394/2007, which has been in effect since December 2008 and applies directly to the national law of member states (Regulation (EC) no 1394/2007 of the European Parliament and of the Council of November 13, 2007 on ATMPs and amending Directive 2001/83/EC and Regulation (EC) no 726/2004).[8]
According to the abovementioned regulation, licensing and supervision procedures of the specific medicines to be used on humans are regulated and the Committee for Advanced Therapies (CAT) is founded. Regulation (EC) no 1394/2007 has relatively recently been modified by Regulation (EU) 1235/2010, mainly in relation to the regulation of pharmacovigilance issues.
More specifically, products/medicines for advanced or innovative cell therapies, which pertain to Regulation (EC) no 1394/2007 modified by Regulation (EU) 1235/2010, follow a registration procedure that takes place at a European level. Responsibilities lie with the European Medicine Agency (EMA) and the CAT, and not with the national organizations for medicines of member states.
2. Products of advanced or innovative therapies are exempted and regulated separately, according to article 28 of Regulation (EC) no 1394/2007. Even though the abovementioned cell products fall within the definition of advanced or innovative therapy and can be regarded as medicines, when they are personalized for a unique patient, used by a hospital within a single country and are not industrially produced products, they are regulated at a national level, mainly in relation to licensing. These products, however, are not exempted from the requirements of traceability and pharmacovigilance. Moreover, they pertain to quality regulations equivalent to those provided for medicines regulated by Regulation (EC) no 1394/2007 and Regulation (EC) no 726/2004, as well as Directives 2010/84/EU and 2011/62/EU.
3. Products for cell therapies that do not fall within the definition of advanced or innovative cell therapies are not considered medicines. In this case, the abovementioned products are not legally considered medicines and pertain to the Directive of "tissues and cells" of the EU 2004/23/EC, which is legally incorporated in the internal law of member states. Licensing procedures take place at a national level.
Regulation (EC) no 1394/2007 is enforced and applied directly to all countries. However, a deadline is given to member states of the EU, which have products in their internal markets and fall within the definition of advanced or innovative therapy, in order for them to become "legitimate."
Products, however, manufactured for a single patient, within a single hospital, prepared on a nonroutine basis pertain to article 28 of Regulation (EC) no 1394/2007, and their regulation takes place at a national level with authorization from a national body (even though they are considered medicines).
The main strategy of the EU aims to allow the safe circulation within its grounds for reasons of free circulation and legal protection of the health of civilians only of medicines that have been approved by the EMA. Regarding products that pertain to article 28 of Regulation (EC) no 1394/2007, their license from a national body of a member state is by way of exception still valid. However, after 2012 it is likely that difficult legal issues will arise, especially regarding the issue of

whether the pharmaceutical product has been industrially manufactured or not. Moreover, there is a great risk that the public sector and the nonprofitable sector, which deal with cell therapies and for which "commercialization" of cell products is not an objective, will limit their scientific and research activity to products for cell therapies.

The following table presents differences in the regulatory frame in the EU:

| Legislation | Advanced or Innovative Cell Therapies/ Pharmaceutical Products | Personalized Advanced Cell Therapies | Products from Hematopoietic Stem Cells (HSCs) |
|---|---|---|---|
| | HSCs for the regeneration of another tissue. Advanced therapy medicinal products (ATMPs) in particular are manufactured either industrially or with a method that includes industrial processing:<br>• Gene therapies<br>• Somatic cell therapies<br>• Cell and tissue engineering<br>• Combined medicine ATMPs | Autologous use/personalized production of HSCs for a specific patient, within the hospital and within a single member state for the regeneration of another tissue | Products from HSCs that are not ATMPs and are destined for blood regeneration through transplantation, especially after chemotherapy |
| Directives of "tissues and cells"<br>2004/23/EC<br>2006/17/EC<br>2006/86/EC | Are valid regarding donation, collection, testing, storage, and disposal of tissues and cells | Are valid regarding donation, collection, testing, storage, and disposal of tissues and cells | Pertain only to integrated directives of "tissues and cells" of the EU in the internal law of a member state and to national legislation |
| Directives:<br>2004/23/EC<br>2006/17/EC<br>2006/86/EC +<br>Regulation (EC) 1394/2007 +<br>Directive 2001/83/EC | Compulsory European central procedure by Committee for Advanced Therapies/European Medicine Agency (licensing, approval, etc.) | Directive 2001/83/EC and national legislation | |

## 3. CORD BLOOD BANKING IN THE USA

Over the last decade, we have seen a significant increase in the use of UCB as a source of HSCs for allogeneic transplantation. Each year thousands of patients are diagnosed with life-threatening illnesses for which an HSC or cord blood transplant is the treatment of choice. A successful transplant, however, requires an HLA (Human Leucocyte Antigen)-compatible (tissue type matched) donor.

In response to the recognition of the clinical value of cord blood, public and private cord blood banks were established in the mid-1990s in the United States and internationally.[9] Although the US still lacks a well-funded, well-publicized system of public donor banks, the number of public Cord Blood banking programs in the country is growing. The American Academy of Pediatrics, the American College of Obstetricians and Gynecologists, the American Medical Association, and the American Society for Blood and Marrow Transplantation have recently issued new or revised guidelines that encourage public Cord Blood banking, and cord blood education laws have been passed in 27 states.

Since the Food and Drug Administration first established standards for family banking in 1998, public policy initiatives at both the federal and state level continue to reinforce the importance of stem cells. In 2005, Congress passed the Stem Cell Therapeutic and Research Act 18.[10] The act, reauthorized in 2010,[11] includes the C.W. Bill Young Cell Transplantation Program to encourage cord blood research and build a national public cord blood inventory to address inequalities in the access for minorities to transplant by funding public banking programs that target minority donors.[12] The Stem Cell Acts of 2005 and 2010 are managed by the Health Resources and Services Administration of the US Department of Health and Human Services (HHS). The Stem Cell Acts include:

- the C.W. Bill Young Cell Transplantation Program. Congressman C.W. Bill Young was a long-time supporter of bone marrow transplantation. Congressman Young's devoted efforts helped start the National Bone Marrow Donor Registry. The Program expanded on the previous requirements of the registry to increase the number of marrow donors and cord blood units (CBUs) and continues to serve patients who need a bone marrow or UCB transplant. The Program also established an outcomes database to collect data and perform research.

- the National Cord Blood Inventory (NCBI). The goal of the NCBI cord blood banks is to collect and store at least 150,000 new CBUs to treat patients. The NCBI banks will also provide CBUs for research studies. The Stem Cell Act 2010 also required the US Government Accountability Office to report on efforts to increase CBUs.
- the 3-year demonstration project under which qualified cord blood banks will receive a portion of the funding in order to collect and store CBUs for a family where a first-degree relative has been diagnosed with a condition that will benefit from transplantation (including blood disorders, malignancies, metabolic disorders, hemoglobinopathies, and congenial immunodeficiencies) at no cost to such family.

In 2007 Health Resources and Services Administration (HRSA) has entered into contracts with 13 cord blood banks to contribute cord blood units to the NCBI, with each contract specifying by racial and ethnic group a specific goal of bankable units. Since then, 24 more public cord blood banks have been established, and the demand for and utilization of CBUs have increased substantially. These developments highlight the compelling need to assess existing public banking policies and practices.

## 3.1 Cord Blood Banking by State

### 3.1.1 Arizona

As of January 2007, health-care professionals are required to inform a pregnant patient in her second trimester about her ability to family bank or donate her newborn's cord blood. The Department of Health Services has developed a brochure on all options for expectant mothers, including the benefits of UCB collection to the newborn, biological family, and nonrelated individuals. Arizona is the first state to inform expectant parents about free cord blood collection and storage programs offered by family and sibling donor banks.

### 3.1.2 Arkansas

By June 2008, the Arkansas commission will have developed a voluntary program to educate patients on Cord Blood banking, including an explanation of the differences between public and private banking options relating to availability and cost.

### 3.1.3 California

Consistent with the recommendations of the IOM, the Maternal and Child Health Advancement Act will authorize the Department of Health to create a cord blood awareness campaign that will offer standardized, objective information to expectant mothers about the differences between public and private banking, current and future uses of cord blood, and how medical or family history can impact a family's decision to donate or family bank their newborn's stem cells.

### 3.1.4 Connecticut

In July 2009, Governor M. Jodi Rell signed into law a bill with a provision requiring physicians and other health-care providers who provide health-care services to pregnant women to provide timely, relevant, and appropriate information during the last trimester of pregnancy, sufficient to allow each woman to make an informed and voluntary choice regarding options to bank or donate UCB following the delivery of a newborn.

### 3.1.5 Georgia

Following the governor's executive order last year, Governor Perdue signed into law SB 148, Saving the Cure Act in 2007. The act establishes the Newborn Umbilical Cord Blood Bank and the creation of a commission whose numerous tasks will include developing a program to explain differences between public or private banking, the medical process involved in the collection and storage of postnatal tissue and fluid, the current and potential future medical uses of stored postnatal tissue and fluid, the benefits and risks involved in the banking of postnatal tissue and fluid, and the availability and cost of storing postnatal tissue and fluid in public and private UCB banks. Beginning June 2009, all physicians and hospitals in Georgia have informed pregnant patients of the full range of cord blood options before the third trimester.

### 3.1.6 Illinois

In 2004, the Hospital Licensing Act was amended to add a mandate that hospitals offer pregnant women the option to donate their newborn's cord blood to a public bank.

### 3.1.7 Louisiana

In June 2008, Governor Jindal signed HB 861 requiring Louisiana Department of Health and Hospitals (LDHH) to promote awareness of the potential benefits of Cord Blood banking. The LDHH will develop an outreach campaign via written materials, brochures, the Internet, and public service announcements to promote Cord Blood banking awareness. It educates medical professionals and establishes a toll-free number for information on all cord blood banks serving the State of Louisiana. Implementation of this program was dependent upon monies being allocated in July 2008.

### 3.1.8 Massachusetts

An act enhancing regenerative medicine in the Massachusetts Commonwealth has become law and has been incorporated into the acts that govern the state. The Massachusetts Department of Public Health will establish a program to educate women on all cord blood options, so they can make an informed decision on whether or not to participate in a private or public UCB banking program. Hospitals within the Commonwealth will inform pregnant patients of their ability to donate to a public bank.

### 3.1.9 Michigan

Several bills focused on creating a network of UCB banks were passed in 2006. An education bill was passed, which encouraged health-care professionals, facilities, and agencies to disseminate educational materials on all cord blood options, developed by the Department, to a pregnant woman before her third trimester, including the differences between public and private banking.

### 3.1.10 New Jersey

In January 2008, New Jersey passed cord blood legislation aimed at helping educate expectant families on all their cord blood preservation options, including family banking and public donation. The legislation includes a provision for the Department of Health and Senior Services to create an educational brochure and directs prenatal health-care providers to share the brochure with expectant parents prior to labor and delivery, preferably in the first trimester.

### 3.1.11 New York

In August 2007, New York State amended the state's public health law, establishing a public and private Cord Blood banking program to promote public awareness of the potential benefits of both public or family Cord Blood banking, promote research into the uses of cord blood, and facilitate predelivery arrangements for public or private banking of cord blood donations. The program is charged with providing educational materials and brochures on both cord blood options, to be made available to the general public and potential donors, through local departments of health, health-care practitioners, hospitals, clinics, and other organizations serving pregnant women. In addition, the bill requires the coordination and promotion of educational materials for health-care providers.

### 3.1.12 North Carolina

The HHS shall make online materials available and will encourage health-care professionals to educate all pregnant patients on their cord blood options, to ensure every family has the opportunity to make an informed choice.

### 3.1.13 Oklahoma

Oklahoma governor Brad Henry has signed a public Cord Blood banking bill HB 3060, which discusses the establishment of a public bank/public donation collection program in the state. The bill includes a Department of Health program to educate maternity patients with sufficient information to make an informed decision on whether or not to participate in a private or public UCB banking program.

### 3.1.14 Pennsylvania

In April 2008, the Umbilical Cord Blood Education and Donation Act was passed into law. The law encourages health-care providers to educate expectant parents about their cord blood preservation options prior to the end of the second trimester, so they are equipped to make an informed choice between family (private) banking, public donation, or free family banking in instances where there is an existing medical need within the family. The new law also includes a provision calling for the Department of Health of the Commonwealth to create an online brochure that health-care providers can print and give to their patients. This brochure will outline the current and future medical uses of cord blood stem cells and differences between family banking and public donation.

### 3.1.15 Rhode Island

Enacted June 27, 2008, the law states that health-care professionals shall educate a pregnant patient after her first trimester about her ability to family bank or donate her newborn's cord blood. It also states that hospitals shall facilitate cord blood collections. In addition, physicians will inform expectant parents about free cord blood collection and storage programs offered by family and sibling donor banks.

### 3.1.16 Tennessee

The original cord blood legislation in Tennessee was enacted in 2006 as a donation inform-only bill. It was amended in March 2010 to direct the department of health to create a publication educating pregnant women about all Cord Blood banking options and encourage physicians to provide the information by the end of a woman's second trimester of pregnancy. The effective date of this updated legislation was July 1, 2010.

### 3.1.17 Texas

In May 2007, Governor Perry signed HB 709 requiring Texas Health and Human Services Commission to develop a brochure regarding public donation and family banking, including the free programs where there is an existing medical need. The brochure includes current and potential future uses and benefits of cord blood stem cells to a potential recipient of donated stem cells, such as a biological family member or a nonrelated individual, the medical process to collect cord blood, and any risks and associated costs with Cord Blood banking. The brochure will be provided by maternal health-care professionals to pregnant women before the third trimester, or as soon as reasonably feasible.

### 3.1.18 Virginia

The department of health posted information on their Web site developed by Parent's Guide to Cord Blood Banking foundation outlining the value of cord blood stem cells, the differences between related and unrelated stem cell sources, and the difference between public and private banking. Physicians make this information available to pregnant patients in their third trimester of pregnancy. The effective date of this updated legislation was July 1, 2010.

### 3.1.19 Wisconsin

The prenatal provider must offer information on public donation by the 35th week of pregnancy. The effective date of this legislation was July 1, 2010.

### 3.1.20 Washington

In 2008, the Washington state legislature signed into law House Bill 2431, which encourages health-care providers to better educate their patients about the value of cord blood stem cells and the options for preserving them. The law ensures that expectant parents have greater access to cord blood information prior to the third trimester of pregnancy to help them more readily arrange for private cord stem cell storage or public donation as alternatives to discarding cord blood as medical waste. The new law also required cord blood banks to have licenses, accreditations, and other authorizations required under federal and Washington State law.

## 4. CORD BLOOD BANKING IN ASIA, AFRICA, AND OCEANIA

Cord Blood banking in other parts of the world does not follow a uniform model or regulations as is the case in Europe or North America. In Asia, Oceania, and Africa, there are great disparities due to the coexistence in the same geographical area of developed and developing countries, and different political and economic systems.[1,2]

Public and private cord blood banks coexist and in some cases support each other. Regarding private cord blood banks—they are present in most countries. Sometimes, they are local subsidiaries of international corporations with a presence in several continents. They usually have an operating license, but in a lot of countries the absence of concrete regulations on Cord Blood banking means that the license is for a medical establishment that does not necessarily meet any specific requirements regarding Cord Blood banking. These organizations are usually self-regulated, by submitting to accreditation by organizations like AABB or FACT. Public cord blood banks need stricter and preferably government-enforced regulations that ensure the safety of the offered product. In any case, there is a disparity between regulations of different countries.

There have been some efforts to present unified standards for cord blood banks to follow. AsiaCORD is one such example: it was founded in 1999 in order to promote Cord Blood banking in Asia with members in China, Japan, Thailand, Korea, Taiwan, and Vietnam, and to supply high-quality CB units for transplantation. It originally had its own standards and granted accreditation to banks. It has evolved to an organization that promotes Cord Blood banking and related research in the field of cellular therapy (http://www.asiacord.umin.jp/). Another organization is the Asia Pacific Cord Blood Bank Consortium (http://www.apcbbc.org), uniting private cord blood banks in Asia, Africa, and the Pacific.

## 4.1 Japan

Public Cord Blood banking in Japan was established by parent organizations. There are 11 public cord blood banks in Japan, partly funded by government, forming the Japan Cord Blood Banking Network (which started in 1999).[13] UCB transplantation follows the same regulations as bone marrow transplantation. It is defined as a medical procedure, and CB is not considered a blood or tissue product that would fall under the Pharmaceutical Affairs Law (with mandatory Good Manufacturing Practices (GMP) and Good Tissue Practices (GTP)) and is not supervised by the Ministry of Health, Labour and Welfare as such.[14] So, the issues of quality of CB are self-regulated (JCBBN Guidance for UCB Collection and SOPs for UCB Collection), as this product is not covered by the "Notice for quality and safety of cellular and tissue-based pharmaceuticals and devices" issued in 2008 by the MHLW. Only the Tokyo Cord Blood Bank is a member of international organizations Netcord, BMDW, and AsiaCORD, and the government does not allow export or exchange of cord blood units.[15]

## 4.2 China

In China, the first cord blood bank was founded in Beijing in 1996, and guidelines and administrative measures concerning Cord Blood banking were set by the Ministry of Health in 1999.[16] The following documents have been implemented in order to regulate the operation of cord blood banks in China: the (Trial) Administrative Measures for Cord Blood Hematopoietic Stem Cell Banks ("CBB Measures"); the Administrative Measures for Blood Banks ("Blood Bank Measures"); the (Trial) Technical Norms on Cord Blood Hematopoietic Stem Cells; the (Trial) Administrative Norms on Facilities of the Cord Blood Hematopoietic Stem Cells; and the Principles of Guideline for Plans of Establishment of Institutes for Blood Collection and Provision.[16]

The Ministry of Health has decreed that 10 cord blood banks should operate in China in order to provide CBUs for the Chinese population. Nevertheless, the cord blood banks are not financed by the Ministry and are expected to raise the funds needed for their operation. Although according to Chinese law a cord blood bank cannot be a for-profit organization, a mixed system has arisen, with corporations owning at least partly the existing cord blood banks that have both public and private storage facilities. The profits from the private cord blood banks finance the public cord blood banks. There is speculation that not all licensed cord blood banks promote the donation procedures and instead focus on the family storage of cord blood.

The license required for Cord Blood banking operation is granted (since 2006) by the health administration of the province where a cord blood bank is based, that is also responsible for the supervision of the Cord Blood banking. There are a limited number of licenses per province, but an organization may hold more than one license. The private (and public) banks are held by three corporations (http://chinacordbloodcorp.com, www.shanghaicordblood.org, http://www.chinastemcell.com). The Ministry of Health still inspects cord blood banks and issues the requirements a cord blood bank must meet.

At the moment, seven public cord blood banks operate in China. Six of them cooperate with the CMDP (www.cmdp.com.cn) for the listing and distribution of the CBUs (WMDA, Shanghai CB). In total, approximately 54,000 units are available. There are no data on the total number of CBUs for family use in China, but they are estimated to be much higher.

On November 13, 2009, China's Ministry of Health issued the Technology Management Norms for Umbilical Cord Blood Hematopoietic Stem Cell Therapy in order to standardize clinical use of such therapy.[17]

## 4.3 Hong Kong

There is one public cord blood bank operating in Hong Kong and several banks offering private storage. Although cord blood is regarded as an "organ" under the Human Organ Transplant Ordinance (CAP 465 Human Organ Transplant Ordinance), the Administrative Guidelines for the Human Organ Transplant Ordinance (CAP.465) ed. of 2011 do not address the issues of the collection and storage of cord blood in Hong Kong. The only general outlines were given in 2002,[18] but some legislators have urged for regulation of Cord Blood banking.[19]

## 4.4 Taiwan

There are four public cord blood banks in Taiwan and eight family banks, some of which have a public donations inventory. Cord blood banks must register with the Department of Health and get a 3-year permit in order to operate in Taiwan, according to "Measures on the Administration of Human Organ Banks," issued in 2009, which follow the Human Organ Transplant Regulations (last amended 2003).[20]

## 4.5 Singapore

Singapore is home to one public cord blood bank and at least two private ones. There seem to be no regulations in place governing the operation of Cord Blood banking, as it is not covered either by the 2004 Regulation regarding licensing for tissue banking,[21] or by the 2008 Human Organ Transplant Act.[22]

## 4.6 Australia

Cord blood banks in Australia are licensed by the Australian Therapeutic Goods Act 1989.[23] The standards for the collection, testing, processing, storage, and release must follow the Therapeutic Goods Order 75 standards for hematopoietic progenitor cells derived from cord blood,[24] which are based on the 3rd Ed. FACT-Netcord Standards. They must comply to the Therapeutic Goods (Manufacturing Principles) Determination No.1 of 2013 for human blood and blood components, human tissues, and human cellular therapy products 2013 (2013 code of GMP),[25] and to the Therapeutic Goods Order No.88 (TGO 88) for the minimization of infectious disease transmission.[26] All Australian cord blood banks are FACT-accredited, and there are three private banks and three public banks. The public banks form the AusCord Netcord.

## 4.7 South Korea

In South Korea, there are both private and public cord blood banks. Some of the private banks also have public CB banking programs. Since 2011, the "Cord Blood Management and Research Act" has been implemented and all banks are approved by the Ministry of Health.[27] It addresses issues concerning the operation of both private and public Cord Blood banking, funding for public banks, as well as research-related activities. It is accompanied by ordinances and regulations based on the FACT Standards, 4th ed. and by updates on relevant legislation (regarding ethics, transplantation, health insurance, etc.).[28]

## 4.8 Saudi Arabia, Jordan, Egypt, Qatar, UAE

Cord blood stem cell banking is permitted under Islamic law and there are several fatwas supporting Cord Blood banking in Arab countries.[29] The policies regarding public or private Cord Blood banking differ from country to country. Saudi Arabia and the UAE already have a government-funded public Cord Blood banking program and the relevant legislation. Qatar is trying to implement its own program with hybrid characteristics, while Egypt and Jordan have no policy regarding public banking. With the exception of the UAE, which prohibits private banking, private banks operate in all of these countries.

## 4.9 India

In India, the 2011 amendment of the 1945 Drugs and Cosmetics Acts and the ICMR-DBT Guidelines for Stem Cell Research give the framework for Cord Blood banking operation. They have been in effect since 2012 and concern both private and public banking. They outline all aspects of banking, from collection to release for transplantation. Cord Blood banking requires registration and license from the government regulatory agency, namely Drug Controller General of India.[30–32]

## 4.10 Malaysia

In 2008, the "National Standards for Cord Blood Banking and Transplantation" (Ministry of Health) were established in Malaysia. One public cord blood bank and several private ones operate in Malaysia and are licensed by the Ministry of Health.[33]

## 4.11 Israel

In Israel, the Cord Blood Law was passed by the Knesset in 2007[34] and implemented in 2010.[35] It regulates the activities of both public and private Cord Blood banking, defines the protocols from collection to transplantation, and provides state funding for the three public cord blood banks. In 2012, further regulation concerning the operation of private Cord Blood banking was implemented.[36]

## 4.12 Turkey

In Turkey, since 2005 there is a bylaw on cord blood—the "Regulation for Umbilical Cord Blood Banking." In Turkey, private cord blood banks are allowed, but 25% of their stored CBUs must be offered to the public for allogeneic transplantation.[37]

In some countries (Thailand, Pakistan, South Africa, Vietnam, Indonesia, Philippines), there is no clear regulatory environment for the operating Cord Blood banking (whether public or private), while sometimes cord blood banks are part of blood banks and transfusion services that are government regulated as such (e.g., Iran).[38]

# 5. CONCLUSION

Across the world, not all countries have established legislation to regulate Cord Blood banking and banks. Moreover, there is a pluralism of laws governing cord blood. Harmonization of regulations in the EU, including efforts by WHO and the Council of Europe, as well as North America and a few other countries, remains a goal to be achieved.

Consequently, the enactment of uniform or, at least, converging rules at a national level is necessary for the safety, sufficiency, and promotion of UCB stem cells or processed products across the globe. Respect of human rights, biovigilance, traceability, and the exclusion of persons from high-risk groups constitute fundamental guarantees for public health. The abovementioned rules relate not only to UCB for allogeneic or autologous use for transplantation from private or public banks, regarding setting standards of quality and safety for the donation, procurement, testing, processing, preservation, storage, and distribution of human tissues and cells, but also to a "healthy" development of cord blood industry relating to their industrial processing (HSCs, mesenchymal stem cells, and iPS) for use in regenerative medicine.

# REFERENCES


1. Sniecinski I, MD. Cord blood banking in developing countries. In: Broxmeyer HE, editor. *Cord blood: biology, transplantation, banking, and regulation*. Bethesda, MD: AABB Press; 2011.
2. Reagan DM. Cord Blood Banking: the development and application of cord blood banking processes, standards, and regulations. In: Broxmeyer HE, editor. *Cord blood: biology, transplantation, banking, and regulation*. Bethesda, MD: AABB Press; 2011.
3. Boo M, JD, Welte K, Confer D, MD. Accreditation and regulation of cord blood banking. In: Broxmeyer HE, editor. *Cord blood: biology, transplantation, banking, and regulation*. Bethesda, MD: AABB Press; 2011.
4. Mitrossili M. *Health law*. 2009 [Athens Papassizi (in Greek)].
5. Directive 2004/23/EC of the European Parliament and of the Council of 31 March 2004 on setting standards of quality and safety for the donation, procurement, testing, processing, preservation, storage and distribution of human tissues and cells (OJ L 102, 7.4.2008, p. 48).
6. Commission Directive 2006/17/EC of 8 February 2006 implementing Directive 2004/23/EC of the European Parliament and of the Council as regards certain technical requirements for the donation, procurement and testing of human tissues and cells (OJ L 38, 9.2.2006, p. 40).
7. Commission Directive 2006/86/EC of 24 October 2006 implementing Directive 2004/23/EC of the European Parliament and of the Council as regards traceability requirements, notification of serious adverse reactions and events and certain technical requirements for the coding, processing, preservation, storage and distribution of human tissues and cells (OJ L 294, 25.10.2006, p. 32).
8. Regulation (EC) No 1394/2007 of the European Parliament and of the Council of 13 November 2007 on advanced therapy medicinal products and amending Directive 2001/83/EC and Regulation (EC) No 726/2004 (OJ L 324, 10.12.2007, p. 121).
9. Martin P, Brown N, Turner A. Capitalizing hope: the commercial development of umbilical cord blood stem cell banking. *New Genet Soc* 2008;**27**(2):127–43.
10. Stem Cell Therapeutic and Research Act of 2005, rev. 42 U.S.C. 274k.
11. Stem Cell Therapeutic and Research Reauthoritzation Act of 2010, rev. 42 U.S.C. 274k.
12. Meyer EA, Hanna K, Gebbie K. *Cord blood: establishing a national hematopoietic stem cell bank program*. Washington, DC: National Academies Press; 2005.

13. Atsuta Y, Suzuki R, Yoshimi A, Gondo H, Tanaka J, Hiraoka A, et al. Unification of hematopoietic stem cell transplantation registries in Japan and establishment of the TRUMP system. *Int J Hematol* 2007;**86**(3):269–74.
14. Tada N, Hinotsu S, Urushihara H, Kita F, Kai S, Takahashi TA, et al. The current status of umbilical cord blood collection in Japanese medical centers: survey of obstetricians. *Transfus Apher Sci* June 2011;**44**(3):263–8.
15. www.worldmarrow.org/fileadmin/Committees/ASIA/Appendix_D.pdf [last accessed on 12.09.14].
16. Liu Y. Investigation of the immature stage of the cord blood banks and their regulation in China. *J Int Bioethique* 2008;**19**(4):105–16, 164.
17. http://www.loc.gov/lawweb/servlet/lloc_news?disp3_l205401696_text [last accessed on 12.09.14].
18. http://www.legco.gov.hk/yr02-03/english/counmtg/hansard/cm1016ti-translate-e.pdf [last accessed on 12.09.14].
19. http://www.scmp.com/article/557922/lawmaker-demands-regulation [last accessed on 12.09.14].
20. http://www.loc.gov/lawweb/servlet/lloc_news?disp3_l205401251_text. [last accessed on 12.09.14].
21. https://elis.moh.gov.sg/elis/info.do?task=guidelines§ion=GuidePHMCTnC [last accessed on 12.09.14].
22. http://statutes.agc.gov.sg/aol/search/display/view.w3p;page=0;query=CompId%3A018ae65c-7438-4c49-9da4-93594f0e5431;rec=0 [last accessed on 12.09.14].
23. http://www.comlaw.gov.au/Series/C2004A03952 [last accessed on 12.09.14].
24. http://www.comlaw.gov.au/Details/F2007L03598 [last accessed on 12.09.14].
25. http://www.comlaw.gov.au/Details/F2013L00855 [last accessed on 12.09.14].
26. http://www.comlaw.gov.au/Details/F2013L00854 [last accessed on 12.09.14].
27. Lee Y-H, Yoon Kim J, Mun Y-C, Koo HH. A proposal for improvement in the utilization rate of banked cord blood. *Blood Res* 2013;**48**(1):5–7.
28. Lee Y-H. The prospect of the government management for cord blood in Korea - at the time of enactment of the Cord blood management and research act. *Korean J Hematol* 2010;**45**(1):1–2.
29. http://bakerinstitute.org/research/more-than-oil-and-uprisings-current-developments-in-cord-blood-banking-in-the-arab-world/[last accessed on 12.09.14].
30. *Ethical guidelines for biomedical research on human participants*. New Delhi: Indian Council of Medical Research; 2007.
31. Committee for medicinal product for human use [CHMP]. *Guidelines on human cell based products*. September 2008, European Medicines Agency.
32. Viswanathan C, Kabra P, Nazareth V, Kulkarni M, Roy A. India's first public cord blood repository — looking back and moving forward. *Indian J Hematol Blood Transfus* 2009;**25**(3):111–7.
33. http://www.moh.gov.my/images/gallery/Garispanduan/Stem_Cell/guideline_national_cord_blood_banking.pdf [last accessed on 12.09.14].
34. http://www.icbb.org.il/?categoryId=8537 [last accessed on 12.09.14].
35. http://www.knesset.gov.il/mmm/data/pdf/me02735.pdf [last accessed on 12.09.14].
36. http://www.israelnationalnews.com/News/News.aspx/152020#.UkBYbT8rhU4 [last accessed on 12.09.14].
37. O'Connor MAC, Samuel G, Jordens CFC, Kerridge IH. Umbilical cord blood banking: beyond the public-private divide. *J Law Med* 2012;**19**:512.
38. Cheraghali A, Amini-Kafiabad S, Amirizadeh N, Jamali M, Maghsudlu M, Manshadi M, et al. Iran national blood transfusion policy; goals. objectives and milestones for 2011–2015. 2011;**3**(2):35–42.

Chapter 19

# International Development and Import/Export—WMDA

Lydia Foeken and Daniela Orsini

*World Marrow Donor Association, WMDA Office, Leiden, The Netherlands*

## Chapter Outline



## 1. INTRODUCTION

The world's first cord blood banks were established in New York (United States) and Düsseldorf (Germany) in 1991 and 1992, respectively. To date (2014), an estimated 700,000 cord blood units (CBUs) for unrelated transplantation are stored in 160 public cord blood banks, established in 36 different countries.

In addition, over 34,000 CBUs have been distributed worldwide for both adults and children with life-threatening malignant and nonmalignant diseases. These data are retrieved from the annual reports of the World Marrow Donor Association (WMDA),[1] which show that 36% of all CBUs have been shipped to North America, 34% to Asia, and 23% to Europe. The international exchange of cord blood products has grown from 129 (1999) to 1162 (2013)[1] cord blood products annually, accounting for nearly 30% of all CBUs provided for unrelated hematopoietic progenitor cell (HPC) transplantation (Figure 1). As 30% of the CBUs are passing an international border, when transported from one country to another, rules and regulations are important to ensure a high-quality cord blood product for a patient in need of an HPC transplant. Therefore, international strategies have been developed governing the collection, processing, storage, and use of cord blood (see Chapter 18 for Regulation Across the Globe).

In this chapter, we will first discuss factors to be considered when setting up a cord blood bank and the possibility of collaboration with an HPC donor registry. Next, we will describe the challenges when providing a cord blood product internationally. An HPC donor registry is an organization responsible for the search for HPCs from donors (including cord blood) unrelated to any potential recipients. An HPC donor registry can assist a starting cord blood bank with listing their cord blood products in international databases, like Bone Marrow Donors Worldwide (BMDW), to make these CBUs available for transplant units. Next, we will focus on the issues transplant units face when a cord blood transplant search is initiated and on what needs to be done if the best-matched CBU has to be imported from another country. Finally, we will give an analysis of the field based upon the WMDA annual reports and future plans for cord blood banks.

Cord Blood Stem Cells Medicine. http://dx.doi.org/10.1016/B978-0-12-407785-0.00019-0

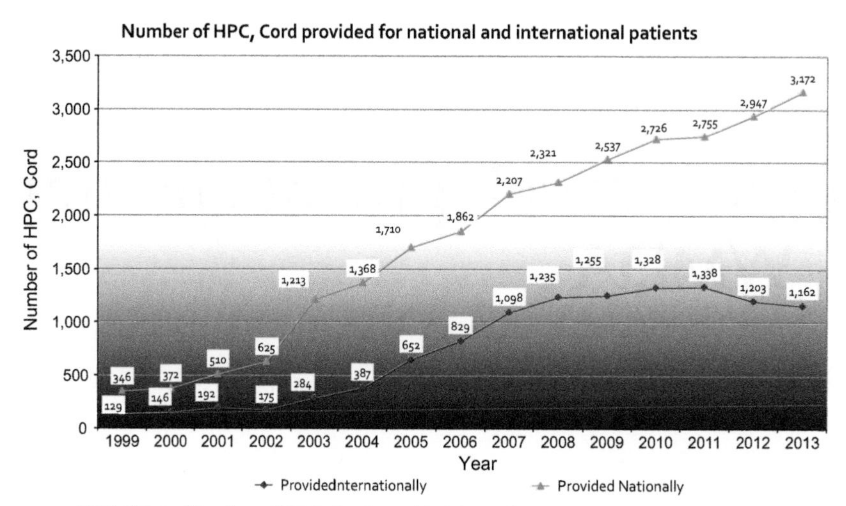


**FIGURE 1** Number of HPC, Cord provided for national and international patients.

## 2. ISSUES TO CONSIDER WHEN STARTING A CORD BLOOD BANK

Establishing a cord blood bank is a very large and complex endeavor, which requires extensive resources. It also requires commitment, motivation, funding, and competence in collecting and processing of cord blood. The motivation to start a cord blood bank is often a patient in need of a transplant.

Setting up a cord blood bank and taking part in the international exchange of cord blood products starts in your own country, as the primary goal of a cord blood bank is to help patients in need of a transplant in their own country. Therefore, it is important to assess the status of and need for unrelated donor transplantations in your country. This concerns questions such as:

- Which country or registry or cord blood bank is currently providing the cord blood products for patients in your country?
- Will the import of these products with acceptable human leukcocyte antigen (HLA) matches and at acceptable cost be sustainable in the future?
- What percentage of patients have no matches among their siblings or family members?
- Are there large family sizes or consanguineous marriages in your country that minimize the need for unrelated donors? Or is the need increased because of genetic heterogeneity, which causes hybrids of rare haplotypes?

Also consider whether in your country haploidentical transplants, where HLA type is half-matched, are preferred to unrelated donor transplants. Many transplant units are successfully using haploidentical graft sources as an alternative to unrelated donor sources due to logistical and financial limitations. These haploidentical donors are usually parents or children of the patient and thus, in general, available to the patient. However, posttransplantation, the patient's immune reconstitution may be poorer and relapse rates may be higher.[2]

Once you have assessed the status of unrelated donor transplantation in your country, the next thing to determine is if you would like to start an organization, recruiting adult volunteer donors (the so-called donor center or donor registry), or a cord blood bank. The WMDA developed a handbook with answers to any questions that come to mind when you consider establishing a cord blood bank in your country.[3]

When you would also like to ship CBUs to other countries to help patients abroad, it is highly recommended that representatives of your cord blood bank become familiar with the WMDA and BMDW to learn more about the international exchange, and with the American Association of Blood Banks (AABB) and NetCord Foundation to learn more about quality standards (see below for more information about these organizations). Being a member of professional societies gives you the opportunity to learn more from the experience of colleagues in the field and to share your experience with others. Professional societies have developed educational materials in order to help you start up a cord blood bank.

| Organization | Description | Educational Resources |
|---|---|---|
| AABB www.aabb.org | AABB, formerly known as the American Association of Blood Banks, is an international nonprofit association committed to advancing the practice and standards of transfusion medicine and cellular therapies to optimize patient and donor care and safety. | The AABB Web site presents a wide variety of resources: http://www.aabb.org/resources |

| Organization | Description | Educational Resources |
|---|---|---|
| BMDW www.bmdw.org | BMDW is a voluntary collaborative effort of stem cell donor registries and cord blood banks whose goal is to provide centralized information on the human leukocyte antigen (HLA) phenotypes and other relevant data of unrelated stem cell donors and cord blood units and to make this information easily accessible. | The BMDW users guide provides you information about how to work with the matching programs of BMDW, see link: http://www.bmdw.org/uploads/media/BMDW_User_Guide_03.pdf |
| NetCord www.netcord.org | The International NetCord Foundation is a nonprofit foundation of umbilical cord blood banks whose members supply the largest source of high-quality cord blood grafts for patients in need of HPC transplant. | The accreditation program of NetCord is facilitated by the Foundation for the Accreditation of Cellular Therapy (FACT) office. The Web site of FACT provides information on aspects of this accreditation program, see link: www.fact-website.org |
| WMDA www.wmda.info | WMDA is a global association whose mission is to ensure that high-quality stem cell products are available for all patients in need, while maintaining the health and safety of the volunteer donors. | WMDA handbook[3] Home page of the cord blood working group on www.wmda.info |

These professional societies are part of the umbrella organization Worldwide Network for Blood and Marrow Transplantation (WBMT), a nonprofit scientific organization that promotes excellence in stem cell transplantation, stem cell donation, and cellular therapy. One of the standing committees of the WBMT, the Alliance for Harmonisation Cellular Therapy Accreditation (AHCTA), developed a document that is used as a template for new or developing transplant programs.[4] The document is a helpful guide describing how a transplant unit can comply with regulation and accreditation requirements. The document lists requirements, followed by examples of how a transplant unit can comply to this specific requirement. For example: a requirement is that a transplant unit must have procedures in place for the administration of cellular products. An example of how to meet this requirement is that a transplant unit must describe a procedure when a CBU is used that is not red blood cell reduced. In that case the transplant unit must require post-thaw dilution or wash techniques to dilute or remove red blood cell. The document does not contain the full requirements of standards, but seeks to provide clear examples of compliance and additional detail to support basic quality system elements.

WMBT organizes scientific and educational conferences to educate regulators and physicians on how to set up transplant programs. The first conference was organized in Vietnam (2011), the second one in Brazil (2013), and a third one is planned for South Africa (2014). The presentations from these conferences are available on the WBMT Web site (www.wbmt.org).

One of the main activities of the WBMT is to analyze data on global transplant outcomes. Based on these data, in December 2012, the WBMT announced that the 1 millionth transplant was performed.[5] In these transplants, 53% were autologous and 47% allogeneic. An estimated 250,000 transplants originated from an unrelated adult volunteer donor or cord blood. The WMDA, BMDW, NetCord, and AABB are partners in the WBMT.

One of the reasons to start a cord blood bank is that for some patient populations a fully-matched adult donor cannot be found. Fortunately, for the majority of Caucasian patients, matched unrelated donors can be found among the 25 million potential adult volunteers registered worldwide, but this is not the case for most blacks, Hispanics, Asians, and Native Americans,[6] due to greater HLA polymorphism and underrepresentation in the international database of potential donors. Therefore, some cord blood banks aim to increase their pool with uncommon haplotypes by recruiting ethnic/minority donors, to maximize chances of finding a matched donor or CBU for every patient. These cord blood banks collaborate with birth clinics where many mothers from ethnic minorities give birth in order to collect cord blood that may result in enrichment of the current donor database. Examples of countries with a relatively high percentage of unique HLA-A, -B, -DR split phenotypes of CBUs are: Thailand (21.14%), Mexico (13.13%), Greece (9.36%), France (8.76%), and the United States (8.56%).[7]

Funding typically is the biggest obstacle for the establishment of a cord blood bank, as storage of cells requires a guarantee of continuous and indefinite monitoring. In order to get funding, you need to set out a mission, vision, and goals for your cord blood bank. For example, you need to determine if you will only provide CBUs nationally, internationally, or both. If you decide to ship CBUs internationally, it is important to estimate the number of CBUs your cord blood bank might be able to ship for patients in need of a transplant.

In 2013, the WMDA indicated that 30% of all cord blood products were provided to a patient residing in a different country than the CBU. In 2013, more than 4334 CBUs were

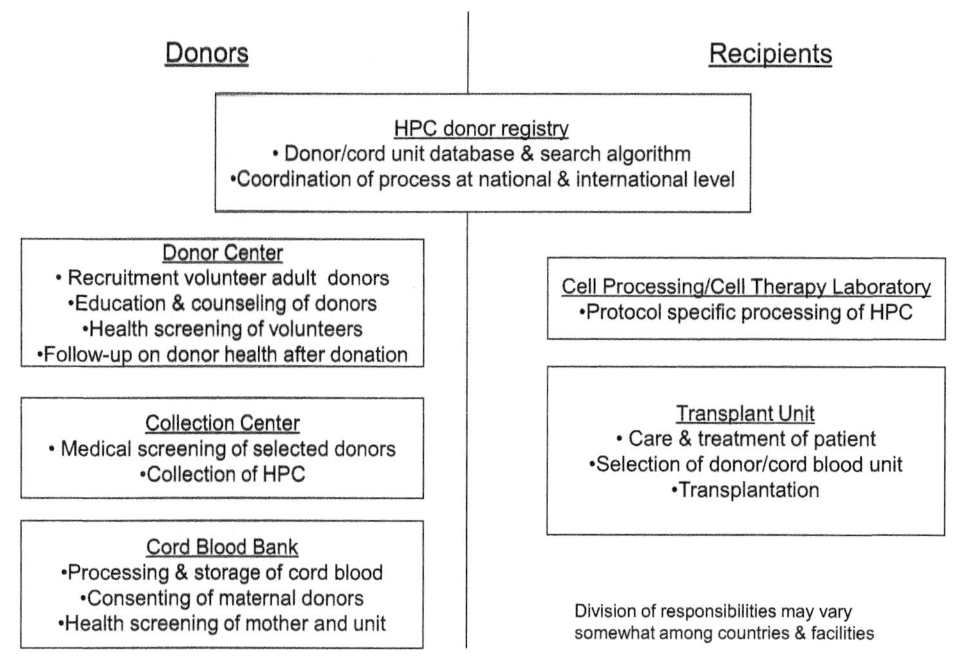


**FIGURE 2** Areas of responsibility related to hematopoietic stem cell transplantation with unrelated donors and cord blood units.

provided for unrelated transplantation globally. From these CBUs, 1402 were provided for patients in the United States, followed by 1159 CBUs for Japanese patients. Nearly 59% of all cord blood transplants took place in only these two countries.[1] The Japanese population is an HLA-homogenous population, so almost 100% of the cord blood products for Japanese patients come from Japanese cord blood banks. The population of the United States is less homogenous and, in 2013, 20 countries outside the United States provided cord blood products for US patients.

If you decide to provide CBUs internationally, it is crucial that your CBUs are accessible to transplant units who are searching for a CBU. One possibility is to collaborate with an HPC donor registry in your country. Usually, HPC donor registries are already listing adult volunteer donors and are responsible for the coordination of the search for HPC products from peripheral blood or bone marrow/CBUs unrelated to the potential recipient (Figure 2). As a starting cord blood bank, you should decide if you will be collaborating with an HPC donor registry or if you will be handling all communication with transplant units directly. In general, the development of one single entity that coordinates all activities of cord blood banks and donor centers is preferred, as it is generally easier for the transplantation community, funding agencies, potential donors, and the general public to work with one single partner only. If a new cord blood bank chooses to work independently from an existing HPC donor registry, it is important to distinguish itself and communicate to all parties involved what the underlying reasons are. If an HPC donor registry has not yet been established in your country, the cord blood bank will be contacted directly by transplant units. Examples are the Chile Cord Blood Bank and the Vietnam Hochiminh City Cord Blood Bank. In these countries an HPC donor registry has not yet been set up.

Over 85% of the public cord blood banks have established a relationship with a national HPC donor registry, which oversees the facilitation of CBUs. Usually, the cord blood bank and the HPC donor registry have a written agreement or contract describing the nature of their relationship, the allocation of responsibilities, and the division of tasks. As part of the agreement, the HPC donor registry may audit the cord blood bank to inspect their counseling of maternal donors, medical evaluation, testing, collection, labeling, documentation, or transport.

## 3. LISTING OF CORD BLOOD UNITS—MAKING THEM AVAILABLE TO TRANSPLANT UNITS

In Chapter 15, you have read about the cord blood banking process. When the CBU is processed, tested, and released for listing in international databases, an infrastructure needs to be developed to facilitate the search requests. In this section you can read more about the procedures that start with the listing of a CBU in international database, and end with the infusion of the product into a patient.

CBUs are prepared for listing after performing a comprehensive review of details or characteristics for each unit, including maternal donor selection and maternal and infant donor evaluation for specific medical requirements, as well as testing to ensure the product meets international requirements, and at least has been tested for the following information: gender and race of the infant donor, total nucleated cell (TNC) count, HLA typing, infectious disease marker testing done on the mother and on the CBU, collection date, processing method, sample inventory, and the type of bags used for cryopreservation of the unit. All information about the CBU is summarized on a form, a so-called CBU report (Figure 3), which can be provided to transplant units on request.

Beside this comprehensive report the CBUs needs to be listed in an international database, which is accessible for transplant units searching for a cord blood product. An example of an international database is BMDW. BMDW is a centralized database in which adult volunteer donors and CBUs are listed from 74 HPC donor registries and 49 cord blood registries from 53 different countries.[8] An organization listing CBUs in BMDW has to comply to the following criteria:

- The cord blood bank must have at least 50 HLA-typed CBUs.
- The cord blood bank must be operational.
- All CBUs registered in the BMDW database must be available for donation under the legislation of the individual donors' country.
- The registry must be willing and able to provide hematopoietic stem cell products from unrelated CBUs to transplant units abroad.
- The registry must provide BMDW a complete data file at least twice per year, but preferably more often.
- The registry must keep the BMDW office informed about the registry's contact information.

A cord blood bank that fulfills these criteria can contact the BMDW office at bmdw@europdonor.nl. The next step will be to send the file with the HLA data of the CBUs to the BMDW office in order to list the CBUs in the database and make them available for transplant units worldwide.

BMDW currently lists 619,051 CBUs (October 2014).[8] It is known that there are cord blood banks together containing at least 45,175 CBUs that are not listed in international databases. These cord blood banks only list their units locally.[1] Transplant units can access these CBUs by contacting the cord blood banks directly. Examples of cord blood banks that have not yet listed their CBUs in international databases are:

- Japanese Cord Blood Bank Network (JCBBN), Japan
- Sichuan Cord Blood Bank, China
- Guangzhou Cord Blood Bank, China
- Beijing Cord Blood Bank, China
- CNTS Cord Blood Bank, Mexico
- La Raza Cord Blood Bank, Mexico

## 4. REGISTRY—GATEWAY TO THE WORLD

As soon as transplant units are aware that your cord blood bank has stored CBUs that are available for unrelated patients they can contact you and send a search request.

Some cord blood banks have developed their own infrastructure to facilitate the search requests coming from national and international transplant units. It requires investments in terms of staff and communication technology to set up an infrastructure to facilitate all search requests from transplant units. Beside staff with knowledge of collection and processing of cord blood products, these cord blood banks have appointed staff with expertise in human histocompatibility and hematopoietic stem cell transplantation. These individuals have a basic understanding of diseases that can be treated by hematopoietic stem cell transplantation, comprehend alternative therapies and donor search problems associated with these diseases, understand HLA specificities (serologic and DNA-based) and haplotypes, and possess knowledge of (inter)national transplant unit and registry protocols.[9] Next to the investment in staff, the cord blood banks also invested in adequate communication technology to facilitate searches for CBUs, like telephone, fax and international telematics links, and access to search databases.

One benefit of working with an HPC donor registry is their expertise in listing data in international databases and their communication technology to facilitate search requests from the transplant units. In addition, HPC donor registries have systems in place to coordinate data transactions including demographic, clinical, and genetic information and lab samples and results for CBUs. One of the most complicated components, which requires HLA expertise, is creating and maintaining a search algorithm to match potential donors/CBUs with patients.

HPC donor registries also offer knowledge of import and export requirements per country. As regulations differ from country to country, excellent communication and cooperation is essential to ensure that all appropriate donor screening questions and infectious disease tests have been performed, according to the importing country's requirements.

HPC donor registries have HLA expertise and knowledge about searching for and listing of adult volunteer donors. The establishment of cord blood banks brought a change to the global infrastructure providing HPC products. The unique properties of cord blood cells and differences in the search processes between cord blood and adult unrelated donors have made HPC donor registries review their current practice and develop new processes to provide cord blood products (Figure 4).

Until 1993, only freshly collected HPC products from adult volunteer donors had been provided internationally through HPC donor registries. These donations were

**CB21**

## CORD BLOOD UNIT REPORT

Page 1 of 2

| CORD BLOOD UNIT DATA | |
|---|---|
| Cord Blood Bank/Registry: | Unit collection date: (YYYY-MM-DD) |
| Registry CBU ID: | Local CBU ID: |
| ID on CBU bag 1: | Gender: |
| ID on CBU bag 2: | Baby's race: |
| ID on CBU bag 3: | OR: Mother's race: |
| ID on CBU bag 4: | Father's race: |

| CORD BLOOD UNIT HLA TYPING | | | | | |
|---|---|---|---|---|---|
| Locus: | A | B | C | DRB1 | DQB1 |
| First antigen: | | | | | |
| Second antigen: | | | | | |
| Is this typing confirmed? ☐ Yes ☐ No | | | If yes, was a contiguous sample (segment) used? ☐ Yes ☐ No | | |
| Was viability testing performed? ☐ Yes ☐ No | | | Viability testing results: % viable | | |

| CBU POST-PROCESSING COUNT, CULTURES & IDMs | | | | |
|---|---|---|---|---|
| Processing date: (YYYY-MM-DD) | TNC: x $10^7$ | CD34+: x $10^6$ | CFU: x $10^5$ | nRBC: x $10^6$ |
| Bacterial Cultures: | ☐ Positive | ☐ Negative | ☐ Not Done | |
| Fungal Cultures: | ☐ Positive | ☐ Negative | ☐ Not Done | |

**IDM RESULTS** N = Non-reactive/Negative, R = Reactive/Positive, NA = Not tested

| Maternal IDM Results | Sample collection date: (YYYY-MM-DD) | | | Test date: (YYYY-MM-DD) | | | |
|---|---|---|---|---|---|---|---|
| HBsAg | ☐N | ☐R | ☐NA | Anti-HTLV I/II | ☐N | ☐R | ☐NA |
| Anti-HBc | ☐N | ☐R | ☐NA | Syphilis | ☐N | ☐R | ☐NA |
| Anti-HCV | ☐N | ☐R | ☐NA | EBV | ☐N | ☐R | ☐NA |
| Anti-HIV 1/2 | ☐N | ☐R | ☐NA | Toxoplasmosis | ☐N | ☐R | ☐NA |
| HIV-1 NAT | ☐N | ☐R | ☐NA | CMV | ☐N | ☐R | ☐NA |
| HCV NAT | ☐N | ☐R | ☐NA | ☐ IgG ☐Total | | | |
| HBV NAT | ☐N | ☐R | ☐NA | | | | |
| **Additional tests:** | | | | | | | |
| | ☐N | ☐R | ☐NA | | ☐N | ☐R | ☐NA |

| CBU IDM Results (optional) | Sample collection date: (YYYY-MM-DD) | | | Test date: (YYYY-MM-DD) | | | |
|---|---|---|---|---|---|---|---|
| HBsAg | ☐N | ☐R | ☐NA | Anti-HTLV I/II | ☐N | ☐R | ☐NA |
| Anti-HBc | ☐N | ☐R | ☐NA | Syphilis | ☐N | ☐R | ☐NA |
| Anti-HCV | ☐N | ☐R | ☐NA | EBV | ☐N | ☐R | ☐NA |
| Anti-HIV 1/2 | ☐N | ☐R | ☐NA | Toxoplasmosis | ☐N | ☐R | ☐NA |
| HIV-1 NAT | ☐N | ☐R | ☐NA | CMV | ☐N | ☐R | ☐NA |
| HCV NAT | ☐N | ☐R | ☐NA | ☐ IgG ☐Total | | | |
| HBV NAT | ☐N | ☐R | ☐NA | | | | |
| **Additional tests:** | | | | | | | |
| | ☐N | ☐R | ☐NA | | ☐N | ☐R | ☐NA |

Reset Form

20140724-CBWG-FORM-CB21

**CB21**

## CORD BLOOD UNIT REPORT

Page 2 of 2

| Cord Blood Bank/Registry: | Registry CBU ID: | Local CBU ID: |
|---|---|---|

| HEALTH HISTORY | |
|---|---|
| Did the mother participate in any behavior that put her at high risk for contracting HIV or hepatitis or did she travel to an area or areas that put her at risk for disease transmission? | ☐Yes ☐No |
| If yes, please describe: | |
| Were any familial or genetic disease risks identified, including cancer, blood disorders, enzyme deficiencies, metabolic/storage diseases or autoimmune disorders? | ☐Yes ☐No |
| If yes, please describe: | |
| Hemoglobinopathy screening: ▾ | Blood type: |
| If trait or disease, please specify: | |

| PROCESSING, SAMPLE INVENTORY AND MATERNAL FOLLOW UP | | | |
|---|---|---|---|
| **Unit Processing** | **Available Samples** | | |
| Collection volume: | **CBU Samples** | Quantity | Volume |
| Total volume frozen: | Aliquots for DNA | | ml |
| Processing method: ▾ | Viable cell aliquots | | Mio |
| If manual, please specify; If automatic, what type of system? | Plasma aliquots | | ml |
| | Serum aliquots | | ml |
| Type of bag used: ▾ | Attached segments | | |
| | **Maternal Samples** | Quantity | Volume |
| Other: | Aliquots for DNA | | ml |
| Product modification: ▾ | Viable cell aliquots | | Mio |
| | Plasma aliquots | | ml |
| Other: | Serum aliquots | | ml |

| MATERNAL FOLLOW UP | | |
|---|---|---|
| Was maternal follow up done? | ☐Yes ☐No | Date of contact (YYYY-MM-DD) |
| Any problems? | | |
| Was maternal IDM testing repeated? | ☐Yes ☐No | If yes, please enter results below: |

**IDM RESULTS** N = Non-reactive/Negative, R = Reactive/Positive, NA = Not tested

| **Maternal IDM Results** | Sample collection date: (YYYY-MM-DD) | | Test date: (YYYY-MM-DD) |
|---|---|---|---|
| HBsAg | ☐N ☐R ☐NA | Anti-HTLV I/II | ☐N ☐R ☐NA |
| Anti-HBc | ☐N ☐R ☐NA | Syphilis | ☐N ☐R ☐NA |
| Anti-HCV | ☐N ☐R ☐NA | EBV | ☐N ☐R ☐NA |
| Anti-HIV 1/2 | ☐N ☐R ☐NA | Toxoplasmosis | ☐N ☐R ☐NA |
| HIV-1 NAT | ☐N ☐R ☐NA | CMV | ☐N ☐R ☐NA |
| HCV NAT | ☐N ☐R ☐NA | ☐ IgG ☐Total | |
| HBV NAT | ☐N ☐R ☐NA | | |
| **Additional tests:** | | | |
| | ☐N ☐R ☐NA | | ☐N ☐R ☐NA |

| Person completing form: | Date (YYYY-MM-DD): |
|---|---|

Reset Form

20140724-CBWG-FORM-CB21

FIGURE 3 Cord blood unit report form.

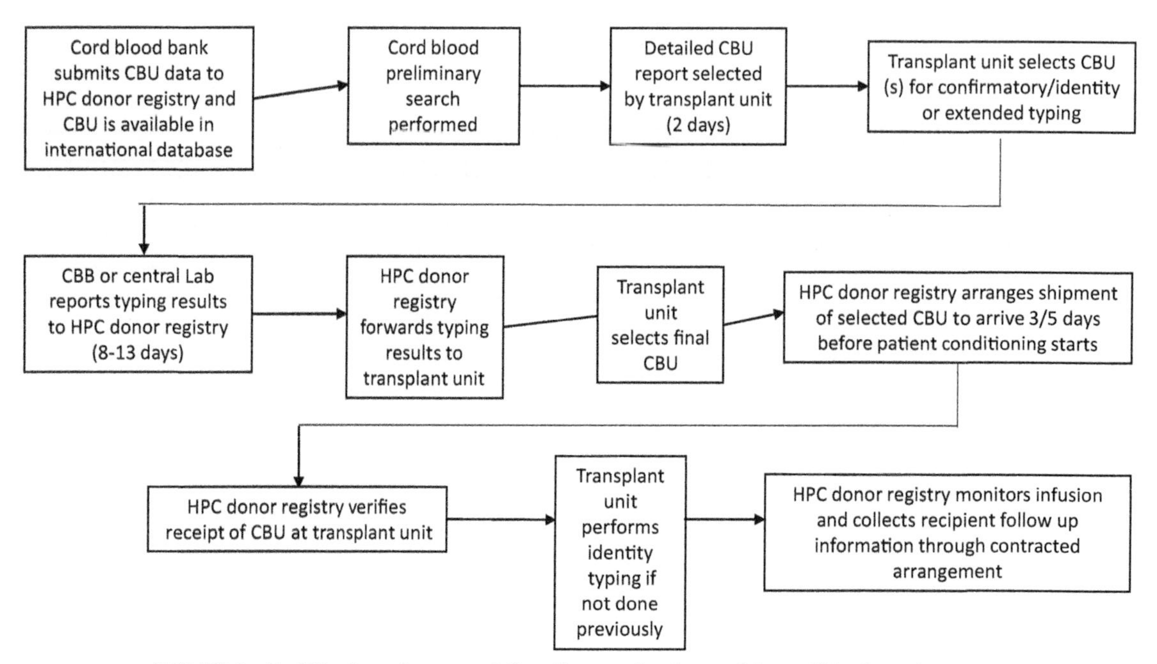


**FIGURE 4** Cord blood search process. A flow diagram of each step of the cord blood search process.

collected either in a marrow collection center or in an apheresis collection center and were immediately transported to the transplant unit. Usually, the courier was someone working for the transplant unit who transported the product personally, carried by hand, from one country to another. For cord blood the process is completely different. Cord blood is collected directly after the baby is born, processed in a specialized facility, and stored in a freezer. Cord blood products are shipped in dry shippers by specialized courier companies.

One of the disadvantages of CBUs is their low cell count, which means that they are mainly used for juvenile patients as they do not contain enough cells for adult patients. To overcome the low cell content of single CBUs, various alternatives have been tried. Infusing two CBUs instead of one is now frequently used.[10] In 2013, at least 1300 CBUs provided for unrelated transplantation were used in double cord transplantations.[1] These double cord blood transplants required adaptation of the matching programs and also logistical solutions, as sometimes the CBUs come from different cord blood banks.

In the next paragraphs you can read how HPC donor registries have organized the search process for a cord blood transplant and which challenges HPC donor registries have faced over time.

## 5. SEARCH FOR CORD BLOOD

Cord blood as a hematopoietic cell source has practical, biological, and clinical advantages for donors and recipients: prompt availability upon request (a few weeks for cord blood versus a number of months for adult blood or bone marrow); provision for ethnic minorities, who are underrepresented in HPC donor registries; no risk of anesthesia for donors or stimulation with growth factors; lower risk of transmissible infectious diseases; partially compatible transplants may be performed; and lower immunogenicity and lower risk of severe posttransplant immune responses, such as Graft-versus-Host Disease, which is the main cause of high posttransplantation mortality. Search strategies differ between countries and between transplant units. Some clinicians only transplant CBUs, others only transplant HPCs from adult volunteer donors, and other clinicians decide case-by-case. When HPC donor registries also started to facilitate cord blood transplants, they were confronted with an entirely different process from that which they were used to. A search process for cord blood differs from a search process for adult volunteer donors: there are fewer steps in the release of CBUs, because these are fully tested and the HLA typing is done at the time of storage and listing for searches. Therefore, the process is faster, because it does not involve the donor work-up activities required for marrow and apheresis collection. Figure 4 shows the cord blood search process with recommended time frames.[11]

Once the CBUs are HLA-typed and have been medically qualified, they can be listed and become accessible to transplant physicians/units. As described earlier the majority of the CBUs are listed in the international database BMDW.

Transplant units and search coordination units have access to databases like BMDW, when searching for a CBU. Running a BMDW search report is particularly helpful to set an optimal but realistic target for an international donor search. However, when considering the overall search strategy and the usefulness of an extended international search, one must also take into account the variation of allele and haplotype frequencies in different geographic, racial, or ethnic groups. HPC donor registries can help transplant units with requests to worldwide HPC donor registries and cord blood banks, but the selection of a unit is the final responsibility of the transplant unit. An important parameter to indicate the quality of a CBU is the TNC count measured postprocessing, prior to cryopreservation and once again determined on a segment, prior to the release of the CBU.

Although the specific requirements of the transplant programs vary accordingly, published data from both Europe and the United States underline the importance of the TNC count per kilogram of patient weight in reduced intensity and myeloablative conditioning regimens for both single and double cord blood transplantation. Secondly, with regard to the HLA-matching status, it has been found that a single (or no) mismatch is preferred to a double mismatch, and that three or four mismatches should be avoided for almost all patients. The third important factor is to identify whether the donor is selected for a patient with malignant or nonmalignant disease since the effect of the TNC and HLA may differ.[12]

An initial search report contains information about the CBUs that have the highest HLA-matching grade with the patient; usually, the class I phenotype of cord blood cells is determined by antibody-binding analysis and the class II phenotype by DNA analysis of the alleles encoding the HLA–DRB1 chain, after which the CBUs are sorted according to the reported TNC count before freezing. International recommendations from Eurocord and other data registries advise a minimum of $2 \times 10^7$ TNC/kg and $2 \times 10^5$ CD34+cells/kg of recipient body weight.[13]

By contacting the HPC donor registry/cord blood bank and asking for a "Cord Blood Unit Report" (Figure 3), a transplant unit can obtain detailed information about the CBU. This "Cord Blood Unit Report" contains information about the gender and race of the donor, TNC, HLA typing, testing done on the mother and on the CBU, collection date, processing method, sample inventory, and the type of bags used for cryopreserving the unit. This first interaction between the transplant unit and the HPC donor registry/cord blood bank is called a preliminary search.

Based on the information shown on the Cord Blood Unit Report, the transplant unit decides whether to move to the next step in the process, called the "formal search." A formal search means that the transplant unit asks for more detailed HLA typing of the unit and sometimes also for testing for additional local infectious disease markers, such as human T-lymphotropic virus and West Nile virus, to be performed on maternal blood samples. The transplant unit contacts HPC donor registries by filling in a "Cord Blood Unit—information and typing request." (Figure 5)

When a transplant unit orders a CBU by filling in a CBU shipment request (Figure 6), the shipment date is agreed upon between the transplant unit, the HPC donor registry, and the cord blood bank. Although CBUs can be shipped within 1 or 2 days for emergency orders, most cord blood banks prefer at least 3–5 days notice. Cord blood banks that perform additional prerelease viability or potency tests like a colony-forming unit assay may require up to 15 days notice. CBUs are transported in a dry shipper, which contains the CBU in vapor phase liquid nitrogen. The dry shipper is collected from the cord blood bank by a courier company with expertise in international shipping and should not be irradiated for inspection.

Professional societies, like NetCord-FACT and the WMDA, developed guidelines in their standards for the transport of cord blood products. An important aspect is that procedures must be in place to protect the integrity of the CBU, next to a plan for alternative transportation in case of an emergency. The transport of a CBU starts with the dry shipper: a vessel containing hydrophobic absorbent material, fixed in the inner capacity. After complete saturation of the material with liquid nitrogen, the vessels are ready to transport samples at cryogenic temperatures, without the risk of liquid nitrogen spilling, should the container be overturned. The temperature needs to be validated to guarantee it will stay below −150 °C for a period of at least 48 h. To ensure and to check this, the dry shipper needs a device that continuously monitors the temperature, as the older minimum temperature single use indicators do not meet the current standards. Documentation with the correct address and emergency contact information needs to be shipped together with the CBU. The label also needs to show the date and time of distribution; the name and contact details of the cord blood bank providing the unit; the name of the receiving transplant unit; the identity of the person responsible for the receipt of the CBU; statements like: "do not X-ray," "medical specimen," "handle with care," "cord blood for transplantation," and shipper handling instructions. Finally, a cord blood bank needs to ensure that the dry shipper will not be opened during transport.

CBUs are ordered before the patient is given the conditioning regimen used in the transplant procedure, to ensure that the shipment arrives safely and in time. When the CBU arrives, the transplant unit is responsible for checking the accompanying unit records, unloading the shipper, checking the integrity of the unit, and placing it in a liquid nitrogen freezer for temporary storage until the transplant date. The dry shipper is returned to the cord blood bank to inspect for any damage and to read the data from the continuous temperature monitor in order to ensure the desired temperature was maintained during shipment. One of the advantages of using a cord blood donor is that the CBUs are sent frozen, which enables the transplant unit to obtain the unit before starting the conditioning regimen.

The last step is to issue an invoice for services requested and rendered (e.g., shipping of cord blood product) for the transplant unit. If a cord blood bank collaborates with

# CB10 CORD BLOOD UNIT - INFORMATION AND TYPING REQUEST

Page 1 of 1

☐ Urgent request

| PATIENT DATA | | | |
|---|---|---|---|
| Patient name: | | Patient ID: (assigned by patient registry) | |
| Patient registry: | | Patient ID: (assigned by donor registry) | |
| Transplant center: | | | |
| Date of Birth (YYYY-MM-DD): | Gender: ▾ | Weight in kg: | Blood group/RhD: |
| Diagnosis: | | Estimated transplant date (YYYY-MM-DD): | |

| PATIENT HLA | | | | | |
|---|---|---|---|---|---|
| **Locus:** | **A** | **B** | **C** | **DRB1** | **DQB1** |
| **First antigen:** | | | | | |
| **Second antigen:** | | | | | |

| CORD BLOOD UNIT EXTENDED HLA TYPING REQUEST | | | | | | |
|---|---|---|---|---|---|---|
| **CBU ID:** | | | | | | |
| **A** | **B** | **C** | **DRB1** | **DQB1** | **Other:** | **Other:** |
| ☐ ▾ | ☐ ▾ | ☐ ▾ | ☐ ▾ | ☐ ▾ | ☐ ▾ | ☐ ▾ |

| ADDITIONAL CORD BLOOD UNIT DATA | | |
|---|---|---|
| The requesting institution requests the following details: | Cord Blood Bank representative answers: | |
| ☐ Was red cell reduction performed prior to cryopreservation? | ☐Yes ☐No | |
| ☐ Please give the **total** erythrocytes of the unit: | x $10^9$ | |
| ☐ Was viability testing performed on **post**-cryopreserved material? | ☐Yes ☐No | Testing results: % viable |
| ☐ Was colony testing (e.g. CFU-GM) performed on **post**-cryopreserved material? | ☐Yes ☐No | |
| ☐ Was HLA verified on segment of the unit? | ☐Yes ☐No | If yes, test date: (YYYY-MM-DD) |
| ☐ Is maternal HLA typing available? | ☐Yes ☐No | |
| ☐ What type of bag is used? | ▾ | |
| ☐ Please provide a detailed unit report. | | |
| ☐ Additional questions: | | |

| **Requesting institution:** | **Invoice address:** |
|---|---|
| Institution: | Institution: |
| Address: | Address: |
| Attention: | Attention: |
| Phone: | Phone: |
| Fax: | Fax: |
| E-mail: | E-mail: |

| Cord Blood Bank representative: | Date (YYYY-MM-DD): | Signature: |
|---|---|---|

Reset Form

20140724-CBWG-FORM-CB10

FIGURE 5 Cord blood unit—information and typing request form.

# CB30 CORD BLOOD UNIT SHIPMENT REQUEST

Page 1 of 1

| PATIENT DATA | |
|---|---|
| Patient name: | Patient ID: (assigned by patient registry) |
| Patient registry: | |
| Transplant center: | Patient ID: (assigned by donor registry) |
| Date of Birth (YYYY-MM-DD): Gender: | Weight in kg: Blood group/RhD: |
| Diagnosis: | Estimated transplant date (YYYY-MM-DD): |

| PATIENT HLA | | | | | |
|---|---|---|---|---|---|
| **Locus:** | **A** | **B** | **C** | **DRB1** | **DQB1** |
| **First antigen:** | | | | | |
| **Second antigen:** | | | | | |
| ☐ Initial typing ☐ Verification typing | | | Typing date (YYYY-MM-DD): | | |
| Cord Blood Unit ID: | | | | | |

**ADDITIONAL PRE-RELEASE CHECKS**

The transplant center requests the following tests to be done on CBU at time of release and/or additional information:

Please test the following on post-cryopreservation attached segment of CBU at time of release:

- ☐ Viability test
- ☐ Colony testing (e.g. CFU-GM)
- ☐ CD34 pos test
- ☐ HLA verification test
- ☐ Additional IDM tests, please specify:
- ☐ Blood or other sample shipment, please specify:
- ☐ Maternal health questionnaire or summary statement
- ☐ Other tests:

| PROPOSED TIME FRAME FOR CORD BLOOD UNIT SHIPMENT | |
|---|---|
| Preferred date: (YYYY-MM-DD) | Preferred delivery time: (HH:MM + local time zone) |
| Start of conditioning: (YYYY-MM-DD) | Conditioning regimen: ☐ Myeloablative ☐ Non-myeloablative |
| Transplant type: ☐ Single cord ☐ Double cord ☐ Multiple cord ☐ Single cord in combination with haplo-donor ☐ Ex-vivo expansion transplant ☐ Other: | |
| Transplant date: (YYYY-MM-DD) | Comments: |
| Transport to be organised by: <br> Dry shipper to be provided by: | Preferred courier: |

| **Cord blood unit** to be shipped to: | **Invoice(s)** to be sent to: |
|---|---|
| Institution: | Institution: |
| Address: | Address: |
| Attention: | Attention: |
| Phone: | Phone: |
| Fax: | Fax: |
| E-mail: | E-mail: |

| Person completing form: | Date (YYYY-MM-DD): | Signature: |
|---|---|---|
| | | |

Reset Form

20140724-CBWG-FORM-CB30

FIGURE 6 Cord blood unit shipment request form.

an HPC donor registry, the HPC donor registry is fully responsible for the settlement of charges. Some HPC donor registries/cord blood banks invoice when the CBU leaves the cord blood bank, while other HPC donor registries/cord blood banks invoice when the CBU has safely arrived at the transplant unit. It is important to establish an agreement with the courier company in which the responsibilities are described in situations where the CBU does not arrive as agreed upon.

As part of the quality management system, it is mandatory to collect recipient follow-up medical data to safeguard the quality control of the collection, processing, storage, and distribution of the CBU. In Europe, the Eurocord registry facilitates as a service provider to collect clinical outcome data and produce an annual activity report for the participating cord blood banks with a statistical analysis of the cord blood transplant outcomes. Other organizations that collect outcome data are the European Group for Blood and Marrow Transplantation (EBMT), the Centre for International Blood and Marrow Research (CIBMTR), the Asia-Pacific Blood and Marrow Transplantation Group, and the Eastern Mediterranean Blood and Marrow Transplantation. The following information should be obtained to evaluate the processes: the number of TNC and CD34+ cells, viability after thawing, method of thawing, and adverse events at the time of infusion. Overall results are published every year by Eurocord–EBMT and National Marrow Donor Program (NMDP–CIBMTR). Collecting outcome data helps formulating important guidelines on cord blood selection, indications, role of HLA, prognostic factors, and comparison with other stem cell sources.

In addition to the patient outcome registries, the WMDA has developed a centralized database, in which adverse events are recorded related to the procurement, processing, storage, and distribution. You will read more about this database later in this chapter (Challenges in the provision of cord blood—regulation).

To summarize, HPC donor registries adopted cord blood as another possibility to help a patient in need of a transplant in the past 20 years. An infrastructure was set up to perform searches and to provide CBUs internationally. Despite these developments, we still face some challenges concerning the provision of cord blood, which will be discussed in the next section.

## 6. CHALLENGES RELATED TO THE PROVISION OF CORD BLOOD—SELECTION

In 1995, when the New York Blood Center's Placental Blood Project published that they had provided 24 cord blood products to unrelated recipients, the way was opened to alternative sources of HPC beside bone marrow and peripheral blood donated by adult volunteer donors.[14] By that time, only 50% of the searches resulted in identification and availability of an acceptably matched donor.[15] Over time, the technology to collect, process, and store cord blood has changed. As transplant units are always looking out for CBUs of high quality for their patients, how do you make sure that a CBU is the best available for the patient?

One parameter for the quality of CBUs is determined by whether a CBU is derived from an accredited cord blood bank or not. Over time, standards have been developed based on events faced by transplant units, HPC donor registries, or cord blood banks. Three international organizations offer accreditation programs and actively collaborate to promote high-quality CBUs: the WMDA, NetCord Foundation in collaboration with the FACT, and the AABB. All these organizations are involved in the AHCTA and aim to harmonize standards internationally. Each accreditation program covers different aspects of unrelated cord blood transplantation. The main objective of the NetCord-FACT[16] accreditation is to evaluate all aspects of cord blood banking including donor selection, methods of collection, processing, testing, and storage of the units and release and transportation of the CBUs to the transplant units. AABB[17] developed high-quality standards in the area of cord blood collection, storage, and distribution, and the WMDA Standards focus on HPC donor registry operations but do not cover aspects of unrelated transplantation that are included in the standards of other organizations (Figure 7). For example, the activities of collection centers collecting either marrow or peripheral blood stem cells, cord blood banking and tissue typing are covered by accreditation standards from entities such as the Joint Accreditation Committee – International Society for Cellular Therapy, EBMT, NetCord, and the European Federation for Immunogenetics (EFI), respectively.

These accreditation schemes address different aspects of cord blood banking, listing, and transplantation, and they all refer to each other for complementary requirements.[18] For example, the WMDA Standards[9] state "If a registry relies on an independent donor centre or cord blood bank to recruit and characterise donors, the registry must ensure that the donor centre/cord blood bank complies with relevant WMDA standards." In this case, requiring accreditation of the cord blood bank by an organization such as NetCord-FACT might be one possibility by which the registry ensures that this WMDA standard can be met. The NetCord-FACT standards describe the requirement in another way: "If a cord blood bank utilises a registry to deliver services related to the listing, search and selection, reservation, and/or release of a CBU, the registry should be accredited by the WMDA." A list of accredited cord blood banks and cord blood registries can be found on the Web sites of the FACT Office (www.factwebsite.org), AABB (www.aabb.org), and WMDA (www.wmda.info).

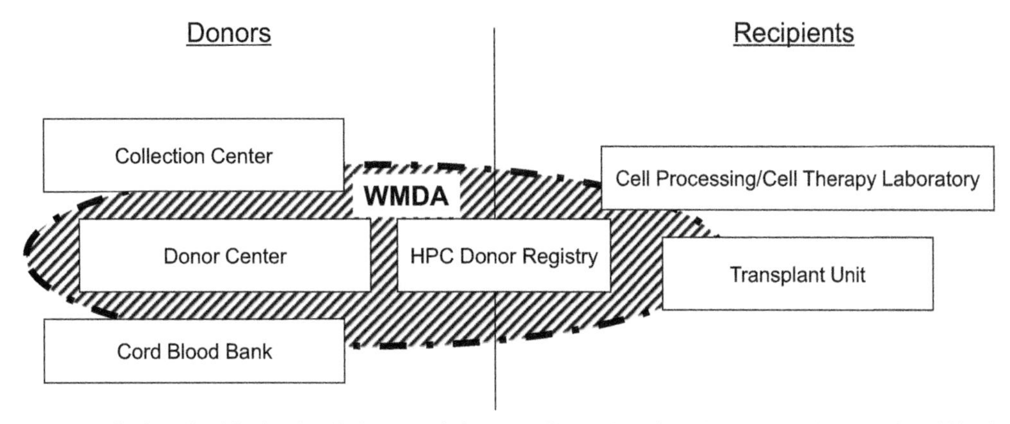


**FIGURE 7** WMDA accreditation of entities involved in hematopoietic stem cell transplantation with unrelated donors and cord blood units.

Despite the existence of these voluntary accreditation programs, a significant number of cord blood banks are not yet accredited, and for transplant units it is difficult to find out if a CBU originates from an accredited cord blood bank or not, as accredited CBUs are not clearly marked in the preliminary search reports.

## 7. CHALLENGES RELATED TO THE PROVISION OF CORD BLOOD—SERVICE PROVIDER

A second important challenge is the registry/cord blood bank's role as service provider, since transplant units need timely and accurate information about a CBU. The WMDA accreditation program states that a preliminary search request needs to be responded to within 24 h. The WMDA has also developed recommendations for cord blood banks about how fast verification typing requests need to be reported back to the requesting transplant units. Verification typing is HLA-typing performed on an independent sample with the purpose of verifying concordance of that typing assignment with the initial HLA-typing assignment. Concordance does not require identical levels of resolution for the two sets of typing but requires the two assignments to be consistent with one another. The WMDA recommendations for verification typing are:

- **For urgent cord blood verification typing requests:** time from request received to results reported to the transplant unit should be a maximum seven calendar days for 80% of the requests received.
- **For standard cord blood verification typing requests:** time from request received to results reported to the transplant unit should be a maximum 13 calendar days for 80% of requests received.

A unique aspect of CBUs is the limited availability of additional samples to perform HLA-typing when requested by a transplant unit. The most valuable sample type is the small segment attached to the CBU, called a contiguous segment. This is the only sample type that can definitively confirm that the HLA typing reported for a CBU really belongs to the CBU actually listed. The number of contiguous segments is limited (one to three small volume segments for one CBU), so when all contiguous segments of a CBU are exhausted, the likelihood that this unit will be ordered in the future is low.

In 2009, HPC donor registries[19] were faced with a situation in which transplant units asked cord blood banks to fully work-up CBUs and then persistently "delayed" the date of transplant—eventually "canceling" the order altogether. Although all parties recognized that this is sometimes unavoidable, some transplant units appeared to be working-up many CBUs simultaneously, while making "final" selection decisions relatively late in the process. Since the number of attached/contiguous segments is limited, the consequence, apart from the (sometimes unpaid) work-up costs, is the decrease in available CBUs with the required testing samples for subsequent searches.

When a request for additional sample testing arrives, three major concerns for transplant units and HPC donor registries/cord blood banks must be addressed:

1. Transplant units need to update and expand the initial HLA typing.
2. Transplant units and HPC donor registries need to be sure about the HLA "identity" of the CBU, i.e., that the CBU is the one with the HLA type listed on the search report. NetCord-FACT standards require that, before a CBU is released, a sample obtained from a contiguous segment of that CBU shall be tested to verify HLA type and, if possible, cell viability (Standard E 3.2, Fifth Edition).
3. If the first transplant unit that requests additional typing of the CBU decides not to order the unit, the cord blood bank or HPC donor registry should have a process in place to prevent loss of value of that unit for use by a future patient.

The professional societies worked out definitions to clarify the critical issue of the sample source used for

HLA typing and to standardize international communication between cord blood banks, HLA labs, transplant units, and HPC donor registries. As historically the term "confirmatory typing" was used, which is commonly used in the selection process of an adult donor, it was agreed to replace all references to "confirmatory" typing with "verification" typing and to implement a new term "extended" typing.

1. Verification (to replace the term "confirmatory") typing must be performed on cells derived from an attached segment and must meet the requirements for the level of resolution and loci as established by the registry. The purpose of typing is to confirm HLA and identity of the CBU selected, i.e., that the unit was the one selected based on its HLA typing shown on the search report.
2. Extended typing is defined as any additional HLA typing to obtain higher or allele-level resolution at HLA-A, -B, -DRB1, or to identify assignments at other loci such as -C or -DQB1. When typing is performed on cells from an attached segment, it can also serve as verification (confirmatory) typing.

If there is no attached segment for the unit, the fifth edition of the NetCord-FACT[16] International Standards for Cord Blood Collection, Banking, and Release for Administration guidance statement for standard E3.2.2 requires the cord blood bank to define and validate an alternative method to confirm CBU identity prior to release and to notify the transplant unit well before release of a CBU.

Issuing financial penalties to transplant units as a general practice should be avoided. Cord blood banks must keep control of available additional cord blood samples and now generally charge transplant units for requested HLA testing. Segment typing only needs to be performed once, as the cord blood bank should keep a historical record of all HLA typing performed on a given unit's samples, and this information should be made available upon request by a transplant unit or search coordination department of an HPC donor registry. HLA-typing data, forwarded to international databases (e.g., BMDW), should always be updated and thus contain the most useful (highest level of HLA typing) information available. A cord blood bank should also have a contract with a quality-controlled central laboratory to do verification or extended typing, rather than releasing samples to individual transplant units. These rules have been accepted by transplant units, aware that they can only get samples for testing in their facility of units they decide to order. Find below a summary of the conclusions:

- It is important that the transplant unit knows what type of sample was used for typing and, in particular, at what point typing can be requested from a contiguous segment.
- Centralized laboratories need to be accredited by professional societies, such as EFI or American Society for Histocompatibility and Immunogenetics, and preferably these laboratories differ from the one that performed the initial typing.
- Results are reported to the transplant unit and the cord blood bank and are available to any other transplant unit that may request unit information in the future.
- If an HPC donor registry or cord blood bank has the policy that segment and viability testing (if required) will only be done after a transplant unit has ordered a unit, results should be available within 48 h.
- If the CBU has two segments, the last segment should stay attached to the unit for additional testing by the transplant unit, if desired.

HPC donor registries and cord blood banks can establish a reservation policy that ensures units be put back into the general available inventory if a transplant unit reserves a unit for a longer period of time. Such a policy should be communicated clearly to a transplant unit at the time of CBU reservation.

The HPC donor registry serves as an intermediate between the cord blood bank and transplant unit. It is important that adequate information technology support is available to ensure that the exchange of information is validated and that the information provided is accurate. For example, if a CBU is removed from the cord blood bank's inventory, this change is directly reported to the HPC donor registry. Chapter 6 of the WMDA handbook[3] describes the listing and search process from the HPC donor registry's side.

## 8. CHALLENGES RELATED TO THE PROVISION OF CORD BLOOD—REGULATION

The third challenge has to do with the regulatory landscape. Apart from submission of clinical data to patient outcome registries and accreditation by professional societies like AABB and NetCord-FACT, national health authorities have also developed regulations.

The last decade, the European Committee and the US Food and Drug Administration (FDA) developed regulations to control the import and export of cord blood products in their region. The EU Directive 2004/23/EC[20] (article 9) describes: *Member States shall take all necessary measures to ensure that all imports of tissues and cells from third countries are undertaken by tissue establishments accredited, designated, authorised or licensed for the purpose of those activities, and that imported tissues and cells can be traced from the donor to the recipient and vice versa in accordance with the procedures referred to in Article 8. Member States and tissue establishments that receive such imports from third countries shall ensure that they meet standards of quality and safety equivalent to the ones laid down in this Directive.* The implementation of this article changes among the EU member states. Theoretically, all EU member states will have to ensure that an HPC product originating from a third country is coming from an authorized cord blood bank. If you realize that over 70 cord blood banks are established in third countries, it is easy to calculate that this will result in nearly 2000 inspections for authorization if all EU countries independently authorize these 70 cord blood banks. The professional societies have established tools to monitor the activities in third countries by sharing information about licensing by national health authorities and the accreditation status of cord blood banks in third countries.

A second development in Europe is the implementation of a European coding system. According to Directive 2004/23/EC[20] on tissues and cells, EU Member States shall establish a system for the identification of all human tissues and cells procured, processed, stored, or distributed in their territory, in order to ensure their traceability from donor to recipient and vice versa. This Directive also requires designing a single European coding system to provide information on the main characteristics and properties of tissues and cells. A consortium, called Eurocet128, is working on the implementation of such a system. More information about this project can be found at www.eurocet128.org.

The American FDA announced that manufacturers of cord blood products are required to have an approved Biologics License Application (BLA) or Investigational New Drug Application (IND) in effect for unrelated CBUs shipped after October 20, 2011.[21] There are a variety of reasons why a cord blood product may not be BLA licensed. One reason may be that the CBU is collected and stored before the cord blood bank received a biologics license. Another reason may be that the CBU is coming from a cord blood bank outside the United States. All CBUs are now categorized as:

- Licensed—collected by a cord blood bank with an FDA-approved BLA and meeting licensure requirements.
- Unlicensed—collected before regulations were in place, collected by a cord blood bank without a BLA, or do not meet licensure requirements of the FDA.

Prior to the licensure regulations, cord blood banks followed industry-accepted quality standards to collect and store CBU for use in transplantation. It is important to understand that although a cord blood bank's units may be categorized as "unlicensed" by the FDA, the bank may still meet stringent quality standards and may have been successfully providing units for transplantation during the past 20 years. More importantly, these units may be the best and/or only available match for a patient. The FDA recognizes the importance of unlicensed CBUs and, therefore, is allowing access to unlicensed CBUs for use in transplant through IND clinical research protocols. The National Marrow Donor Program-Be The Match Registry has assisted the international cord blood banks in complying with these IND requirements.

One of the requirements to comply with the IND requirements is to register adverse events. Registration of adverse events has become part of the regulatory requirements (e.g., WHO, FDA, and EU) as well as part of the requirements for accreditation by professional societies, like the WMDA.

CBUs are increasingly requested as part of life-saving procedures for patients in need. Annually, over 4000 allogeneic cord blood products are shipped for patients with blood disorders worldwide. To safeguard the system of global HPCs exchange (including CBUs), donor health and safety is of critical importance. Since almost 30% of the CBUs cross international borders when shipped for transplantation, optimal donor safety, and quality of the cord blood products require global strategies. The World Health Organization has published the Guiding Principles. The WHO Guiding Principle 10[22] established the need for safe and efficacious procedures for living donors. It highlights the need for quality and vigilance systems for all cells, tissues, and organs used for transplantation, nationally and across international borders, as well as requiring adverse event and reaction reporting. World Health Assembly Resolution WHA63.22, adopted in May 2010, extended these principles, and introduced the concept that the collection and analysis of all serious adverse events/reactions should be collaborative and global. This enhances the likelihood of recognizing relatively rare adverse events, which may occur at such a low frequency, or not at all, in a single HPC donor registry or cord blood bank. It also enables the tracking of trends or hazards, which again may not be recognized within a single country. The analysis of adverse events can lead to new insights into underlying system failures and may be the basis for new recommendations and corrective/preventative actions. Another advantage is that global expertise and experience can be developed within formal structures, resulting in continuity of analysis of adverse events and a global "institutional memory," allowing for adverse events and reactions to be collected for several years and be rapidly accessible, again advancing the recognition of rare events.

There are challenges associated with global data reporting such as inconsistency in definitions, duplication of, or contradiction with, national reporting systems, cost, infrastructure issues, availability of labor resources for data capture and processing, and fear of reprisals or punishment. This is why there are several requirements for a reporting system to be successful. Important aspects for reporting serious events are that the reporter remains anonymous and that the report is confidential. The system should be independent and nonpunitive and thus clearly separated from any regulatory or funding implications. In 2002, the WMDA started a system to report serious adverse events/reactions in bone marrow donors/peripheral blood stem cell donors and patients receiving an unrelated transplant. At that time, only donor reactions were reported. In 2006, the European Directive[23] required the Member States to report adverse events in article 11:

*Member States shall ensure that there is a system in place to report, investigate, register and transmit information about serious adverse events and reactions* ***which may influence the quality and safety of tissues and cells and which may be attributed to the procurement, testing, processing, storage and distribution of tissues and cells,*** *as well as any serious adverse reaction observed during or after clinical application which may be linked to the quality and safety of tissues and cells*

An adverse event could be attributed to incidents during procurement, testing, processing, storage, and distribution. Therefore, the WMDA implemented a complementary system to collect product events. In 2014, the system has grown and four types of serious incidents are reported to the WMDA:

- Serious adverse event/reactions related to donation
- Serious adverse event/reactions related to infusion
- Serious events related to the product
- Serious events related to transport of the product

WMDA-qualified and -accredited HPC donor registries are required to report their incidents by filling in an online questionnaire to report incidents in the central database: S(P)EAR. Each WMDA HPC donor registry has appointed one individual who serves as the designated individual for S(P)EAR reporting and for receiving S(P)EAR communications from the WMDA. Each report received by the WMDA is evaluated by an independent committee. The committee determines the imputability of the reported incident, but does not take on a legal reporting role for S(P)EAR, and the system is in no way a replacement for individual HPC donor registries to comply with their national/competent authorities' regulations. In some cases, the committee can decide to send out a rapid alert to inform the WMDA HPC donor registries and transplant community about incidents. This first occurred when the WMDA received several reports describing the development of Takotsubo syndrome ("shocked heart syndrome") in recipients of double CBUs for transplantation. The reporting HPC donor registry determined in retrospect if similar events had occurred. Interestingly, five or more of these events had been reported. Even though it is not within the jurisdiction of WMDA to investigate S(P)EAR, pertinent information resulting from the investigation conducted and shared by the involved HPC donor registry is disseminated to WMDA's membership, representing the vast majority of cord blood banks shipping CBUs nationally or internationally. Therefore, a letter was prepared for dissemination. The rapid alert led to the recognition and reporting of at least two more cases, from countries different from the first reporting one. Following a formal investigation by the first reporting HPC donor registry, the WMDA was able to circulate their findings, which suggested that the problem had arisen due to procedural issues when preparing the cord blood for transplantation. Recommendations based on these assessments were then disseminated to all relevant parties. A second WMDA rapid alert related to cord blood was issued in 2013, when two large volume RBC-depleted CBUs, given by the thaw and infuse method to a patient with a previous history of cardiac disease, resulted in a fatal outcome. The WMDA recommended transplant units to consider this adverse reaction in the context of patients' cases when deciding on the optimal method for cord blood administration. More examples that underline the importance to move to a global reporting system can be found in a recent publication of the WMDA.[24]

Other incidents reported to the WMDA involved different aspects of shipment of cord blood products. In 0.5–1% of the CBU shipments, a serious event was reported.

Examples of incidents from January 2011 till August 2012 are depicted in this table:

| | # | Negative Influence on Patient |
|---|---|---|
| Leak/loss of integrity of bag | 8 | 6 |
| CBU arrived thawed | 11 | 9 |
| Decreased viability | 7 | 4 |
| Low TNC | 3 | 1 |
| Unusual appearance of CBU | 3 | 3 |

In total, 40 incidents related to cord blood shipment were reported to the WMDA S(P)EAR system. Other incidents concerned documentation mistakes, dry shipper alarm flashing upon arrival, technical problems during thawing, dry shipper being X-rayed, minor physical damage, positive bacterial culture, and coagulated cord blood upon thawing.

A task force has been formed to analyze the shipments of cord blood products in dry shippers to find out if global recommendations can be developed to reduce the number of S(P)EAR incidents related to these shipments.

Beside the WMDA S(P)EAR system, other initiatives have been developed to work toward and to create awareness for a global biovigilance reporting system. The first initiative is the Notify library (www.notifylibrary.org). The Notify library focuses on the rare occasions when unforeseen complications or errors result in negative outcomes. Although such incidents are unusual, they present opportunities for people working in the field to learn and improve, so that these services can be made safer and more effective for future donors and patients. A second initiative is the European Union funded project SOHO (http://www.sohovs.org/soho/), where competent authorities are trained in analyzing serious adverse events and reactions, and how to communicate rapid alerts between the European Member

States. Biovigilance becomes increasingly important when trying to get the best quality products available for patients worldwide.

Beside the challenges described earlier (to ensure the quality of the CBU, the arising regulations and the registry as service provider), another challenge is locating the right unit if a transplant unit is searching for cord blood. At the moment CBUs are listed in different databases and not all information related to the CBU is available online.

Over time, the HPC donor registries and cord blood banks have been and still are working to improve the framework for search and provision of cord blood products. The cord blood search process is complex and should be streamlined, so that the global inventory can easily be accessed with all relevant information needed by transplant units.[25] Initiatives have been launched to generate one global database. Complementary to the BMDW database, the EMDIS (European Marrow Donor Information System) has been developed. BMDW is merely an information system easy to access and use by any transplant unit and HPC donor registry, whereas EMDIS provides the communication system for the exchange between HPC donor registries of data from adult volunteer donors and CBUs on five continents. At the moment, 40% of the HPC donor registries are connected via EMDIS, and over 80% of international transplant activities take place between these partners, using these systems. A milestone in the development of EMDIS was the development of EMDIScord, which is a collaboration between NetCord and EMDIScord, a first step to harmonize the data of the NetCord cord blood banks, National Marrow Donor Program-Be The Match Registry cord blood banks, and the cord blood banks connected to EMDIS. Currently, the following countries are connected through EMDIScord: Belgium, Cyprus, Czech Republic, France, Germany, Italy, Spain, Switzerland, and the United States (NMDP cord blood banks only). Chapter 7 of the WMDA handbook[3] provides detailed information on how to set up an EMDIS architecture in an HPC donor registry or cord blood bank. Also, the Web site of NetCord provides information on how cord blood banks can participate.

The next paragraph summarizes where public cord blood banks are located worldwide.

## 9. CORD BLOOD BANKS WORLDWIDE

The New York Blood Center's Placental Blood Project and the José Carreras Cord Blood Bank Düsseldorf were the first two cord blood banks established in the early 1990s. From 1994 onward, initiatives were launched to establish cord blood banks in Asia, Europe, and the United States. In 2014, the number of cord blood banks worldwide has grown to 160 and their collective inventory provides CBUs for any patient in the world. Each country has different regulations regarding cord blood banking; below are some examples organized by region.

As described earlier, some cord blood banks collaborate with an HPC donor registry, while other cord blood banks work directly with transplant units.

### 9.1 North America

One of the largest networks of cord blood banks is the network of the United States Registry, the Be The Match Registry. The NMDP, which operates Be The Match, establishes relationships with public cord blood banks both throughout the United States as well as internationally. To ensure the quality of each unit listed on the Be The Match Registry, every bank in the network meets and maintains strict criteria and standards.[26] One of the criteria is that the cord blood bank must maintain accreditation by either AABB or NetCord-FACT. The network consists of 24 different cord blood banks, from which 19 cord blood banks are US cord blood banks.[27] The non-US cord blood banks are: StemCyte Taiwan, Singapore Cord Blood bank, Sheba Israel, Healthbanks Taiwan, and Duesseldorf Cord Blood Bank.

In Canada, Canadian Blood Services opened a national cord blood bank in Ottawa on September 30, 2013. Another Canadian Cord Blood Bank is the Hema Quebec Cord Blood Bank.

In Mexico, three cord blood banks can be found that are operating independently: La Raza Cord Blood Bank, CNTS Cord Blood Bank, and Bacecu Cord Blood Bank. The Bacecu Cord Blood Bank is the only one listing its CBUs in international databases.

### 9.2 South America

In South America, three countries have set up a cord blood bank: Argentina, Brazil, and Chile. In Chile, there is no HPC donor registry overseeing the recruitment of adult volunteer donors. This cord blood bank lists its CBUs directly in the international databases. The Argentine cord blood bank is AABB-accredited and collaborates with the national registry. In Brazil, a network of cord blood banks will be established.

### 9.3 Asia

In Japan, a large network of eight cord blood banks has been established: the JCBBN. These cord blood banks are currently not listed in international databases. In China, at least four public cord blood banks have been established. These cord blood banks (Shanghai, Sichuan, Beijing, and Guangzhou cord blood bank) have an estimated inventory of over 50,000 CBUs available for unrelated patients.

In Korea, cord blood banking projects have been developed since 1996. Seven cord blood banks, licensed by the national government, are listing their CBUs in international databases.

In Taiwan, five cord blood banks are active: Healthbanks Taiwan (through the NMDP), StemCyte Taiwan (through the NMDP), BIONET cord blood bank, and the cord blood banks affiliated with the Tzu Chi Registry.

In Singapore, a cord blood bank was established in 2007. The Singapore cord blood bank is listing its units in the database of the NMDP.

Other cord blood banks in Asia are the Thai cord blood bank, and the cord blood bank of the Hong Kong Marrow Program.

Vietnam is another country where a cord blood bank has been established, even though a registry has not been set up till now. This Hochiminh City cord blood bank is in the process of starting up.

### 9.3.1 Europe

The largest European network of cord blood banks can be found in Italy, comprising 18 cord blood banks.[1] The Italian Bone Marrow Donor Registry has coordinated and facilitated the search requests of these cord blood banks since 2007, based upon an agreement between the Health Department and Regions (October 5, 2006-2.637). The Italian cord blood banks are ISO9001 certified and licensed by the National Competent Authority.

Other large networks of cord blood banks can be found in Spain, France, Belgium, and Germany.

The European countries: Cyprus, Austria, Czech Republic, Finland, the Netherlands, Sweden, Slovakia, Slovenia, and Turkey have a one single cord blood bank. Greece, Switzerland, and the United Kingdom have a network of two/three cord blood banks.

### 9.3.2 Australia

In Australia three cord blood banks are collaborating in AusCord. The three cord blood banks: Melbourne cord blood bank, Sydney cord blood bank, and Queensland cord blood bank are all NetCord-FACT accredited. The search requests are facilitated by the Australian Bone Marrow Donor Registry.

### 9.3.3 Middle East

Israel and Iran both have three public cord blood banks in their country.

## 10. ANALYZING THE FIELD—NUMBER OF CORD BLOOD SHIPMENTS

The annual reports of the WMDA have become an established instrument to describe and observe the current status and trends for the number and types of HPC products provided worldwide and to monitor the exchange of products between countries and continents. The WMDA uses these data to recommend benchmarks for standardized practice and to identify barriers in donor exchange. This paragraph describes the trends during 14 years of collecting data and the results of the 2013 WMDA annual report, summarizing the status of activities per January 1, 2014.

Stem cell donor registries and cord blood banks worldwide are requested to report the number of products provided for unrelated stem cell transplantation by filling in a questionnaire. Cord blood registries/banks receive a questionnaire asking for the number of CBUs available, the numbers provided for transplantation, the countries to which CBUs have been shipped, and the degree of matching (-A, -B serological split level, -DRB1 allele). The project started with 19 cord blood registries/banks (1 from the West Pacific, 14 from Europe, and 4 from the Americas) in 1999 and has grown to 56 cord blood registries/banks (13 from the West Pacific, 26 from Europe, 12 from the Americas, 2 from Southeast Asia, and 3 from the Eastern Mediterranean) in 2013. In 2013, almost 63,000 CBUs were added to the inventory, increasing the total number of CBUs available to any patient in need to over 700,000, stored in 160 cord blood banks worldwide. In 2013, a total of 4334 cord blood products were provided from 31 countries.

Nearly all CBUs (98.5%) are registered in only three WHO regions: 28.8% in the West Pacific Region, 35.5% in the Americas, and 34.2% in Europe. The size of cord blood registries varies. The largest cord blood registries are located in the United States. The NMDP lists over 191,071 CBUs followed by the New York Blood Center (United States) with over 58,000 CBUs. However, more than 75% of all cord blood registries have an inventory of less than 10,000 CBUs.

Cord blood banks must continue to collect CBUs to replace the units used for transplantation. In addition, as quality standards for CBUs have evolved, many stored CBUs are no longer acceptable for transplantation. Older units may not have attached segments of blood required for sample testing, may not have been volume depleted, or may have low cell counts, which make them less likely to be selected by a transplant unit for infusion.

In 1997, 11,115 searches for a stem cell donor were initiated and 2965 patients received a transplant. In 2013, over 23,000 searches were initiated for a volunteer adult stem cell donor and over 3500 searches for a cord blood product. The stem cell donor registries reported that 3500 patients received a cord blood transplant in 2013.

In the same period, the use of cord blood products increased from 270 to 4334. A large increase occurred in 2006 and 2007, when the first reports on double CBU transplantation were published[10] and more CBUs were infused into adult patients. Figure 8 shows that the number of units provided for adult patients increased from 447 to 3000 in 2013. The United States NMDP Network provides 52% of the total number of CBUs shipped. The number of CBUs shipped from other countries is much smaller: Japan 28%, Spain 7%, France 5%, Australia 2%, and Italy 2%.

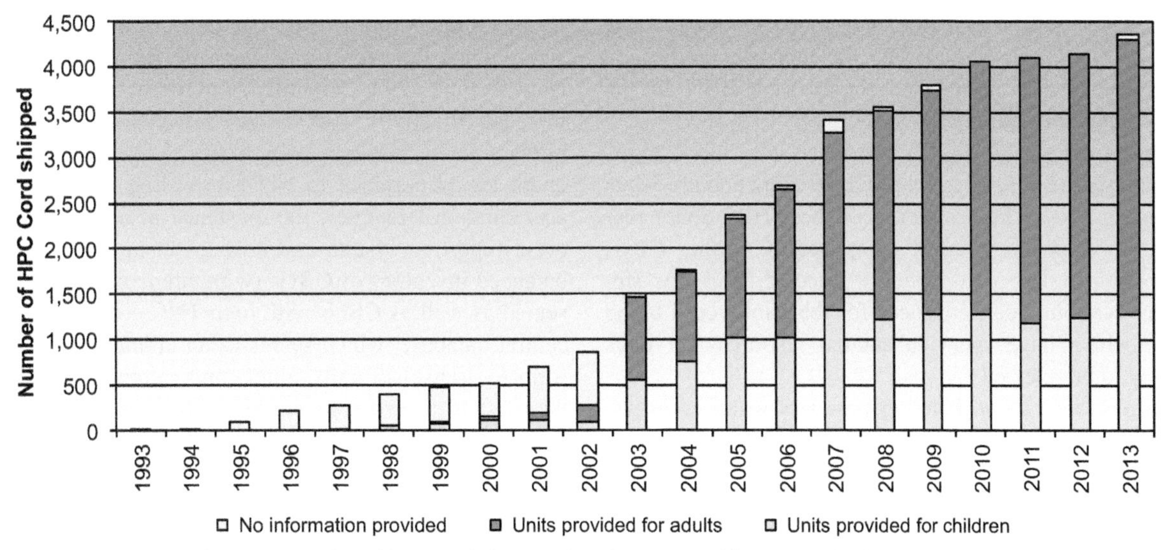


**FIGURE 8** Number of hematopoietic progenitor, Cord provided for unrelated patients over time.

The criteria used to select a CBU for transplantation have changed over time. Two important parameters used to select a CBU nowadays are the TNC count and the degree of HLA matching. In 2013, the median TNC of a CBU provided for children was $155 \times 10^7$, compared to $212 \times 10^7$ for units destined for adult patients. Many cord blood banks use a minimum TNC count as a prerequisite to store a CBU for transplantation. Thirty-five of the 160 cord blood banks only store units with TNC counts over $120 \times 10^7$. The second parameter used to select a CBU is the degree of matching. Matching criteria for cord blood products were also included in the WMDA questionnaire. For patients younger than 16 years old, 47% of the CBUs provided had one mismatch (defined at the antigen level for HLA-A, -B, and allele level for DRB1). For adult patients, 33% of the CBUs provided had two mismatches. Over the years, the distribution of the degrees of matching has not significantly changed: 12% 0-mismatches, 48% 1-mismatch, 40% 2-mismatches, and 0.6% with 3 or more mismatches.

For many patients there is no suitable stem cell donor available in their own country's registry. The percentage of products crossing international borders is nearly 30% (Figure 1). The WMDA was founded to facilitate this extensive international exchange of HPC products, and through their annual reports, the WMDA continues to monitor this exchange. The data collected are used by the organization to establish guidelines for a globally standardized practice.

## 11. FUTURE PLANS FOR CORD BLOOD BANKS

The WMDA reports show that the growth of the number of CBUs provided is flattening. This has urged cord blood banks to reevaluate their strategies and to work toward an efficient system to store cord blood products for unrelated patients.

Some banks have implemented a hybrid model, with CBUs stored for both private and public use. A public cord blood bank with a private banking operation needs to take into account that the CBUs stored for public use have to meet all regulatory and accreditation standards, including informed consent at the time of collection. CBUs collected for private use should not be stored together with CBUs for public use. Also, the marketing communication must clearly explain the differences between public and private use, including a statement concerning potential future use of private units, which lack supporting scientific data.[28]

Other cord blood banks have invested in either better HLA typing of the CBU or in HLA typing of the maternal donor. Since in recent publications,[29] matching for HLA-C was recommended beside HLA-A, -B, and -DRB1, HLA-C typing is now considered as an important piece of information in order to maximize possibilities of an 8/8 matched transplant.

Another interesting development, published in 2009, is the fact that the HLA type of the donor's mother may influence unit selection in the future.[30] A recent analysis of 1121 CBT recipients demonstrated lower transplant-related mortality and improved survival rates for the 79 recipients of units in which the donor–recipient mismatched locus was matched to the donor's noninherited maternal HLA antigens (NIMA). This suggests that exposure of the fetal donor to NIMA results in more permissive mismatches between donor and patient.

As mentioned earlier in this chapter, many cord blood banks do not store every collected cord blood product. Only the larger CBUs collected with a TNC count of over $120 \times 10^7$ are stored. At the moment, 31% of the CBUs listed in the international database BMDW have a TNC count over $120 \times 10^7$. The number of CBUs provided for transplantation for adult patients in 2013, show that 92%

have a TNC count over $120 \times 10^7$ versus 62% for juvenile patients. The CBUs provided by the Japanese cord blood banks are excluded in this analysis, as the JCBBN did not supply this information.

The experience of the US and Swiss cord blood registries[31] shows a similar pattern of CB banking and use when comparing TNC. The inventories of these HPC donor registries are disproportionately composed of smaller CBUs when compared to units actually selected. This has significant economic consequences for the public cord blood banks, which have spent substantial resources on units unlikely to be selected.

Current cord blood banking practices using comparatively low minimum TNC numbers as a cutoff for storage, appear not to be sustainable. Increasing the cutoff rate will result in a sustainable CB banking model, as long as the commitment of current CB banks for collection and banking continues, ensuring that the units preferred for transplant will continue to be banked at the same or higher rates.

For a successful implementation of this fundamental change in the CB banking strategy, it is necessary, in addition to purely technical adjustments of locally relevant banking regulations, to include all key stakeholders (especially harvesting/collection units, gynecology departments, but also expectant mothers) in integrated communication and information measures, to explain the rationale and the background of the changes to come.

Beside storing units with high TNC counts and performing HLA-C typing, a cord blood bank can also develop a strategy to recruit maternal donors from ethnic minorities in order to get a mixed inventory. This has been the strategy of the London Cord Blood Bank,[32] where, following a standardized, validated procedure during a period of 7 years, CBUs have been stored from hospitals targeted for their ethnically diverse populations, resulting in a cord blood bank with nearly half of the units originating from ethnic minorities within the United Kingdom. In 2005, the London Cord Blood Bank reported that 31% of the CB units issued for transplant were of non-European Caucasoid origin, thereby fulfilling one of the London Cord Blood Bank primary aims. Interestingly, this approach has resulted in a stable number of cord blood shipments over time for this particular cord blood bank, while the number of cord blood products provided by European cord blood banks has decreased.

Another service is the delivery of double cords to the transplant unit. The growth in provision of two CBUs for one patient and the publications showing very encouraging long-term results after double umbilical cord blood transplantation, make this procedure a promising treatment strategy for patients in need of a transplant. The big disadvantage is that a double cord blood transplant is twice as expensive as a single one. Discussions are therefore emerging on how to keep cord blood products competitive in relation to HPC donations from adult donors.[33]

In summary, cord blood banking has developed extensively since the pioneering work by Rubinstein et al. in New York, in 1993. Certainly, there has been an impressive increase in activity over the last few years; however, significant quality issues remain. In this chapter, you have read about the importance to work according to international accreditation strategies, the existence of a global adverse event reporting system and how to set up a global, well-balanced inventory of CBUs (with ethnic minorities represented as well as CBUs with high TNC counts) listed in a central database, which serves as an efficient service platform for transplant units. Safety and identity issues remain the most critical responsibilities of the cord blood banks in functioning as a reliable provider of CBUs for transplant units. The main priority of cord blood banks should be to focus on building quality rather than quantity and thus improve the outcome of cord blood transplantation.[34]

The mission of the WMDA allows us all to talk to each other and solve problems together. We are continuously evaluating incoming results to better understand what is working well and what is not. We described one example about the way guidelines were developed regarding the contiguous segments and the way rapid alerts are issued to the global community. Another role for the WMDA is educating and helping to start up new cord blood banks. To this end, a handbook has been published with tools and recommendations, serving as an elaborate source of information for starting cord blood banks.[3]

The WMDA is grateful to all members of the Cord Blood Working Group and the participating cord blood banks/registries for helping to develop the guidelines and recommendations referenced in this chapter and for providing data that enabled us to get a total overview of global shipments.

## 12. LIST OF CORD BLOOD REGISTRIES/BANKS

We would like to thank all the following registries and cord blood banks for their support and cooperation.

| Country | HPC Donor Registry—Cord Blood Registry |
|---|---|
| Argentina | Argentine CPH Donors Registry |
| Australia | AusCord—Australian Bone Marrow Donor Registry |
| Austria | Austrian Cord Blood Registry |
| Belgium | Marrow Donor Program Belgium |
| Brazil | BrasilCord/REDOME |
| Canada | Héma-Québec Public Cord Blood Bank |
| Canada | Victoria Angel Registry of Hope |
| Chile | Chile Cord Blood Bank |

| Country | HPC Donor Registry—Cord Blood Registry |
|---|---|
| China | Beijing Cord Blood Bank , Shanghai Cord Blood Bank |
| China | Guangzhou Cord Blood Bank |
| China | Sichuan Cord Blood Bank |
| China Hong Kong | Local Cord Blood Bank of HKBMDR |
| Croatia | Croatian Cord Blood Bank |
| Cyprus | The Cyprus Rotary Cord Blood Registry (CYCORD) |
| Czech Republic | Czech Stem Cells Registry |
| Finland | Finnish Cord Blood Bank |
| France | France Greffe de Moelle |
| Germany | ZKRD Zentrales Knochenmarkspender Register |
| Germany | José Carreras Stammzellbank |
| Greece | Hellenic Cord Blood Bank (HCBB) |
| Greece | Thessaloniki Cord Blood Bank |
| Iran | Iranian National Cord Blood Bank (INCBB) |
| Iran | Iranian Stem Cell Donor Program |
| Iran | Royan Cord Blood Bank |
| Israel | Hadassah Unrelated Donor and Cord Blood Registry |
| Israel | Sheba Cord Blood Bank |
| Italy | Italian BMDR and Cord Blood Bank Network |
| Japan | Japan Cord Blood Bank Network |
| Korea | Catholic Hematopoietic stem Cell Bank |
| Korea | KoreaCORD |
| Mexico | Banco Central de Sangre Centro Medico Nacional La Raza |
| Mexico | Banco de Sangre de Cordon—CNTS Mexico |
| Mexico | Mexican Cord Blood Bank and Registry BACECU |
| Netherlands, The | Europdonor Foundation |
| Poland | POLTRANSPLANT |
| Russia | HPC Registry |
| Singapore | Singapore Cord Blood Bank |
| Slovakia | Eurocord Slovakia |
| Slovenia | Slovene Cord Blood Bank—ESPOK |
| Spain | REDMO—Spanish Cord Blood Registry |
| Sweden | SWEDCORD—Swedish National Cord Blood Bank |
| Switzerland | Swiss Blood Stem Cells/Swiss Transfusion SRC |
| Taiwan | Bionet Babybanks Corp. |
| Taiwan | Buddhist Tzu Chi Stem Cells Center (BTCSCC) |
| Taiwan | Healthbanks Biotech Co. Ltd. |
| Taiwan | StemCyte Taiwan National Cord Blood Center |
| Thailand | Thai National Stem Cell Donor Registry |
| Turkey | Ankara Cord Blood Registry |
| United Kingdom | Anthony Nolan Cord Blood Bank |
| United Kingdom | British Bone Marrow Registry |
| United States of America | Celgene Cord Blood Bank Cedar Knolls |
| United States of America | National Marrow Donor Program (NMDP) |
| Unites States of America | New York Blood Center—National Cord Blood Program |
| Vietnam | Blood and Transfusion & Hematology HospitalHochiminh City |

## REFERENCES


1. *WMDA annual report cord blood Banks/Registries*. 2013. Available on request at mail@wmda.info.
2. Alshemmari S, Ameen R, Gaziev J. Haploidentical haematopoietic stem cell transplantation in adults. *Bone Marrow Res* 2011;2011: 303487. http://dx.doi.org/10.1155/2011/303487.
3. A gift for life – WMDA handbook for blood stem cell donation, ISBN 978-90-821221-0-7. Available on request at mail@wmda.info.
4. Available from: http://www.ahcta.org/documents.html.
5. Available from: http://www.wbmt.org/fileadmin/pdf/01_General/Press_release_final.pdf.
6. Confer D Draft inventory and adult donor models In: *Presented at a meeting of the department of health and human services Advisory Council on blood stem cell transplantation.* Retrieved on May 5 2010 at: http://bloodcell.transplant.hrsa.gov/about/advisory_council/meetings/notes/files/acbsctsummarynotesmay2010.pdf.
7. *Bone marrow donors worldwide – annual report.* 2012. Available on request at: bmdw@europdonor.nl.
8. Bone Marrow Donors Worldwide. Number of donors/CB per registry in BMDW. Cited 2014, October 23, Available from: http://www.bmdw.org/.
9. *World marrow donor association international standards for hematopoietic progenitor cell registries.* 2014. Available from: www.wmda.info.

10. Barker JN, Weisdorf DJ, DeFor TE, Blazar BR, McGlave PB, Miller JS, et al. Transplantation of 2 partially HLA-matched umbilical cord blood units to enhance engraftment in adults with hematologic malignancy. *Blood* 2005;**105**:1343–7.
11. Welte K, Foeken L, Gluckman E, Navarrete C. International exchange of cord blood units: the registry aspects. *Bone Marrow Transplant* 2010;**45**:825–31.
12. Rocha V, Gluckman E. Improving outcomes of cord blood transplantation: HLA matching, cell dose and other graft- and transplantation-related factors. *Br J Haematol* 2009;**147**:262–74.
13. Gluckman E, Ruggeri A, Volt F, Cunha R, Boudjedir K, Rocha V. Milestones in umbilical cord blood transplantation. *Br J Haematol* 2011;**154**:441–7.
14. Rubinstein P, Adamson JW, Stevens C. The Placental/Umbilical cord blood program of the New York blood Center. A progress report. *Ann N Y Acad Sci* 1999;**872**:328–5.
15. Confer DL. Unrelated marrow donor registries. *Curr Opin Hematol* 1997;**4**:408–12.
16. NetCord-FACT international standards for cord blood collection, banking, and release for administration, Available from: www.fact-website.org/Standards.
17. AABB standards for cellular therapy product services. Available from: www.aabb.org.
18. Hurley CK, Foeken L, Horowitz M, Lindberg B, McGregor M, Sacchi N. Standards, regulations and accreditation for registries involved in the worldwide exchange of hematopoietic stem cell donors and products. *Bone Marrow Transplant* 2010;**45**:819–24.
19. Available from: http://www.worldmarrow.org/fileadmin/Committees/CBWG/20100707-CBWG-RECO-Samples.pdf. www.wmda.info under professionals.
20. Available from: http://eur-lex.europa.eu/LexUriServ/LexUriServ.do?uri=OJ: L:2004:091:0025:0039:EN: PDF.
21. Available from: http://www.fda.gov/BiologicsBloodVaccines/Tissue TissueProducts/default.htm.
22. Available from: http://www.who.int/transplantation/Guiding_Principles Transplantation_WHA63.22en.pd.
23. Available from: http://eur-lex.europa.eu/LexUriServ/site/en/oj/2006/l_294/l_29420061025en00320050.pdf.
24. Shaw BE, Chapman J, Fechter M, Foeken L, Greinix H, Hwang W, et al. Towards a global system of vigilance and surveillance in unrelated donors of haematopoietic progenitor cells for transplantation. *Bone Marrow Transplant* 2013;**48**(12):1506–9.
25. Barker J, Byan C, Scaradavou A. How I treat: the selection and acquisition of unrelated cord blood grafts. *Blood* 2011;**117**:2332–9.
26. Available from: http://bethematch.org/About-Us/Global-transplant-network/Standards/.
27. Available from: http://bethematch.org/About-Us/Global-transplant-network/Cord-blood-banks/.
28. Available from: http://www.worldmarrow.org/fileadmin/Committees/CBWG/20120328-CBWG-PPR-Hybrid.pdf. www.wmda.info under professionals.
29. Spellman SR, Eapen M, Logan BR, Mueller C, Rubinstein P, Setterholm MI, et al. A perspective on the selection of unrelated donors and cord blood units for transplantation. *Blood* 2012;**120**:259–65.
30. Van Rood JJ, Stevens CE, Smits J, Carrier C, Carpenter C, Scaradavou A, Reexposure of cord blood to non inherited maternal HLA antigens improves transplant outcome in hematological malignancies. *Proc Natl Acad Sci* 2009;**106**:19952–7.
31. Bart T, Boo M, Balabanova S, Fischer Y, Nicoloso G, Foeken L, et al. Impact of selection of cord blood units from the United States and Swiss registries on the cost of banking operations. *Transfus Med Hemother* 2013;**40**:14–20.
32. Davey S, Armitage S, Rocha V, Garnier F, Brown J, Brown CJ, et al. The London Cord Blood Bank. *Br J Haematol* 2004;**125**:358–65.
33. Querol S. In: *Confronting the cost of cord blood transplants – abstract at cord blood Symposium June 8$^{th}$*. 2013. Available from: http://www.cordbloodsymposium.org/pdf/2013_CBS_ABSTRACT_web.pdf.
34. Querol S, Gomez SG, Pagliuca A, Torrabadella M, Madrigal JA, Quality rather than quantity: the cord blood bank dilemma. *Bone Marrow Transplant* 2010;**45**:970–8.

Section VI

# Cord Blood Banking: Current and Future Outlooks

Chapter 20

# Allogeneic and Autologous Cord Blood Banks

Paolo Rebulla

*Foundation Ca' Granda Ospedale Maggiore Policlinico, Milano, Italy*

## Chapter Outline



## 1. A "PERFECT" MATCH?

The outcome of allogeneic cell, tissue, and organ transplantation is positively associated to the degree of genetic match between donor and recipient. Although minor degrees of genetic mismatch do not necessarily translate into statistically significant and clinically relevant negative outcomes in all cases and conditions, sound scientific evidence supports the current standard medical practice to pursue the best possible match. A potential unfortunate consequence of this correct medical principle is a generalization that "autologous is better than allogeneic" under any circumstances, while evidence supports a positive outcome associated with minor degrees of mismatch between donor and recipient, which can generate antitumor effects in the transplant recipient.

The appeal of the "perfect autologous match," in association with bombastic media representations of stem cell discoveries, has contributed to support the development of the commercial autologous cord blood banking industry—a profitable business totaling US$ 4.5 billion in 2010[1]—as opposed to and in competition with solidaristic cord blood donation programs aimed at benefitting any patient in need of transplantation.[2] The latter were implemented two decades ago with the main purpose of supporting allogeneic hemopoietic stem cell transplant therapy in conditions including leukemia, lymphoma, thalassemia, immunological defects, and some metabolic disorders.[3,4]

The aim of this chapter is to report contrasting opinions on the current value of commercial autologous cord blood storage versus solidaristic donation to public banks,[5–18] and to challenge the view that routine conservation of umbilical cord blood with the perspective of a future autologous transplant is a choice made in the best interest of the newborn.

## 2. STEM CELL TRANS-DIFFERENTIATION AND TISSUE REPAIR?

During the last two decades, a classical dogma in biology—the inability for stem cells to cross boundaries of the mesoderm, ectoderm, and endoderm embryonic germ layers—has been challenged with conflicting evidence by a large number of investigators. Early provocative data from these studies[19] gained top visibility not only in specialistic scientific journals, but also through editorials and commentaries published in highly respected journals of general medicine,[20,21] which were promptly reported in the grey and lay press.

An oversimplified view of complex and only partially understood biochemical and molecular pathways was that any cell, bearing the whole genomic information of the human organism, could be instructed by an appropriate molecular environment to activate the necessary gene makeup to develop any cell and tissue, independently from its original embryonic germ layer. If confirmed, an immediate consequence of this possibility would be to develop laboratory procedures carried out according to good manufacturing practices for large-scale expansion and trans-differentiation of easily procured stem cells—like hemopoietic stem cells from bone marrow, peripheral blood, or cord blood—with the ultimate aim of developing a large array of cellular therapies for autologous and allogeneic tissue and organ repair.

**Cord Blood Stem Cells Medicine. http://dx.doi.org/10.1016/B978-0-12-407785-0.00020-7**

Highly attractive interpretations of these early data suggested, for example, that hematopoietic stem cells (which belong to the mesoderm) from a male bone marrow donor could home in the brain of a female bone marrow transplant recipient and trans-differentiate into neurons (which belong to the ectoderm).[20] Another discussed example was the finding of male cells in the parenchyma of a heart from a female cadaver donor transplanted into a male recipient, which was interpreted as supportive of the possibility that circulating cells from the recipient's bone marrow (possibly hemopoietic or mesenchymal stem cells) could be attracted by mediators of inflammation, migrate to a site of lesion, and differentiate into the host tissue (belonging in this case to the same germ layer, the mesoderm), thus contributing to its repair.[21]

A detailed discussion of data supporting or contrasting the early views on adult stem cell differentiation and trans-differentiation exceeds the scope of this chapter. Alternative interpretations of the observed phenomena have been proposed, including cell fusion.[22,23] Suffice to say that, although possibly supported in principle by some evidence, trans-differentiation with currently available technology is at best a very inefficient phenomenon, and formidable hurdles need to be overcome before real therapeutic prospects of human applications can be foreseen.[24–26]

The full understanding of the above findings and the uncertainties surrounding their interpretation are particularly critical not only for the scientific community, but also for families considering the alternative options of cord blood donation to a public bank versus autologous conservation in a commercial program.

## 3. SOURCE AND QUALITY OF INFORMATION

Information on public and private cord blood banking programs is readily accessible through current electronic communication systems. National and international networks of public cord blood banks, international registries of transplanted patients, and scientific societies involved in hemopoietic cell transplants regularly report information on the number of active banks, size of cord blood inventory, and number of cord blood transplants performed in different countries and institutions. In the private sector, carefully designed and highly attractive Web sites display information on commercial programs offering autologous cord blood conservation for a variable fee including the initial cost of cord blood collection, transportation and processing, and a regular annual fee for frozen storage.

An important difference between public and private programs is that public banks include in their frozen inventory only a small proportion of collected units (in several banks less than 20%), as clinical experience has shown that storage of units with low cell numbers is not cost effective, because the latter have a very small chance of being used for transplantation. On the contrary, most private programs have much less stringent criteria for unit cellularity, as they offer a service primarily based on the hope that future in vitro (pretransplant) or in vivo (posttransplant) cell expansion procedures could overcome the limitation due to a small original cell number or even that very few cells could provide effective treatments for "miniature" transplant procedures of tissue or organ repair.

Another difference regards the quality of information reported by the public and private sectors, as the former usually follows the pattern of peer-reviewed publication in scientific journals, whereas the latter generally presents information through Web sites which may operate independently from stringent regulatory oversight and accreditation and through grey literature.

A careful evaluation of the quality of information on stem cell medicine displayed at 19 direct-to-consumer Web sites was reported in 2008 by Lau et al.[27] These authors investigated three critical parameters of information: determinacy, i.e., "the extent to which indications were well bounded and specific, as opposed to open ended or vague;" relevance, defined as "the extent to which the risk or benefit is likely in frequency or dramatic in magnitude;" and the readiness of therapies described at the Web site for public access. The study showed that "(s)eventy-nine percent of websites portrayed benefits as somewhat or very relevant. In contrast, websites' risk portrayals scored as very irrelevant in the majority (14; 74%) of sites. This asymmetric portrayal of relevance of risks and benefits contributed to an overall impression that therapies were safe and effective. (...) Ten (53%) sites scored at the 'very' positive end of the five point scale for readiness." Based on data available at the time of their publication, Lau et al. concluded that "patients should be wary of claims made by stem cell clinics on the Internet."[27]

The very low probability for a newborn to require the use of his/her own cord blood unit for autologous transplant before the first two decades of life was extensively discussed by Sullivan[28] in 2008, who concluded that "in the absence of any published transplant evidence to support autologous and non-directed family banking, commercial cord banks currently offer a superfluous service." This investigation stirred a number of scientific letters contrasting in some cases the above conclusions, mostly based on the expectation that future technical developments could overcome the current difficulties and new, successful clinical studies could pave the way to effective autologous cell therapy treatments.[29–31]

With regard to the latter, specific clinical studies have been recently started or completed in cerebral palsy, type 1 diabetes, and severe aplastic anemia.

A large placebo-controlled trial of autologous cord blood infusion, which was designed after the collection of improved outcomes in animal models, has been started at Duke University in neonates with hypoxic-ischemic encephalopathy and in children with cerebral palsy.[32] In a similar condition of cerebral palsy occurring after cardiac arrest in a normally developed 2.5-year-old boy, the infusion of autologous cord blood 9 weeks after the cardiac arrest was followed by remarkable neurological improvement at 40-month follow-up, which was interpreted by the authors as "difficult to explain by intense rehabilitation alone."[33]

Great hopes were raised by preliminary investigations in diabetes, but a recently published open-label, randomized study of autologous cord blood infusion followed by daily supplementation with vitamin D and oral docosahexaenoic acid for 1 year in 10 children with type 1 diabetes, and five controls treated with intense diabetes management alone, failed to show preservation of C-peptide.[34–36]

Successful autologous cord blood transplantation following immunoablative chemotherapy with fludarabine and cyclophosphamide has been reported in a 9-year-old child with acquired severe aplastic anemia, a condition annually occurring in over 400 pediatric cases in the United States.[37] Another four cases of autologous cord blood transplant treated with different conditioning regimens showed variable results.[38,39]

Another issue relevant for the choice of pregnant women between solidaristic public donation and autologous commercial conservation is the quality of the banking program. While public banks undergo regulatory oversight from governments, which provides a standard, or at least a harmonized framework for quality, a standard approach to define the quality of commercial programs has not been developed. This is particularly important for any current and future plan to use cord blood units stored in commercial programs for clinical trials of autologous transplant.

A comparative evaluation of the quality of cord blood units stored in public versus private, commercial banks is found in the study published by Sun et al.,[40] as a part of the experimental clinical study on autologous cord blood infusion in children with acquired neurologic disorders carried out at Duke University. This study found that precryopreservation volume, total nucleated cell count, and $CD34^+$ cell count were significantly lower in units from commercial programs as compared to publicly stored cord blood units. Moreover, post-thaw sterility cultures were positive in 7.6% of the infused autologous cord blood units obtained from private banks, as compared to 0.5% in a control sample of public units used for allogeneic transplantation.[40] Another study assessing the characteristics of thawed autologous cord blood units as part of the previously described experimental protocol for the treatment of early pediatric type 1 diabetes found high variability and low cell counts in autologous cord blood units, suggesting the need for further standardization of characterization, collection, and processing procedures performed in private banks.[41]

Within the domain of public banks, McCullough et al.[42] reported the outcomes of a careful quality assessment of 268 cord blood units provided by cord blood banks in the United States and Europe to the Clinical Cell Therapy Laboratory and the Blood and Marrow Transplant Program of the University of Minnesota. Quality issues were detected in 151 (56%) of the 268 units, with likely risks to patients in 10% of the cases. In another study,[43] the same group reported that 2 of 871 (0.2%) cord blood units provided for allogeneic transplantation during 6 years had been originally mislabeled. Although small in frequency, this error related to the identity of cord blood units, if undetected, could have had potentially devastating clinical consequences.

The above data stress the importance of careful oversight of both private and public cord blood banking programs.

Moreover, families considering the options of solidaristic donation versus autologous commercial conservation should include independent documentation of high quality—such as a certificate of accreditation by internationally recognized organizations—among the criteria used to select a specific bank. General information on national cord blood banking programs has been reported in peer-reviewed journals from Italy[44]—where private banking is illegal—France,[45] and Germany.[46] Additional information on accredited banks can be found at the Web sites of NetCord, the Foundation for the Accreditation of Cellular Therapy, the National Marrow Donor Program, and Bone Marrow Donors Worldwide.

## 4. MATERNAL AND PATERNAL KNOWLEDGE AND PREFERENCES

The psychological vulnerability of expectant families toward the appeal of commercial proposals of autologous cord blood storage "in the best interest of the newborn" prompted the development of studies aimed at assessing stem cell knowledge and storage preferences in parents in different countries.

Thirty-seven percent of 425 pregnant women completing a survey performed in 2006 by Perlow[47] in Phoenix, Arizona, had no knowledge of cord blood banking. Of those indicating familiarity with cord blood banking, 74% felt "minimally informed."

Fox et al.[48] evaluated the understanding about cord blood banking in a cohort of 325 pregnant women attending the antepartum testing unit of a department of obstetrics and gynecology in the United States. The study, which was published in 2007, disclosed a "strikingly poor understanding regarding the current uses for cord blood therapy." More specifically, when asked if cord blood had been used successfully to treat Alzheimer's disease, Parkinson's disease, and spinal cord injuries, only 28%, 24%, and 24%, respectively, of the pregnant women correctly responded that it had not.[48]

A more recent survey carried out in 1001 highly educated pregnant Korean women receiving active prenatal care showed that a high proportion of them (n=863; 86%) had heard of cord blood.[49] However, 56% of them underestimated the probability of identifying a matched cord blood unit in the public bank, and correct answers on current usefulness were given by just 57.4% of respondents.

An analysis of decision-making in cord blood donation was performed through a participatory approach in a large maternity unit in Milan, Italy.[50] The attitudes of midwives practicing in hospital and outside hospital, obstetricians/gynecologists, pregnant women informed about and unaware of cord blood, future parents, and blood donors were collected by seven focus groups led by two psychologists. The study outlined "large support to altruistic cord blood donation and need for better health professionals education in this field."[50]

A larger study on pregnant women's awareness of cord blood stem cells and attitude on banking options was performed in France, Germany, Italy, Spain, and the United Kingdom by Katz et al.[51] Among the 89% of women who would opt to store cord blood, 76%, 12%, and 12% would choose a public bank, a mixed (public-private) bank, and a private bank, respectively. The study showed "a strong preference for public banking in all five countries, based on converging values such as solidarity."[51]

Emerging hopes of cure with autologous cord blood were outlined in a study performed in 2010–2011 by Parco and Vascotto[52] in a Northern Italian region, where the law prohibits private banking but allows families to export cord blood for storage in a foreign private commercial bank. These authors reviewed 129 questionnaires filled as part of the authorization to exportation and found that 75% of the families had opted for public donation. In the remaining 25% families opting for autologous or family-directed conservation in a foreign private bank, the main motivation by disease was treatment of diabetes (22.4%) and celiac disease (19.7%), both conditions in which therapy with autologous cord blood is still not routinely available.

Finally, particular attention to the attitude of actual and potential cord blood donors toward the different models of banking (public, private, and hybrid) was given in the studies performed in Switzerland by Manegold et al.[53] and by Wagner et al.[54] In the former study, a standardized, anonymous questionnaire was sent to the most recent 621 public cord blood donors. The returned questionnaires (48%) showed the will to donate again to the public bank in 95% of the cases, whereas 27% of donors had never heard about private banking. Among those informed about private banks, concern was expressed for the high cost of commercial programs. This investigation also outlined difficulties in mail contact with the former donors shortly after donation, an issue that, in the opinion of the authors, "might be relevant in any sequential hybrid banking."[53] The more recent study performed by Wagner et al.[54] specifically investigated the attitude of parents and pregnant women with or without children toward a hybrid allogeneic-autologous model of cord blood banking. The study documented a fundamental agreement with cord blood donation and overall acceptance of private banking from 47% of study participants. Furthermore, it showed that if a public–private, hybrid model would become available, 49% would opt for this option and the choice of private banking alone would decrease to 13% only.

## 5. CONCLUSIONS: WHAT IS THE CURRENT "CHILD'S BEST INTEREST"?

Responsible parental protection and guidance is a fundamental element for survival during development into adulthood. Several studies discussed in this chapter demonstrate the commitment of expectant parents to make responsible decisions on cord blood donation versus autologous conservation. They also identify areas of improvement for healthcare professionals in relation to their duties on unbiased communication with expectant families and discussion of scientific data on stem cell medicine. Hopes for future possible treatments supporting the choice of autologous cord blood storage should be balanced with demonstrated benefits from current treatments of proved clinical effectiveness, which require a large availability of solidaristically donated cord blood.

Besides any other scientific consideration, the ultimate question that a mother-to-be should answer is: How could I ask another mother, in the unfortunate event that my child develops leukemia and he/she needs an allogeneic hemopoietic transplant, to give up a potentially precious source of future therapies for her child, if I am not available to donate my child's cord blood to the community for the current benefit of any of us who could need it?

Attempts to predict the future can be very disappointing in several fields and stem cell medicine is no exception. For the current times this author shares the opinion of those who do not consider autologous cord blood commercial storage cost effective.[55] The possibility of short-term autologous directed storage for prompt treatment of neonatal cerebral ischemia, followed by crossover into the general solidaristic inventory for hemopoietic transplantation, could be considered based on the outcomes of the currently ongoing clinical studies in this condition.[4]

## CONFLICT OF INTEREST

Paolo Rebulla has been the medical director of a large public allogeneic cord blood bank during 2001–2013.

## REFERENCES

1. Webb S. Banking on cord blood stem cells. *Nat Biotechnol* 2013;**31**:585–8.
2. Gunning J. Umbilical cord cell banking: an issue of self-interest versus altruism. *Med Law* 2007;**26**:769–80.
3. Gluckman E, Ruggeri A, Volt F, Cunha R, Boudjedir K, Rocha V. Milestones in umbilical cord blood transplantation. *Br J Haematol* 2011;**154**:441–7.
4. Gluckman E, Ruggeri A, Rocha V, Baudoux E, Boo M, Kurtzberg J, et al. Eurocord, Netcord, World Marrow Donor Association and National Marrow Donor Program; Family-directed umbilical cord blood banking. *Haematologica* 2011;**96**:1700–7.
5. Cord blood banking for potential future transplantation: subject review American Academy of Pediatrics. Work Group on Cord Blood Banking. *Pediatrics* 1999;**104**:116–8.

6. Dalle JH. Cord blood banking: public versus private banks–facts to ponder and consider. *Arch Pediatr* 2005;**12**:298–304.
7. Ecker JL, Greene MF. The case against private umbilical cord blood banking. *Obstet Gynecol* 2005;**105**:1282–4.
8. Carpenter Jr RJ. Commercial cord blood banking: public cord blood banking should be more widely adopted. *BMJ* 2006;**333**:919.
9. Edozien LC. NHS maternity units should not encourage commercial banking of umbilical cord blood. *BMJ* 2006;**333**:801–4.
10. American Academy of Pediatrics Section on Hematology/Oncology; American Academy of Pediatrics Section on Allergy/Immunology, Lubin BH, Shearer WT. Cord blood banking for potential future transplantation. *Pediatrics* 2007;**119**:165–70.
11. Committee on Obstetric Practice; Committee on Genetics. ACOG committee opinion number 399, February 2008: Umbilical cord blood banking. *Obstet Gynecol* 2008;**111**:475–7.
12. Samuel GN, Kerridge IH, O'Brien TA. Umbilical cord blood banking: public good or private benefit? *Med J Aust* 2008;**188**:533–5.
13. Harris DT. Non-haematological uses of cord blood stem cells. *Br J Haematol* 2009;**147**:177–84.
14. Hollands P, McCauley C. Private cord blood banking: current use and clinical future. *Stem Cell Rev* 2009;**5**:195–203.
15. Thornley I, Eapen M, Sung L, Lee SJ, Davies SM, Joffe S. Private cord blood banking: experiences and views of pediatric hematopoietic cell transplantation physicians. *Pediatrics* 2009;**123**:1011–7.
16. Vidalis T. A matter of health? Legal aspects of private umbilical cord blood banking. *Eur J Health Law* 2011;**18**:119–26.
17. O'Connor MA, Samuel G, Jordens CF, Kerridge IH. Umbilical cord blood banking: beyond the public-private divide. *J Law Med* 2012;**19**:512–6.
18. Weisbrot D. The ethical, legal and social implications of umbilical cord blood banking: learning important lessons from the protection of human genetic information. *J Law Med* 2012;**19**:525–49.
19. Mezey E, Key S, Vogelsang G, Szalayova I, Lange GD, Crain B. Transplanted bone marrow generates new neurons in human brains. *Proc Natl Acad Sci U S A* 2003;**100**:1364–9.
20. Vastag B. Health agencies update. From blood cells to neurons. *JAMA* 2003;**289**:833.
21. Schwartz RS, Curfman GD. Can the heart repair itself? *N Engl J Med* 2002;**346**:2–4.
22. Song YH, Pinkernell K, Alt E. Stem cell induced cardiac regeneration: fusion/mitochondrial exchange and/or transdifferentiation? *Cell Cycle* 2011;**10**:2281–6.
23. Sanges D, Lluis F, Cosma MP. Cell-fusion-mediated reprogramming: pluripotency or transdifferentiation? Implications for regenerative medicine. *Adv Exp Med Biol* 2011;**713**:137–59.
24. Bianco P. Don't market stem-cell products ahead of proof. *Nature* 2013;**499**:255.
25. Abbott A. Doubt cast over tiny stem cells. *Nature* 2013;**499**:390.
26. Shen H. Stricter standards sought to curb stem-cell confusion. *Nature* 2013;**499**:389.
27. Lau D, Ogbogu U, Taylor B, Stafinski T, Menon D, Caulfield T. Stem cell clinics online: the direct-to-consumer portrayal of stem cell medicine. *Cell Stem Cell* 2008;**3**:591–4.
28. Sullivan MJ. Banking on cord blood stem cells. *Nat Rev Cancer* 2008;**8**:555–63.
29. Polymenidis Z, Patrinos GP. Towards a hybrid model for the cryopreservation of umbilical cord blood stem cells. *Nat Rev Cancer* 2008;**8**:823.
30. Nietfeld JJ. Opinions regarding cord blood use need an update. *Nat Rev Cancer* 2008;**8**:823.
31. Harris DT. Cord blood stem cells: worth the investment. *Nat Rev Cancer* 2008;**8**:823.
32. Liao Y, Cotten M, Tan S, Kurtzberg J, Cairo MS. Rescuing the neonatal brain from hypoxic injury with autologous cord blood. *Bone Marrow Transplant* 2013;**48**:890–900.
33. Jensen A, Hamelmann E. First autologous cell therapy of cerebral palsy caused by hypoxic-ischemic brain damage in a child after cardiac arrest-individual treatment with cord blood. *Case Rep Transplant* 2013:1–6. article ID 951827.
34. Haller MJ, Wasserfall CH, McGrail KM, Cintron M, Brusko TM, Wingard JR, et al. Autologous umbilical cord blood transfusion in very young children with type 1 diabetes. *Diabetes Care* 2009;**32**:2041–6.
35. Haller MJ, Wasserfall CH, Hulme MA, Cintron M, Brusko TM, McGrail KM, et al. Autologous umbilical cord blood transfusion in young children with type 1 diabetes fails to preserve C-peptide. *Diabetes Care* 2011;**34**:2567–9.
36. Haller MJ, Wasserfall CH, Hulme MA, Cintron M, Brusko TM, McGrail KM, et al. Autologous umbilical cord blood infusion followed by oral docosahexaenoic acid and vitamin D supplementation for C-peptide preservation in children with type 1 diabetes. *Biol Blood Marrow Transplant* 2013;**19**:1126–9.
37. Buchbinder D, Hsieh L, Puthenveetil G, Soni A, Stites J, Huynh V, et al. Successful autologous cord blood transplantation in a child with acquired severe aplastic anemia. *Pediatr Transplant* 2013;**17**:E104–7.
38. Fruchtman SM, Hurlet A, Dracker R, Isola L, Goldman B, Schneider BL, et al. The successful treatment of severe aplastic anemia with autologous cord blood transplantation. *Biol Blood Marrow Transplant* 2004;**10**:741–2.
39. Rosenthal J, Woolfrey AE, Pawlowska A, Thomas SH, Appelbaum F, Forman S. Hematopoietic cell transplantation with autologous cord blood in patients with severe aplastic anemia: an opportunity to revisit the controversy regarding cord blood banking for private use. *Pediatr Blood Cancer* 2011;**56**:1009–12.
40. Sun J, Allison J, McLaughlin C, Sledge L, Waters-Pick B, Wease S, et al. Differences in quality between privately and publicly banked umbilical cord blood units: a pilot study of autologous cord blood infusion in children with acquired neurologic disorders. *Transfusion* 2010;**50**:1980–7.
41. Rosenau EH, Sugrue MW, Haller M, Fisk D, Kelly SS, Chang M, et al. Characteristics of thawed autologous umbilical cord blood. *Transfusion* 2012;**52**:2234–42.
42. McCullough J, McKenna D, Kadidlo D, Schierman T, Wagner J. Issues in the quality of umbilical cord blood stem cells for transplantation. *Transfusion* 2005;**45**:832–41.
43. McCullough J, McKenna D, Kadidlo D, Maurer D, Noreen HJ, French K, et al. Mislabeled units of umbilical cord blood detected by a quality assurance program at the transplantation center. *Blood* 2009;**114**:1684–8.
44. Capone F, Lombardini L, Pupella S, Grazzini G, Costa AN, Migliaccio G. Cord blood stem cell banking: a snapshot of the Italian situation. *Transfusion* 2011;**51**:1985–94.
45. Katz G, Mills A. Cord blood banking in France: reorganising the national network. *Transfus Apher Sci* 2010;**42**:307–16.
46. Virt G. Ethical aspects of human embryonic stem cell use and commercial umbilical cord blood stem cell banking. Ethical reflections on the occasion of the regulation of the European Council and Parliament on advanced therapy medicinal products. *Bundesgesundheitsblatt Gesundheitsforsch Gesundheitsschutz* 2010;**53**:63–7.
47. Perlow JH. Patients' knowledge of umbilical cord blood banking. *J Reprod Med* 2006;**51**:642–8.

48. Fox NS, Stevens C, Ciubotariu R, Rubinstein P, McCullough LB, Chervenak FA. Umbilical cord blood collection: do patients really understand? *J Perinat Med* 2007;**35**:314–21.
49. Shin S, Yoon JH, Lee HR, Kim BJ, Roh EY. Perspectives of potential donors on cord blood and cord blood cryopreservation: a survey of highly educated, pregnant Korean women receiving active prenatal care. *Transfusion* 2011;**51**:277–83.
50. Salvaterra E, Casati S, Bottardi S, Brizzolara A, Calistri D, Cofano R, et al. An analysis of decision making in cord blood donation through a participatory approach. *Transfus Apher Sci* 2010;**42**:299–305.
51. Katz G, Mills A, Garcia J, Hooper K, McGuckin C, Platz A, et al. Banking cord blood stem cells: attitude and knowledge of pregnant women in five European countries. *Transfusion* 2011;**51**:578–86.
52. Parco S, Vascotto F. Autologous cord blood harvesting in North Eastern Italy: ethical questions and emerging hopes for curing diabetes and celiac disease. *Int J Gen Med* 2012;**5**:511–6.
53. Manegold G, Meyer-Monard S, Tichelli A, Granado C, Hösli I, Troeger C. Controversies in hybrid banking: attitudes of Swiss public umbilical cord blood donors toward private and public banking. *Arch Gynecol Obstet* 2011;**284**:99–104.
54. Wagner AM, Krenger W, Suter E, Ben Hassem D, Surbek DV. High acceptance rate of hybrid allogeneic-autologous umbilical cord blood banking among actual and potential Swiss donors. *Transfusion* 2013;**53**:1510–9.
55. Kaimal AJ, Smith CC, Laros Jr RK, Caughey AB, Cheng YW. Cost-effectiveness of private umbilical cord blood banking. *Obstet Gynecol* 2009;**114**:848–55.

Chapter 21

# The Future of Cord Blood Banks

**Catherine Stavropoulos-Giokas, Theofanis K. Chatzistamatiou, Efstathios Michalopoulos and Andreas Papassavas**
*Hellenic Cord Blood Bank, Biomedical Research Foundation Academy of Athens, Athens, Greece*

## Chapter Outline



## 1. INTRODUCTION

In the modern era of genomics, systemic and molecular biology, great advances have enhanced knowledge and understanding of disease pathophysiology. New biological markers help early diagnosis and play an important role in decision-making regarding the treatment. Additionally, new technologies allow to better understand the differences between individual cases, a fact that could ultimately lead into personalized medicine. This, however, has increased the scientific demand for access to high-quality material and information, in order to further advance research and therapeutic applications. Therefore, Biobanks have been created to assist medical research, clinical, and translational medicine. Depending on their purpose and specialty, they are classified into different types. Disease oriented, population based, case control are but some examples of existing Biobanks.[1]

A special category of Biobanks is the "Tissue and Cell Banks" that focus on providing cells and tissues, including umbilical cord blood (CB), corneas, skin, bone fragments, and other, for clinical applications. The particularity of those banks resides in the fact that they are patient oriented ("banks for patients"). They can be anything from a small collection of vials of frozen cells to a large laboratory facility with dedicated storage, testing, and distribution systems, supplying high-quality and controlled cells for international users. Recent advances in stem cell biology have led to enormous interest in the use of stem cells in translational and clinical approaches. In the 1990s, it was demonstrated that embryonic and several other categories of adult stem cells have the capacity to give rise to any cell type. More recently, induced pluripotent stem cells (iPSCs) "created" from adult human somatic cells,[2] have been considered for use in several medical research areas such as drug development, drug screening, disease models, etc. More importantly, iPS opened a new field of stem cell-based therapeutic strategies for a large number of human pathologies such as neurological diseases, heart disorders, liver failures, diabetes, etc.[1] All these novel, developing activities, rely on the procurement of a starting biological material that "Tissue and Cell Banks" could easily supply.

In this context, cord blood banks (CBBs) could greatly enhance their activities, providing biological material for new applications. Their main advantages over other

Cord Blood Stem Cells Medicine. http://dx.doi.org/10.1016/B978-0-12-407785-0.00021-9

Biobanks are the existing infrastructures, designed under stringent standards to deliver high-quality products, as well as the easy access to a rich source of different kinds of cells, namely the CB. Indeed, the quality of stored biomaterial is a key factor that could determine the success of its usage and working protocols, conceived, validated, and routinely used by CBBs for the collection, processing, and storage of clinical grade biological material, could give them a significant edge. Regarding the various types of cells encountered in the CB, several research teams have reported studies performed mainly in animal models, suggesting that some categories could repair tissues other than blood and be applied in diseases ranging from heart attacks to strokes. The cell populations implicated include nonhematopoietic stem and progenitor cells like mesenchymal stem/stromal cells (MSCs), endothelial progenitor cells (EPCs),[3] or cells that could derive from them, namely the iPSCs.[4,5] The regenerative ability, if verified, might be either due to a trophic effect, exerted by the cells, that helps the body to repair damaged tissues, or to direct differentiation of stem cells into the damaged tissue cell type. Either way, their contribution in tissue repair and regeneration would be significant. With the focus shifting on these "other" cell populations of CB, the stakes for CBB are also increasing.

## 2. CB AND FETAL ANNEX TISSUES ARE AN ABUNDANT SOURCE OF "YOUNG" STEM CELLS

CB initially came into focus as a rich source of hematopoietic stem cells (HSCs) with practical and ethical advantages making it an attractive graft choice for HSC transplantation. In the last 25 years, numerous studies have confirmed its feasibility and safety for the treatment of malignant as well as nonmalignant diseases. Although cell dose imposes a significant challenge for the treatment of adults, innovative strategies such as ex vivo expansion and nonmyeloablative conditioning regimens promise to improve outcomes and also increase the flexibility of treatment options available to the physicians. Of the above strategies, ex vivo expansion strongly relies on the ability of the cells to proliferate. It has been demonstrated that CB HSCs, when compared to their adult bone marrow (BM) counterparts, have distinctive proliferative advantages including increased cell cycle rate, production of autocrine growth factors, and increased telomere length.[6] Taken together, these characteristics indicate that HSCs in CB are more naïve and more potent than adult cells.

In addition to the well-studied and routinely used HSCs, other stem/progenitor categories have recently been isolated from CB and studied. Similar to the HSCs, these stem/progenitor cell populations present particularities that distinguish them from the respective adult populations. First, several CB originating stem populations, such as MSCs and unrestricted somatic stem cells (USSCs), are immunologically immature, which allows for less stringent matching requirements between donor and recipient. Another example of CB-derived stem cells immaturity refers to their telomeres, which, as mentioned before, correlate with proliferation capacity.[7] Indeed, it has been determined that aging shortens telomeres and impairs self-renewal potential of cells.[8] However, CB MSCs, USSCs, and also EPCs, originating from this relatively naïve and immature source, sustained the lowest level of changes in their stemness and self-renewal characteristics.[8,9] This property suggests that, CB MSCs, USSCs, and EPCs might be more efficient than adults' stem cells. Especially for EPCs, Naruse et al. provided evidence that CB-derived EPCs could promote neovascularization in diabetic mice when BM-derived EPCs could not, reinforcing this hypothesis.[10]

Finally, MSC populations isolated from other perinatal tissues, such as amniotic fluid,[11] amnion,[12] placenta,[13] or umbilical cord's (UC) Wharton's Jelly, exhibit the same characteristics of immaturity as CB MSCs. For this reason, MSCs from perinatal tissues are already being considered for clinical applications, in particular Wharton's Jelly MSCs,[14] due to the ease of procurement of the cell source that involves no invasive procedures and the fact that Wharton's Jelly appears to be immune privileged.[15]

## 3. CB-DERIVED EPCs

Tissue regeneration, wound healing, and tissue engineering repair of damaged tissues rely to a large extent on the ability of the human organism to create new blood vessels.[16] This phenomenon, called neovascularization, is crucial, since failure of the organism to channel blood to the regenerating tissue will lead to cell death and necrosis of the area. In many cases neovascularization is very slow and results in the failure of regenerative medicine techniques. Various attempts to facilitate and accelerate the creation of vessels have been proposed that rely on the use of cells of endothelial progeny. Mature endothelial cells (ECs) have been successfully isolated from vessels but, even though they have been shown to spontaneously organize into tubes, they rapidly reached proliferative arrest and subsequently, could not be expanded in vitro. Thus, scientific focus shifted to the more immature EPCs. Indeed, EPCs have been detected in human peripheral blood as well as in human CB and could prove to be promising candidates for the development of future cellular therapy strategies.[17]

Recently, a combination of immunophenotypic and cell culture characteristics has helped to define three distinct EPC populations: the colony-forming unit endothelial cells (CFU-ECs), the circulating angiogenic cells (CACs), and the endothelial colony-forming cells (ECFCs). The first two types, previously indistinguishable and termed early outgrowth or pro-angiogenic cells appear to be heterogeneous populations of hematopoietic progeny, while the ECFCs, also termed late outgrowth cells, are a nonhematopoietic population.[18] Studies have shown that out of the three cell types, the pro-angiogenic cells, when implanted into animal models, did not form tubes and did not differentiate into ECs, but significantly promoted tubulogenesis of mature ECs by releasing paracrine factors such as interleukin 8 and monocyte chemoattractant protein 1.[19] On the contrary, human ECFCs demonstrated de novo tubulogenesis capacity and were capable of interacting with mature ECs through surface proteins to form capillary structures in vitro but most importantly to form human vessels in mice that integrated the murine circulatory system.[20] Thus, ECFCs display all the properties of a real EPC.

## 3.1 Therapeutic Potential of CB EPCs

With their work, Prater et al. have shown that one unit of CB can easily give rise to between $10^8$ and $10^9$ ECFCs, which are still proliferating and functional.[18] This yield can certainly be improved, thus increasing numbers of clonogenic endothelial progenitors within primary colonies. When CB EPCs' expression of endothelial markers was compared to that of peripheral blood EPCs or even to differentiated ECs such as human umbilical vein endothelial cells (HUVECs) and human umbilical cord artery-derived endothelial cells (HUCACs) they have been found to express higher levels of KDR (kinase insert domain receptor), an early vasculogenesis and progenitor marker. Consequently, CB EPCs are more sensitive to angiogenic factors than HUVECs and HUCACs and give rise to a higher number of colonies than PB EPCs.[17] This fact indicates that although CB EPCs are not multipotent and have undergone some degree of differentiation, they have properties of immature cells and probably greater tissue repair capabilities.

Various experimental data demonstrate that there is a significant improvement in reperfusion in NOD/SCID mice with hind limb ischemia after injection of human CB EPCs-derived ECs, compared to noninjected animals.[21] It has also been shown that adult PB EPCs form blood vessels that are unstable and regress within 3 weeks, while CB EPCs form normal functional blood vessels that last for more than 4 months. These vessels have similar functional abilities to normal vessels.[22] Therefore, CB represents a valuable source for isolation of EPCs to be used in the treatment of a wide range of cardiovascular diseases.

## 3.2 Regenerative Medicine and Tissue Engineering Applications of EPCs

Regenerative medicine and tissue engineering are two rapidly evolving disciplines focused on the repair of damaged tissues and organs. The first involves the use of cells, growth factors, or other biological substances in order to induce natural healing processes. Tissue engineering uses the same therapeutic factors adding an extra component, the scaffold. Scaffolds are biomaterials, or synthetic materials, designed to simulate human tissues. When they are loaded with a patient's cells, a copy of the damaged tissue is created in the laboratory and is used to replace the defect. With the aid of this technique, simple organs such as bladder[23] and trachea[24] have been constructed and implanted to patients, while technological advances such as 3D printing could further increase tissue engineering's potential.

Therapeutic angiogenesis is a new and exciting approach to the treatment of cardiovascular diseases. It is a regenerative medicine application that involves the promotion of generation of new vessels from existing ones in order to treat myocardial ischemia or peripheral ischemia diseases. Inser and his team were among the first to attempt revascularization in a rabbit hind limb ischemia model with the injection of a single growth factor, the vascular endothelial growth factor (VEGF) isoform 165. They demonstrated that the angiogenic activity of VEGF was sufficiently potent to achieve therapeutic benefit.[25] Subsequently, in the last 20 years many research groups have utilized various angiogenic factors, in an attempt to promote angiogenesis. In parallel, novel strategies have been proposed that include gene delivery, protein delivery, and cell delivery. Compared with single-factor therapies, cell therapy is believed to have a more comprehensive and extensive effect.[26] Very good candidates for clinical applications are early outgrowth EPCs since they secrete pro-angiogenic factors. Applied in the infarcted heart, the infusion of cells either intravenously or directly into the infarct zone, aiming to restore organ vascularization and function is called cellular cardiomioplasty. This technique is effective in patients with transmural infarction where, however, the early injection of cells is critical since the therapeutic benefit declines with the time post-infarct.[27] Moreover, clinical trials in myocardial infarction patients and in patients with chronic ischemic heart failure using either BM stem/progenitor cells or circulating EPCs have shown promising results.[28] In general, therapeutic angiogenesis has been studied in the treatment of various diseases, as well as in processes that rely on blood supply such as tissue regeneration and wound healing.

In the field of tissue engineering, the use of EPCs relies on their ability to differentiate into mature ECs, thus participating in the formation of vessels. The main goals of EPCs' use in tissue engineering are the creation of vascular grafts and the in vitro vascularization of tissue-engineered organs. The cells to be used in these applications must have high proliferation and differentiation potentials. For this reason, ECFCs, endothelial cells deriving from embryonic stem cells (ESC-EC) or, as will be discussed later, iPS-derived endothelial cells (iPS-ECs) could be considered. Since the use of ESCs is limited due to legal and ethical issues and iPS technology still lacks efficiency, is time-consuming, and costly, it appears that ECFCs are, at least in the short-term, the most viable solution.

In summary, in the damaged vessels or in cardiovascular disease, there is a reduction in EC number and their function is impaired. Possible cell-based therapeutic approaches are the use of BM-derived progenitors, including EPCs, in order to mobilize cells and stimulate endogenous repair. Also ECFCs, human ESC-ECs, or human iPS-ECs could be expanded in vitro, and be used to engineer vessels for grafting at the site of damage.

## 3.3 Immunogenicity of EPCs

Since EPCs are being considered for the in vitro creation of artificial grafts, avoidance or reduction of rejection due to a natural immune response of the human recipient is a highly critical parameter.[29] Obviously, utilization of autologous cells remains the ideal scenario for cellular therapy applications, provided that such cells are available and retain their therapeutic potential whenever the need arises. Indeed, the rarity of EPCs in circulation and the concept of reduced, with age, potency of circulating EPCs[30,31] have prompted the investigation of CB as an allogeneic source.[32,33] However, allogeneic CB EPCs would be immunogenic to patients. It is therefore important to study allo-immunity of these cells as well as allo-immunity of cells from alternate fetal sources, namely the HUVECs. In a study by Suarez et al., CB EPCs were compared to same donor HUVECs. The results showed that both cell lines were comparable regarding the expression of proteins relevant to allo-immunity, including major histocompatibility complex (MHC) molecules, costimulators, adhesion molecules, cytokines, chemokines, and in their ability to initiate allogeneic CD4 and CD8 memory T cell response both in vitro and in vivo.[34] These data indicate that arterial grafts, created by EPCs and used in a human leukocyte antigen (HLA)-mismatched allogeneic setting would be rejected and destroyed by the host organism. Thus, the use of engineered arterial grafts should follow the rules and HLA-matching requirements of solid organ transplantation.

In addition to the HLA-mediated immune response, there is a growing interest in the role of non-HLA antibodies in transplantation immunology. These non-HLA antigens are classified either as allo-antigens, such as the MHC class I chain-related genes A (MICA) and B (MICB), or tissue-specific auto-antigens such as vimentin, cardiac myosin (CM), collagen V (Col V) agrin, and angiotensin II receptor type I (AT1).[35] While many of the non-HLA antigens still remain poorly defined, it is known that the principal antigenic targets are expressed, among others, on cells of the allograft's endothelium, and that the immune responses to these non-HLA antigens are critical in the pathogenesis of the allografts. The existence of polymorphisms or the presence of antibodies with different specificities that could target endothelial antigenic systems indicate that, in addition to the HLA-matching requirements, detection and specificity of non-HLA antibodies should also be considered for the use of tissue-engineered arterial grafts. Undoubtedly, this is going to be one of the important areas of future interest for both organ and cell transplantation but also in therapeutic approaches to promote vasculogenesis in patients failing conventional revascularization therapies.[36]

## 3.4 EPCs Banking

In modern, Western societies, ischemic diseases such as coronary artery disease (CAD), cerebrovascular disease (stroke), and peripheral vascular thrombosis have high socioeconomic implications. CAD alone caused approximately 1 of every 6 deaths in the United States in 2009, i.e., 386,324 persons died of CAD. Each year, an estimated 635,000 Americans have a new coronary attack (defined as first hospitalized myocardial infarction or coronary heart disease death) and approximately 280,000 have a recurrent attack. It is estimated that an additional 150,000 silent first myocardial infarctions occur each year. Approximately every 34 s, one American has a coronary event, and approximately every 1 min, an American will die of one. From 1999 to 2009, the relative rate of stroke death fell by 36.9% and the actual number of stroke deaths declined by 23.0%. Yet each year, approximately 795,000 people continue to experience a new or recurrent stroke (ischemic or hemorrhagic). Approximately 610,000 of these are first attacks and 185,000 are recurrent attacks. In 2009, stroke caused 1 of every 19 deaths in the United States. On average, every 40 s, someone in the United States has a stroke and one dies of it approximately every 4 min. The total direct and indirect cost of cardiovascular disease and stroke in the United States for 2009 was estimated to be $312.6 billion. This figure included health expenditures (direct costs, which include the cost of physicians and other professionals, hospital services, prescribed medications, home health-care, and other medical durables) and lost productivity that results from morbidity and premature mortality (indirect costs).[37] Similar epidemiological characteristics have also been reported in Europe, confirming CAD as the leading cause of death in the developed countries.

As seen in the aforementioned examples, novel approaches based on EPCs can provide effective treatment in a variety of pathologies, either through stimulation of endogenous repair (therapeutic angiogenesis) or with the in vitro creation of ready-to-use arterial grafts (tissue engineering). In all of these cases timing is a critical factor since application of EPCs early in the disease onset, is correlated with best results.[27] Although autologous cells should be, ideally, used in most patients, the time required for their isolation, expansion, and overall preparation could prove detrimental to the therapeutic potential. On the other hand, allogeneic use of EPCs is limited by donor–recipient histocompatibility restrictions, based on HLA, and possibly also on non-HLA antigen, matching criteria. For these two reasons, the creation of EPCs banks could provide the necessary, ready-to-use biological material for the propagation of cell-based therapeutic approaches.

In this setting, CB appears to be the source of choice for the isolation of EPCs, as it presents several advantages. First, the immaturity of CB progenitor cells increases, as discussed before, their proliferative and therapeutic potential. Additionally, there is a great availability of CB units (CBUs) which are unsuitable for CB banking, but could be used for the isolation of other cell populations such as EPCs. These CBUs have a great diversity in their HLA–antigen expression and could help in the creation of EPCs banks where both common and rare HLA haplotypes would be represented, thus covering the needs of the majority of patients. However, several issues need addressing before the creation of such CB-derived EPCs banks is possible. Presently, proper characterization of EPCs relies, as stated before, on a combination of immunophenotypic and cell culture characteristics. The absence of clear and well-defined cell markers complicates not only the study, but also the large-scale isolation of the cells. Thus, simple techniques must be developed to measure EPC numbers and function accurately and quickly as a first step toward EPCs banking. The other important step is the development of effective cryopreservation techniques for EPCs. While the impact of freezing on HSCs has been extensively studied over the years, there is no clear evidence whether or not this process maintains viability and functionality of other stem and progenitor cells within the CBU. A recent study by Broxmeyer[3] showed efficient recovery of functional HSCs up to 21–23.5 years after freezing, as well as recovery of responsive T cells and detection of ECFCs. However, ECFCs colony numbers recovered from thawed CB mononuclear cells were only 1/5 to 1/10 of the corresponding numbers in fresh CB. Therefore, it would be safe to assume that the freezing procedure that works well for the efficient recovery of HSCs may not be optimal for storage of ECFCs. Nevertheless, the study has shown that ECFCs can be cryopreserved and recovered, and with the optimization of storage techniques EPCs banking is feasible.

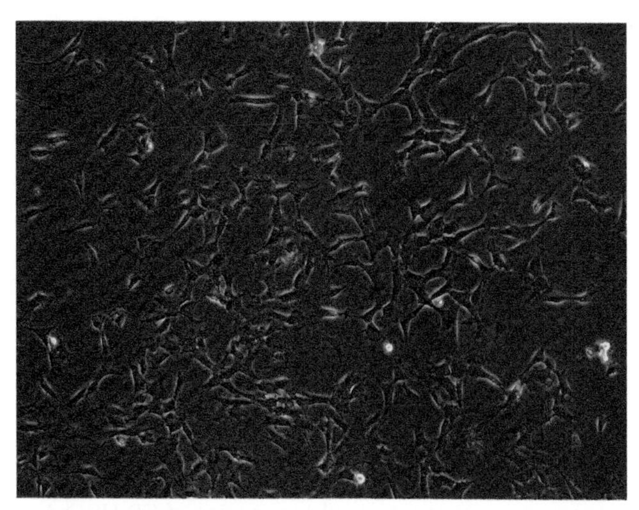

FIGURE 1 **Wharton's Jelly-derived mesenchymal stem/stromal cells in culture (objective 10×).**

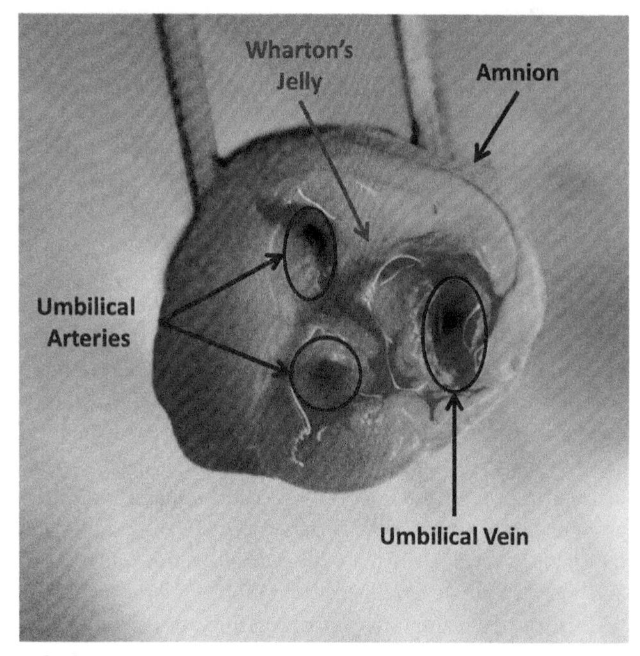


FIGURE 2 **Transverse section of the umbilical cord: Wharton's Jelly is the connective tissue surrounding the two umbilical arteries and the umbilical vein and is delimited externally by the amnion.**

## 4. UC WHARTON'S JELLY MSCs

MSCs (see Figure 1) have been initially known for forming the hematopoietic supportive stroma of the BM[38,39] but can also be isolated from different sources such as adipose tissue,[40] amniotic fluid,[41,42] CB,[43–45] the umbilical vein,[46] and from the connective tissue composing the UC, commonly known as Wharton's Jelly[47,48] (see Figure 2). Initially, MSCs were in focus due to their differentiation potential and their capacity to migrate and engraft into injured tissues, characteristics that made them a potential regenerative medicine tool. However, the discovery of the alternative sources in the last years revolutionized the study of the MSC population. With the increase of cell availability, this field evolved rapidly and many potential clinical applications for MSCs have arisen.

### 4.1 MSCs to Enhance CBB Banking

Currently, one of the major factors limiting CB transplantation is the late engraftment of the HSCs that can ultimately lead to graft failure and patient loss. The main reason for late engraftment being the low number of HSCs present in CB, and in order to overcome this hurdle, strict criteria have been established regarding cell dosage. It is now commonly accepted that the minimal required cell dose to make CB transplantation possible is $2.5 \times 10^7$ total nucleated cells[49] or $2 \times 10^5$ HSCs/kg of patient's body weight.[50] These criteria translate to a loss of 65–75% of all collected CBUs for any CBB aiming at the storage of high-quality CBUs (units that could be administered to patients with an average body weight of 50–65 kg). In an example set by Querol et al., it is shown that if a CBB accepted only CBUs containing $12.5 \times 10^8$ total nucleated cells or more, it would lead to a rejection rate of 62%.[51] The drawbacks of such practices are the elevated financial cost, due to the discarded CBUs, as well as the great loss of biological material. In this context, it is easy to understand why the expansion of CB HSCs has ever been the holy grail of CB banking. The initial attempts dated to 1993[52] and led to the first clinical trial using expanded HSCs in 2002.[53] Since then, various efforts have been made to improve HSC expansion's outcome utilizing diverse techniques.[54–59] The overall experience resulted in the consensus that better HSC expansion is achieved when the cells are cultured on a feeder layer of MSCs.[60] Thus, availability of MSCs could allow CBBs to develop HSC expansion protocols taking advantage of CBUs with low cell content, which would otherwise be discarded.

Another factor limiting hematopoietic stem cell transplantation (HSCT) is post-transplant complications that lead to increased morbidity and mortality of transplanted patients. Graft-versus-Host Disease (GvHD) is a common complication after HSCT and consists of immune cells, produced by the graft, attacking the patient's (host) tissues and organs. It usually causes lesions on the skin, the gastrointestinal tract, and the liver. GvHD can be classified either as acute Graft-versus-Host Disease (aGvHD), characterized by an early onset (less than 100 days after transplantation) and very rapid development, or chronic. Chronic GvHD (cGvHD) can occur after a successfully treated aGvHD, can be an extension of aGvHD, or occurs de novo in transplanted patients. The frequency of aGvHD incidence is 30% in HLA-matched sibling transplantations and can reach up to 70–90% in high-risk patient groups[61] (BM transplantation between mismatched siblings or unrelated transplantations), while cGvHD incidence ranges from 33% to 80%.[62] There are various factors affecting both the frequency and the severity of the disease, the source of the transplant (BM, CB, or peripheral blood stem cells), the age of the recipient, the degree of HLA match and the relation between donor and recipient, being the most important of them. With the exception of aGvHD in autologous transplantations (can occur in 5–20% of the cases) that can be relatively easily treated,[63] acute and chronic GvHD require strong immunosuppressive or immunomodulating medication which in turn can lead to severe infections. Post transplant mortality rates due to the disease can be as low as 25% for patients with aGvHD that respond to treatment, 40% in cGvHD patients, and 75% in nonresponding aGvHD patients.[61,64] All of the above numbers demonstrate the need for the development of novel, alternative treatments in order to provide more effective care to the transplanted patients. As was the case with HSC expansion, it appears that MSCs can also provide a solution to GvHD. First, Le Blanc and her team reported the durable remission of steroid-resistant aGvHD after MSCs infusion in a pediatric patient in 2004.[65] Since then, multiple phase II clinical trials[66–69] have provided a strong rationale for what now tends to become a routine medical practice in many European countries[70] and, most importantly, have provided a strong rationale to MSC banking for potential use in GvHD treatment.

## 4.2 MSCs for Regenerative Medicine/Tissue Engineering Applications

Similarly to the previously described EPCs, MSCs appear to be one of the most valuable cellular agents for both regenerative medicine and tissue engineering applications. They have been shown to exert a trophic effect,[71,72] when located at the site of injury, and to accelerate wound healing.[73,74] Additionally to their regenerative medicine applications, differentiated[24] or undifferentiated MSCs[75–77] can be loaded in scaffolds for the creation of artificial tissues. Since 2010, our team in the Hellenic Cord Blood Bank, in cooperation with the 2nd Orthopedics department of "Papageorgiou" Gen. Hospital of Thessaloniki, is participating in a phase II clinical trial providing MSCs for tissue engineering of knee cartilage defects (see Figure 3). More than 30 patients have already been successfully treated with this method, showing substantial recovery of their joint functionality and pain-free movement. All these advances in regenerative medicine and tissue engineering rely on the availability of controlled, ready-to-use MSCs; a demand which could easily be met by MSC banks.

## 4.3 MSCs for Advanced Cellular Therapies

As stated in the previous examples, MSCs attract scientific interest due to two main properties they possess. They have an immunomodulatory activity, which is used in GvHD treatment, and they exert a trophic effect, which is important for regenerative medicine applications. However, there are various other pathological situations where a combination of both anti-inflammatory and reparative effects is required. All these pathologies are, for the time being, incurable and often lethal. Through the years, drug-based therapies have been developed aiming, primarily, at providing relief from symptoms, improving quality of life, and elongating patients' life expectancy. Cells' infusion represents a novel, alternate strategy, called "advanced cellular therapies," in

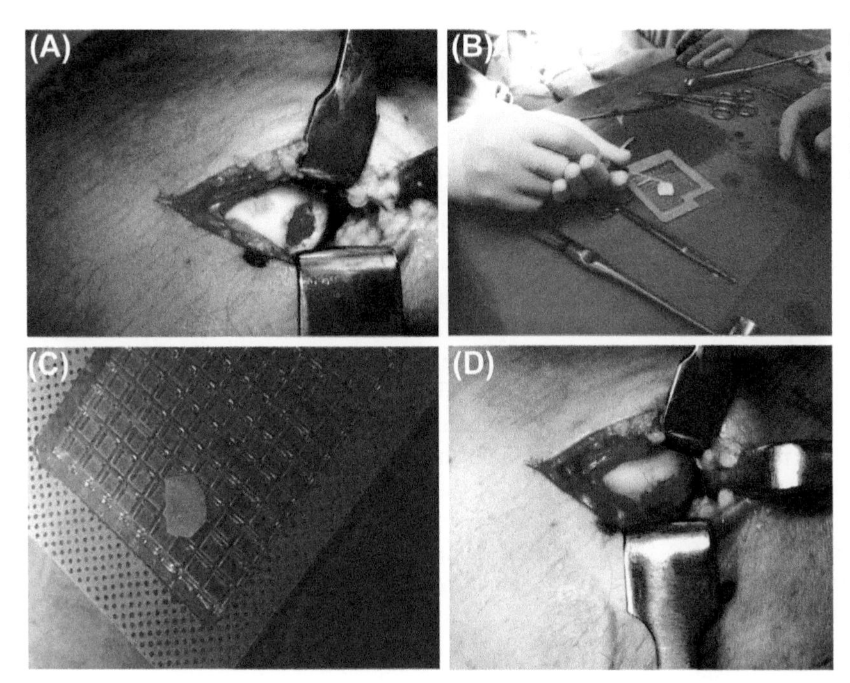


**FIGURE 3** Cartilage defect repair with tissue engineering: To repair a delimited cartilage defect (A), MSCs are injected on a hyaluronic acid-based synthetic scaffold (B). Thus, an artificial "graft" is created (C) and placed surgically in the area of defect (D). MSCs, mesenchymal stem/stromal cells.

the arsenal of physicians for the care of patients suffering from diseases with no effective therapy. Autoimmunities that cause extensive tissue damage are one category of such diseases. Rheumatoid arthritis, an inflammatory disease that causes articular deterioration; multiple sclerosis, an immune-mediated disease causing neuron degeneration; systemic lupus erythematosus, a chronic inflammatory disease affecting various organs; and type 1 autoimmune diabetes are some illustrative examples of autoimmunity where MSC treatment is being considered.[78–82] Degenerative diseases constitute another major group of pathologies that could benefit from MSCs' applications. For example, Parkinson's disease is one of the most common neurodegenerative disorders, affecting approximately 1% of individuals older than 60 years. However, preliminary studies of differentiated (to neuronal cells) MSCs infusion in Parkinson's animal models held promising results.[83] MSCs can help alleviate another major neurodegenerative disease, amyotrophic lateral sclerosis (ALS). Patients with ALS have a median life expectancy of 3 years from the onset of weakness[84] but already phase I[85] and II (FDA-approved clinical trial NCT01051882) clinical trials aim to achieve regression of symptoms and survival prolongation with the infusion of MSCs. Other degenerative diseases, where preclinical and/or clinical trials suggest that MSCs use could be warranted, include osteoarthritis,[86] osteoporosis, and type 2 diabetes mellitus.[87,88]

It still remains unclear what the future contribution of MSCs in advanced cellular therapies will be and if they will find their way to bedside, complementing or even replacing traditional, drug-based treatments. However, current data are promising and MSC banking activities could increase availability of cells for future studies and possibly provide cells for novel applications. Following this paradigm, the Hellenic Cord Blood Bank, participating in a phase II clinical trial, stores autologous BM-derived MSCs which are then infused to patients suffering from multiple sclerosis and ALS.

## 4.4 Actual Status of MSC Banking

All of the above developing applications, even though they demonstrate the utility of MSC banking, also lead to new, unanswered questions. The choice of the cell source to be used for clinical grade (and large-scale) MSC isolation is one of them. To answer this question, the accessibility and the availability of the biological material, as well as the quality and characteristics of isolated cells should be taken into account. Many of the original studies denoted that the MSCs obtained from different tissues have the same characteristics as their BM-isolated counterparts.[40,41,89] However, more recent ones indicate that there might be differences on the proliferative capacity[90] or even the therapeutic potential of the cells[90,91] depending on their origin. The clear advantages of Wharton's Jelly, compared to other sources of MSCs, are the ease of procurement of the cell source, that involves no invasive procedures, the fact that Wharton's Jelly appears to be immune privileged[15] and could, therefore, be used in an allogeneic setting, and finally their greater expansion potential.[92] Additionally, UC is routinely collected and cryopreserved, by many CBBs, as a source of backup biological material for testing. In the Hellenic Cord Blood Bank, our team validated an isolation and cryopreservation protocol for Wharton's Jelly MSCs[93] which can be used for the safe banking of clinical grade MSCs starting from those UCs.

The next important question that needs answering is whether MSCs should be banked for allogeneic or autologous use. In a study for the Joint Survey Committee of the Tissue Engineering and Regenerative Medicine International Society (TERMIS)-Europe, the International Cartilage Repair Society (ICRS), the European League against Rheumatism (EULAR), the International Society for Cellular Therapy (ISCT)-Europe, and the European Group for Blood and Marrow Transplantation (EBMT), use of MSCs throughout Europe was analyzed.[94] According to this survey, in 2011 alone, MSCs were used either for regenerative medicine or for the treatment of autoimmune diseases in 679 patients. In 97% of these cases (658 patients) the cells were autologous. The same study also showed that MSCs were used as an adjuvant in HSC transplantation, to either improve engraftment, or treat GvHD. Virtually all of the MSCs used in transplantation (299 out of 300) were allogeneic. Under this scope, the development of both autologous and allogeneic banks seems justified. However, the creation of such banks should be highly regulated by standards ensuring quality and safety of the MSCs. In the European Union, MSCs are considered (in most cases) advanced therapy medicinal products (ATMPs). Since the ATMPs are at the forefront of scientific innovation in medicine, a specific regulatory framework has been developed implemented in 2009. The Committee for Advanced Therapies (CAT) has been established at the European Medicines Agency (EMA) for centralized classification, certification, and evaluation procedures, and other ATMP-related tasks. Guidance documents, initiatives, and interaction platforms are available to make the new framework more accessible for small- and medium-sized enterprises, academia, hospitals, and foundations.[95]

In parallel, emerged the idea of "off-the-shelf" industrial MSC products. However, attempts for industrial-scale production of MSCs in the United States were hindered when an industry sponsored, FDA-licensed phase III clinical trial of an "off-the-shelf" MSC product for the treatment of GvHD failed to meet endpoints. These results contradicted the European experience of academia and hospital-based trials registering significant progress in the field. The reasons for this contradiction will not be investigated here but it is important to bear in mind, that in the field of MSCs, scientific knowledge is not on par with clinical applications. In truth, scientists do not understand to their full extent the biological mechanisms and biochemical pathways through which MSCs exert their therapeutic action. Although life-threatening situations and patients suffering can legitimize the exceptional use of a not fully understood therapeutic agent, only well-characterized and thoroughly comprehended products can be in the basis of industrial ATMPs. Obviously, this was not the case for the failed clinical trial. Academic and health institutions, on the other hand, have a more patient-focused approach. Each case is considered individually not only for the selection of the optimal therapeutic strategy, but also for the preparation of the cells, the selection of the cell source, and the cells donor. In this context, MSCs appear to be a personalized medicine tool, rather than an "off-the-shelf" product, a fact that gives academia and hospitals a clear short- and midterm advantage over industry in leading the field.

In conclusion, all the necessary ingredients required for MSC banking are already present; the scientific basis, the know-how, and the regulatory framework. Even though we are still, in scientific reckoning, at the first steps of MSCs' clinical applications, the creation of MSC banks could benefit greatly both the scientific community and patients in need for novel treatments. The CBBs, having access to rich sources of MSCs and a long-standing expertise in cell banking, are ideally suited for establishing the first of such banks in order to enhance, as well as diversify their services.

## 5. THE IPSCs

The iPSC technology is an emerging field that has created new pathways in basic research, in vitro models of disease, drug discovery, pathological studies, and toxicology studies.[2] In addition to their use in modeling diseases and drug screening, the most promising purpose of iPSCs will be their potential application in regenerative medicine. The ultimate goal of this technology would be the transplantation of progenitor cells that do not trigger any immune response or promote tumor formation, since they are derived from patient-specific iPSCs, and may assist in the recovery of damaged tissue. The great promise offered by iPSCs is that studying human cells, which may reflect a disease phenotype more accurately than previous cellular models or animal models, will make therapeutic drug discovery faster, more efficient, and eventually, customizable to individual patients. However, several practical issues must be addressed to increase the impact of these stem cells on disease-based research.

### 5.1 Ideal Sources for the Generation of iPSCs

Selection of a starting material is important because each type of somatic cell possesses a unique epigenetic profile that could make the cells differentially liable to clinical application subsequent to reprogramming.[96] Most of the studies to date have used fibroblasts as the target population for reprogramming, but results with alternative targets including peripheral blood and CB have also been reported.[97] However, the cost and time required to produce and validate individual iPSC lines make use of autologous iPSCs unlikely in the near future. Instead, the establishment of a set of standardized iPSC lines that can be used to derive cells and tissues for allogeneic transplantation to the majority of the population seems a more viable approach.[98]

## 5.2 CB for the Generation of iPSCs

CB may be the ideal source for the generation of iPSCs, since all samples are collected with the appropriate consent and the cells are well characterized and HLA-typed. In addition, if medical history of the mother-donor along with the consent is available, it would be easier to gather data for a further follow-up. The efficiency of generating iPSCs from CB cells is higher and faster, since the final number of cells that can be obtained from a fraction of the CB aliquot exceeds what is required for the generation of a cell line.[99,100] Use of CB cells as a cell source for iPSCs could pave the way to the establishment of "healthy" allogeneic iPSC banks through association with existing CB bank networks.[98]

CB is considered to be superior to cells isolated from adult individuals because nuclear and mitochondrial mutations tend to accumulate in adult stem cells and differentiated somatic lineages over an organism's lifetime, and have been suggested to contribute to aging and cancer formation.[101,102]

## 5.3 iPSCs Therapy from HLA-homozygous CB Donors

Differentiated derivatives of such iPSCs should be transplantable to the majority of the population with three-locus HLA (i.e., HLA-A, HLA-B, HLA-DR) matches if iPSCs are generated from donors that are homozygous at each of the three HLA-loci.[103] The size of many existing CB banks should allow identification of all required homozygous HLA combinations.

It has been stated that human pluripotent stem cells may show clonal variations in lineage bias; therefore, cells best suited for the purpose of each application may be chosen.[104] Judging by previous experience with the transplantation of BM, matching the three major types of HLA-loci between the recipient and donor is anticipated to cause less immune rejection after transplantation. Therefore, it could be stated that creation of iPSC stocks shall derive from HLA-homozygous donors for an iPSC therapy.[105]

Banking of CB iPSC lines and large-scale production of these lines, representing a wide panel of HLA haplotypes organized from a public network could represent an alternative approach for future clinical applications.[4] All public CB banks can provide qualified sources of cells and advice to help researchers and clinicians meet these demands and avoid wasted time due to receipt of unsuitable cells.[106] Moreover, by using banked CBUs, the selection of donors homozygous for common HLA haplotypes could be accomplished; therefore, that could reduce significantly the number of CB iPSC lines needed to provide a perfect HLA match for a greater percentage of the population.[4] In addition, based on the medical history of donors, it is inevitable to define whether single nucleotide and copy number variations found in iPSC clones already exist in the donor or whether these are newly acquired during iPSC formation.[106] A further advantage would be the availability of multiple clones from the same donor.[107]

Furthermore, CB presents the advantage of being particularly rich in immature progenitors, which demonstrate a great expansion capacity. In the foreseeable future, it should be possible to efficiently stimulate the proliferation of human pluripotent stem cells, that is, human embryonic stem cells (hESCs) and human-induced pluripotent stem cells (hiPSCs), so as to provide an inexhaustible source of human HSCs to generate high volumes of in vitro-produced red blood cells for transfusion purposes.[107]

## 5.4 Strategies to Generate iPSCs

Subsequently several groups have used alternative strategies to generate iPSCs[108,109] by using a single vector to express all factors,[110] or using only two factors,[111] or using transfection of two expression plasmids, one containing the complementary DNAs (cDNAs) of Oct3/4, Sox2, and Klf4 while the other contains the c-Myc cDNA.[112] Although the retroviral introduction of reprogramming factors remains the most commonly used method, it has the obvious drawback of genetically modifying the host cells, which can have deleterious effects after re-differentiation.[113] Recently, several attempts to overcome the problems associated with modifying the host genome have been made using nonintegrating adenoviruses,[114] transient transfection of plasmid expression vectors,[112] the piggyBac transposition system,[115,116] cre-excisable viruses,[117] and the oriP/Epstein–Barr virus nuclear antigen 1 episomal expression system,[118] xeno-free culture to generate iPSCs that either were produced without modifications to the host genome or where those modifications have been excised. To date, all of these approaches are limited by low efficiency of reprogramming[115,116,118] and logistical considerations related to the need to generate sufficient quantities of recombinant proteins used in protein-based reprogramming protocols.[119] A recent protocol by Warren et al. using a simple nonintegrating method with synthesized RNA to generate in vitro transcription templates (IVT) for each of the pluripotency genes, achieves reprogramming to levels that are much higher than previously reported.[120]

While the issues listed above may obstruct applications on a large industrial-scale production, there may be benefit for patients individually, particularly those in need of customized blood cells, for instance, patients requiring lifelong erythrocyte substitution therapies that over time develop allo-antibodies that highly restrict the pool of matching blood cell donors. Generating patient-specific blood cells from CB iPSCs will possibly benefit and increase treatment therapies in these patients.[121] The ability to generate transplantable hematopoietic cells from human pluripotent stem cells has the potential to have a huge impact on the fields of transfusion medicine and HSCT.

## 5.5 Banking of iPSCs

CB as source for the generation of an iPSC line could be the best option by the public CB banks, since CB banks are well organized and validated. CB is considered as one of the youngest sources of stem cells one may obtain, and it has been stated by Kaufmann et al.[122] "*The challenge, is to integrate pluripotency induction in the highly regulated existing workflow of a Cord Blood Bank*" (see Figure 4). Synoptically, a segment from an existing stored buffy coat may be removed and then processed to generate an iPSC line for potential therapeutic use, or when a CBU is shipped for use, a small sample is retained and iPSC lines are made to provide replenishment for the used unit.

Therefore, careful documentation of iPSC lines must be maintained in order to track biological characteristics such as cell and patient phenotypes, technical issues such as the method used to generate a line or the number of times the line has been passaged in culture, and legal aspects such as the degree of patient consent for the use of each line.

Since iPSC technology is still at the beginning, in order to be considered as a therapeutic tool all different aspects of the processes will have to be optimized. All protocols have to be compared systematically such as culturing, expansion, and storage, as well as the cellular characterization just to mention a few. That may be accomplished through the creation and establishment of iPSC banking, where collaboration between research centers, clinicians, and biotechnology industry may maximize the use of iPSC lines. Therefore, existence of iPSC banks would facilitate the commercial banking of iPSC lines that are already being generated in research laboratories. Banks would also generate high-quality, comparable cell lines available for the study of particular diseases that are not being adequately served through existing research efforts.

So far, two models of iPSC bank have been suggested by the scientific community such as an iPSC bank as a repository and distribution mechanism for cell lines generated by investigators in the field, and an iPSC facility that would recruit patients and generate cell lines for certain targeted disorders.

## 5.6 Benefits of the Creation of iPSC Banks

- Collection of specific-disease cell lines generated in research laboratories, then characterization of these lines as well as optimization of their growth conditions.
- Standardized specific-disease iPSC lines will be available to the clinicians, academic research centers, and biotechnology industry.
- All the information regarding the extensive characterization of iPSC lines could be collected in an economical way by a centralized facility, additionally, all that background information collected for the iPSC lines could be available on database engines.

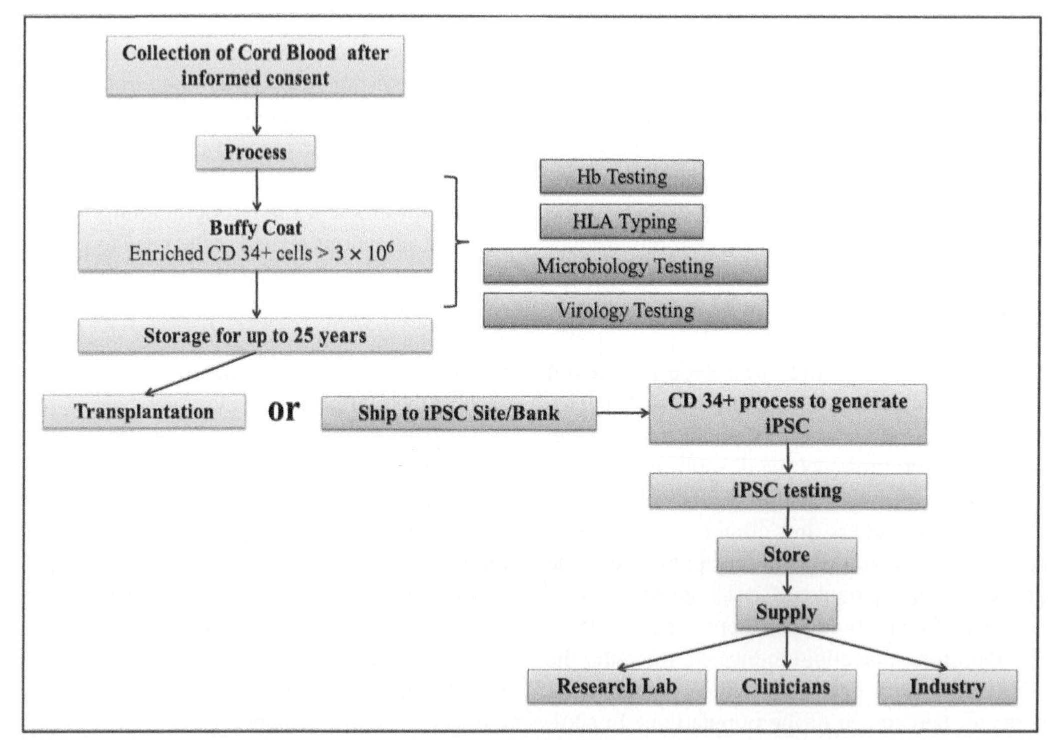


**FIGURE 4 Integration of CBB banking into a classic CBB model. CBB, cord blood bank.** *Adapted from Rao M et al.*[121]

- A centralized cell bank could dedicate resources toward developing more consistent cell lines and dedicate resources to creating new iPSC lines from an appropriate sample size of patients that could more usefully represent the heterogeneity of the human disease than any other approach. Furthermore, it could focus on specific diseases as a priority for iPSC research and assist in the generation of new iPSC lines and control lines to study these diseases.
- Another function of an iPSC bank would be the storage and distribution of iPSC lines derived from healthy donors that could be used for therapeutic reasons.

To sum up, the use of iPS-derived cells has been proposed in the field of regenerative medicine,[123] HSCT,[124] blood transfusion,[125] and neurology.[126] Various tissue types have been proposed for the generation of iPSCs for therapeutic purposes, including CB. The HLA haplotypes of the banked iPSC lines are crucial for these applications. Several studies address the issue of histocompatibility in regenerative medicine and propose different strategies to addressing the issue of the haplotypic content required for maximum coverage of a population by the banked cell lines.[127]

In conclusion, the existence of iPSC banks would facilitate the commercial banking of iPSC lines that are already being generated in research laboratories and would also generate high-quality cell lines available for the study of particular disorders that cannot be investigated through current research protocols.

## 6. CB AND UC COMPONENTS

While scientific research and translational medicine focus on cell populations that can be isolated from CB and UC, there are also other, noncellular components that could be of great clinical relevance. Human plasma, growth factors, and scaffold constituents gained, in the last decade, an increasing scientific interest. These components, which could be used for the production of culture media, cell culture supplements, or scaffolds, are routinely collected by the CBBs and might prove to be their link with industry.

### 6.1 CB Plasma

A large portion of the CBU is considered to be waste, namely the CB plasma. Plasma is consistently being removed during volume reduction, i.e., the preparation of CB for storage. More importantly CBBs reject and discard a large number of CBUs that do not meet acceptance criteria for transplantation. All these CBUs also contain plasma that could be recovered for purposes other than HSCT. Indeed the plasma contains platelets, growth factors, and nutrients that could be of use.

A major issue of cells cultured for clinical applications is that they have been dependent, for many years, in animal sera. Fetal calf serum (FCS) and fetal bovine serum (FBS) were used (and still are to a great extent) as media supplements, in order to provide cultured cells with growth factors and nutrients necessary for their survival and proliferation. The problem that arises with animal sera is that they are unsuitable for the culture of clinical grade cells destined to be used in human recipients. The main reasons are: (1) the inability to effectively screen them for pathological agents, and particularly prions that cause the human variant of bovine spongiform encephalopathy (BSE), the variant Creutzfeldt–Jakob disease (vCJD); and (2) they contain uncharacterized substances that could be immunogenic in human recipients. There have been many attempts at the creation of xeno-free media that would not rely on sera supplements to promote cell growth. However, their success has been limited or at the very best, they seem to be effective in the culture of only a few cell populations. To overcome the lack of consistent results, different attempts have been made to produce a human equivalent of animal sera. For this purpose, several cell culture techniques have been developed where culture media were supplemented with human platelet-rich plasma (PRP), platelet lysates (PL), platelet-poor plasma (PPP), and human autologous serum. The first attempts at the use of human plasma (in the form of PRP, PL, or PPP) dated to the early 1980s.[128,129] However, it is in the last decade that great progress has been made in the field. A recent comparative study by Pawitan which demonstrated the beneficial effects of activated PRP and PL, produced various protocols in stem cell cultures.[130] Similarly, the potential use of human serum or PPP has been demonstrated.[131] In view of these data, many scientists consider CB as a source of the necessary human plasma and/or serum for the execution of these novel techniques.[132]

Fibrin is another protein component present in activated plasma with potential applications in tissue engineering and in regenerative medicine. This protein is normally formed when fibrinogen is cleaved by a protease called thrombin and is the main clotting factor of human blood. The interest in fibrin arose when it was used for the creation of biodegradable scaffolds[133] and it has been amplified with the advances in tissue engineering. Similar scaffolds have been considered for bone,[134] blood vessels,[135] and cartilage[136] repair, but also for non-tissue engineering applications.[137] Moreover, Anitua and colleagues suggested that fibrin matrices can improve wound healing, and tested their theory in the reconstitution of tendons in animal models.[138]

CBBs are in a unique position to satisfy the demand for high-quality, screened, and controlled human plasma and plasma components. Indeed, CBBs not only collect CBUs, but additionally a complete donor's family medical history and samples for serological and if necessary genetic disease testing. Thus, the material processed and provided by CBBs meets very high quality standards for efficacy and safety. Consequently, the plasma derived from CBU processing and the plasma of CBUs that do not meet banking requirements but meet safety criteria could be routinely processed and made available to industry.

## 6.2 UC Arteries

Similarly to CB, the UC is also routinely collected by CBBs. However, besides the Wharton's Jelly that can be used for MSCs isolation, other parts could be of use, especially the blood vessels. Arterial bypass is the primary therapeutic strategy in patients with cardiovascular diseases. With more than 570,000 arterial bypasses being performed each year there is, understandably, a great need for arterial transplants.[139] Several synthetic materials such as Dacron and ePTFE have been tested as arterial grafts and failed, due to very serious adverse reactions, such as clot development, rejection, and chronic inflammation.[140] However, vessel development with tissue engineering techniques is a rapidly evolving field. As mentioned before, EPCs can be used, along with scaffolds, to create artificial blood vessels. Since the early 2000s decellularized tissues have been proposed as substitutes for synthetic scaffolds.[141–143] The technique consists of removal of all cellular components from a tissue; the remaining extracellular matrix, i.e., the skeleton of the tissue, can then be repopulated by patient-derived cells. The human UC reaches a length of over 50cm, from which two artery segments of 20–30cm each can be easily isolated. There are no branches along this artery, which is also characterized by a uniform diameter throughout the whole length.[144] Native tissues, including human saphenous veins, have been decellularized and have shown potential use as small-diameter vascular grafts.[145] The human umbilical artery might represent a more attractive source because it is widely available and easily isolated. In the Hellenic Cord Blood Bank, our team has successfully established protocols for the decellularization of umbilical arteries (see Figure 5) for tissue engineering applications (in press).

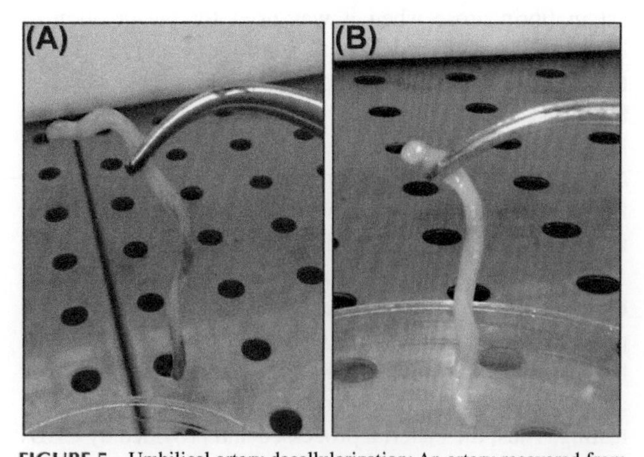


**FIGURE 5** Umbilical artery decellularization: An artery recovered from the umbilical cord before (A) and after undergoing decellularization (B).

## 7. CONCLUSION

The terms, regenerative medicine, tissue engineering, and advanced cellular therapies have been added to the medical glossary toward the end of the twentieth century. What all these disciplines have in common is their dependence on human stem cells, the use of which could provide clinicians with a valuable arsenal for the improvement of health-care quality. As the potency and therapeutic potential of stem cells decline with aging, sources of fetal origin could supply a sufficient number of high quality cells.

CB is considered an abundant source of such cells. CB stem cells have a unique biologic potential and they may provide a powerful research and clinical tool for pharmacology, genomics, cell therapy, and tissue engineering. The future of CB stem cell therapy extends beyond conventional transplantation into regenerative medicine, as CB stem cells may become an important resource for regeneration of tissue and organs. Research is currently under way in a variety of new areas of stem cell applications. Recently there have been many new discoveries in the field of stem cells and heart disease. In fact, EPCs have recently been isolated from CB to augment collateral vessel growth to ischemic tissue and even help regenerate damaged heart muscle.

In addition to the EPCs, CB can be a good source for the generation of iPSC lines, since CB sourcing is well organized and validated. All stored CBUs are HLA-typed and donors' medical history is available. Therefore, all existing samples can be used without compromising their ultimate use. Finally, it should be mentioned that CB is considered as one of the youngest and most naïve sources of stem cells one may obtain, ensuring the highest quality of iPSCs.

Another source of fetal stem cells is the UC. Wharton's Jelly is rich in MSCs, which can be isolated and stored for future use. Their unique properties make MSCs a valuable agent for advanced cellular therapies, regenerative medicine, and tissue engineering applications, and they are already being used in a variety of applications, ranging from the treatment of neurological disorders to bone and cartilage regeneration.

A CB bank possessing a strategic plan that integrates the bank's approach to these issues by developing reliable procedures to make useful cell lines available to the research community, would eventually become self-sustaining. However, this attempt should be promoted and funded since it will have a great impact on public health.

If the focus is on increasing the availability and therapeutic utility of high-quality stem cell lines for research on human diseases, particular thought should be given to issues of patient consent, ethics, and intellectual property that could impact the use of these cells for commercial development. Also, the bank must fit into the international space by establishing connections with other banks, research organizations, patient organizations, and commercial entities. In conclusion, it will be necessary for clinical grade stem cells to be isolated and maintained in conditions that are established under good manufacturing practices in order to improve, not only survival, but also the therapeutic potency of the cells.[146] In conclusion, in an era where stem cells and the growing field of regenerative medicine elicit excitement and anticipation across the scientific world, it seems appropriate to reconsider the potential future of CB biology, transplantation, and banking.

## LIST OF ABBREVIATIONS

**aGvHD** Acute Graft-versus-Host Disease
**ALS** Amyotrophic lateral sclerosis
**AT I** Angiotensin II receptor type I
**ATMPs** Advanced therapy medicinal products
**BM** Bone marrow
**BSE** Bovine spongiform encephalopathy
**CACs** Circulating angiogenic cells
**CAD** Coronary artery disease
**CAT** Committee for Advanced Therapies
**CB** Cord blood
**CBB** Cord blood bank
**CBU** Cord blood unit
**CD** Cluster of differentiation
**cDNA** Complementary deoxyribonucleic acid
**CFU-ECs** Colony-forming unit endothelial cells
**CM** Cardiac myosin
**Col V** Collagen V
**EBMT** European Bone Marrow Transplantation
**ECFCs** Endothelial colony-forming cells
**ECs** Endothelial cells
**EMA** European Medicines Agency
**EPCs** Endothelial progenitor cells
**ePTFE** Expanded polytetrafluoroethylene
**ESC-EC** Endothelial cells deriving from embryonic stem cells
**EULAR** European League against Rheumatism
**FBS** Fetal bovine serum
**FCS** Fetal calf serum
**FDA** Food and Drug Administration
**GvHD** Graft-versus-Host Disease
**hESCs** Human embryonic Stem Cells
**hiPSCs** Human-induced pluripotent stem cells
**HLA** Human leukocyte antigen
**HSCs** Hematopoietic stem cells
**HSCT** Hematopoietic Stem Cell Transplantation
**HUCACs** Human umbilical cord artery-derived endothelial cells
**HUVECs** Human umbilical vein endothelial cells
**ICRS** International Cartilage Repair Society
**iPSCs** Induced pluripotent stem cells
**IVT** In vitro transcription templates
**KDR** Kinase insert domain receptor
**MHC** Major histocompatibility complex
**MICA** Major histocompatibility complex class I chain-related genes A
**MICB** Major histocompatibility complex class I chain-related genes B
**MS** Multiple sclerosis
**MSCs** Mesenchymal stem/stromal cells
**NOD** Non-obese diabetic
**PL** Platelet lysates
**PPP** Platelet-Poor Plasma
**PRP** Platelet-rich plasma
**SCID** Severe combined immune deficient
**TERMIS** Tissue Engineering and Regenerative Medicine International Society
**USSCs** Unrestricted somatic stem cells
**vCJD** Variant Creutzfeldt–Jakob Disease
**VEGF** Vascular endothelial growth factor

## REFERENCES


1. Gottweis HKJ, Bignami F, Rial-Sebbag E, Lattanzi R, Macek M, Biobanks for Europe. *A challenge for governance. Report of the expert group on dealing with ethical and regulatory challenges of international biobank research.* Luxembourg: Publications Office of the European Union; 2012, ISBN: 978-92-79-22858-2. p. 63.
2. Yamanaka S. A fresh look at iPS cells. *Cell* 2009;**137**(1):13–7.
3. Broxmeyer HE, Lee MR, Hangoc G, Cooper S, Prasain N, Kim YJ, et al. Hematopoietic stem/progenitor cells, generation of induced pluripotent stem cells, and isolation of endothelial progenitors from 21- to 23.5-year cryopreserved cord blood. *Blood* 2011;**117**(18):4773–7.
4. Giorgetti A, Montserrat N, Rodriguez-Piza I, Azqueta C, Veiga A, Izpisua Belmonte JC. Generation of induced pluripotent stem cells from human cord blood cells with only two factors: Oct4 and Sox2. *Nat Protoc* 2010;**5**(4):811–20.
5. Broxmeyer HE. Will iPS cells enhance therapeutic applicability of cord blood cells and banking? *Cell Stem Cell* 2010;**6**(1):21–4.
6. de La Selle V, Gluckman E, Bruley-Rosset M. Newborn blood can engraft adult mice without inducing graft-versus-host disease across non H-2 antigens. *Blood* 1996;**87**(9):3977–83.
7. Tondreau T, Meuleman N, Delforge A, Dejeneffe M, Leroy R, Massy M, et al. Mesenchymal stem cells derived from CD133-positive cells in mobilized peripheral blood and cord blood: proliferation, Oct4 expression, and plasticity. *Stem Cells* 2005;**23**(8):1105–12.
8. He S, Nakada D, Morrison SJ. Mechanisms of stem cell self-renewal. *Annu Rev Cell Dev Biol* 2009;**25**:377–406.
9. Can A, Karahuseyinoglu S. Concise review: human umbilical cord stroma with regard to the source of fetus-derived stem cells. *Stem Cells* 2007;**25**(11):2886–95.
10. Naruse K, Hamada Y, Nakashima E, Kato K, Mizubayashi R, Kamiya H, et al. Therapeutic neovascularization using cord blood-derived endothelial progenitor cells for diabetic neuropathy. *Diabetes* 2005;**54**(6):1823–8.
11. Kaviani A, Perry TE, Dzakovic A, Jennings RW, Ziegler MM, Fauza DO. The amniotic fluid as a source of cells for fetal tissue engineering. *J Pediatr Surg* 2001;**36**(11):1662–5.
12. Alviano F, Fossati V, Marchionni C, Arpinati M, Bonsi L, Franchina M, et al. Term Amniotic membrane is a high throughput source for multipotent mesenchymal stem cells with the ability to differentiate into endothelial cells in vitro. *BMC Dev Biol* 2007;**7**:11.
13. Li CD, Zhang WY, Li HL, Jiang XX, Zhang Y, Tang PH, et al. Mesenchymal stem cells derived from human placenta suppress allogeneic umbilical cord blood lymphocyte proliferation. *Cell Res* 2005;**15**(7):539–47.

14. Sun L, Wang D, Liang J, Zhang H, Feng X, Wang H, et al. Umbilical cord mesenchymal stem cell transplantation in severe and refractory systemic lupus erythematosus. *Arthritis Rheum* 2010;**62**(8):2467–75.
15. Weiss ML, Anderson C, Medicetty S, Seshareddy KB, Weiss RJ, VanderWerff I, et al. Immune properties of human umbilical cord Wharton's jelly-derived cells. *Stem Cells* 2008;**26**(11):2865–74.
16. Carmeliet P. Angiogenesis in life, disease and medicine. *Nature* 2005;**438**(7070):932–6.
17. Ingram DA, Mead LE, Tanaka H, Meade V, Fenoglio A, Mortell K, et al. Identification of a novel hierarchy of endothelial progenitor cells using human peripheral and umbilical cord blood. *Blood* 2004;**104**(9):2752–60.
18. Prater DN, Case J, Ingram DA, Yoder MC. Working hypothesis to redefine endothelial progenitor cells. *Leukemia* 2007;**21**(6):1141–9.
19. McAllister SE, Medina R, O'Neill C, Stitt AW. Characterisation and therapeutic potential of endothelial progenitor cells. *Lancet* 2013;**381**(Suppl. 1):S73 (0).
20. Yoder MC, Mead LE, Prater D, Krier TR, Mroueh KN, Li F, et al. Redefining endothelial progenitor cells via clonal analysis and hematopoietic stem/progenitor cell principals. *Blood* 2007;**109**(5):1801–9.
21. Lavergne M, Vanneaux V, Delmau C, Gluckman E, Rodde-Astier I, Larghero J, et al. Cord blood-circulating endothelial progenitors for treatment of vascular diseases. *Cell Prolif* 2011;**44**(Suppl. 1): 44–7.
22. Au P, Daheron LM, Duda DG, Cohen KS, Tyrrell JA, Lanning RM, et al. Differential in vivo potential of endothelial progenitor cells from human umbilical cord blood and adult peripheral blood to form functional long-lasting vessels. *Blood* 2008;**111**(3):1302–5.
23. Atala A, Bauer SB, Soker S, Yoo JJ, Retik AB. Tissue-engineered autologous bladders for patients needing cystoplasty. *Lancet* 2006;**367**(9518):1241–6.
24. Macchiarini P, Jungebluth P, Go T, Asnaghi MA, Rees LE, Cogan TA, et al. Clinical transplantation of a tissue-engineered airway. *Lancet* 2008;**372**(9655):2023–30.
25. Takeshita S, Zheng LP, Brogi E, Kearney M, Pu LQ, Bunting S, et al. Therapeutic angiogenesis. A single intraarterial bolus of vascular endothelial growth factor augments revascularization in a rabbit ischemic hind limb model. *J Clin Invest* 1994;**93**(2): 662–70.
26. Chu H, Wang Y. Therapeutic angiogenesis: controlled delivery of angiogenic factors. *Ther Deliv* 2012;**3**(6):693–714.
27. Strauer BE, Brehm M, Zeus T, Kostering M, Hernandez A, Sorg RV, et al. Repair of infarcted myocardium by autologous intracoronary mononuclear bone marrow cell transplantation in humans. *Circulation* 2002;**106**(15):1913–8.
28. Losordo DW, Dimmeler S. Therapeutic angiogenesis and vasculogenesis for ischemic disease: part II: cell-based therapies. *Circulation* 2004;**109**(22):2692–7.
29. Liras A. Future research and therapeutic applications of human stem cells: general, regulatory, and bioethical aspects. *J Transl Med* 2010;**8**:131.
30. Chauhan A, More RS, Mullins PA, Taylor G, Petch C, Schofield PM. Aging-associated endothelial dysfunction in humans is reversed by L-arginine. *J Am Coll Cardiol* 1996;**28**(7):1796–804.
31. Tschudi MR, Barton M, Bersinger NA, Moreau P, Cosentino F, Noll G, et al. Effect of age on kinetics of nitric oxide release in rat aorta and pulmonary artery. *J Clin Invest* 1996;**98**(4):899–905.
32. Finney MR, Greco NJ, Haynesworth SE, Martin JM, Hedrick DP, Swan JZ, et al. Direct comparison of umbilical cord blood versus bone marrow-derived endothelial precursor cells in mediating neovascularization in response to vascular ischemia. *Biol Blood Marrow Transplant* 2006;**12**(5):585–93.
33. Murohara T, Ikeda H, Duan J, Shintani S, Sasaki K, Eguchi H, et al. Transplanted cord blood-derived endothelial precursor cells augment postnatal neovascularization. *J Clin Invest* 2000;**105**(11):1527–36.
34. Suarez Y, Shepherd BR, Rao DA, Pober JS. Alloimmunity to human endothelial cells derived from cord blood progenitors. *J Immunol* 2007;**179**(11):7488–96.
35. Zhang Q, Reed EF. Non-MHC antigenic targets of the humoral immune response in transplantation. *Curr Opin Immunol* 2010; **22**(5):682–8.
36. Sumitran-Holgersson S. Relevance of MICA and other non-HLA antibodies in clinical transplantation. *Curr Opin Immunol* 2008;**20**(5):607–13.
37. Go AS, Mozaffarian D, Roger VL, Benjamin EJ, Berry JD, Borden WB, American Heart Association Statistics Committee, Stroke Statistics Subcommittee, et al. Heart disease and stroke statistics–2013 update: a report from the American Heart Association. *Circulation* 2013;**127**(1):e6–245.
38. Till JE, Mc CE. A direct measurement of the radiation sensitivity of normal mouse bone marrow cells. *Radiat Res* 1961;**14**:213–22.
39. Friedenstein AJ, Petrakova KV, Kurolesova AI, Frolova GP. Heterotopic of bone marrow. Analysis of precursor cells for osteogenic and hematopoietic tissues. *Transplantation* 1968;**6**(2):230–47.
40. De Ugarte DA, Morizono K, Elbarbary A, Alfonso Z, Zuk PA, Zhu M, et al. Comparison of multi-lineage cells from human adipose tissue and bone marrow. *Cells Tissues Organs* 2003;**174**(3):101–9.
41. In 't Anker PS, Scherjon SA, Kleijburg-van der Keur C, de Groot-Swings GM, Claas FH, Fibbe WE, et al. Isolation of mesenchymal stem cells of fetal or maternal origin from human placenta. *Stem Cells* 2004;**22**(7):1338–45.
42. Tsai MS, Lee JL, Chang YJ, Hwang SM. Isolation of human multipotent mesenchymal stem cells from second-trimester amniotic fluid using a novel two-stage culture protocol. *Hum Reprod* 2004;**19**(6):1450–6.
43. Erices A, Conget P, Minguell JJ. Mesenchymal progenitor cells in human umbilical cord blood. *Br J Haematol* 2000;**109**(1):235–42.
44. Lee OK, Kuo TK, Chen WM, Lee KD, Hsieh SL, Chen TH. Isolation of multipotent mesenchymal stem cells from umbilical cord blood. *Blood* 2004;**103**(5):1669–75.
45. Yang SE, Ha CW, Jung M, Jin HJ, Lee M, Song H, et al. Mesenchymal stem/progenitor cells developed in cultures from UC blood. *Cytotherapy* 2004;**6**(5):476–86.
46. Romanov YA, Svintsitskaya VA, Smirnov VN. Searching for alternative sources of postnatal human mesenchymal stem cells: candidate MSC-like cells from umbilical cord. *Stem Cells* 2003;**21**(1):105–10.
47. Wang HS, Hung SC, Peng ST, Huang CC, Wei HM, Guo YJ, et al. Mesenchymal stem cells in the Wharton's jelly of the human umbilical cord. *Stem Cells* 2004;**22**(7):1330–7.
48. Lu LL, Liu YJ, Yang SG, Zhao QJ, Wang X, Gong W, et al. Isolation and characterization of human umbilical cord mesenchymal stem cells with hematopoiesis-supportive function and other potentials. *Haematologica* 2006;**91**(8):1017–26.
49. Scaradavou A, Brunstein CG, Eapen M, Le-Rademacher J, Barker JN, Chao N, et al. Double unit grafts successfully extend the application of umbilical cord blood transplantation in adults with acute leukemia. *Blood* 2013;**121**(5):752–8.

50. Gluckman E. Ten years of cord blood transplantation: from bench to bedside. *Br J Haematol* 2009;**147**(2):192–9.
51. Querol S, Rubinstein P, Marsh SG, Goldman J, Madrigal JA. Cord blood banking: 'providing cord blood banking for a nation'. *Br J Haematol* 2009;**147**(2):227–35.
52. Gabutti V, Timeus F, Ramenghi U, Crescenzio N, Marranca D, Miniero R, et al. Expansion of cord blood progenitors and use for hemopoietic reconstitution. *Stem Cells* 1993;**11**(Suppl. 2):105–12.
53. Shpall EJ, Quinones R, Giller R, Zeng C, Baron AE, Jones RB, et al. Transplantation of ex vivo expanded cord blood. *Biol Blood Marrow Transplant* 2002;**8**(7):368–76.
54. Bornhauser M. Ex vivo expansion of umbilical cord blood cells on feeder layers. *Methods Mol Biol* 2003;**215**:341–9.
55. Jaroscak J, Goltry K, Smith A, Waters-Pick B, Martin PL, Driscoll TA, et al. Augmentation of umbilical cord blood (UCB) transplantation with ex vivo-expanded UCB cells: results of a phase 1 trial using the AastromReplicell System. *Blood* 2003;**101**(12):5061–7.
56. Kashiwakura I, Takahashi TA. Basic fibroblast growth factor-stimulated ex vivo expansion of haematopoietic progenitor cells from human placental and umbilical cord blood. *Br J Haematol* 2003;**122**(3):479–88.
57. Rosu-Myles M, Bhatia M. SDF-1 enhances the expansion and maintenance of highly purified human hematopoietic progenitors. *Hematol J* 2003;**4**(2):137–45.
58. Peled T, Mandel J, Goudsmid RN, Landor C, Hasson N, Harati D, et al. Pre-clinical development of cord blood-derived progenitor cell graft expanded ex vivo with cytokines and the polyamine copper chelator tetraethylenepentamine. *Cytotherapy* 2004;**6**(4):344–55.
59. Huang GP, Pan ZJ, Jia BB, Zheng Q, Xie CG, Gu JH, et al. Ex vivo expansion and transplantation of hematopoietic stem/progenitor cells supported by mesenchymal stem cells from human umbilical cord blood. *Cell Transplant* 2007;**16**(6):579–85.
60. Robinson SN, Ng J, Niu T, Yang H, McMannis JD, Karandish S, et al. Superior ex vivo cord blood expansion following co-culture with bone marrow-derived mesenchymal stem cells. *Bone Marrow Transplant* 2006;**37**(4):359–66.
61. Deeg HJ, Henslee-Downey PJ. Management of acute graft-versus-host disease. *Bone Marrow Transplant* 1990;**6**(1):1–8.
62. Atkinson K. Chronic graft-versus-host disease. *Bone Marrow Transplant* 1990;**5**(2):69–82.
63. Holmberg L, Kikuchi K, Gooley TA, Adams KM, Hockenbery DM, Flowers ME, et al. Gastrointestinal graft-versus-host disease in recipients of autologous hematopoietic stem cells: incidence, risk factors, and outcome. *Biol Blood Marrow* 2006;**12**(2):226–34.
64. Jacobsohn DA, Vogelsang GB. Acute graft versus host disease. *Orphanet J Rare Dis* 2007;**2**:35.
65. Le Blanc K, Rasmusson I, Sundberg B, Gotherstrom C, Hassan M, Uzunel M, et al. Treatment of severe acute graft-versus-host disease with third party haploidentical mesenchymal stem cells. *Lancet* 2004;**363**(9419):1439–41.
66. Le Blanc K, Frassoni F, Ball L, Locatelli F, Roelofs H, Lewis I, Developmental Committee of the European Group for Blood and Marrow Transplantation, et al. Mesenchymal stem cells for treatment of steroid-resistant, severe, acute graft-versus-host disease: a phase II study. *Lancet* 2008;**371**(9624):1579–86.
67. Ringden O, Uzunel M, Rasmusson I, Remberger M, Sundberg B, Lonnies H, et al. Mesenchymal stem cells for treatment of therapy-resistant graft-versus-host disease. *Transplantation* 2006; **81**(10):1390–7.
68. Bernardo ME, Ball LM, Cometa AM, Roelofs H, Zecca M, Avanzini MA, et al. Co-infusion of ex vivo-expanded, parental MSCs prevents life-threatening acute GVHD, but does not reduce the risk of graft failure in pediatric patients undergoing allogeneic umbilical cord blood transplantation. *Bone Marrow Transplant* 2011;**46**(2):200–7.
69. Kebriaei P, Isola L, Bahceci E, Holland K, Rowley S, McGuirk J, et al. Adult human mesenchymal stem cells added to corticosteroid therapy for the treatment of acute graft-versus-host disease. *Biol Blood Marrow Transplant* 2009;**15**(7):804–11.
70. Galipeau J. The mesenchymal stromal cells dilemma–does a negative phase III trial of random donor mesenchymal stromal cells in steroid-resistant graft-versus-host disease represent a death knell or a bump in the road? *Cytotherapy* 2013;**15**(1):2–8.
71. Watanabe S, Arimura Y, Nagaishi K, Isshiki H, Onodera K, Nasuno M, et al. Conditioned mesenchymal stem cells produce pleiotropic gut trophic factors. *J Gastroenterol* 2014;**49**(2):270–82.
72. Tamari M, Nishino Y, Yamamoto N, Ueda M. Acceleration of wound healing with stem cell-derived growth factors. *Int J Oral Maxillofac Implants* 2013;**28**(6):e369–75.
73. Dash SN, Dash NR, Guru B, Mohapatra PC. Towards reaching the target: clinical application of mesenchymal stem cells for diabetic foot ulcers. *Rejuvenation Res* 2014;**17**(1):40–53.
74. Zografou A, Papadopoulos O, Tsigris C, Kavantzas N, Michalopoulos E, Chatzistamatiou T, et al. Autologous transplantation of adipose-derived stem cells enhances skin graft survival and wound healing in diabetic rats. *Ann Plastic Surg* 2013;**71**(2):225–32.
75. Bray LJ, George KA, Hutmacher DW, Chirila TV, Harkin DG. A dual-layer silk fibroin scaffold for reconstructing the human corneal limbus. *Biomaterials* 2012;**33**(13):3529–38.
76. Reichert JC, Cipitria A, Epari DR, Saifzadeh S, Krishnakanth P, Berner A, et al. A tissue engineering solution for segmental defect regeneration in load-bearing long bones. *Sci Transl Med* 2012;**4**(141):141ra93.
77. Mendez JJ, Ghaedi M, Steinbacher D, Niklason L. Epithelial cell differentiation of human mesenchymal stromal cells in decellularized lung scaffolds. *Tissue Eng Part A* 2014;**20**(11-12):1735–46.
78. Papadopoulou A, Yiangou M, Athanasiou E, Zogas N, Kaloyannidis P, Batsis I, et al. Mesenchymal stem cells are conditionally therapeutic in preclinical models of rheumatoid arthritis. *Ann Rheum Dis* 2012;**71**(10):1733–40.
79. Wang L, Wang L, Cong X, Liu G, Zhou J, Bai B, et al. Human umbilical cord mesenchymal stem cell therapy for patients with active rheumatoid arthritis: safety and efficacy. *Stem Cells Dev* 2013;**22**(24):3192–202.
80. Karussis D, Karageorgiou C, Vaknin-Dembinsky A, Gowda-Kurkalli B, Gomori JM, Kassis I, et al. Safety and immunological effects of mesenchymal stem cell transplantation in patients with multiple sclerosis and amyotrophic lateral sclerosis. *Arch Neurol* 2010;**67**(10):1187–94.
81. Liang J, Zhang H, Hua B, Wang H, Lu L, Shi S, et al. Allogenic mesenchymal stem cells transplantation in refractory systemic lupus erythematosus: a pilot clinical study. *Ann Rheum Dis* 2010;**69**(8):1423–9.
82. Mabed M, Shahin M. Mesenchymal stem cell-based therapy for the treatment of type 1 diabetes mellitus. *Curr Stem Cell Res Ther* 2012;**7**(3):179–90.
83. Hayashi T, Wakao S, Kitada M, Ose T, Watabe H, Kuroda Y, et al. Autologous mesenchymal stem cell-derived dopaminergic neurons function in parkinsonian macaques. *J Clin Invest* 2013;**123**(1):272–84.

84. Hardiman O, van den Berg LH, Kiernan MC. Clinical diagnosis and management of amyotrophic lateral sclerosis. *Nat Rev Neurol* 2011;**7**(11):639–49.
85. Mazzini L, Ferrero I, Luparello V, Rustichelli D, Gunetti M, Mareschi K, et al. Mesenchymal stem cell transplantation in amyotrophic lateral sclerosis: a Phase I clinical trial. *Exp Neurol* 2010; **223**(1):229–37.
86. Noth U, Steinert AF, Tuan RS. Technology insight: adult mesenchymal stem cells for osteoarthritis therapy. *Nat Clin Pract Rheumatol* 2008;**4**(7):371–80.
87. Ende N, Chen R, Reddi AS. Transplantation of human umbilical cord blood cells improves glycemia and glomerular hypertrophy in type 2 diabetic mice. *Biochem Biophys Res Commun* 2004;**321**(1):168–71.
88. Si Y, Zhao Y, Hao H, Liu J, Guo Y, Mu Y, et al. Infusion of mesenchymal stem cells ameliorates hyperglycemia in type 2 diabetic rats: identification of a novel role in improving insulin sensitivity. *Diabetes* 2012;**61**(6):1616–25.
89. Wagner W, Wein F, Seckinger A, Frankhauser M, Wirkner U, Krause U, et al. Comparative characteristics of mesenchymal stem cells from human bone marrow, adipose tissue, and umbilical cord blood. *Exp Hematol* 2005;**33**(11):1402–16.
90. Jin HJ, Bae YK, Kim M, Kwon SJ, Jeon HB, Choi SJ, et al. Comparative analysis of human mesenchymal stem cells from bone marrow, adipose tissue, and umbilical cord blood as sources of cell therapy. *Int J Mol Sci* 2013;**14**(9):17986–8001.
91. Ryu HH, Kang BJ, Park SS, Kim Y, Sung GJ, Woo HM, et al. Comparison of mesenchymal stem cells derived from fat, bone marrow, Wharton's jelly, and umbilical cord blood for treating spinal cord injuries in dogs. *J Vet Med Sci/Jpn Soc Vet Sci* 2012;**74**(12):1617–30.
92. Prasanna SJ, Gopalakrishnan D, Shankar SR, Vasandan AB. Pro-inflammatory cytokines, IFNgamma and TNFalpha, influence immune properties of human bone marrow and Wharton jelly mesenchymal stem cells differentially. *PloS one* 2010;**5**(2):e9016.
93. Chatzistamatiou TK, Papassavas AC, Michalopoulos E, Gamaloutsos C, Mallis P, Gontika I, et al. Optimizing isolation culture and freezing methods to preserve Wharton's jelly's mesenchymal stem cell (MSC) properties: an MSC banking protocol validation for the Hellenic Cord Blood Bank. *Transfusion* 2014 doi: 10.1111/trf.12743. [Epub ahead of print].
94. Martin I, Baldomero H, Bocelli-Tyndall C, Emmert MY, Hoerstrup SP, Ireland H, et al. The survey on cellular and engineered tissue therapies in europe in 2011. *Tissue Eng Part A* 2014;**20**(3–4):842–53.
95. Ancans J. Cell therapy medicinal product regulatory framework in Europe and its application for MSC-based therapy development. *Front Immunol* 2012;**3**:253.
96. Lowry WE, Quan WL. Roadblocks en route to the clinical application of induced pluripotent stem cells. *J Cell Sci* 2010;**123**(Pt 5): 643–51.
97. Dravid GG, Crooks GM. The challenges and promises of blood engineered from human pluripotent stem cells. *Adv Drug Deliv Rev* 2011;**63**(4–5):331–41.
98. Takenaka C, Nishishita N, Takada N, Jakt LM, Kawamata S. Effective generation of iPS cells from CD34+ cord blood cells by inhibition of p53. *Exp Hematol* 2010;**38**(2):154–62.
99. Chou BK, Mali P, Huang X, Ye Z, Dowey SN, Resar LM, et al. Efficient human iPS cell derivation by a non-integrating plasmid from blood cells with unique epigenetic and gene expression signatures. *Cell Res* 2011;**21**(3):518–29.
100. Hu K, Yu J, Suknuntha K, Tian S, Montgomery K, Choi KD, et al. Efficient generation of transgene-free induced pluripotent stem cells from normal and neoplastic bone marrow and cord blood mononuclear cells. *Blood* 2011;**117**(14):e109–19.
101. Trifunovic A, Larsson NG. Mitochondrial dysfunction as a cause of ageing. *J Intern Med* 2008;**263**(2):167–78.
102. Marion RM, Strati K, Li H, Tejera A, Schoeftner S, Ortega S, et al. Telomeres acquire embryonic stem cell characteristics in induced pluripotent stem cells. *Cell Stem Cell* 2009;**4**(2):141–54.
103. Nakatsuji N, Nakajima F, Tokunaga K. HLA-haplotype banking and iPS cells. *Nat Biotechnol* 2008;**26**(7):739–40.
104. Kajiwara M, Aoi T, Okita K, Takahashi R, Inoue H, Takayama N, et al. Donor-dependent variations in hepatic differentiation from human-induced pluripotent stem cells. *Proc Natl Acad Sci USA* 2012;**109**(31):12538–43.
105. Okita K, Yamakawa T, Matsumura Y, Sato Y, Amano N, Watanabe A, et al. An efficient nonviral method to generate integration-free human-induced pluripotent stem cells from cord blood and peripheral blood cells. *Stem Cells* 2013;**31**(3):458–66.
106. Stacey G. Banking stem cells for research and clinical applications. *Prog Brain Res* 2012;**200**:41–58.
107. Takahashi K, Yamanaka S. Induced pluripotent stem cells in medicine and biology. *Development* 2013;**140**(12):2457–61.
108. Hochedlinger K, Plath K. Epigenetic reprogramming and induced pluripotency. *Development* 2009;**136**(4):509–23.
109. Nishikawa S, Goldstein RA, Nierras CR. The promise of human induced pluripotent stem cells for research and therapy. *Nat Rev Mol Cell Biol* 2008;**9**(9):725–9.
110. Carey BW, Markoulaki S, Hanna J, Saha K, Gao Q, Mitalipova M, et al. Reprogramming of murine and human somatic cells using a single polycistronic vector. *Proc Natl Acad Sci USA* 2009;**106**(1): 157–62.
111. Huangfu D, Osafune K, Maehr R, Guo W, Eijkelenboom A, Chen S, et al. Induction of pluripotent stem cells from primary human fibroblasts with only Oct4 and Sox2. *Nat Biotechnol* 2008;**26**(11): 1269–75.
112. Okita K, Nakagawa M, Hyenjong H, Ichisaka T, Yamanaka S. Generation of mouse induced pluripotent stem cells without viral vectors. *Science* 2008;**322**(5903):949–53.
113. Okita K, Ichisaka T, Yamanaka S. Generation of germline-competent induced pluripotent stem cells. *Nature* 2007;**448**(7151):313–7.
114. Stadtfeld M, Nagaya M, Utikal J, Weir G, Hochedlinger K. Induced pluripotent stem cells generated without viral integration. *Science* 2008;**322**(5903):945–9.
115. Woltjen K, Michael IP, Mohseni P, Desai R, Mileikovsky M, Hamalainen R, et al. piggyBac transposition reprograms fibroblasts to induced pluripotent stem cells. *Nature* 2009;**458**(7239):766–70.
116. Kaji K, Norrby K, Paca A, Mileikovsky M, Mohseni P, Woltjen K. Virus-free induction of pluripotency and subsequent excision of reprogramming factors. *Nature* 2009;**458**(7239):771–5.
117. Soldner F, Hockemeyer D, Beard C, Gao Q, Bell GW, Cook EG, et al. Parkinson's disease patient-derived induced pluripotent stem cells free of viral reprogramming factors. *Cell* 2009;**136**(5):964–77.
118. Yu J, Hu K, Smuga-Otto K, Tian S, Stewart R, Slukvin II, et al. Human induced pluripotent stem cells free of vector and transgene sequences. *Science* 2009;**324**(5928):797–801.
119. Zhou H, Wu S, Joo JY, Zhu S, Han DW, Lin T, et al. Generation of induced pluripotent stem cells using recombinant proteins. *Cell Stem Cell* 2009;**4**(5):381–4.

120. Warren L, Manos PD, Ahfeldt T, Loh YH, Li H, Lau F, et al. Highly efficient reprogramming to pluripotency and directed differentiation of human cells with synthetic modified mRNA. *Cell Stem Cell* 2010;**7**(5):618–30.
121. Lengerke C, Daley GQ. Autologous blood cell therapies from pluripotent stem cells. *Blood Rev* 2010;**24**(1):27–37.
122. Rao M, Ahrlund-Richter L, Kaufman DS. Concise review: cord blood banking, transplantation and induced pluripotent stem cell: success and opportunities. *Stem Cells* 2012;**30**(1):55–60.
123. Wu SM, Hochedlinger K. Harnessing the potential of induced pluripotent stem cells for regenerative medicine. *Nat Cell Biol* 2011;**13**(5):497–505.
124. Togarrati PP, Suknuntha K. Generation of mature hematopoietic cells from human pluripotent stem cells. *Int J Hematol* 2012;**95**(6):617–23.
125. Peyrard T, Bardiaux L, Krause C, Kobari L, Lapillonne H, Andreu G, et al. Banking of pluripotent adult stem cells as an unlimited source for red blood cell production: potential applications for alloimmunized patients and rare blood challenges. *Transfus Med Rev* 2011;**25**(3):206–16.
126. Jung YW, Hysolli E, Kim KY, Tanaka Y, Park IH. Human induced pluripotent stem cells and neurodegenerative disease: prospects for novel therapies. *Curr Opin Neurol* 2012;**25**(2):125–30.
127. de Almeida PE, Ransohoff JD, Nahid A, Wu JC. Immunogenicity of pluripotent stem cells and their derivatives. *Circ Res* 2013;**112**(3):549–61.
128. Fischer-Dzoga K, Kuo YF, Wissler RW. The proliferative effect of platelets and hyperlipidemic serum on stationary primary cultures. *Atherosclerosis* 1983;**47**(1):35–45.
129. Delwiche F, Raines E, Powell J, Ross R, Adamson J. Platelet-derived growth factor enhances in vitro erythropoiesis via stimulation of mesenchymal cells. *J Clin Invest* 1985;**76**(1):137–42.
130. Pawitan JA. Platelet rich plasma in xeno-free stem cell culture: the impact of platelet count and processing method. *Curr Stem Cell Res Ther* 2012;**7**(5):329–35.
131. Koellensperger E, von Heimburg D, Markowicz M, Pallua N. Human serum from platelet-poor plasma for the culture of primary human preadipocytes. *Stem Cells* 2006;**24**(5):1218–25.
132. Murphy MB, Blashki D, Buchanan RM, Yazdi IK, Ferrari M, Simmons PJ, et al. Adult and umbilical cord blood-derived platelet-rich plasma for mesenchymal stem cell proliferation, chemotaxis, and cryo-preservation. *Biomaterials* 2012;**33**(21):5308–16.
133. Bensaid W, Triffitt JT, Blanchat C, Oudina K, Sedel L, Petite H. A biodegradable fibrin scaffold for mesenchymal stem cell transplantation. *Biomaterials* 2003;**24**(14):2497–502.
134. Osathanon T, Linnes ML, Rajachar RM, Ratner BD, Somerman MJ, Giachelli CM. Microporous nanofibrous fibrin-based scaffolds for bone tissue engineering. *Biomaterials* 2008;**29**(30):4091–9.
135. Yao L, Liu J, Andreadis ST. Composite fibrin scaffolds increase mechanical strength and preserve contractility of tissue engineered blood vessels. *Pharm Res* 2008;**25**(5):1212–21.
136. Sage A, Chang AA, Schumacher BL, Sah RL, Watson D. Cartilage outgrowth in fibrin scaffolds. *Am J Rhinol Allergy* 2009;**23**(5):486–91.
137. Ferreira MS, Jahnen-Dechent W, Labude N, Bovi M, Hieronymus T, Zenke M, et al. Cord blood-hematopoietic stem cell expansion in 3D fibrin scaffolds with stromal support. *Biomaterials* 2012;**33**(29):6987–97.
138. Anitua E, Sanchez M, Nurden AT, Zalduendo M, de la Fuente M, Orive G, et al. Autologous fibrin matrices: a potential source of biological mediators that modulate tendon cell activities. *J Biomed Mater Res Part A* 2006;**77**(2):285–93.
139. Dahl SL, Koh J, Prabhakar V, Niklason LE. Decellularized native and engineered arterial scaffolds for transplantation. *Cell Transplant* 2003;**12**(6):659–66.
140. Teebken OE, Haverich A. Tissue engineering of small diameter vascular grafts. European journal of vascular and endovascular surgery. *Off J Eur Soc Vasc Surg* 2002;**23**(6):475–85.
141. Zeltinger J, Landeen LK, Alexander HG, Kidd ID, Sibanda B. Development and characterization of tissue-engineered aortic valves. *Tissue Eng* 2001;**7**(1):9–22.
142. Kaushal S, Amiel GE, Guleserian KJ, Shapira OM, Perry T, Sutherland FW, et al. Functional small-diameter neovessels created using endothelial progenitor cells expanded ex vivo. *Nat Med* 2001;**7**(9):1035–40.
143. Herson MR, Mathor MB, Altran S, Capelozzi VL, Ferreira MC. In vitro construction of a potential skin substitute through direct human keratinocyte plating onto decellularized glycerol-preserved allodermis. *Artif Organs* 2001;**25**(11):901–6.
144. Stehbens WE, Wakefield JS, Gilbert-Barness E, Zuccollo JM. Histopathology and ultrastructure of human umbilical blood vessels. *Fetal Pediatr Pathol* 2005;**24**(6):297–315.
145. Schaner PJ, Martin ND, Tulenko TN, Shapiro IM, Tarola NA, Leichter RF, et al. Decellularized vein as a potential scaffold for vascular tissue engineering. *J Vasc Surg* 2004;**40**(1):146–53.
146. Li Y, Ma T. Bioprocessing of cryopreservation for large-scale banking of human pluripotent stem cells. *Biores Open Access* 2012;**1**(5):205–14.

Section VII

# The Viewpoint of Society

Chapter 22

# An Introductory Note to the Cord Blood Banking Issues in a European and International Environment

**Marietta Giannakou**
*MEP, Head of the Greek EPP Parliamentary Delegation, Former Minister of National Education and Religious Affairs, Former Minister of Health, Welfare and Social Security*

The European Union (EU) constitutes one of the greatest political achievements in European history. The people of Europe achieved peace and pursued a mutually beneficiary collaboration by establishing institutions, procedures, and instruments that rely on state parity, democratic governance, and citizen freedom.

The results for peace, democracy, and prosperity, a few decades after the end of two World Wars are of historic significance. These developments are confirmed by the participation of states that now had freedom of choice, after the fall of the socialist states which barely respected the freedom of citizens. It is not an accident that the EU constitutes today an international light for world peace and prosperity. Its institutions and targets are embraced and supported by 28 governments and peoples, and several countries are under a preaccession process.

We certainly do not witness the end of history, as new challenges question the achievements and create the necessity to draw new action plans. There is no doubt that the Euro zone crisis constitutes a serious challenge for European unification, while for some people, the Greeks in particular, it is a painful process.

We must learn from our mistakes, acquire knowledge and turn our weaknesses to advantages. It would be wrong if the economic crises carried away the fundamentals and led to a retreat. Some interests seem to pursue the return to the era of ethnocentrism, intolerance, and the entrenchment of the people. We need to create a strong alliance in order to support everything we consider as acquired and for everything that would not exist without a united Europe.

The multiple positive dimensions of the EU for the life of the citizens must be fully understood. Such is the case in public health, which constitutes a primary index of the citizen's welfare. In this field the EU timely acquired an active role in the formulation of common rules and objects, ensuring high quality protection for the citizens' health. Health policies still fall within the competence of the national authorities, but the role of the EU is often necessary to effectively face problems of international character, like pandemics and bioterrorism.

An EU role is obviously needed to ensure that other European policies, like the movement of goods and services, will promote the objectives of public health. We must point out that the EU does not seek a common policy in the field of health and the abolishment of national competence. On the contrary it should be understood that national policies and common European targets are closely related.

Consequently, the EU has an important auxiliary role regarding national policies, that is irreplaceable in cases like illness prevention, food and nutritional safety, the safety of pharmaceutical products, the curbing of smoking, legislation for blood, tissues, cells, and human organs, quality of the water and the air, and the formation of relevant committees to oversee the respect of common rules. Most of the problems faced by European societies, such as demographic evolution and the aging population, are common.

Article 68 of the EU treaty provides the EU with a complete health strategy, the most recent example being the 2007 initiative titled "Together for Health."

Some of the most important principles concerning public health guiding the actions of the EU are common values for health, health as social wealth, health in all policies, the strengthening of the voice of the EU in international health institutions, the promotion of good health for the European people, the support of dynamic health systems, and the use of new technologies.

One typical example of the role of the EU regarding health matters is the 2004/23 EC Directive of the European Parliament (EP) and the European Council (EC), setting quality and safety standards for the donation, provision, controlling, processing, storage, and distribution of human tissues and cells.

**Cord Blood Stem Cells Medicine. http://dx.doi.org/10.1016/B978-0-12-407785-0.00022-0**

The Directive sets quality and safety standards for human tissues and cells that are intended to be for human application, so as to ensure a high level of safety for human health. The provisions of the Directive deal with donation, provision, controlling, processing, maintenance, and distribution of human tissues and cells that are intended to be applied to humans, as well as on processed products coming from humans and intended to be used on humans.

Thus a supervision system has been established, with common rules and practices, for the provision of human tissues and cells. The member states responsibility is to ensure that the provision and control of the tissues and cells is conducted by properly trained and experienced personnel, and takes place under conditions that have been certified, defined, approved, or licensed for this reason by a competent authority.

The first cord blood transplantation was successfully performed by E. Gluckman in Paris in 1998 on a child suffering from Fanconi anemia. Since then more than 20,000 allergenic transplantations of cord blood have been performed to patients that needed primeval hematopoietic cells. More than 2000 allergenic transplantations using cord blood cells are performed globally each year and more than 600,000 units are available for clinical usage.

The wide use of cord blood cells became possible with the creation of the first cord blood bank (CBB) in New York in 1993. The role of the CBB is to collect, process, freeze, and dispose of the units to the global banks.

The advantage of developing a big public bank is collective because a large part of the population will benefit from it. Furthermore, experts maintain that only a public bank can ensure a good transplant, since one of the drawbacks of private storage is the fact that a graft may involve a mutation that could cause leukemia in a child. So in cases where the disease we want to be healed has a genetic base (the disease precedes the birth so the stem cells also have the disease) it is better for a sick child to seek cure through a compatible donor.

Moreover, private banking concerns autologous and allergenic transplantation, but within the family. The storage cost of stem cells in a private bank is not negligible. Depending on the company the cost varies between 1500 and 3000€ for a 20-year storage period; finally, private banks do not take histocompatibility into consideration and therefore they do not participate in the international transplant search network NetCord.

Within public banks, the provision of cord blood from the parents is a donation at no cost, aiming to serve any patient who needs it. So cord blood donated to the bank by One can be used by an Other (Nonrelative). The banks are called nonrelated cord blood banks where one donates for all and all donate for the one (allogeneic transplantation). By donating his child's stem cells one does not acquire automatically priority to a future transplantation, since one will not get his transplant and furthermore histocompatibility is the first high priority. The time needed in order to locate a matching transplant cannot be estimated.

According to the above, banks with "public" character have a very strict selection procedure regarding the units that will be frozen in the final stage. Therefore, 80% of the cord blood cells that are collected on a daily basis are rejected. The main rejection reason concerns the total number of primeval hematopoietic cells that are contained in the unit. The cells that correspond to the standards set by the international organization NetCord-FACT are processed and, if they are found suitable, then and only then these cells will be stored at −196 °C in liquid nitrogen. The transplantation cost is covered by the insurance fund of the recipient.

All the above described in brief are some important developments in the area of medicine that concern every family. At the same time it constitutes a presumption about the important role of the EU in the everyday life of its citizens. Regarding public health, the EU plays a subsidiary role offering a framework of common rules and objects that are irreplaceable in the confrontation of some common challenges and dangers of the European citizen. Finally, the EU with its multidimensional actions, policies, and means promotes in practice the prosperity of the people, for the sake of whom it was formed initially.

Chapter 23

# Ethical and Legal Issues in Cord Blood Stem Cells and Biobanking

Maria Mitrossili, Marcos Sarris and Yannis Nikolados
*Health Law of Technological Educational Institute of Athens, Greece*

## Chapter Outline



## 1. INTRODUCTION

Over the last decade, great developments in biology have given a new boost to research on stem cells, their manipulation and banking, while the applications of biology in the field of biomedical technology have created challenges regarding the treatments of human illnesses and disabilities. More specifically, contemporary regenerative medicine constitutes a privileged field which, through cell treatments, gene technology, and tissue engineering, offers a perspective in the use of stem cells as therapies for a wide range of health problems which are considered difficult or impossible to treat, to the benefit of the human species. It seems that human beings today, at a symbolic and imaginary level, are less dependent on nature and probably more enslaved by their ethical positions.

The above-mentioned scientific advances have generated many bioethical and legal issues, which cause intense debate and controversy among scientists from different fields or even scientists from the same field. Specifically, "Regenerative medicine goes to the very core of the moral and the metaphysical understandings that ask what it is to be human. If one is to remake what it is to be human, one should know what goals are appropriate and what constraints should apply."[1]

Some of these issues focus on a well-known ethical and metaphysical question; that of the definition of human beings and the restrictions on the interventions which can be performed to them.[2] This dilemma causes a greater interest in embryonic stem cells than in somatic ones, to which hematopoietic stem cells (HSCs) and other cells that derive from the placenta, the umbilical cord, and cord blood can be included.[2–5] Other issues refer to doubts regarding the therapeutic results from the use of cell treatments, while some issues pose questions about the manipulation and modification of cells by researchers and doctors, the extent and limits of informed consent, and so forth.[6] Specifically in the case of biobanking, informed consent includes not only the removal of biological material, its collection, and storage in the bank, but also the collection and accumulation of information and personal data regarding the donor's personal and family medical history, as well as information about their lifestyle. The latter pieces of information are collected along with the biological material, while it is important that they be coded or encoded and identified with the donor for medical or scientific reasons, abiding by the principle of confidentiality. Respect of confidentiality ensures the protection of the donors whether that is in an electronic environment or not.

**Cord Blood Stem Cells Medicine. http://dx.doi.org/10.1016/B978-0-12-407785-0.00023-2**

However, the above-mentioned indicative bioethical and legal issues do not only differentiate when examined in the light of theory or practice, but also show different intensity depending on whether they refer to stem cells that come from umbilical cord blood (HSCs), a part of the umbilical cord and the placenta (mesenchymal stem cells), adult tissues, or embryos and fetuses. Various ethical arguments underline the diversity of views which surround the question: is it ethically acceptable to collect and manipulate stem cells from embryos? Different ethical opinions are inevitable in a pluralist word, but, in general, from a bioethical standpoint, approaches converge today at a level of principles regarding respect for the person and the primacy of the human being in relation to research benefits. Moreover, the necessity to find balance among scientific progress, economy, society, protection of fundamental rights and freedoms of the individual, and respect for the integrity of the human body is important. These principles are valuable to the extent that biobanking and biobanks are relatively new social phenomena and there is great institutional and legal diversity, worldwide, in the internal law of the countries. However, there is a lack of a strong accord or regulatory uniformity as far as international and European law are concerned. Exceptions to this context are some specialized provisions in documents of hard or soft laws at the international and European levels. These are the Convention of Human Rights and Biomedicine[7] of the European Council, also known as the Convention of Oviedo (1997) with the Additional Protocols,[8] the Charter of Fundamental Rights[9] of the European Union of 2000/2009, the Directives of the European Union (2004/23/EC, 2006/86/EC etc.),[10] and the Regulation (EC) no 1394/2007[11] of the European Parliament, as well as nonbinding texts, such as the Recommendation Rec(2006)4[12] of the European Council on research which uses biological material of human origin.

## 2. SCIENTIFIC BACKGROUND: STEM CELLS FROM CORD BLOOD, THE UMBILICAL CORD, AND THE PLACENTA

Umbilical cord blood is an important source of HSCs, which once collected, processed, cryopreserved, and stored can be used for the treatment of many illnesses and disabilities. This therapeutic wealth provides three basic potentials, some of which are already being applied.

The first one involves the treatment of blood diseases and blood malignancies. Already in the 1980s, Professor Eliane Gluckman and her team from Saint-Louis Hospital in Paris, France, replaced bone marrow transplantation with transplantation of cord blood HSCs and hematopoietic progenitor cells (HPCs) for the treatment of Fanconi anemia.[13,14] Since then many blood diseases and malignancies, which are successfully treated with cells collected from umbilical cord blood, have been recorded.[15]

The second potential involves skin restoration in the case of burns with the use of mesenchymal cells, which have the ability to be directly used, in contrast to cells that derive from bone marrow, as well as the promotion of testing of new drugs on human tissues, without the need to resort to living organisms.[15–17]

The third type focuses on the development of cell treatments that restore or regenerate different organs, such as neurons, cardiac muscle, liver, etc.[15]

It is scientifically known that cord blood consists of HPCs and it also contains HSCs, both of which are multipotent stem cells able to give rise to a limited number of specialized cell types.[18] HPCs and HSCs exist throughout a person's life.[19,20] It is important to highlight that, at an experimental level, the generation of induced pluripotent stem cells (iPSCs) has many cell sources, including cord blood, the umbilical cord, and the placenta.[19–22] They are pluripotent cells and as such probably have a potential like embryonic stem cells and are able to give rise to most cell types.[21–23] Moreover, it contains specialized stem cells that have the potential to produce a tissue in the transplanted organ, which is different from the original tissue. The "plasticity" of these cells opens new paths to research, with therapeutic results which are likely to surpass the original predictions, cultivating new expectations to the society regarding the expansion of the number of diseases and disabilities that can be treated.[15,23,24] The umbilical cord and the placenta contain non-hematopoietic somatic stem cells with the ability to generate cells, etc.[6,15,17]

Even though umbilical cord blood poses the fewest obstacles regarding not only ethics,[2,21] but also accessibility, quantity, or quality, it is sometimes unsafe or unsuitable for transplantation and cell treatment.[15] For this reason, the above-mentioned collections, which cannot be used for therapeutic purposes, should be available for research after the donor has consented. However, apart from blood collection, stem cell research can be expanded to the umbilical cord and the placenta, thus providing greater potential for therapeutic use.[15] The above-mentioned scientific advances with their therapeutic potentials are not known to people in all their dimensions. It is necessary that individuals—citizens and foreigners—be informed about the best possible therapeutic and research use of cord blood, the umbilical cord, and the placenta. It is well known that not all diseases and disabilities can be treated, in contrast to what some so-called scientific articles in the media claim, cultivating within the people a fantasy of omnipotence of biomedical technology, which is often a pretext for profiteering.[6] However, it is very important that they know that the therapeutic results do not cease to be significant for a capable number of sick and vulnerable people, as well as that their removal is not particularly dangerous and does not pose, at least in the majority of cases, major ethical dilemmas to the mother or the parents of the newborn, as well as to health professionals and researchers. Balancing the advantages and disadvantages of the removal of stem cells and other cells from umbilical cord blood, the umbilical cord and the placenta, according to the principle of proportionality, the advantages seem to outweigh the disadvantages.

## 3. FUNDAMENTAL ETHICAL AND LEGAL ISSUES

The presentation of the research and therapeutic potentials of the stem cells contained in umbilical cord blood, as well as in parts of the umbilical cord and the placenta, does not raise serious ethical issues.

To begin with, expressing an opinion on issues regarding the beginning of life, the definition of the person (as "qualified life")[25] and the human being before birth, "excluded from the community of people"[25] and with the potentiality to develop as a person, as well as on the manipulation of stem cells is of no interest, as this basically concerns somatic stem cells, which do not have the ability to wholly create a new complete and functioning organism. The concept of "potential person" which preoccupied the Committees of Bioethics in the 1990s at an international, European, and national level, has recently been replaced by the expression "human person becoming" connected with new projects of research (iPS) and Regenerative Medicine. It is a fact that there was minimum consensus on official documents (hard and soft laws or Opinions of Bioethics Committees) regarding the definition of an embryo. Philosophers' points of view diverge[26–29] and the ethical positions of institutions are different, because States, Administrations, or Bodies adopt opinions which are unacceptable to others. In general, many important international and European texts proclaim the right to life, but only a few texts state more specifically the right to life of the conceived child. The American Convention on Human Rights[30] of 1969 proclaims that "every person has the right to have his life respected. This right shall be protected by law, and in general, from the moment of conception" (art. 4). Moreover, the Council of Europe states in certain documents the primacy and the protection of the human being.

However, there is a latent consensus, which is based on values, principles, and rights. Apart from the right to life, the principle of human dignity is a basic reference point. Moreover, the following are of great importance: (1) the principles of autonomy and personal freedom which, respectively, lay the foundations of consent in every medical or research action and the principle of respect of private life; (2) the principles of justice and beneficence or charity which refer to the protection of health, safety, security, quality of products and services; (3) the quality of solidarity, which stands for the protection of collective interest and particularly of vulnerable people; (4) the principle of equality which on the one hand prohibits discriminations in the field of health, and on the other hand allows access to health services and provisions; and (5) the principle of freedom of thought which lays the foundation of respect of freedom of research by public authorities.[31]

In Western societies, cord blood, the umbilical cord, and the placenta were for many years considered to be merely useless remains just like surgical medical waste, while there was no need for consent on behalf of the person they came from for their use in immunoglobulin products and cosmetic medicine.[17] This practice acquired its meaning on the one hand in the context of a paternalistic practice and organization of medicine and on the other hand in the context of the Hippocratic Oath and morality based on the principles of utilitarianism, distant from bioethical ideas of virtue, deontology, and human rights approaches. The usual destination of this waste for many years was their destruction or burial. Even though in Western societies no one was interested in the umbilical cord and the placenta and they were considered useless, infected, and unwanted material, anthropological studies demonstrate cultural habits involving rituals for the burial of the umbilical cord and the placenta.[32]

However, over the last 20 years, from the moment that the placenta, the umbilical cord, and the cord blood, acquired great significance for medical treatment and research, they have changed status and have become human biological material and products for transplantation and cell therapy. Their collection takes place under specific rules and procedures by specialized trained staff in maternity hospitals. The above-mentioned collection changes from an unlawful act to a lawful one due to informed consent. Informed consent becomes the cornerstone of ethical and legal legalization of such an "intervention," according to the terminology of the Convention on Human Rights and Biomedicine[7] of the Council of Europe, which was signed in Oviedo, Spain, in 1997. Moreover, the collection of cord blood, the umbilical cord, and the placenta is morally justified because it is based on the autonomy of the individual and legally because it is based on the voluntary act of the donor. Even though it is provided to the benefit of specific groups of the population, patients and disabled people, whether it involves allogeneic nondirected use or autologous directed use, it does not cease to serve public health and the good of humanity, in the light of the international community which was formed on the principle of the universal value of respect and protection of human life, as well as international solidarity. This happens in many countries of the European Union and in the United States with regard to allogeneic use and this is confirmed by the fact that regulations and standards have mainly an international impact.

## 4. THE STATUS OF THE UMBILICAL CORD AND THE PLACENTA

Even though the umbilical cord and the placenta come from the fertilized ovum and include living cells, they do not have the potential capacity to grow as a human embryo, as they do not include any totipotent cells. In this way, the arguments regarding the "potential person" or "human person becoming" are easily refuted. Stem cells from the umbilical cord and the placenta cannot develop into a human being who acts in a moral or social manner and is responsible for his or her own actions. According to Western ethical and legal thought, the umbilical cord and the placenta are "things," just like every other organ or tissue, in the same category as the biological material from abortions and miscarriages or even corpses.[15,33] They are connected to the fetus till they are detached from the mother's body. However, they are considered "res derelictae" and "res extra commercium."[17,34] The latter category was invented to show their existing relation with the human body, their symbolic relation to the human person, and in some respect regarding their use.

As far as the law is concerned, the legal nature of the umbilical cord and the placenta does not seem to be specifically defined in legislation. However, as they bear the characteristics of a thing, they are consistent with the concept of a res extra commercium (noncontractual/noncommercial), while the legislations regarding the removal and transplantation of organs, tissues, and cells refer to the concepts of donation, the person's own will, informed consent, etc. The question that arises is "Who has ownership rights?" It is known that today there is no ownership over human beings, but only over things and immaterial goods. As mentioned above, the umbilical cord, the placenta, and the fetus's cord blood were of no interest. Hospitals were responsible for them, just like they were for other surgical remains, biopsies, biological material from abortions, following rules of invariable hospital practice and more specifically best clinical practice.[17,33]

Nonetheless, three individuals could have ownership rights of cord blood, the umbilical cord, and the placenta, even though their rights could not be equated with a classic form of ownership rights: the newborn, the mother, and the father, each one for different reasons. The newborn is the only person who can genetically identify with it, as thanks to the embryo there is umbilical blood. The child, however, does not have will because it is under age. The child's parents or mother can legally replace its will.[33] The mother is the one who contributed genetically to the creation of the embryo, was responsible for its nutrition and development through the placenta and the umbilical cord for 9 months, and moreover has to undergo tests regarding the object of donation. Finally, the father is the person who contributed genetically and socially to the whole parental process, as far as the typical form of family is concerned.[33,35]

At an ethical and legal level there is a conflict of views. The legislations regarding organ, tissue, and cell transplantation do not have the same regulations regarding the issue of the person who makes a donation valid and gives their consent. Some predict that both parents should give their consent, while others define that only the mother should give her consent.[18,36]

One could argue that both parents have the right of co-ownership on biological materials, and even if it is accepted that the right belongs to the newborn, it is required that the child be represented by both parents. In this case, in order for the donation, which is a type of contract, to be valid, both parents' accord is required and the informed consent given by the parents legalizes the donation. The criterion is whether the removal of the biological material takes place before or after the detachment of the placenta.[18] Once the donation has been completed, the parents, the mother, and the newborn donor no longer have ownership rights over the cells of the umbilical cord blood and the other human biological materials. Especially in the case of allogeneic nondirected donation, the identities of the donor and the recipient are consistent with the principle of anonymity of the donation. The only link is research carried out by the Bank of Umbilical Cord Blood and the provision of information to the parties regarding health issues.[33,35]

Others would argue that the mother has the right of ownership, because the placenta and the umbilical cord are parts of her body until they are detached from it. This is the opinion of NetCord–FACT (Foundation for the Accreditation of Cellular Therapy).[37] In order for the donation to be valid, the mother's accord is required and informed consent given by the mother legalizes the collection and the donation. Even if it is accepted that the right belongs to the newborn, it is required that the mother represent him. In this way, greater significance is given to the rights of the pregnant woman and her importance in the couple's procreation plans. Today, most banks adopt this position and obtain informed consent from the mother.[18,36,38]

## 5. LEGISLATIVE CHARACTERIZATIONS OF STEM CELLS FROM CORD BLOOD, THE UMBILICAL CORD, AND THE PLACENTA

Today, depending on national legislation, umbilical cord blood cells have acquired a particular status and are sometimes regarded as a medicine, sometimes as a tissue, and other times as a human biological product.[15,16]

1. *Medicine*. In Germany, umbilical cord blood is considered a special category of medicine and is subsumed under the Act of 1976 on medicine, as it stands today.[39] The Act characterizes blood products as medicine (Arzneimittel). Moreover, it classifies umbilical cord blood that has been processed in advance, is destined for allogeneic use, and pertains to stricter licensing rules by the Institute Paul-Ehrlisch, at a Federal level, in a different category of medicine (Fertigarzneimittel). In addition, it differentiates cord blood for allogeneic use

from cord blood for autologous use.[39] Moreover, in the Medicinal Products Act[40] of 2005 it is predicted that: "Advanced therapy medicinal products are gene therapy medicinal products, somatic cell therapy medicinal products or tissue engineered products pursuant to Article 2, paragraph 1, letter a of Regulation (EC) No. 1394/2007 of the European Parliament and the Council of November 13, 2007 on advanced therapy medicinal products and amending Directive 2001/83/EC and Regulation (EC) No. 726/2004."[40,41]

2. *Tissue or product of the human body.* In the United Kingdom, according to the Human Tissue Act[42] 2004 regarding tissues of human origin, umbilical cord blood is regarded as the donor of tissue and it is specified that the law should be implemented on every product of the human body which consists of human cells, with the gametes being the only exception. In Spain, according to the Royal Decree[43] of November 10, 2006, umbilical cord blood is also regarded as tissue. The same approach of equation with tissues stands for the Belgian legislation,[44] as a tissue is regarded as "every part of the human body consisted of cells." As far as the removal of stem cells is concerned, bone marrow, peripheral blood, umbilical cord blood, embryonic stem cells, etc., belong to the same categorization. Organs to be transplanted are the only exception.

Beyond the Atlantic, according to the anticipated government regulations of the Food and Drug Administration[45] in the United States, umbilical cord blood is referred to as "human cells, tissues, and cellular and tissue-based products"(HCT/Ps) whose purpose is to be transplanted to candidate patients and to be given for cellular therapy and research: "(d)*Human cells, tissues, or cellular or tissue-based products (HCT/Ps)* means articles containing or consisting of human cells or tissues that are intended for implantation, transplantation, infusion, or transfer into a human recipient."[45] Consequently, they are considered products of the human body to be transplanted, while those being stored for allogeneic transplantation belong to the category of biological products.[46] While for autologous use, cord blood is under section 361[264] of the Public Health Service Act. Moreover, in Canada they are characterized products of the human body with the aim of being transplanted, according to Regulation 7/6/2007 regarding the safety of cells, tissues, and organs to be transplanted (article 2).[47] This regulation is under the Regulation regarding food and drugs in which provisions about drugs are included.

## 6. INFORMED CONSENT

Respect of the autonomy of the individual constitutes one of the fundamental principles of bioethics and influences today, to a great extent, the rules of the game among scientists and individuals during the removal of cells, tissues, and pieces or products of the human body, as well as the administration of medical, biological, and other data. Lawful Western orders have converging views, regarding consent as a result of the respect of the will of the individual, which protects the freedom of their decision on matters related to their health and life.[48,49]

The Oviedo Convention,[7] a text of international prestige, outlines about consent that "An intervention in the health field may only be carried out after the person concerned has given free and informed consent to it. This person shall beforehand be given appropriate information as to the purpose and nature of the intervention as well as on its consequences and risks. The person concerned may freely withdraw consent at any time." At the same time, the Charter of Fundamental Rights[9] of the European Union predicts that "In the fields of medicine and biology, the following must be respected in particular: the free and informed consent of the person concerned, according to the procedures laid down by the law," (article 3 Section 2). Moreover, in the United States, the Patient Self Determination Act[50] of 1991 defines the principle of former consent in every medical act, using as a source the principles of common law, which the historic decisions of the Supreme Court invoke. Also, the organized in nations societies should not forget the Nuremberg Code (1947) and the mandate of consent in every research procedure.

More specifically, the removal, collection, processing, administration of information, cryopreservation, storage, and disposal, according to the mandate of consent, for therapeutic or research purposes, of the placenta, the umbilical cord, and the cord blood it contains are or should be a chain of responsibilities on the part of maternity hospitals and banks, whether they are public or private bodies. A pregnant woman is in a very sensitive period of her life during her pregnancy. This is why valid and timely provision of information constitutes an important step in the whole procedure. The rights of the pregnant woman and the fetus must be respected and under no circumstances is it justified from a bioethical standpoint to treat the pregnant woman and the fetus as instruments, even if a higher purpose is to be served.

The whole procedure of the donation takes place in a context of mutual trust and according to the standing rules of science and art (lege artis), in order for it to be given to the recipients as the best possible treatment, provided it has been tested for suitability and no danger to their health can be foreseen in the future. Moreover, provided that the biological material is not used for therapeutic purposes and there is consent, it can be used for research.

The above-mentioned are general principles which are accepted worldwide. In practice, however, there is a wide range of types of consent in relation to the legislation of the national states and each bank.

Opt-in consent is explicitly given after provision of information and opt-out consent is presumed to have been given as the individual does not object or refuse to consent and shows a passive attitude.[51] The first one usually applies to the removal and collection of biological material, while the second one applies to the collection and administration of the data related to the biological material, mainly with regard to processing, filing, etc., always with respect to the existing legislation about personal data. However, there is great dispute about opt-out consent from a bioethical standpoint, as this type of consent applies more to anonymous research procedures, nonpersonally identifiable, such as epidemiologic research.[51]

Moreover, wide consent is given to the bank only once for all procedures that will follow the collection and banking, while special consent is given for each separate act or research. However, in any case, the donor should be acknowledged with the right to retreat from consent, which will specifically define the content of the retreat from the consent as following: (1) "no further contact" with the donor; (2) "no further access" to the biological material and the data of the donor; (3) "no further use" by the bank and destruction of the biological material and its data.[52]

Two more types of consent related to the individuals who consent seem to appear: the consent of the mother and the consent of the parents. In both cases consent is given in a very specific written form, in which the likely dangers, the consequences, the benefits, and the negative points of the collection and banking are described in detail. Even the guarantee of the confidentiality of the privacy that the bank provides, the security for the protection of personal and sensitive data (health, lifestyle, etc.), the legitimate, legal, and authorized use and, why not, likely benefits from the research for the individual are defined. If for the collection of cord blood and parts of the umbilical cord or the placenta informed consent is required, just like in any other medical act, then storage in banks requires enhancement of the terms regarding their removal and the whole content of consent. For example, the consent documents of New York Blood Center-National Cord Blood Program of New York (NYBC) or the Hellenic Cord Blood Bank of BRFAA of the Academy of Athens include in detail on the one hand the content of the full information in order for the validation of the consent regarding collection, manipulation, banking, disposal to be established, and on the other hand separate consent for research purposes. In this way, the person who consents is enabled to choose whether they desire their biological material to be given to the bank for therapeutic or even research purposes. This position creates conditions of transparency, trust, and respect for the individual and their privacy.[51–53] The information must be clear, comprehensive, and free from any form of pressure or coercion, and must show the mother or the parents the complexity of this deposit on behalf of the newborn.[51,53,54]

However, major ethical dilemmas and legal issues arise in relation to the control the individual who has consented can have, particularly regarding research results. Who could establish rights in an economy of the market over the results of medical and biological research: the donor, the scientist and the bank, the researcher and the research institute, or the pharmaceutical industry? The answer to this question is difficult, as the donor in fact provides cellular material, or to be more precise part of their biological material unprocessed along with other donors. This material, however, without scientific process or research procedure would be useless.

Apart from this, conflicted rights appear between on the one hand the recognition of possible ownership of the individual who has consented over his biological material, which, in this instance, is classified and could easily be accessible by the individual himself, with the aim of observing the completion of the research procedure for any positive research results and on the other hand the demand for the safety of personal and sensitive data which creates conditions of nonidentification of the individual and their genetic data. The safety and anonymity of the individuals who have consented outweigh the right of ownership, a fact that is consistent with sound procedures and good practices in the development of the bank and creates a climate of trust to the citizens. Moreover, this bind tends to be demarcated today through regulations concerning the protection of the patent of the invention for a specific amount of time. Apart from that, the specific individual biological material is removed from the specific person and is used for the good of society, supporting the public character of banks and underlining a turn an orientation toward the social purpose of private banks.[53,54]

## 7. COMMUNICATION OF INFORMATION TO THE DONOR OF CORD BLOOD, UMBILICAL CORD, AND PLACENTA REGARDING LIKELY DISEASES

It is a fact that in the consent documents given in biobanks, the ability to give information to the donor of cord blood, umbilical cord, or placenta and specifically to the mother through the family doctor or the attending physician regarding any health problems of the child and its mother or likely health issues that become known and come to the surface from the processing of the biological material and involve them is provided. This provision of information presupposes the consent or refusal of consent on behalf of the mother and is morally established on the right to knowledge or ignorance regarding sensitive health issues or issues of a different nature which involve their life history. To the extent that the diagnostic capability of medicine evolves through cell tests, the above-mentioned individuals could find therapeutic solutions to pathological problems they suffer from, or are likely to be afflicted by in the near future and are not known to them. Sometimes, however, they are exposed to knowledge that can influence the rest of their lives, especially when there is a therapeutic impasse. For this reason, symmetrically to the right to knowledge the right to ignorance is developed, which is defined in the Oviedo Convention (article 10 Section 2)[7] as following: "Everyone is entitled to know any information collected about his or her health. However, the wishes of individuals not to be so informed shall be observed." Consequently, it is both lawful and legal that the provision of information should belong to the sphere of decision-making of the donor.

## 8. THE DONATION OF CORD BLOOD, UMBILICAL CORD, AND PLACENTA AND NONCOMMERCIALIZATION

At the international level, the Oviedo Convention[7] recognizing the precedence of human beings over society and science as the ultimate moral principle defines that "The human body and its parts shall not, as such, give rise to financial gain or comparable advantage." According to the Additional Protocol[8,55] of the Oviedo Convention regarding the transplantation of organs and tissues of human origin, which are removed in order for them to be used for therapeutic reasons, its predictions involve human cells, among which HSCs are included. However, this Protocol does not apply to the reproduction of organs, tissues, organs and tissues of embryos and fetuses, as well as blood and blood products. Moreover, the Charter of the Fundamental Rights[9] of the European Union is introduced with the prohibition of profit from the human body or parts of it: "The prohibition on making the human body and its parts as such a source of financial gain."

Even in national legislations, commercialization, marketing and making profit, or financial benefit is prohibited and the concept of donation and, in particular, free donation is introduced. Donation is legally an accord or a type of contract. For example, in the Swiss federal legislation, whose policy is clearly oriented toward public banks for allogeneic use, the prohibition of the sale of organs, tissues, and cells, as such, is explicitly stated. The donors, the mother or the parents, who consent to the free donation, do so on the one hand for altruistic or solidarity reasons, and on the other hand for reasons of nondiscrimination and exploitation of vulnerable groups of the population.[56] Only a small compensation to the mother or the parents for any likely expenses could be accepted. The above-mentioned concern the donation for allogeneic and not for autologous aims, because in the case of autologous use there is a special agreement regarding banking, which is in fact contracted with the bank separately from the mother's consent. Autologous directed use for specific, proven reasons regarding the treatment of a member of the family is part of the context of the donation.

Human beings are virtually selfish, who become social persons through speech and learning, passing part of their instinctive life to civilization and its products, to which the fight against diseases and research on new treatments belong.

As with every gift there is an antigift,[57] in this case the antigift is the satisfaction that derives from the applications of medicine and biology for the treatment of serious diseases and disabilities to the benefit of everyone; the individual and his generation. This can happen when there is sufficient umbilical cord blood, as well as other products of biological material and all banks, private or public, follow international rules and standards regarding their collection, processing, quality, and volume. As for the removal and banking of umbilical cord blood, the umbilical cord, or the placenta, which take place in private banks in order for them to be used for a future disease of the child or any members of its family, this choice is based on the principle of egoism. According to some thinkers, ethical egoism adopts that it is necessary and sufficient for an action to be ethical right when it maximizes one's self-interest;[58,59] this is so in relation to autologous banking in commercial banks.

It is important, due to the development of regenerative medicine, that the status of private banks evolve and they turn to cooperation with public institutions for the benefit of the common good.

Even though it is morally and legally right for transplantation to continue to be based on free donation as res extra commercium, the characterization of cord blood, tissues, and other biological products of human beings will be equated with the statute of medicine and consequently their commercialization will become acceptable.[33,54] This is one of the greatest dilemmas that the applications of regenerative medicine, which, as mentioned above, does not identify with transplantation but follows its own rules, will have to face.

## 9. ANONYMITY—ANONYMIZATION OF DONATION AND PROTECTION OF DATA

In banks of umbilical cord blood, which have been established to have mainly a therapeutic purpose, the principle of the anonymity of the donor and the recipient, as well as the principle of communication of information to them by a doctor involved in the therapeutic process only in case of medical reason take effect, as it is with every transplantation of organs, tissues or cells. Consequently, there is no bond between the donor and the recipient, which from a moral standpoint would cultivate relations of dependence or relations of informal purchase and sale, while from a legal standpoint it would possibly lead to courts with lawsuits for compensation.

Nonetheless, some employees in banks classify in files, adopt a procedure of anonymization or often "coded" and/or "encoded" information and data, which concern the donor, and they can have access to them at any time.[53,54] These people, however, are bound to protect the anonymity of the donor and the people in their family for a long period of time. There is confidentiality and protection of personal and sensitive data according to the legislation of each country, as well as respect for the inviolability of private and family life. "Everyone has the right to respect for private life in relation to information about his or her health" (article 10 Section 1).[7]

One can find examples of deviation from the principle of anonymity in the case of "targeted banking" or directed donation of stem cells from the umbilical cord, cells destined to treat a sick child, when in the same family a brother or a sister is born with identical human leukocyte antigen (HLA). Even though the practice of giving birth to a "child-medicine" is common, especially in assisted in vitro reproduction, it does not cease to pose major ethical dilemmas regarding the fact that the second child is treated as an instrument, not only from a psychological standpoint but also in relation to the destruction of the embryos which do not have identical HLA, violating the principle of protection of every human being.[33,60]

In the field of transplantation, where banks have been established to serve a therapeutic purpose, the following rules shall also apply to HSCs and other cells, including mesenchymal stem cells from umbilical cord and the placenta: "All personal data relating to the person from whom organs or tissues have been removed and those relating to the recipient shall be considered to be confidential. Such data may only be collected, processed and communicated according to the rules relating to professional confidentiality and personal data protection" (article 23 Section 1).[8]

As far as the research field is concerned, the principle of anonymization of personal data is also in effect as a necessary condition for their collection, processing, circulation among researchers-doctors, filing, preservation, and disposal or destruction, as arguments regarding either the control of citizens' personal data by the state authority or the commercial use of individuals' biological material and data at a global level by private companies are being developed.[54] Principles and regulations are foreseen at the international and European level[9,61–63] regarding their protection "Everyone has the right to respect for private life in relation to information about his or her health" (article 10 Section 1).[7] In the light of the Additional Protocol[9] for biomedical research of the Oviedo Convention (article 25) it is stated that "Any information of a personal nature collected during biomedical research shall be considered as confidential and treated according to the rules relating to the protection of private life." In the following article there is provision for the recognition of the right of the individuals who participate in the research to be informed, prior to authorization (article 26): "(1) Research participants shall be entitled to know any information collected on their health in conformity with the provisions of Article 10 of the Convention. (2) Other personal information collected for a research project will be accessible to them in conformity with the law on the protection of individuals with regard to processing of personal data."

In general, the legal protection of personal data provides for prior authorization, security measures of confidentiality, use of methodologies of research and informatics, as well as anonymization or codification and/or encoding, particularly during their transfer to research or pharmaceutical units and prohibition of their exploitation whether that is commercial or any other form.[53,54]

## 10. "BIOLOGICAL SAFETY" OF THE DONOR AND BIOBANKING

The removal and banking of cells from the umbilical cord blood of the donor in order for them to be used by the donor him/herself for therapeutic reasons if needed (autologous transplantation) or by a third party who belongs to the close family of the donor has led to the foundation of private commercial banks of umbilical cord blood. This choice is based on the ethical theories of egoism or individualism. Egoism is the opposite position of altruism and individualism the opposite position of solidarity.

Many arguments against the collection and banking of individuals' cells for autologous transplantation for mainly medical but also financial reasons have been expressed by institutional bodies and scientists from the European Union. There are many countries, such as France and Switzerland, which implement voluntary, anonymous, and free donation of umbilical cord blood for allogeneic use as a public policy, giving emphasis to the principles of altruism, solidarity, and social interest, proportionally to organ transplantation and blood donation.[33] The collection and banking of umbilical cord blood in public banks or nonprofitable private banks is integrated in this context.[15,53,54] While, at the same time, there is the ability for purely medical reasons and under strict conditions, to deviate from the principle of anonymity and to allow the removal and preservation of umbilical cord blood cells for transplantation to a close member of the donor's family who has been inflicted by serious, yet curable, disease or disability.[53,63–66]

It seems that in an environment of universalization and competition of the economy (financial competition), biomedical research and treatment could not stay uninfluenced and as a result the activities of all types of biobanks in general are incorporated in some rules of the market and turn their attention to the scientific and research policies of the future.

As far as umbilical cord blood banks specifically are concerned, whose main purpose since their establishment has been treatment, the great challenge today, along with serving public health, is their turn toward regenerative medicine, which will once again serve therapeutic purposes.

If umbilical cord blood banks are classified according to their statute, one can see that there is no uniform regulation. There are countries, such as Germany, Denmark, the Netherlands, Greece, Poland, the United Kingdom, Canada, and the United States, whose legal documents do not foresee a specific statute. For this reason and to the extent that there were no legal restrictions, public, private, and mixed banks were founded. In other countries, such as Belgium, Spain, and Italy, the legislation originally forbade the foundation of private banks while, today, it is being modified in stages and allows it under specific conditions. France and Switzerland should be classified under the latter category, even though their public character is more intense.[53,54,66]

As far as the first group of countries is concerned, umbilical cord blood banks can operate after they have acquired administrative license, which is not mentioned in their statute. In Germany, for example, there are four public banks, among them the Bank of Dusseldorf, while in Denmark umbilical cord blood banking takes place in private banks in cooperation with the big hospitals of the country, as well as the Public Bank of Finland, which covers all the needs of the Scandinavian countries.[66]

The emergence of mixed banks is of great interest. The first mixed bank with both a public and private character, which enables banking of two samples of umbilical cord blood, one for autologous use and one for allogeneic use, in any case available to any third applicant, was founded in the United Kingdom, in 2007. At the same time, a private bank in Canada is developing a donation program for allogeneic use.[66] Symmetrically, the Hellenic Cord Blood Bank of BRFAA of the Academy of Athens, a legal nonprofitable entity governed by private law, is by legislation a public bank and operates as such, while it develops collaborations with private maternity hospitals.

Regarding the second group of countries, which are modifying their legislation in stages and are allowing the establishment of private banks of umbilical cord blood, the example of Spain is important. In the Royal Decree[43] of November 10, 2006 about quality and safety rules regarding actions in cells and tissues of human origin emphasis is put on common good, nonprofitable purpose, and cases of emergency. When, for instance, private banks cease to operate they are obliged to give their stock to the public network.

It is obvious today that at a national level there is a tendency to converge public and private banks, not only to prevent medical tourism,[67] but also because from an ethical standpoint and mainly in private commercial banks, banking for autologous use creates stocks of biological material, which are not being used. Moreover, there is no consensus within the scientific community regarding their suitability. Finally, they are expensive to store and for this reason lead to discrimination in the population.[68] Consequently, there are questions regarding the usefulness of these stocks: should they be given as a donation for allogeneic use? Should they be given for research purposes for free or in exchange for money? Should they be given to pharmaceutical companies and if so, who should benefit (the donor, the bank, the doctor or the researcher, the pharmaceutical industry, etc.)? Should they be destroyed or stored and if they are stored, how long should that be for? Should public health, common good or the best interest of the individual come first? What is of great importance regarding the above-mentioned dilemmas is the mother's or parents' consent, as well as the convention between the individuals and the banks. Nonetheless, these issues do not fall only within the regulatory sphere of the autonomy of the individual, but constitute major problems of politics at international and national level, to the extent that the value of the "biological safety" of the individual is more symbolic than real. So, the dilemma of public or private banks is ethically rightful, while the contemporary tendency from a legal standpoint for the cooperation of public and private banks seems to take it into consideration, solving a lot of issues, while questioning the myth of "biological safety."

## 11. REGENERATIVE MEDICINE AND MESENCHYMAL CELLS OF THE UMBILICAL CORD AND THE PLACENTA

Regenerative medicine, whose name is attributed to L. Kaiser in an article about hospital administration, appeared in 1992.[69] It promises to answer the problem of organ, tissue, and cell shortage with their replacement or the regeneration of ailing parts and their return to normal function.[1,70,71,] It involves cell treatments and focuses its research interests on stem cells that come not only from cord blood, but also from the walls of the umbilical cord (Wharton's Jelly) and the placenta.[17,72] The placenta and that part of the umbilical cord contain non-hematopoietic cells, mesenchymal stem cells, and endothelial cells, which are today collected by biobanks for research or preclinical purposes, with clinical trial, their use in pharmacology, and regenerative medicine being the ultimate purpose.[2,72] At this stage of the research, the collections concern autologous banking without, however, excluding allogeneic banking, they are important for treatments with comparable phenotype HLA, and perhaps in the near future, they will be of clinical use. Even though they raise issues regarding the nature of banks, their storage or their use in research, they constitute a challenge to public banks or nonprofitable banks with a collective purpose to develop banking sections with mesenchymal or endothelial stem cells. Moreover, autologous private profiteering banks need to acquire a mixed character or to form partnerships, on condition that they abide by national regulations and standards regarding qualitative and quantitative criteria, the collection procedure, the mother's consent, testing, information, etc.

Scientific debates about publications at an experimental stage are wide, in relation to the advantages of mesenchymal and endothelial stem cells. These cells not only have multipotent potential, but may even have pluripotent potential. It is underlined that according to scientists their potential is comparable to that of iPSCs, while they have the advantage of being young, self-renewing, and able to differentiate.[17,72]

The collection of the placenta and the umbilical cord is easy. It can be done with the removal of umbilical cord blood or it can even be collected separately with the expulsion of the placenta. This has the advantage that the procedure of the removal can take place in a room other than the delivery room and then the biological material can be cultivated. This ability emphasizes the respect toward the mother and the newborn.

The above references are important in relation to likely additional or differentiated ethical or legal issues that may occur.[2] The placenta and the connected part of the umbilical cord are part of the mother's body, as already mentioned, until their expulsion. As such, the consent of the mother is necessary. In our opinion, consent obtained from the mother in the specific case of donation of the placenta and the umbilical cord must be informed, explicitly expressed, written, and signed by the mother, with all the characteristics that are required in the mother's informed consent for the donation of her child's umbilical cord blood. Obtained informed consent is an ethical duty and a legal obligation. As already mentioned in Section 4, this kind of consent originates from the principle of autonomy, one of the four basic principles of bioethics. According to Beauchamp and Childress,[73] autonomy, beneficence, nonmaleficence, and equality are the most important principles. Touching the mother, selecting and banking hers or her fetus's biological material without permission could be assault or battery under criminal and civil law. For this reason and in order for the consent to be valid an individual who has the capacity and has been fully informed prior to the act must give their voluntary accord without coercion. Consent is stronger when it is given in writing and written consent is evidence that consent has been given. Many consent forms which are being used by banks include sections in which it is recorded that the mother has been informed.

According to another point of view it is sufficient that the consent be nonrefusal for specific uses. This type of consent involves not only the removal of biological material, but also, and mainly, the processing of data and specific uses (for example commercial use). Such an opt-out type of consent is used by legislations regarding postmortem organ donation.[54] However, medical professionals must evaluate the mother's capacity to consent. Then her wishes will be respected. If nonrefusal consent is to be adopted, it should at least follow some principles: the principle of nonharm, the principle of suitability, the principle of charity, and the principle of justice. Moreover, it can be given along with the mother's consent for the selection and banking of her child's umbilical cord blood.

The reasoning behind nonrefusal consent is that the placenta and the umbilical cord were until recently considered biological material meant to be destroyed due to the risk of contamination. Despite this provision, sometimes instead of being destroyed they were given irregularly, apparently for a financial exchange, by maternity hospitals and companies to be used for immunoglobulin products as well as in cosmetic medicine, posing major ethical and legal dilemmas regarding the issue of who has the authority for such actions over the biological material of the individual or who has ownership rights. From a moral standpoint, the rights of human beings have put up barriers to the utilitarian perception of medical and research power, as we must not forget the delicate balance between respect for the dignity and autonomy of human beings and freedom of research. The latter should place the ethics of research in an advantageous position for the welfare of patients in relation to its financial or utilitarian aspects.

## 12. CONCLUSION

At an ethical and legal level, the collection, testing, processing, storage, and disposal for therapeutic or research use of stem cells that come from cord blood, the umbilical cord, and the placenta open a new chapter of dilemmas in relation to the development of regenerative medicine. This chapter of ethical dilemmas is, at the same time, accompanied by important financial issues.

People want to benefit from the achievements and future plans of regenerative medicine and biotechnology. Serious diseases and disabilities need to be treated and desperate individuals await for "a scientific revolution" through gene and cell treatments.

Nonetheless, one cannot bypass fundamental moral values, principles, and rules of contemporary civilization, the most important of which are respect for the dignity, freedom, and life of man.

Stem cells that come from cord blood, the umbilical cord, and the placenta and their collection, processing, and use do not seem to pose major ethical questions in relation to the definition of the person and the human being.

However, many moral values, principles, and rules, as well as delicate legal issues are involved worldwide and in every society separately, are often in conflict and need to be answered.

Striking a balance between the principles of freedom of man and freedom of science creates the necessary and sufficient context for the resolution of the conflict between the principles of altruism and ethical egoism, the values of charity and individualism, in other words, between allogeneic and autologous therapeutic use.

The choice of banking stem cells from cord blood, the umbilical cord, and the placenta in a public or commercial bank is also an important issue which can lead to future cooperations or to a change in their legal character.

Finally, the legal nature and character of stem cells from cord blood, the umbilical cord, and the placenta also pose major questions that need to be answered internationally. Do these specific stem cells belong to the field of transplantation, in the same category as tissues? Must they be regarded as cells and be classified with other stem cells or are they medicine? In any case, will cellular treatment or processed cellular body products become available by banks or will they be the object of pharmaceutical companies?

## REFERENCES


1. King Tak I. The bioethics of regenerative medicine. 3 ed. New York: Springer; 2009, as cited in Lysaght T, Campbell AV. Broadening the scope of debates around stem cell research. *Bioethics* 2013; **27**(5):251–6.
2. Lysaght T, Campbell AV. Broadening the scope of debates around stem cell research. *Bioethics* 2013;**27**(5):251–6.
3. Havstad JC. Human reproductive cloning: a conflict of liberties. *Bioethics* 2010;**24**(2):71–7.
4. Gabolde M, Hors J. Utilisation aux fins de greffe de cellules et tissus humains d'origine fœtale ou embryonnaire. *Med Droit* 2000;**44**:1–5.
5. *International declaration on genetic data of Unesco, Paris*; 2003. Available at: www.unesdoc.unesco.org (July 25 2013).
6. Kahn A. Cellules couches et médecine régénérative. Réalités, promesses et lobbies. *S.E.R Etudes* 2006;**4**(404):474–86.
7. Convention for the Protection of Human Rights and Dignity of the Human Being with regard to the Application of Biology and Medicine. Convention on human rights and biomedicine of human rights and biomedicine of the european Council. *Oviedo* 1997;**4**(iv). Available at: www.coe.int/t/dg3/healthbioethic/texts_and_documents. March 10, 2013.
8. Additional Protocols to the Convention on Human Rights and Biomedicine, on the prohibition of cloning human beings, on transplantation of organs and tissues of human origin, on biomedical research, on genetic testing for health purposes. [Available at: www.coe.int/t/dg3/healthbioethic/texts_and_documents (May 14, 2013)].
9. *Charter of fundamental rights of European Union*. 2000/2009. Available at: ec.europa.eu/justice/fundamental-rights/Charter (May 14, 2013).
10. Directives of the European Union: Directive 2004/23/EC of the European Parliament and of the Council of 31 March 2004, JO of European Union; Commission Directive 2006/86/EC of 24 October 2006, JO of European Union.
11. Regulation (EC) no 1394/2007 of the European Parliament and of the Council of 2007 [Available at: eur.lex.europa.eu/Lex (August 24 2013)].
12. Recommendation Rec(2006)4 of the European Council and Draft Recommendation of European Council Steering Committee on Bioethics (CDBI) on research on biological materials of human origin. [Available at: www.coe.int/t/dg3/healthbioethic/texts_and_documents (September 1 2013)].
13. Cluckman E, Broxmeyer HE, Auerbach AD, Freidman HS, Douglas GW, Devergie A, et al. Hematopoeietic reconstitution in a patient with Fanconi' s anemia by means of umbilical-cord blood from an HLA-identical sibling. *N Engl J Med* 1989;**321**:1174–8.
14. Ong SY, Hwang WYK. Banking of cord blood. In: Kadareit S, Udolph G, editors. *Umbilical cord blood: a future for regenerative medicine*. Singapore: World Scientific; 2011. p. 291–320.
15. *Senat de la République Française, Sur le potentiel thérapeutique des cellules souches extraites du sang de cordon ombilical. Paris*. 2008. Rapport d'Information No 79.
16. Teskrat F. Nouveau paysage réglementaire français dans le domaine des tissus et cellules. *Médecine Droit* 2012;**112**:16–21.
17. *Comité Consultatif National d' Éthique pour les Sciences de la Vie et de la Sante, sur l'utilisation des cellules souches issues du sang de cordon ombilical, du cordon lui-même et du placenta et leur conservation en biobanques. Questionnement éthique. Paris*. 2012. Avis No 117.
18. Committee on Establishing a National Cord Blood Stem Cell Bank Program. In: Ann Meyer E, Hanna K, Gebbie K, editors. *Board on health sciences policy*. Washington DC: Academie of Sciences Press; 2005. Available at: http://www.nap.edu/catalog/11269.html. September 2, 2013, 2013.
19. Broxmeyer HE. Preface. In: Broxmeyer HE, editor. *Cord blood: biology, transplantation, banking and regulation*. Bethesda, MD: AABB Press; 2011. p. xxviii.
20. Notta F, Doulatov S, Laurenti E, Poeppl A, Jurisica I, Dick JE. Isolation of single human hematopoietic stem cells capable of long-term multilineage engraftment. *Science* 2011;**333**:218–21.
21. Yamanaka KT. Induction of pluripotent stem cells from mouse embryonic and adult fibroblast cultures by defined factors. *Cell* 2006;**126**:663–76.
22. Yu J, Vodyanik MA, Smuga–Otto K, Antosiewicz–Bourget J, Frane JL, Tian S, et al. Induced pluripotent stem cell lines derived from human somatic cells. *Science* 2007;**318**:1917–20.
23. Fung RKF, Kerridge IH. Uncertain translation, uncertain benefit and uncertain risk: ethical challenges facing first–in–human trials of induced pluripotent stem (iPS) cells. *Bioethics* 2013;**27**(2): 89–96.
24. Turhan A. Plasticité des cellules souches adultes. *Hématologie* 2003;**9**(2):105–16.
25. Agamben G. *Homo Sacer. Il potere sovrano e la nuda vita*. Torino: Einaudi; 1995.
26. Engelhardt Jr HT. *The foundations of bioethics*. 2nd ed. Oxford: Oxford University Press; 1996.
27. Singler P. *Practical ethics*. 3 ed: Cambridge: Cambridge University Press; 2011.
28. Jonas H. *Le principe responsabilité*. Paris: Flammarion, coll; 1990. « Champs ».
29. Ricœur P. *Lectures II*. Paris: Seuil; 1992.
30. *American convention on human rights, Costa Rica*: 1969. Available at: www.cidh.org/basicos/english/basic3.american%20convention.htm (June 30 2013).
31. Altavilla A. La recherche sur les cellules souches, Quels enjeux pour l'Europe? Paris: Le Harmattan; 2012.
32. Saura B. Le Placenta en Polynésie française un choix de santé confronté à des questions identitaires. *Sci Sociales Santé* 2000;**18**(3): 5–27.
33. *Conseil d' Éthique Clinique HUG-Bank de cellules souches issues du cordon ombilical Genevacord*. 2002. Avis consultatif.
34. Gailloux JC. Réflexions sur la catégorie des choses hors du commerce: l'exemple des éléments et des produits du corps humain en droit français. *Les Cah droit* 1989;**30**(4):1011–32. http://dx.doi.org/10.7202/042991ar. Available at: http://id.erudit.org/iderudit/042991ar.

35. Petrini C, Farisco M. Informed consent for cord blood donation. A theoretical and empirical study. *Blood Transfus* July 2011;**9**(3):292–300.
36. Nuffield Council on Bioethics. *Human bodies: donation for medicine and research*. London: Nuffield Council on Bioethics; 2011. Available at: www.nuffielddbioethics.org/sites/default/filesdonation_full_report.pdf (June 30, 2013).
37. NetCord-FACT. *International standards for cord blood collection, banking, and release for administration*. Accreditation Manual DRAFT. 5th ed. 2012.
38. Guindi ES, Moss TF, Ernst MT. Public and private cord blood banking. In: *Cord blood: biology, transplantation, banking and regulation*. Bethesda, MD: AABB Press; 2011. p. 600–1.
39. Gesetz über den Verkehr mit Arzneimitteln (Arzneimittelgesetz - AMG) of 24/8/1976 German: Ein Service des Bundesministeriums der Justiz in Zusammenarbeit mit der juris GmbH. Available at: www.juris.de.
40. Medicinal Products Act. *German: federal law Gazette [BGBl.]*2005. [Part I].
41. Regulation (EC) No 1394/2007 of the European Parliament and of the Council of November 13, 2007 on advanced therapy medicinal products and amending Directive 2001/83/EC and Regulation (EC) No 726/2004, OJ L 324 of 10.12.2007.
42. *Human tissue act*. UK: Stationery Office Limited; 2004.
43. Royal decree of November 10, 2006, Spain: BOE num. 270.
44. *Loi relative à l'obtention et à l'utilisation de matériel corporel humain destiné à des applications médicales humaines ou à des fins de recherche scientifique*. Belgique: Moniteur Belge; 30/12/2008.
45. Food and Drug Administration, Code of federal regulations Title 21, Volume 8 Revised as April 1, 2013, CITE: 21CFR1271.3 [Available at: http://www.accessdata.fda.gov/scripts/cdrh/cfdocs/cfcfr/cfrsearch.cfm?fr=1271.3 (August 24, 2013)].
46. United States Code, Title 42, section 361[264], public health service act, Washington DC: US Government Printing Office, 2010 [Available at: www.house.gov/legcoun/Comps/PHSA_CMD pdf, September 1 2013].
47. Regulation of 7/6/2007 Canada: Minister of Justice [Available at: http://laws-lois.justice.gc.ca (August 20 2013)].
48. Hennette-Vauchez S. Le droit de la bioéthique. *La Découverte* 2009;**34**.
49. Nerson R. *Le respect par le médecin de la volonté du malade in Mélanges offerts à Gabriel Marty*. Toulouse: Publications de l'Université des sciences sociales de Toulouse; 1978. p. 853.
50. United States patient self-determination act of 1991, in the Omnibus Budget Reconciliation act of 1990, P.L.I01-50S, sections 4206, 4751, enacted November 5, 1990.
51. Kurosu M. Ethical issues of presumed consent in the use of patient materials for medical research and the organ donation for transplantation. *J Philosophy Ethics Health Care Med* July 2008;**64**(3):p64–85.
52. UK Biobank. *Ethics and governance framework version 3.0*. October 2007.
53. Comité Consultatif National d' Éthique. Avis relatif aux problèmes éthiques poses par les collections de matériel biologique et les données d'information associées. *Biobanques, "Biothèques"* 2003;**77**.
54. Bellivier F, Noiville C. *Les Biobanques*. Paris: Puf; 2009.
55. Blumberg-Mokri M. Convention de biomédecine et transplantation d'organes. *Médecine Droit* 2002;**56**:20–5.
56. Rial-Sebbag E. Vulnérabilité, enfant et recherche médicale. *Médecine Droit* 2011;**111**:231–4.
57. Mauss M. *Essai sur le don. Sociologie et anthropologie*. 9th ed. Paris: Puf; 2001.
58. Norman R. *The moral philosophers*. Oxford: Oxford University Press; 1983.
59. Egoism SR. *The Stanford Encyclopedia of Philosophy*. Stanford, Ca; 2010.
60. Amy TY, Lai PhD JD. To be or not to be my sister's keeper? *J Leg Med* 2011;**32**(3):261–93. http://dx.doi.org/10.1080/01947648.2011.600169.
61. *Convention 108 of the Council of Europe for the protection of individuals with regard to automatic processing of personal data, Strasbourg*. 28.1.1981. Available at: www.coe.int/dataprotection. 15 April 2013.
62. Directive 95/46/EC of the European Parliament and of the Council of 24 October 1995 on the protection of individuals with regard to the processing of personal data and on the free movement of such data, OJ L281, 23/11/1995.
63. Technopolis group. *Cord blood banking in the UK*. 2009. Available at: www.technopolis-group.com (August 10 2013).
64. Marville L, Haye I, Torre RM, Katz G. Quel statut pour les banques de sang de cordon ombilical? *Médecine Droit* 2010;**102**:81–5.
65. Petrini C. Ethical issues in umbilical cord blood banking: a comparative analysis of documents from national and international institutions. *Transfusion* 2013;**53**:902–10.
66. *Senat sur la conservation du sang placentaire, Les documents de travail du sénat série législation comparée*. 2008. No LC 187.
67. Murdoch CE, Scott CT. Stem cell tourism and the power of hope. *Am J Bioeth* 2010;**10**(5):16–23.
68. Steinsbekk KS, Ursin LØ, Skolbekken JA, Solberg B. We're not in it for the money—lay people's moral intuitions on commercial use of 'their' biobank. *Med Health Care Philosophy* 2011:1–12.
69. Kaiser L. The future of multihospital systems. Top health care finance, as cited in Bhattacharya N, Stubblefield editors. *Regenerative medicine using pregnancy-specific biological substances*. (preface). Springer; 2011.
70. Mason C, Dunnill P. A brief definition of regenerative medicine (editorial). *Regen Med* 2008;**3**(1):1–5.
71. Aditya J, Ramta B. Biological solutions to biological problems. *Indian J Med Specialities* 2013;**4**(1):41–6.
72. Forraz N, MacGuckin CP. The umbilical cord: a rich and ethical stem cell source to advance regenerative medicine. *Cell Prolif* April 2011;**44**(Suppl. 1):60–9. http://dx.doi.org/10.1111/j.1365-2184.2010.00729.x. [pubmed].
73. Beauchamp TL, Childress JF. *Principles of biomedical ethics*. 6th ed. Oxford: Oxford University Press; 2009.

Chapter 24

# Industrial Economics of Cord Blood Banks

Gregory Katz
*ESSEC Business School, Chair of Therapeutic Innovation; Fondation Générale de Santé, Paris, France*

## Chapter Outline



To analyze the economic issues raised by cord blood banks, it is important to clearly define their activities and their objectives. There are two basic types of banks: (1) public banks are not-for-profit organizations; and (2) private banks are for-profit companies. In between these two models, hybrid banks have emerged at the interface of public and private banking: they also operate as for-profit corporations. From a terminology standpoint, the "public" and "private" labels can be misleading since several public banks depend on private not-for-profit organizations (Anthony Nolan Trust, Jose Carreras Foundation, etc.). "Public" or "private" refers less to the bank's legal status than to the final usage of the stored units, which can be either available to everyone (public bank) or reserved for a single family (private bank). Beyond the various qualifiers commonly used (such as "autologous," "allogeneic," "community," "family," "directed," or "altruistic") the "for-profit" or "not-for-profit" criterion is what truly differentiates all these models from an economic standpoint. Whether free or for a fee, the main difference between all these banks lies in their objectives, organization, and the economic model that will determine their future development.

## 1. TYPES OF CORD BLOOD BANKS

Public banks store cord blood units (CBUs) for the general population. The units are collected anonymously, and if they are required to meet specific biological criteria they are mainly used for allogeneic transplants where the donor and the recipient are two different people. For certain specific indications concerning high-risk families, storage for family members is free. These banks are supported by charity organizations and financed by public health, but their limited funds hinder their growth. They are inspected periodically by accreditation bodies and generally meet strict quality assurance criteria that lead them to approve and keep just 25% of their collected units. Allogeneic banks collaborate through a global network (NetCord) and share their inventory online via an interface that provides the human leukocyte antigen (HLA) typing of available grafts for all transplant centers worldwide (Bone Marrow Donor Worldwide, BMDW).[1]

Cord Blood Stem Cells Medicine. http://dx.doi.org/10.1016/B978-0-12-407785-0.00024-4

Commercial banks are for-profit enterprises: they bill the parents-to-be an upfront payment of roughly 2.000 USD and store the newborn's cord blood—and sometimes placental tissue—for 20 years on average, with an annual storage fee of 125 USD. The units are stored for autologous use or for a recipient in the family. Virtually all units collected at birth are kept, regardless of their biological quality and stem cell concentration. Stored units cannot be used for patients outside the family. Aside from a minority of banks accredited, the quality assurance systems of commercial banks are rather opaque.[2] Their operating procedures and the clinical results of transplants performed using the CBUs they store are not reported or analyzed in the scientific literature.

Hybrid banks store CBUs for either allogeneic or family member usage. They run a for-profit business, but they also develop a special storage channel that makes the units available to any patient in the general population (not-for-profit activity). These banks are generally required to meet the same accreditation criteria as public banks. Several types of hybrid banks have developed. Some split the CBU volume in two, allocating 20% of the CBU for family usage while making the other 80% available to public health services. Others bill the parents for CBU storage for family usage but perform HLA typing on the units so that the entire graft is available to the health authorities at all times, in order to allow allogeneic transplants in the general population. If a graft is requisitioned, the parents' outlays are reimbursed. In some cases, parents cannot contest the requisition (the case in Spain), while in others they can reject it and retain exclusive usage of the CBU. "Binary" banks offer yet another model; they provide two separate channels under the same brand: one that is allogeneic and free, and the other that is for family members, for a fee (Figure 1).

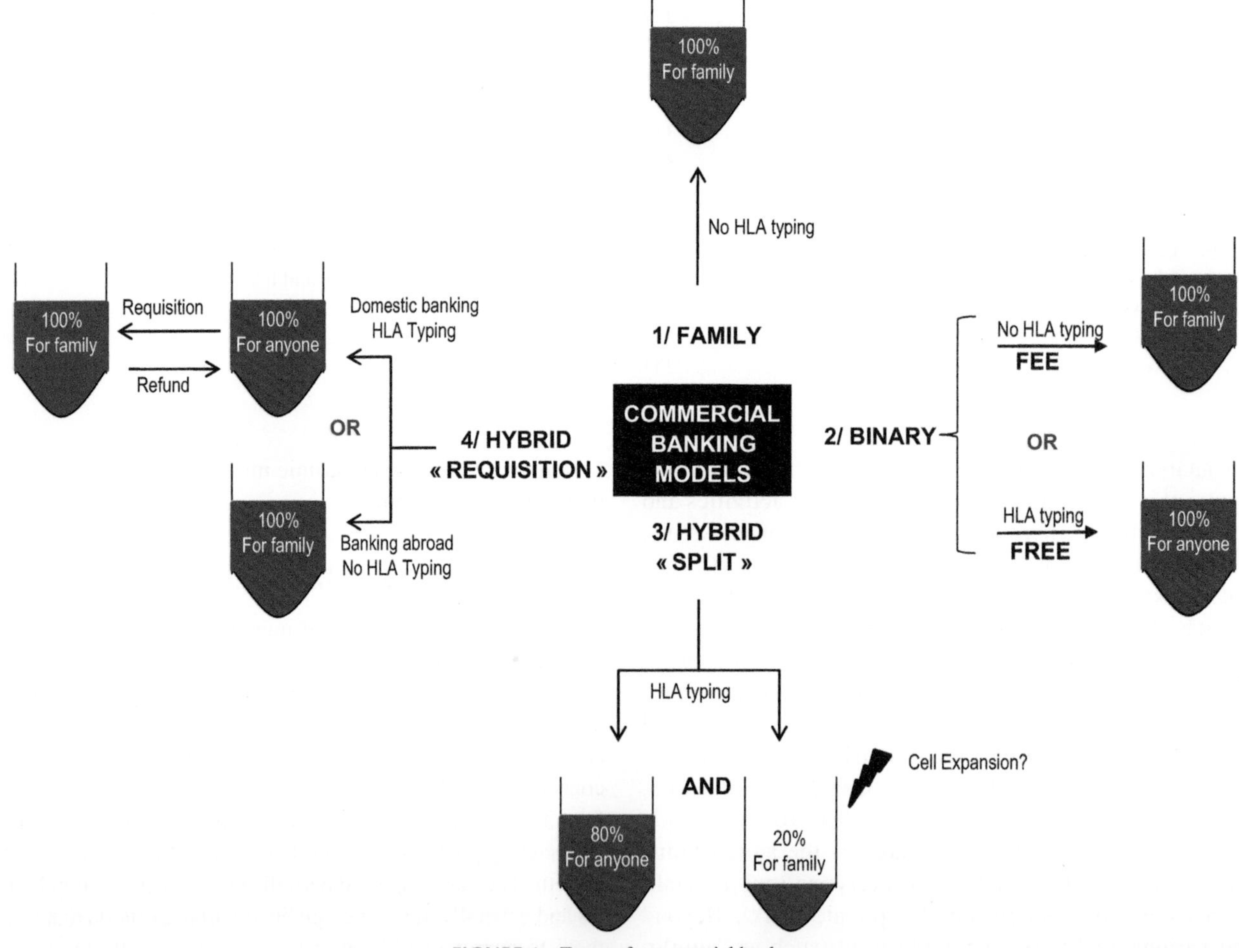


FIGURE 1 Types of commercial banks.

## 2. EMERGENCE OF HYBRID MODELS

In theory, the public/private channel used in the hybrid model seeks to reconcile private funding and public health, i.e., exclusive paid storage combined with altruistic free usage. This channel differs from the commercial family channel in that, since allogeneic usage is a possibility, the operating procedures must follow those of public banks. In most mixed banks, collections must be performed through a network of accredited maternity hospitals, by specially trained medical staff. Standard operational procedures are audited periodically by the health authorities and accreditation agencies. CBU approval leads to mixed cryopreservation on the one hand, and billing of the parents on the other. If the CBU is not approved for allogeneic usage, the parents are not debited and can choose to either have the CBU destroyed, or donate it to science. Even if the CBU does not meet the eligibility criteria for therapeutic use applied by public banks, some parents still opt for paid storage strictly for family usage in the hope that the biological quality of the graft will be improved in the future through techniques that are currently still at the experimental stage.

### 2.1 Split Storage and Community Banking

In 2006, Richard Branson, the founder of Virgin, in partnership with the Excalibur-Merlin investment fund and in collaboration with the British health authorities, launched a hybrid bank with a pioneering model: community banking. The Virgin Health Bank (VHB) model allows future parents to keep 80% of the cord blood volume for allogeneic usage and 20% for family usage. The CBU is stored in a single bag with the two aliquot fractions split in sealed compartments. The ethical justification for this model is based on graft usage probabilities. The focus is clearly on allogeneic usage whose probability is much higher than family usage, for which clinical indications are exceptional. The 20% fraction is reserved for the family, in the hope of regenerating 100% of the initial volume through cell expansion, a technology that is still in its experimental stage. Without cell expansion, the family unit will be useless because its volume and cell dosage are insufficient for therapeutic applications. Some critics have also pointed out that, much like the 20% fraction, the 80% fraction of the CBU also contains a concentration of stem cells that is insufficient to perform a transplant in a patient older than 10 years.[3]

When VHB opened, its goal was to store 4000 units per year, i.e., 0.5% of the 690,000 annual births in Great Britain in 2007. With the authorization of the UK Human Tissue Authority, and in collaboration with the public banks of the National Health Service, VHB vowed from the outset to avoid any dichotomy or commercial relations with physicians and midwives. Branson also pledged to reinvest his share of the profits in other initiatives dedicated to developing stem cell research. With its ethical commitment, the hybrid model developed by Virgin sought to support the development of public banks through private funding. From a marketing standpoint, the idea of community banking implies that when a family decides to keep cord blood for their child, it benefits the community. Hailed by the British authorities, patient associations, and the Royal College of Obstetrics and Gynaecology,[4] VHB quickly ran into major financial difficulties. By 2009, the business targets had not been reached and the bank showed a large deficit, bringing it to the brink of bankruptcy. The market was not responding to the community banking offer because future parents could not understand why they should pay a price of 1195 GBP for just 20% of the volume, while 80% of the stem cells were allocated to the community. For almost the same price, competing commercial banks guarantee exclusivity for 100% of the volume.

With the British market saturated by a number of commercial banks that have been in business for several years, VHB was forced to radically rethink its strategy in order to become competitive. In 2011, the firm opened a subsidiary in Qatar, offering a family model for 12,950 QR (2670 EUR) in a country with no allogeneic banks, while taking advantage of the world's highest GDP per capita. VHB continues to operate in Great Britain, offering future parents a choice between two options with a small price difference: classical family storage (2626 USD), or community banking (1852 USD), in partnership with the hospitals of the Cambridge University NHS Trust Foundation. VHB has not publicly disclosed the breakdown of their inventory between the two options, but given that the price difference is not huge, it is likely that a large majority of customers would prefer family storage, which guarantees they get 100% of the volume. The community banking option probably only accounts for a very small fraction of VHB's inventory, obviously providing a strong marketing impact for the firm, but with a very small benefit for public health (Figure 2).

| Bank (country) | Initial processing (USD) | Annual storage (USD) | Total for 20 years (USD) | Total for 25 years (USD) |
|---|---|---|---|---|
| CORD BLOOD REGISTRY (USA) | 1,695 | 130 | 4,295 | 4,945 |
| CORD:USE family (USA) | 1,87 | 125 | 4,37 | 4,995 |
| CORDLIFE (Singapore) | 1,685 | 196 | 5,603 | 5,016* |
| CRYO-CELL (USA) | 1,924 | 125 | 4,299 | 3,799* |
| CRYO-SAVE (Netherlands) | NA | NA | NA | 2,835 |
| STEMCYTE family (USA) | 1,500 | 125 | 3,875 | 4,500 |
| VIACORD (USA) | 1,670 | 125 | 4,170 | 3,970 |
| VIDACORD (SPAIN) | 2,366 | NA | NA | 2,829 |
| VIRGIN HEALTH BANK - Community (UK) | 1,852 | NA | NA | 1,852 |
| VIRGIN HEALTH BANK - family (UK) | 2,626 | NA | NA | 2,626 |
| VITA34 (Germany) | 2,630 | 64 | 3,902 | 3,424 |

*Price is for 21 years of storage.

All prices are in US dollars (USD). Conversion rate at Sept 2, 2013
(Only classic pricing and payment plans were taken into account)

**To obtain the pricing, the following websites were accessed on September 2, 2013 :**

CordLife : http://www.cordlife.com/sg/en/price-plan
Cryo-Cell http://www.cryo-cell.com/cord-blood-banking-costs
ViaCord http://www.viacord.com/pricing/cord-blood-banking/
Cord Blood Registry http://www.cordblood.com/cord-blood-banking-cost/cord-blood-stem-cells
Virgin Health Bank http://www.virginhealthbank.com/our-services/family-banking
Vita34 http://www.vita34.de/en/products-prices/offering/
Cryo-Save http://nld.cryo-save.com/en/details/d5fpr/prices/costs-fees
StemCyte https://www.stemcyte.com/pricing/payment-calculator
Cord:Use https://cordbloodbank.corduse.com/enrollment-cord-use-pricing.php

**FIGURE 2** Pricing for storage in commercial banks.

## 2.2 Requisition or Export

In Spain, the hybrid bank VidaCord has developed another mixed banking model based on two different options: requisition or export. The two paid options are offered at the same price (2366 EUR). The requisition option works as follows: at birth, the parents pay the bank to keep the CBU on Spanish territory. If required, national law allows the health authorities to requisition the CBU for a transplant in a nonfamily member. The parents cannot oppose this requisition: they lose 100% of the CBU. In this case, their outlays are reimbursed. The other option consists in paying the bank to export the unit at birth to a location outside Spain, in order to keep it strictly for family members. The two options of this hybrid model have developed asymmetrically: on the one hand, the export and family option has been extremely successful, and on the other, the requisition option has failed, with a negligible number of grafts available to the general population. VidaCord shares the HLA typing of all stored CBUs with the Spanish bone marrow registry (REDMO) so the CBUs can be requisitioned at any time for an allogeneic transplant.[5] Yet according to Enric Carreras, director of REDMO, 3 years after the launch of this Spanish hybrid bank, "no grafts have been made available for public health."[6]

There are two main reasons for this disparity: the pricing policy established by the commercial bank, which charges the same price for both options, and a poorly designed regulatory system that seeks to strengthen national cohesion around a common public health project, but paradoxically pushes parents-to-be to export their bioresources abroad. Until 2005, Spanish law prohibited commercial banks from storing cord blood exclusively for family usage. At the birth of the Infante Eleonor, Prince Philip, heir to the Spanish throne, decided to export the unit of royal blood for storage in Arizona,[7] in the American bank, Cord Blood Registry. Since this export was illegal with regard to Spanish law, the press immediately interpreted the prince's decision as a refusal on moral grounds to mix royal blood with the blood of the Spanish people. As a result of the popular uproar, and given the political stakes, a royal decree was finally issued to change the way cord blood banks are regulated.[8] Since 2006, anyone can either export cord blood abroad exclusively for family usage, or store it in Spain in a hybrid bank, with the chance that the health authorities will requisition it in case of need. As soon as this reform was adopted, Vida-Cord set itself up in Spain and proposed this mixed model, and was followed by other competing banks such as Ivida.

## 2.3 Imbalance of Mixed Channels

After several years under this system, the Spanish health authorities have observed that virtually all customers prefer the export solution in order to avoid requisitions and ensure exclusive use of their graft. On December 31, 2012, after operating for 7 years, VidaCord stated on its Web site that it held more than 15,000 CBUs—while REDMO only totaled 68 units available for the general population.[9] In other words, when parents are presented with two options, the fact is that 99.6% of CBUs are stored abroad for family usage. The mixed bank only contributes 0.4% of its inventory to public health needs.

While it is true that parents are theoretically free to choose between the two options, in practice, why should they hesitate between total or partial exclusivity, given that the price is virtually the same for both options? The causes and effects are the same for VHB, VidaCord, Vita34, and Ivida: whenever a hybrid bank charges approximately the same price for both options, it creates a tremendous imbalance between the two options, with the altruistic model consistently on the losing end. For Ivida, only four units were registered in the national registry as of December 31, 2012.[9] The pricing policy automatically causes rational economic actors to turn away from the altruistic model and therefore opt for a family bank. Price distribution is the key issue. Unless the community option is significantly less expensive, parents will choose the family option. Price setting is a sensitive issue and commercial banks do not want their hands tied in this area, but for hybrid models to work in a balanced way, the public authorities need to require that commercial banks establish a price difference that is sufficient to incite parents to carefully consider the two options available to them.

From a marketing strategy standpoint, the hybrid model offers several advantages. It allows banks to enhance their reputation, facilitate accreditation, stand out from the competition, and accelerate business development, all at a relatively low cost. This hybrid model is being adopted by a growing number of private banks who use it as a marketing showcase. For instance, when Prince Philip's second daughter was born in 2007, it would have been logical for the Prince to decide to store the cord blood in Madrid, in a public or hybrid bank, in order to set an example and check the flow of bioresources crossing the Spanish border. Yet despite the new regulations allowing hybrid storage, he again preferred to export the cord blood to Vita43 to ensure the exclusive usage of his daughter's blood.

In Germany, commercial banks such as Vita34 and Eticur offer a hybrid model where a privately stored CBU can be requisitioned for nonfamily members, but unlike the Spanish decree, the family can refuse to comply with the requisition and keep the unit for strictly private use.[10] In 2012, Vita34 opened an online portal allowing real-time searches for allelic typing in its allogeneic CBU registry.[11] On the other hand, in Turkey, commercial banks are authorized to store CBUs on condition that 25% of stored units are made available to public health authorities to meet the population's needs.[12] Basically, without government intervention designed to balance the pricing policies of hybrid banks, mixed models may remain an ethical illusion with no real impact on public health.

## 2.4 Single Marketing and Binary Channels

In addition to the "split" and "requisition" mixed models, some commercial banks such as StemCyte or Cord:Use have developed two separate channels under the same brand. The parents choose between two different types of banks: paid storage for exclusive family usage ("StemCyte family" or "Cord:Use family"), or free storage of an anonymous donation in a public bank ("StemCyte" or "Cord:Use public bank"). This model has the advantage of clarifying the offer by clearly differentiating the two channels, how they work, and the conditions for accessing the grafts. The pricing policy of the binary model does not create confusion in the minds of parents and it promotes a more balanced situation in terms of the choices actually made. Developed more generally in the United States, the binary model contributes to the harmonization of quality standards through a stringent accreditation process (Foundation for the Accreditation of Cellular Therapy, FACT), collaboration with the best hospital maternity staff, and cooperation with the National Marrow Donor Program (NMDP) in the United States.

# 3. ECONOMIC MODEL OF PUBLIC BANKS

Public and commercial banks differ not only in their objectives; their economic models are based on opposing approaches to each phase of the production chain: (1) at entry into the bank immediately following collection; (2) during storage in the bank; and (3) at release to a transplant center. Each unit added to the inventory of a commercial bank generates approximately 2000 USD in revenue, which are paid by the parents (family model). For public banks, each unit added generates an expense of approximately 2000 USD (anonymous, free model).[13] A stored unit also generates an annual cost of roughly 125 USD (consumables, maintenance, etc.). In a public bank, the full cost is borne by the bank, whereas in a commercial bank it is funded entirely by the parents. However, an outgoing unit does not generate any revenue for a commercial bank, whereas for a public bank an international release can generate close to 30,000 USD.[13] Some banks that are affiliated with the NMDP in the United States charge close to 50,000 USD per CBU, including shipping costs. International prices of allogeneic CBUs for 2012 are listed in Figure 3.

The release prices are generally set by each public bank, but in some countries they are determined by the health authorities. They are billed to the transplant centers where the unit is transplanted. Prices vary depending on the biological quality and the genetic characteristics of the graft, and on whether the transplant requires one graft or two. The average release rate for established public banks is between 1% and 2%. For a public bank to finance close to 98% of its inventory with a 2% release rate, the price of each CBU needs to cover the collection and storage costs of approximately 20 new units. This break-even formula reflects the pricing practiced by allogeneic banks. Though these banks are not-for-profit, they follow the logic of a competitive market where transplant physicians compare the biological quality of grafts before selecting them. Public banks are repositioning their development strategy by raising their selectivity standards (cell concentration, rare haplotypes), increasing their release rate, growing their revenue, then offsetting the resulting additional costs through higher prices.[13]

Several public banks with release rates of less than 2%, such as the US Memorial Medical Center (Worcester, USA), filed for bankruptcy. Created in 1997, the bank's annual operating costs (750,000 USD) were deemed too high by the university hospital, which decided to shut it down. The bank was rescued from bankruptcy by the American Red Cross in 2003 and acquired by Cryobanks International in 2006. Since then, Cryobanks International has been managing it like an accredited allogeneic bank, in parallel with its commercial family bank. Another case is the public bank of the Saint-Louis Hospital in Paris, located on the campus where the first umbilical cord blood transplant worldwide was performed in 1988. A pioneer in cord blood storage as far back as 1989, the bank was forced to completely halt its activities in 2002 due to lack of funding from Assistance Publique-Hôpitaux de Paris (AP-HP). Hundreds of grafts were transferred to the Etablissement Français du Sang. In 2008, the Saint-Louis bank was reopened by AP-HP and is now one of the leading banks in the French National Cord Blood Bank Network (RFSP). To avoid the risk of bankruptcy, some public banks such as Alberta Cord Blood Bank (Canada), created in 1996, have been

| COUNTRY | Rate per CBU |
|---|---|
| GERMANY | |
| Düsseldorf | 21,000 EUR |
| ZKRD | - |
| BELGIUM | 22,450 EUR |
| SPAIN | |
| Redmo | 23,500 EUR |
| BCB Barcelona | 23,000 EUR |
| FINLAND | 22,491 EUR |
| ITALY | 17,461 EUR |
| NETHERLANDS | 22,450 EUR |
| SWIZERLAND | 27,049 EUR |
| UNITED KINGDOM | |
| Nolan Trust | 21,000 GBP |
| BBMR | 19,623 GBP |
| AUSTRALIA | 39,000 AUD |
| ISRAEL | |
| Hadassah | 26,000 USD |
| Sheba | 22,000 USD |
| TAIWAN | |
| Healthbanks Biotech | 10,000 USD |
| Bionet Corp | 4,500 USD |

| NMDP Banks (USA) | City | Rateper CBU (USD) |
|---|---|---|
| LifeCord | Gainesville | 43,250 |
| Belle Bonfils Memorial Blood Center | Denver | 32,750 |
| San Diego Blood Bank | San Diego | 38,000 |
| St Louis Cord Blood Bank | St Louis | 37,160 |
| ITxM Clinical Services Cord Blood Lab | Rosemont | 34,850 |
| Texas Cord Blood Bank | San Antonio | 34,850 |
| Children's Hospital of Orange County | Orange | 32,750 |
| New Jersey Cord Blood Bank | Camden | 43,250 |
| Michigan Cord Blood Bank | Grand Rapids | 38,000 |
| Puget Sound Blood Center & Program | Seattle | 38,000 |
| StemCyte International Cord Blood Center | Covina | 37,615 |
| J.P. McCarthy Cord Stem Cell Bank | Detroit | 38,000 |
| Carolinas Cord Blood Bank | Durham | 40,100 |
| COBLT Units – Carolinas Cord Blood Bank | Durham | 21,335 |
| Lifeforce Cryobanks | Altamonte Springs | 37,475 |
| New Jersey Cord Blood Bank | Allendale | 43,250 |
| M.D. Anderson Cord Blood Bank | Houston | 36,425 |
| Puget Sound Blood Center (ARC Units) | Seattle | 38,000 |
| ITxM Cord Blood Services (ARC Units) | Glenview | 34,850 |
| Carolinas Cord Blood Bank (ARC Units) | Durham | 40,100 |
| CORD:USE Cord Blood Bank | Orlando | 40,100 |
| Gift of Life Bone Marrow Foundation | Boca Raton | 24,350 |
| University of Colorado Cord Blood Bank | Aurora | 41,150 |
| SemCyte Taïwan National Cord Blood Center | Taipei – Taïwan | 34,150 |
| New York Blood Center | New-York | 41,190 |
| Cleveland Cord Blood Center | Warrensville Heights | 41,845 |
| Celgene Cellular Therapeutics / Lifebank USA | Cedar Knolls | 47,450 |

NB: an additional charge of 1,950 USD for shipment of CBU from NMDP banks

**FIGURE 3** Prices of allogeneic CBUs applicable to french transplant centers in 2013.

offering commercial banking since 2005 to help finance the growth of their not-for-profit public banking.[14]

To summarize, in economic terms, public and commercial banks follow diametrically opposed approaches, leading to vastly different financial results. For public banks, 1 or 2% of stored units generate more than 70% of their revenue, whereas for commercial banks, 100% of stored units generate 100% of their revenue. This explains the financial difficulties that allogeneic banks face and their dependency on the limited public funding available, which results in hindered growth. It also explains the strong financial appeal of commercial banks, whose economic model is at once stable and highly profitable.

## 4. ECONOMIC MODEL OF COMMERCIAL BANKS

The economic analysis of commercial banks remains unexplored in the literature because of a lack of available data on their management practices and their financial performance. In 2013, out of 310 commercial banks identified in 86 countries,[15] six banks were listed on European, Asian, and American stock exchanges: Cryo-Save (Euronext); China Cord Blood Corp. (NYSE); Vita34 (FWB); Cordlife (SGX); CryoCell; and Cord Blood America (Nasdaq). We analyzed the annual report and financial statements of each of these six banks for the years 2009–2012. These reports were subject to an external audit: they comply with international accounting standards (IFRS) and underwent stock market authority oversight before being disclosed to the markets. Therefore, they represent a relatively reliable source of information, in a context where the other private banks do not publish their business and financial records. This analysis only covers six commercial banks and does not reflect the entire private banking market. However, in the absence of alternative transparent sources, it highlights several economic trends (Figure 4).

### 4.1 Analysis from Six Listed Cord Blood Banks

The listed banks are small companies (between 15 and 890 employees) that were created between 1989 and 2007. They total 973,982 stored CBUs. Only three of the banks publish the number of CBUs used since their creation, without specifying therapeutic or scientific usage, type of transplant, therapeutic indications, or clinical outcomes. For those three banks, when comparing the total number of units used (44) with the total number of units stored (360,000), we find an overall turnover rate of 0.012%. Turnover is a key indicator that reflects the usefulness of a commercial bank to its customers and patients. It is surprising to note that half the listed banks prefer to not publish this indicator in their annual reports.

Over the 2009–2012 period, the compound annual revenue growth rate for the six banks was 9.18% despite a global economic crisis. Exceptionally, in 2012, Cyro-Save posted negative growth of −12%. Its share price fell sharply that year when, after a three-year dispute with the French health authorities, the Conseil Constitutionnel finally rejected Cryo-Save's request to operate in France, requiring it to transfer its technical platforms outside the French borders.[16] We are unaware whether shipping the units illegally stored in France since 2010, resulted in the deterioration or loss of CBUs and the reimbursement of the parents (Figure 5).

For these six banks, the average gross profit margin was 67.3%: this is much higher than the average observed for biotech firms of similar size, which—unlike cord blood banks—are required to invest massively in R&D, with no guarantee of a result. Yet in 2012, these listed banks had invested on average 1.1% of their revenue in

| | CRYO-CELL | VITA34 | CRYO-SAVE | CORDLIFE | CORD BLOOD AMERICA | CHINA CORD BLOOD | TOTAL* |
|---|---|---|---|---|---|---|---|
| Year of creation | 1989 | 1997 | 2000 | 2001 | 2003 | 2007 | NA |
| Nb of employees 2012 | 80 | 101 | 270 | 121 | 15 | 890 | 1,477 |
| Total inventory 2012 | 275,000 | 95,000 | 225,000 | 40,000 | 27,000 | 311,982 | 973,982 |
| Nb of CBU released 2012 | NA | 24 | 12 | 8 | NA | NA | NA |
| Stock market | NASDAQ | FWB | EURONEXT | SGX | NASDAQ | NYSE | NA |
| Revenue 2012 (growth) | $ 17,969,855 **(+0,3%)** | € 13,603,000 **(-15,0%)** | € 36,842,000 **(-12,0%)** | $ 28,775,000 **(+12,1%)** | $ 5,992,948 **(+18,1%)** | CNY 526,123,000 **(+38,3%)** | $ 197,005,303 **(+7,0 %)** |
| Revenue 2011 (growth) | $ 17,918,270 **(+0,9%)** | € 16,001,000 **(-5,7%)** | € 41,853,000 **(+3,6%)** | $ 25,673,000 **(-9,0%)** | $ 5,075,292 **(+22,9%)** | CNY 380,490,000 **(+12,1%)** | $ 178,830,962 **(+4,1 %)** |
| Revenue 2010 (growth) | $ 17,760,000 **(+0,6%)** | € 16,963,000 **(+12,4%)** | € 40,400,000 **(+5,2%)** | $ 28,200,000 **(+24,8%)** | $ 4,130,000 **(+27,5%)** | CNY 339,532,000 **(+29,8%)** | $ 169,744,900 **(+16,7 %)** |
| Revenue 2009 | $ 17,660,000 | € 15,097,000 | € 38,400,000 | $ 22,600,000 | $ 3,240,000 | CNY 261,540,000 | $ 151,371,300 |
| Gross profit margin 2012 | **+72,8 %** | **+59,1 %** | **+64.7%** | **+69,6 %** | **+70.0 %** | **+65.0 %** | **+67,3 %** |
| R&D expenses 2012 | $ 109,640 | € 429,000 | € 378,000 | $ 0 | $ 0 | CNY 8,459,000 | $ 2,515,344 |
| R&D budget vs revenue 2012 | **0,6%** | **3,2%** | **1,0%** | **0,0%** | **0,0%** | **1,6%** | **1,1%** |

*Annual reports cover the period from Jan 1st to Dec 31 st , except for Cryo- Cell (Dec 1st Nov 30th), China Cord Blood (April 1st - March 31st) and Cordlife (July 1st - June 30th).
Totals in EUR, CNY and SGD are converted into USD at the corresponding date exchange rate

FIGURE 4 Economic performance of six listed banks based on their annual report.

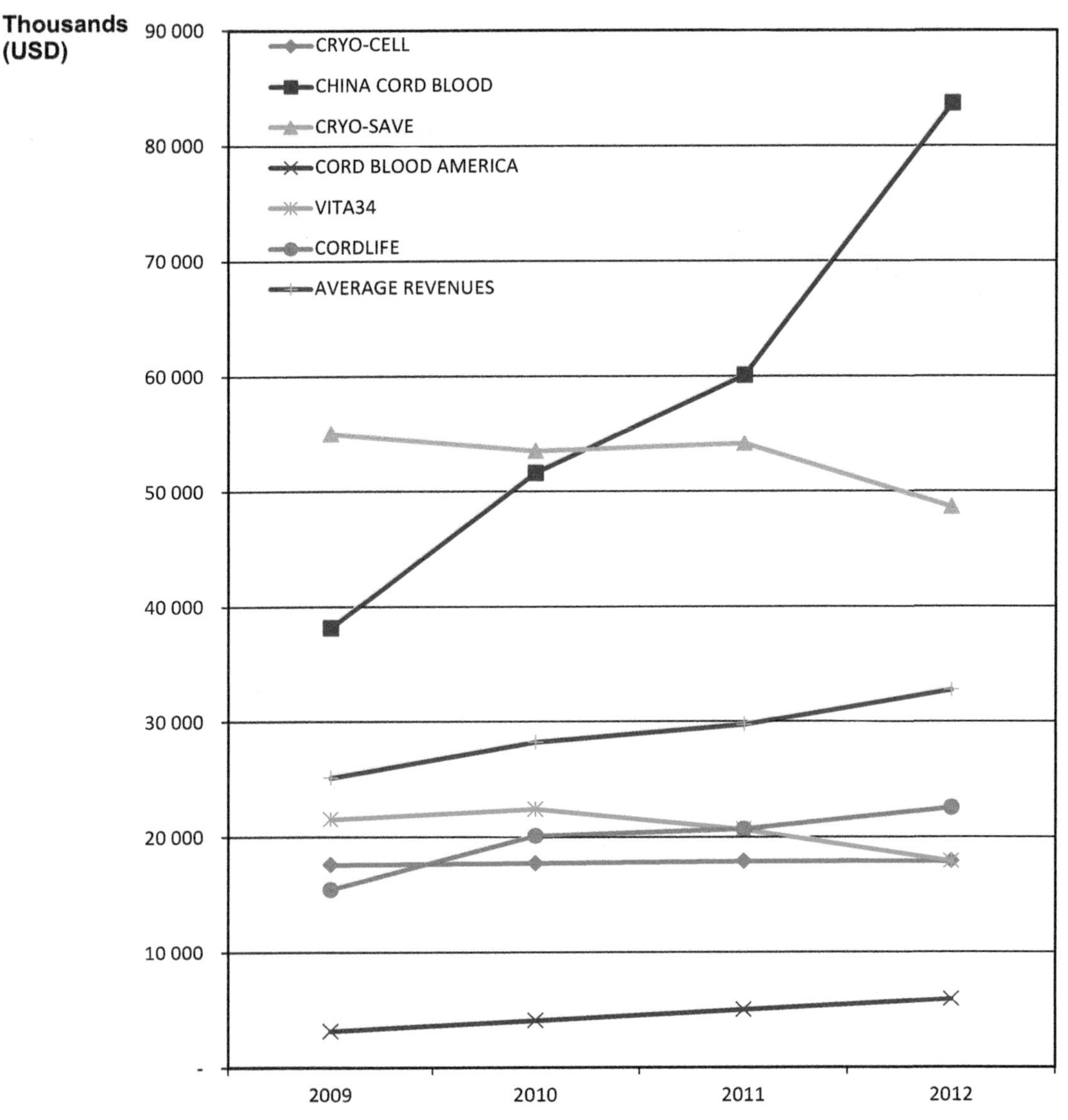


**FIGURE 5** Evolution of revenues of six listed cord blood banks (2009–2012).

R&D. This is especially surprising given that these firms operate in the fields of cell therapy and regenerative medicine, two areas of scientific research recognized as exceptionally innovative. On the one hand, they refuse to invest in R&D, but on the other hand, these commercial banks boast to their customers of the therapeutic promise held by this scientific research. To make a comparison, the R&D investment levels of these banks are the same as those found in businesses where high-tech innovation is not a major issue, such as the paper, tobacco, and beverage industries, where levels range from 0% to 2%. This is also true of the shipping/storage industry, which is in fact the economic sector that has the most similarities with the everyday operation of commercial banks.[17] Compared to the erratic fluctuations in biotech firm share prices, private banks operate in a financial environment with low volatility, without R&D investment risks, and with a gross profit margin that is much higher than those observed in the health industries.[18]

## 4.2 Industrial Investors

Beyond the purely financial investments, the high return of commercial banks has attracted a number of manufacturers in the health sector who seek to diversify their lines of business. Drug companies and medical device firms are investing in the capital of private banks such as LifebankUSA, owned by Celgene Cellular Therapeutics. In 2003, Amgen acquired shares in the firm ViaCell,[19] whose ViaCord family bank is now a subsidiary of PerkinElmer.

For biopharmaceutical companies, such investments are crucial because they provide direct access to large collections of human biological samples. These collections play a key role in epidemiological and population genomics studies in order to identify biomarkers correlated with pathologies and targeted treatments. They also allow the use of biospecimens obtained from the banks to perform high-throughput molecular screening of candidate drugs. These biobanks also allow predictive toxicology tests to be performed on human cells before beginning tests on animals or humans, which help to humanize the preclinical phases before investing large budgets on clinical trials. They also provide raw material for experimental protocols in the field of cell expansion and differentiation, before patenting new devices or culture media that are essential for tissue regeneration. Lastly, biobanks allow large-scale testing of procedures and new medical devices designed for the collection, miniaturization, and preservation of stem cell grafts.

## 4.3 Scientific Banking

However, the creation and development of scientific biobanks is sometimes slowed by the wariness of potential donors, who focus on the confidentiality issues tied to genetic data and intellectual property. The famous cases of the atypical cells of Henrietta Lachs and John Moore raise other critical ethical questions which, due to a lack of appropriate legal frameworks, generate mistrust and hinder the research and financing on which they depend.[20] For CBUs that are not approved for therapeutic use, 92% of pregnant women are ready to offer them to research for the advancement of cell therapies, though they remain concerned regarding the protection of genomic data and possible commercial applications.[21]

The issue of access to bioresources for scientific use is directly related to the issue of donor consent, but in the case of commercial banking, clients pay a fee precisely to guarantee exclusive rights over the cells and prevent researchers from gaining access to them. Private banks do not have a title of ownership for the units they store, unlike public banks which, through the consent mechanism, legally transform the donations into property.[22] However, other channels may emerge for CBU collection with donor consent for strictly scientific usage, while ensuring both data anonymity and trade secrecy relative to the industrial applications resulting from research.[23] The units are collected by the commercial banks, which prepare them on their technical platforms and store them in tanks that are separate from those reserved for therapeutic use. Allogeneic banks are also developing scientific channels based on disqualified CBUs that cannot be used for therapeutic purposes. Some banks use these scientific units to conduct their own research, or sell them to other research teams. In 2012, the price for a bag of CD133+ stem cells extracted from cord blood was 2304 EUR.[24]

Very few public banks in the world actually have systems for managing the distribution of scientific CBUs to make them available for research. In France, the 114 research teams working in 2011 on human stem cells had a hard time obtaining biospecimens for their research protocols under proper ethical framework (donor consent) and biological conditions (bacteriological and virological analysis of samples). Public–private partnerships have been created since 2009 to help academic researchers obtain quality bioresources free of charge, in particular the partnership between the Fondation Générale de Santé and the public cord blood bank of the Assistance Publique—Hôpitaux de Paris.[25] Some 2000 selected CBUs are offered each year to 25 research teams working on cell expansion, amplification, differentiation and isolation, biomarker identification, and improvements in transplant protocols.

The development of scientific banks is essential for the emergence of regenerative medicine, but it raises major economic, organizational, and ethical challenges. In Germany, 43% of pregnant women are concerned about the confidentiality of the genetic data in the samples entrusted to researchers.[21] Confidentiality should be managed through the adoption of international standards to ensure that consent forms are harmonized. This would reassure candidates for scientific donations, which would in turn help accelerate the development of biological sample collections.[26]

# 5. ATTITUDES AND KNOWLEDGE OF PREGNANT WOMEN

To understand the tissue economy of cord blood, it is necessary to analyze the supply and demand mechanisms affecting the attitude of pregnant women, particularly their vision of the issues at stake in cord blood banking. In 2011, a large-scale study was conducted in five European countries: a questionnaire was distributed to 1620 women in maternity units in the United Kingdom, Germany, France, Spain, and Italy.[21] The results were globally homogeneous in the five countries. On average, 79% of pregnant women had very low awareness of the possibility of storing cord blood. In terms of banking choices, 76% preferred to give their baby's cord blood to a public bank and 24% preferred to preserve it in a commercial bank, of which 12% in a hybrid bank (see Figure 6). Among the reasons given, 59% stated that their choice was mainly driven by a sense of solidarity (only 26% mentioned the fact that the donation was free). Moreover, 24% of pregnant women would be ready to switch maternities to make a cord blood donation if their own maternity facility was not accredited to collect the unit. The study also showed that the choices of pregnant women are not correlated with household income or socioeconomic category. In addition, 91% of pregnant women feel that the father should be involved in choosing the bank type (Figure 6).

In the five countries, 30% of pregnant women on average feel that human body parts are not a marketable commodity like any other consumer good—with a significant difference between France (42%) and the United

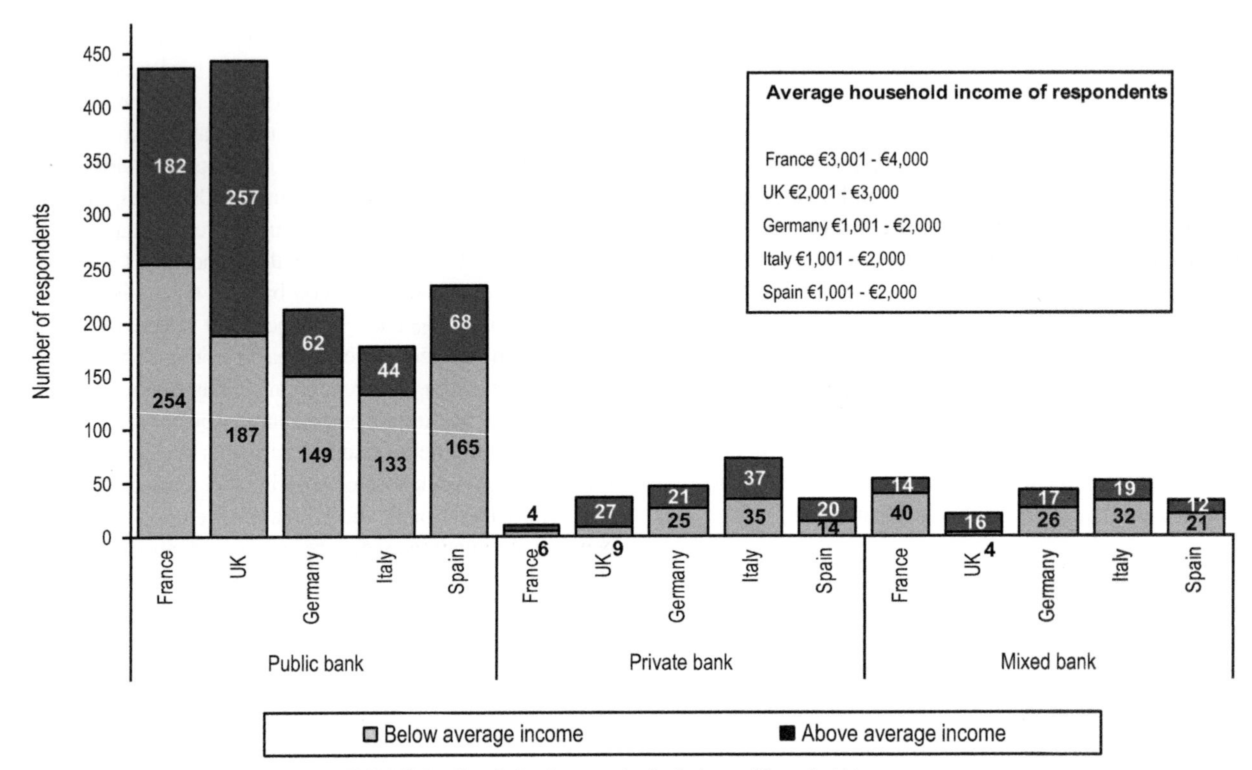


FIGURE 6 Correlation between bank choice and household income.

Kingdom (17%, p <0.001). This result echoes the French tenets regarding the noncommercialization and nonownership of human body parts, a humanistic principle inscribed at the core of the French Civil Code and the Oviedo Convention adopted by the Council of Europe in 1997.[27] In contrast, the UK results reflect more utilitarian values and liberal economic principles illustrated by the thriving cord blood banking business over the past decade.[28]

In the five countries, on average, only 21% of pregnant women received information from their obstetrician or midwife on the reasons for donating cord blood. The fact that information is not provided to future parents is at the root of the success of commercial banks, which use advertising campaigns to target ill-informed parents-to-be. In the five countries, the average rate for inadequate information provided is so high (79%) that it is legitimate to look for its causes among the medical profession. Could it be lack of interest or lack of time? Perhaps they do not share the information which they themselves did not receive during their university training. More in-depth research is needed to determine the number of course hours dedicated to cord blood collection in medical and midwifery schools. Childbirth professionals would probably be more committed to informing future parents if they had better training. Obstetricians and midwives could reassure parents that public banks can offer a compatible graft for each patient in need. They could also present the clinical results of hematopoietic stem cell transplants, the very low utilization probability of autologous units, the free banking system, and the essential value of solidarity in public health.

## 6. COST ANALYSIS FOR PUBLIC BANKS

Public banks are faced with the challenge of building a registry of high-quality transplantable CBUs with few resources. Their economic vulnerability requires that selective criteria be established when deciding which units are eligible for therapeutic usage. A composite scoring system is being developed to measure the biological quality of the collected units before storage or after thawing in order to help transplant physicians obtain the best possible graft for their patients.[29] The parameters making up this score have a direct influence on costs and prices according to the banks' level of selectivity.[30] Basically, the more stringent the CBU approval criteria, the fewer units qualified for therapeutic use, the higher the final cost of an approved graft.

### 6.1 Correlation Between Costs and Release Rates

"Final cost" refers to the total expenditure required to produce a graft, i.e., to list it in the international registry (BMDW) so it is available for a transplant. To assess the production cost of an allogeneic graft, each step is taken into account, starting with informing the parturient woman in the maternity unit, obtaining consent, identifying contraindications, collecting the unit, shipping it from maternity hospital to bank, performing bacteriological and serologic tests, typing CBUs' HLA, miniaturizing the sample, and cryopreserving each unit in tanks. Depending on the calculation method, this cost may also include collected CBUs that were rejected and therefore not stored, but for which resources were consumed.[31] The rejection rate depends on the stringency of the criteria used by the bank. In 2012, established allogeneic banks reported a CBU approval rate of between 20% and 30%, for release rates ranging from 1% to 2% of their total inventory.

Financed by public funds, allogeneic banks have been comparing costs for listing a CBU in the international registries since 1999.[32] In 2012, Prat Arrojo et al. published a total cost of 720.41 EUR per CBU approved for therapeutic use.[33] The calculation method used in the study is debatable since it does not take disqualified units into account, i.e., those collected at a cost but finally not banked. In the United States, the NMDP analyzed the costs of four allogeneic banks. The findings show a total cost of 1524 USD per unit registered, which includes 206 USD for collection and 886 USD for preparation.[13] The cost for preserving a CBU in a tank over a period of 1 year is 27 USD. Salaries account for 23.08% of the final cost. With an average release rate of 1.16% and an average release price per graft of 30,358 USD, the four NMDP public banks generated total annual revenue of 51,062,396 USD for a total expenditure of 64,130,176 USD. Therefore, without public funding, these banks would post an annual deficit of 13,067,780 USD, which makes them highly vulnerable. To break even, they need to review their growth strategies by raising the quality standards for CBU storage in order to optimize their utilization rate.

On the one hand, this strategy increases costs, but, on the other hand, there is a direct correlation between graft quality and the clinical outcome of a child or adult transplant.[34] CBU quality depends mainly on the number of total nucleated cells (TNC) and CD34+ cells.[35] The more a bank applies selective criteria in order to only store units with a high cell count, the lower the collected CBU approval rates, the higher the final production cost of the graft, and, as a consequence, the higher the probability of a release.[13] The facts support this cause and effect relationship. According to NMDP data, the cost for collecting and preparing a unit with a minimum TNC count of $90 \times 10^7$ is estimated at 206 USD. The cost rises to 2768 USD for a minimum TNC count of $150 \times 10^7$.[13] In 2012, Querol concluded that banks should focus on selective storage of CBUs with a high TNC count rather than expanding their registries without targeting them.[36] The cellular richness of the new stored CBUs represents a strategic criterion the guarantees the bank's financial stability and economic sustainability.

In 2010, the Mannheim Cord Blood Bank found a median TNC count of $119 \times 10^7$ for released grafts, whereas the median TNC count of the total inventory was only $65 \times 10^7$.[31] This result is consistent with the observations of other allogeneic banks worldwide, which conclude that cell count improves not only therapeutic effectiveness for the patient, but the bank's release rate and financial robustness as well. In 2013, in countries such as France and Spain, allogeneic banks increased their minimum counts for storing a unit (a TNC count of $100 \times 10^7$ and a CD34+ cell count of $1.8 \times 10^6$).[37] Based on these criteria, the Agence de la Biomédecine estimated that the total cost of a CBU listed in the French bone marrow registry is 2674 EUR. This amount is paid to the bank for each new registered CBU in order to offset its operating costs, in particular expenses generated by collected CBUs that were rejected before storage.

## 6.2 Economic Vulnerability of Public Banks

New CBU registrations and CBU releases represent the two main sources of income through which public banks seek to achieve financial equilibrium. However, given the economic crisis and the very low level of public funding available, public banks are struggling to ensure their growth. It is likely that some of them will close down in the coming years. In practice, most public banks—including competitive ones—are unable to reach the break-even point without public funding or philanthropic donations. There is positive emulation among allogeneic banks who seek to optimize their international releases to save patients, help transplant physicians find the best graft, and increase their financial independence and growth. This medico-economic and organizational pressure raises three challenges: (1) the quantity challenge, i.e., the need to reach optimal inventory size; (2) the quality challenge, because the overall survival of transplanted patients is directly correlated with graft quality; (3) the diversity challenge, i.e., to increase the representation of ethnic minorities in banks and registries. Over the past few years, the issue of quantity has gradually been replaced by those of quality and diversity. This dynamic captures the challenging mission of public banks: to offer the most patients the best graft at the lowest cost.

## 6.3 Ethnicity and Selection of Collecting Maternity Units

The fact that ethnic minorities are underrepresented in international registries raises strategic questions for allogeneic banks.[38] Some propose loosening the restrictions on TNC counts, especially for banks located in geographic areas with high antigenic variability due to the population mix. A minimum TNC count of more than $4 \times 10^8$ would allow a greater number of underrepresented patients to receive transplants.[33] In addition to the cellular richness criteria (TNC, CD34+) that is generally used to make banking decisions, new CBU selection strategies are being applied before the collection stage. They consist in selecting collecting maternity units located in geographic areas in which the local population features greater ethnic diversity. Indeed, ethnic monitories are underrepresented in the registries and therefore overexposed to the risk of not finding a compatible match. To assess the contribution of each newly banked CBU toward increasing registries' HLA diversity, the Marseille bank used an algorithm to deduce information on the putative ethnic origin of the newborn's parents. Whereas 93% of CBUs identified with one or two non-Caucasian haplotypes enrich the international registries, the proportion drops to 42% for CBUs from families with two Caucasian haplotypes.[39] Other studies are needed to demonstrate that a collection strategy specifically targeting maternity units receiving these non-Caucasian ethnic profiles improves the release rate and provides a dual benefit: therapeutic—for patients—and financial—for the bank.

When it comes to non-Caucasian ethnic minorities with immigrant backgrounds, the language barrier plays a role in reducing their chances of making a cord blood donation. The lack of translated consent forms results in misunderstandings and refusals to donate, which further increases the underrepresentivity of these ethnic profiles in banks, overexposing them to the risk of not finding a compatible graft. A leading French player in human stem cells collection, the Fondation Générale de Santé, has translated the standard consent forms into seven languages (including Mandarin, Arabic, Tamil, and Turkish) in order to remove the language barrier for non-French speakers, thereby increasing their genetic representation in public banks. In collecting maternity units, midwives specifically target non-Caucasian women for cord blood donations, providing them with a form translated in a language they understand. These translated forms are available online and can be downloaded in any maternity unit. This cost-effective, easy-to-implement solution allows allogeneic banks to overcome the language issue, take in more ethnic profiles in their registry, increase their release rate, and strengthen their financial situation. The reorganization of collection channels is the new challenge facing allogeneic banks. It implies stringent selection of collecting maternity units according to population pools and the haplotypes of their potential donors. Maternity units are the pillars on which banks rest. Selection of collecting maternity units is a pivotal step toward optimizing allogeneic inventories, improving their utilization rate, which results in greater economic independence and financial sustainability.

## 7. COST ANALYSIS FOR PRIVATE BANKS

Private banks that publish precise data on their production channels are rare, and virtually none of them disclose detailed information on their procedure costs,[40] Though unpublished, the costs of CBU storage for a commercial bank should logically be much lower than the costs for an allogeneic bank, because they require fewer biological analyses before freezing, and generally no HLA typing when they are banked.[41] Furthermore, unlike public banks, the number of disqualified CBUs in private banks is virtually nil, since close to 100% of collected units are stored, even if the CBU is not suitable for therapeutic usage.[2]

While no precise data is available, one can safely make the assumption that, like any corporation, most commercial banks seek to optimize their net profit margin and minimize their operating costs. At minimum, these low costs include the collection kit, shipping from the maternity unit to the bank, the basic cord blood tests (count, TNC), the miniaturization procedure, the nitrogen, tank depreciation, salaries, and administrative costs. Serological and bacteriological analyses may also be performed in some cases, whereas HLA typing is only performed in those very rare cases where the CBU is for therapeutic usage,[2] i.e., a probability of roughly 0.00044% (see section 8.1).

By applying the costs of NMDP public banks[13] to each of these stages, and by adding the salaries, administrative and marketing expenses, total cost ranges from 1050 USD to 1190 USD per stored CBU in a private bank. The average price billed to the parents is 2000 USD. Annual storage prices average 125 USD in a private bank, whereas the actual cost in a public bank is 27 USD (NMDP).[13] These assumptions on commercial bank operating expenses need to be verified using real and transparent data. However, these estimates are consistent with the net profit margins published in the financial statements of banks listed on the stock exchange (between 64.7% and 72.8%).

### 7.1 Quality Assurance in Private Banks

The quality assurance of a bank is essential to guarantee operational procedures that allow production of a graft that is usable for a transplant. These procedures concern every stage of the production chain. A quality management plan must include the staff organization chart, the tasks and duties of each member, a map of the location, the safety rules, and a periodic audit plan. The plan must describe the procedures for sample labeling, test validation, paper and electronic file archiving, personnel training, and the methods for stock inventory and the periodic analysis of results. An operational backup plan must be applied in case of unexpected incidents or interruption of tank operation in order to guarantee the integrity of each cryopreserved graft. The bank's quality assurance manager must ensure strict application of these rules and improvement of the standard operating procedures. These procedures receive accreditations delivered following auditing by agencies such as the American Association of Blood Banks[42] and FACT.[43]

Of the 310 private banks operating in 2012, only 10 had received FACT accreditation.[44] While public banks work with a small network of accredited collecting maternity units with trained staff, most commercial banks operate through an unstructured network of collecting maternity units that are not accredited, and whose standard operating procedures for cord blood collection are not inspected. In the vast majority of these maternity units, midwives are not specially trained to collect cord blood, to optimize volume and asepsis, or to screen potential donors based on medical contraindications. The quality of stored units in commercial banks is generally below the level found in public banks.[2] With nearly 2 million stored units in commercial banks, one would expect the latter to publish scientific results on the number of autologous transplants performed, their indications and clinical results, and the quality assurance procedures implemented. In the absence of published data, and given the opacity surrounding their operating procedures, the media has conducted several investigations into their practices.

## 7.2 Investigative Reports on Private Banks

In 2008, a journalist from the television channel SkyNews reported on the quality of service offered by one of the leading commercial banks in the United Kingdom. Pretending to be a customer, he received a collection kit from the bank, which he then filled with his own whole blood instead of cord blood. The commercial bank did not detect any issues and cryopreserved the blood bag. A few days later, the reporter received a storage certificate with an invoice for 1500 GBP. Asked to explain himself live on the SkyNews show, the bank's Chairman and CEO stated that this kind of incident was exceptional. In 2011, the firm demanded that the report be taken off-line.

On June 5, 2010, ABC News devoted two special reports to the marketing techniques used by commercial banks to find new customers.[45] Obstetricians have admitted to being paid by private banks to speak at conference dinners with future parents in order to sell them family storage, in contradiction with the ethics of the medical profession. In its opinion on private cord blood banks, the American Medical Association stated "Physicians shall not accept financial or other inducements for providing samples to cord blood banks."[46] As a result of this report, the American Congress established a committee to investigate the practices of commercial cord blood banks.

Another investigative report conducted in 2012 by the CBS News channel described the case of a commercial bank based in Monterrey, California, which declared bankruptcy in 2010 and, without informing its customers, continued to collect their annual fees the following year.[47] The stored units were transferred to another Californian commercial bank which, after analyzing the units in the transferred tank, found that the registered CBUs had disappeared, despite the fact that the parents paid their bills regularly. After interviewing the Food and Drug Administration (FDA), the CBS News report indicated that the FDA had closed down four private banks, and that seven others had filed for bankruptcy in 2011.[45]

With private banks promising their customers a "bioinsurance" against a large number of diseases, a few voices are pointing out that there is probably more "assurance" in storing the cord blood of your child in an accredited public bank, free of charge, that is available to all, including the donor and their relatives, than in paying a high price for some highly improbable future use, with technical conditions that often remain opaque.[48]

# 8. THE EMERGENCE OF BIOINSURANCE

Private banks target future parents by seeking to sell "biological insurance" for their newborn. Their advertisements, which address consumers directly, present cord blood and cord tissue as an "all-purpose" emergency kit without which it would be risky to start out in life.[49] Since birth is a unique opportunity to collect these cells, future parents may regret failing to invest in preserving them. As for classical insurance products, whose sale is based on the perception of low risks, the advertising messages of private banks make a subtle reference to the probability of requiring a stem cell transplant at some point in life. Without specifying what types of stem cells these probabilities refer to, the advertisements praise the upcoming advances in regenerative medicine in order to incite parents-to-be to invest immediately in a sort of "universal remedy" or "guarantee of health."

The industrial economy of private banks does not exactly create the kind of relationship that exists between an owner and a disposable good that is negotiable on the market. The relationship is closer to that found between an investor and their investment. The private bank offers a new kind of bank account where the biological investment is intended to protect the health of the investor. The account owners—i.e., the parents—invest in the biological capital of their child and are swayed by the feeling of living in a dual biological moment. Indeed, the child's body will grow and gradually lose its regenerative capabilities, with the eventual appearance of a disease. However, the cryopreserved cord blood fragment will retain its "autopoietic capacities," and thanks to its "negentropic properties," will allow the blood, immune system, and other vital organs to be regenerated.[22]

This biological bank account is presented by commercial banks as a farsighted investment that follows the same prudential logic as the products offered by private insurance companies. In their marketing message, private banks cast doubt on the capacity of public banks to cover the future needs of the population, predicting shortages in public registries due to the hypothetical growth of transplant indications. Their Web sites "play up the uncertainties of the public system, the possibility that no donation is exactly right for you, and liken private banking to a form of insurance."[50] They emphasize the theoretically unlimited possibilities of stem cell research and remind future parents that failing to preserve the cord blood means losing a unique opportunity.

While often presented as life insurance, cord blood stem cell preservation differs from conventional insurance in several ways. Its principle consists in covering a low risk—appearance of a disease that is treatable by an autologous transplantation—in exchange for a fee that both generates a profit for the firm and compensates the policy-holder if the risk materializes. This biological insurance is based on the idea that storing cord blood offers a guaranteed therapeutic solution for numerous diseases. In fact, the banks are unable to guarantee the therapeutic effectiveness of the CBU. They only claim to guarantee storage quality, despite the fact that 97% of them do not have FACT accreditation and continue to operate without conforming to stringent international standards.[44]

## 8.1 Probability of Use

Whether it is the risk of fire, burglary, a traffic accident, disability or death, the purchase of an insurance policy is determined by the assessment of this risk. However, in the case of cord blood banks, the method used to calculate the probability of utilizing a CBU for a transplantation is controversial for methodological reasons. Some epidemiological studies combine the prevalence and incidence of diseases for which autologous or family CBUs are indicated.[51] Other models evaluate the release rate of CBUs stored for at-risk families in public banks, free of charge, for a limited number of indications.[52] For instance, of the 4030 family units stored in several public banks, 192 were used in directed transplants, i.e., a probability of utilization of 4.76%.[53] However, this probability only applies specifically to public banks which, unlike commercial banks, screen at-risk families before collection and destroy units whose biological quality is inadequate for therapeutic use before cryopreservation. In no way can this probability be extended to private banks.

Other studies assess the probability of receiving a hematopoietic stem cell transplant in the course of a lifetime by adding up the data collected by accredited public institutions for bone marrow, peripheral blood, and cord blood transplants. This probability ranges between 0.23% and 0.98%, depending on age and the simulations used.[54] However, these probabilities are published on general consumer Web sites such as *Parent's Guide to Cord Blood*, though nowhere is it stated that these figures are not specific to cord blood.[55] This site, which is sponsored by commercial banks, claims to provide future parents with "accurate and balanced"[56] information on their cord blood storage options—in particular on the risk that their child will require a "stem cell" transplant. In the section dedicated to probability of use, the site publishes a probability of "1 in 5000 (i.e., 0.02%) for blood diseases and hereditary disorders" and a probability of "1 in 10,000 (i.e., 0.01%) for solid tumors," without telling the future parents the types of stem cells used, the types of transplants performed, or the types of banks in which the grafts were stored.[55] Without critical analysis, these figures may cause concern for future parents who—a few weeks before the birth of their child—were swayed into opting for commercial banking through one of the companies listed on this same portal.

The validity of these probabilities is limited, given that they are based on the assumption that the number of clinical indications for cord blood transplants will remain stable. In fact, it is impossible to predict whether new therapeutic indications for cord blood transplants will be validated in the future. The prevalence and incidence of these new indications will necessarily impact the probabilities of CBU use in public and private banks. Several clinical tests are currently being conducted on the autologous usage of CBUs to treat disorders such as type 1 diabetes or cerebral palsy.[57] Encouraging studies offer hope that one day we will be able to use the mesenchymal stem cells contained in cord blood to isolate cell lines and treat heart and neurodegenerative diseases. The clinical results of these trials will demonstrate whether autologous transplantation holds promise for the future, for both commercial and public banks.

Aside from these predictions, there is empirical data available from the public registries. Between 1998 and 2007, a Eurocord study found that, among 3372 cord blood transplants performed in the public banks of 43 countries, only three were autologous.[58] Over a 9-year period, the probability of autologous usage of a graft stored in a public bank was 0.089%. In 2008, a survey conducted with private banks by the American Society for Blood and Marrow Transplantation showed that out of roughly 460,000 stored CBUs, only 99 were used for a transplant.[59]

In its 2012 annual report, Cryo-Save declared 12 transplants out of the 225,000 CBUs stored since its creation in 2000.[60] For this 12-year period, the probability of use of a stored CBU was 0.0053%, which means an average annual probability of 0.00044%. This very low figure is roughly of the same magnitude as those published in 2008 by two of the largest commercial banks in the United States. In their annual reports, they announced an inventory of 210,000 units and 145,000 units, respectively.[52] Of these 355,000 total units, only 84 units were transplanted, of which 77 were for allogeneic usage and only 7 were for autologous usage—the latter being the main stated purpose of their business. For these two large commercial banks, the probability that a customer will use a stored unit is 0.002% over 16 years of storage, i.e., an average annual probability of 0.000125%.[61] Along the same lines, Smith emphasizes that the probability of CBU use for an autologous transplant lies somewhere between 1 in 10,000 and 1 in 250,000. These probabilities are equivalent to those of drowning (1 in 9000), a plane crash

(1 in 20,000), or death due to an asteroid hitting the earth (1 in 200,000).[62]

### 8.2 "Bioinsurance" Usefulness

More seriously, the real issue raised by private banks is not so much that they are selling an insurance policy to cover a risk whose probability of occurrence is extremely low, but rather that they are selling a second insurance policy to cover a risk that is already covered by public banks free of charge. In 2010 and 2011, investigative reports on television channels such as ABC News and CBS News described cases of parents who, a few months after having paid 2000 USD for the storage of cord blood in a private bank, learn that their child has osteopetrosis, a hereditary disease for which a cord blood transplant is indicated.[63] Expecting to be covered by the CBU bioinsurance in which they invested, they find out that in fact their child cannot receive a transplant from his or her own cord blood because the cells it contains carry the very genetic mutations that caused the disease. To avoid infusing the patient with the same disease, the transplant physicians preferred to use allogeneic CBU stringently selected through public registries, with a higher post-transplant overall survival rate. On seeing their children receive transplants from compatible grafts obtained from free, anonymous donations, customers of private banks understand the absurdity of the so-called "bioinsurance policy" they paid for. They realize that this "bioinsurance" works neither as supplemental insurance nor reinsurance for residual risks: it is useless because all the risks tied to the search for compatible grafts are already covered by the international network of public banks.

Though useless, this investment has a significant cost for both the parents and society. By comparing the prices charged by 27 commercial banks in the United States, we found that in 2010, the average fee for the first year of storage was 1450 USD, with an additional 120 USD in annual expenses.[21] Commercial banks point out that the total storage cost is 4000 USD for 20 years, 6400 USD for 40 years, and 10,000 USD for 70 years. These figures may seem reasonable, assuming that the collected sample remains viable for an entire life. However, from a strictly economic standpoint, the value of "life insurance" is measured in terms of opportunity cost, in other words by evaluating the profitability of the same amount placed in other types of investments, according to their level of risk. An investment of 1450 USD with an annual payment of 120 USD generates 118,000 USD after 70 years at a rate of 5%, and 2,180,000 USD at a rate of 10%. Given the virtually nil probability of using a CBU stored in a commercial bank, and the free and full availability of hemotopoïetic matches of superior quality in public banks, the "bioinsurance" concept can only mislead the ill-informed.

### 8.3 Partnerships With Insurance Companies

With the commercial success of the "biological insurance" concept among future parents, a number of partnerships have been concluded between insurance companies and private cord blood banks. In the United States, insurance companies such as Blue Cross Blue Shield, AmeriHealth, LifeWise Health Plan, and some 20 other health insurance companies have entered into partnerships with private banks such as Cord Blood America and its subsidiary CorCell, bearing a share of the costs for collecting and storing cord blood. In Europe, the Spanish commercial bank VidaCord has created partnerships with five insurance companies including AXA, which agrees to reimburse 50% of the storage costs—providing that the parents and the newborn take out a health insurance policy with them. From an actuarial standpoint, this kind of partnership is based on adverse selection: it allows the insurer to target a preselected population according to its preventive behavior with regard to very low morbid risks.

The private sector is not the only one to be interested in bioinsurance. The Singaporean government subsidizes half the cord blood storage cost of commercial banks listed on the stock exchange, such as Cordlife. This measure is part of the government program coined *Baby Bonus Scheme*, which encourages Singaporeans to have more children by providing financial assistance.[64] The *Child Development Account* is a bank account opened by a family at the birth of their child. Each dollar deposited by the family is matched by one government dollar. Financial caps have been set: 6000 SGD for the first child, 12,000 SGD for the second, and 18,000 SGD for subsequent children. Through this bank account, the CBU storage costs of a family bank are cofinanced using public money.[65] This model materializes the emergence of a true tissue economy where financial banks and insurers diversify their products by building alliances with stem cell banks, with a business model that targets a young and fairly affluent clientele potentially interested in purchasing a wide range of insurance products.

## 9. COST-UTILITY FOR PUBLIC HEALTH

Private banks rarely publish their utilization rates of CBUs. In their 2012 annual reports and on their Web sites, four leading commercial banks—Cord Blood Registry,[66] Cryo-Save,[67] Vita34,[68] and ViaCord[69]—report inventories of more than 400,000, 225,000, 95,000, and 300,000 CBUs, respectively, i.e., a total of more than 1.02 million units stored. Between 1992 and 2012, the total number of their CBUs used was 486. This translates into a turnover for the four inventories combined of 0.048% over 20 years, i.e., an average probability of use of 0.0024% per year. In 2012, the combined CBU inventories of all public banks worldwide totaled 646,772 grafts: in 2012, 4150 transplants were performed,[70] i.e., a 0.64% turnover for that year alone. In short, in 2012, the total inventory of these four leading commercial banks was 1.58 times larger than that of the combined inventories of all 137 public banks worldwide, yet their turnover—i.e., their therapeutic usefulness for patients—was 269 times lower.

In terms of cost-utility, in other words the medical service provided for the price paid, the customers of these four banks paid a total of 2.04 billion USD to use only 486 units over 20 years, i.e., an average of 24 units used per year (for an average market price of 2000 USD per unit for 20 years of storage). If those 2.04 billion USD had been invested in public banks, they could have stored 1,338,583 new allogeneic grafts (with a real production cost observed in NMDP public banks of 1524 USD per CBU). For a single year, this stock would have allowed the use of 15,528 grafts for patient transplants worldwide, including in the donor families (actual turnover of 1.16% observed by NMDP).[13] Given the financial instability of public banks, and keeping in mind that each CBU is used to treat a life-threatening disease, the imbalance in the number of CBUs used each year (24 from private banks vs. 15,528 from public banks) sheds a glaring light on the economic and moral impasse that commercial banks represent for public health.

## 9.1 Cost–benefit Analysis of Private Banks

In 2009, Kaimal et al. used a decision model to compare the cost-effectiveness ratio of cord blood storage in a private bank, with the absence of storage. The assumptions made were a cost of 3620 USD for 20 years of storage, a 0.04% probability the stored unit would be used for an autologous transplant, and a 0.07% probability it would be used for an allogeneic transplant for siblings. The results show that cord blood storage in a private bank is not cost-effective because it generates an additional cost of 1,374,246 USD per year of life gained.[71] To be cost-effective the storage cost in a private bank should be less than 262 USD and the probability that a child would require a stem cell transplant should be greater than 1 in 110. However, both of these numbers are very different from the real-life figures. A Monte Carlo simulation was performed to test the robustness of the model, modifying several parameters and probabilities simultaneously. In 99.2% of the cases, private banks were not cost-effective because they imply a cost that is far above the 100,000 USD limit per year of life gained.[72,73]

## 9.2 Overestimating Very Low Risks

Storing cord blood in a commercial bank is not cost-effective because the expenditures made greatly outweigh the benefits. Yet despite this, and even though the American College of Obstetricians and Gynecologists and the American Association of Pediatrics have come out against storage in private banks, future parents continue to turn to commercial banks. What is the logic behind this phenomenon? One explanation of this behavior could be the fact that parents vastly overestimate the probability of use of CBUs stored in private banks. This behavior is undoubtedly influenced by the Direct-to-Consumer (DTC) advertising used by these banks. This kind of marketing has developed swiftly in the United States since 1997, when the FDA loosened regulations on the advertising of pharmaceutical drugs. Some believe that DTC marketing plays a preventive role because it helps educate the general public on certain medical risks. Others observe that aside from increased drug sales, no concrete results have demonstrated any advantages for patients or public health.[74]

In the case of cord blood banks, several studies have demonstrated that pregnant woman have very little knowledge of the therapeutic indications requiring a hematopoietic stem cell transplant.[21] They are also unaware what the odds are that their child will require an autologous cord blood transplant at some point in the future.[75] It is difficult to assess whether the 2000 USD expenditure by future parents to store cord blood is actually a way to control their anxieties, buy themselves peace of mind, and alleviate their guilt.[62] Another possibility is that parents are willing to spend that amount of money due to insufficient information provided through the pregnancy by obstetrician–gynecologists and midwives. A few weeks before birth, providing factual information on the advantages of preserving cord blood, the diseases concerned, and the probability of using a privately banked unit would probably help future parents hone their critical thinking, alleviate their concerns, and prevent pointless expenditures. The dearth of information from the medical profession tilts the balance in favor of commercial banks.

Some believe that the decision to preserve a child's cord blood is a personal choice that should not be based on statistical or medico-economic analyses.[76] This argument does not take account of the difficulties inherent in each individual properly assessing the risks they face. Behavioral studies on decision-making show that in the presence of extremely rare events, individuals tend to significantly overestimate their probability.[77] Furthermore, we know that the most recent information has a strong influence when assessing probable events and may cause more likely events to be minimized or neglected.[78] The DTC marketing of commercial banks precisely leads parents to overestimate the probability of needing a cord blood transplant for their child, in particular by amplifying the results in the still experimental field of regenerative medicine.

Without providing explanations on the probability of utilization, and given the difficulty individuals experience in conceptualizing very low risks, it is difficult for future parents to rationally assess the advantages of this expenditure compared with the other immediate needs of their newborn child.[71] In addition the financial burden for families, the commercial development of private banks generates costs for public health by funneling away bioresources from free high-quality storage that would have been available for the entire population, including the donor's family. Commercial banks will continue to grow as long as they are profitable. However, the imbalance between the profitability of private banks and the financial vulnerability of public banks is prompting the emergence of new tissue economies that are based, not on values such as solidarity and public good, but on opposing values such as exclusivity and private property.

## 10. TISSUE ECONOMIES

Cord blood banks represent a major step in the advent of tissue economies in the twenty-first century.[79] In their principle, these economies follow the dialectic of supply and demand. CBUs are becoming biological assets imported and exported worldwide, and valued according to a pricing logic based on their level of quality and compatibility. The controversies between public and private banks are rooted in the history of economic theory and the polarization between altruism and self-interest—the "solidary ethos" and the "atomization of the self."[80] Esposito analyzes this dichotomy through the word *munus* (Lat. duty, donation, service, obligation), which is at the root of the words *immunitas* and *communitas*.[81] In its principle, *immunitas*—from which the word *immunity* is derived—exemplifies a sophisticated interaction between the "self" and the "nonself." Immunitas guarantees the protection, integrity, and exclusivity of the "self" by deploying defense mechanisms designed to prevent intrusion of the "nonself." Inversely, *communitas*—the Latin root of *community*—connects the "selves" and "nonselves" through solidary mechanisms, moral obligations, and altruistic reciprocities.

Transposed onto the cord blood bank debate, we could say that commercial banks defend *immunitas*: they propose a model where each family is centered on its identity of "self" and seeks to remove its immunological identity from the community of "nonselves" to avoid losing it, or to become alienated through community obligations. In the medical and legal discourse, immunity expresses a form of "exemption and untouchability."[80] Inversely, public banks defend the *communitas*: they offer a model where "nonselves" come together in a network. They share their donations and mutualize the available immunities in order to promote the compatibility—and therefore survival—of the greatest number of individuals.

### 10.1 Public Good and Privatization of "Self"

Organ donation laws are generally based on the idea of presumed consent: each individual must expressly state their refusal to donate their tissue or organs in order to be excluded from the registries of the *communitas*. However, there are two ways to apply this logic. In a country such as Singapore, only those who accept donation of their organs are eligible to receive organs in their lifetime. In other words, those who refuse to give exclude themselves from the *communitas* and are disqualified from receiving its bioresources. Reciprocity is applied in a mechanical logic: renouncing the "self" is the necessary condition for benefiting from the "nonself." Inversely, in countries such as France, you can receive donations from the *communitas* even if you have expressly chosen to remove yourself from it. In 2012, the 33.7% of the French population that expressly refuse to donate organs postmortem[82] would still receive organs from the community, if in need. This generous freedom has a high societal cost for France, which must meet the needs of the entire population, including nondonors. In Singapore, however, less than 3% of the population refuses postmortem organ donation. More than 97% join the *communitas* and share their bioresources.[83]

Much like blood transfusion centers, public cord blood banks uphold the idea that grafts are a common good, a life-saving resource that should be shielded from the market and entrusted to the *communitas*. On the contrary, commercial banks advocate the appropriation and privatization of the "self." In this sense, *immunitas* represents an obstacle to exchange circulation with any type of obligation toward the *communitas*. Whether public, private, or hybrid, cord blood bank inventories are like "stockpiles of immunity."[80]

### 10.2 Community versus Commodity

From a medical standpoint, the reasoning promoted by private banks isolates families in an immunologically closed model: they believe that by paying for the preservation of "self," they guarantee the compatibility and therefore the survival of their members, but they are unaware that they remain insidiously dependent on community grafts in case of incompatibility. It is important to emphasize the asymmetry of this relationship, because all units stored in the *communitas* are available at all times to the proponents of *immunitas*, while the opposite is not true. According to Esposito, *communitas* expresses mutuality of ties and reciprocity; *immunitas* represents a "negative resistance to reciprocity," a defense against obligations.[80]

The hybrid bank model seeks to reconcile *immunitas* and *communitas* by reconnecting with the polysemy of *munus*: a donation, a duty that obligates. Beyond their good intentions, the results of the first hybrid banks highlight a very strong imbalance, where the "self" escapes from the *communitas*: by charging virtually identical prices, these firms encourage their customers to shirk their altruistic obligations toward the community. Other hybrid models are being developed that could eventually balance the focus on the "self" with its obligations toward the "nonself."

## 10.3 Self-sufficiency and Dependency

A new trade of immunitary profiles is developing through a global economy where 29% of CBUs transplanted in 2012 crossed the borders of dozens of countries across five continents.[70] Nations belonging to the *communitas* network cannot reasonably claim to be self-sufficient: their level of immunitary dependence can be assessed by their CBU import rates. Japan is an exception to this rule on two counts: firstly, due to its insular geography and the haplotypical homogeneity of its population; secondly, through the nuclear catastrophes the country has had to overcome throughout its history, causing it to support cord blood banks to ensure that its population has permanent access to nonirradiated hematopoietic cells.

Without allogeneic banks or transplanted cord blood grafts in 2012, the countries of sub-Saharan Africa remain outside of the *immunitas/communitas* equation, being both isolated and completely dependent on foreign bioresources. Yet equal access to treatment and equal protection against disease are recognized in Article 25 of the Universal Declaration of Human Rights.[84] Inspired by this moral obligation (Lat. *munus*), allogeneic banks have become diasporic centers for "immunity capitalization;"[80] they collect and store the immunotypes of a variety of ethnic profiles and transfer them to diasporas, or offer them to groups that are underrepresented in the registries, and therefore overexposed to the risk of not finding immunological matches.[38]

## 10.4 Biological Capitalization

As a direct effect of globalization, immigration, and interethnic marriages, the panmixia of genomes establishes immunological compatibility as the cornerstone of tissue economies. In the global cord blood marketplace, each unit represents a full-fledged immune system that carries power to save a life. Like a biological currency, CBUs have become life units that are capitalized in "biobanks" and sold as "bioinsurance." The tissue economies are leading to an economy of blood, which until the twentieth century was considered a gift, shared property categorized as a public good.[85] In the twenty-first century, this concept is being reversed. A growing number of private banks are diversifying beyond cord blood services, marketing the storage of menstrual blood,[86] fat cells,[87] and dental tissue.[88] Prices range from 1000 to 2500 USD, depending on the firm. Other banks such as Scéil are already marketing the banking of adult skin cells in order to reprogram them as immature cells (IPS) from which it may be possible to generate many organs and tissue types in the future. Though this research is still experimental, Scéil is already charging 60,000 USD to bank a skin sample plus 500 USD per year in storage costs.[89] These examples illustrate the gradual replacement of the donation economy by a market economy. Though the scientific validity of these experimental approaches has yet to be demonstrated, these new forms of biological capitalization create both a demand and a business based on marketing claims that are insidiously eroding the legitimacy of scientific research in regenerative medicine.[90]

## 10.5 Guaranteeing the Storage, Not the Graft

The controversy over public and private banking is based on a confusion between access to storage and access to a graft. These are two clearly distinct notions. Public banks guarantee free access to a compatible hematopoietic graft for anyone in need, however, they cannot offer CBU storage for every pregnancy due to limited funding and a small local network of collecting maternity units. On the other hand, commercial banks provide access to storage, for a fee, performing collections through a decentralized network that can include virtually any maternity unit. However, most of these banks fail to inform their customers that they are unable to guarantee access to the graft—of all the CBUs they collect, only a minority will be transplantable, and just a tiny fraction will actually be transplanted. At some point, parents-to-be should probably ponder what they really want: paid storage or a free life-saving treatment?

"Graft" and "storage" cannot be placed on the same level. Private banks' marketing practices illustrate a caveat emptor approach: it is up to the consumer to obtain information on the product or service that is being sold. The Web site home pages of several leading commercial banks advertise testimonials from celebrities, sport legends, and Olympic athletes—but are evasive when it comes to the actual therapeutic usefulness of the services they offer. These marketing practices should be subject to stricter controls by the regulatory authorities and accreditation agencies.

## 10.6 Impasses and Dilemmas

It has now been established that, in addition to hematopoietic progenitors, cord blood cells also contain stem cells with embryonic properties. A large number of cell lines have already been generated from cord blood, including mesenchymal, endothelial, neuronal, liver, muscle, cardiac, pancreatic, bone, and skin cells.[91] These results are very promising but require more in-depth research before they can be used clinically as standard treatments. Given the availability of human cord blood stem cells and the fact that they do not raise ethical controversies, umbilical cord blood could become the preferred source of cells for regenerative medicine in the future.

Are hopes for the future sufficient to justify that today, each individual should pay to preserve their cord blood at birth? There are two possibilities: (1) either the therapeutic promise of cord blood is confirmed, requiring public banks to meet the needs of populations in order to ensure equal access to care; (2) or the promise is not confirmed, in which case continuing to sell useless storage would be unethical.

Either way, commercial banks serve little purpose with regard to health systems. For this reason, academic institutions, ethics committees, and professional associations widely denounce the fact that these banks are accelerating the merchandising of human body parts, that their prices exclude the disadvantaged, and that their marketing claims discredit the legitimacy of scientific research in regenerative medicine. More generally, the groups criticize the fact that these banks are undermining the altruistic spirit of the social body. Given this impasse, commercial banks have no choice but to rethink their business model. They should move away from therapeutic storage and develop into scientific banks.

## 11. RETHINKING THE BUSINESS MODEL OF PRIVATE BANKS

Induced reprogramming of somatic cells (IPS) was first discovered in 2006 using murine models,[92] then confirmed by many other teams[93] before being applied to humans in 2008. Induced reprogramming allows the cell aging process to be reversed by bringing mature differentiated cells back to the embryonic stage and then redirecting them toward all types of cell lines. Rewarded by the Nobel Prize in Medicine in 2012, this technique represents a true revolution for cell therapy and regenerative medicine. Applied until now to human fibroblasts, induced reprogramming can also be applied to stem cells from umbilical cord blood, with better results in terms of effectiveness and safety.[94] The results are still preliminary and must be confirmed by further research.[95] However, cord blood could represent an excellent cell source for induced reprogramming because the cells it contains are still naïve and their immunological history is still immature. These characteristics greatly facilitate their reprogramming into a quasi-embryonic state.

### 11.1 Cord Blood Stem Cells and IPS

Cord blood cells, which are 8 months older than embryonic cells, have the advantages of the latter but not their drawbacks: they offer high plasticity without inducing tumors when they are transplanted in adults or children. With close to 30,000 transplants performed since 1988, significant clinical experience and large existing storage infrastructures, cord blood stem cells are easily available at birth. Without risk to the donor, their therapeutic use does not raise the ethical dilemmas that human embryonic cells do. Given all these advantages, cord blood could serve as a universal biological material to which induced reprogramming could be applied, like a virtually blank computer disk on which one might reprogram almost any type of tissue lines used to regenerate diseased organs. This approach has already enabled the production of many organs and cell lines including red blood cells, and cardiac, intestinal, urinary, and lung tissue.[96]

### 11.2 IPS Scientific Banking

Since 2012, IPS cell banks have been developing worldwide. At Oxford, an IPS cell bank—StemBancc—was created in partnership with academic institutions and large drug manufacturers such as Pfizer, Lilly, Merck, Sanofi, Roche, Novo Nordisk, Abbott, Orion, and Boehringer. This bank should enable access to large collections of characterized cell lines with a wide genetic diversity. These lines can be used to create cell phenotypes that will serve as models for the study of new treatments for diabetes, Alzheimer's disease, Parkinson's disease, and disorders such as autism and schizophrenia. This IPS cell collection could also allow the development of predictive toxicology tests and serve as a molecular screening platform for future drugs. Given its size, StemBancc could also help to optimize the protocols for cell culture, differentiation, and expansion that will be used for tissue regeneration. Lastly, the statistical depth of the data collected could help strengthen the organization of biobanks, in particular their maintenance and their archiving procedures.[97]

Cord blood banks could play a decisive role in accelerating the development of IPS banks. One possible solution would consist in encouraging private banks to share the CBUs stored in their tanks so that they can be used for scientific purposes. The fact is, of the 2 million CBUs stored in 2012, in 310 commercial banks worldwide, more than 99% will serve no therapeutic purpose. These biological samples would probably be more useful if they were used to advance scientific projects which the parents have in fact chosen to support by investing in CBU storage. These underutilized bioresources have substantial scientific value. Following the model developed by StemBancc, drug companies and research institutions could cofinance the purchase of these collections of biological samples. Parents would be reassured that they could still obtain a compatible graft through the international network of public banks. The commercial banks would refund those parents who gave their consent for the CBUs to be used for scientific purposes. This strategy would make it possible to draw the full benefits of CBUs collected and stored according to standardized procedures. It would make it possible to produce large quantities of IPS stem cells on demand and in record time.[98]

Great industrial revolutions are not driven by scientific discovery alone. They require the mobilization of well-known existing tools and infrastructures, thereby enabling technological leaps. Commercial cord blood banks could play this key role of technology accelerator. Despite their very small investments in R&D, they could unexpectedly become true enablers of regenerative medicine research. This strategic development would allow them to invent a new business model, at the frontier of science and the biopharmaceutical industry. This business model would also offer them a way out of their ethical impasse—by reconciling *immunitas* and *communitas*.

## REFERENCES


1. http://www.bmdw.org/.
2. Sun J, Allison J, McLaughlin C, et al. Differences in quality between privately and publicly banked umbilical cord blood units: a pilot study of autologous cord blood infusion in children with acquired neurologic disorders. *Transfusion* 2010;**50**:1980–7.
3. Fisk NM, Atun R. Public-private partnership in cord blood banking. *BMJ* 2008;**336**:642–4.
4. Royal College of Obstetricians and Gynaecologists, Scientific Advisory Committee, Opinion Paper 2, revised June 2006, Umbilical cord blood banking. http://www.rcog.org.uk/files/rcog-corp/uploaded-files/SAC2UmbilicalCordBanking2006.pdf [accessed 04.04.11].
5. http://www.vidacord.es/experto/ventajas-banco-mixto/ventajas-de-ser-un-banco-mixto// [accessed 21.08.13].
6. Quoted in El Mundo, 17th October 2009, page 2.
7. http://www.elmundo.es/elmundo/2006/02/26/sociedad/1140968206.html [accessed 22.08.13].
8. Real Decreto 1301/2006.
9. http://www.fcarreras.org/en/spanish-bone-marrow-donors-registry_4768 [accessed 21.08.13].
10. http://www.eticur.de/preis-stammzell-aufbewahrung-nabelschnurbluteinlagerung-eticur/stammzellen-aus-nabelschnurblut-spenden-eticur.html [accessed 23.08.13].
11. http://www.stemcellsearch.org/ [accessed 26.08.13].
12. Brand A, Rebulla P, Engelfriet CP, et al. Cord blood banking. *Vox Sang* 2008;**95**:335.
13. Bart T, Boo M, Balabanova S, et al. Impact of selection of cord blood units from the United States and Swiss registries on the cost of banking operations. *Transfus Med Hemotherapy* 2013;**40**:14–20.
14. O'Connor M, Samuel G, Jordens C, et al. Umbilical cord blood banking: beyond the public-private divide. *J Law Med* 2012;**19**:512–6.
15. http://parentsguidecordblood.org/findabank.php?country=233 [accessed 22.08.13].
16. Haye I, Marville L, Katz G. Quel statut pour les banques de sang de cordon ombilical? *Med Droit* 2010:81–5.
17. Hernández H, Tübke A, Hervás Soriano F, et al. The 2012 EU industrial R&D investment scoreboard, joint research centre institute for prospective technological studies. *Eur Comm* 2013, http://iri.jrc.ec.europa.eu/scoreboard12.html.
18. Goodman M. Pharma industry strategic performance: 2008–2013E. *Nat Rev Drug Discov* May 2009;**8**(5):348.
19. http://wwwext.amgen.com/media/media_pr_detail.jsp?year=2004&releaseID=487340.
20. Weir R, Olick R. *The stored tissue issue*. Oxford, UK: Oxford University Press; 2004.
21. Katz G, Mills A, Garcia J, et al. Banking cord blood stem cells: attitude and knowledge of pregnant women in five European countries. *Transfusion* 2011;**51**:578–86.
22. Waldby C, Mitchell R. *Tissue economies: blood, organs and cell lines in late capitalism*. London: Duke University Press; 2006. p. 71.
23. Mitchell R, Conley JM, Davis AM, et al. Genomics, biobanks, and the trade-secret model.Science. *15* 2011;**332**(6027):309–10.
24. http://www.stemcell.com/ [accessed 21.04.12].
25. Katz G, Mills A. Cord blood banking in France: reorganizing the national network. *Transfus Apher Sci* 2010;**42**:307–16.
26. Salvaterra E, Lecchi L, Giovanelli S, et al. Banking together: a unified model of informed consent for biobanking. *Eur Mol Biol Organ Reports* 2008;**9**(4):307–13.
27. Convention for the Protection of Human Rights and Dignity of the Human Being with regard to the Application of Biology and Medicine: Convention on human rights and biomedicine, Oviedo, 4.IV. 1997. Article 21 – Prohibition of financial gain. "The human body and its parts shall not, as such, give rise to financial gain."
28. Morrell K. Governance, Ethics and the national health service. *Public Money & Manag* January 2006;**26**(No. 1):55–62.
29. Page KM, Zhang L, Mendizabal A, et al. The cord blood apgar: a novel scoring system to optimize selection of banked cord blood grafts for transplantation (CME). *Transfusion* 2012;**52**(2):272–83.
30. Allan D, Petraszko T, Elmoazzen H, et al. A review of factors influencing the banking of collected. *Umbilical Cord Blood Units, Stem Cells Int* 2013;**2013**:1–7.
31. Lauber S, Latta M, Kluter H, et al. The Mannheim cord blood bank: experiences and perspectives for the future. *Transfus Med Hemother* 2010;**37**:90–7.
32. Sirchia G, Rebulla P, Tibaldi S, et al. Cost of umbilical cord blood units released for transplantation. *Transfusion* 1999;**39**: 645–50.
33. Prat Arrojo IP, Lamas Mdel C, Ponce Verdugo L, et al. Trends in cord blood banking. *Blood Transfus* 2012;**10**:95–100.
34. Rodrigues CA, Sanz G, Brunstein CG, et al. Analysis of risk factors for outcomes after unrelated cord blood transplantation in adult lymphoid malignancies: a study by the Eurocord-Netcord and lymphoma working party of the European group for blood and marrow transplantation. *J Clin Oncol* 2009;(27):256–263.
35. Peffault de Latour R, Purtill D, Ruggeri A, et al. Influence of nucleated cell dose on overall survival of unrelated cord blood transplant for patients with severe acquired aplastic anemia: a study by Eurocord and the aplastic anemia working party of the EBMT. *Biol Blood Marrow Transpl* 2010;**17**:78–85.
36. Querol S. Cord blood banking: current status. *Hematology* 2012; **17**(1):187–8.
37. Rouard H, Birebent B, Vaquer G, et al. Cord blood banking: from theory to an application. *Transfus Clin Biol* 2013; **20**(2):95–8.
38. Barker J, Byam C, Kernan N, et al. Availability of cord blood extends allogeneic hematopoietic stem cell transplant access to racial and ethnic minorities. *Biol Blood Marrow Transpl* 2010;**16**:1541–8.
39. Magalon J, Billard-Daufresne LM, Gilbertas C, et al. Assessing the HLA diversity of cord blood units collected from a birth clinic caring for pregnant women in an ethnically diverse metropolitan area. *Transfusion* 2014;**54**(4):1046–54.
40. Badowski M, Harris D. Collection, processing, and banking of umbilical cord blood stem cells for transplantation and regenerative medicine. [Chapter 16]. In: Singh SR, editor. *Somatic stem cells: methods and protocols. Methode in Molecular Biology*, vol. 878. 2012 Springer online; p. 279–90.
41. Butler M, Menitove J. Umbilical cord blood banking: an update. *J Assist Reprod Genet* 2011;**28**:669–76.
42. http://www.aabb.org/sa/facilities/bbts/Pages/default.aspx [accessed 14.08.13].
43. http://www.factwebsite.org/cbstandards/ [accessed 02.09.2013].
44. http://www.factwebsite.org/CordSearch.aspx?&type=CordBloodBank&country=All&state=Select+State+or+Province [accessed 14.08.13].

45. http://abcnews.go.com/WNT/video/big-business-banking-cord-blood-10579005, [accessed 09.07.13].
46. http://www.ama-assn.org/ama/pub/physician-resources/medical-ethics/code-medical-ethics/opinion2165.page [accessed 13.08.13].
47. http://www.youtube.com/watch?v=knoKQsCKbp8 [accessed 15.08.13].
48. Rebulla P, Lecchi L. Towards responsible cord blood banking models. *Cell Prolif* 2011;**44**:30–4.
49. http://www.cliniqueovo.com/ovo-clinique/index.asp?page=historique [accessed 09.07.13].
50. Waldby C, Mitchell R. *Tissue economies: blood, organs and cell lines in late capitalism.* London: Duke University Press; 2006. p. 125.
51. Thornley I, Eapen M, Sung L, et al. Private cord blood banking: experience and views of pediatric hematopoietic cell transplantation physicians. *Pediatrics* 2009;**123**(3):1001–7.
52. Rosenthal J, Woolfrey A, Pawlowska A, et al. Hematopoietic cell transplantation without cord blood patients with severe aplastic anemia : an opportunity to revisit the controversy regarding cord blood banking for private use. *Pediatr Blood Cancer* 2011;**56**:1009–12.
53. Gluckman E, Ruggeri A, Rocha V, et al. Family-directed umbilical cord blood banking. *Haematologica* 2011;**96**(11):1700–7.
54. Nietfeld J, Pasquini M, Logan B, et al. Lifetime probabilities of hematopoietic stem cell transplantation in the US. *Biol Blood Marrow Transplant* 2008;**14**:316–22.
55. http://parentsguidecordblood.org/odds.php [accessed 15.08.13].
56. http://parentsguidecordblood.org/about.php [accessed 21.08.13].
57. Ilic D, Miere C, Lazic E. Umbilical cord blood stem cells: clinical trials in non-hematological disorders. *Br Med Bull* 2012:1–15.
58. Cairo MS, Rocha V, Gluckman E, et al. Alternative allogenic donor sources for transplantation for childhood diseases: unrelated cord blood and haploidentical family donors. *Biol Blood Marrow Transpl* 2008;**14**(Suppl. 1):44–53.
59. Ballen KK, Barker JN, Stewart SK, et al. American Society of Blood and Marrow Transplantation. Collection and preservation of cord blood for personal use. *Biol Blood Marrow Transplant* 2008;**11**:356–63.
60. Cryo-Save Group N.V., Annual report 2012, p. 4 (http://www.cryo-save.com/cms/bib/files/2893_cryosavear12final3413.pdf), [accessed 11.06.13].
61. NetCord inventory and usage March 2008, www.netcord.org/inventory.html, 2008.
62. Smith FS. Why do parents engage in private cord blood banking: Fear, realistic hope or a sense of control? *Pediatr Blood Cancer* 2011;**56**(7):1003–4.
63. http://abcnews.go.com/WNT/video/banking-cord-blood-10567759 [accessed 01.09.13].
64. www.babybonus.gov.sg [accessed 28.08.13].
65. http://www.cordlife.com/sg/en/child-development-account, [accessed 17.06.13].
66. http://www.cordblood.com/best-cord-blood-bank/best-stem-cell-bank [accessed 26.08.13].
67. http://www.cryo-save.com/group/news__reports.html [accessed 26.08.13].
68. http://www.vita34.de/en/about-vita-34/ [accessed 26.08.13].
69. http://www.viacord.com/cord-banking/viacords-treatment-experience/ [accessed 26.08.13].
70. World Marrow Donor Association, annual report 2012.
71. Kaimal AJ, Smith CC, Laros Jr RK, et al. *Cost-effectiveness private umbilical cord blood Bank* 2009;**114**(4):848–55.
72. Owens DK. Interpretation of cost-effectiveness analyses. *J Gen Intern Med* 1998;**13**:716–7.
73. Gold MR, Siegel JE, Russell LB, et al. *Cost-effectiveness in health and medicine.* New York (NY): Oxford University Press; 1996.
74. Donohue JM, Cevasco M, Rosenthal MB. A decade of direct-to-consumer adverstising of prescription drugs. *N Eng J Med* 2007;**357**:673–81.
75. Fox NS, Stevens C, Ciubotariu R, et al. Umbilical cord blood collection: do patients really understand? *J Perinat Med* 2007;**35**:314–21.
76. Nietfeld JJ. Opinion regarding cord blood use need an update. *Nat Rev Cancer* 2008;**8**:823.
77. Kahneman D, Tversky A. Prospect theory: an analysis of decision under risk. *Econometrica* 1979:263–91.
78. Hertwig R, Barron G, Weber EU, et al. Decisions from experience and the effect of rare events in risky choice. *Psychol Sci* 2004;**15**:534–9.
79. Waldby C, Mitchell R. *Tissue economies: blood, organs and cell lines in late capitalism.* London: Duke University Press; 2006.
80. Brown N, Machin L, McLeod D. Immunitary bioeconomy: the economization of life in the international cord blood market. *Soc Sci Med* 2011;**72**(7):1115–22.
81. Esposito R. *Bios: Biopolitics and philosophy.* Minnesota, MN: University of Minnesota Press; 2008.
82. http://www.france-adot.org/chiffres-cles-don-organe/ [accessed 01.09.13].
83. Kwek TK, Lew TW, Tan HL, et al. The transplantable organ shortage in Singapore: has implementation of presumed consent to organ donation made a difference? *Ann Acad Med Singap* April 2009;**38**(4):346–8.
84. http://www.un.org/en/documents/udhr/ [accessed 01.09.13].
85. Titmuss R. *The gift relationship: from human blood to social policy.* London: LSE Books; 1970.
86. http://www.lifecellfemme.com/ [accessed 01.09.13].
87. http://www.biolifecellbank.com/ [accessed 01.09.13].
88. http://www.store-a-tooth.com/ [accessed 01.09.13].
89. http://www.sceil.com/ [accessed 02.09.13].
90. Ledford H. Stem-cell scientists grapple with clinics. *Nature* 2011;**474**.
91. McKenna DH, Brunstein CG. Umbilical cord blood: current status and future directions. *Vox Sang* 2011;**100**:150–62.
92. Okita K, Ichisaka T, Yamanaka S. Generation of germline-competent induced pluripotent stem cells. *Nature* 2007;**448**:313–37.
93. Wernig M, Meissner A, Foreman R, et al. In vitro reprogramming of fibroblasts into a pluripotent ES-cell-like state. *Nature* 2007;**448**:318–24.
94. Yamanaka S. *Keynote speech, conference on tissue regeneration.* Paris: Académie des sciences; November 11, 2012.
95. Blanpain C, Daley GQ, Hochedlinger K, et al. Stem cells assessed. *Nat Rev Mol Cell Biol* 2012;**13**(7):471–6.
96. Giorgetti A, Montserrat N, Aasen T, et al. Generation of induced pluripotent stem cells from human cord blood using OCT4 and SOX2. *Cell Stem Cell* 2009;**5**:353–7.
97. Neitfeld JJ, Sugarman J, Litton J-E. The Bio-PIN: a concept to improve biobanking. *Nat Rev* 2011;**11**:303–8.
98. Rao M, Ahrlund-Richter L, Kaufman D. Concise review: cord blood banking, transplantation and induced pluripotent stem cell: success and opportunities. *Stem Cells* 2012;**30**:55–60.

Chapter 25

# Public Health Policies in European Union: An Innovation Strategy—Horizon 2020

**Sotiris Soulis[1], Marcos Sarris[2], George Pierrakos[3], Aspasia Goula[4], George Koutitsas[5] and Vasiliki Gkioka[6]**

*[1]Health Economics and Social Protection, Technological Educational Institute of Athens, Greece; [2]Health and Sociology and Quality of Life, Technological Educational Institute of Athens, Greece; [3]Primary Health Management, Technological Educational Institute of Athens, Greece; [4]Organizational Culture in Health Services, Technological Educational Institute of Athens, Greece; [5]Process Analysis and Strategy Implementation Expert, National Insurance, Athens, Greece; [6]Evaluation Expert, Hellenic Transplant Organization, Athens, Greece*

## Chapter Outline



**Cord Blood Stem Cells Medicine. http://dx.doi.org/10.1016/B978-0-12-407785-0.00025-6**

## 1. INTRODUCTION

The human condition has been the focus of philosophical thought and quest throughout the history of societies. Nowadays, the significance that is attributed not only in maintaining the health of the individual, but primarily in promoting it as well as improving the quality of life of the modern European citizen, can and should be a fundamental feature of the new European social and cultural reality. With this goal in mind, we can sustain within the fabric of the European Union (EU) the eternal individual and social rights of continuous improvement of the quality of life and the state of health of the population while, at the same time, being aware of the risks as they escalate due to the globalization of economies and societies.

Since 1958, there has been a 56-year-old continuous effort to establish economic cooperation within Europe and transform the European Economic Community from an economic to a political union. The current geopolitical formation of the EU is a powerful factor on a global scale:

- 28 countries, 509 million people representing 7% of the planet
- The third largest population and the world's top economic power
- 25% of global GDP (US$ 16 trillion), 20% of world trade
- 16.5% of world imports (the United States 15.5%, China 12.0%) and 15.5% of world exports (the United States 13.5%, China 10.5%)
- Life expectancy of over 80 years, 24 years more than countries with low human development index (80 vs 56 years)
- Of the first 25 positions of the human development index, 14 belong to the EU in 2010.

However, the big dilemma inherent in individual and collective European thought is the one between economic optimization and real human development where the latter embodies in a single framework and model both in economic growth and social welfare. The battle between fiscal discipline and public social-health interest in recent years is uneven under the pressure of competition and the prevalence of markets.[1]

This dominant view does not take into account the negative fiscal conditions that may be brought about by an epidemic or pandemic shock, such as the reappearance of the Ebola virus. The loss of life and disruption of economic relations can create a larger fiscal gap than the short-term savings in public spending that may result, for example, from denying coverage to the uninsured and illegal immigrants.

By extension, the theoretical field of health policies should therefore be enriched with the economic theory of externalities, so as to clarify how to resolve the major dilemmas we face when trying to secure minimum social rights.

The fears and serious concerns arising from modern public health risks as well as the strong external impact from transmitted diseases that cause endemics and pandemics make the need to secure and protect health on an international level a top priority. These risks relate to:

- transmitted diseases
- the spread of addictive substances
- environmental pollution and ionizing radiation
- the disturbance of ecological balance and the depletion of natural resources
- industrial risks, nuclear waste, and bioterrorism
- food safety and the cultivation of genetically modified produce
- an unhealthy diet
- hygiene and safety at work,
- migration, poverty, and socioeconomic disparities
- lawlessness, modern way of life, and stressful life events.[2]

The type, the extent, and the complexity of diseases involving the entire population require the creation of a system that will meet health needs and promote health in general. This system must be based on scientific criteria and principles, must take into account developments in the sciences of medicine, biogenetics, biotechnology, economic developments and the requirements of the citizens for protection and promotion of their health.

This whole field of knowledge requires systematic research and study in order to produce innovative responses and best practices, to investigate and verify the assumptions of a structural analysis of the health system and at the same time to highlight the attitudes, perceptions, and behaviors that contribute to disease prevention combined with the protection of the environment and the support of sustainable development.

This combined approach is made possible by processing data showing on one hand the current situation, and on the other the projections of health needs that determine public health policies in the EU. These policies are supported by and are documented in the form of data which, after similar processing, allow both the comparative mapping and the differences of health needs, which help formulate distinct health policies among EU Member States.

In the introductory part of this chapter, we attempt a high-level approach to: (1) the fundamental principles of health policies; (2) the methodological assessment of health needs in conjunction with the investigation of trends in nosology and epidemiology; and (3) the classification of priorities when formulating health policies.

In the first part of the chapter, we attempt to investigate the state of health of the population in the EU with: (1) the theoretical approach to quantitative and qualitative indicators of health and well-being; (2) the analysis of the rate of change in life expectancy; (3) the quantification of the reduction in mortality and the differences in the causes of mortality; and (4) the estimation of possible years of life lost as well as the expected years of life in good health.

In the second part of the chapter, the redesign of health policies in the EU during the current period is defined through: (1) the approach of the nosology spectrum as a multidimensional and multivariate social phenomenon; (2) the assessment of the new trends in health indicators for defining health policies; (3) the formulation of the new strategy regarding health policies; and (4) the general and specific objectives and proposed actions for 2014–2020.

Finally, in the third part of the chapter, we present the specific actions of the health policies in the field of Cord Blood Stem Cells Banking through: (1) the development of an integrated policy for the sufficiency and the represented collection of cord blood units; (2) the management of the serious adverse reactions (SAR) in tissues and cells quality assurance; (3) the development of vigilance and surveillance across border networks for disease prevention and control; and (4) the promotion of public campaign programs for informing and raising awareness among the general population and among specific/target social groups.

# 2. THE CONFIGURATION FRAMEWORK OF GUIDING PRINCIPLES OF HEALTH POLICIES

## 2.1 The Fundamental Principles of Health Policies

Health is a fundamental human right, the exercise of which ensures and protects the lives of people and therefore each country through its constitution protects the health of its citizens.[3]

Health:

1. has the highest multifactor, multidimensional usefulness and the maximum contribution to Human Development because the fulfillment of needs, the psychosocial balance, and work are dependent on the state of health of individuals,
2. is affected externally and therefore is subject to the most serious cross-border threats, and
3. establishes and develops Human Capital.[1]

The lack of basic coverage and provision of health services, even for only one citizen, can cause on one hand loss of the right to life and on the other various problems to the rest of the citizens either at a national or international level. It is therefore an obvious necessity to provide health-care services, a necessity that creates the principle of universal coverage of the population in an inclusive society that makes sure it is covered.

Because health is presented in modern economic relationships of markets both as a public and as a private consumer good, access to it sometimes depends on: (1) the level of income of citizens; (2) the magnitude of the risk and the nature and the severity of the disease, i.e., the size of the resulting expense; (3) the availability of health services in some areas; (4) the nature of human mobility due to age or disability in conjunction with formal or informal social or family structures; and (5) the type of insurance coverage.

It is therefore obvious that many times, if even one of the above factors comes into play or more than one in combination, there is lack of access to health goods and services causing inequalities because it results in inability to earn income from work and hence the inability to meet basic needs and sociocultural goods, losing the right to equal opportunities. Therefore, equitable access to health services and goods is the second basic principle of health policy.

The type and severity of diseases have a serious external impact in a world of continuous mobility due to: (1) national and international trade relations; (2) tourism exchanges; and (3) migration, and as a result they cause epidemics and pandemics and affect the health of citizens through transmitted diseases.

The maximization of the social cost from transmitted diseases as opposed to the private cost, the loss of human capital in the basic economic activity which is work, the disruption of family and social life, but also the loss of national power, make the protection of public health at national and international level the third basic principle of health policy.

## 2.2 Methodological Assessment of Health Needs and Investigation of the Nosology and Epidemiology Spectrum

Health planning and policy seek to create a group of mutually interacting bodies to produce goods and services to meet the health needs of a population. The investigation of the state of health of citizens often coincides with the field of health planning and policy because it tries to determine the qualitative and quantitative characteristics of the output of these goods and services. This is achieved through the theory of opportunity cost which aims at not wasting resources in order to allocate these resources to meet other socioeconomic needs. The investigation, therefore, of the nosology and epidemiology spectrum and population and demographic data highlights and continuously reflects on the type and quantity of goods and health services that must be produced.

Many times, however, despite the scientific approach of this investigation, the market of health-care, but also the failures of planning and health system development, produce either services, such as medical services, far in excess of what is necessary, or pharmaceutical products or diagnostic centers or private clinics, creating strong induced demand

and imbalance in demand and supply of the health-care market. In any case, if we do not investigate systematically the size and age structure of the population, its regional distribution, and especially the types of diseases and causes of death, the prevalence and incidence of disease, it is not possible to specify the necessary actions of health policies on a preventive and therapeutic level.

Social planning and health policy explore, on the one hand, the type and quantity of health services and goods needed by the population and on the other, the human, material, and financial resources, i.e., the necessary inputs for the system to operate as well as the outputs and outcomes. This is indispensable to the availability of resources and to the use of innovative methods to ensure: (1) the rationalization of health services; and (2) the planning of appropriate and specialized output for the operation of the health system and the equilibrium in the market of health-care.

The establishment of the criteria of the mode of distribution of produced goods and services takes into account: (1) the strong external influence that health causes; (2) income inequality; (3) the form of the system of social security and thus the way of insurance coverage for citizens; (4) regional inequalities; and (5) the market types prevailing in an economy and the restrictions brought about by the development of private initiatives.

In all countries, even those in which the free market insurance model prevails, a minimum of state intervention exists to protect public health through the delivery of health services and goods to categories of the population who are in a state of absolute poverty and deprivation. The more health is considered a public good, the stronger is the criterion of redistributive justice when it comes to health-care distribution and care is provided irrespective of the financial contributions of the citizens either through taxes or through insurance premiums.

## 2.3 The Classification of Priorities When Formulating Health Policies

The common condition of coverage of health needs under the restriction of the scarcity of financial resources brings forth the theoretical interest regarding the classification of priorities and choices, both for the design of a system for the prevention, preservation, and promotion of health and for other fiscal intervention policies in various areas of socioeconomic activity of individuals. It is also known that these policies affect and shape the state of health of the population to a much greater extent than the health-care system itself. The actual numbers regarding health-care needs, the severity of diseases, the degree of external adverse effects, and the financial resources available determine the hierarchy of options in combination with the analysis of the factors as mentioned above.

Recent developments in the state of health of the population and in the health-care systems are characterized by:[4–6]

- continuous improvement in life expectancy
- chronic diseases which absorb 90% of total health-care expenditure
- the reappearance of transmitted diseases
- more than 70% mortality from cancer, heart and brain, and external causes
- increase in the number of elderly people
- increase in demand for health-care services
- falling rate of patient satisfaction with the quality of health services.

The health-care system enhances the health of a population, but to a much lesser extent than other socioeconomic factors—determinants of health status. Avoidable mortality is significantly influenced by lifestyle and the environment which are the key determinants of population health. Even if the entire budget was allocated to health-care, the health-care system could only determine the state of health of the population by a small proportion. Changing the consumption patterns and behaviors regarding health is the main focus of the new policy for the prevention of disease and promotion of health.

The health-care system must be sustainable in order to be able to offer services on the one hand, and on the other hand to be sufficient to meet the health-care needs of the population. The health system must be efficient and its operations must: (1) abide by the fundamental economic principle of scarcity of resources; and (2) take into account the documented finding that the health system affects the state of health of the population much less than expected. Consequently, resources can be allocated to other public policies that indirectly benefit health and quality of life for the population under an unwavering long-term perspective.

This means an increase in public spending to improve other social sectors that have a positive effect to address the underlying causative factors and key determinants of health. The transition from the perception of further spending increases that simply result in an increase in numbers of the health system to a way of thinking aiming at continuous rationalization and restructuring of the system in order to improve the quality of service, should be the modern scientific guarantee for the development of the health-care system.

Therefore, the key priorities when formulating health policies are:

1. *The prevention of disease and the improvement of life in good health*: The health-care system must prevent disease, promote health, and prolong a healthy and active life.
2. *Accessibility*: The health system should be accessible to all citizens, should offer and distribute in an equitable way health-care services and goods to ensure the basic

human rights such as the fundamental right to life, and at the same time prevent the external adverse effects such as the spread of diseases.

3. *The safety and quality of services*: The system must provide safety and quality of service; it must protect the dignity of the patient.
4. *Readiness to manage crises and health risks*: The system must be prepared to manage crises and major risks from cross-border threats through mechanisms of information and data processing and be in-sync on a cross-European and international level to protect people from the spread of transmitted diseases. The serious cross-border threats that can cause epidemics and pandemics have mobilized the international community to prevent the spread of disease and reduce social cost.

Europe and the global community have created the institutional framework for implementing health policies to combat major health risks at both preventive and interventional level. To address the prevalence and incidence of diseases, international organizations coordinate, monitor disease classification developments, and set the priorities according to the assessment of risk.

# 3. THE HEALTH STATUS OF THE POPULATION IN THE EU

## 3.1 Current Approach Using Qualitative Indices of Health and Well-being

Demographic changes such as population aging and migration affect the health sector both in terms of the need for increased spending, i.e., the financial viability of the system and in terms of the responsiveness of the system to meet health-care needs. The state of health of the population of the EU follows some general, common trends among the countries of the EU-28 (all the countries in the EU); however, some variations can be observed due to the particular socioeconomic and cultural characteristics but also the relative slowing progression of certain health indicators. This is evident when considering the comparative data of the old EU of 15 ("EU-15") members and the 13 new member countries ("EU-13").

Positive indicators approach in an integrated fashion the state of health of the population and are particularly related to the demographic indicators and can be regarded as indicators of well-being. Some of the positive indicators are the birth rate, fertility, and life expectancy. The life expectancy index can represent life expectancy without restriction, disease-free life expectancy, or life expectancy in good health (self-perceived).

The most common negative indicators are the indicators of morbidity, mortality, and disability, expressing the frequency of diseases and serving as the means of comparison and evaluation of the state of health between different countries.

The Crude Mortality Rate approximates the percentage of the population of a territory that dies during a specified period or within a year. It is affected by the age distribution of the population and does not allow for reliable comparisons of international data. For this reason, the standardized mortality rate is used.

To investigate the impact of the large number of deaths at older ages compared to the much lower number in younger age-groups, we use the standardized mortality rate, the weighted average rate per age-group. The weighting factor depends on the age distribution of the fixed reference population. A commonly used indicator is age-specific early mortality at various ages before 65 years. The standard European reference population does not include the age-groups under 5 years of age and over 95 years.

The index of potential years of life lost (PYLLs) to early mortality takes into account the age at death, for example, 65 or 75 years, and records the years that the deceased could have lived until the age X. The index is calculated per age-group among 100,000 inhabitants. The importance of the indicator rests in the way by which the health-care system affects the health of the population or in the determinants that could affect mortality to a lesser extent. For example, according to Halley des Fontaines (2007),[7] prevention policy can reduce the consumption of tobacco and alcohol and reduce accidents, suicides, and AIDS at an early age, while early diagnosis within a reliable health-care system can significantly reduce mortality from circulatory diseases and cancer.

The concept of years of life lost could be extended to those years where a person cannot engage in creative activities, is going through periods of severe symptoms of chronic disease and disability and generally perceives his or her health to be less than good or less than very good. PYLLs are the years that a person could have lived up to a certain age if he or she had not died, including the years lived with disability or chronic illness, i.e., the concept of PYLLs enables the mathematical calculation of the probability for a person to live longer and in good health.

PYLLs and disability-adjusted life years (DALYs) reflect the mortality and morbidity that could have been avoided.[8] To this end, we study the total avoidable mortality. Total avoidable mortality is defined as that part of mortality that could have been curable and could have been avoided had science, technology, and innovation been effectively applied, which means making effective use of all the available human, economic, and material resources to prevent the incident. Total avoidable mortality consists of preventable and treatable avoidable mortality.

We use PYLLs to explain early mortality due to the increase at younger ages of mortality from malignant neoplasms and circulatory diseases, while we use the DALYs to explain mortality due to causes such as malignant neoplasms, circulatory diseases, external causes, and mental or behavioral disorders.

The PYLLs are directly linked to early mortality, while the DALYs are linked with morbidity. For example, years of life lost per 100,000 inhabitants from malignant neoplasms amount to 918, while the years of life adjusted for disability resulting from malignant neoplasms per 100,000 inhabitants amount to 210.[9,10] In any case, these two indicators reflect the years of active life lost either due to death or disability.

The health-care system and health policies (prevention, preservation, and health promotion) should aim at reducing early mortality, reducing the PYLLs, reducing the avoidable preventable and treatable mortality, and reducing life years with disability, with chronic diseases, and without normal creative activity.

Modern quantitative approaches and qualitative goals such as the increase in life without disability, without problems in daily activities, without chronic diseases, with good social and family life examine the need to identify, besides the classical documentation of the frequency of diseases—the prevalence and incidence of disease—and the number of deaths, new indicators such as:

- life expectancy without disability,
- without chronic diseases,
- early mortality before a certain age,
- life expectancy at age over 65 years,
- and so on.

## 3.2 Rate of Development in Life Expectancy

The population of the EU, as calculated in accordance with the projections for 2060, will not show significant quantitative changes in the EU-28 (from 507,162,571 in 2013 to 525,845,522 in 2060) except for the very significant structural evolution by age-group and cultural composition.

The slight increase in physical mobility of the population, in births, and the fertility index shape the overall developments and highlight the phenomenon of aging of the population. It is estimated that according to recent projections, the number of Europeans aged 65 years and over will almost double over the next 50 years, from 86.7 million (17%) in 2010 to 149 million (29%) in 2060. Also in 2010 in the EU-27, people aged over 80 years amounted to 4.3%. In 2020, this figure is predicted to rise to 5.7% and in 2060 to 12.1%.[11]

It is easily understood that the fifth age-group, those over 80 years of age, will be the main concern of the health system in terms of coverage of multiple health needs. The coverage of the health needs of people aged 65 years and older and especially those over 80 years old will lead to subsequent changes in the health system which should be designed so as to address more specialized health needs. This is the main reason why we expect health spending to increase by 3% by 2060 due to the aging population.[4]

Eurostat data focus on the increase of pensioners by 39.8% in 2060 in the countries of EU-27, from 119,265,000 in 2010 to 166,683,000 in 2060. This is a historic challenge for health economists because the combination of: (1) multiple needs; (2) available free time; and (3) demands for elongation of healthy and fulfilling life for older people will send health expenditures skyrocketing.

Europe shows continuous improvement in life expectancy, which is confirmed by the growth rate of the index in the EU-15 for both sexes in the last four decades (1970–2010) with an average per decade of 2.39 years. Especially in the decade 2000–2010, the growth rate was the highest with 2.62 years in the EU-15 and 2.71 years in the EU-28. However, it should be noted that the rate of increase in life expectancy shows a tendency to slow down if we look at all countries (EU-28) because from 2.71 years in 2000 it has decreased to 2.18 in 2010.

## APPENDIX 1: EVOLUTION RATE (%) OF LIFE EXPECTANCY FOR BOTH SEXES PER DECADE (1970–2010)

This relative slowdown is explained by the variations presented in the rate of progression in life expectancy between countries in the EU. In the EU-28, life expectancy at birth for men in 2010 is 77.2 years and for women 83.1 years. Men live more in Cyprus with 78.48 years, while the highest life expectancy for women is noted in France and Spain with 85.4 years. Fewer years of life expectancy are recorded in Lithuania with 73.57 years for men, and Bulgaria with 77.42 years for women. The EU-13 shows the fewest years of life expectancy at 79.6 for women and 71.8 for men in 2010. On the contrary, in the EU-15, life expectancy at birth is comparatively high in 2010 with 84.0 years for women and 81.8 years for men.[10,1]

## APPENDIX 2: LIFE EXPECTANCY AT BIRTH, MALES–FEMALES (1970–2010)

Similar differences between the sexes are also observed in life expectancy at the age of 65 years where in the EU-28 in 2010 for women it was 21.2 years and for men 17.6 years. Also significant differences between the sexes can be observed in the EU-13 and EU-15. In the EU-13, life expectancy at the age of 65 years appears significantly reduced to 18.7 years for women and 14.7 years for men, whereas in the EU-15 it is has increased to 21.9 years for women and 18.4 for men.

## APPENDIX 3: LIFE EXPECTANCY AT THE AGE OF 65YEARS, MALES–FEMALES (1980–2010)

### 3.3 Reduction in Mortality and Difference in the Causes of Mortality

The socioeconomic and cultural changes that have occurred with the entry of new Member States to the EU since 2004 and the overall economic growth observed over the past decades, significantly affected the reduction of mortality rates and causes of variation in mortality of the European population.

The trend of mortality in Europe from 1970 onward shows continued reduction (Appendix 4) when of course mortality is weighted by age. The crude mortality rate can be increased, but this is due to the accumulation of most of the morbid cases aged over 65 years.

## APPENDIX 4: AGE-STANDARDIZED MORTALITY RATE (SDR-ICD-10-DISEASES OF ALL CAUSES, ALL AGES), PER 100,000 INHABITANTS

The weighted mortality rate in 2010 for both sexes in the EU-28 decreased to 595 deaths from 965 deaths in 1980 per 100,000 people. However, there is a marked difference between countries in the EU-15 and the EU-13 with 530 and 835 deaths per 100,000 persons, respectively. The lowest index is noted in Spain (478 deaths/100,000 people).[10]

In the comparative mortality between the sexes, men have a higher mortality rate from all causes. Also analysis by cause of mortality reflects the same variations. Mortality by age between the age-groups of 0–64 years and 65 and older has clear negative characteristics for people over 65 years even in the case of suicide, self-inflicted injuries, and traffic accidents.

In the EU-28 in 2010, the mortality rate for men was 762 whereas for women it was just 462 per 100,000. Italy has the lowest index in men with 598 deaths, and Cyprus in women with 361 per 100,000. The index value of 1118 deaths per 100,000 men in the countries of the EU-13 is considered very high.[10]

Cause-specific mortality follows the same trend since the middle of the interwar period. The causes have now shifted from diseases which are transmitted, infectious-parasitic such as malaria, typhoid fever, tuberculosis, pneumonia, etc., to diseases of the modern way of life, social isolation, disruption of social relations, and the environment, such as circulatory diseases, cancer, and mental disorders. The nosological profile of Europeans with 8% deaths from transmitted diseases compared to 80% of deaths in less-developed countries portrays the relationship between the state of health of a population and the socioeconomic and cultural context within which the people live and work.

In the EU-28 in 2010, 74% of Europeans died from circulatory diseases and neoplasms, while in the EU-13 this rises to 86%. If to these percentages we add deaths from external causes (injuries, poisoning) and deaths from mental and neurological disorders then the percentage of the fundamental causes of mortality in the EU-28 in 2010 increases to 85% and to 94% in the EU-13, respectively.[1,10,12–15]

## APPENDIX 5: BASIC CAUSES OF MORTALITY IN THE EU-28, EU-15, EU-13 (2010)

Diseases of the circulatory system are regarded as the leading cause of mortality with 201 cases per 100,000 in the EU-15 and 417 per 100,000 in the EU-13 in 2010. This percentage is equivalent to the one the EU-15 had back in the 1980s. This means that we need to intensify the policy of prevention for cardiovascular diseases in the new countries of the EU.

Malignancies showed a slight decrease since 1980 in the EU-28 (200–160/100,000) while in the EU-13, a small increase was seen (189–193/100,000), with men having a very high rate with 269 deaths in 100,000 people.

Deaths from external causes (injuries, poisoning) have shown a clear decline in the last three decades with the countries of the EU-13 recording almost twice the rate of the corresponding rate for EU-15 (55–30), and men recording four times the rate for women (90–23) in the EU-13 and more than double in the EU-15 (44–18). These large disparities between men and women are due to work conditions and the participation of men in manual and technological activities.

For suicides and self-inflicted injuries, despite the reduction achieved mainly for women, the rate for the EU-13 remains quite high at 15 deaths/100,000, which corresponds to the rate for the EU-15 in the 1980s. It should also be noted that the rate for men in the EU-13 is five times that of women (26–5).

Diabetes registers a slight decrease for the EU-15 (16–12) from 1980 to 2010 and stabilization in the countries of EU-13 (12–12).

## APPENDIX 6: CAUSES OF DEATH—STANDARDIZED DEATH RATE PER 100,000 INHABITANTS IN THE EU-28, EU-15, EU-13 (1980–2010)

The cultural progress, the continuous effort for economic advancement, consumerism, the prevalence of poorly imitated social standards, the architectural remodeling of cities, and the increase in individualistic relations and solitary-living individuals, are reflected in the increase in mental illness with the EU-15 exhibiting double the rate of that for the EU-13 (35–15) and double the rate since 1980 (18–35).[10,1]

Infant and maternal mortality in 2010 with four infant deaths in 1000 live births and six cases of maternal deaths in 100,000 live births are considered of lesser importance in the modern European nosological spectrum due to advancements in pediatrics, in the respective health services, and due to income growth and education of European citizens. A serious effort is still required, however, to reduce these indicators in countries like Romania with the highest infant mortality rate (9.79/1000 births) or Latvia with the highest maternal mortality (26.02/100,000 live births).[10,1]

The general evaluative assessment of the state of health of the population is considered satisfactory, showing a trend toward a significantly halting pace of diseases such as ischemic heart disease or mortality from external causes such as traffic accidents or accidents at the workplace. Cancers and vascular diseases of the brain remain at high levels with very low rates of decline.

## 3.4 PYLLs and Life Expectancy in Good Health

In the EU-28 over the last 30 years, 2.9 years have been gained for men before the age of 65 and 1.9 years for women hoping to avoid a deadly event, i.e., how many lost years of life a person could gain if he or she had not died (early mortality). Nevertheless, men still have a greater risk with 5.4 as opposed to 3.1 years for women.[10,1]

## APPENDIX 7: LOSS IN LIFE EXPECTANCY YEARS FOR MEN AND WOMEN FROM DEATH BEFORE THE AGE OF 65 YEARS (1980–2010)

PYLLs per 100,000 in the last 50 years (1961–2010) for men and women have fallen significantly, contributing to the increase in life expectancy. PYLLs from all causes for ages 0–69 years per 100,000 people show significant decreasing trends for both women and men during the period 1961–2010. In women, the greatest reductions are noted in Spain at 78% (from 8238 in 1961 to 1787 in 2010) and Greece at 73% (from 7576 in 1961 to 2036 in 2010). For men, the percentages of reduction are 68% in Spain and 57% in Greece. Therefore, both the falling rate of the indicator and the sharp differences between the sexes can be clearly seen.

In the United States, the corresponding rate of decline of PYLLs during the same period was 53% for women and men equally. In Japan, for women it was 79.9% and for men 76.5%. In 1961, the years of life lost for women was at 8004 years for the OECD countries, and in 2010 only 2457 years, respectively. The years of life lost for men was 12,210 in 1961 for the OECD countries, while in 2010 the corresponding figures stood at 4798.[12,13]

## APPENDIX 8: PYLLS FROM ALL CAUSES WITH COMPARATIVE REFERENCE TO LIFE EXPECTANCY (100,000 MEN–WOMEN AGED 0–69), (1961, 2010)

Health-adjusted life expectancy (HALE) is a measurement developed by the World Health Organization that attempts to capture a more complete estimate of health than standard life expectancy rates (note that HALE was previously referred to as disability-adjusted life expectancy or DALE). HALE estimates the number of healthy years an individual is expected to live at birth by subtracting the years of ill health—weighted according to severity—from overall life expectancy. HALE is also calculated at age 65 to provide a measurement of the quality of life of seniors. By moving beyond mortality data, HALE is meant to measure not just how long people live, but the quality of their health through their lives.

The interest regarding changes in life expectancy centers on the number of years in good health the European citizen hopes to live. The average number of years of life for the European citizen is 80 and by the 65th year, a basic period of life and work comes to an end and he or she usually retires. It is argued that the age of 65 years is a milestone and a comparative point of reference, a benchmark so to speak, as to how many years in good health the European citizen could live beyond this age limit or what he or she would not want to experience in terms of morbidity or disability.

In the EU-28 at age 0, men expect to live in good health 61.3 years and women 61.9 years, a difference of only 0.6 years. At age 65, the difference is even smaller (0.1) with life expectancy for men at 8.4 and women at 8.5 years.

The highest at age 0 is observed in Malta (72.4 for women and 71.8 for men) and the lowest in Slovakia and Estonia (53.1 equally for women and men). Similarly, at age 65 life expectance is the highest in Sweden (15.4 for women and 14 for men) and the lowest in Slovakia (3.1 for women and 3.5 for men).[16]

## APPENDIX 9: HEALTH-ADJUSTED LIFE EXPECTANCY

# 4. REDESIGNING HEALTH POLICIES IN EUROPE

## 4.1 New Trends in Health Indicators in the Multidimensional Context of Changing Health Needs

The new trends observed in health indicators, namely the decline in the rate of growth of life expectancy and the differences in the causes of mortality, the new population data with the rapid growth of people aged over 65 years, and the strong migration flows force us to further investigate phenomena such as the reappearance of tuberculosis, syphilis, and hepatitis which also represents a significant cross-border threat because of the heavy external impact.

These new trends in health indicators define the current health policies and lead to the necessity of adapting the health system so as to meet the changing health needs of the population. These trends are identified by the persistent findings observed in the study of four categories of indicators, namely the state of health, the determinants of health, the efficiency–effectiveness of the health system, demographic trends and changes in the composition of the population, and socioeconomic development.

### *4.1.1 State of Health of the Population*

The increase in people aged over 65 years and the new way of life contribute to the increase in chronic diseases and consequently to an increase in health-care spending. Thus, the systematic recording of morbidity and mortality of chronic diseases (asthma, arthritis, migraine, chronic bronchitis, musculoskeletal spine problems, diabetes, stomach ulcer, cerebrovascular problems after injury, Alzheimer's disease, etc.) is a priority for health indicators. Also the increase in mental disorders makes the need for the documentation of relevant indicators the highest priority.

### *4.1.2 Sociocultural Determinants of Health*

The growth in cancers, cardiovascular and cerebral diseases, and chronic diseases has given rise to a systematic investigative field for the explanation of a population's state of health, placing a greater emphasis on noniatrogenic health determinants such as:

1. the correlation between free time and systematic exercise,
2. the enrichment of the diet with fruit and vegetable consumption, and
3. the degree of autonomy of individuals in their teens or older within the local community, which has proved to be a strong factor in physical and mental health.

We can also investigate determinants such as the correlation between the absence of sleep disorders and good health, the degree to which people feel healthy in general, or the correlation between the degree of introversion and social and professional advancement on the one hand, and physical or mental disorders on the other.

### *4.1.3 Effectiveness of Health Services*

It has been found that the way by which we cover health needs, the speed of the offered services, the specialized scientific skills, and organizational-administrative innovation in the health system receives considerable contribution from health education, from primary care all the way to tertiary therapeutic intervention aiming at preventing the disease or at achieving full therapeutic rehabilitation. It is therefore reasonable to investigate whether the health system could contribute: (1) to a decline in mortality and reduction of years of life lost at a late age (65 or 75 years), i.e., to reduce early mortality; (2) to rehabilitation without complications; and (3) to safety and patient satisfaction. Thus, potentially avoidable mortality has been introduced in the statistics of health indicators, waiting time for specialized surgery, as for example in the case of hip fracture, and the rate of readmissions within 30 days of discharge for a particular diagnostic category.

Mortality before the age, say, of 65 or 75 years could have been avoided if, thanks to primary prevention and care, the disease had never occurred or if through secondary and tertiary treatment the disease had been dealt with effectively (potentially avoidable mortality). The indicator of PYLLs is now considered a key indicator used to assess the degree of influence of the health system on the population's state of health.

### *4.1.4 Efficiency of the Health-care System*

So far the study of the classic health indicators such as the aging index, addiction, unemployment rate, per capita income, square meters per person, public health expenditure per capita, ratio of human, or technological resources has enabled assessment of the adequacy of resources to meet the health needs of the population. Now, however, within the context of curtailing expenditure, the need emerges for investigating the impact of a rational use of health services and especially the flexible forms such as one-day surgeries and to this end, corresponding indicators are being developed. The purpose of these indicators is to highlight the degree of utilization of resources and services, examining the admissions in general or for specific cases comparatively and proportionately to the cases meeting the demand for services in other areas.

### *4.1.5 Demand and Supply of Health Services*

The new trend in the investigation of how supply meets demand in health services is in two directions. The first has to do with inequalities depending on the place of residence in combination with income inequality in an area of reference where a strong correlation between the two variables has been observed. The second relates to the disparities in the use of services or in the results of a therapeutic process. In other words, an attempt is made to investigate if the health system treats all citizens the same way, whether they have a very low or very high income, and to measure the relative variations.

The first case of inequalities can be attributed to behavioral and quality of life factors that depend on the socioeconomic level (unemployment, housing conditions, level of education, crime, alcohol, tobacco, food or work conditions affected by the social and professional status of the inhabitants). In the second case, the disparities should be attributed to the allocation of health-care resources that affect health service use (high costs, inability to participate in the costs, waiting time for a medical operation, availability of expert medical opinion doctors with specific experience and expertise, new technologies in micro-robotics or biotechnology, specialized staff for home health-care, complete and fast diagnostic procedures with subsequent effective treatment, etc.).

## APPENDIX 10: THE MULTIDIMENSIONAL FIELD OF POPULATION-APPRECIATED HEALTH NEEDS

### 4.2 The Need for Redesigning Health Policies in the Context of Changes in the Multifactor Nosological Spectrum

In order for a proposal that pinpoints the necessity for redesigning the guidelines of EU health policies to be considered sufficient, it must demonstrate a high degree of scientific credibility and be based on thorough and informed analysis. The initial proposal must move within the limits and the general guidelines of the European policy for health and social care and take into account the changes that have occurred in the macroeconomic, legal, and organizational-administrative level of the health systems. Then it must integrate within the framework of analysis all developments in technology and medical science, to investigate and determine the resulting health needs of the population taking into account socioeconomic, cultural, and environmental factors and health determinants.

More specifically, the proposal must provide an investigative methodological framework for defining the individual objectives taking into account the accelerated aging of the population, economic recession, unemployment, population at risk of poverty, the slowdown in the rate of growth in life expectancy for some countries of the EU, health costs and their significant containment, control of pharmaceutical expenditure, the high private costs, satisfaction of consumer health in relation to the offered services, the level of service quality, and the difficulties in implementing the restructuring of health services.

The proposal should also analyze the correlation between environmental and climatic factors in shaping the state of health of the population and consider in a systematic way the causative factor "working conditions" as a basic cause for shaping the state of the population's health and as a variable affecting direct and indirect costs of workplace accidents on GDP, on the system of social security, and on the insured themselves.

This approach, although it highlights goal setting in health-care policies, can be further enhanced by systematic analysis of:

- the epidemiological and nosological spectrum of supply of health goods and services,
- health spending relative to the amount and distribution, and
- policies for promoting biomedical hightech or sociocultural determinants of health,

so as to better specify the objectives relating to policy on cancer, cardiocerebral accidents, transmitted diseases, tobacco policy, and alcohol or dietary beliefs.

The analysis at the micro-level may further specify the proposed measures and prescribe actions for:

- the digital modernization of the system and the introduction of e-health,
- the improvement of skills and competencies of human resources,
- the development of modern procurement systems, analytical cost accounting, compensation methods, methods for preparing and monitoring the budget of hospitals, and
- the introduction of internal control systems, quality management systems, one-day surgery, home health-care, distant care.

The analysis can also focus on interventions regarding a particularly critical field, that of the management of hospital waste and nuclear contaminants.

The redesign of the guiding health policies defines in a rational and documented manner and takes into account the wider European social protection policies. Specifically, the EU has pledged to modernize the social protection systems and improve its social model in the light of shared values of social justice and the active participation of citizens in economic and social life. For this purpose and in order to develop integrated policies and coordinated monitoring methods, the open method of coordination (OMC) was adopted.

In 2005 three areas of coordination were defined: (1) social integration; (2) adequate and sustainable pensions; and (3) high-quality sustainable health-care and long-term care, which merged into a single social OMC.[17] In 2008, the EU further enhanced the OMC as a fundamental tool of social development and revised the term to "Social OMC."

Member States have defined common objectives emanating from the Lisbon Strategy as well as monitoring indicators and analytical tools through the Progress Program to improve the capabilities of comparative analysis, knowledge transfer, mutual learning, better coordination, and strategic planning. The directions of the OMC for social protection and the basic indicators of an ongoing pan-European research should be expanded and revised regularly as for example, the indicators of material deprivation (digital exclusion), of affluent households, life expectancy, and the transition from work to retirement.[18]

The directions of the OMC as an intergovernmental cooperation method on accessible, quality and sustainable health-care, and long-term care, focus on the following:

- Access for all to adequate health-care and long-term care, ensuring that the need for care does not lead to poverty and financial dependency while at the same time addressing inequalities related to access to care and the results of health services.
- Quality in health-care and long-term health care and adjustment, including preventive care, to the changing needs and preferences of society and individuals, primarily through the development of quality standards that meet international best practices and through the development of a higher sense of responsibility among health professionals and patients and recipients of care.
- Sustainability of adequate and quality health-care and long-term care through the rational use of resources and mainly through appropriate incentives for users and providers, good governance and coordination between care systems and public and private entities. Long-term sustainability and quality also require the promotion of a healthy and active lifestyle as well as the existence of good human resources for the health-care sector.

## 4.3 The Strategic Context of Health Policies for the Period 2014–2020

### *4.3.1 From the Innovation of the Health System to the Promotion of Health and Improvement of the Quality of Life in Good Health*

The fundamental principle of health policies lies in the documented argument that the principal objective of health policies is not solely to improve a health system but primarily to improve the state of health of the population.[1] The success of the guidelines of health policies is an essential prerequisite for the control of health-care spending, for the improvement of satisfaction of health services consumers, and for greater work output due to an increase in productivity from the application of innovation and planning at all operating levels of the health system.

The strategic objectives must specify the design of a strategic framework that centers on ensuring the transition from health system innovation to the promotion of health. The focus on applying best practices in innovation as far as the organization and management of the health system is concerned ensures sustainability, access, safety, and quality in the provision of health-care services, essential components for the prevention, maintenance, and promotion of the population's health.

The best practices in innovation refer to the methods and techniques of human, financial, and material resources of the health system, the use of technology, the organizational and structural models of health-care services as well as their funding. Innovative e-health applications have a high added value because they have a wide range of applications from complete and personalized care to independent living and distant care.

The transition from innovation in the health system to the promotion of health is a prerequisite for ensuring the long-term supply of health services on an equal basis and with equal distribution of all produced results. However, the dilemma of attaining maximalist objectives—the effective development of the health system—at the expense of the minimalist implementation of coverage for the socially disadvantaged, particularly in times of crisis, is of particular interest and guides the analysis of health policies with regards to the allocation of available resources.

Within the context of the fundamental principles of civil and social rights and health policies the key objectives of the EU regarding health for the period 2014–2020 emerge:[4]

1. Prolonging healthy life by two (2) years,
2. Healthy and active aging (preventing early retirement),
3. Reducing absenteeism from work for health reasons,
4. Reduction of the number of people who are at risk of poverty and those who are socially excluded by 4%.

Especially in the context of the changes in the nosological spectrum and the variability in the indices of morbidity and mortality, the focus is not only to further prolong life expectancy but also to improve life expectancy without chronic diseases and without disabilities and to reduce the avoidable mortality, the PYLLs before the age of 65 years (active aging).

## APPENDIX 11: MAIN OBJECTIVES OF HEALTH POLICIES: PROLONGING LIFE IN GOOD HEALTH AND WITH QUALITY YEARS

According to this approach to the redesign of health policies in conjunction with the quantitative and qualitative results of the positive and negative health indicators and the changing health needs of the population, the following are the strategic priorities of health policy in the EU for the period 2014–2020.[19,20]

### *4.3.2 Strategic Objective 1. Viability of the Health System*

The lack of resources and the need for innovation in the health system are dealt with through a series of policy measures for comprehensive EU-based long-term health-care, for active and healthy aging, for information, awareness, prevention, early diagnosis, cost containment through organizational reforms, product innovation, and the use of information and communication technologies (ICT).

### *4.3.3 Strategic Objective 2. Patient Access and Safety*

Access for all patients to medical expertise and quality of care is addressed by a number of policy measures for their safety, support for low-income patients and the socially excluded, citizens' access to health services across Europe, implementation of best practices, fighting antimicrobial resistance, high safety standards in the blood, organs, tissues, and cells for transplantation and generally reducing and eliminating disparities in health.

### *4.3.4 Strategic Objective 3. Prevention and Health Promotion*

Addressing premature mortality and low growth of healthy years of life requires the development of measures for the prevention and health education that relate, for example, to smoking, alcoholism, poor diet and obesity, lack of exercise, sexually transmitted diseases, cancer screening, and generally changing health behavior that is affected by modern lifestyle, socioeconomic, and cultural factors.

### *4.3.5 Strategic Objective 4. Protection from Cross-border Threats*

Improving preparedness for emergencies and cross-border threats such as pandemic influenza (H1N1) virus, bovine spongiform encephalopathy, sexually transmitted diseases, the Severe Acute Respiratory Syndrome (SARS) pandemic, and *Vibrio* cholera can be managed through a series of policy measures and actions concerning effective coordination, the development of crisis management systems, documented health risk assessments, early warning, monitoring and collection of reliable data and information, and finally, the development of general and specific vigilance and surveillance systems.

The focus, therefore, of the European health policies is to ensure:

1. The adequacy and viability of the health system in order to meet health needs despite of potential shortages of resources (human, material, financial) and the absence of innovative organizational, management, and operation methods.
2. The promotion of accessibility to quality and safe health services and the equitable distribution of health resources (regardless of income, social status, place of residence, ethnicity).
3. Address the critical risk factors for prevention and health promotion.
4. The protection of public health from external impact caused by the spread of diseases such as epidemics or pandemics.
5. The adequacy of efforts to inform and update the population, to create a mechanism to protect public health which is in constant readiness through a permanent cross-European body.

Consequently the up-to-date objectives of health policies are briefly:

1. Coverage of health needs, adequacy, and viability of the health system
2. Mitigation of inequalities and access to health services
3. Prevention, promotion, and elongation of a healthy life
4. Information, preparedness, coordination, and protection from cross-border threats

## 4.4 Proposed Actions for the Achievement of the Strategic Objectives of the European Health Policy

The guidelines and policy measures aggregate and enrich the proposed objectives of health policy. Also, the specialized reconfiguration of health policies leads to specific policy measures and actions in the following areas:

- Meet the needs of a rapidly evolving epidemiological spectrum due to the modern lifestyle and environmental changes
- Prevent diseases, maintain and promote health in the new conditions of the nosological spectrum
- Address the strong external impact from cross-border threats
- Develop bio-vigilance and surveillance systems and control blood derivatives
- Ensure the supply of health services for the population of the third, fourth, and fifth age
- Develop combined measures for mental health to address contemporary psychosocial–pathological phenomena as well as the mentally ill with comprehensive, friendly, and efficient mental health services
- Patient safety
- Examine accessibility and equality in health services
- Improve and expand the primary care and the interconnection between primary and secondary care
- Continuous improvement of quality of health services
- Sustainability, accessibility, effectiveness, efficiency, and preparedness of the health system
- Improve working conditions
- Protect the environment

Following the above analysis and documentation of the redesign of the broad health policies and strategic objectives, we present the proposed interventions describing the specific objectives, the problems to be addressed, and suggested actions to achieve the objectives.

### 4.4.1 Viability

*Specific objectives*: To address the lack of resources by adopting innovation at the organizational and administrative level.

*Problems to be addressed*: Excessive increase in costs, low level of organization of health-care, low level of EU care, lack of individualized and comprehensive health-care, lack of use of tools to assess health-care technology and health-care with fewer resources, shortage of health professionals (generally, specifically), low cross-European cooperation, insufficient use of ICT, lack of interconnection functionality for the electronic patient records.

*Actions and measures to achieve the objectives*: Implement innovation (in prevention, in early diagnosis, in treatment, in integrated care, active aging, and independent living), and ensure the adequacy and quality of human resources, of frugal and efficient use of medical technologies. Implement integrated management and administration systems for: (1) the economic efficiency of medical technologies; (2) decision-making processes; (3) organizational–economic management; (4) the performance of health-care units based on measurable quantitative and qualitative targets; (5) the utilization of ICT; (6) the restructuring of resources in the smaller scale within health-care units; and (7) the coordinated and interconnected operations of health services and general or specialized medical departments/clinics within each health district. Respond to increased demand due to an aging population (home health-care, distant care for the elderly and implementation of long-term health-care systems at the local and regional level).

### 4.4.2 Accessibility

*Specific objectives*: Address the inequalities of access to medical expertise and the problem of low quality and safety of health-care.

*Problems to be addressed*: Increase in poverty and the number of people with low incomes, increase in inequalities and in the number of those who are socially excluded, life in deprived areas or in regions with lack of resources, elimination of regional disparities—inadequate interconnection between service networks at the European level. Increase in antimicrobial resistance, lack of actual rights of patients to cross-border health-care, absence of uniform rules and regulations for the safety of blood, organs, tissues, cells, pharmaceuticals.

*Actions and measures to achieve the objectives*: Access to diagnosis using reliable people with sufficient medical expertise, improve the quality and safety of health-care, support for European reference networks (access regardless of where one resides), support for low-income and socially excluded people, improvement of the specifications for safety and quality in the substances of human origin such as blood and its derivatives.

### 4.4.3 Prevention and Health Promotion

*Specific objectives*: Combat diseases associated with the modern way of life, the so-called diseases of civilization, by implementing policies of health education and promotion.

*Problems to be addressed*: Insufficient planning and provision of preventive services, maintenance and promotion of health, absence of a mechanism for systematic intervention for prevention and screening of diseases affecting public health and the modern way of life (behavioral health model).

*Actions and measures to achieve the objectives*: Develop specific and targeted health education programs based on methods and techniques of experiential awareness and psychoemotional empowerment to deal with the impact of the main determinants of health (smoking, alcohol, diet, physical exercise). Also develop awareness programs to improve health and safety conditions at work, traffic education, environmental education, and environmental control regarding air pollution, subsoil and sea, recycling, low energy consumption, promoting alternative energy, sex education and the use of condoms to reduce sexually transmitted diseases, the phytosanitary control, and the protection of public health. Additionally, create a permanent observatory for the promotion of health and quality of life, develop models for long-term care and home health-care, enable the creative involvement of people of the third, fourth, and fifth age, active participation in the local community—these interventions have a positive impact for the design of prevention (not creating new health needs) and for treatment (less treatment in closed-type services).

### 4.4.4 Protection from Cross-border Threats

*Specific objectives*: To address the external impact from cross-border relations, combating epidemics, pandemics, and contamination from chemical incidents.

*Problems to be addressed*: Pandemic influenza, bird flu, SARS, bovine spongiform encephalopathy, transmitted diseases, sexually transmitted diseases.

*Actions and measures to achieve the objectives of health policy*: Readiness—early warning, coordination of services—public health centers of response to emergencies, development of a specialized information system for collecting and analyzing data. Improve capacity for risk assessment and crisis management, services for the control of immigration, development of common methods for combating cross-border threats, and dissemination and awareness regarding condom use for sexually transmitted diseases.

# 5. HEALTH POLICIES ON CORD BLOOD STEM CELLS BANKING

## 5.1 Development of an Integrated Policy for the Sufficiency and the Represented Collection of Cord Blood Units

Transplantation of hematopoietic cells is the standard therapeutic method and often the only treatment for a range of diseases characterized by malfunction of the bone marrow and the cells it produces. Nowadays, blood derived from an infant's umbilical cord is the appropriate source of hematopoietic progenitor stem cells for bone marrow transplantation.

The increase in the offered therapeutic applications using tissues and cells during the last years, the legal differences among the Member States, and the related risks of the imprudent use of these substances have led the EU lawmakers and the Member States to establish Community standards which ensure the quality and safety of human tissues and cells, and also the quality of the procedures related to donation, procurement, processing, and use of them.[21]

The use of tissues and cells for application on the human body involves the risk of transmission of infectious diseases and other possible side effects on the recipients. The EU takes under serious consideration the transmission of genetic diseases mostly originated from reproductive cells and for this reason has made adjustments to the European Directives for tissues and cells which vary from the general principles referred to in Directive 2004/23/EC for the setting of quality and safety standards for donation, procurement, control, processing, preservation, storage, and distribution of human tissues and cells to the Directives 2006/17/EC and 2006/86/EC which contain detailed technical requirements on processes of standardization and traceability, which means the list of procedures performed in order to prevent serious adverse events (SAE) and SAR.[22–24]

We should also refer to the 2011/2307 (INI) European Parliament report (A7-0223/29-6-2012) related to the voluntary and unpaid donation of tissues and cells, where for the first time it is established that Member States should implement policies for the promotion of the sufficiency of cord blood units. It also urges the States to develop programs which will encourage the general population and especially some specific groups (minorities) to donate cord blood, since it is proved that the cells can be used for the treatment of diseases like childhood diseases, in order to deal with the problem of lack of compatible donors.

In order to optimize the sufficiency of cord blood units in each Member State, the competent authorities (CA) should develop public campaigns in order to raise awareness mostly targeted to expectant mothers and their families more than other groups from the general and special groups of population. In this framework, the development of an integrated policy in the EU Member States is firstly necessary in order to ensure the sufficiency of genetically varied cord blood units concerning HLA histocompatibility antigens, and secondly to optimize the prediction and provision of predisposing factors for the health restoration of hematological patients who need hematopoietic cell transplantation.

This integrated policy is developed in accordance with the following strategies:

1. Enforcement and support of health services and of the organizations involved in promotion of cord blood units, in order to develop public awareness campaigns targeted to the general and special groups of population.
2. Coordinated health services, through the person-oriented (expectant mother) cord blood donation system and through the management of the most important health risk factors of cord blood donors in order to ensure the suitability of the donor and the good health state of the recipient.

In the context of the above strategies, the main goals are formed into two main courses of management actions:

- Networking and involvement of the organizations, health services, and units for the represented cord blood collection concerning the HLA histocompatibility antigens.
- Creation of HLA-oriented registries so as to map the population's genes.

Furthermore, "Member States take all the necessary measures to ensure that every promotion and advertising activity for the support of the human cells and tissues donation is in accordance with the guidelines or the legislative provisions in force. These guidelines or legislative provisions include appropriate constraints or prohibitions of advertising the need or the availability of human tissues and cells which aim to the offering or the seeking of financial gain or anything similar. Member States try to assure that the procurement of tissues and cells is performed on a nonprofit basis."

## 5.2 Management of SAE and SAR: Quality Assurance and Safety of Tissues and Cells

Health policies established by the EU do not deal with the legislative framework for the promotion and the implementation of strict quality regulations. The Directorate General for Health and Consumer Affairs (SANCO),[25] whose aim is to promote a healthier and safer place for European citizens, has financed a number of surveys which were initially expected to help Member States to understand and to conform with the existent legislation and later help them with the detection and dealing with potential problems.

With tools like monitoring, consultation, supporting, and implementation, the EU initially funded the 3-year program Eustite in order to promote the standardization of procedures for tissues and cells among the Member States through common control guidelines. The Eustite program involved a consortium of organizations from 10 Member States in cooperation with the WHO[26] and coordinated by the National Transplant Center in Italy. Other partners of the project were Austria, Bulgaria, Denmark, France, Ireland, Italy, Poland, Spain, Slovakia, and the United Kingdom.

The aims of the Eustite program were:

1. the promotion of good practice standards for the inspection of tissue establishments, and
2. the development of optimized systems for the reporting and management of adverse events and reactions related to quality and safety of tissues and cells implanted on patients within the EU, regardless if the tissues and cells come from inside the EU or from other countries outside the EU.

Later, the knowledge acquired by the Eustite program was transferred to the 3-year program "Soho Vigilance & Surveillance" which continued one step forward involving issues like the possible relation between the adverse events with tissues and cell donation or if the application on humans has caused events which could be evaluated and investigated. One of the most important Work Packages was responsible for a survey carried out by the Spanish Transplant Organization concerning the development of biovigilance systems for tissues and cells and also for assisted reproduction in Member States in the year 2010.

In September 2010, "Soho V&S" cooperated with WHO and with the Italian Transplant Organization (Centro Nazionale Trapianti) to take a global initiative which aimed in the enforcement of promoting substances of human origin. This initiative was called "The Notify Project."[27] Ten international groups of experts worked together through a Google Web site, where more than 100 participants (clinical doctors, health professionals, representatives of civil societies, experts) collaborated to gather documented cases of adverse reactions and events using published articles and vigilance system reports as their sources. More than 1700 published references were introduced to the database.

The cases were used as the basis for developing draft guidance on detection and confirmation of reactions and events, with an emphasis on the key role of the treating physician, and on July 03, 2013, the European Council approved the proposal of the Commission for issuing the decision concerning serious cross-border health threats. The issue of this decision is a big step forward for the safeguard of health safety in EU and the protection of citizens from a wide range of health risks.

A recently funded program has the acronym "ARTHIQS" (ART and Hsc Improvements for Quality and Safety throughout Europe) and aims to develop guidelines for key aspects of service provision and regulation in ART and HSCT.

The specific goals of the program are:

1. Institutional and organizational guidelines for the enhancement of an ART-specific expertise at official level in each Member State, leading to a more appropriate and consistent level of national organizations.
2. Set up of shared specifications regarding the main characteristics for HSC donor (related and nonrelated) follow-up registries to be implemented locally and/or nationally.
3. Minimum requirements for authorizing/reauthorizing cord blood banks (CBB) and the minimum quality and safety standards for CBB of all sorts (public/private, allogeneic/autologous), in accordance with European Union Tissue and Cell Directives (EUTCD) and existing standards.

ARTHIQS should bring solutions to some current issues related to EUTCD implementation: lack of consistent institutional competencies at CA and EU; lack of defined requirements for safety and quality both for HSC donors and CBB. Vademecum and curriculum for CBB and ART inspectors will also be prepared.

For the first time, in this project the policy followed by the CBB is separated. Since 2002, when EU started legislating for tissues and cells had never separated before the health policy of CBB from the policy followed by the Hematopoietic Stem Cell Volunteer Donor Registries. This happens because tissues and cell applications have the same target group, which is the patients suffering from hematological malignancies, despite the fact that the processing of the cord blood unit until it reaches the recipient has nothing to do with the processing performed until the graft of the volunteer donor reaches its destination.

## 5.3 Development of Bio-vigilance Systems: An Across-border Network for Disease Prevention and Control

More than 500,000 patients in Europe receive annually tissues and cells of human origin. These substances represent, on the one hand "traditional grafts" whose therapeutic interest has been recognized a long time ago (corneas, bones, skin, blood vessels, heart valves) and on the other hand, state-of-the-art biotechnologies (hematopoietic stem cells of cord blood after special processing for their application on humans). Like all activities which involve products of human origin, these techniques include the risk of adverse events or reactions which are related to the transmission of communicable diseases. The risk of infection is bigger than it is in blood transfusions, since tissues and cells can travel across borders.[28,29]

According to article 7 of European Legislation 2006/86/EC "Member States shall submit to the European Commission, through their Competent Authorities, annual reports on the notification of serious adverse reactions and events concerning the transplantations of tissues and cells including the centers of assisted reproduction." The above guidelines promote the application of strict quality requirements in order to reassure the safety of tissues and cells. Traceability and bio-vigilance are the basic tools of quality and safety of tissues and cells intended for transplantation.

Traceability refers to all relevant data relating to products and materials coming into contact with tissues and cells. In this framework, tissues and cell establishments have to implement a system which will be able to locate and identify the donor and the donated tissues and cells and to generate one unique number for every donation and for every product related to this donation in order to safeguard the quality control and the recipient's safety.[30] In this way, there is the possibility to withdraw from the availability and distribution procedure any product that may cause an SAE or SAR.

Bio-vigilance concerns the timely and valid reporting of all SAE and SAR and it works at three levels:[27,31]

1. At the European level, the European Commission supports preventive action through the early warning system for tissues and cells and also the European Center for Disease Prevention and Control has the mission to identify, assess, and communicate current and emerging threats to human health posed by infectious diseases.
2. At the national level, CA ensures that countries adhere to the requirements coming from the European Directives.
3. At the local level, tissues and cell establishments meet all the quality and safety requirements.

The field of bio-vigilance application is extended from the procurement of tissues and cells to the monitoring of patients after transplantation. The general scope of bio-vigilance involves activities like controlling, processing, transportation, preservation, import, export, and distribution of grafts. For example, the French network of bio-vigilance is running at a national level under the Agence Française de Sécurité Sanitaire des Produits de Santé (AFSSaPS) which is responsible for classification of information from local professionals. At a local level, the network consists of health centers, tissues and cell establishments, and private health units and services which are related to this field.

## 5.4 Promotion of Awareness Campaign Programs in Target Groups

Taking into consideration the current percentage of cord blood stem cells stored nowadays, it represents only 1% of the total number of births in the EU. Because it is necessary to create an integrated policy to raise public awareness, the EU stresses the importance of donating cord blood to cord banks which meet the strict requirements and also emphasizes the fact that public awareness and public opinion play a very important role on the increase of hematopoietic cell donation figures. In this framework, it is necessary to develop a specialized action plan concerning cord blood donation and specifically how to inform the public and future parents in cooperation with CA at a local community level (public health organizations, primary healthcare units, etc.).[32]

Moreover, the EU estimates that donations of allogeneic character should be promoted apart from the members of the family, regardless of whether the CBB is public or private, in a way that stored units are available to *Bone Marrow Donors Worldwide* and thus to every hematological patient who needs transplantation. For this reason it is expected that Member States will establish at least one public CBB. For the achievement of the main goal, which is the safeguard and the availability of cord blood units, each EU Member State is invited to establish a regional network of maternity hospitals which will have the authorization to receive samples in order to supply all population centers.

Civil Societies Organizations in collaboration with CA can also play a special role. Institutions like associations of volunteer blood donors or/and organ donors, organizations which promote the information, the procurement, and the distribution of cord blood and do not run only as organized and legal structures but also have the ability to intervene in local communities. After conducting a comprehensive strategic social marketing plan, these organizations can be the core through which cord blood units can be collected, and therefore a representative sample of the most frequent or rare HLA haplotypes of the population can be formed. Finally, it is understandable that in this procedure there are evaluation stages of the actions taken and also strategy redefinition when required, in order to form an applicable, gradually developing, flexible, and efficient plan of investigating the Member States' registries.

### 5.4.1 Epilogue

The successful implementation of a program for the formulation of health policies depends, to a great extent, on the adoption of a system of ongoing evaluation and feedback. This system can ensure the successful implementation and operation of the proposed program by monitoring the effects resulting from these actions, with continuous modification in the light of the experiences gained from the evaluation itself and from continuous system adjustments. Finally, it should be noted that we need to list the proposed measures in a hierarchy and clarify the priorities mainly with regards to the feasibility of their implementation under the financial constraints of each time period.

Therefore, the knowledge that supports designers of health policies in order for them to ensure the collective public interest and competitiveness of Europe is particularly complex and maximizes their scientific responsibility.

We hope that politics and science will take a common course so that political analysis will incorporate the innovative scientific knowledge and added value that is produced, if we really want to link Europe's mission in history with the vision of its citizens and the future of the global community.

## REFERENCES


1. Soulis S. In: Sarris M, editor. *Social policy and health planning*. Athens: Papazisis; 2014.
2. Sarris M. In: *Health sociology and quality of life*. Athens: Papazisis; 2001.
3. European Committee of Social Rights Activity Report 2011, Council of Europe, 2012.
4. European Commission 2012, Communication from the commission to the european parliament and the Council, taking forward the strategic implementation plan of the european innovation partnership on active and healthy ageing, COM 2012; 83 [Final, Brussels].
5. Turner K. *The impact of chronic diseases on health care system*. 2013. www.voices.yahoo.com/the-impact-chronic-diseases-healthcare-system-12288873.1. Centers for Disease Control and Prevention.
6. Dusse R, Blumel M, Scheller-Kreinsen D, Zentner A. Tackling chronic disease in europe, strategies interventions and challenges. *Observatory studies series, no 20*, WHO; 2010.
7. Halley des Fontaines V. *Introduction a la Santé Publique, Analyse des etats de santé et indicateurs de santé, 1ere annee Psychomotricite Sante Publique 2006–2007, Faculte de medecine, Pierre et Marie Curie*, 2007.
8. Institut Canadien d'information sur la Sante. *Indicateurs de Santé, ICIS 2013*, Statistique Canada, 2013.
9. OECD. *Complete data bases available via OECD'S, library, OECD stat extracts**, Organization of Economic Cooperation and Development, 2013.
10. WHO. *European health for all data base*. ECHI– European Community Health Indicators; 2013. http://www.euro.who.int/en/data-and-evidence/databases/european-health-for-all-database-hfa-db.
11. Eurostat, L'Europe en chiffres – L'Annuaire d' Eurostat, 2010.
12. OECD, Health data, 2012.
13. European Commission. *Population projections updated ECHI-5 with EUROPOP 2013 data*, Directorate General Health and Consumer, 2013.
14. Eurostat. *Statistics in focus, 72/2008, population and social conditions, Eurostat, statistics explained*. European Commission, 2008.
15. O.M.S. *Statistiques sanitaires mondiales*, 2010. Organisation Mondiale de la Sante, 2010, France.
16. Eurostat. *HEIDI data tool, data on HLY in the EU*. European Commission, 2013.
17. Commission of the European Communities. *Working together, working better: a new framework for the open coordination of social protection and inclusion policies in the European Union, 22.12.2005 COM*2005. 706 final. Commission of the European Communities, 2005, Brussels.
18. Commission of the European Communities, Sec(2008)2153, 2169, 2170, 2179, Bruxelles 8 July 2008 (11.07), (OR.fr), 11560/08, soc 414, ecofin 292.
19. European Commission. *Regulation of the European parliament and of the council on establishing a health for growth programme, the third multi-annual programme of EU action in the field of health for the period 2014–2020*, 2013. Brussels, COM(2011)709 final. European Commission, 09-11-2011.
20. World Health Organization. *The European health report 2012, charting the way to well-being*. World Health Organization, Regional Office for Europe; 2013.
21. Opinion of the European economic and social committee on the 'Proposal for a directive of the European parliament and of the council on setting standards of quality and safety for the donation, procurement, testing, processing, storage and distribution of human tissues and cells' (COM) (2002) 319final-2002/0128 (COD) (2003/C 85/14).
22. Directive 2004/23/EC of the European Parliament and of the Council of 31 March 2004 on setting standards of quality and safety for the donation, procurement, testing, processing, preservation, storage and distribution of human tissues and cells.
23. Directive 2006/17/EC of 8 February 2006 on the implementation of Directive 2004/23/EC of the European Parliament and of the Council as regards certain technical requirements for the donation, procurement and testing of human tissues and cells.
24. Directive 2006/86/EC of 24 October 2006 implementing Directive 2004/23/EC of the European Parliament and of the Council as regards traceability requirements, notification of SAR and SAE and certain technical requirements for the coding, processing, preservation, storage and distribution of human tissues and cells.
25. European Commission Directorate General of Health and Consumers Public health and risk assessments. Legislation on health and international issues, Inspection Notice Tissues and Cells-Operating Manual for competent authorities (SANCO-2010–10703),Version 1.0 (19/10/2009).
26. Second global consultation on Regulatory requirements for human cells and tissues for transplantation: towards global Harmonization through Graduated standards. WHO Geneva June 2006; 7–9.
27. NOTIFY exploring vigilance notification for organs, Tissues and Cells A Global Consultation Organised by CNT with the co-sponsorship of WHO and the participation of the EU-funded SOHO V&S Project February 7–9, 2011.
28. Guidance for Industry Eligibility Determination for Donors of Human Cells, Tissues, and Cellular and tissue-based products (HCT/Ps) U.S. Department of health and human services food and Drug administration Center for Biologics evaluation and Research, August 2007.
29. Guidance for Industry Minimally Manipulated, Unrelated allogeneic Placental/Umbilical cord blood intended for hematopoietic Reconstitution for specified Indications U.S. Department of health and human services food and drug administration center for Biologics Evaluation and Research October 2009.
30. Regulation (EC) No1394/2007of the European Parliament and of the Council of13 November 2007 on advanced therapy medicinal productsandamending Directive 2001/83/EC and Regulation(EC) No726/2004.
31. Working group on the European Coding system for tissues and cells established by the tissues and Cells Regulatory Committee 31/01-01/02/2011.
32. Report on voluntary and unpaid donation of tissue and cells (2011/2193) (INI) Committee of the environment, Public Health and Food Safety, European Parliament.

# Appendix

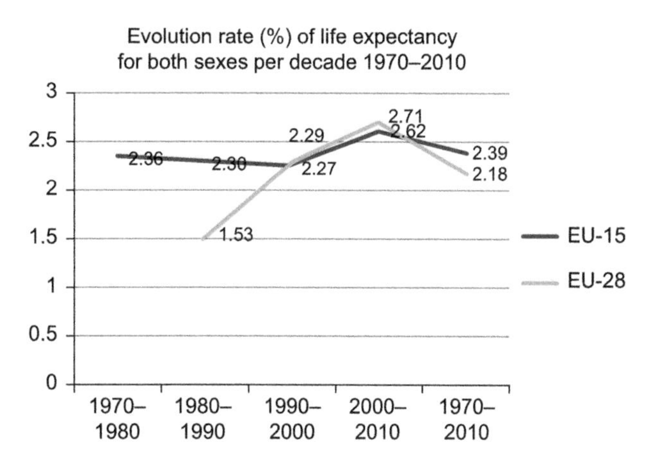


**APPENDIX 1:** Evolution rate (%) of life expectancy for both sexes per decade (1970–2010). *Source: Soulis 2014*[1]

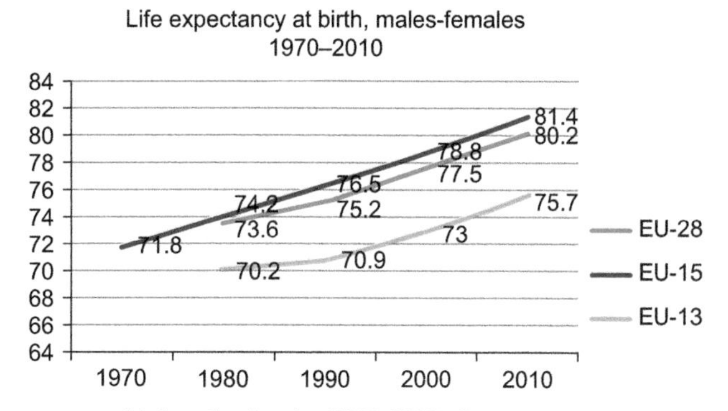


**APPENDIX 2:** Life expectancy at birth, males–females (1970–2010). *Source: WHO European Health for all data base*[10]

| Years | 1980 | | 1990 | | 2000 | | 2010 | |
|---|---|---|---|---|---|---|---|---|
| EU | males | females | males | females | males | females | males | females |
| EE-28 | 13.5 | 17.1 | 14.3 | 18.1 | 15.6 | 19.4 | 17.6 | 21.2 |
| EE-15 | 13.6 | 17.4 | 14.8 | 18.8 | 16.3 | 20.1 | 18.4 | 21.9 |
| EE-13 | 12.3 | 15.2 | 12.7 | 15.9 | 13.4 | 16.9 | 14.7 | 18.7 |

**APPENDIX 3:** Life expectancy at the age of 65 years, males–females (1980–2010). *Source: WHO European Health for all data base*[10]

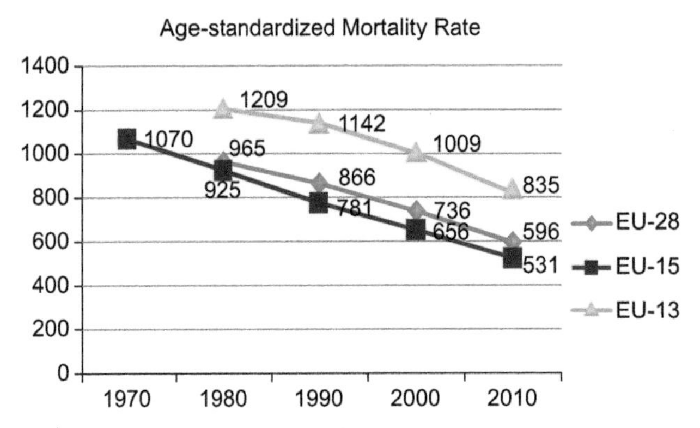


**APPENDIX 4:** Age-standardized mortality rate (SDR-ICD-10-diseases of all causes, all ages), per 100,000 inhabitants. *Source: WHO, European Health for all data base*[10], *Soulis 2014*[1]

| Fundamental Causes of Mortality in the EU-28, EU-15, EU-13 (2010) | | | | | | |
|---|---|---|---|---|---|---|
| Causes<br>Years | SDR - I00 - I99 - Diseases of circulatory system and SDR - C00 - C97 - malignant neoplasms 1 | % a+b causes/Total | + SDR - V00 - V99, W00 - W99, X00 - X99, Y00 - Y99 - Diseases of external causes of injury and poisoning 2 | % a+b+c causes/Total | + SDR - F00 - F99, G00 - G99, H00 - H95 - Diseases of Mental Disorders, nervous system and sense organs 3 | % a+b+c+d causes/Total |
| | | | | | | |
| EE - 28 | 271+167=438/595 | 74% | 438+36=474/595 | 80% | 474+31=505/595 | 85% |
| EE - 15 | 201+160=361/530 | 68% | 361+30=391/530 | 74% | 391+35=426/530 | 80% |
| EE - 13 | 526+193=719/835 | 86% | 719+55=774/835 | 93% | 774+15=789/835 | 94% |

1= a + b causes / Total, 2= a + b+c causes / Total, 3= a + b+c+d causes / Total

| a | | b | | c | | d | |
|---|---|---|---|---|---|---|---|
| | Circulatory system | | Malignant neoplasms | | External causes of injury and poisoning | | Nervous system and sense organs |

**APPENDIX 5:** Basic causes of mortality in the EU-28, EU-15, EU-13 (2010). *Source: Soulis 2014*[1]

| STANDARDIZED SPECIFIC MORTALITY RATE BY SEX<br>ALL AGES, PER 100.000<br>CAUSES OF DEATH | EU-28 (1980/2010) | | | EU-15 (1980/2010) | | | EU-13 (1980/2010) | | |
|---|---|---|---|---|---|---|---|---|---|
| | T | M | F | T | M | F | T | M | F |
| SDR- I00-I99- Diseases of all causes | 965 | 1.248 | 758 | 925 | 1.204 | 725 | 1209 | 1.531 | 697 |
| | 595 | 762 | 462 | 530 | 666 | 419 | 835 | 1.118 | 621 |
| SDR- I00-I99- Diseases of circulatory system | 460 | 571 | 380 | 427 | 538 | 347 | 649 | 777 | 556 |
| | 218 | 271 | 176 | 164 | 201 | 133 | 417 | 526 | 336 |
| SDR- C00-C97-malignant neoplasms | 196 | 265 | 150 | 200 | 269 | 153 | 189 | 251 | 145 |
| | 167 | 261 | 128 | 160 | 208 | 125 | 193 | 269 | 142 |
| SDR- V00-V99, W00-W99, X00-X99, Y00-Y99- external causes of injury and poisoning | 63 | 91 | 37 | 58 | 81 | 37 | 85 | 130 | 43 |
| | 36 | 54 | 19 | 30 | 44 | 18 | 55 | 90 | 23 |
| SDR- V02-V04, V09, V12-V14, V20-V79, V82-V87, V89- Motor vehicle traffic accidents | 16 | 25 | 8 | 17 | 26 | 8 | 15 | 24 | 8 |
| | 6 | 9 | 5 | 5 | 8 | 3 | 8 | 13 | 2 |
| SDR- X60-X84-Suicide and Self-inflicted Injury | 14 | 21 | 8 | 13 | 19 | 8 | 19 | 31 | 9 |
| | 10 | 17 | 4 | 9 | 14 | 4 | 15 | 26 | 5 |
| (SDR- A00-A99, B00-B99-Infectious and Parasitic Diseases | 8 | 12 | 6 | 7 | 10 | 5 | 12 | 18 | 7 |
| | 8 | 11 | 7 | 9 | 11 | 7 | 7 | 10 | 4 |
| SDR- E10-E14-Diabetes | 15 | 14 | 16 | 16 | 15 | 17 | 12 | 11 | 13 |
| | 12 | 14 | 10 | 12 | 14 | 10 | 12 | 14 | 11 |
| SDR- F00-F99, G00-G99, H00-H95-MentalDisorders, Diseases of nervous system and sense organs, | 18 | 23 | 14 | 18 | 22 | 15 | 14 | 18 | 11 |
| | 31 | 34 | 27 | 35 | 38 | 32 | 15 | 19 | 12 |
| SDR-A15-A19, B90-Tuberculoses | 4 | 6 | 2 | 3 | 5 | 1 | 7 | 12 | 3 |
| | 1 | 1 | ≤1 | ≤1 | 1 | ≤1 | 2 | 4 | 1 |

1980 2010 T= Total M=Males F=Females

**APPENDIX 6:** Causes of death-standardized death rate per 100,000 inhabitants in the EU-28, EU-15, EU-13 (1980–2010). *Source: WHO, European Health for all data base,[10] Soulis 2014[1]*

| Years | 1980 | | 1990 | | 2000 | | 2010 | |
|---|---|---|---|---|---|---|---|---|
| | Males | Females | Males | Females | Males | Females | Males | Females |
| EE-28 | 8.3 | 5 | 7.8 | 4.4 | 6.5 | 3.7 | 5.4 | 3.1 |

**APPENDIX 7:** Loss in life expectancy years for men and women from death before the age of 65 years (1980–2010). *Source: WHO, European Health for all data base,[10] Soulis 2014[1]*

| Countries | 1961 | | | | 2010 | | | |
|---|---|---|---|---|---|---|---|---|
| | Females | | Males | | Females | | Females | |
| | PYLLs | LE | PYLLs | LE | PYLLs | LE | PYLLs | LE |
| **Greece** | **7.576** | **73.4** | **10.189** | **70.2** | **2.036** | **82.8** | **4.340** | **78.4** |
| **Netherlands** | **4.864** | **75.9** | **7.931** | **71.6** | **2.276** | **83.0** | **3.193** | **78.9** |
| **USA** | **7.310** | **73.6** | **12.328** | **67.1** | **3.447** | **81.1** | **5.814** | **76.2** |
| **Portugal** | **6.354** | **65.4** | **21.694** | **59.9** | **2.105** | **82.8** | **4.841** | **76.7** |
| **Spain** | **8.328** | **72.4** | **11.712** | **67.4** | **1.787** | **85.3** | **3.657** | **79.1** |
| **Japan** | **8.901** | **70.8** | **12.798** | **66.0** | **1.795** | **86.3** | **3.525** | **79.6** |

**APPENDIX 8:** Potential years of life lost-PYLLs from all causes with comparative reference to life expectancy-LE (100,000 men-women aged 0–69), (1961, 2010). *Source: OECD, Health Data, 2012*[12] *European Community Health Indicators (ECHI)*[13]

| EU-28 | Health-adjusted life expectancy at birth | | | Health-adjusted life expectancyat 65 years | | |
|---|---|---|---|---|---|---|
| | Females | Males | Difference | Females | Males | Difference |
| EU-28 | 62.7 | 61.9 | 0.7 | 8.9 | 8.7 | 0.2 |
| Max. | Malta 72.4 | Malta 71.8 | 0.6 | Sweden 15.4 | Sweden 14 | 1.4 |
| Min. | Slovakia 53.1 | Estonia 53.1 | | Slovakia 3.1 | Slovakia 3.5 | -0.4 |

**APPENDIX 9:** Health-adjusted life expectancy (HALE). *Source: Eurostat (2013), HEIDI data tool, Data on HLY in the EU*[16]

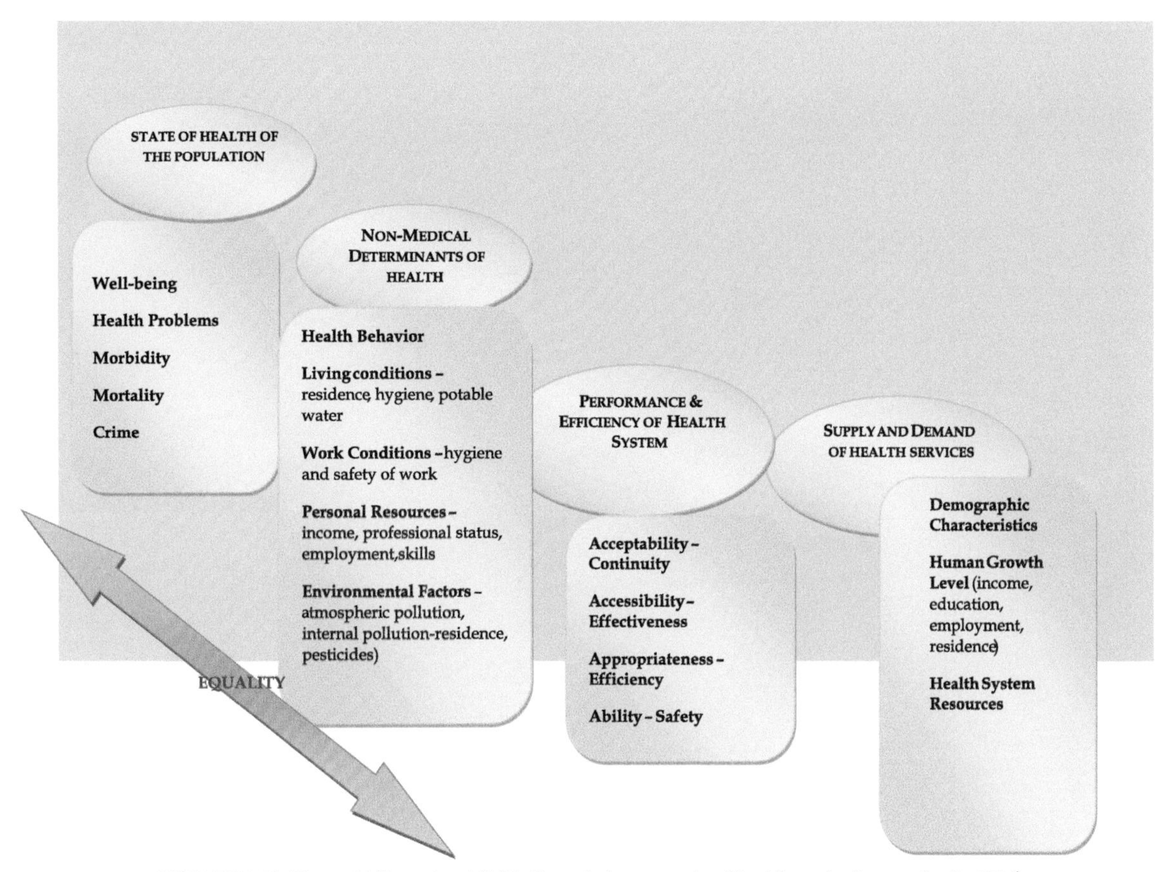


**APPENDIX 10:** The multidimensional field of population appreciated health needs. *Source: Soulis 2014*[1]

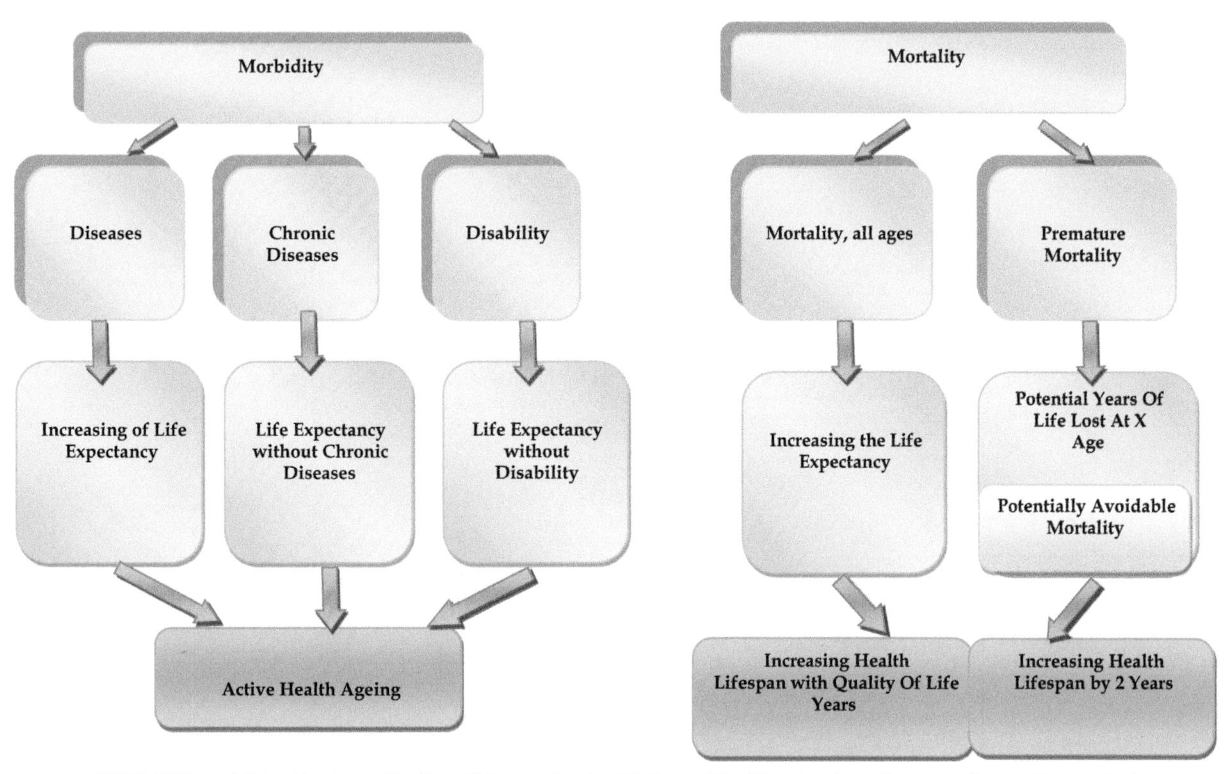


**APPENDIX 11:** Main objectives of health policies: prolonging life in good health and with quality years. *Source: Soulis 2014*[1]

# Index

*Note*: Page numbers followed by "f" and "t" indicate boxes, figures and tables respectively.

## G



## H



## I

## J



## K



## L



## M

## Q



## R



## S

## T



## U

## V



## W

Printed and bound by CPI Group (UK) Ltd, Croydon, CR0 4YY
08/05/2025
01865026-0003